Energy and protein metabolism and nutrition in sustainable animal production

EAAP – European Federation of Animal Science

The University of California

Department of Animal Science

The European Association for Animal Production wishes to express its appreciation to the *Ministero per le Politiche Agricole e Forestali* and the *Associazione Italiana Allevatori* for their valuable support of its activities

Energy and protein metabolism and nutrition in sustainable animal production

4th International Symposium on Energy and Protein Metabolism and Nutrition

Sacramento, California, USA

9-12 September 2013

EAAP publication No. 134

edited by:

James W. Oltjen

Ermias Kebreab

Hélène Lapierre

Wageningen Academic Publishers

Buy a print copy of this book at
www.WageningenAcademic.com/eaap134

ISBN: 978-90-8686-227-6
e-ISBN: 978-90-8686-781-3
DOI: 10.3920/978-90-8686-781-3

ISSN 0071-2477

Photo cover: Jim Oltjen and Romeo Capell

First published, 2013

The 4th EAAP International Symposium on Energy and Protein Metabolism and Nutrition was organized in Sacramento, California (USA) on 9-12 September 2013. A special symposium as tribute to the late Professor R. Lee Baldwin was also organized on 12 September 2013.

The symposium was chaired by:
- James Oltjen

The scientific programme of the symposium was co-chaired by:
- Ermias Kebreab
- Hélène Lapierre

In consultation with the International Scientific Committee:
- Mario Luis Chizzoti (Brazil)
- Matteo Crovetto (Italy)
- Kristen Johnson (Alternate, USA)
- Michael Kreuzer (Secretary, Switzerland)
- Hélène Lapierre (Canada)
- Jan Erik Lindberg (President, Sweden)
- Toyomizu Masaaki (Japan)
- Cornelia Metges (Outgoing, Germany)
- James Oltjen (USA)
- Isabelle Ortigues-Marty (France)

and liaising with the National Scientific Committee:
- Ransom Baldwin
- Karen Beauchemin
- Brian Bequette
- John Cant
- Theresa Davis
- Kees de Lange
- Lorraine Doepel
- Jeff Firkins
- Harvey Freetly
- Mark Hanigan (Chair, Baldwin Symposium)
- Kris Johnson
- Kyle McLeod
- John McNamara
- Phil Miller
- Gordon Murdoch
- Daniel Ouellet (Editorial)
- Luis Tedeschi

Table of contents

Part 6. Products and health

Part 7. Tissue metabolism

Part 9. Baldwin symposium

Preface

The 4th EAAP International Symposium on Energy and Protein Metabolism and Nutrition (ISEP) was held at the Sheraton Grand Hotel in Sacramento, California, USA from September 9th to September 12th, 2013. This followed the 3rd ISEP in Parma (2010), the 2nd ISEP in Vichy (2007), and the 1st ISEP in Rostock-Warneműnde (2003). These follow the previous Energy Symposium and Protein Symposium which were held separately, all under the auspices of the European Association of Animal Production and the Commission on Animal Nutrition; it is also the first time the symposium had been held in North America.

As world population increases, demand for food, particularly animal products, is expected to grow substantially. Because of limited area for expansion of animal agriculture and increased consumer concern for the environmental impact of animal production, gains in animal efficiency will have to be part of the solution. The 4th International Symposium on Energy and Protein Metabolism and Nutrition addressed key issues of how energy and protein are utilized and interact in farm animals from the molecular to the whole animal and even to the herd or group level of organization. Key issues addressed include energy/protein interactions, methodology such as *in vitro* and *in vivo* techniques, regulation including pre-natal programming and endocrine regulation, modeling/systems biology, products and health of animals, tissue metabolism, and environmental sustainability in agriculture. ISEP also included a tribute to the late Professor R. Lee Baldwin of the University of California, Davis, a leader in the field.

The 4th Symposium began with the premise that improved understanding of animal energetics and protein metabolism will be required for sustainable animal production. Over 200 participants, from 27 countries, made theatre and poster presentations; two-page abstracts for contributed papers and full length papers by invited speakers are contained herein.

Attendees at the ISEP heard significant new research, with invited speakers and oral and poster communications by participants. The Symposium combined fundamental research with applied research and practical applications. Because energy and protein metabolism and nutrition cannot be addressed separately, a better and deeper understanding of nutrient metabolism and nutrition can be achieved only by integrating the outcomes of scientists conducting research on different aspects.

Participants and accompanying persons were housed in one hotel and shared common meals in an effort to increase networking possibilities and stimulate interactions; they were also treated to a showcase of California agriculture and hospitality.

We thank all those who helped make this Symposium successful, especially the sponsors and the International, North American, and Local Organizing Committees. Finally, we would also like to thank all the participants for making possible a meeting with a great deal of interaction; we hope this meeting has given us more tools to address questions that need to be answered for a real sustainable agriculture scientifically.

James W. Oltjen

Keynotes

Feeding the planet: key challenges

M. Herrero
Commonwealth Scientific and Industrial Research Organisation, 306 Carmody Road, St Lucia 4067, QLD, Australia; m.herrero@cgiar.org

Abstract

The notion of feeding 9 billion people sustainably in the next forty years presents considerable challenges. Population growth and human dietary changes remain the key drivers of the large increases in future food demand. While the world food system can respond to meet this demand, there are challenges to ensure this happens sustainably, equitably and within our conceived limits of a safe environmental space. This paper discusses the key trends in food production and consumption, and the key challenges for feeding the planet. Especial attention is given to livestock, a key part of the puzzle for ensuring sustainable nutritional security and environmental sustainability for future generations.

Introduction

The global food system is experiencing profound changes as a result of anthropogenic pressures. The ever-increasing human population (to reach 9 billion by 2050) together with changes in consumption patterns (i.e. increasing demand for livestock products) caused by urbanization, increasing incomes, and nutritional and environmental concerns, are shaping what we eat, who eats, and how much, more than ever. The double burden of nutrition (overconsumption and under-nutrition) is defining research agendas, policies and conceptions about food in different ways around the world.

Against this background is a global food system that will have to improve its resource use efficiency and environmental performance significantly in order to ensure the sustainability of global food production and consumption within established planetary boundaries of greenhouse gas emissions, and water and nutrient use amongst others.

Livestock, the largest land use sector on Earth is an important part of this puzzle and many solutions to the challenges facing how to feed the world sustainably lie in how we manage this sector. This brief paper aims to discuss some of the key ways in which we could increase the sustainability of the world food system, and some of the challenges to overcome.

Key challenges: the future demand for food

Table 1 shows the FAO projections of global food consumption to 2050 (Alexandratos and Bruinsma, 2012). The main conclusions from these projections, as well as others (ie. IAASTD 2009), is that a shift to diets with more animal products and fats, is likely to happen, mostly in the developing world as a result of increased incomes and urbanization. While the consumption per capita of cereals is likely to stabilize, population growth will increase the total quantities of both meat (almost doubling) and cereals (50%) needed to feed the world in 2050.

The supply response of the global agriculture and livestock sectors is likely to be able to accommodate these demand increases (Alexandratos and Bruinsma, 2012). All recent projections have important common features: (1) Local production under current yield trends in many parts of the world, like Sub-Saharan Africa (SSA) and parts of Asia, will not be able to meet local food demand. Hence increases in food trade are projected to increase in the future in some parts of the world. This is a key aspect of balancing the food supply and demand equation. (2) While increases in the yields of crops and livestock have occurred in most regions of the world (apart from SSA), all projections show a variable increase in cropland and grassland expansion to meet demand (Smith *et al.*, 2010)

Table 1. Projections of global demand for food to 2050 (Alexandratos and Bruinsma, 2012).

	2005/2007	2050
Population (millions)	6,584	9,306
Cereals for food (kg per capita)	158	160
Cereals for all uses (kg per capita)	314	330
Meat consumption (kg per capita)	38.7	49.4
Oil crops for food (kg per capita)	12.1	16.2
Oil crops for all uses (kg per capita)	21.9	30.5
Meat production (million tonnes)	258	455
Cereal yields, rice paddy (t/ha)	3.32	4.30
Arable land area (million ha)	1,592	1,661

and (3) also an increase in animal numbers, but with monogastric production (pork and poultry) growing at faster rates than ruminants (meat especially, and less so for milk). (4) These factors lead to net increases in greenhouse gas emissions (GHG) from the agricultural and livestock sectors, but a diminishing trend in the emissions intensities across commodities (GHG per unit of product). (5) projections of water use show increased pressure on total fresh water resources, notably on blue water (irrigation), and moderate increases in the efficiency of green water use (CA, 2007). Other studies have also demonstrated large quantities of reactive nitrogen used and a potential depletion of phosphorus stocks in the future (Bouwman *et al.*, 2011). Hence, food production can be attained under current productivity and demand trends, but not necessarily making inroads in the improvement of our environmental goals.

Several authors (Foley *et al.*, 2011; Garnett and Godfray, 2011; Godfray *et al.*, 2010; Herrero *et al.*, 2010) have suggested different mechanisms for improving the sustainability of the world food system. The three most often mentioned are:

1. Increasing productivity (managing the supply side): Increasing agricultural productivity and overall food production have been the pillar for designing strategies for feeding the world since the industrial revolution. Notable gains have been made in many parts of the world (developed countries and Latin America and Asia). There is significant ongoing research on how to sustainably intensify global food production, how to bridge yield gaps of crops and livestock and how to improve value chains so that both producers and consumers benefit from potential yield increases, while using less, the same, or slightly more inputs.
2. Reducing waste in food value chains. This subject has received attention recently (Godfray *et al.*, 2010), and it has been estimated that food waste can account for up to 40% of losses relative to food production. Figure 1 (Godfray *et al.*, 2010) shows that in the developing world these losses occur mostly due to post-harvest activities like deficient harvesting and storage methods, pests, export regulations and others. In the developed world this occurs mostly at the post-consumption stage, due to poor management of product sell-out dates in the value chain and direct food disposal by consumers (i.e. discarding food from fridges).
3. Consuming more sustainable diets (managing the demand for food): There is evidence that modifying what we eat could have significant impacts on the use of resources like land and water, it could reduce GHG emissions and it could have important health and nutritional benefits. A lot of emphasis has been put in the potential benefits of reducing red meat consumption and the promotion of 'healthy' diets (Stehfest *et al.*, 2009) (Table 2). These studies have shown that reductions in livestock consumption could lead to reduced land use change, directly from less land clearing for raising animals or for producing feed crops. These land sparing gains, in turn lead to lower GHG emissions in general.

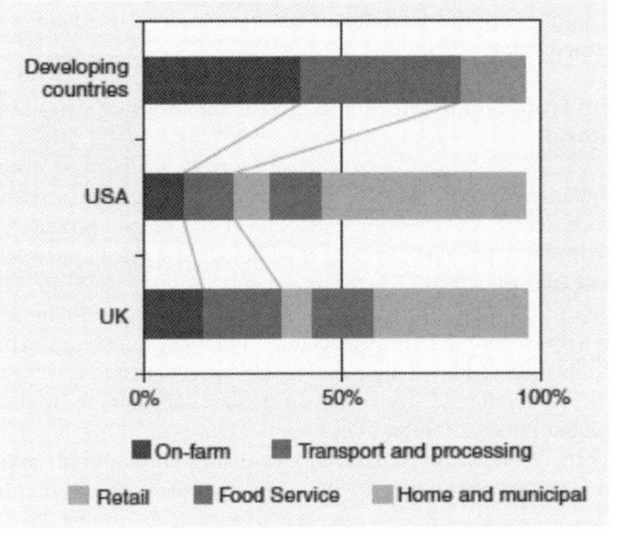

Figure 1. Make up of total food waste in developed and developing countries (Godfray et al., *2010).*

This is a significant finding. However, this space has not been explored sufficiently to provide suitable, practical, regional/country guidance for consumers and for policy makers to effect the necessary changes in local food systems, and to modify consumer behavior. At the global level, this concept has only been studied superficially, and without considering important dimensions such as dietary diversity and cultural preferences extending beyond measures of kilocalorie consumption, and what would be the social and economic impacts of reducing the size of the livestock sector. These additional dimensions are essential for understanding the biological and socio-economic implications of diet sustainability on the global food system. We need to go beyond simplistic recommendations like 'stop eating meat' to make this area of research useful, and provide alternatives and practical guidelines for achieving these kinds of gains. This is of particular importance for the developing world, where livestock product demand projections demonstrate that even with significant consumption growth, consumption per capita will remain significantly lower than in the developed world (IAASTD, 2009).

The implementation of these strategies is not straightforward. There are many challenges and trade-offs and they are complex because there are competing economic, social and environmental claims to their implementation. Additionally, human nature, the single biggest ingredient in this mix, plays a key role in defining our choices. These choices are not always in favor of long-term sustainability. More often the attainment of shorter term gains prevails, especially in a world of

Table 2. Land use emissions in 2000 and 2050 for the reference scenario and four variants of dietary composition (Stehfest et al., *2009).*

	GtCeq
2000	3.0
2000 – reference	3.3
2050 – no red meat	1.7
2050 – no meat	1.5
2050 – no animal products	1.1
2050 – healthy diet	2.1

increasing resource limitations and with significant pressures on livelihoods. Some of these key challenges are explained below.

Closing yield gaps of crops and livestock and increasing adoption rates of key technologies in the developing world

Closing yield gaps, mainly of crops, has been the strategy of choice for increasing food production worldwide (Foley *et al.*, 2010; Van Ittersum *et al.*, 2013). Significant productivity gains have been observed in the last 40 years, however in many smallholder systems in the world, low productivity still prevails (Tittonell and Giller, 2013). In terms of livestock yield gaps, there has been continuous productivity increases in the developed world or in parts of the developing world where market orientation has been a thrust for the livestock sector. A notable example, presented by Capper *et al.* (2009) demonstrates that milk yields per animal in the US have increased several-fold since the 1940s, while at the same time reducing the size of the dairy herd considerably in the process. However, we still lack solid global information on livestock yield gaps to inform the potential productivity gains for different parts of the world. This is an area that requires significant research, especially for guiding research in the developing world, where most of the potential for increasing yields lies.

In livestock systems, attaining yield gaps has essential pre-requisites. The availability of inputs (high quality feeds, fertilizers, etc.), services (veterinarians, extension) and in many cases, the development of markets and their associated value chains need to be developed (McDermott *et al.*, 2010), as these are key incentives for systems to intensify (Herrero *et al.*, 2010; McDermott *et al.*, 2012). Currently, adoption of better feeding practices, like improved forages, have shown low adoption rates. For example Thornton and Herrero (2010) found adoption rates of dual purpose crops, agroforestry practices and improved pastures in the order of 15-25% of farmers in selected developing regions, over a 10-15 year horizon. Increasing adoption rates will require significant public and private investment and institutional change to be able to not only increase the % of farmer adopting, but also for reducing the adoption lag times that are often large.

Achieving sustainable intensification without its potentially perverse incentives

The concept of sustainable intensification sounds to many as a win/win strategy to increase resource use efficiencies. From a livestock perspective, most well managed intensification practices in the past have led also to improved systems profitability (i.e. pasture intensification and supplementation in the tropics has significantly improved milk and meat production). As a result, farmers have often increased the size of the operation (more animals, more land use changes) in order to increase even further the economic returns. This growth in turn has led to increased environmental problems (more deforestation, increased GHG emissions, land degradation). A critical challenge ahead is how to regulate intensification so that it is truly sustainable and operates within limits of production growth, protects biodiversity and other ecosystems services, and attains net or near net reductions in the use of resources. This is of particular importance, as having less animals, but of higher productivity, seems to be essential to maximize the environmental benefits (i.e. reductions in GHG emissions and land use) of productivity growth in livestock systems (Thornton and Herrero, 2010).

What we eat matters: how far can we go towards modifying human diets in different parts of the world?

The debate on human diets is dominated by the concerns of the developed world and the middle classes of middle income countries, on the negative health impacts of livestock product over-consumption (McMichael *et al.*, 2007), and by the global integrated assessment community interested in reducing GHG emissions from the agriculture, food and land use sectors (Smith *et al.* 2013; Stehfest *et al.*, 2009). At the same time, the discourse in the developing world is changing towards the recognition that achieving nutritional security might be a more important target than just promoting a higher kilocalorie consumption (Herrero *et al.*, 2013). According to these authors, this does not mean that we should take as given the projected trajectories of animal consumption proposed by the so called 'livestock revolution'. They are not inevitable. Part of our responsibility is to challenge these future trajectories, and ensure that we identify consumption levels for different parts of the world that will achieve the best compromise between a culturally-appropriate, healthy diet that includes livestock products (or not), economic growth, livelihoods and livestock's impacts on the environment. The other part of the agenda is ensuring that policies geared towards the promotion of healthy sustainable diets occur throughout the food systems (from the regulatory bodies, the food industry, all the way to the different value chain actors.

Structural change and competitiveness in the livestock sector

Large parts of the food systems of the developing world have as a starting point the smallholder mixed crop-livestock farmer or the vulnerable pastoralist. Some of these systems produce a significant amount of food, mostly for local consumption, they have significant, exploitable, yield gaps in crops and livestock, and in many cases they have low opportunity costs of labor. If livestock is to be used as an engine for poverty reduction, it is essential that these producers become market-orientated. The degree of competitiveness of smallholders against imports from countries that can produce vast amounts of animal products, at lower production costs, will be a crucial factor to determine the success of many men and women livestock farmers in the developing world, especially as the volume of traded livestock products increases due to trade liberalization. Formal and informal markets will need to ensure the supply of cheaper, locally produced, safe livestock products to adequately compete. This implies a significant reduction in transaction costs for the provision of inputs, increased resource use efficiencies, and very responsive, innovative and supporting institutions for the livestock sector in developing countries (FAO, 2009). Hence investment in developing efficient value chains (including market development, service provision, adequate institutional support, etc.) should be high in the development agenda, to create incentives for smallholders to integrate in the market economy, formal or informal.

Recently land consolidation has occurred in many parts of Africa, as foreign investors buy large blocks of land for developing them into large farms. Advocates of large scale farming argue in favor of the higher efficiencies of resource use often found in these systems and how simple it is to disseminate technology and effect technological change. However, we lack comprehensive information on the impacts of different farming avenues and their future evolutionary pathways on water cycles, biodiversity, social aspects (nutrition, incomes, employment), coping with production risks (i.e. climate variability and change, commodity prices) and others.

A thriving environmentally-responsible, diversified, commercial smallholder sector, that helps feed the world and lifts people out of poverty; together with a large scale efficient livestock sector that efficiently produces food while generating enough employment for rural people should co-exist. The balance between these ways of farming is likely to be different in different regions of the world, but understanding where the balance should lie for achieving socially and environmentally goals is still the question that merits significant research.

The success of paying for environmental services (PES)

PES schemes as an income diversification strategy, for mitigating climate change or for protecting important regional or global goods that regulate essential biogeochemical cycles have received a lot of attention recently. However, not many successful examples exist with smallholder livestock producers or pastoralists (Herrero *et al.*, 2013). Proofs of concept that test how these schemes could operate in very fragmented systems, with multiple users of the land or in communal pastoral areas are necessary. Research on fair, equitable and robust monitoring and evaluation frameworks and mechanisms for effecting payments schemes that work under these conditions is necessary. The promise of PES schemes as a means of deliver the income diversification for the poor, and natural resource protection necessary to produce food while protecting the world's ecosystems is yet to be seen on a large scale.

Institutional and market mechanisms for reaching smallholders

The reality is that livestock production in the developing world is largely fragmented and disorganized. Under-investment in extension systems and other support services have rendered poor producers disenfranchised to access key support systems necessary for increasing productivity and efficiency, or in cases important safety nets for reducing vulnerability (i.e. to drought/famines). The poorer farmers are unlikely to be able to respond sustainably to the increased demands for animal products without increased public investment in innovation and support platforms, as these are essential to foster the technological change required to increase productivity and link them to markets (McDermott *et al.*, 2010). More advanced farmers or larger farmers in the developing world are likely to rely on the private sector for these support services. It is essential that the roles of women in production and trading of livestock products and in controlling livestock assets are taken into consideration when designing these institutional mechanisms.

Animal health and food safety: regulation and surveillance in an era of more animals, higher volumes of food trade and more diseases

Security of animal source foods, health threats to animal assets, and food safety are inextricably linked. Animal disease threatens livelihoods and economies. For intensive livestock farmers, animal health costs while a small part of overall farm enterprise costs are a large part of avoidable costs, and hence a leverage point for increasing productivity. For poor livestock-keepers, animal health typically represents one of the largest single costs and epidemic animal disease one of the biggest and most feared risks to livestock-keeping.

Animal disease is also the main obstacle to trade in animals and animal products. Despite recent attempts at liberalization, sanitary and phytosanitary regulations still allow importing countries to take a precautionary 'if in doubt, keep it out' approach. This denies people-poor but livestock-rich countries an opportunity to trade their way out of poverty while imposing unpredictable shocks on all countries for which livestock trade is important (for example the Rift Valley Fever pandemic has periodically interrupted the lucrative trade in live sheep and goats from the horn of Africa to the Arabian Peninsula).

We are living in an era of unprecedented ecosystem change and the current upsurge in emerging disease (75% of which are zoonotic) is predictable. An ecosystems perspective sees farming and natural systems as containing pools of pathogens with circulate among hosts, vectors and the environment. In systems which are stable, connected and diversified, hosts and pathogen co-evolution favors lowered pathogenicity and less disease. However, anthropogenic incursion into ecosystems and ecosystem alteration allows pathogens to encounter new hosts and new diseases to emerge. An example of ecosystem provision of disease regulation services is the flare-ups of human sleeping sickness seen in West Africa after African swine fever killed village pigs leading the tsetse vector of sleeping sickness to shift closer to houses and bite more people.

Food safety is increasingly viewed as inseparable from food security and while intensive agriculture can produce cheap products it also introduces new health risks for both animals and people. This is of essential importance to consider how far we go in intensifying livestock production, and the food system in general. In particular, intensification selects pathogens hard to detect in animal populations or (such as *Campylobacter* spp. in poultry or diarrhoeagenic *Escherichia coli* in cattle) or survive conventional treatment (such as antibiotic resistance). The wide geographic scale and large volume of consolidated food distribution systems means that food-borne diseases can spread rapidly and affect large numbers of people greatly removed from the point of production.

Can the food system adapt to climate change and mitigate GHG emissions at a fast enough pace?

Climate change is likely to cause severe impacts on livestock systems and on poor vulnerable producers. According to Herrero *et al.* (2013), the capacity and speed of adaptation of smallholders will play and important role in defining the contribution of livestock to livelihoods under climate change. At the same time, in a low carbon economy, it will be essential that the sector mitigates GHG effectively in relation to other sectors. Demonstrating that these options are real with tangible examples is essential to generate the evidence for increasing the investments in climate change adaptation and mitigation for the livestock sector. This becomes more imperative as the global food system prepares to become part of the climate change negotiations.

Conclusions

Feeding 9 billion people is a formidable but attainable task. Doing it sustainably will depend on several things, including our capacity to reach more equitable and healthy levels of consumption, our ability to invest in key food production systems of the world, on maintaining well regulated and economically viable production systems, and on achieving a balance between environmental and social goals.

The livestock sector, the largest land user on Earth, holds a large stake on how to achieve the balance between food production, livelihoods and environmental objectives. A mixture of sustainable intensification, protecting biodiversity and ecosystems services, together with strategies and policies to reduce animal numbers simultaneously, may yield a suitable compromise for achieving these goals.

References

Alexandratos, N. and Bruinsma, J. 2012. World agriculture to 2030/2050. The 2012 revision. ESA Working Paper 12-03. Food and Agriculture Organization of the United Nations, Rome, Italy, 160 p.

Bouwman, A.F., Klein Goldewijk, K., Van der Hoek, K.W., Beusen, A.H.W., Van Vuuren, D.P., Willems, J., Rufino, M.C. and Stehfest, E. 2011. Exploring global changes in nitrogen and phosphorus cycles in agriculture induced by livestock production for the period 1900-2050. PNAS, doi/10.1073/pnas.1012878108.

CA. 2007. Comprehensive Assessment of Water Management in Agriculture (D. Molden, Editor). Earthscan, London UK.

Capper, J.L., Cady, R.A. and Bauman, D.E. 2009. The environmental impact of dairy production: 1944 compared with 2007. Journal of Animal Science 87, 2160-2167.

FAO 2009. The State of Food and Agriculture. Livestock in the Balance. Food and Agriculture Organization of the United Nations, Rome, Italy.

Foley, J.A., Ramankutty, N., Brauman, K. A., Cassidy, E. S., Gerber, J. S., Johnston, M., Mueller, N. D., O'Connell, C., Ray, D. K., West, P. C., Balzer, C., Bennett, E. M., Carpenter, S. R., Hill, J., Monfreda, C., Polasky, S., Rockstrom, J., Sheehan, J., Siebert, S., Tilman, D. and Zak, D. P. M. 2011. Solutions for a cultivated planet. Nature 478, 337-342.

Garnett, T. and Godfray, H.C.J. 2011. Sustainable intensification: navigating through a course of complexities. Food and Climate Research Network, University of Oxford, UK.

Godfray H.C.J., Beddington, J.R., Crute, I.R., Haddad, L., Lawrence, D., Muir, J.F., Pretty, J., Robinson, S., Thomas, S.M. and Toulmin, C. 2010. Food security: the challenge of feeding 9 billion people. Science 327, 812-818.

Herrero, M., Grace, D., Johnson, N., Njuki, J., Enahoro, D., Silvestri, S and Rufino, M. 2013. The roles of livestock in developing countries. Animal 7 (1), 3-18.

Herrero, M., Thornton PK., Notenbaert AM., Wood S., Msangi S., Freeman HA., Bossio D., Dixon J., Peters M., van de Steeg J., Lynam J., Parthasarathy Rao P., Macmillan S., Gerard B., McDermott J, Seré C. and Rosegrant M. 2010. Smart investments in sustainable food production: revisiting mixed crop-livestock systems. Science 327, 822-825.

IAASTD. 2009. Agriculture at a Crossroads. International Assessment of Agriculture, Science and Technology Development. Island Press, 590 p.

McDermott, J.J., Staal, S., Freeman, H.A., Herrero, M., van de Steeg, J. 2010. Sustaining intensification of smallholder systems in the tropics. Livestock Science 130, 95-109.

McMichael, A.J., Powles, J.W., Butler, C.D. and Uauy, R. 2007. Food, livestock production, energy, climate change, and health. The Lancet 370: 1253-1263

Smith, P., Gregory, P.J, van Vuuren D.P., Obersteiner, M., Havlik, P.,Rounsevell, M., Woods, J., Stehfest, E. and Bellarby, J. 2010. Competition for land. Philosophical Transactions of the Royal Society B, 365: 2941-2957.

Smith, P., Haberl, H., Popp, A., Erb, K., Lauk, C., Harper, R., Tubiello, F., Siqueira Pinto, A., Jafari, M., Sohi, S., Masera, O., Böttcher, H., Berndes, G., Bustamante, M., Ahammad, H., Clark, H., Dong, H., Elsiddig, E.A., Mbow, C., Ravindranath, N.H., Rice, C.W., Robledo-Abad, C., Romanovskaya, A., Sperling, F., Herrero, M., House, J.I. and Rose, S. 2013. How much land based greenhouse gas mitigation can be achieved without compromising food security and environmental goals? Global Change Biology (DOI: 10.1111/gcb.12160, in press).

Stehfest, E., Bouwman, L. van Vuuren, D.P., den Elzen, M. G. J., Eickhout, B. and Kabat, P. 2009. Climate benefits of changing diet. Climatic Change 95, 83-102.

Thornton P.K. and Herrero M. 2010. The potential for reduced methane and carbon dioxide emissions from livestock and pasture management in the tropics. PNAS 107, 19667-19672.

Tittonell, P. and Giller, K. 2013. When yield gaps are poverty traps. The paradigm of ecological intensification in African agriculture. Field Crops Research 143, 76-90.

Van Ittersum, M., Cassman, K., Grassini, P, Wolf, J., Tittonell, P. and Hochman, Z. 2013. Yield gap analysis with local to global relevance – a review. Field Crops Research 143, 4-17.

Role of animal products in feeding the planet

J.E. Hermansen, M.T. Knudsen and T. Kristensen
Aarhus University, Blichers Allé 20, 8830 Tjele, Denmark; john.hermansen@agrsci.dk

Abstract

Livestock products contribute today by approximately 13% of calories and 28% of protein to the global human population with a huge difference between developed, transition and developing countries. As a consequence of increased consumption per capita and a larger population, it is projected that global meat consumption will increase by 40% in 2030 and 70% in 2050. This has large resource use implications and furthermore the foreseen increased production is to happen within the context of growing scarcity of natural resources and challenges posed by climate change. All this put challenges on the livestock production systems. The major growth in livestock production will take place in intensive industrialized systems very much focused on poultry and pork. It will be important that the feeds can be based on very high yielding new crops that serve both as feed and other purposes in order to reduce land use requirements. This asks for new knowledge and technology that can transform the biomass from the high yielding crops to high quality feed. Also, it will be very important that the energy left in the manure is recovered to substitute fossil energy or energy produced from new biomass. Still, a major part of dairy and beef production will take place in mixed systems, which in some parts of the world are rather inefficient in their use of natural resources. Such systems have to increase efficiency having more of the feed consumed transformed into food instead of feed requirements for animal maintenance. Food from extensive grazing systems will in particular be important to support the nutritional needs of populations in food-insecure areas. The major challenge here is to stop land degradation and restoring the functioning of the grassland, including grassland role to sequester carbon and contribute to other eco-system services.

Introduction

Although the expressions 'feeding the planet' and 'food security' may be perceived as two distinct challenges, it is clear that when considering animal products, they are two sides of the same coin. Animal products clearly have a role in ensuring food security – food security including the dimensions: availability, access, stability and utilization for all human communities. At the same time, however, it is beyond arguing that the ecological costs of animal products typically are much higher than for plant products. While in a typical Northern European diet it is estimated, that animal products covers approximately 25% of the calorie coverage of the diet, the animal products are responsible for more than 60% of the carbon footprint of the typical diet (Hermansen and Olesen, 2009). At the same time it is well recognized that production of feed for livestock is a major driver of land use change (Steinfeld *et al.*, 2006) putting pressure on global biodiversity, and enhance emissions of carbon dioxide to the atmosphere as a consequences of transforming grassland, forest, and savanna to arable land. E.g. UNEP (2011) estimates an increase in soybean production from 6250 square km in 1992 to almost 10.000 square km in 2009. Thus, there are good reasons to focus on the role of animal products in our diets and in relation to the overall environmental concerns of today.

Several developments in addition tend to aggravate these environmental concerns. Thus, besides the growth in global population, during the latest 20 years the overall food production has increased considerable more than the growth in population to ensure a better food supply and this trend is expected to be maintained (UNEP, 2011). Further, the consumption of meat has increased considerable more than the population growth. Figure 1 shows the global dietary change towards higher per capita meat consumption in the period from 1961 to 2009 (FAOSTAT, 2013). The meat consumption per capita in Europe and North America, though flattening out, is still considerably higher than in Asia and Africa. Recently, the meat consumption per person has shown a larger increase in Asia than in

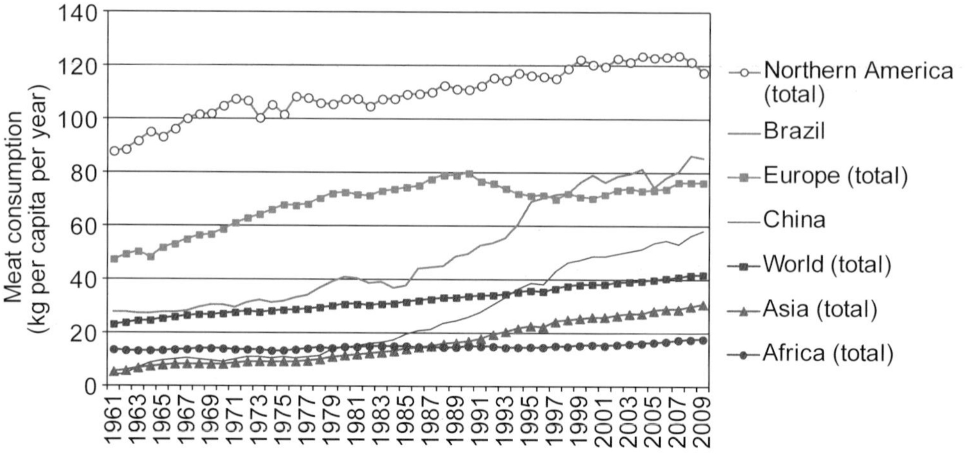

Figure 1. Meat consumption per capita per year on average in the world, in Europe, North America, Asia, Africa and in China and Brazil (FAOSTAT, 2013).

Africa and especially China has shown a remarkable increase during the last 30 years. Projections of future meat consumption predict that China by 2030 will have a per capita meat consumption of approx. 75 kg/person/year and Brazil is predicted to have a per capita meat consumption of almost 90 kg/person/year by 2030 (Msangi and Rosegrant, 2011). Thus, the Internal Food Policy Research Institute (IFPRI) predicts that over the next decades, virtually all growth in demand for meat will come from the transition economies and the developing world (IFPRI, 2012).

At the same time, there is competition for land for biomass production for other purposes than food, in particular energy, which contributes to the overall pressure on land resources. The overall transformation of land to agricultural land ranged between 5-10 million ha per year during the latest 40 years with a particular increase in the latest years.

All this being said, livestock contribute with around 13% of global calories for human consumption, but approximately 28% of protein through provision of meat, milk and eggs (FAO, 2011). In the developed world the numbers are 20 and 48%, respectively. While in the wealthiest part of the world, the protein coverage is easily met, and actually can be met for the major part by animal protein, this is not the case in the poorest part of the world. In these regions lack of calories and protein seems to go hand in hand and here livestock products often play a significant role in ensuring protein as well as other essential nutrient. Even in small amounts livestock products contribute to ensure the proper nutrition of poor families due to the content and bioavailability of important micro-nutrients. In addition, livestock has a role in supporting food security in vulnerable small-holder populations, through delivering stability of food and income. FAO (2012) estimates that close to one billion of the world's poorest people rely on livestock for their livelihoods.

Thus, animal products have a significant role in feeding the planet, but the production is challenged by ecological side-effects which need to be addressed. These side-effects naturally depend on type of animal production system and how they are managed.

Global livestock production and livestock production systems

Livestock are produced in different ways in different parts of the world. Thus, the contribution to food supply and food security as well as the ecological impact of production differs markedly among different livestock products and livestock systems. In Table 1 is shown the global production and the

Table 1. Changes in global livestock production total and per person 1967 to 2007 (after FAO, 2011).

	Production (million tonnes)			Production per person (kg)		
	1967	2007	2007/1967	1967	2007	2007/1967
Pig meat	33.9	100.0	294%	9.8	14.9	152%
Beef and buffalo meat	36.5	65.6	180%	10.6	9.8	93%
Eggs, primary	18.2	64.0	353%	5.3	9.6	183%
Milk, total	381.8	680.7	178%	110.3	102.0	92%
Poultry meat	12.4	88.0	711%	3.6	13.2	369%
Sheep and goat meat	6.5	13.1	202%	1.9	2.0	105%

supply per capita, inclusive the changes in the 40 year period from 1967 to 2007. For all products the production increased by a factor 2-7 in that period, most pronounced for poultry meat but also very pronounced for pigs and eggs. Also per capita production increased except for beef and milk, which were almost unchanged. Thus, from a food supply point of view, poultry- and pig products became much more important during that period. Projections estimate total meat consumption in 2030 of 376 million ton – an increase of 40% from the numbers given in Table 1.

The increasing demand for meat on the world markets will mainly be supplied by Brazil, USA, Oceania and Europe. In an international trade perspective, Brazil is expected to play a major role being a strong net exporter of livestock products (FAO, 2012) and has currently the world's largest export of both beef and chicken meat (FAOSTAT, 2013).

Data on what type of production from different production systems are less updated, but Table 2 gives a good picture of the relative importance of the different productions systems for each product. Grazing systems include both extensive grazing system and intensive grazing systems. The extensive systems are prevailing in dry areas that are marginal for crop production and sparsely populated (Southern Africa, central and western Asia, Australia and western North America) and these systems provides for about 7% of beef production and 12% of sheep and goat meat production. The intensive grazing systems appear in the temperate zones where high quality forages can be grown (most of Europe, North America and South America) and account for about 17% of total beef and sheep and goat meat production.

The mixed farming systems are systems where cropping and livestock production are interlinked activities and where at least 10% of the total value of production comes from non-livestock activities. Most of the milk and beef produced comes from such systems – the rain fed systems prevailing

Table 2. Global livestock production average by production system 2001 to 2003, million ton (after FAO, 2011).

	Grazing	Rainfed mixed	Irrigated mixed	Landless/industrial	Total
Beef	14.6	29.3	12.9	3.9	60.7
Mutton	3.8	4.0	4.0	0.1	11.9
Pork	0.8	12.5	29.1	52.8	95.2
Poultry meat	1.2	8.0	11.7	52.8	73.7
Milk	72	319	204	-	594
Eggs	0.5	5.6	17.1	35.7	58.9

in temperate Europe and North America as well as sub-humid regions of tropical Africa and Latin America, and the irrigated systems prevailing in East and South Asia. It is worth mentioning also, that approximately 50% of the world's cereal production in developing countries comes from mixed systems (Herrero *et al.*, 2010), and as highlighted by Smith *et al.* (2013) by providing manure and – in the case of many smallholder systems – draught power for field operations, thereby supporting the production of this staple food.

The industrial systems are defined as those systems that receive at least 90% of the feed from other enterprises. The major part of poultry are produced in such systems often found near large urban centres, and thus relying on global feed resources, and slightly more than half of pork production takes place in such systems.

Ecological impacts related to livestock production

The livestock production is related to a range of ecological impacts affected by the production system and its efficiency. While not designed for it, the carbon footprint of livestock products can be a good indicator of the overall impact, since often a good correlation exist between the carbon footprint and impacts like eutrophication, fossil energy use and acidification. The reason is that the carbon footprint aggregates the CO_2 emission related to use of fossil energy, the global warming effect of emission of nitrogenous substances from feeds and animals, and the methane emissions related to enteric digestion as well as from manure management, both of which are related to overall feed use. Thus, feed use is a very important aspect when considering the carbon foot print of livestock products (Hermansen and Kristensen, 2011). This is not to say that the carbon foot print is the only relevant indicator, but it helps to get an overview in the broad sense. Furthermore, the carbon footprint is a highly relevant indicator in policy for consideration of combating climate change.

Table 3 shows typical numbers for carbon footprint and land requirements mainly representing relatively productive Western European systems. Red meat clearly has a higher carbon footprint

Table 3. Land requirement (m^2) and carbon footprint (kg CO_2eq) of products, per kg product. Sources for carbon footprint: De Vries and De Boer (2010); Nguyen et al. *(2010, 2011). Importance of iLUC after Audsley* et al. *(2009) (~0.15 kg CO_2/m^2) and Schmidt* et al. *(2012) (0.75 CO_2/m^2)*

	m^2	Carbon footprint without and with accounting for indirect land use changes (iLUC)	
		Without iLUC	To be added with iLUC
Milk	1-2	~1	+0.2-1.5
Egg	4-7	4-5	+0.6-5
Broiler	5-7	2-4	+0.8-5
Pork	6-8	3-5	+0.9-5
Beef	15-45		
– Intensive bulls 12 mo.	~17	~12	+3-13
– Steers 24 mo.	~23	~22	+2-10
'Productive land'	~13		
'Less productive land'	~10		
– Suckler system	~43	~25	+2-10
'Productive land'	~13		
'Non productive land'	~30		

Meat is given per kg weight of carcass.

than 'industrialized' poultry and pork. The carbon footprint of the red meat, however, is very must dependent on how the beef is produced – intensive beef from one year old calves being produced with the lowest carbon footprint whereas the suckler system results in the highest carbon footprint of around 27 kg CO_2eq per kg carcass. For reference, Desjardins *et al.* (2012) in a review of carbon footprint of beef found that beef cattle from Canada, US, EU, and Southern Australia had similar carbon footprints on average but with large regional differences, whereas the carbon footprint from cattle grazing under extensive conditions, such as Northern Australia and Brazil, were likely to have a higher carbon footprint.

In the numbers given above no consideration was given to how the production affects land use changes. As already mentioned requirements for feed is a main driver of land use changes, which in turn is a considerable source of CO_2 emissions. Thus, it is estimated that such continues land use changes are responsible for 12% of the world's yearly CO_2 emissions (World Resources Institute, 2009), because huge stocks of carbon in vegetation and soil are released to the atmosphere. Some attempts have been made to include indirect land use changes in the assessment of the carbon footprint of the land based products. Eg. Audsley *et al.* (2009) argue that all land use results in a pressure on the global (limited) land resources, and consequently that all crops have to carry this burden proportionally. Audsley *et al.* (2009) found that on average every cultivated ha of land resulted in an emission of 1.43 ton of CO_2 as a results of indirect land use changes following that occupation. This corresponds to a load of 140 g CO_2 per occupied m^2 globally. In contrast, Schmidt *et al.* (2012) modeled the marginal effect of including 1 extra ha land for cropping, and found that this effect results in on average 7.83 t CO_2eq per occupied ha (depending on the productivity of the land used) or as a global average 783 g CO_2eq per m^2.

In Table 3 we illustrate the impact on the carbon footprint of including this effect of indirect land use changes using the value from Audsley *et al.* (2009) as the most conservative estimate and the results from Schmidt *et al.* (2012) as the highest estimate. Using the lowest estimate means that the carbon footprint of most products are increased by 25-30%, while using the highest value increase the carbon foot print by more than 100% for the non-grazing based systems, and for the grass based systems the increase is about 50%. The estimates by Schmidt *et al.* (2012) seems close to the values found by Laborde (2011), which are used to evaluate the consequences of increased use of land for energy purpose as recommended for use in the EU. Thus, in a situation where the meat production does expand rapidly globally, it seems most reasonable to include this high value in the overall assessment.

In the assessments given in Table 3, we distinguished between use of productive cropland (including cultivated grassland) and less productive grass land that in fact does not represent a resource for other biomass production purposes. This separation of land use in the different systems is of course not universal, but was specific for the case investigated. However, this distinction has a huge impact on the assessment of the environmental impact as illustrated in Table 3, when indirect land use changes are included.

Like there is a close connection between land use and impacts on global warming, when taking indirect land use changes into account, the same is true when considering impacts on biodiversity. Impacts on biodiversity are challenging to describe satisfactorily. However, some concepts are present to support assessment of land use related impacts on biodiversity. Thus, De Schryver *et al.* (2010) established a framework where the damage to biodiversity was expressed as Potential Disappeared Fraction (PDF) of vascular species compared to a baseline that would be the natural vegetation of the area in question. This characterization factor is unit-less and can be aggregated according to different types of m^2 of land occupied for a given product. Based on UK conditions they established the PDF's as given in Table 4.

It is clear from Table 4 that while use of land for intensive cultivation has a huge negative impact on biodiversity, use of land for grazing, and in particular grazing on low fertilized land has a much lower negative impact on biodiversity and may even enhance biodiversity compared to a semi natural forest. Thus, such aspects need to be brought into the picture when assessing land use requirement of different livestock products, but as also mentioned the methodology is only in its virgin stage. It is obvious, that a huge reduction in complexity takes place when reducing the biodiversity aspect to the species richness of vascular plants, but on the other hand this indicator clearly catches important elements of biodiversity, since most often there is a correlation between species richness of vascular plants and species richness of insects and mammals (e.g. birds).

While no doubt the industrial livestock systems to a wide extent is based on feed that compete with foods for humans directly or indirectly, and in this way is in a land competition situation and causing land use changes, this is not always the case for other systems. Thus, systems that primarily depend on grazing compete less with resources that could be used for human food, and such systems produce about 12% of global milk and 9% of global meat production (FAO, 2011). Considering the contribution of livestock products to the protein supply, FAO estimated the edible protein output/input ration for different countries. In countries with huge landless production of monogastrics, like USA and Germany, this ration was less than 1 (0.5-0.6), in Brazil and China approximately 1, in India around 4 and in Mongolia and East African countries beyond 10 (FAO, 2011). In a huge country like India this positive protein balance supplied by smallholder livestock systems amount to 3.4 mill ton which can be compared to a protein 'deficit' of livestock in USA and Germany of approximately 8.9 mill ton.

The estimations as mentioned above are related with huge uncertainties, but it is important to have this overall picture in mind when considering the efficiency of livestock production in feeding the planet. Also, as pointed out by Smith *et al.* (2013) in most mixed crop-livestock system used by small holders the main animal feed consists of crop residues. When considering the challenges and prospects in the future, it is important that these features of such systems are not deteriorated, but in fact enhanced.

Table 4. Damage to biodiversity expressed as Potential Disappeared Fraction (PDF) of vascular species for different types of land use (modified after De Schryver et al.*, 2010).*

Type of land use	Median	Min (2.5%)	Max (2.5%)
Baseline (semi-natural forest)	0.00		
Grazing organic infertile grassland	-0.33	-0.56	0.13
Grazing organic fertile grassland	-0.01	-0.18	0.15
Grazing less intensive fertile grassland	0.36	0.14	0.52
Organic arable land	0.36	0.15	0.51
Less intensive arable land	0.44	0.31	0.54
Intensive woodland	0.55	0.44	0.65
Grazing intensive fertile grassland	0.65	0.56	0.72
Intensive tall grassland	0.70	0.61	0.77
Intensive arable land	0.79	0.73	0.83

Challenges and prospects for different livestock production systems

Taking the point of departure in the different livestock systems presented in Table 2, the challenges and opportunities seem to differ substantially as also highlighted in the 'Global Agenda for Action in support of sustainable livestock sector development' (FAO, 2013).

Landless/industrial systems

Looking at the trend in production of different types of livestock, the increased demand for meat globally seems to a wide extent to be covered by pork and poultry products. Today these production systems are by large based on feed that heavily compete with human food (or produced on land which potential could be used for human food). A further expansion is foreseen to have significant negative consequences in relation to land use changes and the connected effects such as global warming and reduction of global biodiversity, and to be in sharp competition with the request for biomass for other purposes. On the other hand given the expected growth in human wealth in the long term in transition and developing countries, this increased demand will no doubt substantiate, and thus this form of livestock production must be foreseen to get a steady more significant role in supplying the growing population's nutritional needs.

In addressing this challenge, the production efficiency needs to be considered further. Looking at the livestock part in itself, substantial improvements have been obtained in reducing feed use per kg of meat produced during the last decades. This is still an important issue, but two issues should be given more focus: the feed stock supply and the waste treatment.

Regarding feed stock, at the moment significant efforts are put how best to use the byproducts from the biofuel industry as feed, like Dried Distillers Grains with Solubles (DDGS), which constitute a significant feed source. This is very important in aiming for lowering the environmental impact of the food produced, but the main feed needed still seems to be cereals and soybean, both crops with a relatively low yield of biomass per unit of land occupied. In temperate areas of the world mixtures of grass and legumes typically can produce 50% or more of both dry biomass and protein per ha compared to cereals, and – if legumes are included – typically with less resource use like fertilizer and pesticides. This biomass it not directly suited for monogastrics, but given the demand for biomass for a range of purposes, the introduction of bio-refinery processes may transform part of this biomass to high quality feed, leaving other parts of the biomass for other high quality products or biofuel. It is a challenge for animal nutritionists to contribute with insights and options for new feeds based on very high yielding crops that are not immediately suitable as feed.

Regarding waste treatment, still in many places the manure nutrients (50-70% of the input) are not efficiently used for plant production but represent an environmental overload. Further, approximately 30% of the energy input is present in the manure. Efficient use of this resource for biogas constitutes a major potential for counteracting the emission of greenhouses gasses of producing livestock products. Thus, we estimated that utilizing manure in the form of slurry from the pig production for biogas potentially reduced the carbon footprint of the pork meat by 30%, partly as a result of energy recovery and partly because less a greenhouse gas emissions were released from the manure (Nguyen *et al.*, 2010). Such an efficient utilization requires a good infra-structure. However, in particular for the systems in consideration with bigger and bigger enterprises, and mainly systems based on liquid manure, this should be possible to establish.

Mixed systems

A very huge part of the global food supply is produced in mixed systems, not least for milk and beef. One can distinguish between large scale mixed systems in temperate areas of the developed world and small scale mixed systems in the developing world. The large scale mixed systems in fact also often rely to a wide extent on feed that compete with human food, and in the development of such system to further fulfill their role in global food supply, the same aspects as described for land less systems should be addressed.

Regarding small scale mixed systems, the main challenge is with the words from the Global Agenda of Action in support of sustainable livestock development (FAO, 2013) to 'Closing the efficiency gap'. It is recognized that a number of technological possibilities or production concepts are yet not fully exploited in such systems. Mixed systems in the developing world is estimated to supply 65% of the beef, 75% of the milk and 55% of the lamb meat (Tarawali *et al.*, 2011) in these regions, where the demand for animal products (per capita and not least in total due to increased population) are expected to increase markedly. Thus, it is very important that the production is better optimized. Taraweli *et al.* (2011) pointed out a number of issues to be dealt with to allow a better optimization of such systems. Main issues are to allow for keeping fewer animals with higher productivity, thus turning more feed energy into production instead of maintenance. This includes better feeding and understanding of importance of small feed supplements to local feed sources, use of better genotypes, and improved multipurpose crops, including legumes. While realizing that such changes are not easy since livestock has other important roles than as direct food in such communities, it is found imperative that intensification has to happen which results in more food produced without compromising use of natural resources.

Grazing systems

Approximately 50% of the land used for livestock production is extensively used grassland for grazing. While the supply in absolute terms to global animal based foods are limited, these systems support nutrition and livelihood of many vulnerable and food insecure people. In fact livestock provide more food security in arid regions than do crop production (Kratli *et al.*, 2013). Due to climatic conditions, it is not likely that the biomass yield from such systems can be increased, except in cases where the land was been partly degraded. It is estimated that at least 20% of such land are degraded due to inappropriate management resulting in lower biomass yield and high erosion. Thus the challenge for the extensive grassland based systems is on the one hand to improve grazing management by adapting the stocking rate to the carrying capacity, and hereby avoid further degradation. In fact better grazing practice may as a results have such areas to contribute significantly to global carbon sequestration (Wilkes *et al.*, 2012).

Conclusion

While the consumption of livestock products seems to have reached a plateau per person in developed countries, it is increasing rapidly in transition and developing countries. As a consequence of increased consumption per capita and a larger population, it is forecasted that global meat consumption will increase considerably. While the livestock sector provides high value food and other social functions, its resource use implications are however large. Livestock uses the major part of the agricultural land through grazing and consumption of feed crops, and plays a major role in emissions related to climate change and reduced biodiversity, which are all major concerns of today.

To overcome the challenges mentioned, it will be very important that the feeds can be based on very high yielding new crops that serve both as feed and other purposes in order to reduce land use requirement, and that animal nutritionist are active in exploring these possibilities. For dairy and beef production taking place in mixed systems, some of which are rather inefficient in their use if natural resources, it is imperative to increase efficiency having more of the feed consumed transformed into food instead of feed requirements for animal maintenance. Also extensive grazing systems need to be adapted to ensure stop of land degradation and enhance the role of grassland to sequester carbon and contribute to other eco-system services.

References

Audsley, E., Brander, M., Chatterton, J., Murphy-Bokern, D., Webster, C. and A. Williams, 2009. How low can we go? An assessment of greenhouse gas emissions from the UK food system and the scope for reducing them by 2050. WWF-UK.

De Schryver, A.M., Goedkoop, M.J., Leuven, R.S.E.W. and M.A.J. Huijbregts, 2010. Uncertainties in the application of the species area relationship for characterisation factors of land occupation in life cycle assessment. International Journal of Life Cycle Assessment 15, 682-691.

Desjardins, R.L., Worth, D.E., Verge, X.P.C., Maxime, D., Dyer, J. and D. Cerkowniak, 2012. Carbon Footprint of Beef Cattle. Sustainability 4, 3279-3301.

De Vries, M., and I.J.M. de Boer, 2010. Comparing environmental impacts for livestock products: A review of life cycle assessments. Livestock Science, 128, 1-11.

EC, 2012. EC (draft) Commission staff working document – Impact Assessment on indirect land use change related to biofuels and bioliquids, 125 p.

FAO, 2011. World livestock 2011 – livestock in food security. Rome, FAO 115 pp.

FAO, 2012. Livestock sector development for poverty reduction: An economic and policy perspective –Livestock's many virtues. FAO, Rome.

FAO, 2012. World food and agriculture in review. In: The State of Food and Agriculture 2012, p. 97-106.

FAO, 2013. Global Agenda for Action http://www.livestockdialogue.org/.

FAOSTAT, 2013. FAOSTAT. Food and Agriculture Organization of the United Nations (FAO). Online at http://faostat.fao.org/site/610/default.aspx#ancor.

Hermansen, J.E. and Kristensen, T., 2011. Management options to reduce the carbon foot print of livestock products. Animal Frontiers 1, 13-19.

Hermansen, J.E. and Olsen, J., 2009. (Climate impacts from Danish agriculture and food production (In Danish)) Vand og Jord 16,137-139.

Herrero, M., Thornton, P.K., Notenbaert, A.M., Wood, S., Msangi, S., Freeman, H.A., Bossio, D., Dixon, J., Peters, M., van de Steeg, J., Lynam, J., Parthasarathy Rao, P., MacMillan, S., Gerard, B., McDermott, J., Seré, C. and M. Rosegrant, 2010. Smart investments in sustainable food production: Revisiting mixed crop-livestock systems. Science 327, 822-825.

IFPRI, 2012. The meat of the issue. IFPRI Insights Magazine, Magazine of the International Food Policy Research Institute, October 26, 2012. Online at http://insights.ifpri.info/2012/10/the-meat-of-the-issue-2/.

Krätli, S., Huelsebusch, C., Brooks, S. and B. Kaufmann, 2013. Pastoralism: A critical asset for food security under global climate change. Animal Frontiers 3, 42-50.

Laborde, D., 2011. Assessing the Land Use Change Consequences of European Biofuel Policies. Final Report, October 2011, IFPRI, 111 pp.

Msangi S. and M.W. Rosegrant, 2011. Feeding the future's changing diets. Implications for agriculture markets, nutrition, and policy. 2020 Conference: Leveraging Agriculture for Improving Nutrition and Health. February 10-12, 2011, New Delhi, India. 12 p.

Nguyen, T.L.T., J.E. Hermansen, Horsted, K., 2011. Design of innovative pork chains with low environmental load. Deliverable submitted in the project: Improving the quality of pork and pork products for the consumer. 36 pp.

Nguyen, T.L.T., J.E. Hermansen, and L. Mogensen, 2010. Environmental consequences of different beef production systems in the EU. Journal of Cleaner Production 18, 756-766.

Nguyen, T.L.T., Hermansen, J.E. and L. Mogensen, 2010. Fossil energy and GHG saving potentials of pig farming in the EU. Energy Policy 38, 2561-2571.

Nguyen, T.L.T., Hermansen, J.E. and L. Mogensen, 2011. Environmental assessment of Danish Pork. Internal Report, Faculty of Agricultural Sciences, Aarhus University, Aarhus, Denmark. 34 pp.

Schmidt, J.H., Reinhard, J. and B. Weidema, 2012. A model of indirect land use change. 8th international conference on LCA in the agri-foodsector, Rennes, France, 2-4 October 2012.

Smith, J., Sones, K., Grace, D., MacMillan, S., Tarawali, S. and M. Herrero, 2013. Beyond milk, meat, and eggs: Role of livestock in food and nutrition security. Animal Frontiers 3, 6-13.

Steinfeld, H., Gerber, P., Wassenaar, T., Castel, V., Rosales, M., and C. de Haan, 2006. Livestock's long shadow – environmental issues and options. FAO. 390 pp.

Tarawali, S., Herrero, M., Descheemaeker, K., Grings, E. and M. Blummel, 2011. Pathways for sustainable development of mixed crop livestock systems: Taking a livestock and pro-poor approach. Livestock Science 139, 11-21.

UNEP, 2011. Keeping track of our changing environment. From Rio to Rio +20. United Nations Environmental Programme, Nairobi, Kenya 96 pp. www.unep.org.

Wilkes, A., Solymosi, K. and T. Tennigkeit, 2012. Options to support grassland restoration in the context of climate change mitigation. Report commissioned by FAO. Freiburg, Germany 56pp.

World Resources Institute, 2009. World Greenhouse Gas Emissions: 2005. Online på http://www.wri.org/chart/world-greenhouse-gas-emissions-2005.

Part 1. Energy and protein interactions, ruminants

Challenges in ruminant nutrition: towards minimal nitrogen losses in cattle

J. Dijkstra[1], C.K. Reynolds[2], E. Kebreab[3], A. Bannink[4], J.L. Ellis[1], J. France[5] and A.M. van Vuuren[4]
[1]Animal Nutrition Group, Wageningen University, 6700 AH Wageningen, the Netherlands; jan.dijkstra@wur.nl
[2]Animal Science Research Group, School of Agriculture, Policy and Development, University of Reading, Reading RG6 6AR, United Kingdom
[3]Department of Animal Science, University of California, Davis, CA 95616, USA
[4]Wageningen UR Livestock Research, P.O. Box 65, 8200 AB Lelystad, the Netherlands
[5]Department of Animal and Poultry Science, University of Guelph, Guelph ON, N1G 2W1, Canada

Abstract

Ruminants play a key role in human food production by converting fiber-rich plant resources that humans cannot (or choose not to) consume into high-quality food that humans can eat. However, this conversion causes unavoidable losses of nitrogen (N) in feces and urine from ruminants that may become an environmental burden, in particular nitrate (NO_3^-) leaching, ammonia (NH_3) volatilization and nitrous oxide (N_2O) emissions. The aim of this paper is to identify the maximal theoretical N efficiency at the animal level, and the challenges and opportunities to achieve this maximal N efficiency. This is done via striving for the lowest possible N excretion in urine and feces, and for purposes here, with a focus on dairy cattle. Inevitable N losses in dairy cattle include losses associated with urinary excretion of urea synthesized from ammonia produced in the rumen; undigested microbial protein excreted in feces; microbial nucleic acids synthesized in the rumen and excreted mainly in urine; fecal and urinary excretion resulting from endogenous secretions; and urinary excretion related to maintenance and milk protein synthesis. The theoretical upper limit of N use efficiency in a dairy cow producing 40 kg fat and protein corrected milk/d is 0.43. Higher efficiencies may be achieved, but these require major inputs of human edible resources. The present analysis demonstrates there is little or no scope to reduce N losses related to microbial nucleic acid synthesis, recycling of N to the rumen, intestinal digestion of microbial protein, and animal maintenance requirements. Strategies to reduce N losses and improve N efficiency should focus on an optimal supply of rumen degradable N and optimal efficiency of utilization of absorbed amino acids for milk protein synthesis. To improve N efficiency, integration between protein and energy metabolism is essential, and energy and protein should be considered together rather than as two distinct entities. A major challenge in strategies to optimize high-fiber diets for high milk N efficiency will be to avoid increases in enteric methane production associated with these dietary strategies.

Introduction

Global consumption of dairy products and beef is projected to rise by well over 50% by 2050 (FAO, 2011). In the face of competing demands for resources, ruminants play a key role in human food production in converting plant resources that humans cannot (or choose not to) consume into high-quality food that humans can eat. For dairy cattle, the return on human edible protein inputs (calculated as the output of human edible protein in products compared with human edible protein input with feed) is larger than 1 (range: 1.4 to infinite; infinite if diet contains no human edible protein) (reviewed by Dijkstra *et al.*, 2013a), indicating that dairy cattle add to the total human food supply. For beef cattle, protein efficiencies on a human-edible basis are also often larger than 1 but are more variable (range: 0.33 to infinite) than for dairy cattle. The ability of ruminants to turn fibrous feed resources into edible animal food of high biological value is likely to become of greater significance in terms of global human food production as the population of the planet and demand for human-edible plant resources increases rapidly.

Despite the positive contribution of ruminants in terms of food security, there is a widespread perception that the production of milk or meat is an inefficient way to use natural resources. In cattle production systems, simple overall efficiency (amount of product produced versus amount of feed consumed) is well below 1. The general reliance of cattle on forages and fibrous co-products as feeds and the role of fermentation in their digestion, results in lower overall efficiencies than in pig or poultry systems (Reynolds *et al.*, 2011). The low overall efficiency of feed utilization in cattle is also a major determinant of their environmental impact, and the need for improved efficiency in milk or meat production to minimize environmental impact is evident.

Among the main environmental concerns is the emission of nitrogen (N) to the environment. Major losses of N occur in cattle production systems, including nitrate (NO_3^-) leaching, ammonia (NH_3) volatilization, and nitrous oxide (N_2O) emissions (Steinfeld *et al.*, 2006). The principal driver of N losses from cattle is N intake. Variation in dietary N supply will affect urinary N output in particular and urine N is more susceptible to leaching and volatile losses than fecal N. Thus, reducing urinary N excretion will reduce environmental impact significantly. The efficiency of conversion of feed N into milk or meat N in cattle varies widely (Figure 1). Such large variation suggests that major improvements in reducing N excretion in feces and urine are possible. The aim of this paper is to identify maximal N efficiency possible at the animal level, and the challenges and opportunities to achieve this maximal N efficiency through lowest possible N excretion in urine and feces, keeping in mind the role of cattle in utilizing human inedible feed resources. The present paper focuses on dairy cattle, but the main concepts apply to beef cattle and other ruminants as well.

Theoretical maximum in nitrogen efficiency

Production of milk or meat causes some avoidable and some inevitable losses of N in feces and urine. The N use efficiency of a dairy cow is the quantity of feed N consumed that is captured in milk, also called the milk N efficiency (MNE). Van Vuuren and Meijs (1987) calculated an MNE of 0.40 to 0.45 in an ideal situation of a 600-kg cow producing 25 kg milk/d with 33 g protein/kg milk and fed on a well-balanced diet, with inevitable losses in feces and urine being 115 and 55 g N/d, respectively. Assuming a daily DM intake of ~18 kg/d, the dietary crude protein (CP) content in this optimal situation is ~105 g/kg DM. Van Vuuren and Meijs (1987) did not include inevitable inefficiency in rumen microbial protein synthesis (MPS) though, and assumed all microbial protein is digested. Therefore the actual maximal MNE may be lower than 0.40 to 0.45.

An updated calculation of the inevitable N losses associated with milk protein production in dairy cattle is presented in Table 1. These unavoidable losses are calculated for a reference cow of 650 kg live weight producing 40 kg/d of fat and protein corrected milk (FPCM) with a true protein content of 31.5 g/kg. Dry matter intake (DMI) is 24.1 kg/d and diet net energy content 6.9 MJ/kg DM. Total tract diet digestibility is 0.80 and dietary content of rumen fermentable organic matter (RFOM) 0.55 kg/kg DM. The estimated minimal N losses in feces and urine are 89 and 174 g/d, respectively, and the theoretical upper limit of MNE is 0.43 (Table 1). Usually, MNE is well below this maximal attainable value. Calsamiglia *et al.* (2010) reported MNE in the lowest and highest quartile being 0.22 and 0.33 in a US dataset (n=167) and 0.21 and 0.32 in a European dataset (n=287), respectively. The mean MNE in the large database presented in Figure 1 is 0.26 (SD, 0.072). The challenge is thus to improve MNE to be closer to the theoretical maximum efficiency of 0.43. Although an increase in N intake level increases milk N output, it is not related (P>0.05) to MNE (Figure 1). However, dietary N content is related to MNE, with a slope of -0.00466 (SE, 0.00145). These results indicate the significance of level of feed intake on N efficiency. Occasionally, MNE values larger than 0.43 have been reported. These high efficiencies tend to be associated with body protein mobilization, as indicated by the relationship between milk N efficiency and N balance (Figure 1). A negative N balance may occur for several days or weeks, in particular in early lactation (Van Knegsel *et al.*,

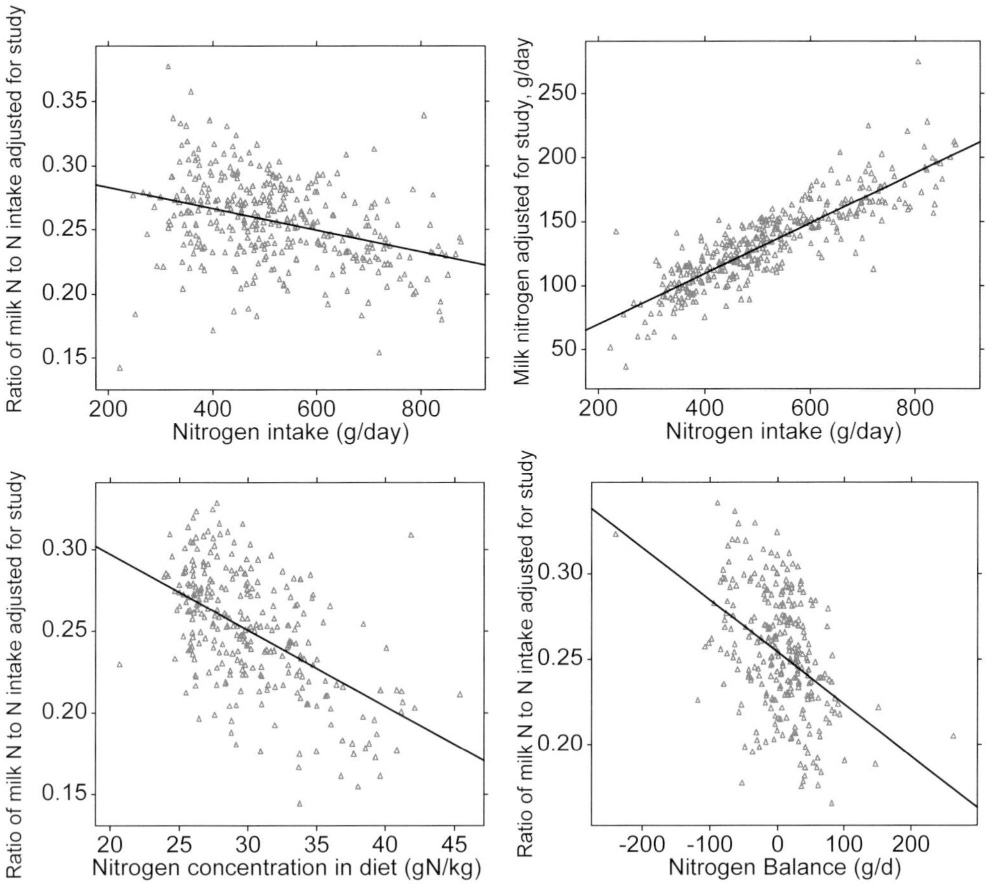

Figure 1. Univariate relationships between N intake (g/d) or N content in the diet (g/kg DM) and N output in milk (g/d) and milk N efficiency (N output in milk divided by N input in feed). The responses have been adjusted for study effect. Nitrogen balance data of individual cows (n=470) from Kebreab et al. *(2010). The equations describing the relationships were: Milk N = 30 (SE, 10) + 0.20 (SE=0.03) N intake; Ratio of milk N to N intake = 0.30 (SE=0.033) – 0.000075 (SE=0.000062) N intake; Ratio of milk N to N intake = 0.39 (SE=0.048) – 0.0047 (SE=0.0015) N concentration; Ratio of milk N to N intake = 0.25 (SE=0.0097) – 0.00031 (SE=0.000064) N balance.*

2007), but has to be replenished later in the lactation cycle. In the present approach, zero body N balance is assumed.

The main areas in which N losses in dairy cattle occur include urinary excretion of urea synthesized from ammonia lost from the rumen; undigested microbial true protein excreted in feces; microbial nucleic acids synthesized in the rumen and excreted mainly in urine; fecal and urinary excretion resulting from endogenous secretions; and urinary excretion related to maintenance, milk protein synthesis and amino acids (AA) absorbed in excess of requirement. These losses are discussed below.

Table 1. Inevitable N losses (g N/day), N output in milk (g N/day) and theoretical maximum milk N efficiency in a dairy cow producing 40 kg fat and protein corrected milk per day and a true protein content of 31.5 g/kg.

	N in feces	N in urine	N in milk
Inefficiency rumen microbial protein synthesis		35	
Rumen microbial nucleic acids	13	71	
Undigested microbial protein	37		
Endogenous secretions	39	19	
Maintenance		13	
Milk		36	198
Total	89	174	198
Maximum milk N efficiency			0.43

Nitrogen losses related to rumen microbial protein synthesis

Sources of N for microbial protein synthesis

Ammonia is the most important source of N for protein synthesis in the rumen. Satter and Slyter (1974) observed that ammonia-N concentration of ~3 mM is sufficient to achieve maximal MPS in continuous cultures. In vivo, higher ammonia-N concentrations (between 6 to 18 mM) are required to maximize MPS (Reynal and Broderick, 2005). Because ammonia is washed out from the rumen with the fluid phase and is absorbed through the rumen wall, such levels of ammonia-N to achieve optimal MPS will always result in some losses of N. Assuming in the reference dairy cow a rumen volume of 80 l, a fractional liquid passage rate of 0.15/h and a fractional ammonia absorption rate of 0.50/h, the minimal ammonia concentration of 6 mM gives rise to a loss of 105 g/d ammonia-N from the rumen. If 75 g urea-N/d is recycled to the rumen (discussed later), the net unavoidable N-loss is at least 30 g N/d which is excreted as urea-N in the urine.

Although ammonia-N may be the sole source of N for MPS, benefits of supplying preformed AA and peptides on MPS and MPS efficiency are well established. Based on stoichiometric principles, Dijkstra *et al.* (1992) calculated that the efficiency of synthesis of microbial CP with preformed AA and peptides, compared with ammonia, is ~20% higher when expressed per unit OM utilized and ~75% higher when expressed per unit carbohydrate utilized. Russell and Sniffen (1984) observed in vitro that addition of trypticase to a medium with ammonia as sole N source increased MPS efficiency per unit carbohydrate according to a pattern of diminishing returns, with efficiency being 54% higher at the highest level of trypticase addition. However, when expressed per unit OM utilized, microbial efficiency improved by 19% up to a level of trypticase being 0.14 of total OM, whereas efficiency declined with further trypticase additions. Maximal MPS efficiencies per unit OM when peptide supply is balanced with requirement have been confirmed in other in vitro experiments (e.g. Brooks *et al.*, 2012). Thus, supply of preformed AA or peptides to rumen microbes in addition to supply of ammonia initially increases microbial efficiency, but after reaching an optimum further AA and peptide additions decrease efficiency. At the highest microbial efficiency level, a part of the AA and peptides supplied are fermented to ammonia, increasing rumen fluid ammonia concentrations and adding to the net unavoidable N-losses from the rumen of 30 g N/d calculated above.

In a different approach, Bach *et al.* (2005) analyzed the relationship between MPS and efficiency of N use in the rumen (amount of microbial N produced per unit available N) in continuous culture fermenters using mixed model analysis adjusting for study effect. They observed a quadratic relationship between N use efficiency and efficiency of MPS, with an optimum efficiency of growth obtained at efficiency of 29 g microbial N/kg RFOM and rumen N use efficiency of 0.69 (Figure 2). In various continuous culture studies higher N use efficiencies (~0.80) were obtained. To calculate the lowest level of rumen degradable N (RDN) required and minimal N losses related to microbial growth, the N use efficiency relationship presented by Bach *et al.* (2005) was arbitrarily increased by 10%. Recycling of urea to the rumen contributes to the supply of RDN, and the relationship between N intake and urea-N recycled to the gastro-intestinal tract of dairy cattle derived by Reynolds and Kristensen (2008) was adopted to calculate the contribution of dietary RDN to total RDN required. Up to 75% of urea-N recycled to the gastro-intestinal tract enters the forestomachs (Lobley *et al.*, 2000). For the reference cow, the lowest loss of N calculated based on these assumptions was 35 g/d at an efficiency of 25.2 g microbial N/kg RFOM and a calculated apparent efficiency of use of dietary N to microbial N of 0.90. This N loss is in broad agreement with the net unavoidable N loss from the rumen (>30 g/d) calculated before. For the reference cow, to avoid such losses dietary RDN content should only be 15 g/kg DM or 94 g CP/kg DM. This dietary RDN level is lower than levels recommended in various evaluation systems [e.g. the NRC (2001) recommendation on rumen degradable protein is ~95 to 105 g/kg DM] and lower than used in practice. Surplus rumen degradable protein-N is excreted as urea-N in urine. Urea-N is the most variable component in urine, contributing from 50 to well over 90% of all N in urine (Dijkstra *et al.*, 2013b). Current protein evaluation systems have only limited value in providing guidance to reduce these losses, and drawbacks of these systems are well recognized (e.g. Dijkstra *et al.*, 2007; Hanigan, 2005). Single protein values for a feed quote static conditions, whereas the protein supply depends on a number of interrelated factors, including site of digestion and level of energy substrates available. Energy and protein systems have been developed independently, despite vast evidence of the interactions between energy- and protein-yielding nutrients. Protein evaluation systems do not predict how performance, in terms of production and excretion, will change in response to deliberate changes in feeding strategy. To improve upon current prediction schemes, new feeding systems need to be based on mechanisms that govern the response of animals to nutrients, by quantitatively describing metabolite supply at a more detailed level than the current aggregated components.

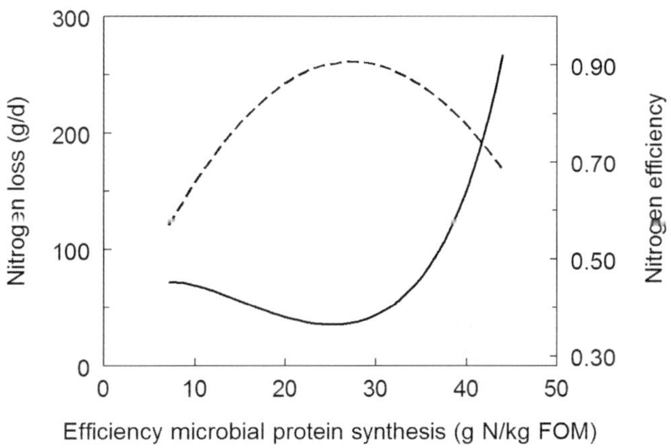

Figure 2. Relationship between efficiency of microbial crude protein synthesis (g microbial N/kg fermented organic matter) and N loss from the rumen (g/d) (solid line) and N efficiency (g microbial N/g dietary rumen available N) (dashed line) in a reference cow producing 40 kg fat and protein corrected milk/d. The relationship between microbial synthesis and N efficiency (g microbial N/g total available N) was based on Bach et al. *(2005). Dietary rumen available N is the difference between total available N and urea-N recycled to the rumen, with the latter based on Reynolds and Kristensen (2008). See text for further details on calculations.*

Nitrogen recycling to the rumen

The use of non-protein N (NPN) sources, including ammonia originating from recycled urea, for MPS in the rumen enables ruminants to survive and produce milk and meat on diets that are based solely on NPN sources (Virtanen, 1966). The ability of rumen microorganisms to synthesize protein from urea recycled to the rumen could, in theory, compensate for some of the potential loss of N that occurs through degradation of protein in the rumen and metabolism of AA absorbed from the gut. In order to take advantage of this potential, endogenous N sources must be transferred to the forestomachs and microbial protein synthesized, digested and absorbed. On diets with a relatively high ratio of rumen fermentable energy to rumen degradable protein, the flow of non-ammonia N into the duodenum can be higher than dietary N intake, reducing the amount of N excreted in urine through reduced ammonia absorption and urea synthesis. This high duodenal non-ammonia N flow is possible when microbes in the rumen use more ammonia from recycled urea to synthesize microbial CP, than ammonia produced from fermentation of dietary N sources. Thus the recycling of N to the rumen and subsequent incorporation of recycled N into microbial protein may well be central to the success of diets of low N content in dairy production (Calsamiglia *et al.*, 2010).

Cattle are particularly adept at using as much intake N as possible. Depending on dietary N concentration, between 30 and almost 100% of all urea entering the blood pool is returned to the gut (Reynolds and Kristensen, 2008). Urinary N excretion still occurs even when 100% of urea is returned to the gut, but the N is then excreted primarily in forms other than urea, including purine derivatives (that may act as a marker of MPS), hippuric acid, creatine and creatinine (review Dijkstra *et al.*, 2013b). The transfer of urea-N from blood to the gut is via saliva and across the gut epithelia. In general, this transfer is positively related to blood urea concentration and rumen fermentable energy supply, and negatively to rumen ammonia concentration (Reynolds and Kristensen, 2008). Although salivary and epithelial urea transport is adaptable to the N-status of the ruminant, when expressed in absolute amounts, urea-N transport to the gut is little affected by changes in dietary N content. Røjen *et al.* (2011a,b) reported that various urea transporters are expressed in bovine rumen papillae. However, none of the investigated transcripts or proteins correlated with the increased rumen epithelial urea permeability observed with low dietary N concentration. Reduced N supply increased the relative extraction of arterial urea-N across the rumen and portal-drained viscera with an overall shift towards the rumen. Adaptation of epithelial urea-N extraction with decreasing N supply compensated for decreasing arterial concentration of urea-N, and thus a rather constant transport of urea across treatments was observed. Thus, cows appear largely unable to up-regulate urea-N transport sufficiently to fully compensate for the N removed from the diet and to substantially increase MNE. In the present calculations, a linear relationship between N intake and urea-N transfer to the gut with a slope of 0.12 g urea-N per g dietary N intake (Reynolds and Kristensen, 2008) was adopted.

Microbial nucleic acids

Microbial N is composed mostly of AA-N and nucleic acid-N. Synthesis of microbial nucleic acids represents an irreversible loss of N excreted largely in urine (Dijkstra *et al.*, 2013b). Pyrimidines of microbial or endogenous origin undergo ring cleavage, and the major end products of catabolism are β-amino acids, ammonia and CO_2. Various derivatives of purine metabolism are present in urine, viz. allantoin, uric acid, xanthine and hypoxanthine. Some 20-25% of rumen microbial N is present as nucleic acid-N (Dijkstra *et al.*, 1992; NRC, 2001). There is virtually no scope to reduce losses of N related to nucleic acid synthesis in the rumen. Nucleic acid contents of rumen microorganisms are smallest at low fractional growth rates (Bates *et al.*, 1985). However, in general high fractional growth rates are required to reduce maintenance cost of microbes as a fraction of total energy expenditure (Pirt, 1965) and consequently to maximize efficiency of MPS. The true ileal digestibility of nucleic acids is 0.81 to 0.87 (Storm *et al.*, 1983). Thus, rumen microbial synthesis of nucleic acids leads to considerable losses of N particularly in urine. For the reference cow, assuming high MPS efficiency and thus a high nucleic acid N content (25% of total microbial N), these inevitable N losses are 71 and 13 g N/d in urine and feces, respectively.

Digestion of microbial true protein and escape protein

The N-containing compounds entering the small intestine comprise mainly microbial protein and nucleic acids produced in the rumen and feed protein that has escaped degradation in the rumen. The digestibility of these fractions in the small intestine determines the amount of N absorbed as well as that excreted with feces. There is little opportunity to improve the intestinal digestibility of microbial and most rumen undegraded protein sources. In various protein evaluation systems used worldwide, true digestibility of microbial true protein in the small intestine is 80-85%. These values are largely based on digestibility of microbial AA in the small intestine of sheep (Storm *et al.*, 1983). In dairy cattle, small intestinal digestibility of microbial AA-N was 75-77% (Larsen *et al.*, 2001). The true digestibility of undegraded escape protein is variable but may be nearly 100% (Vérité and Peyraud, 1989). For the reference cow, assuming complete digestion of rumen escape feed protein and 85% of microbial true protein, fecal N losses are 37 g/d.

N losses related to endogenous secretions

Amino acids required for endogenous protein synthesis are related to replacement of sloughed epithelial cells and release of digestive enzymes (Tamminga, 1992). These losses are not restricted to the excretion of metabolic fecal N. Reabsorption of AA originating from endogenous losses will mask its importance, because resynthesis of protein to replace endogenous losses is associated with significant losses of energy and AA. Endogenous secretions are mainly related to the quantity of DM or OM passing through the intestinal tract or to the undigested DM or OM excreted in feces (NRC, 2001; Van Duinkerken *et al.*, 2011; Vérité and Peyraud, 1989). Endogenous protein losses including microbial protein synthesized from urea transferred into the hindgut may vary between 60 and 90 g/kg of fecal DM output (Tamminga, 1992). In the present approach, efficiency of utilization of absorbed AA for endogenous protein of 0.67 and gross synthesis of 12 g endogenous N/kg fecal DM output are assumed (Van Duinkerken *et al.*, 2011). For the reference cow, inevitable N losses in feces and urine are 39 and 19 g/d, respectively. In general, fiber rich diets give rise to higher endogenous losses than diets with high levels of non-structural carbohydrates. A main challenge in the conversion of human-inedible feed resources into milk or meat in cattle whilst improving N use efficiency, is to use fiber rich yet highly digestible diets. For example, a change of total DM digestibility from 0.80 to 0.90 improves maximal MNE from 0.43 to 0.46.

N losses related to maintenance and milk protein synthesis

Losses of N related to maintenance requirements are largely unavoidable and include scurf, skin secretions and hair, and tissue maintenance requirement losses. In various protein evaluation systems, endogenous losses (already discussed) may or may not be part of maintenance requirements. Except for endogenous N losses, N losses due to maintenance are relatively small (Tamminga *et al.*, 1992). For the reference cow, this N loss is 13 g/d and is excreted mainly in urine, with a minor part being losses in hair, scurf and skin secretions. No substantial improvement in MNE can be achieved by trying to reduce maintenance requirements.

The utilization efficiency of absorbed AA for milk protein synthesis assumed in various protein evaluation systems is 0.64 to 0.68. Even though this is below the theoretical maximum efficiency of 0.85, in experiments much lower efficiencies are usually observed. In the AFRC (1992) approach, the efficiency of AA utilization for milk protein when AA supply limits animal performance is the result of two efficiency factors. The first factor is the efficiency with which an 'ideal' AA mixture is utilized, which is assumed to be an animal characteristic and is set at 0.85. The second factor is the biological value or the extent to which the absorbed AA mixture differs from the ideal one, which is a characteristic determined mainly by diet composition. The default biological value is 0.80 for lactation (AFRC, 1992). The multiplication of both factors results in an efficiency conversion factor of 0.68. For the reference cow, minimal N losses in conversion of absorbed AA to milk protein occur when an ideal AA mixture is assumed, and are 36 g/d.

Usually, factors other than absorbed AA supply limit milk protein production, and the efficiency of absorbed AA utilization is lower than the theoretical maximum. Based on relationships between supply of AA absorbed in the small intestine and milk protein yield across a number of experiments, this efficiency was 0.64 when the lowest AA allowance giving the greatest protein yield was regressed (Cant, 2005). However, when the data were adjusted for trial effects, the marginal efficiency averaged only 0.24. In a meta-analysis, the transfer efficiency of post-ruminal casein or AA with casein profile infusions on milk protein yield showed a quadratic pattern, and maximum marginal efficiency was still only 0.38 (Lapierre *et al.*, 2010). These estimates of AA utilization efficiency reflect the utilization of increments of metabolizable protein (AA) above and below specific requirements (Reynolds, 2001), and have a major impact on overall MNE. An efficiency of absorbed AA utilization for milk protein of 0.85 (reference cow), 0.64 (protein evaluation systems) and 0.38 (maximum in experiments) results in maximal MNE of 0.43, 0.37 and 0.26, respectively.

Intermediary metabolism of AA between the duodenum and the mammary gland explains the decreased efficiency of transfer of absorbed AA into milk protein as maximal yield is approached. On average, 0.65 of estimated digestible AA was recovered in the portal vein, with the loss (0.35) due to endogenous secretions (discussed previously) and oxidation of AA across the gut wall (Lapierre *et al.*, 2006). The magnitude of this loss is not uniform among AA and varies between less than 0.05 for histidine to more than 0.90 for some non-essential AA. Next, the liver removes on average 0.45 of portal absorbed AA, again with considerable variation between individual AA. Liver fractional extraction of AA relative to portal absorption increases at higher AA supply. Finally, the utilization of post-liver available AA by the mammary gland is relatively high but again variable, which implies that improvements can be made. A considerable part of variation in AA use by splanchnic tissues and mammary gland is related to the supply of energy, and thus AA requirements, and further details can be found in various reviews (Hanigan, 2005; Lapierre *et al.*, 2010). Thus, the use of constants (0.64 to 0.68) for absorbed AA utilization for milk protein synthesis in most current protein evaluation systems appears inadequate in view of the decreasing efficiency with increasing supply of absorbed AA observed. Mechanistic models at the organ level that integrate the metabolism of energy substrates and individual AA enable incorporation of the rapidly expanding experimental knowledge on post-absorptive AA dynamics, and help to obtain a better quantitative understanding to further reduces post-absorptive N losses. Ultimately, the efficiency of absorbed AA utilization is determined by their requirements by the mammary gland, which is modulated by stage of lactation, genetics, and energy supply (Reynolds, 2001).

Integration: nitrogen losses, energy substrates and methane production

The present approach has confirmed that there is little or no scope to improve MNE and reduce N losses related to microbial nucleic acids, recycling of N to the rumen, intestinal digestion of microbial protein, and animal maintenance requirements. Strategies to reduce N losses should focus on an optimal supply of RDN and optimal efficiency of absorbed AA utilization for milk protein synthesis. A major challenge here is to achieve such reductions using human inedible feed resources. In theory, if the reference dairy cow is fed a diet without RFOM (digestion to occur in intestine only) whilst maintaining the default total tract digestibility of 0.80, maximal MNE increases from 0.43 to 0.65. Thus, the use of fibrous, human inedible feed resources by microorganisms in the forestomachs occurs at a cost, which is amongst others unavoidable N loss associated with microbial metabolism and a reduced MNE.

In striving for optimal supply of RDN and optimal efficiency of utilization of absorbed AA, the proper supply of energy yielding nutrients is crucial. If the supply of energy to rumen microbes increases whilst RDN supply does not change, less ammonia-N is formed and lost as urea-N in urine. Similarly, increased supply of metabolizable energy (ME) may reduce N losses in post-absorptive tissues. For example, MNE of a low energy, high protein diet improved from 0.20 to 0.24 when either supply of protein was reduced or supply of energy was increased, and improved to 0.28 with both (Rius *et al.*, 2010). Data presented in Figure 1 were subjected to multivariate meta-analysis by Kebreab *et al.* (2010) and prediction of urinary N output, but not fecal N output, improved when metabolizability (ratio of ME to gross energy) or ME intake was added as co-variable to a model that already included N intake. The slopes were 0.56 g urine N/g diet N and −71.4 g urine N/MJ ME, indicating reduced urinary N output when dietary ME concentration increases. Thus, in attempts to improve MNE, changes in energy supply have to be considered. In view of an integrated approach to reduce greenhouse gas emissions on farm, it should be noted that mitigation options aimed at reducing urinary N excretion may result in elevated methane (CH_4) emission depending largely on the type of carbohydrate consumed with grass (Ellis *et al.*, 2012). Methane production declines if starch or digestible nutrients escaping rumen fermentation replace protein in the diet, but rises if dietary fiber levels increase. Dijkstra *et al.* (2011) estimated an increase of on average 0.30 g CH_4 per g urinary N decrease for various nutritional interventions with grass silage based diets aimed to improve MNE. Using standard emission factors for direct and indirect N_2O emissions, the estimated N_2O emission reduction (in CO_2 equivalents) resulting from decreased manure N output was more than offset by a rise in enteric CH_4 production (Dijkstra *et al.*, unpublished). Moraes *et al.* (2012) developed a linear programming model to formulate minimum cost diets when environmental policies are present. In their evaluations, imposing CH_4 restrictions increased N losses from the animal. In view of these results, a major challenge in reducing N losses from dairy cattle is to find an optimal nutritional balance without increasing enteric production of CH_4.

Improvements in nutrition, genetics and technology have dramatically improved the efficiency of milk production in the past decades, particularly in developed countries. Such improvements are due to increased yield per animal which dilutes the nutrient costs of maintenance. Although feed efficiency will significantly improve with increases in production level, maximal achievable MNE may not improve to a similar extent (Figure 3). For example, an increase in FPCM production from 6,000 to 10,000 kg/yr improves feed efficiency by 16% (from 0.78 to 0.65 kg feed DM/kg FPCM), and enteric CH_4 production per kg FPCM is lowered as well (Bannink *et al.*, 2011). However, the maximal MNE improves by only 5% (from 0.40 to 0.42). Thus, to reduce the amount of feed resources used per unit milk, further production gains may be effective, but diet composition options (as discussed previously) rather than further production gains may be preferred to improve MNE.

Conclusions

Reducing N output in urine from cattle is critical to reducing NO_3 leaching, NH_3 volatilization and N_2O emission and achieving an environmentally sustainable production. Large variation in N efficiency presents an opportunity to manipulate diets to improve N efficiency. Given the role of dairy cattle in conversion of human inedible resources into human edible high quality foods, milk N efficiencies higher than some 0.43 are unlikely to be achieved. There is little opportunity to reduce N losses related to incomplete digestion of microbial protein, to synthesis of microbial nucleic acids, and to animal maintenance requirements. Dietary strategies to reduce N losses should focus on an optimal supply of rumen degradable N and optimal efficiency of absorbed amino acid utilization for milk protein synthesis. Integration between protein and energy metabolism is essential, and energy and protein should be considered conjointly rather than as two distinct entities. Current protein evaluation systems predict requirements for dietary N, but these requirements are usually fixed or comprise linear approximations, and the systems do not predict responses in N efficiency to dietary changes. A major challenge in strategies to optimize high-fiber diets for high milk N efficiency will be to avoid increases in enteric CH_4 production associated with these dietary strategies.

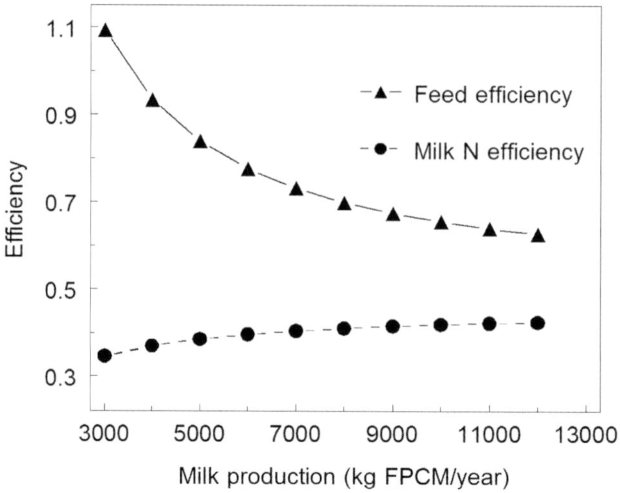

Figure 3. Relationship between milk production level (kg fat and protein corrected milk/year) and feed efficiency (kg feed/kg FPCM) or theoretical maximal milk N efficiency (kg N output in milk/ kg N input in feed). Feed efficiency calculated with the Dutch energy evaluation system and maximal milk N efficiency calculated as described in text.

Acknowledgements

Partially funded by the Commission of the European Communities (Rednex project FP7-KBBE-2007-1).

References

AFRC, 1992. Technical committee on responses to nutrients, report no. 9. Nutritive requirements of ruminant animals: Protein. Nutr. Abstr. Rev., Series B 62, 787-835.

Bach, A., S. Calsamiglia and M. D. Stern, 2005. Nitrogen metabolism in the rumen. J. Dairy Sci. 88, E9-E21.

Bannink, A., M.W. van Schijndel and J. Dijkstra, 2011. A model of enteric fermentation in dairy cows to estimate methane emission for the Dutch National Inventory Report using the IPCC Tier 3 approach. Anim. Feed Sci. Technol. 166-167, 603-618.

Bates, D.B., J.A. Gillet, S.A. Bargo and W.G. Bergen, 1985. The effect of specific growth rate and stage of growth on nucleic acid-protein values of pure cultures and mixed ruminal bacteria. J. Anim. Sci. 61, 713-724.

Brooks, M.A., R.M. Harvey, N.F. Johnson and M.S. Kerley, 2012. Rumen degradable protein supply affects microbial efficiency in continuous culture and growth in steers. J. Anim. Sci. 90, 4985-4994.

Calsamiglia, S., A. Ferret, C.K. Reynolds, N.B. Kristensen and A.M. van Vuuren, 2010. Strategies for optimizing nitrogen use by ruminants. Animal 4, 1184-1196.

Cant, J.P., 2005. Integration of data in feed evaluation systems. In: Dijkstra, J., J.M. Forbes and J. France (editors), Quantitative aspects of ruminant digestion and metabolism, 2nd edition. CABI Publishing, Wallingford, UK: 707-726.

Dijkstra, J., E. Kebreab, J.A.N. Mills, W.F. Pellikaan, S. López, A. Bannink and J. France, 2007. From nutrient requirement to animal response: predicting the profile of nutrients available for absorption in dairy cattle. Animal 1, 99-111.

Dijkstra, J., H.D. St.C. Neal, D.E. Beever and J. France, 1992. Simulation of nutrient digestion, absorption and outflow in the rumen: model description. J. Nutr. 122, 2239-2256.

Dijkstra, J., J. France, J.L. Ellis, A.B. Strathe, E. Kebreab and A. Bannink, 2013a. Production efficiency of ruminants: feed, nitrogen and methane. In: Kebreab, E. (editor), Sustainable animal agriculture. CAB International, Wallingford, UK, in press.

Dijkstra, J., O. Oenema and A. Bannink, 2011. Dietary strategies to reducing N excretion from cattle: implications for methane emissions. Curr. Opin. Environm. Sust. 3, 414-422.
Dijkstra, J., O. Oenema, J.W. van Groenigen, J.W. Spek, A.M. van Vuuren and A. Bannink, 2013b. Diet effects on urine composition of cattle and N_2O emissions. Animal, in press.
Ellis, J.L., J. Dijkstra, J. France, A.J. Parsons, G.R. Edwards, S. Rasmussen, E. Kebreab and A. Bannink, 2012. Effect of high-sugar grasses on methane emissions simulated using a dynamic model. J. Dairy Sci. 95, 272-285.
FAO, 2011. World Livestock 2011 – Livestock in Food Security. FAO, Rome.
Hanigan, M.D., 2005. Quantitative aspects of ruminant splanchnic metabolism as related to predicting animal performance. Anim. Sci. 80, 23-32.
Kebreab, E., A.B. Strathe, J. Dijkstra, J.A.N. Mills, C.K. Reynolds, L.A. Crompton, T. Yan and J. France, 2010. Energy and protein interactions and their effect on nitrogen excretion in dairy cows. In: Crovetto, G.M. (ed.), 3rd EAAP international symposium on energy and protein metabolism and nutrition. Wageningen Academic Publishers, Netherlands: 417-425.
Lapierre, H., C.E. Galindo, S. Lemosquet, I. Ortigues-Marty, L. Doepel and D.R. Ouellet, 2010. Protein supply, glucose kinetics and milk yield in dairy cows. In: Crovetto, G.M. (editor), 3rd EAAP international symposium on energy and protein metabolism and nutrition. Wageningen Academic Publishers, Netherlands: 275-285.
Lapierre, H., D. Pacheco, R. Berthiaume, D.R. Ouellet, C.G. Schwab, P. Dubreuil, G. Holtrop and G.E. Lobley, 2006. What is the true supply of amino acids for a dairy cow? J. Dairy Sci. 89, E1-E14.
Larsen, M., T.G. Madsen, M.R. Weisbjerg, T. Hvelplund and J. Madsen, 2001. Small intestinal digestibility of microbial and endogenous amino acids in dairy cows. J. Anim. Physiol. Anim. Nutr. 85, 9-21.
Lobley,G.E., D.M. Bremmer and G. Zuur, 2000. Effects of diet quality on urea fates in sheep as assessed by refined, non-invasive [$^{15}N^{15}N$] urea kinetics. Br. J. Nutr. 84, 459-468.
Moraes, L.E., J.E. Wilen, P.H. Robinson and J.G. Fadel, 2012. A linear programming model to optimize diets in environmental policy scenarios. J. Dairy Sci. 95, 1267-1282
National Research Council, 2001. Nutrient requirements of dairy cattle, 7th revised ed. National Academy Press, Washington DC, USA.
Pirt, S.J., 1965. The maintenance energy of bacteria in growing cultures. Proc. R. Soc. London Ser. B 163:224-231.
Reynal, S.M. and G.A. Broderick, 2005. Effect of dietary level of rumen-degraded protein on production and nitrogen metabolism in lactating dairy cows. J. Dairy Sci. 88, 4045-4064.
Reynolds, C.K., 2001. Economics of visceral energy metabolism in ruminants: Toll keeping or internal revenue service? J. Anim. Sci. 80, E74-E84.
Reynolds, C.K., L.A. Crompton and J.A.N. Mills, 2011. Improving the efficiency of energy utilisation in cattle. Anim. Prod. Sci. 51, 6-12.
Reynolds, C.K. and N.B. Kristensen, 2008. Nitrogen recycling through the gut and the nitrogen economy of ruminants: An asynchronous symbiosis. J. Anim. Sci. 86, E293-E305.
Rius, A.G., M.L. McGilliard, C.A. Umberger and M.D. Hanigan, 2010. Interactions of energy and predicted metabolizable protein in determining nitrogen efficiency in the lactating dairy cow. J. Dairy Sci. 93, 2034-2043.
Røjen, B.A., S.B. Poulsen, P.K. Theil and N.B. Kristensen, 2011b. Effects of dietary nitrogen concentration on messenger RNA expression and protein abundance of urea transporter-B and aquaporins in ruminal papillae from lactating Holstein cows. J. Dairy Sci. 94, 2587-2591.
Røjen, B.A., P.K. Theil and N.B. Kristensen, 2011a. Effects of nitrogen supply on inter-organ fluxes of urea-N and renal urea-N kinetics in lactating Holstein cows. J. Dairy Sci. 94, 2532-2544.
Russell, J.B. and C.J. Sniffen, 1984. Effect of carbon-4 and carbon-5 volatile fatty acids on growth of mixed rumen bacteria in vitro. J. Dairy Sci. 67, 987-994.
Satter, L.D. and L.L. Slyter, 1974. Effect of ammonia concentration on rumen microbial protein production in-vitro. Br. J. Nutr. 32, 199-208.
Steinfeld, H., P. Gerber, T. Wassenaar, V. Castel, M. Rosales and C. De Haan, 2006. Livestock's Long Shadow: Environmental Issues and Options. FAO, Rome.
Storm, E.R., D.S. Brown and E.R. Ørskov, 1983. The nutritive value of rumen micro-organisms in ruminants. 3. The digestion of microbial amino and nucleic acids in, and losses of endogenous nitrogen from, the small intestine of sheep. Brit. J. Nutr. 50, 479-485.
Tamminga, S., 1992. Nutrition management of dairy cows as a contribution to pollution control. J. Dairy Sci. 75, 345-357.

Van Duinkerken, G., M.C. Blok, A. Bannink, J.W. Cone, J. Dijkstra, A.M. van Vuuren and S. Tamminga, 2011. Update of the Dutch protein evaluation system for ruminants: the DVE/OEB2010 system. J. Agric. Sci., Camb. 149, 351-367.

Van Knegsel, A.T.M., H. van den Brand, J. Dijkstra, W.M. van Straalen, M.J.W. Heetkamp, S. Tamminga and B. Kemp, 2007. Dietary energy source in dairy cows in early lactation: energy partitioning and milk composition. J. Dairy Sci. 90, 1467-1476.

Van Vuuren, A.M. and J.A.C. Meijs, 1987. Effects of herbage composition and supplement feeding on the excretion of nitrogen in dung and urine by grazing dairy cows. In: Van der Meer, H.G., R.J. Unwin, T.A. van Dijk and G.C. Ennik (eds.), Animal manure on grassland and fodder crops. Fertilizer or waste? Martinus Nijhoff Publishers, Dordrecht, the Netherlands: 17-25.

Vérité, R. and J.L. Peyraud, 1989. Protein: the PDI system. In: Jarrige, R. (ed.), Ruminant nutrition: recommended allowances and feed tables. John Libbey Eurotext, Paris: 33-47.

Virtanen, A.I., 1966. Milk production of cows on protein-free feed. Science 153, 1603-1614.

Small intestinal fermentation contributes substantially to starch disappearance in milk-fed calves

M.S. Gilbert[1], A.J. Pantophlet[2], J.J.G.C. van den Borne[1], H.A. Schols[3] and W.J.J. Gerrits[1]
[1]Animal Nutrition Group, Wageningen University, P.O. Box 338, 6700 AH Wageningen, the Netherlands; myrthe.gilbert@wur.nl
[2]Centre for Medical Biomics, University Medical Centre Groningen, P.O. Box 30001, 9700 RR Groningen, the Netherlands
[3]Laboratory of Food Chemistry, Wageningen University, P.O. Box 8129, 6700 EV Wageningen, the Netherlands

Introduction

Calf milk replacers commonly contain 40-50% lactose. For economic reasons, starch is of interest as a lactose replacer. Small intestinal disappearance of starch (66%) was lower than that of glucose (85%) when infused in the abomasum of steers (Kreikemeier and Harmon, 1995), indicating that enzyme activity required for the hydrolysis of starch to glucose limits starch digestion. Which enzyme system is limiting starch digestion in milk-fed calves is unknown. Portal glucose appearance was only 57% of small intestinal starch disappearance (Kreikemeier and Harmon, 1995). This gap includes starch fermentation and glucose use by portal drained visceral tissues. In steers, abomasal infusion of a starch hydrolysate resulted in a linear decrease in ileal pH (Branco *et al.*, 1999), illustrating that fermentation may be an important contributor to small intestinal starch disappearance.

The objectives were therefore (1) to determine the rate-limiting enzyme for hydrolysis and disappearance of starch from the intestinal lumen and (2) to quantify starch fermentation in milk-fed calves.

Material and methods

Forty male calves (224±2.0 kg BW) were fed milk replacer containing either 18% lactose (control) or 18% of one of 4 corn starch products. The 4 corn starch products differed in the enzymes required for their complete hydrolysis to glucose: gelatinized starch (α-amylase and maltase); maltodextrin (α-amylase and maltase); maltodextrin with α-1,6-branching (α-amylase, maltase and isomaltase) and maltose (maltase). Calves were adapted to the diets for 15 weeks before the start of the measurements. All diets included Co-EDTA as an indigestible marker. The corn starch products (1.093 atom% ^{13}C) differed in natural ^{13}C enrichment from lactose (1.073 atom% ^{13}C) and the remainder of the diet (1.078 atom% ^{13}C).

Feces were collected quantitatively during 4 days to measure total tract digestibility of the starch products and to calculate total tract starch fermentation based on fecal ^{13}C excretion (Gerrits *et al.*, 2012). Blood samples were taken at -30, 30, 60, 120, 180, 240 and 360 min after feeding to measure ^{13}C enrichment in plasma glucose. On the day of blood sampling only, control calves received ^{13}C enriched lactose (1.092 atom% ^{13}C). Calves were sacrificed 4 h after feeding and ileal digesta were collected to measure ileal digestibility of starch products. Variables were analyzed for treatment effects by ANOVA.

Results and discussion

Apparent total tract (98.6±0.5%) and ileal (60.3±2.8%) starch digestibility did not differ between starch products. Total tract starch fermentation, estimated from increased fecal ^{13}C excretion, was not affected by treatment and averaged 478±28 g/d, corresponding to 101% of starch intake (Table 1). Starch-fed calves produced more feces (+66 g DM/d) than control calves ($P<0.001$), which

Table 1. Fecal characteristics of milk-fed calves fed a milk replacer containing only lactose (CON) or 18% of naturally ^{13}C enriched starch products (GS, gelatinized starch; MD, maltodextrin; MDB, maltodextrin with a high degree of branching; MT, maltose).

Treatment	CON	GS	MD	MDB	MT	*P*-value[1]
	Mean ± SE					
Number of calves	7	8	7	8	7	
Fecal ^{13}C enrichment, atom%	1.0776	1.0829	1.0832	1.0825	1.0828	<0.001
Fecal dry matter output, g/d	168±18	259±15	245±19	222±9	225±12	0.003
Fermentation[2], g/d	-	512±36	530±91	432±20	431±53	0.422
Fermentation[2], % of intake	-	98±7	107±9	97±5	104±13	0.925

[1] *P*-value for differences between treatments. When no value for the CON treatment is shown, the *P*-value applies for differences between the starch product treatments only.

[2] Estimated fermentation of starch products.

consisted of 6.0 g starch, 0 g fat and 0 g ash per day. This leaves 60 g/d unaccounted for, which is hypothesized to be increased undigested microbial mass resulting from starch fermentation. Assuming a ratio of 5 gram starch fermented for each g of fecal microbial output (Lanzas *et al.*, 2007), this would require 300 g/d starch to be fermented, corresponding to 63% of starch intake. In the control calves, ^{13}C enrichment in plasma glucose increased from 1.083 to 1.089 atom% within 3h after feeding ($P<0.001$). In starch-fed calves, ^{13}C enrichment in plasma glucose did not increase relative to baseline (1.085 atom%). Hence, absorption of starch-derived glucose was not sufficient to lead to a measurable increase in ^{13}C enrichment in plasma glucose.

The combination of the 4 starch products would lead us to deduce the rate-limiting enzyme for starch digestion in milk-fed calves. Ileal starch digestibility did not differ between starch products, suggesting that maltase activity limits starch digestion in milk-fed calves. Two methods were used to quantify starch fermentation; one based on increased fecal DM output and the other on increased fecal ^{13}C excretion. Based on these methods, starch fermentation was 63 to 100% of the starch intake in milk-fed calves. This is in agreement with the absence of a postprandial response in ^{13}C enrichment of plasma glucose to feeding corn starch products that are characterized by a relatively high natural ^{13}C enrichment. Nearly 40% of the starch was fermented in the colon. Therefore, an additional 23 to 60% of the starch intake is fermented before the colon. Overall, this study shows that small intestinal fermentation contributes substantially to starch disappearance and that maltase limits starch digestion in milk-fed calves.

References

Branco, A.F., D.L. Harmon, D.W. Bohnert, B.T. Larson and M.L. Bauer, 1999. Estimating true digestibility of nonstructural carbohydrates in the small intestine of steers. J. Anim. Sci. 77, 1889-1895.

Gerrits, W.J.J., M.W. Bosch and J.J.G.C. van den Borne, 2012. Quantifying resistant starch using novel, *in vivo* methodology and the energetic utilization of fermented starch in pigs. J. Nutr. 142, 238-244.

Kreikemeier, K.K. and D.L. Harmon, 1995. Abomasal glucose, maize starch and maize dextrin infusions in cattle – Small intestinal disappearance, net portal glucose flux and ileal oligosaccharide flow. Br. J. Nutr. 73, 763-772.

Lanzas, C., C.J. Sniffen, S. Seo, L.O. Tedeschi and D.G. Fox, 2007. A revised CNCPS feed carbohydrate fractionation scheme for formulating rations for ruminants. Anim. Feed. Sci. Technol. 136, 167-90.

Nutrient digestion by dairy cows fed diets replacing starch with non-forage fiber

S.D. Ranathunga, M.M. Abdelqader and K.F. Kalscheur
Dairy Science Department, South Dakota State University, SD 57007, USA; kenneth.kalscheur@sdstate.edu

Introduction

Corn starch is used as the main energy source in lactating dairy cow diets. Feeding high levels of corn starch may be associated with negative health impacts on lactating dairy cows, such as ruminal acidosis and laminitis along with higher feed costs and lower income from reduced milk components. Dried distillers grains with solubles (DG), a co-product of the ethanol industry, is an excellent source of energy. Ranathunga *et al.* (2010) demonstrated that that incrementally reducing the amount of starch in a ration from a high of 29% to a low of 20% by adding DG resulted in similar milk production and composition by lactating dairy cows. The objective of the study was to evaluate the effect of replacing starch from corn with non-forage fiber from DG and soybean hulls on the nutrient flow to the omasum, ruminal nutrient degradability, total tract nutrient digestibility, and nitrogen partition of lactating dairy cows.

Material and methods

Six Holstein cows with ruminal fistula were assigned to a multiple 3×3 Latin square design. Three diets were formulated: (1) high starch (HS) containing 33% starch with 0% DG; (2) medium starch (MS) containing 25% starch with 12% DG; and (3) low starch (LS) containing 17% starch with 24% DG. Ground corn, soybean meal, expeller soybean meal, and an inert fat were replaced by DG and soybean hulls to formulate diets containing high, medium, and low starch concentrations. Ingredient and nutrient composition is described in Table 1.

Table 1. Ingredient and nutrient composition of the diets.

Item	HS	MS	LS
Alfalfa hay	25.0	25.0	25.0
Corn silage	25.0	25.0	25.0
Ground corn	31.5	20.5	9.5
DG	0	12.0	24.0
Soybean hulls	0	7.5	15.0
Soybean meal, 44%	8.00	4.00	0
Expellers soybean meal	7.25	3.60	0
Ruminally inert fat	1.60	0.80	0
Minerals and vitamins	1.65	1.57	1.50
DM, % of diet	70.8	71.4	71.9
CP, % of DM	17.2	17.2	17.2
NDF, % of DM	25.4	31.5	37.6
Forage NDF, % of DM	20.5	20.5	20.5
Starch, % of DM	32.4	24.6	16.7

Nutrient flow to the omasum was measured using omasal sampling and the triple marker method (CoEDTA, Yb, and indigestible NDF as markers; Reynal and Broderick, 2005). Total tract digestibility of nutrients and nitrogen excretion data were measured by total collection of feces and urine.

Data were analyzed using the MIXED procedure (SAS, 2001). Orthogonal contrasts were used to test linear (L), and quadratic (Q) effects.

Results and discussion

As dietary starch concentration decreased from 31% to 18%, DMI increased linearly (P=0.04). As dietary starch concentration decreased across diets, milk yield (P=0.04), milk and fat percentages (P=0.04) increased linearly, whereas protein percentages (P=0.01) decreased linearly.

Flow of DM (10.5, 11.2, and 11.8 kg/d) and OM (8.2, 8.8, and 9.1 kg/d) to the omasum increased linearly when starch was replaced with non-forage fiber. Average flow of NDF, ADF, CP, and starch to the omasum was not affected by the diets (3.6, 1.6, 3.4, and, 0.4 kg/d, respectively).

Table 2 summarizes the data on nutrient digestion and nitrogen metabolism. Apparent ruminal degradability of DM, OM, and CP was similar across the diets (50.5, 58.6, and 14.8%), whereas the apparent ruminal degradability of NDF (P<0.01) and ADF (P=0.10) increased linearly when starch was replaced with non-forage fiber. There was a tendency to reduce the apparent ruminal degradability of starch (P<0.09) when starch was replaced with non-forage fiber. Total tract digestibility of DM (71.0%), OM (72.5%), CP (70.6%) and starch (96.0%) were similar irrespective of the diet, but total tract digestibility for NDF and ADF increased linearly (P<0.01) with the inclusion of non-forage fiber to replace starch.

As dietary starch concentrations were decreased across diets, fecal N excretion tended to increase linearly (P=0.09) and urinary N excretion increased linearly (P=0.01). However, average N excreted

Table 2. Nutrient digestion and N metabolism.

Item	HS[1]	MS[1]	LS[1]	SEM	P-value[2]
Ruminal degradability, %					
DM	52.0	49.3	50.3	1.51	NS
NDF	33.7	46.9	58.9	3.96	L
ADF	51.5	57.1	68.0	3.32	L
Starch	94.2	91.9	90.8	1.36	LT
Total tract digestibility, %					
DM	70.9	71.8	70.3	0.77	NS
NDF	41.1	52.7	59.5	2.50	L
ADF	41.0	50.9	58.2	3.00	L
Starch	95.6	96.1	96.4	0.67	NS
Nitrogen metabolism					
Fecal N excretion, g/d	176	185	198	20.1	LT
Urine N excretion, g/d	214	218	236	8.86	L
Milk N, g/d	161	156	150	11.5	LT
N efficiency, %	25.7	25.7	22.2	1.82	L

[1] HS = high starch diet; MS = medium starch diet; LS =low starch diet.

[2] L = linear effect (P<0.05); Q = quadratic effect (P<0.05); LT = linear effect (tendency) (P<0.10); NS = non-significant.

via feces and urine as a percentage of total N intake was similar across the diets (28.9 and 35.5% respectively). There was a tendency to reduce N excreted as true milk protein (P=0.06) and N efficiency (P=0.03) decreased linearly across the diets when starch was replaced with non-forage fiber. Although, replacing dietary starch with non-forage fiber did not affect ruminal and total tract digestibility of DM, OM, and CP, it increased ruminal and total tract digestibility of NDF and ADF.

References

Ranathunga, S. D., K. F. Kalscheur, A. R. Hippen, and D. J. Schingoethe, 2010. Replacement of starch from corn with nonforage fiber from distillers grains and soyhulls in diets of lactating dairy cows. J. Dairy Sci. 93, 1086-1097.

Reynal, S. M., and G. A. Broderick, 2005. Effect of dietary level of rumen-degraded protein on production and nitrogen metabolism in lactating dairy cows. J. Dairy Sci. 88, 4045-4064.

SAS, 2002. Version 9.3. SAS Institute Inc., Cary, NC, USA.

Effect of different dietary levels of Quebracho tannin extract on nitrogen and fiber digestibility and post-ruminal microbial protein flow in heifers

S. Ahnert[1], A. Susenbeth[1] and U. Dickhoefer[1,2]
[1]Institute of Animal Nutrition and Physiology, Christian-Albrechts-Universitaet zu Kiel, 24098 Kiel, Germany; ahnert@aninut.uni-kiel.de
[2]Current address: Institute of Animal Production in the Tropics and Subtropics, University of Hohenheim, 70599 Stuttgart, Germany

Introduction

The effect of condensed tannins (CT) on protein digestion in ruminants appears to be dosage-dependent. While low to moderate CT concentrations are considered to increase rumen-escape protein, higher CT intakes may inhibit microbial crude protein (MCP) synthesis and decrease nutrient digestibility (Barry and McNabb, 1999). Our aim was thus to determine how graded dosages of a Quebracho tannin extract (QTE; 75% total phenolic content) influence MCP flow and digestibilities of neutral detergent fiber (NDF), acid detergent fiber (ADF), nitrogen (N), and acid detergent insoluble N (ADIN).

Material and methods

Six rumen-fistulated heifers (491±35 kg body weight) were offered 2.6 kg grass hay, 2.6 kg concentrate feed, and 60 g of a mineral premix per day in two equal meals. The study comprised one period without QTE supplementation (Control I) followed by four periods when all animals received 1, 2, 4, or 6% QTE of their daily dry matter (DM) intake. Half of the QTE dosage was infused intra-ruminally at each feeding. Every period comprised 9 d of adaptation and 6 d of total urine and feces collection. During the collection period, ruminal fluid samples were taken from three animals at 1, 2, 4, and 8 h after morning feeding. Subsequent to period 5, QTE dosing was ceased and urine and feces were collected again for 10 d (Control II) after 14 d of adaptation. Feed and feces samples were analyzed for NDF, ADF, N, and ADIN concentrations and rumen fluid for ammonia-N content. Duodenal MCP flow was estimated from urinary allantoin and uric acid excretions according to Chen and Orskov (2003), assuming a constant proportion of purine-N in microbial-N. Significant ($P<0.05$) treatment effects were determined by ANOVA and regression analyses by R software 2.12.0.

Results and discussion

Urinary purine derivative excretion was highest with 99 and 103 mM/d at 0% and 1% QTE, respectively, and declined with increasing QTE supplementation to 80 mM/d at 6% QTE, reflecting a 36% decrease in MCP flow from 334 (at 1% QTE) to 213 g/d (at 6% QTE; $P<0.001$; Table 1). Apparent total tract digestibility of NDF declined linearly with increasing QTE dosage from 71.8% without QTE to 59.0% at 6% QTE ($R^2=0.58$; $P<0.001$). Digestibility of ADF decreased linearly from 62.6% at 0% QTE to 49.6% at 6% QTE ($R^2=0.54$; $P<0.001$). Besides a direct inhibition of rumen microbes, this lower rumen carbohydrate fermentation may have also caused the decrease in MCP flow. No clear trend was observed for apparent digestibility of ADIN ($R^2=0.15$; $P=0.021$). Apparent N digestibility decreased linearly by 17% from 71.6% without QTE to 59.8% at 6% QTE ($R^2=0.90$; $P<0.001$). This reduction can be associated with an incomplete post-ruminal dissociation of tannin-protein complexes, an impaired protein absorption in the small intestine, reduced digestibility of fiber-bound N, and increased endogenous N losses. Rumen ammonia-N concentrations were unaffected by QTE (4.7-5.8 mM/l; $P=0.171$), most likely due to a lower ammonia-N incorporation in MCP that outweighed the decrease in ammonia-N release from protein degradation. However,

Table 1. Effect of different levels of Quebracho tannin extract (QTE) on urinary purine derivative (PD) excretion, microbial crude protein (MCP) flow, apparent total tract digestibility of NDF, ADF, N, ADIN, and ruminal ammonia-N concentration in heifers (least square means; SEM; n=6).

Parameter, unit	Dietary QTE concentration (% daily dry matter intake)												SEM	*P*-value
	0 (Control I)		1		2		4		6		0 (Control II)			
PD, m*M*/d	99	c,d	103	d	95	b,c	90	b	80	a	98	b,c,d	4	<0.001
MCP, g/d	316	c,d	334	d	291	b,c	267	b	213	a	307	b,c,d	21	<0.001
Apparent total tract digestibility, %														
NDF	71.8	b	70.0	b	70.7	b	62.5	a	59.0	a	71.0	b	1.6	<0.001
ADF	62.6	b	61.0	b,c	63.0	b	54.2	a,c	49.6	a	63.6	b	1.8	<0.001
N	71.6	d	69.4	c,d	67.7	c	62.7	b	59.8	a	71.8	d	0.6	<0.001
ADIN	12.5	a	9.1	a	11.5	a	1.7	a	5.3	a	35.8	b	3.9	<0.001
Ruminal ammonia-N, mM/l														
	5.4		5.7		4.7		5.8		5.2		5.6		0.4	0.171

NDF, neutral detergent fiber; ADF, acid detergent fiber; N, nitrogen; ADIN, acid detergent insoluble nitrogen; SEM, standard error of the mean.

these conclusions are valid only, if the purine content in microbial matter remained constant and thus the accuracy of MCP estimates were unaffected by the diet.

While dietary QTE levels of 4% and 6% markedly reduced nutrient digestibilities, MCP synthesis was already affected at 2% QTE in the diet. In conclusion, the supplementation of QTE at low levels is expected to achieve no negative effects on MCP synthesis and a reduced ruminal protein degradation may therefore improve post-ruminal protein supply. At dietary levels of QTE<2% of DM intake it is unlikely that the decrease in MCP might be compensated by an increase in rumen-escape protein. Furthermore, the decrease in energy supply to the animal and a possible negative effect on feed intake allow for the use of QTE at very low concentrations only. Further studies are needed to identify to what extent ruminal protein degradation is reduced and whether post-ruminal protein digestibility is affected by QTE supplementation.

References

Barry, T.N., and W.C. McNabb, 1999. The implications of condensed tannins on the nutritive value of temperate forages fed to ruminants. Br. J. Nutr. 81: 263-272.

Chen X.B., and E.R. Orskov, 2003. Research on urinary excretion of purine derivatives in ruminants: Past, present and future. International Feed Research Unit, Aberdeen, UK, 34 p.

Effect of fescue toxicosis on nitrogen and energy balance in Holstein steers

A.F. Koontz[1], D.H. Kim[1], A.P. Foote[1], L.P. Bush[2], J.L. Klotz[3], K.R. McLeod[1] and D.L. Harmon[1]
[1]Department of Animal and Food Sciences, University of Kentucky, Lexington, KY, USA; anne.koontz@gmail.com
[2]Department of Plant and Soil Sciences, University of Kentucky, Lexington, KY, USA
[3]USDA-ARS, Forage-Animal Production Research Unit, Lexington, KY, USA

Introduction

Animals consuming endophyte-infected tall fescue exhibit reduced weight gain and feed intake, with the most severe effects occurring during summer months. Heat stress and plane of nutrition have independent effects on metabolic activity (Birkelo *et al.*, 1991; O'Brien *et al.*, 2010) while, ergot alkaloids reduce blood flow to and nutrient absorption from the rumen (Foote *et al.*, 2012). This research was designed to separate the effects of alkaloid consumption from those of reduced intake and environmental temperature.

Material and methods

Two experiments were conducted, each using six ruminally cannulated Holstein steers weight-matched (348±13 and 217±7 kg, respectively) and pair-fed alfalfa cubes to separate the effects of alkaloid ingestion and energy intake. During each period, one steer per pair was ruminally dosed twice daily with ground endophyte-infected tall fescue seed (E+; 0.025 mg ergovaline/kg BW/d), the other with ground endophyte-free tall fescue seed (E-). Fescue seed dosing provided negligible N and E to the animals. Exp. 1 utilized a two period crossover design, with two temperatures, (22 °C and 30 °C) within each period. E+ steers were offered feed at $1.5 \times NE_m$. On d8 of each temperature segment, animals were moved to metabolism stalls with indirect calorimetry head-boxes for fasting heat production (FHP) determination. Rumen contents were removed, the reticulorumen was washed and filled with a buffer (NaCl=96; $NaHCO_3$=24; $KHCO_3$=30; K_2HPO_4=2; $CaCl_2$=1.5; $MgCl_2$=1.5 mmol/kg buffer), to which an E+ or E- fescue seed extract was added at 12h intervals to maintain alkaloid treatment presentation during FHP. After 12 h, heart rate (HR), O_2 consumption, CO_2 production and urinary output were recorded for 16 h. Exp. 2 used a four period crossover design with a 2×2 factorial treatment structure. Factors were endophyte (E+ vs. E-) and energy intake ($1.8 \times NE_m$, H vs. $1.1 \times NE_m$, L). After 8 d of diet adaptation animals were dosed twice daily with ground tall fescue seed for the remainder of each period. On d 17-21 total fecal and urinary output were collected, with animals placed into indirect calorimetry head-boxes during d 20-21. Feed, feces, and urine samples were analyzed for DM, N, and GE. Heat production was determined using the Brouwer (1965) equation. Data were analyzed using the MIXED procedure of SAS, with steer as the experimental unit, animal and period as random effects, and fixed effects of endophyte treatment and temperature (Exp. 1) or intake (Exp. 2).

Results and discussion

There was no difference (P>0.9) in DMI or DMI/kg$^{0.75}$ between endophyte treatments in both experiments, and DMI was different (P<0.01) between H and L in Exp. 2 by design.

During Exp. 1, increased temperature decreased intake (P<0.01), but had no effect on other measurements. There were no interactions of temperature and endophyte treatment. O_2 consumption decreased (P=0.04) and CO_2 production tended to be reduced (P=0.07) during E+ treatment. Fasting heat production (kcal/kg BW$^{0.75}$) was lower (P=0.006; Table 1) in animals receiving E+ treatment at both temperatures.

Table 1. Calorimetric data from steers dosed with endophyte free (E-) and endophyte infected (E+) tall fescue seed at 22 °C and 30 °C (Exp. 1).

Item	Treatment				SEM	P=		
	E-		E+			Main effects		Interaction
	22 °C	30 °C	22 °C	30 °C		Endophyte (E)	temperature (T)	E × T
Heat produced (kJ/kg $BW^{0.75}$/d)	323.6	353.7	288.2	288.9	18.81	0.006	0.332	0.356

In Exp. 2, animals on H feeding had higher (P<0.03) water, N, and energy intakes, energy and N excretion, as well as retained N, DE, ME, and RE. There were no differences (P>0.15) in these values between endophyte treatments. On average, L fed animals were below maintenance requirements, based on an average RE of -111 kJ/kg BW.75 (Table 2). Thus the H and L diets show animals above and below maintenance, respectively. The interaction of intake × endophyte was significant for CO_2 production (P=0.04), and tended to be significant (P≤0.11) for O_2 consumption, CH_4 production, and HP. For each of these measures H was greater than L, and the difference between intakes was greater with E+ treatment.

These data indicate that ingestion of endophyte infected tall fescue seed in pair fed steers reduces basal metabolism, but does not alter total N or E balance. The reduction in FHP may be offset by increased urinary and gaseous energy losses, as well as increased heat production by E+ dosed animals at intakes above maintenance. These results indicate that reduced intake is likely the primary cause of the reduced weight gain associated with fescue toxicosis.

Table 2. Digestible, metabolizable, and retained energy and heat production (kJ/kg $BW^{0.75}$/d) of steers dosed with endophyte free (E-) and endophyte infected (E+) tall fescue seed fed at 1.1xNE_m (L) and 1.8xNE_m (H) (Exp. 2).

	Treatment				SEM	P		
	E-		E+			Main effects		Interaction
	L	H	L	H		Endophyte (E)	Intake (I)	E × I
Digestible energy	572.5	986.6	553.9	950.3	53.63	0.362	<0.0001	0.766
Metabolizable energy	446.0	841.9	437.6	795.1	50.85	0.381	<0.0001	0.540
Retained energy	-123.5	169.8	-100.3	97.9	55.55	0.585	0.0002	0.296
Heat production	574.4	667.2	542.9	692.3	18.47	0.828	<0.0001	0.075

References

Birkelo, C. P., Johnson, D. E. and Phetteplace, H. P. 1991. Maintenance requirements of beef cattle as affected by season on different planes of nutrition. J Anim Sci 69(3):1214-1222.

Brouwer, E. 1965. Report of sub-committee on constants and factors Proc. 3rd Symp Energy Metab, Troon, Scotland.

Foote, A. P., Kristensen, N. B., Klotz, J. L., Kim, D. H., Koontz, A. F., McLeod, K. R., Bush, L. P. and Harmon, D. L. 2012. Ergot alkaloids decrease rumen epithelial blood flow. J Anim Sci 90(Suppl 3):408 (Abstr).

O'Brien, M. D., Rhoads, R. P., Sanders, S. R., Duff, G. C. and Baumgard, L. H. 2010. Metabolic adaptations to heat stress in growing cattle. Domest Anim Endocrin 38(2):86-94.

Performance, efficiency and estimated maintenance energy requirements of *Bos taurus* and *Bos indicus* cattle

R.D. Sainz[1,2]*, G.D. Cruz*[1]*, E. Mendes*[2]*, C.U. Magnabosco*[2]*, Y.B. Farjalla*[3]*, F.R.C. Araujo*[3]*, R.C. Gomes*[3,4] *and P.R. Leme*[4]
[1]*University of California, Davis, CA, USA; rdsainz@ucdavis.edu*
[2]*Embrapa, Brasília, DF, Brazil*
[3]*Aval Serviços Tecnológicos, Goiânia, GO, Brazil*
[4]*Universidade de São Paulo, Pirassununga, SP, Brazil*

Introduction

The NRC (2000) concluded that maintenance requirements of *Bos indicus* cattle were about 10% lower than those of *Bos taurus* cattle, however other reports do not support any difference between breed types (Tedeschi *et al.*, 2002). A recent focus on genetic selection for increased feed efficiency has led to greater efforts to quantify individual feed intakes on farms and research institutes. This study has combined several of those data sets, including *B. indicus* and *B. taurus* cattle, in high and low efficiency categories, in order to revisit the maintenance issue.

Material and methods

Individual data on feed intakes and weight gains from 11 experiments with a mean duration of 79.1±26.5 days, conducted at the University of California – Davis (3 trials, Angus, Hereford and crossbreds), the University of São Paulo – Pirassununga (1 trial, Nelore), and three private ranches in Brazil (Guaporé Pecuária, Pontes e Lacerda – MT, 2 trials, Polled Nelore; Rancho da Matinha, Uberaba – MG, 2 trials, Nelore; Fazenda Perfeita União, Pirajuí – SP, 2 trials, Guzerá) were pooled and analyzed. The combined data set comprised 838 animals (127 *B. taurus* steers and 711 *B. indicus* bulls). In each trial, residual feed intakes (RFI) were calculated as the residual of the regression of dry matter intake (DMI) on average metabolic body weight and average daily gain (ADG). Animals with RFI below mean -0.5 SD, within 0.5 SD of the mean or above mean -0.5 SD were classed as Low, Medium and High RFI groups, respectively. In addition, for each trial the dietary metabolizable and net energy values were estimated, and standard NRC (2000) gain equations used to fit the maintenance energy requirement to the data for each animal (Zinn and Shen, 1996). Data were subjected to analysis of variance, with Breed Type and RFI Group as main effects. The interaction between Breed Type and RFI group was not statistically significant for any variable.

Results and discussion

Differences among RFI groups were as expected, with Low RFI cattle having lower DMI, similar ADG, greater gain:feed, and lower RFI values than High RFI cattle (Table 1). Maintenance energy requirements were lower in Low vs. High RFI cattle. Dry matter intakes were lower in *B. indicus* than in *B. taurus* cattle, both in absolute terms and relative to body weight. This was accompanied by lower weight gains and gain:feed. As expected, RFI values did not differ between types, since the RFI calculation was done within trial. Maintenance energy requirements were lower in *B. indicus* than in *B. taurus* cattle, but this difference was smaller than that among RFI groups of both cattle types.

These results support the conclusion of the NRC (2000) of lower maintenance requirements of *B. indicus* cattle as compared to *B. taurus* cattle. In these data, the difference in maintenance requirements was about 7%, but this is confounded by the fact the all the *B. taurus* animals were steers and all the *B. indicus* animals were bulls. One would expect a greater difference between animals of the same sexual condition. On the other hand, the difference in maintenance requirements between the most and least efficient groups was greater than 30%, indicating that variation in maintenance requirements

Table 1. Performance of Bos taurus *and* Bos indicus *cattle in different efficiency categories.*

Variable	Type		RFI group			SD[2]	*P*-value[1]	
	B. taurus	*B. indicus*	High	Medium	Low		Type	RFI group
N	127	711	233	364	241			
Dry matter intake, kg/d	10.34	9.50	10.97	9.63	9.16	1.77	<0.001	<0.001
Dry matter intake, % of BW	2.51	2.33	2.62	2.43	2.22	0.34	<0.001	<0.001
Average daily gain, kg/d	1.511	1.222	1.384	1.349	1.367	0.325	<0.001	0.428
Gain:feed	0.147	0.130	0.126	0.140	0.149	0.027	<0.001	<0.001
Residual feed intake, kg/d	0.032	0.009	0.882	0.032	-0.853	0.374	0.534	<0.001
Maintenance coefficient, Mcal/kg$^{0.75}$/d	0.0750	0.0699	0.0867	0.0710	0.0597	0.020	0.012	<0.001

[1] Probability of a Type I error.

[2] SD, pooled standard deviation.

within breed types is much greater than the variation among breed types. We conclude that there is great scope for genetic selection for increased feed efficiency in all beef cattle breeds, and that RFI may be used as an indicator trait for maintenance requirement.

References

National Research Council, 2000. Nutrient Requirements of Beef Cattle, 7th Rev. ed., National Academy Press, Washington, DC.

Tedeschi, L.O., C. Boin, D.G. Fox, P.R. Leme, G.F. Alleoni and D.P.D. Lanna, 2002. Energy requirement for maintenance and growth of Nellore bulls and steers fed high-forage diets. J. Anim. Sci. 80, 1671-1682.

Zinn, R.A. and Y. Shen, 1996. Interaction of dietary calcium and supplemental fat on digestive function and growth performance in feedlot steers. J. Anim. Sci. 74, 2303-2309.

Energy for maintaining liveweight: an indicator of adaptive abilities of beef cows?

A. De La Torre[1,2], *F. Blanc*[2,1], *E. Recoules*[1,2], *D. Egal*[3] *and J. Agabriel*[1,2]
[1]*INRA, UMR 1213 Herbivores, 63122 Saint-Genès-Champanelle, France; anne.delatorre@clermont.inra.fr*
[2]*VetAgro Sup, UMR 1213 Herbivores, 63000 Clermont-Ferrand, France*
[3]*INRA, UE1296 Monts d'Auvergne, 63210 Orcival, France*

Introduction

Productivity of low input livestock systems partly relies on animals' ability to cope with changing environments while achieving productive and reproductive performances. In such conditions, individuals' robustness could be estimated by indicators of animal adaptive abilities which account for energy variations across the productive cycle. In mature suckler cows, the net energy requirements for production are low (30%) compared to those for maintenance (70%) which complicates the evaluation of adaptive abilities. The later could be approached by estimating the net energy required for maintaining liveweight constant (Em). The objective of this study was to (1) estimate in beef cows having different body reserves at calving the partition of net energy between net energy outputs (Em, Emilk) and net energy inputs (Eintake + Etissues) and (2) test the relevance of Em as an indicator of adaptive abilities of beef cows.

Material and methods

Fourty multiparous charolais cows varying in body condition at calving (BCS scale from 0 to 5): 20 thin (T, BCS=2.0±0.04) vs. 20 fat (F, BCS=2.8±0.08) were submitted to a nutritional challenge consisting in a sequence of two contrasted feeding periods. During period 1 (P1), from calving to 120 days *postpartum*, cows were reared indoors and were fed the same basal diet according to two energy levels (Control, C vs. Low, L). Intake expressed in NE for lactation (NE_L) averaged 0.57 and 0.36 MJ/d/kg$^{0.75}$ for C and L cows. At the end of P1, all cows were turned out to a permanent pasture for a 76 days period (P2). Individual intakes at grazing were estimated from fill unit system (INRA, 2007). During P1 and P2, body weight, BCS and milk production were regularly measured. Body lipids were assessed by measuring subcutaneous adipose cell diameter at calving and at the end of P1 and P2. Em was calculated over P1 and P2. It was defined as follow Em = Eintake – (Emilk + Etissues), all terms expressed in Net Energy for lactation. Energy in milk was taken at 3.2 MJ/kg assuming a standard milk composition for beef cows (INRA, 2007). The tissue net energy conversion rate to milk production was taken as 0.8. We assumed that the net energy value of a kilogram of body mass change is equal to 66.7 MJ × % of lipids + 39 MJ x% of fat-free mass (Garcia and Agabriel, 2007). Data were analyzed using PROC MIXED (SAS) including BCS at calving, *postpartum* energy levels and days as fixed effects and individual as random effect.

Results and discussion

Over the whole experimental period, average milk production and calf body weight gain were similar between the four groups. Over P1, FL and TL cows lost 43±13 and 25±17 kg compared to FC and TC cows, respectively (Table 1). L treatment resulted mainly in a fat mass loss, whereas C treatment induced no (FC) or only a slight (TC) weight gain which was mostly due to an increase in fat mass (99 MJ). These discrepancies resulted in differences in net energy balance (Figure 1). Decreasing E intake led to a decrease in Em, which was 25% lower in L vs. C level of energy intake ($P<0.05$). At the end of P2, FL and TL cows weighed 20 and 10 kg less than FC and TC cows, respectively. Over P2, Em remained lower in cows previously submitted to energy restriction in comparison to C cows ($P<0.05$). This result is related to a compensatory rebound which is associated to lower

Table 1. Effect of body condition at calving and postpartum *energy level on milk production, average daily gain of calves (ADG), liveweight (LW) and adipose cells diameter (ACD).*

	Period 1 (indoors)					Period 2 (pasture)				
	FC	FL	TC	TL	SEM	FC	FL	TC	TL	SEM
Milk (kg/day)	8.3	8.7	9.2	7.0	0.6	6.0	6.8	7.6	7.1	0.5
ADG (kg/day)	0.8	0.9	0.9	0.7	0.09	1.1	1.2	1.1	1.1	0.10
LW start(kg)	846[ab]	852[a]	761[ab]	753[b]	17.4	827[a]	790[ab]	734[ab]	716[b]	17.8
LW end (kg)	834[a]	810[ab]	762[ab]	719[b]	17.9	805	824	767	776	18.2
ACD change	-4.5[a]	-9.6[b]	2.6[a]	-10.0[b]	2.0	7.5	11.2	0.93	3.4	1.6

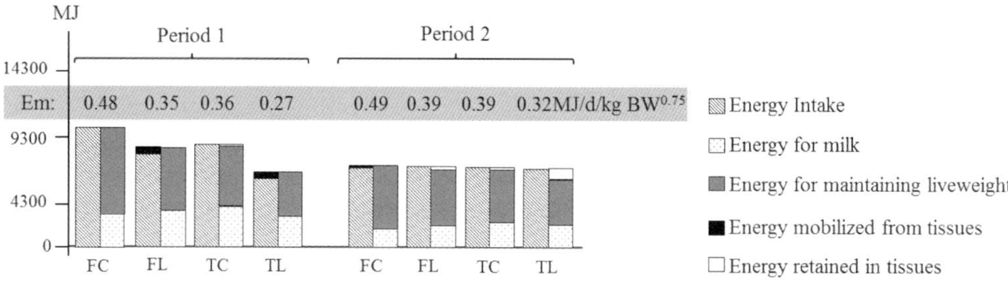

Figure 1. Effect of body condition at calving and postpartum *energy level on energy balance and partition during the initial indoors (1) and pasture (2) periods.*

Em after a period of nutritional restriction (Hornick, 2000). It is the lag time in adaptation of Em before its gradual increase which gives to the restricted cows the ability to recover weight loss and condition without affecting milk production. Much of Em variation with energy level is related to rapid changes in visceral mass and energy expenditures (Freetly *et al.*, 1998; Ortigues *et al.*, 1993).

Thus, the ability of cows to mobilize and recover body reserves under restriction/refeeding periods provides them the adaptive ability to low net energy for maintaining liveweight constant. The lag time in adaptation of Em as a result of the feeding restriction is partly the support of cows' adaptation at least at short and medium term. The long term effects of adaptation should be further investigated. Variations in Em can be so interpreted as an indicator of the ability of mature producing cows to face nutritional constraints.

References

Freetly H., J. Nienaber and T. Brown-Brandl, 1998. Efficiency of energy and nitrogen loss and gain in mature cows. Journal of Animal Science, 76, 896-905.

Garcia F. and J. Agabriel, 2007. Update of feeding recommendations for fattening of culled cows. Development of a model for the estimation of the composition of weight gain and associated requirements. Productions Animales, 20, 137-149.

Hornick J., C. Van Eename, A. Clinquart, O. Gerard and L. Istasse, 2000. Mecanism of reduced and compensatory growth. Domestic Animal Endocrinology, 19, 121-132.

INRA 2007. Alimentation des bovins, ovins et caprins. Besoins des animaux – valeurs des aliments, Quae (Ed), Versailles

Ortigues I., M. Petit, J. Agabriel and M. Vermorel, 1993. Maintenance requirements in metabolizable energy of adult non pregnant non lactating Charolais Cows. Journal of Animal Science, 71, 1947-1956.

SAS, 2009. Release 9.2. SAS Inst. Inc, cary, NC, USA.

Response to high altitude grazing in metabolic traits and performance by yak crossbreds and yaks

S. Barsila[1], N.R. Devkota[2], M. Kreuzer[1] and S. Marquardt[1]
[1]ETH Zurich, Institute of Agricultural Sciences, 8092 Zurich, Switzerland; michael.kreuzer@inw.agrl.ethz.ch
[2]Institute of Agriculture and Animal Science, Rampur, Chitwan, Nepal

Introduction

Yaks (*Bos grunniens*) and yak crossbreds are kept in transhumant systems using different pastures along an altitudinal gradient in the Nepalese Himalayan Mountains. Yaks are known to be well adapted to cope with low oxygen partial pressure, low temperatures and the harsh mountain environment (Wiener *et al.*, 2003), and the resulting energy deficiency. Less information is available on the adaptive capabilities of yak crossbreds. Crossbreeding is a strategy to obtain higher milk yields especially by heterosis. In the Eastern Nepalese Himalayan Mountains, two different local breeds are used for crossbreeding. *Bhelang* bulls from near Tibet (*Bos taurus* genotype) are crossed with female yaks, and yak bulls are crossed with female Nepalese hill cattle (*Bos indicus*). Female crosses of cattle bulls × yak cows are locally called Dimjo chauries, while those which are produced using cows and yak bulls are called Urang chauries (Joshi, 1982). In order to compare the adaptive capacity of these two crossbred types with that of yaks, an experiment was conducted at two altitudes in the Taplejung District of Nepal.

Material and methods

Six adult lactating females per genotype that calved in 1-2 week intervals from the beginning of April onwards (*B. grunniens* × *B. indicus* (n=6 Y×I), *B. taurus* × *B. grunniens* (n=6 T×Y) and *B. grunniens* (n=6 Y)) were selected. Different from the crossbreds, the yaks were accompanied by their calves and only milked in the morning. The crossbreds were milked twice a day. The experiment was conducted at 4,700 m (August) and 3,000 m (October) during 2 weeks each. The first days were used as adaptation period and the last 6 days were used for measurements and data sampling. Daily measured data included milk yield and milk composition. Milk fat, protein and lactose were analysed with a portable milk analyser (Lactoscan SA-L, Milkotronic Limited, Nova Zagora, Bulgaria) at every milking. For the crossbreds, the aliquot was calculated from morning and evening measurements. In yaks, the amount suckled by the calves was determined by the weigh-suckle-weigh method. Heart rate was measured in one animal per genotype attaching the device to a different animal every day. Polar Equine CS600X Trotting instruments (Polar Electro Oy, Kempele, Finland) and the corresponding software were used for measurement and analysis of the cardiac data. Only short events of 5 consecutive minutes where the animals were standing ($\leq$3 steps/min) were used for data analysis by calculating a mean value from 1-4 measurements/animal. Respiration rate and rectal temperature were measured in the morning and evening after milking by using a stethoscope and a thermometer. Ear blood samples were taken on the first day of adaptation and the last day of the sampling period and were immediately analysed for haemoglobin, glucose and lactate using point-of-care devices designed for human application (HemoCue Hb 201+, HemoCue AB, Ångelholm, Sweden; Accu-Chek Aviva, Roche Diagnostics, Mannheim, Germany and Accutrend Plus, Roche Diagnostics, Mannheim, Germany).

The Mixed procedure of the SAS program (SAS Institute Inc., Cary, USA, 2009 version 9.3) was used for statistical analysis. Data on heart rate and milk yield and composition was analyzed with altitude, genotype and its interaction as fixed effects, altitude as repeated factor and animal nested within genotype as subject. For respiration rate and rectal temperature the model included additionally the effect of daytime and a random effect was estimated here to account for the two repeated within-subject factors time and altitude.

Results and discussion

Blood haemoglobin was affected by genotype and altitude ($P<0.001$). The highest levels were recorded at the end of the stay at 4,700 m (161, 153 and 140 g/l in Y, T×Y and Y×I, respectively). Y already started from higher levels (154 g/l) than TxY (145 g/l) and even more than Y×I (119 g/l) as measured at arrival at 4,700 m. Haemoglobin values decreased ($P<0.05$) during the stay at 3,000 m for both crossbreds, but were unchanged ($P>0.05$) for Y. At first sampling at 4,700 m, blood glucose level was higher in TxY (7.3 mmol/l) than in Y×I (6.5 mmol/l), with Y being intermediate ($P>0.05$); this had been levelled out until the second sampling at 4,700 m. Across sampling times and genotypes, glucose levels were higher at 3,000 m as compared to 4,700 m (7.6 and 6.2 mmol/l, respectively). In blood lactate, differences between the genotypes were large at 4,700 m being low in Y (2.6 mmol/l), intermediate in T×Y (3.9 mmol/l) and high in Y×I (5.5 mmol/l). The values decreased during the stay at 4,700 m in all three genotypes by keeping the same order. At 3,000 m, no differences were found between the crossbreds and times of measurement ($P>0.05$). Only the values measured for yaks were higher (5.5 mmol/l, $P<0.05$) at the first measurement. Heart rate was overall higher at 4,700 m as compared to 3,000 m (78 and 50 beats/min, respectively). Also respiration rate and rectal temperature were higher at 4,700 m as compared to 3,000 m. Elevated levels of these three traits and of blood haemoglobin as response to altitude were also found elsewhere (Bianca and Näf, 1979). At 4,700 m, milk yield (kg/day) was 2.5 (T×Y), 2.1 (Y) and 2.1 (Y×I). At 3,000 m, Y and Y×I only had a milk yield of 0.8 kg/day while T×Y still yielded 1.2 kg/day. Milk fat content was higher at 3,000 m being highest in Y (8.2%), followed by T×Y (6.8%), and lowest in Y×I (6.4%). At 4,700 m, milk protein content was generally higher (4.1, 4.0 and 3.9% in Y, T×Y and Y×I, respectively, $P>0.05$) than at 3,000 m.

The initially very high levels of haemoglobin and low levels of lactate and the comparably high yield and milk protein content illustrate that Y are better adapted to high altitude than any of the crossbreds. However, T×Y not only showed the best performance among the three genotypes, but was also close to Y in the variables indicative of adaptation. Having the choice between Y×I and T×Y, the latter therefore seems to be a better alternative for farmers which have access to very high altitude pastures.

References

Bianca W. and F. Näf, 1979. Responses of cattle to the combined exposure, to diurnal temperature rhythm (-5 to 25 °C) and to simulated high-altitude (4,000 m). Int. J. Biometeor 23, 299-310.

Joshi, D.D., 1982. Yak and chauri husbandry in Nepal. K. D. Joshi (Publisher), Kathmandu, Nepal. 145 pp.

Wiener, G., H. Jianlin and R.J. Long (eds.), 2003. The Yak. 2. Edition. RAP Publication 2003/06. FAO Regional Office for Asia and the Pacific. Bangkok, Thailand, 460 pp.

Estimates of nutritional requirement of sheep, goats and cattle in tropical and warm countries: a meta-analysis study

N. Salah[1], D. Sauvant[2] and H. Archimède[1]
[1]INRA UR143, Unité de Recherches Zootechniques, 97170 Petit-bourg, Guadeloupe, France; harry.archimede@antilles.inra.fr
[2]INRA, UMR791, Modélisation Systémique Appliquée aux Ruminants, 16 rue Claude Bernard, 75005 Paris, France

Introduction

Until now, feeding recommendations for livestock in tropical and warm areas are largely based on standards produced in temperate area (ARC, 1984; INRA, 1989; NRC, 1985). Nevertheless, energy and protein requirements of livestock in warm area could differ from those of temperate area. The objective of this study was to estimate energy and protein requirements of ruminants in tropical and warm area using a meta-analysis.

Material and methods

Published studies performed with growing sheep, goat and cattle were used for this analysis. To be included in this meta-analysis, the papers were selected on some criteria. At a minimum, trials should report data on animal weight gain, chemical composition of diets, intake (organic matter, OM and crude protein, CP), total tract digestibility (OM and CP) and nitrogen balance. Metabolisable energy intake (kcal MEI/kg BW) was predicted from digestible OM intake per BW (g DOMI/kg BW) on the basis of a data base:

$$MEI/BW = -2.03 + 4.03\ DOMI/BW\ (n=975,\ R^2=0.99,\ RSD=11.3)$$

In total, 590 publications representing 2,225 dietary treatments were pooled to be used in the present study. There were 325 and 1,287; 145 and 544; 119 and 394 publications and treatments for sheep, goat and cattle respectively. Genotype animals from warm regions were differentiated from those of temperate regions. The meta-analysis has been performed following recommendations of Sauvant *et al.* (2008). Inter-publications regressions of nutrient intake (as explained variables) on average daily gain (ADG) were performed to estimate requirements for maintenance and growth. The intercept and the slope of the regression are the estimation of the maintenance and growth requirements respectively. Moreover, to test simultaneously the influences of species and genotypes on either the intercept or the slope of the regressions, analyses of variance-covariance were applied on the parameters. The study effect was considered random. ME has been used as energy unit. Protein requirements were estimated regressing Digestible CP intake (DCPI) on ADG or retained nitrogen (Nr). Two units of BW (BW and $BW^{0.75}$) were tested because hierarchy between species could differ depending on unit. Normal distribution of data was observed within and across species and BW units.

Results and discussion

Equation 1 indicates that ME for maintenance (MEm) of cattle was significantly higher than those of sheep and goat according to their $BW^{0.75}$. No differences were noted between genotypes. Concerning ME requirement for growth (MEg), differences were not significant between species nor genotypes. When requirements were estimated on the basis of the BW, values were lower with cattle compared with small ruminants (Equation 2). Moreover, MEm was higher for tropical small ruminants compared to the temperate ones (Equation 2).

Table 1. Inter-publications regressions of Energy intake (Kcal metabolic energy intake, MEI/kg body weight, BW or MEI/kg metabolic body weight, MBW) or digestible crude protein intake (g, DCPI) on average daily gains (ADG, g) performed to estimate requirements for maintenance and growth of warm genotype livestock.

Equation	Equation regression
1	MEI/MBW = Ei + 5.78 ADG/MBW, (n=362, species=2, R^2=0.41, RSD=42.2) Ei = 129.2±4.2 kcal ME /kg MBW for small ruminants, 150.3±4.2 kcal ME/kg MBW for cattle.
2	MEI/BW = Ei + 5.39 ADG/BW, (n=362, species=2, R^2=0.51, RSD=16.9) Ei = 60.8±1.58 kcal ME/kg BW for small ruminants, 43.7±1.58 kcal ME/k gBW for cattle.
3	DCPI/MBW = Ei + 6.47 Nr/MBW (n=168, species=3, R^2=0.54, RSD=1.66) Ei = 3.36± 0.27 for sheep, 2.38±0.27 for goats and 2.81±0.27 for cattle.
4	DCPI/MBW = 3.53 + 0.446 ADG/MBW – 0.0058 ADG/MBW^2 (n=331, R^2=0.41, RSD=2.1)
5	ADG/MBW = Ei + 9.85 Nr/MBW (n=182, species=3, R^2=0.41, RSD=3.74) Ei = 3.60±0.57 for sheep, 0.12±0.57 for goats, 3.10±0.57 for cattle

The regression of DCPI against Nr indicated significantly higher protein maintenance requirement for sheep compared to goats with an intermediate value for cattle (Equation 3). No differences were noted between species and genotype for protein growth requirement assimilated as Nr. However, it is known that Nr leads to an overestimation of growth requirements. The regression of DCPI against ADG was curvilinear (Equation 4). No differences were noted between species and genotype for protein maintenance requirements. As the regression is not linear, the marginal DCP requirement/kg ADG decreases, from 0.446 g DCP/kg ADG when ADG is close to 0.0 to 0.326 g DCP/kg ADG when ADG=10 g/kg MBW and equal to 0.206 g DCP/kg ADG when ADG=20 g/kg MBW (corresponding to the highest values recorded for growth in the data base).

As a conclusion, in this study we hypothesized that MEm corresponds to ME requirements for absence of weight gain rather than absence of energy gain. Moreover we have not taken into account changes in the energy content of BW gain. This approach is different from that used to establish some international standards (INRA, ARC or NRC) and could partly explain some of the observed differences. However, the comparisons made in this study, with the same methodology, indicate that the energy and protein requirement for maintenance and growth estimated for tropical livestock are higher than values for temperate livestock.

References

ARC, 1984. Nutrient requirements of ruminant livestock.

INRA, 1989. Ruminant nutrition. Recommended Allowances and Feed Tables. Jarrige R. (ed.) Paris, France, 373 p.

NRC, 1985. Nutrient requirements of sheep, 6th Revised ed. Washington, DC, 2-53. National Academy of Sciences.

Sauvant, D., P. Schmidely, J.J. Daudina and N.R. St-Pierre, 2008. Meta-analyses of experimental data in animal nutrition. Animal 2, 1203-1214.

Net energy and protein requirements for growth of goats kids

I.A.M.A. Teixeira[1], N. St-Pierre[2], A. Cannas[3], A.P. Souza[1], M.H.R. Fernandes[1] and K.T. Resende[1]
[1]UNESP, Univ. Est. Paulista, Dep. de Zootecnia, 14884-900 Jaboticabal, SP, Brazil; izabelle@fcav.unesp.br
[2]The Ohio State University, Department of Animal Sciences, Columbus, OH, 43210, USA
[3]Universitá Degli Studi di Sassari, Dipartimento di Scienze Zootecniche, 07100 Sassari, Italy

Introduction

Knowledge of nutritional requirements is decisive for a successful nutrition system, since animals fed properly convert more efficiently the nutrients ingested in products. Current feeding systems for goats still comprise information extrapolated from sheep and/or cattle. However, the data obtained with these species should not be used for goats, because of differences between these species, such as dietary habits, physical activity, milk and carcass composition, different adaptation mechanisms, among others (NRC, 2007). Therefore, the objective of this study was to determine net energy and protein requirements for growth of goat kids, using meta-analysis as a statistical tool.

Material and methods

A total of 412 goat kids from nine studies carried out at UNESP (Jaboticabal, Brazil) and UFV (Viçosa; Brazil) were used. These studies comprise goats of 4 genotypes, 3 genders and in different composition of mature body weight (BW). The main characteristics of each study are presented in Table 1. In five of the studies (studies 2, 3, 6, 7 and 8) goat kids were unweaned in part of the trial. All studies used diet with roughage to concentrate ratio of approximately 50:50. In all studies individual measurements of intake, digestibility, crude protein and metabolizable energy intake and body composition (based on comparative slaughter) were available.

Net energy and protein requirements for gain were estimated using animals fed ad libitum. Logarithmized allometric equations were used to obtain prediction models of protein and energy

Table 1. Description of the studies used in the meta-analysis.

Study	N	Gender	Genotype	Body weight[1]	Body composition[2]
1	34	Male	1/2 Boer, 1/2 Saanen	15 to 25	18.5; 9.7; 1.9
2	30	Male	1/2 Boer, 1/2 Saanen	5 to15	19.0; 5.4; 1.5
3	41	Male	Saanen	5 to 20	17.8; 7.4; 1.7
4	45	Castrated male	Saanen	20 to 35	17.9; 13.6; 2.2
5	33	Male	3/4 Boer, 1/4 Saanen	20 to 35	18.9; 14.9; 2.3
6	18	Male	1/2 local, 1/2 dairy breed	5 to 25	19.4; 6.6; 1.8
7	30	Male	1/2 local, 1/2 dairy breed	5 to 15	18.0; 8.9; 2.5
8	26	Castrated male			
	23	Female		5 to 15	25.0; 8.2; 2.1
	31	Male	Saanen		
9	36	Castrated male			
	33	Female		30 to 45	16.9; 21.9; 2.9
	32	Male	Saanen		

[1] Body weight (kg) range of the animals from the initial to the slaughter body weight.
[2] Average of protein (% empty body weight – EBW), fat (% EBW) and energy (Mcal/kg EBW) body composition of the animals, respectively.

concentration of empty body weight (EBW), based on the amount of protein and energy in the empty body. And the derivative of these equations were used to estimate net protein and energy requirements for gain, according to ARC (1980). Net protein and energy maintenance requirements were estimated by regressing protein and energy retention on crude protein and metabolizable energy intake, respectively. The data were analyzed using MIXED procedure of SAS and orthogonal contrasts were used to test the intercepts and slope estimates in the regression equations among the main evaluated factors (genotype, gender and body weight range) across studies.

Results and discussion

The parameters of the equation to predict energy and protein concentration of empty body weight are presented in Table 2. Net protein requirement for gain decreased as body weight increased ($P<0.05$), in Saanen goat kids. An opposite pattern was observed with net energy requirement for gain, in which net energy requirement increased with increasing of body weight ($P<0.05$). It was also observed a significant genotype effect in the net energy requirements for gain ($P<0.05$). On the other hand, there was no gender effect in net protein and energy requirement for gain of goat kids ($P>0.05$).
At the beginning of the growth stage, net protein requirements for maintenance of Boer crossbred goat kids were greater compared to Saanen goat kids ($P<0.05$). It was not observed significant gender effect in the net energy and protein requirement for maintenance in Saanen goat kids ($P>0.05$).
In conclusion, net requirements for growth of goat kids is not affected by gender, on the other hand genotype influenced net requirements for growth.

Table 2. Parameters of the regression equations of protein and energy concentration on empty body weight (EBW) of goat kids.

Study	Protein[1]		Energy[2]		P-value[3]
	Intercept	Slope	Intercept	Slope	
1	2.38±0.12	0.90±0.10	2.84±0.11	1.37±0.10	<0.0001
2	2.26±0.06	1.03±0.06	2.94±0.05	1.28±0.06	<0.0001
3	2.20±0.04	1.04±0.04	3.07±0.04	1.16±0.04	<0.0001
4	2.32±0.12	0.94±0.09	2.52±0.11	1.60±0.08	<0.0001
5	2.16±0.12	1.08±0.09	3.04±0.11	1.23±0.08	<0.0001
6	2.18±0.04	1.10±0.04	3.05±0.04	1.19±0.04	<0.0001
7	2.22±0.04	1.04±0.04	2.86±0.04	1.45±0.04	<0.0001
8 castrated male	2.43±0.05	0.96±0.06	3.14±0.05	1.20±0.06	<0.0001
8 female	2.46±0.05	0.92±0.05	3.07±0.05	1.30±0.05	<0.0001
8 male	2.40±0.05	1.00±0.05	3.12±0.05	1.22±0.05	<0.0001
9 castrated male	2.07±0.16	1.11±0.11	2.67±0.16	1.53±0.11	<0.0001
9 female	2.03±0.16	1.14±0.11	2.72±0.16	1.55±0.11	<0.0001
9 male	2.33±0.19	0.93±0.13	2.50±0.18	1.60±0.12	<0.0001

[1] Log protein (g) = b0 + b1 Log EBW (kg).
[2] Log energy (kcal) = b0 + b1 Log EBW (kg).
[3] *P*-value for intercept and slope on protein and energy.

References

ARC. Agricultural Research Council, 1980. The nutrient requirements of ruminant livestock. The Gresham Press, London, England, 351 pp.

NRC. National Research Council. 2007. Nutrient requirements of small ruminants. Sheep, Goats, Cervids and New World Camelids. National Academy Press, Washington, USA, 292 pp.

Low protein solid feed enhances nitrogen utilization by urea-N recycling in veal calves

H. Berends[1], J.J.G.C. van den Borne[1], B.A. Røjen[2], J. van Baal[1] and W.J.J. Gerrits[1]
[1]Animal Nutrition Group, Wageningen University, Wageningen, the Netherlands; harma.berends@wur.nl
[2]Department of Animal Health and Bioscience, Aarhus University, Tjele, Denmark

Introduction

Veal calves fed merely milk replacer (MR) are typically known for their low nitrogen (N) utilization for growth and high urea-N excretion (Gerrits *et al.*, 1996). Urea-N recycling from blood to the gastrointestinal tract is expected to play a minor role in these preruminant calves, as shown by the 80% recovery of an intravenous pulse dose of [^{13}C]urea in 48-h urine. We have recently shown that provision of low-protein solid feed (SF) increases the efficiency of N utilization for protein gain in veal calves, particularly towards the end of the fattening period (Berends *et al.*, 2012). We expect that in these calves, urea-N recycling could be stimulated by a low protein-to-energy ratio in rumen contents, and by high blood urea-N concentrations (Reynolds and Kristensen, 2008). Urea-N entry into the gastrointestinal tract may be mediated through urea-N transporters and aquaglyceroporins and is potentially influenced by dietary factors (Røjen *et al.*, 2011; Simmons *et al.*, 2009). The current study was designed to quantify the effect of low-protein SF intake, provided in addition to MR, on urea-N recycling and expression of genes associated with urea transport in veal calves. We hypothesized that low protein solid feed provision in addition to MR would contribute substantially to urea-N recycling and expression of genes associated with urea transport.

Material and methods

Forty-eight Holstein Friesian male calves, 55±0.3 kg of bodyweight (BW), were fed 41 g of MR/kg BW$^{0.75}$/d, and assigned to 1 of 4 levels of SF intake: 0, 9, 18, or 27 g of DM of SF/kg BW$^{0.75}$/d. The SF mixture was 25% chopped straw, 25% chopped corn silage and 50% concentrate on a DM basis. Urea-N recycling was quantified at 164±1.6 kg BW by intravenous administration of [^{15}N^{15}N] urea for 24 h after a priming dose to enrich the body urea-N pool to approximately 0.15 atom% excess. Cumulative 68-h urine was analysed for urea isotopomers and feces for total ^{15}N enrichment. Complete N-balance was performed over a 4 d period. At slaughter, rumen tissue samples were collected from the ventral caudal site to determine the absolute mRNA levels of urea transporter-B (UT-B), and aquaporins 3 (AQP3) and 7 (AQP7) by real-time quantitative PCR. Each expression level was normalized to that of importin 8 (IPO8). RT-qPCR data were analysed by ANOVA with fixed effect SF intake. Urea-N recycling and RT-qPCR data were analyzed by regression with SF intake as regressor.

Results and discussion

Urea entry rate, assumed equal to total urea synthesis, averaged 28.7 g urea-N/d for all calves and was not affected by SF intake (P>0.05; Table 1). Urea entry to the gut was estimated 6.80 g urea-N/d for calves without SF, and increased with SF intake (0.28±0.09 g urea-N/g DM from SF/kg$^{0.75/}$d; P<0.01). Subsequently, the estimated return of urea-N to the ornithine cycle was 1.08 g urea-N/d for calves without SF and increased with SF intake (0.10±0.02 g urea-N/g DM from SF/kg$^{0.75/}$d; P<0.001). Estimated fecal excretion was 0.15 g urea-N/d for calves without SF and increased with SF intake (0.08±0.01 g urea-N/g DM from SF/kg$^{0.75}$/d; P<0.001), which is likely undigested microbial N. As a net result, urea-N used for anabolic purposes was estimated 5.57 g urea-N/d for calves without SF and increased with SF intake (0.10±0.08 g urea-N/g DM from SF/kg$^{0.75}$/d; P<0.01).

Table 1. Effects of incremental SF intake (in g DM/kg$^{0.75}$/d), fed in addition to a MR diet, on urea-N transfers (in g urea-N/d), based on [$^{15}N^{15}N$]urea infusions in veal calves (164 kg BW). SF intake was 0, 9, 18, or 27 g DM/kg$^{0.75}$/d and MR intake was identical for all calves (41 g DM/kg$^{0.75}$/d).

Response parameter	Covariable	SF0 level	Beta		
			Estimate	SE	*P*-value
Urea entry rate (UER)	SF intake	30.41	-0.12	0.098	NS
Urea entry to the GIT (GER)	SF intake	6.80	0.28	0.088	<0.01
Return to ornithine cycle (ROC)	SF intake	1.08	0.10	0.021	<0.001
Loss to feces (UFE)	SF intake	0.15	0.08	0.008	<0.001
Re-use for anabolism (UUA)	SF intake	5.57	0.10	0.079	<0.01

The increase in urea-N used for anabolic purposes explained 19% of the increase observed in whole body N retention, as reported by Berends *et al.* (2012).

The mRNA expression of UT-B was increased more than five-fold with SF (Figure 1). The mRNA expression of AQP3 increased ($P<0.05$) with SF intake, whereas mRNA expression of AQP7 did not. In conclusion, an increase in low-protein SF intake was accompanied by an increased gut entry rate of urea-N, likely to be a substantial portion of the observed increase in N retention. Furthermore, any SF stimulates expression of mRNA of UT-B in the rumen wall.

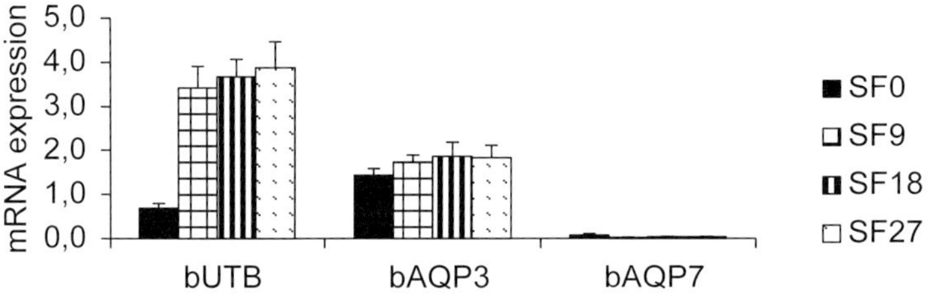

Figure 1. mRNA expression of bUT-B bAQP-3 and bAQP-7 in veal calves (25 wk). SF intake was 0, 9, 18, or 27 g DM/kg BW$^{0.75}$/d and MR intake was identical for all calves (41 g DM/kg BW/d). Data are means ± SEM of 12 calves for each treatment.

References

Berends, H., J.J.G.C. van den Borne, S.J.J. Alferink, C.G. van Reenen, E.A.M. Bokkers and W.J.J. Gerrits. 2012. Low-protein solid feed improves the utilization of milk replacer for protein gain in veal calves. J. Dairy Sci. 95, 6654-6664.

Gerrits, W.J.J., G.H. Tolman, J.W. Schrama, S. Tamminga, M.W. Bosch and M.W.A. Verstegen. 1996. Effect of protein and protein-free energy intake on protein and fat deposition rates in preruminant calves of 80 to 240 kg live weight. J. Anim. Sci. 74, 2129-2139.

Reynolds, C.K. and N.B. Kristensen. 2008. Nitrogen recycling through the gut and the N economy of ruminants: An asynchronous symbiosis. J. Anim. Sci. 86, E293-E305.

Røjen, B.A., S.B. Poulsen, P.K. Theil, R.A. Fenton and N.B. Kristensen. 2011. Effects of dietary nitrogen concentration on messenger RNA expression and protein abundance of urea transporter-B and aquaporins in ruminal papillae from lactating Holstein cows. J. Dairy Sci. 94, 2578-2591.

Simmons, N.L., A.S. Chaudhry, C. Graham, E.S. Scriven, A. Thistlethwaite, C.P. Smith and G.S. Stewart. 2009. Dietary regulation of ruminal bovine UT-B urea transporter expression and localization. J. Anim. Sci. 87, 3288-3299.

Effect of replacing feed grains by food by-product on energy metabolism of lactating cows

K. Higuchi[1], F. Ohtani[1], Y. Kobayashi[1], I. Nonaka[2], K. Yayou[3], M. Sutoh[1] and O. Enishi[1]
[1]National Institute of Livestock and Grassland Sciences, 2 Ikenodai, Tsukuba, Ibaraki 305-0901, Japan; higuchik@affrc.go.jp
[2]National Institute of Kyusyu Okinawa Agricultural Research Center, 2421 Suya, Koshi, Kumamoto 861-1192, Japan
[3]National Institute of Agrobiological Sciences, 2 Ikenodai, Tsukuba, Ibaraki 305-8602, Japan

Introduction

The lactating cow requires a large amount of energy for milk production. Thus, it is important to clarify and satisfy the cow's energy requirement for milk production. Additionally, it is also critical to improve the efficiency of energy utilization for milk production to reduce both cow's metabolic load and feed costs. We have begun work to estimate the impact of roughage to concentrate ratio in the diet on energy metabolism of the lactating cow at the whole-body and mammary gland level. In this experiment, the energy efficiency for milk production from cows fed diets which contained 45% roughage and either feed grains or a by-product were examined.

Material and methods

Three Holstein lactating cows fitted with an ultrasound flow probe around the left external pudic artery were housed in a temperature and humidity controlled-room. Cows were fed 2 diets: a control diet (Control, 40% Italian ryegrass silage, 5% alfalfa hay cube and 55% concentrate mix), or a by-product diet (Bypro, replacing a part of soybean meal, corn and barley of the control diet by beet pulp and soybean curd residue) for 21-days periods according to a cross over design. The crude protein content of the Control and Bypro diets were 17.2 and 18.2%, respectively. The feeds were fed to meet 100% of the total digestible nutrients requirement for a lactating Holstein cow producing 28 kg milk/d. The energy balance trials were conducted using open circuit respiratory chambers with digestion trial apparatus (Higuchi *et al.*, 2010b). The efficiency of metabolizable energy (ME) utilization for lactation (k_l) was calculated assuming ME requirement for maintenance (MEm) as 486.6 kJ/kg $BW^{0.75}$ (NARO, 2007). The energy expenditure of the udder was calculated from their oxygen consumption using arterio-venous difference technique measuring mammary blood flow and blood oxygen concentration (Higuchi *et al.*, 2010a). Blood metabolites were analyzed enzymatically by commercial kits. The data were analyzed according to the following model using GLM procedure of SAS 9.2: $Y_{ijk} = \mu + T_i + F_j + C_k + e_{ijk}$; where μ is the overall mean; T_i, F_j, C_k are the effects of the term, feed, and cow, respectively; and e_{ijk} is the residual error.

Results and discussion

There were no significant differences between treatments in body weight, dry matter intake and milk yield (Table 1). The milk fat content was significantly higher in cows given the Bypro diet. Dry matter and fiber digestibility were same in both diets, but starch digestibility was higher in Control (98.3<95.7%, P<0.05). Gross energy intake tended to be higher in Bypro; however, energy partitioning for feces, urine, methane, heat production, milk and retention were not different between treatments. The k_l did not differ between treatments. The average of metabolizability was calculated as 0.58, and this value was lower than that (0.60~0.61) in the previous study in which cows receiving 30% roughage diets (Higuchi *et al.*, 2010a). In spite of that the k_l would be 0.62~0.63 when metabolizability is 0.60 (ARC, 1980; NARO, 2007), the average of k_l was calculated as 0.67 in this study. Agnew and Yan (2000) suggested that increasing dietary proportion of forage increased MEm but had no significant effect on k_l in lactating dairy cows. Therefore MEm was calculated as

Table 1. Performance, whole body and mammary gland energy metabolism.

	Control	Bypro[1]	SEM
Body weight, kg	596	594	1
Dry matter intake, kg/day	17.5	17.5	0.0
Milk yield, kg/day	24.4	24.7	0.3
Milk fat, %	4.3	4.4[b]	0.0
Gross energy intake, kJ/kg $BW^{0.75}$/day	2,684	2,730[a]	5
Fecal energy, kJ/kg $BW^{0.75}$/day	867	883	4
Urinary energy, kJ/kg $BW^{0.75}$/day	70	77	1
Methane energy, kJ/kg $BW^{0.75}$/day	182	176	4
Heat production, kJ/kg $BW^{0.75}$/day	862	858	15
Milk energy, kJ/kg $BW^{0.75}$/day	641	662	7
Retained energy, kJ/kg $BW^{0.75}$/day	61	74	4
ME intake, kJ/kg $BW^{0.75}$/day	1,564	1,594	12
k_l	0.66	0.67	0.01
Mammary blood flow, l/min	4.3	3.7[a]	0.1
Mammary energy expenditure, kJ/kg $BW^{0.75}$/day	64.9	61.0	5.6
Mammary energy uptake, kJ/kg $BW^{0.75}$/day	705	723[a]	2
Mammary energy efficiency	0.91	0.92	0.01
Mammary glycerol uptake, mmol/h	10.9	17.1[b]	1.5

[1] Superscript a=P<0.10; b=P<0.05.

fixed value, thus the k_l was seemed to show apparently higher value in this study. Mammary energy expenditure did not differ between treatments. Despite the decrease in mammary blood flow in Bypro, mammary energy uptake tended to be higher in the Bypro supplemented cows. The amounts of nutrient supplies toward the mammary gland for milk production were not different between treatments; however, mammary glycerol uptake was higher in Bypro. The higher milk fat content and higher energy uptake of mammary gland in cows given the Bypro diet were due to higher uptake of nutrients, more than the decline in mammary blood flow, for milk production at mammary gland.

References

Agnew, R.E., and Yan, T. 2000. Impact of recent research on energy feeding systems for dairy cattle. Livest. Prod. Sci., 66, 197-215

Agricultural Research Council, 1980. The Nutrient Requirements of Ruminant Livestock. CAB INTERNATIONAL, UK.

Higuchi, K., Y. Kobayashi, I. Nonaka, and O. Enishi, 2010b. A new data acquisition system for respiration trial system on metabolism laboratory in National Institute of Livestock and Grassland Science. Bull. Natl. Inst. Livest. Grassl. Sci. (Jpn.), 10, 15-27.

Higuchi, K., F. Ohtani, Y. Kobayashi, I. Nonaka, K. Yayou, M. Sutoh and O. Enishi, 2010a. Mammary energy consumption and its relation to whole body energy metabolism in lactating cows fed high-concentrate diets. In: Crovetto G.M.(eds), Energy and protein metabolism and nutrition. Wageningen Academic Publishers, Wageningen, the Netherlands, EAAP publication No. 127: 231-232.

National Agriculture and Food Research Organization (NARO), 2007. Japanese Feeding Standard for Dairy Cattle, Japan Livestock Industry Association, Japan.

Source of carbohydrate and protein in the diet of recently weaned dairy calves

T.M. Hill, J.D. Quigley H.G. Bateman, II, J.M. Aldrich and R.L. Schlotterbeck
Nurture Research Center, Provimi North America, Brookville, OH 45309, USA; mhill@provimi-na.com

Introduction

The Dairy NRC (2001) suggests that digestible fiber is needed in the diet of calves. However, in our recent research, live body weight average daily gain (ADG) decreased when starch from corn was replaced with sources of digestible fiber; i.e. soybean hulls, wheat middlings, or distiller's dried grains with soluble (Hill *et al.*, 2008; Suarez-Mena *et al.*, 2012). Additionally, limited recent research has not demonstrated consistent ADG responses to manipulating a calf's diet for metabolizable protein compared to control diets that were predominately corn and soybean meal (Hill *et al.*, 2007). The objective was to measure diet digestibility as a potential means to explain differences in performance when different carbohydrate and protein sources were fed to calves.

Material and methods

We conducted 2, 56-d trials with weaned Holstein dairy calves (initially 72±1.8 kg BW, 58 to 60 d of age) fed a 95% concentrate and 5% chopped grass hay diets. Each trial used 96 calves (4 calves/ pen). During 15 of the last 21 d of the first trial and 10 of 14 d of the second and third week of the second trial, fecal samples were taken from the ground of half of the pens. Care was taken to not sample non-fecal material. Fecal samples were composited by pen. The diet was sampled on the same days and composited by diet. Acid insoluble ash was used as an internal marker of digestion. Digestibility estimates along with 56-d performance were analyzed.

In Trial 1, a textured diet (19% CP) with high starch (52% starch, 13% NDF) based on whole corn and oats or a pelleted low starch (20% starch, 35% NDF), high digestible fiber diet were used (6 pens/ diet). Within starch level, diets were formulated from all supplemental soybean meal or soybean meal with blood meal and Alimet® to provide 2 metabolizable protein levels (1 and 1.07% metabolizable lysine plus methionine). The 4 treatments were analyzed as a completely randomized design with a 2×2 factorial arrangement with pen as experimental unit. Differences were declared at $P<0.05$. Factors were starch, protein, and their interaction.

In Trial 2, all ground, pelleted diets (19% CP) were fed. Diets were based on soybean hulls, wheat middlings, or corn, which contained increasing concentrations of starch (13, 27, and 42% starch; 42, 23, 16% NDF), respectively (8 pens/diet). Contrast statements were constructed to separate differences in the means (soybean hulls plus wheat middlings vs. corn; soybean hulls vs. wheat middlings). Differences were declared at $P<0.05$.

Results and discussion

In Trial 1, intake of OM did not differ (3.0±0.08 kg/d). Digestibility of OM was greater in calves fed high (0.85) vs. low (0.79) starch diets (Table 1). Digestibility of NDF (0.59 vs. 0.67) and starch (0.95 vs. 0.99) were less in calves fed the high vs. low starch diets. Calf ADG and hip width change were greater for high vs. low starch diets. Source of protein did not influence digestibility or ADG.

In Trial 2, intake of OM was not different (2.3±0.10 kg/d). Digestibility of OM was greater in calves fed corn (0.85) vs. other (0.78) diets (Table 2). Digestibility of NDF was greater for calves fed soybean hulls vs. wheat middlings. Starch digestibility averaged 0.98 and was not different among treatments. Calf ADG and hip width change were greater in calves fed corn vs. other diets.

Table 1. Digestibility of nutrients, average daily gain (ADG), and hip width change in Trial 1.

Starch (ST)	High		Low		SEM	*P* value		
Metabolizable protein (MP)	Low	High	Low	High		ST	MP	ST×MP
OM digestion, fraction	0.849	0.850	0.802	0.789	0.006	0.01	0.34	0.28
NDF digestion, fraction	0.568	0.628	0.694	0.661	0.016	0.01	0.54	0.12
Starch digestion, fraction	0.951	0.957	0.990	0.987	0.003	0.01	0.66	0.18
ADG, kg/d	1.07	1.09	1.00	1.00	0.032	0.02	0.79	0.72
Hip width change, cm	6.0	5.9	5.4	5.2	0.22	0.01	0.62	0.65

Table 2. Digestibility of nutrients, average daily gain (ADG), and hip width change in Trial 2.

Nutrient	Soyhulls (S)	Middlings (M)	Corn (C)	SEM	*P*-value	
					S+M vs. C	S vs. M
OM digestion fraction	0.775	0.796	0.858	0.0156	0.01	0.21
NDF digestion fraction	0.707	0.561	0.662	0.0313	0.34	0.01
Starch digestion, fraction	0.976	0.989	0.970	0.0057	0.13	0.15
ADG, kg/d	0.95	0.93	1.11	0.049	0.01	0.57
Hip width change, cm	5.1	5.0	5.6	0.23	0.01	0.44

High starch diets were more digestible and supported more growth in 2 to 4 month old dairy calves than replacing starch with digestible fiber. Manipulating metabolizable protein compared to a control diet that was predominately corn and soybean meal did not alter growth or digestibility.

References

Hill, T.M., J.M. Aldrich, R.L. Schlotterbeck and H.G. Bateman, II, 2007. Protein concentration for starters fed to transported neonatal calves. Prof. Anim. Sci. 23, 123-134.

Hill, T.M., H. G. Bateman, J.M. Aldrich and R.L. Schlotterbeck, 2008. Effect of feeding different carbohydrate sources and amounts to young calves. J. Dairy Sci. 91, 3128-3137.

National Research Council, 2001. Nutrient Requirements of Dairy Cattle. 7th rev. ed. Natl. Acad. Sci., Washington, DC.

Suarez-Mena, F.X., T.M. Hill, A.J. Heinrichs, H.G. Bateman, II, J.M. Aldrich and R.L. Schlotterbeck, 2011. Effects of including corn distillers dried grains with solubles in dairy calf feeds. J. Dairy Sci. 94, 3037-3044.

Substituting barley by sorghum enhances efficiency of starch and protein utilization in lambs

M. Yahaghi[1], J.B. Liang[2], J. Balcells[3], R. Valizadeh[4] and M.F. Jahromi[2]
[1]Northeast Animal Breeding Station, Mashhad, Iran
[2]Institute of Tropical Agriculture, Universiti Putra Malaysia, 43400 UPM, Malaysia; jbliang@putra.upm.edu.my
[3]Department of Animal Production, ETSEA, Lleida University, 25198 Lleida, Spain
[4]Faculty of Agriculture, Ferdowsi University of Mashhad, Mashhad, Iran

Introduction

Use of cereal grains with starch of different degradation rates allows for a more synchronized release of energy and N to prevent a rapid drop in rumen pH and increase outflow of ruminal starch for more efficient digestion in the small intestine (Abramson *et al.*, 2005). Yahaghi *et al.* (2012) reported that partially substituting barley by sorghum significantly increased growth rate of lambs because of increased outflow of starch to the intestine and higher microbial-N (MN) yield. However, when barley was completely replaced by sorghum, high proportion of the intestinal starch was undigested (unpublished data). This follow-up study examined whether extrusion can increase the digestibility of sorghum starch in the intestine to allow for a higher substitution of barley by sorghum in lambs. The optimal extrusion condition for sorghum grain (Bicolor *L. Moench*) was prior-determined as 150 °C/55 bars pressure.

Material and methods

Eighteen newly weaned male Iranian Baluchi lambs (approx. 32 kg) were randomly assigned to three treatments in a complete randomized design (CRD). Three iso-caloric (10.25 MJ/kg DM) and iso-nitrogenous (CP: 140 g/kg) diets consisted of 33: 67 roughage (alfalfa) to concentrate were used. Barley (B) was the main energy grain for the control and was substituted partially (50%, BS_E) or fully (S_E) with extruded sorghum. The study consisted of 10-wk feeding, 7-d digestibility and 2-d gut absorption kinetics trials. Ytterbium (Yb) acetate, as flow marker, was orally administered 4 times a day in the digestibility trial. At end of trial, lambs were euthanized to determine starch and N disappearances through the gastro-intestinal tract (GIT) as in Yahaghi *et al.* (unpublished data). Briefly, beside reticulo-rumen and abomasum, the small intestine was sectioned into 1m-segments. Digesta samples from the different segments were taken. Yb in the digesta samples (Hart and Polan, 1984), rumen MN (Chen and Gomes, 1995) and starch (Horadagoda *et al.*, 2008) were determined. The post-rumen flow of nutrients (*Nf*) (starch and N) in different parts (i) of the GIT was calculated using the nutrient and Yb concentrations in post rumen samples: Nf_i = [N (i, g) × Yb (i, mg)/ Daily Yb doses (mg/d)]. Data were analysed in CRD using GLM procedure (SAS, 2003) and differences in LSD were declared significant at $P<0.05$.

Results and discussion

Our results show significantly higher outflow of starch from rumen in the S_E lambs compared to the other two treatments (Figure 1, left). And 84.9% of the intestinal starch in S_E lambs was digested while only 57.8% of the intestinal starch from untreated sorghum was digested (Figure 1, right), suggesting higher digestibility of the extruded sorghum starch in the intestine.

Lambs in S_E diet had higher ruminal total N, MN and dietary-N (Table 1) outflow, suggesting that S_E diet increased MN synthesis, resulted in higher total N outflow. The higher starch and N flows and their higher digestion rates in the S_E lambs had resulted in an additional 80 g/d BW gain, equivalent to 38% improvement, compared to lambs in the control barley diet.

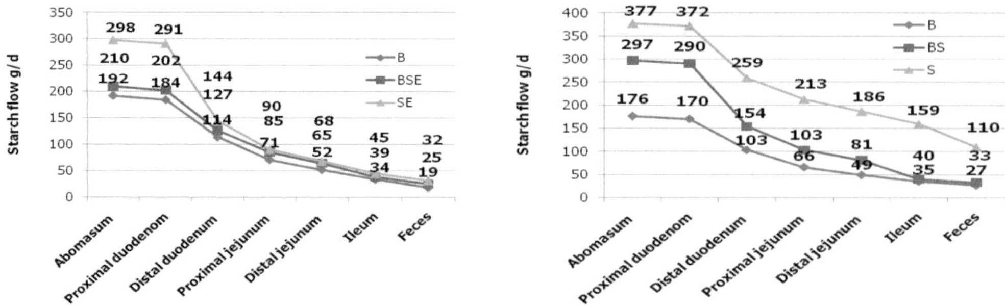

Figure 1. (left) Post-ruminal starch flow in lambs fed diet with barley (B), barley/extruded sorghum (BS_E, 50:50) and extruded sorghum (S_E); (right) Post-ruminal starch flow in lambs feed diets with barley (B), barley-sorghum (BS) and sorghum (S) (unpublished data).

Table 1. Ruminal N flow in lambs fed barley (B), barley/extruded sorghum (50:50 BS_E) and extruded sorghum (S_E) diets.

Item (g/day)	B	BS_E	S_E	SEM*	P<
N intake	45.0	44.0	44.0	0.81	0.38
Duodenal N-flow	45.5^b	46.9^b	53.1^a	1.13	0.01
Microbial N-flow	19.6^b	18.9^b	24.1^a	1.04	0.01
Non-microbial N-flow	25.9^b	26.9^b	29.0^a	0.67	0.02

^{a,b} Means within row with different superscripts differ (P<0.05).

References

Abramson, S., I. Bruckental, L. Lipshitz, U. Moalem, S, Zamwel and A. Arieli, 2005. Starch digestion site: influence of ruminal and abomasal starch infusion on starch digestion and utilization in dairy cows. J. Anim. Sci. 80, 201-207.

Chen, X.B. and M.J. Gomes, 1995. Estimation of Microbial Protein Supply to Sheep and Cattle Based on Urinary Excretion of Purine Derivatives – An Overview of the Technical Details. International Feed Resources Unit. Rowett Research Institute, Aberdeen, Scotland, UK, 21 P.

Hart, S. P. and C.E. Polan, 1984. Simultaneous Extraction and Determination of Ytterbium and Cobalt Ethylenediaminetetraacetate Complex in Feces. J. Dairy. Sci. 67, 888-892.

Horadagoda, A., W. Fulkerson, I. Barchia, R. Dobos and K. Nandra, 2008. The effect of grain species, processing and time of feeding on the efficiency of feed utilization and microbial protein synthesis in sheep. J. Livest. Sci. 114, 117-126.

SAS, 2003. SAS version 9.1: SAS institute Cary, NC.

Yahaghi, M., J.B. Liang, J. Balcells, R. Valizadeh, A.R., Alimon and Y.W. Ho, 2012. Effect of replacing barley with corn or sorghum grain on rumen fermentation characteristics and performance of Iranian Baluchi lamb fed high concentrate rations. J. Anim. Prod. Sci. 52, 263-268.

Influence of different grassland vegetation types on ruminal protozoa and ammonia in beef cattle

I.D.M. Gangnat[1], J.O. Zeitz[1], D. Warner[1,2], M. Kreuzer[1] and F. Leiber[1]
[1]ETH Zurich, Institute of Agricultural Sciences, Animal Nutrition Group, Zurich, Switzerland; isabelle.gangnat@usys.ethz.ch
[2]Wageningen University, Animal Nutrition Group, Wageningen, the Netherlands

Introduction

Grassland vegetation types, which vary in concentration of nutrients and plant secondary compounds, may influence ruminal metabolism differently. Ruminal protozoa play a major role in ruminal protein degradation to ammonia and are involved in fibre digestion. Their role in, and response to, concentrate-based diets is well known. However, the effects of extensive grass-based diets on the composition of the protozoa population in the rumen are poorly understood. The present experiment investigated whether feeding grass or hay harvested from different vegetation types influences ruminal nitrogen metabolism and protozoa populations in beef heifers.

Material and methods

The experiment was conducted with 18 Limousin heifers (466±41 kg live weight at slaughter). During the 2 months prior to slaughter, animals were fed *ad libitum* either fresh grass originating from a lowland ley (*Poa pratensis-Lolium perenne*; CG), or fresh biodiverse alpine grass (*Crepido aureae-Festucetum rubrae*, *Deschampsio cespitosae-Poetum alpinae* and Calthion meadows; AG), or a mixture (1:1 in dry matter (DM); AH) of AG and hay from CG. The dorsal rumen contents were sampled immediately after slaughter and kept on ice for about 2 h until analysis. Strained ruminal fluid was analysed for pH and ammonia (NH_3) concentration using a potentiometer equipped with the respective electrodes. Protozoa (*Isotrichidae*, *Entodininae*, *Diplodininae*, *Ophryoscolecinae* and *Blepharocorythidae*) were counted microscopically in the strained rumen fluid. Nitrogen (N) and ash were determined in freeze-dried rumen content. Grass samples were taken every week per vegetation type to be representative of the area to be harvested in the following 7 days. The oven-dried grass and hay samples were analysed for N and ash (according to AOAC, 1990), neutral detergent fibre (NDF), acid detergent fibre (ADF) and acid detergent lignin (ADL) (according to Van Soest *et al.*, 1991). The values obtained were pooled to obtain the average diet composition over the last 4 weeks before slaughter. Data were subjected to analysis of variance testing for the vegetation type effect.

Results and discussion

Over the last 4 weeks of the trial, the N content of the diet was higher in CG than in AG and AH (29, 18 and 17 g/kg, respectively; $P<0.01$), whereas ash content was lower in AG and AH than in CG (75, 74 and 106 g/kg respectively; $P<0.01$). The NDF content was similar among groups (523 g/kg DM), whereas ADF was higher in AG than CG (323 and 297 g/kg DM, respectively; $P<0.05$) and tended to be higher in AG than AH (299 g/kg DM; $P=0.07$). The ADL content was higher in AG than AH (47 and 38 g/kg DM; $P<0.05$), with CG being intermediate (41 g/kg DM). The addition of hay, which had a lower content of ADF and ADL than the fresh grass due to harvest at an earlier vegetation stage, lowered the ADF and ADL contents of the AH diet. The most contrasting difference between CG and AG, based on the present analysis, was their N content. Additionally, the proportion of herbs (as opposed to legumes and grasses) was higher in AG (40% in the total fresh weight; mainly *Taraxacum officinale*, *Alchemilla xanthochlora*, *Crepis aurea*) than in CG (4% herbs). The ruminal fluid pH was similar among groups (6.1±0.3). Concentrations of N and ash in ruminal content differed between groups in accordance with the N and ash concentrations in the corresponding diets ($R^2=0.81$ for both N and ash). Indeed, the N concentration in ruminal content was higher for CG than AH and AG (31.3, 24.4 and 22.6 g/kg, respectively; $P<0.001$). Ruminal fluid NH_3 concentration of CG differed from that of AH and AG (19.9 vs. 10.5 and 7.6 mmol/l, respectively; SEM=1.30; $P<0.001$), with the highest values found for CG consistent with the highest feed N content. Total protozoal counts were higher in CG than AG (3.7×10^5 and 2.0×10^5/ml, respectively; SEM=0.32×10^4; $P<0.05$), and was intermediate in the AH group (2.7×10^5/ml). A treatment effect on the proportion of some protozoa (sub)families in the entire population was found (Figure 1). The proportion of the Blepharocorythidae was higher in CG than AH (7.9% and 2.1% of total, respectively; SEM=1.46; $P<0.05$) and tended to be higher in CG than AG (2.9%; $P=0.07$). The proportion of Entodininae was lower in CG than AG (59.9% and 78.0%, respectively; SEM=4.80; $P<0.05$), and was intermediate in the AH group (72.8%). The diet had no effect on the proportions of *Diplodininae*, *Ophryoscolecinae* and *Isotrichidae* on total protozoa ($P>0.1$).

The present results suggest that an alpine-grass based diet (AG) low in N and containing a high amount of herbs presumed to be rich in plant secondary compounds decreases both counts of total rumen protozoa and NH_3 concentration compared to a lowland-grass based diet (CG) high in N and composed of a few cultivated grass species. However, this did not influence the growth performance of the animals (data not shown) suggesting that N was excessively present in CG rather than it was deficient with AG. Furthermore, the alpine grass influenced the composition of the protozoal population in a way that the main species Entodininae was represented in a higher proportion at the cost of the less dominant Blepharocorythidae.

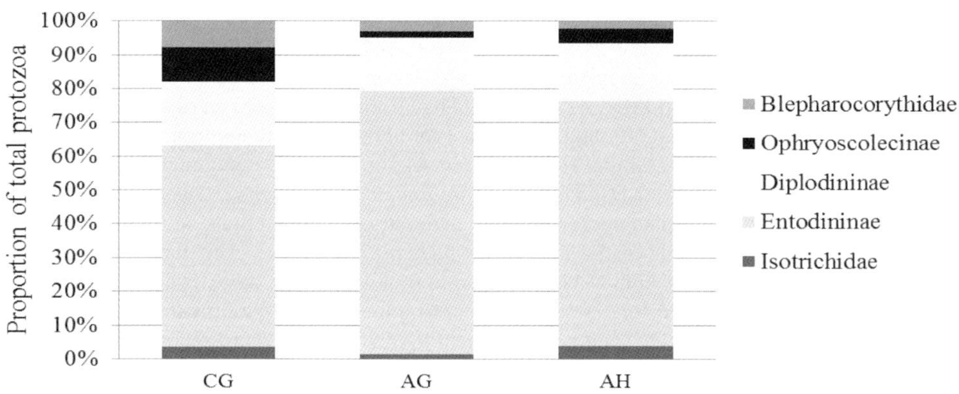

Figure 1. Proportion of protozoa (sub)families in rumen fluid of beef heifers fed fresh lowland cultivated grass (CG), fresh highland alpine grass (AG), or a mixture (1:1 in DM; AH) of AG and hay from CG.

References

Association of Official Analytical Chemists (AOAC), 1990. Official methods of analysis, 15th edition. Arlington, VA: AOAC Inc.

Van Soest, P.J., J.B. Robertson and B.A. Lewis, 1991. Methods for dietary fiber, neutral detergent fiber, and nonstarch polysaccharides in relation to animal nutrition. J Dairy Sci. 74, 3583-3597.

Nutritional evaluation of bulls receiving supplements with different protein: carbohydrate ratios

E.E.L. Valente[1], M.F. Paulino[2], E. Detmann[2], S.C. Valadares Filho[2] and M.L. Chizzotti[1]
[1]Animal Science Departament, Universidade Federal de Lavras, Lavras, Brazil; mariochizzotti@dzo.ufla.br
[2]Animal Science Departament, Universidade Federal de Viçosa, Viçosa, Brazil

Introduction

Modern systems of beef cattle production on tropical pasture aim for continuous growth and slaughter of young animals. Thus strategic supplementation, where supplying limiting nutrients to increase pasture use, is an important way to increase weight gain across the year (Tonello *et al.*, 2011). However, the relationship between protein:carbohydrate in a supplement determines the interactive effects between intake and diet digestibility (Souza *et al.*, 2010). Additionally, the intensity of these effects is determined by composition of basal diet (pasture) that varies through the year. Thus, the aim of present work was to evaluate the intake and digestibility of young bulls supplemented with different protein:carbohydrates ratios grazing tropical pastures from 4 until 18 months of age.

Material and methods

Fifty five beef calves with 138.3±3.4 kg of BW and 90-150 days of age grazing Signal grass (*Brachiaria decumbens*) pasture were used. The experimental period lasted 430 days encompassing 4 seasons (phases): phase 1 = suckling phase in rainy-dry transition season (112 days); phase 2 = post-weaning in dry season (84 days); phase 3 = post-weaning in the dry-rainy transition season (84 days); phase 4 = fattening phase in the rainy season (150 days). The treatments consisted of 5 nutritional plans: control = animals received mineral mixture only; HPHC = high protein and high carbohydrate supplement; HPLC = high protein and low carbohydrate supplement; LPHC = low protein and high carbohydrate supplement; LPLC = low protein and low carbohydrate supplement (Table 1). Approximately 50 and 25% of protein requirement were supplied in high and low protein supplement, respectively, and about 30 and 15% of total digestible nutrients (TDN) requirement was supplied in high and low carbohydrate supplement, respectively. Half of the stipulated requirements were supplied by supplement in the suckling phase due to the milk intake. Every 28 days, the amount of supplement was adjusted by using the estimated protein and energy requirement by BR-CORTE (Valadares *et al.*, 2010). The forage intake and digestibility was measured on day 50 of the experimental period and every 100 days using oxide chrome and indigestible neutral detergent fiber as markers. Milk intake was estimated on days 28, 56 and 84 of the experimental period by milking the cows after an injection of 2 ml of oxytocin (10 IU/ml; Ocitovet®, Brazil). This study was carried on using a completely randomized design using a 2×2 + 1 factorial arrangement to evaluate the nutritional plans (two protein levels, two carbohydrate levels and one control). The variables were evaluated according to a complete random design in repeated measures over time design by using mixed models method.

Results and discussion

Dry matter (DM) intake was higher (*P*<0.05) in dry to rain transitions and rain seasons for all nutritional plans (Table 2). Non-supplemented animals had lower (*P*<0.05) intake of DM and TDN than supplemented animals in all seasons. Although differences on DM intake was not observed between supplemented animals, the supplements with high carbohydrate (HPHC and LPHC) had lower (*P*<0.05) forage intake (*P*<0.05) in the suckling phase (rain to dry transition season) and in the rain season. However, the HPHC plan had higher (*P*<0.05) intake and digestibility of neutral detergent fiber. It can be concluded that supplements with high protein level (supplying 50% CP requirement) increase intake and digestibility of diet, particularly when associated with high carbohydrate level (supplying 30% TDN requirement).

Table 1. Chemical composition of supplement and pasture

	Supplement				Pasture[1]			
	Control	HPHC	HPLC	LPHC	Phase 1	Phase 2	Phase 3	Phase 4
Dry matter	87.1	89.5	85.8	87.0	29.6	42.5	28.0	21.3
Organic matter	89.3	87.4	88.4	85.8	91.4	92.4	92.4	91.5
Crude protein	29.2	55.3	15.4	29.5	8.8	5.5	12.1	10.7
apNDF[2]	8.7	10.2	7.4	9.2	65.3	65.0	61.5	61.6
Ether Extract	2.6	1.5	3.0	2.4	1.2	1.2	1.5	1.2
CNF[3]	46.2	23.3	57.2	43.6	16.1	20.7	17.3	17.9

[1] Obtained by handle plucked sampling.
[2] Neutral detergent fiber corrected for ash and protein.
[3] Non-fibrous carbohydrates.

Table 2. Intake of supplement, total dry matter (DM) and total digestible nutrients (TDN)

Nutritional plan	Production phase				SE
	Phase 1	Phase 2	Phase 3	Phase 4	
DM					
Control	14.68Bb	13.14Bb	21.63Ba	19.63a	0.59
HPHC	17.58Ac	19.30Abc	23.10ABa	20.59ab	0.57
HPLC	18.28Ab	16.20Ab	24.78Aa	21.91a	0.60
LPHC	17.06ABb	17.50Ab	23.69ABa	19.52b	0.60
LPLC	15.66ABc	16.70Abc	23.44ABa	18.87ab	0.58
TDN					
Control	11.04Ba	7.44Cb	12.43Ba	11.41a	0.39
HPHC	13.91Aa	11.88Ab	14.99Aa	13.18a	0.37
HPLC	13.48ABab	9.85ABc	15.77Aa	13.49b	0.40
LPHC	12.11ABab	10.55ABb	14.19ABa	12.05ab	0.40
LPLC	11.87Bb	9.69Bc	14.26ABa	12.52ab	0.38

References

Souza, M.A., E. Detmann, M.F. Paulino, C.B. Sampaio, I. Lazzarini and S.C. Valadares Filho, 2010. Intake, digestibility and rumen dynamics of neutral detergent fiber in cattle fed low-quality tropical forage and supplemented with nitrogen and/or starch. Trop Anim Health Prod 42, 1299-1310.

Tonello, C.L., A.L. Branco, C.Y. Tsutsumi, C.Y. Tsutsumi, L.R. Bueno, R.C. Serrano and S.M. Coneglian, 2011. Suplementação sobre o desempenho de bovinos de corte em pastagens: época do ano. Semin-Cienc agrar 32, 373-382.

Energy value of Tifton-85 (*Cynodon spp.*) for Gir and F1 Holstein × Gir dairy heifers using the respirometric technique

A.L.C.C. Borges[1], H.F. Lage[1], R.R. Silva[1], J.R.M. Ruas[2], A.U. Carvalho[1], E.O.S. Saliba[1] and N.M. Rodriguez[1]
[1]Departament of Animal Science of Veterinary School of Federal University of Minas Gerais, Belo Horizonte, Brazil; analuiza@vet.ufmg.br
[2]Agricultural Research Corporation of Minas Gerais (EPAMIG), Brazil

Introduction

The NRC (2001) proposes the use of net energy system to express the nutritional requirements of the animal and to evaluate the energy value of foods. In tropical regions, however, information at this level of refinement is still scarce, forcing the nutritionists in those areas to use international system's data when formulating diets for dairy cattle.

The aim of this study was to compare the energy value of Tifton-85 (*Cynodon* spp.) hay consumed by Gir and F1 Holstein × Gir dairy heifers using the calorimetric technique.

Material and methods

The study was conducted at the Veterinary School of the Federal University of Minas Gerais, located in Belo Horizonte, Brazil. Six Gir and six crossbred F1 Holstein × Gir heifers with average weight of 450 kg were used. Before the trial, a pre-trial period was conducted in order to adapt the metabolism of the animals to the experimental diet (composed by Tifton-85 hay and mineral nucleus) that was offered in sufficient quantity to allow slight weight gain (200 g/day), keeping the heifers as close as possible to maintenance level. In this period daily food offered and refusals were weighted on a daily basis and animals were weighted every 15 days. Diet digestibility determination and calorimetric measurements began when stability in dry matter intake and minimum daily gain were achieved. In order to express the energy content of hay, energy losses in feces, urine, methane production and heat increment were discounted. The daily amount of feces produced was determined by total collection during five days. The total daily volume of urine was estimated as described by Valadares *et al.* (1999). Gross energy consumption, fecal and urinary energy losses were determined by combustion of these samples in adiabatic bomb calorimeter. The daily methane production and other gases (O_2, CO_2) needed to calculate the heat production of the animals according to Brouwer (1965) equation were obtained by housing the animal for a period of 24h inside a open circuit system calorimetric chamber as described by Rodríguez *et al.* (2007). The heat increment was estimated by difference between the heat production determined from animals fed the experimental diet and after a 72 h fast. The experimental design was completely randomized. The means were subjected to analysis using the statistical package SAEG (2007) and means were compared by Fisher's F test at 5% probability (P<0.05).

Results and discussion

No significant differences were found for evaluating the energy density of the diet expressed in different ways between the two genetic groups (P>0.05) as shown in Table 1.

The average value found in the experimental diet sample combustion (gross energy, GE) was 4.43 Mcal/kg DM, which is very similar to the value suggested by AFRC (1993) of 4.4 to 4.5 Mcal/kg DM to express the gross energy content of ruminant feed.

*Table 1. Gross, digestible, metabolizable and net energy values expressed in megacalories per kilograms of dry matter (Mcal/kg DM) of the experimental diet based on Tifton-85 (*Cynodon *spp.) determined in Gir and F1 Holstein × Gir heifers.*

Item	Genetic group[1]		Standard error
	Gir	F1 H×G	
Gross energy (Mcal/kg DM)	4.44[a]	4.42[a]	0.007
Digestible energy (Mcal/kg DM)	2.69[a]	2.72[a]	0.052
Metabolizable energy (Mcal/kg DM)	2.18[a]	2.19[a]	0.062
Net energy (Mcal/kg DM)	1.41[a]	1.38[a]	0.078

[1] Values followed by different letters in the same row differ by Fisher's test ($P<0.05$).

The digestible energy (DE) had a mean value of 2.7 Mcal/kg DM for both genetic groups, i.e. approximately 39% of gross energy intake was lost in faeces. The value found has great similarity with the DE content of Tifton-85 hay proposed by the NRC (2001) for animals fed at maintenance level, which is 2.5 Mcal/kg DM.

The metabolizable energy of the diet was 2.18 Mcal/kg DM indicating loss of 5.8% and 6% of the GE consumed as urine and methane, respectively. The net energy (NE) had a mean value of 1.39 Mcal/kg DM reflecting the loss of GE intake as heat increment about 18% of the GE intake. The NRC (2001) suggests for animals with intake level three times the maintenance, a ME and NE value of Tifton-85 hay of 1.86 Mcal/kg DM and1.12 Mcal/kg, respectively. The passage rate can explain the difference between the values found in this work and proposed by the NRC (2001) for dairy cows since the increase in consumption is related to increased passage rates of the digesta, leading to reduced dietary energy utilization by the animal.

Conclusion

The energy value of Tifton-85 hay (*Cynodon* spp.) was statistically similar for both Gir and crossbreed F1 Holstein × Gir heifers fed at maintenance level.

Acknowledgements

We would like to thank CNPq, CNPq-INCT, FAPEMIG, CAPES and EPAMIG for their cooperation in carrying out this work.

References

AFRC, 1993. Agricultural and food research council. Energy and requirements of ruminants. Wallingford, Commonwealth Agricultural Bureaux International, 159p.

Brouwer, E., 1965. Report of Sub-Committee on Constants and Factors. Proc 3rd Symp. On Energy Metabolism, EAAP publication no. 11, p. 441-443.

NRC, 2001. National Research Council. Nutrient requirements of dairy cattle. 7 ed., Washington, D.C.: National Academic of Sciences, 381p.

Rodríguez, N.M., W.E. Campos, M.L. Lachica, *et al.*, 2007. A calorimetry system for metabolism trials. Arq. Bras. Med. Vet. Zootec, 59, 495-500.

Valadares, R.F.D., G.A. Broderick, S.C. Valadares Filho, *et al.*, 1999. Effect of replacing alfalfa with high moisture corn on ruminal protein synthesis estimated from excretion of total purine derivatives. J. Dairy Sci., 8, 2686-2696.

Substitution of corn by mesquite pod meal in pellet diets for lambs: nitrogen compounds metabolism

M.L.A. Pereira, T.C.J. Pereira, H.G.O. Silva, J.F. Cruz, P.J.P. Almeida, A.B. Santos, E.J. Santos and C.A.M. Peixoto
State University of Southwest Bahia, Itapetinga, BA, Brazil; marauesb@yahoo.com.br

Introduction

Quantification of total excretion of urine is essential to describe processes such as nitrogen balance, or even to estimate metabolizable energy and microbial protein synthesis by excretion of purine derivatives (PD). Total collection of urine for 24 hours over nine days, is laborious, very uncomfortable for the animals, and can interfere with other variables. For this reason, it is important to develop methodologies that allow the shortest possible time of urine collection, or even to make the total urine collection unnecessary. One opportunity is the use of spot urine collection with estimates based on creatinine excretion. In this context, the present study was conducted to compare the excretion of urinary metabolites using spot urine collection at intervals of 4 hours with those obtained from total urine collection lasting 24 hours in lambs.

Material and methods

We used four lambs Santa Inês x Dorper, with body weight (BW) of 23.0±0.5 kg. The animals were distributed in one 4×4Latin Square, during a four experimental periods of 12 days each. The animals were fed pelleted diets with levels of substitution of milled corn grain by mesquite pod meal (MPM) (0%, 30%, 60% and 90%) in the natural matter (NM) of total diet, which was composed of 30% alfalfa hay and 70% concentrate. The feed was provided twice daily, at 07:00 h and 15:00 h. On the 12th day of each period, urine samples were collected, every 4 hours, after the supply of morning feed, over a 24 hour, period during the spontaneous voiding. The remainder was reserved for obtaining total urine volume each day. At the end of 24 hours, the urine samples collected were also considered for obtaining the total volume of urine.

The effects of treatment, collection time and treatment × time were performed by variance analysis and the probabilities of the effects linear, quadratic and cubic of the contrasts were obtained. Regression analysis was performed according to the level of MPM. All analyzes were performed using the statistical program (PROC MIXED) of SAS statistical package (2006).

Results and discussion

Table 1 shows that there was no treatment effect on total PD excretion, urea and total nitrogen obtained from the total collection of urine. These results differed when using the method of spot urine collection. Based on this methodology, the analysis of variance detected effect of treatment for urinary excretion of total PD, estimating maximum excretion of PD for the diet with corn replacement in 60.4%. Since there was variation in glomerular filtration rate (GFR) (Table 2), the change in the excretion of PD estimated in spot urine samples, as a function of the diet, was not due to the increased synthesis of body PD. It is worth noting that a significant difference was detected between diets for the daily excretion of urea-N in urine when considering all six times of urine collection, which was not observed with the total collection of urine. It is more accurate and precise to the detect differences between the experimental diets when the number of urine samples. For urea-nitrogen it was estimated that minimum excretion happened when the level of MPM was 52.5%, not following the behavior of GFR. Thus, these results corroborate the fact that the increase in GFR may even increase the filtered load of urea, but their subsequent tubular reabsorption was higher at this level of MPM. Based on these results, it is concluded that the GFR ranges, concentrations of creatinine

Table 1. Probability values (P-value) for the effects of treatment, time and treatment × time (Tr x Time) interaction on the urinary excretion of metabolites using total urine and spot urine collection in lambs fed with replacement levels of milled corn grain by mesquite pod meal in pelleted diets.

Item	Total urine (24 hours)	Spot urine (4 in 4 hours)		
	Treatment	Treatment	Time	Treatment × time
Total purine derivatives (g/day)	>0.05	0.008	0.657	0.718
Total nitrogen compounds (g/day)	>0.05	0.736	0.436	0.711
Urea nitrogen (g/day)	>0.05	0.0002	0.437	0.855
PD: C	na	0.304	0.554	0.445
TN: C	na	0.631	0.487	0.800
UN:C	na	<0.0001	0.388	0.838

PD:C = purine derivatives:creatinine ratio; TN:C = total nitrogen:creatinine ratio; NU:C = urea nitrogen:creatinine ratio, calculated from the concentration in mg/dl; na = not analyzed.

Table 2. Means of least squares for the nitrogen metabolism, water intake, glomerular filtration rate and urine volume using 24-hour total urine collection in lambs fed replacement levels of milled corn grain by mesquite pod meal in pelleted diets

Item	Replacement level (% NM)				ASE
	0	30	60	90	
Urea excretion (g/day)[1]	19.42	15.16	17.08	18.06	3.06
Urea nitrogen (g/day)[1]	9.06	7.07	7.97	8.43	1.43
Nitrogen concentration in plasma (ml/dl)[1]	20.59	20.81	20.83	21.09	1.40
Nitrogen balance (g/day)[1]	11.98	14.92	16.50	14.63	4.01
Nitrogen digested (g/day)[1]	27.62	29.80	32.36	28.49	4.08
Losses of nitrogen in urine (g/day)[1]	15.64	14.89	15.86	13.86	0.08
Total purine derivatives (mmol/day)[1]	78.62	79.17	87.37	82.37	16.02
Absorbed purines (mmol/day)[1]	93.60	94.25	104.0	98.06	19.07
Microbial protein synthesis (g/day)[1]	425.3	428.2	472.6	445.6	86.67
Water intake (g/kg BW)[1]	98.13	93.47	103.6	107.9	10.54
Glomerular filtration rate (ml/min BW)[2]	2.07	1.80	2.02	1.76	0.23
Urine volume (g/kg BW)[1]	54.02	43.74	38.45	61.23	21.88

[1]Variance analysis: P>0.05 for linear, quadratic and cubic effects of the contrasts.

[2]GFR = (1.74±0.06)*+ (0.02±0.0003) MPM* – (0.0003±0.0003) MPM^2*; *significance (P<0.0001); ASE=average standard error.

in the urine can vary over the day, so it is recommended that spot urine samples should be collected at different sampling times to constitute representing a 24-hour cycle in experiments with lambs.

References

SAS Institute, 2006. Statistical Analysis System. User's guide. Cary: SAS Institute.

Excretion of purine derivatives and nitrogen compounds in lactating goats fed other protein sources

M.L.A. Pereira, A.B. Santos, A.J. Del Rei, J.F. Cruz, P.J.P. Almeida, T.C.J. Pereira, E.J. Santos and C.A.M. Peixoto
State University of Southwest Bahia, Itapetinga, BA, Brazil; marauesb@yahoo.com.br

Introduction

In the attempt to simplify the collection of urine samples are collected four hours after the feeding in the morning and urinary creatinine has been used as an indicator of the daily urinary production (Leal *et al.*, 2007). However, there are doubts about possible variation in the excretion of purine derivatives and nitrogenous wastes as a function of time, which would derail the obtainment of a spot urine sampling at any hour of the day. The aim of the present experiment was to evaluate the daily excretions of purine derivatives (PD), urea (U) and total nitrogen (TN) and the ratio of those with creatinine (C) obtained using the spot urine collection, at intervals 2 h, in lactating goats.

Material and methods

We used four Alpine goats, 94.0±9.0 days in milk, producing 1.7±0.4 kg of milk and weighing (BW) 42.6±6.1 kg at the beginning of the experiment, distributed in a 4×4 Latin square with 4 periods of 15 days each. The animals were fed with diets containing the same proportions of mesquite pod meal and milled corn grain, as energy sources, associated with different protein sources (soybean meal-SM, cottonseed meal-CM, aerial part hay of cassava-APHC or leucaena hay-LH). Diets consisted of Tifton 85 grass hay (40%) and concentrate (60%), presented 10.2% of CP in DM. Diets were fed *ad libitum* at 07:00 and 16:00, with water constantly available. On the 14th day of each experimental period, urine samples were collected every 2 hours after feeding in the 24 hour period during the spontaneous voiding. The daily excretion of C can be used as indicator of urine volume in dairy goats, whose value was 20.34 mg/kg BW/day. The effects of treatment, collection time and treatment x time interaction were evaluated by variance analysis. Regression analysis was performed for each variable in function of collection time for each treatment. All analyzes were performed using the statistical program (PROC MIXED) of SAS statistical package (2006).

Results and discussion

In diets containing SM, the concentration of C was changed quadratically, reaching a minimum at 11.07 hours after morning feeding. The ratio of the concentration of PD and C ranged over 24 hours, with the peak excretion of PD, which occurred 11.05 hours after morning feeding (Figure 1). However, for CM diet as the C concentration was minimal to 14.26 hours after morning feeding, estimating maximum volume of urine at 13.00 hours and the ratio between the PD and C in 10.62 hours (Figure 1). The APHC diet did not alter any of these metabolic parameters. The LH diet provided variation of U:C ratio to peak at 7.47 hours after the morning feeding, probably due to higher concentrations of blood U and their urinary excretion until this time after morning feeding (Figure 1). Both diets SM and CM affect circadian concentration of C in the urine, being SM diet altered the excretion of PD due to its increased tissue synthesis and CM diet altered the excretion of PD due to the change in the volume of urine produced. Moreover, both the metabolism and excretion of PD were not affected by the use of APHC and LH diets, with the latter affected the liver and renal metabolism of U. Thus, there is need for further studies to identify the time of spot urine sampling which is representative of the total 24 h-urine collection in trials for evaluation of urinary excretion of PD, U and TN.

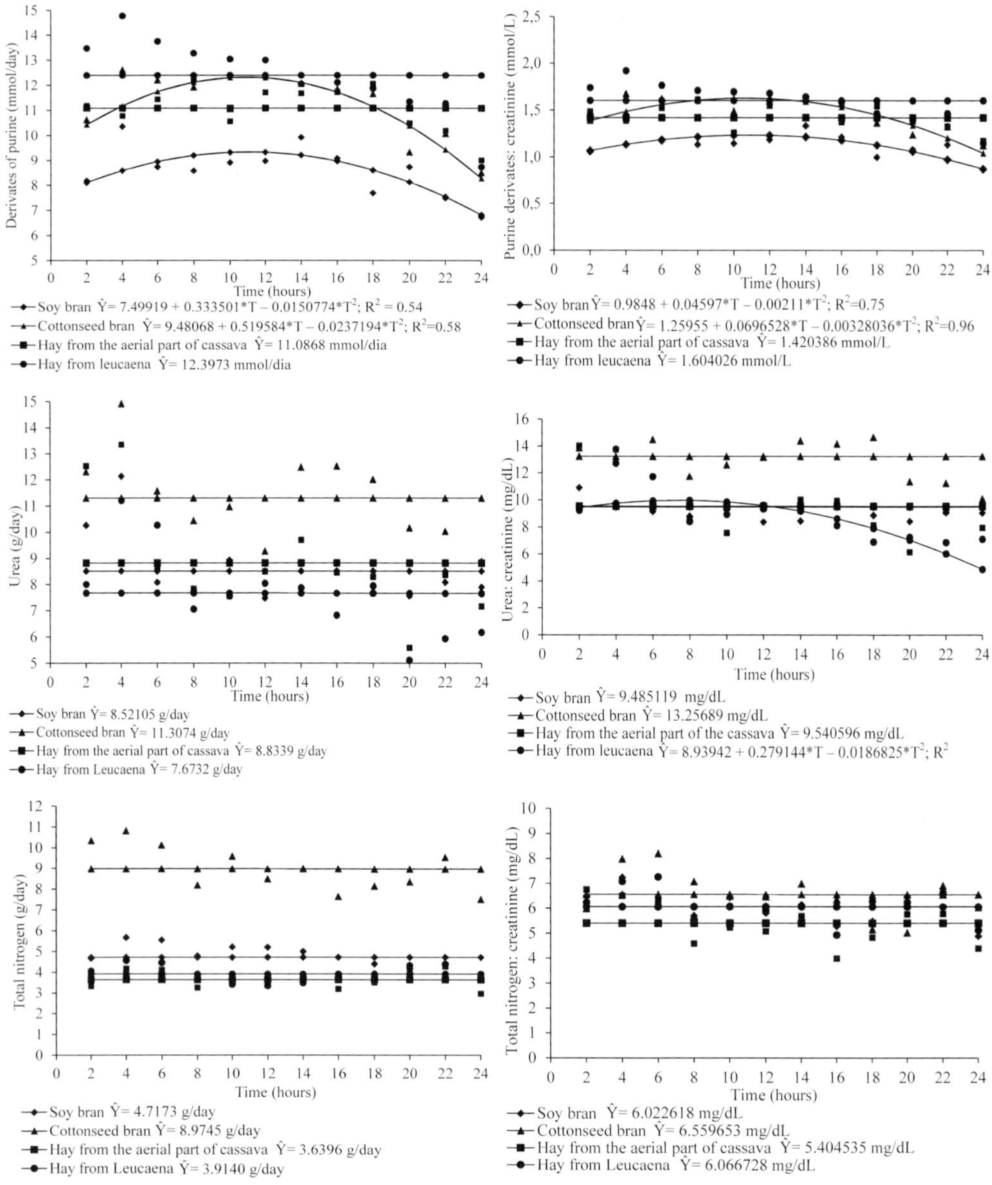

Figure 1. Excretion of purine derivatives (PD), urea (U), total nitrogen (TN) and the ratio of those with creatinine (C) of lactating goats in function of time for different diets (Probability for effect of treatment (Tr), time (T), and time x treatment interaction (T x Tr) for following variables: PD: Tr: 0.0001; T: 0.0096 and T × Tr: 0.6275; Ratio PD:C: Tr: 0.0001; T: 0.0462; T × Tr: 0.7200; U: Tr: 0.0007; T: 0.0353; T × Tr: 0.8663; Ratio U:C: Tr: 0.0183; T: 0,0462; T × Tr: 0.4406; TN: Tr: 0.0001; T: 0.0037; Tr × T: 0.4651; Ratio TN:C: Tr: 0.057; T: 0.1147; T × Tr: 0.6332.

References

Leal, T.L., Valadares, R.F.D., Valadares Filho, S.C., Campos, J.M.S., Detmann, E.,Teixeira, R.M.A., Marcondes, M.I. 2007. Variações diárias nas excreções de creatinina e derivados de purinas em novilhas. R. Bras. Zootec. 36, 905-911.

Nutritional evaluation and performance of beef cattle fed with crude glycerin diets

J.P.I.S. Monnerat[1], *I.M. Oliveira*[1], *P.V.R. Paulino*[1], *R. Mezzomo*[2], *L.H.P. Silva*[1], *P.P. Silva*[1] *and J.P. Ferreira*[1]
[1]*Department of Animal Science, Universidade Federal de Viçosa, Brazil; pveiga@ufv.br*
[2]*Department of Animal Science, Universidade Federal Rural da Amazônia, Brazil*

Introduction

Glycerin is a by-product from the biodiesel industry that results from transesterification of vegetable oils (Crandall, 2004) and presents high concentrations of glycerol (Dasari *et al.*, 2005). Thus, glycerin is a potential ingredient that would be included in ruminant diets as an energy source to replace energetic feedstuffs. However glycerol as a macro ingredient had not have an established estimate of energy values or these values are not available for typical feeding scenarios. Moreover there are few studies that have explored some of the attributes and issues pertinent to glycerol as a feed for beef cattle where the value of glycerol was examined as a replacement for corn grain. Therefore two trials were carried out to evaluate crude glycerin as feedstuff in beef cattle diets. The first experiment was conducted to determine the energy value of crude glycerin as an ingredient in beef cattle diets. The second study was developed with the objective of evaluating the effect of replacing corn with crude glycerin on the intake and total apparent digestibility of diet components and performance of beef cattle under feedlot conditions.

Material and methods

To determine the energy value of glycerin eight crossbred steers with an average initial weight of 334±121.13 kg were used. The animals were divided in four groups of 2 animals and each group was fed diets containing two crude glycerin levels (5 and 15% of diet dry matter) during two experimental periods. The experiment was evaluated according to a two-by-two Latin Square design, with 4 simultaneous squares. Within each square, each crude glycerin level was applied (5 and 15%). Diets with different crude glycerin levels were independently applied to each latin square according to the model: $Y_{ijkl} = \mu + S_i + T_j + A_{(i)k} + P_{(i)l} + ST_{ij} + \varepsilon_{ijkl.}$

Where: μ = general mean; S_i = effect of the Latin Square i (fixed); T_j = effect of the crude glycerin level j (fixed); $A_{(i)k}$ = effect of the animal k nested to the Latin Square I (random); $P_{(i)l}$ = effect of the period l nested to the Latin Square i (random); ST_{ij} = interaction effect of Latin Square i and treatment j (fixed) and ε_{ijkl} = random error

Thirty crossbreed Red Angus × Nellore bulls with an average initial weight of 343.9±16.56 kg were used. The animals were assigned to a complete randomized design with five treatments and six replications per treatments (0, 5, 10, 15 or 20% glycerin inclusion on DM basis) during 84 days. Comparisons between treatments means were performed in accordance with the following orthogonal contrasts: linear and quadratic effects for the substitution level of corn ground with crude glycerin and 0.05 was identified as the probability level.

Results and discussion

The energy values found for the glycerol were 934.46 g/kg TDN,4.01 Mcal/kg DE and 3.64 Mcal/kg of metabolizable energy. These data suggests that crude glycerin could have higher energetic potential than maize (858.3 g/kg TDN), sorghum (788.0 g/kg TDN) and wheat bran (715.4 g/kg TDN) (Valadares Filho *et al.*, 2010). The dry matter intake and nutrient digestibility were unaffected (P>0.05) by treatments (Table 1). Digestibility is the result of the competition between digestion

Table 1. Dry matter intake, total apparent digestibility, levels of total digestible nutrients and performance of beef cattle fed with different levels of crude glycerin.

Item[1]	Crude glycerin, % of DM					SEM	*P*-value	
	0	5	10	15	20		Linear	Quadratic
DMI, kg/d	9.59	10.16	9.01	9.65	9.39	0.9823	0.9385	0.9134
% of DM								
DM	65.14	65.14	65.97	66.72	69.33	4.1063	0.0711	0.4177
OM	66.79	67.09	67.60	69.01	71.04	4.2337	0.0671	0.5086
CP	67.70	71.55	67.39	67.56	69.23	6.1184	0.8947	0.9906
EE	78.58	77.01	81.69	81.19	83.03	4.9859	0.0520	0.8305
NDF	46.94	47.26	47.75	49.52	48.51	7.1839	0.5649	0.9000
NFC	80.62	78.59	81.26	82.38	85.38	5.5301	0.0733	0.3230
TDN	71.57	70.82	73.37	73.26	75.26	4.4098	0.0981	0.6839
ADG, kg/d	1.90	2.12	1.85	1.96	2.07	0.2532	0.8095	0.6824

[1] DMI = Dry matter intake; DM = digestibility of dry matter; OM = digestibility of organic matter; EE = digestibility of ether extract; CP = digestibility of crude protein; NDFap = digestibility of neutral detergent fiber; NFCap = digestibility of non-fiber carbohydrates; TDN= total digestible nutrients; ADG = average daily gain.

and passage rates. Passage rate is positively correlated with the DMI. Thus, the absence of a crude glycerin effect on the intake of DM, OM, and all dietary constituents and similarity of the others ingredients composition in the all diets can explain these results.

In agreement to absence of differences among treatments shown above, animal performance also was not affected by crude glycerin inclusion ($P>0.05$, Table 1) although the diets had been formulated to expected ADG lower than observed. These results are in agreement with results from previous studies evaluating the performance of finishing cattle fed crude glycerin (Mach *et al.*, 2008; Terré *et al.*, 2011). It can be concluded crude glycerin contains as much energy as starch feedstuffs and its inclusion in diets up to 20% of dry matter can be used in diets for finishing beef cattle because it does not lead to detrimental effects on intake and performance.

References

Crandall, L., 2004. Glycerol abundance cause for concern. Inform 15, 146-147.

Dasari, M.A., P.P. Kiatsimkul, W.R. Sutterlin and G.J. Suppes, 2005. Low-pressure hydrogenolysis of glycerol to propylene glycol. Appl. Catalysis A 281, 225-231.

Mach, N., A. Bach and M. Devant, 2008. Effects of crude glycerin supplementation on performance and meat quality of Holstein bulls fed high-concentrate diets. J. Anim. Sci. 87, 632-638.

Terré, M., A. Nudda, P. Casado and A. Bach, 2011. The use of glycerine in rations for light lamb during the fattening period. Animal Feed Science and Technology 164, 262-267.

Valadares Filho, S.C., P.V.R. Paulino, M.I. Marcondes and M.L. Chizzotti, 2010. Exigências nutricionais de zebuínos puros e cruzados. BR-CORTE. 2th ed. UFV: DZO. Viçosa, MG p. 185.

The effect of substituting urea for a commercial slow release urea as supplement to sheep fed a poor quality *Eragrostis curvula* hay

A.M. Mentz, A. Hassen, W.A. Van Niekerk, H. Mynhardt and R. Coertze
Department of Animal and Wildlife Sciences, University of Pretoria, Pretoria 0002, South Africa; abubeker.hassen@up.ac.za

Introduction

The first limiting nutrient for sheep at maintenance receiving poor quality hay is rumen degradable protein (RDP), and addition of urea can substitute part of the RDP (Köster *et al.*, 1997). However, urea is highly soluble and can cause rapid increases in rumen ammonia (NH_3-N) concentrations. The aim of this study was to determine whether a slow release nitrogen (N) source (Optigen II) can be substituted in place of a rapid release N source (urea) without having any negative effects on intake, digestibility, rumen fermentation and microbial protein synthesis when the sheep are fed poor quality roughage.

Material and methods

Five ruminally cannulated wethers were used for the trial in a 5×5 latin square design. The five different treatments were: 100% urea (T1); 75% urea:25% Optigen II (T2); 50% urea:50% Optigen II (T3); 25% urea:75% Optigen II (T4) and 100% Optigen II (T5), with the same inclusion level of starch and a mineral premix between treatments. Supplements were infused twice per day directly into the rumen at 9:00 and 15:00. During the following 9-day experimental period the sheep were transferred to individual metabolic cages. Feed, orts and water intake were measured, sampled and pooled within treatment. Urine was collected, sampled and pooled to determine microbial protein supply (Chen and Gomes, 1992; Chen *et al.*, 1995). During the last three days ruminal fluid was collected at six hour intervals, pooled and analysed for VFA and rumen NH_3-N concentrations, and ruminal fluid pH was measured. The data was analysed using ANOVA (SAS 9.2, 2008) for Latin square design. Treatment differences were detected using the F-test.

Results and discussion

A combination of urea and Optigen II at a proportion of 25:75 appeared to be a preferred level of supplementation due to higher organic matter (OMI), neutral detergent fibre (NDF) and digestible organic matter intakes (DOMI) (Table 1). The intake values recorded for the 100% urea and 100% Optigen II treatment did not differ ($P>0.05$). No differences ($P>0.05$) were recorded for organic matter digestibility (OMD) and neutral detergent fibre (NDF) digestibility between the treatments. However, the 100% Optigen II treatment had a significantly ($P<0.05$) lower apparent nitrogen digestibility, which might have been the result of a slower rumen NH_3-N release and higher nitrogen excretion than the other treatments (data not reported). No differences were observed for pH and VFA between different treatments (Table 1). The effective degradability of both DM and NDF did not differ ($P>0.05$) between treatments (data not reported). Neither were there differences between treatments for the volatile fatty acid (VFA) production (except butyrate) and total microbial crude nitrogen (MCN) production.

Based on the results obtained it seems that it could be beneficial to substitute part (up to 75%) of the urea in a balanced supplement with an slow release urea source (optigen II) without affecting intake, digestibility, rumen fermentation and microbial protein supply to the ruminant receiving poor quality *Eragrostis curvula* hay. However, from an economical point of view, urea might still be the preferred NPN source, because the price of nitrogen from urea is cheaper than Optigen.

Table 1. Effect of slow and rapid release rumen nitrogen on intake, digestibility rumen fermentation parameters and microbial synthesis.

Parameter	Treatments					SE
	T1	T2	T3	T4	T5	
Digestibility (%)						
OM	49.18	46.17	48.47	49.63	48.42	1.16
NDF	57.58	54.15	55.70	56.88	56.91	1.21
N	91.23^{a}	89.29^{a}	92.52^{a}	91.96^{a}	83.75^{b}	1.11
Intake (g/day)						
OM	882.67^{b}	905.11^{ab}	889.96^{b}	973.15^{a}	854.42^{b}	23.45
NDF	777.86^{ab}	781.44^{ab}	768.47^{b}	840.96^{a}	747.21^{b}	20.63
DOMI	450.52^{ab}	438.74^{b}	453.66^{ab}	498.26^{a}	427.90^{b}	15.71
Intake (g/kgBW0:75)						
OM	41.08^{b}	42.88^{ab}	41.91^{ab}	45.33^{a}	40.34^{b}	1.17
NDF	36.20^{ab}	37.01^{ab}	36.17^{ab}	39.16^{a}	35.27^{b}	1.03
DOMI	20.95^{ab}	20.75^{b}	21.30^{ab}	23.21^{a}	20.15^{b}	0.76
Rumen parameter						
pH	6.29	6.28	6.29	6.33	6.34	0.04
NH_3-N (mg/ 100 ml)	5.84	4.93	5.00	5.40	4.55	0.55
Total VFA (mmol/l)	87.50	82.82	84.35	82.01	85.49	1.96
Acetate (%)	76.76	76.50	76.66	77.26	76.00	0.69
Propionate (%)	17.56	17.04	16.92	16.01	16.43	0.67
Butyrate (%)	4.85^{b}	5.46^{a}	5.50^{a}	5.81^{a}	5.60^{a}	0.20
Acetate:Propionate	4.44	4.64	4.59	4.89	4.86	0.22
Microbial protein synthesis						
MCN (g/day)	39.94	41.70	33.18	33.88	44.36	3.94
gMCN/gDOMI	0.0441	0.0418	0.0349	0.0375	0.0444	0.003

[a,b,c]Means within a row with different superscripts differ ($P<0.05$).

T1 = 100% Urea; T2 = 75% Urea : 25% Optigen II; T3 = 50% Urea : 50% Optigen II; T4 = 25% Urea : 75% Optigen II; T5 = 100% Optigen II.

Acknowledgments

The authors would like to acknowledge the national research foundation (NRF) and the international foundation for science (IFS) for providing research grant to cover the running cost of this study.

References

Chen, X. B. and M.J. Gomes, 1992. Estimation of microbial protein supply to sheep and cattle based on urinary excretion of purine derivatives-An overview of the technical details, International Feed Resources Unit, Rowett Research Institute, Bucksburn Aberdeen AB2 9SB, UK Occasional Publication.

Chen, X.B., A.T. Mejia, D.J. Kyle and E.R. Ørskov, 1995. Evaluation of the use of the purine derivative: creatinine ratio in spot urine and plasma samples as an index of microbial protein supply in ruminants: studies in sheep. Journal of Agricultural Science, Cambridge 125, 137-143

Köster, H.H., R.C. Cochran, E.C. Titgemeyer, E.S. Vanzant, T.G. Nagaraja, K.K. Kreikemeier and G. St. Jean, 1997. Effect of increasing proportion of supplemental nitrogen from urea on intake and utilization of low-quality tallgrass-prairie forage by beef steers. J Anim Sci 75, 1393-1399.

Effect of application of fibrolytic enzyme products at different levels on *in vitro* ruminal fermentation of low quality feeds

B. Shenkute[1], *A. Hassen*[1] *and N.E. Odongo*[2]
[1]*Department of Animal and Wildlife Sciences, University of Pretoria, Pretoria 0002, South Africa; abubeker.hassen@up.ac.za*
[2]*Animal Production and Health Section, International Atomic Energy Agency, Vienna, Austria*

Introduction

Plant cell walls typically consist of about 35-50% cellulose, 20-35% hemicelluloses, and 10-25% lignin in the dry mass (Sticklen, 2008). Considering these abundant fibre sources, significant improvement in the cell wall digestibility has been achieved over past decades through various treatment options. Cellulase and xylanase are two major ruminant diet enzyme groups that are able to respectively break down the cellulose and xylans found in the plant cell wall components (Beauchemin *et al.*, 2003), and brought significant effects in improving digestibility. However, their optimal level of inclusion is dependent on the diet under consideration, suggesting the need to determine optimum rate of inclusion of a given enzyme preparation for the various feeds (Yang *et al.*, 1999). This study evaluates the effects of fibrolytic cellulase and xylanase enzymes on in vitro digestibility, gas production and volatile fatty acid production from *Eragrostsis curvula* and maize stover.

Material and methods

Feed samples were collected, dried in forced oven, and ground to pass a 1 mm sieve in a Wiley mill and stored for later analysis. Xylanase activity was assayed using 1% (w/v) birchwood xylan as a substrate following the procedure described by Bailey *et al.* (1992) while Endo-glucanase and Exo-glucanase enzyme activities were assayed following the method described by Wood and Bhat (1988). Ruminal fluid was collected and prepared anaerobically using standard procedures while the culture medium was prepared as described in Goering and Van Soest (1970). A semi-automated system was used to measure gas production as described by Theodorou *et al.* (1994) for 72 h. Enzyme solution was prepared based on required dose rate(s) for specific experimental treatments in order to deliver the desired amount of enzyme in a 1ml aliquot. About 500 mg of the respective feed sample was weighed into 120 ml serum bottle, and 1ml of the appropriate enzyme treatment was directly pipetted onto the substrate and incubated for 24 h. Then 42 ml of ruminal fluid + medium (1:4) was added under a stream of CO_2 to each of the serum bottles. Two replicates per run, and a total of four runs were executed for every sample. Degradability of NDF was measured after terminating the incubation at 48 h. The data were subjected to the general linear model (GLM) procedure of SAS (2001). Analysis of variance was conducted, separately for the each feeds according to the following model: $yijk = \mu + Fi + Rj + eijk$ (where F is the feed and R the experimental run). All multiple comparisons among means were carried out with Duncan's multiple-range test

Result and Discussion

The cumulative gas production recorded at various time intervals, DM and NDF degradability were significantly ($P<0.05$) influenced by addition of both enzymes (Table 1 and 2). Gas production and fiber degradation increased with increasing levels of enzyme application for both test feeds, except the lowest dose rates for the xylanase enzyme that did not differ from the control. There was continuous degradation of the feeds over time, probably due to continuous effect of enzymes. This might be partly due to the pre-incubation effect that resulted in stable enzyme-feed complex, which further increased their resistance for proteolysis and aids during latter fermentation periods (Yang *et al.*, 1999). Among the enzyme treated feeds, the production of acetate and total VFA increased

Table 1. The effects of Cellulase enzymes on gas production (ml/0.5 g DM), NDF degradation (%), and production of VFA.

Cellulase dose rate (mg/g DM)	Poor quality roughage[1]									
	Eragrostis curvula hay					Maize stover				
	GP_{48}	NDF deg	VFA	Acetate	propionate	GP_{48}	NDF	VFA deg	Acetate	propionate
0	21.9^{g}	33.7^{f}	34.7	23.0	7.29	24.7^{f}	39.4^{f}	57.7	36.8	13.9
0.5	22.8^{f}	35.2^{f}	34.9	23.1	7.33	25.3^{f}	40.0^{f}	60.3	38.6	14.0
1	27.3^{e}	39.7^{e}	49.7	33.4	10.1	26.9^{e}	40.9^{e}	76.3	47.3	18.7
2	31.9^{d}	44.6^{d}	50.6	34.1	10.3	33.2^{d}	44.6^{d}	78.2	52.8	15.1
3	32.7^{c}	47.2^{c}	55.3	37.2	10.4	38.2^{c}	49.0^{c}	79.1	55.6	14.2
4	36.1^{b}	48.9^{b}	57.2	38.5	10.6	40.8^{b}	53.8^{b}	80.0	50.0	19.4
5	41.1^{a}	51.0^{a}	34.7	23.0	7.29	42.2^{a}	56.1^{a}	57.7	36.8	13.9

[1] Means with different letters (superscripts) within a column differ significantly at indicated *P*-value, $P<0.05$ for each enzyme.

Table 2. The effects of Xylanase enzymes on gas production (ml/0.5 g DM), NDF degradation (%) and production of VFA.

Xylanase dose rate (mg/ g DM)	Poor quality roughage[1]									
	Eragrostise curvula hay					Maize stover				
	GP_{48}	NDF deg	VFA	Acetate	Propionate	GP_{48}	NDF deg	VFA	Acetate	Propionate
0	21.9^{f}	33.7^{d}	34.7	23.0	7.29	24.7^{f}	39.4^{d}	57.7	36.8	13.9
0.25	22.0^{f}	33.9^{e}	44.2	29.4	9.16	23.9^{f}	40.4^{d}	70.6	50.3	12.0
0.5	22.9^{e}	34.8^{e}	52.1	34.9	10.6	24.6^{f}	41.4^{d}	71.1	43.1	17.9
1	26.1^{d}	37.5^{d}	52.9	35.5	9.99	27.1^{e}	45.7^{c}	71.4	44.1	17.5
2	29.8^{c}	41.2^{c}	57.3	37.7	11.2	30.9^{d}	49.1^{b}	71.9	51.5	11.8
3	33.6^{b}	45.7^{b}	58.5	39.2	10.8	34.6^{b}	53.9^{a}	73.0	52.8	11.6
4	46.9^{a}	48.9^{a}	34.7	23.0	7.29	47.6^{a}	55.9^{a}	57.7	36.8	13.9

[1] Means with different letters (superscripts) within a column differ significantly at indicated *P*-value, $P<0.05$ for each enzyme.

with increasing application level of enzymes. In agreement with our result, many authors (Eun and Beauchemin, 2007), have noticed increasing fibre degradability of diets or feedstuffs with enzyme supplementation. This observed increases in the production of total VFA, acetate and propionate as well as fibre disappearance could increase the flow of microbial-N and microbial colonization of the substrate, resulting in enhanced fibre degradation.

Conclusion

The pre-treatment of these low quality forages with cellulase at 5 mg/g DM and xylanase at 4 mg/g DM improved *in vitro* ruminal fermentation and degradability of NDF.

Reference

Beauchemin, K.A., D. Colombatto, D.P. Morgavi and W.Z. Yang, 2003. Use of exogenous fibrolytic enzymes to improve feed utilization by ruminants. J. of Anim. Sci. 81, E37-E47.

Eun, J.S. and K.A. Beauchemin, 2007. Assessment of the efficacy of varying experimental exogenous fibrolytic enzymes using in vitro fermentation characteristics. Anim. Feed Sci. Technol., 132, 298-315.

SAS Institute Inc., 2001. SAS/STAT Software, Version 8.2. SAS Institute Inc., Cary, NC, USA.

Sticklen, M.B., 2008. Plant genetic engineering for biofuel production:towards affordable cellulosic ethanol. Nature reviews Genetics. 9, 433-443.

Theodorou, M.K., B.A. Williams, M.S. Dhanoa, A.B. McAllen and J. France, 1994. A simple gas production method using pressure transducers to determine the fermentation kinetics of ruminant feed. Anim. Feed Sci. Technol., 48, 185-197.

Van Soest, P.J., 1982. Nutritional ecology of ruminants. O and B books. Inc. Oregon,USA.

Yang, W.Z., K.A. Beauchemin and L.M. Rode, 1999. Effects of an enzyme feed additive on extent of digestion and milk production of lactating dairy cows. J. Dairy Sci. 82,391-403.

Effect of metabolizable energy intake on energy partitioning into muscle and fat in Pelibuey ewes

A.J. Chay-Canul[1], *J.C. Ku-Vera*[2], *A.J. Ayala-Burgos*[2], *M.L. Chizzotti*[3], *J.G. Magaña-Monforte*[2] *and L.O. Tedeschi*[4]
[1]*División Académica de Ciencias Agropecuarias, Universidad Juárez Autónoma de Tabasco, Villahermosa, Tabasco, México; ajchc19@yahoo.com.mx*
[2]*Facultad de Medicina Veterinaria y Zootecnia, Universidad Autónoma de Yucatán, C.P. 97300 Mérida, Yucatán, México*
[3]*Departamento de Zootecnia, Universidade Federal de Lavras, 37200-000 Lavras, MG, Brazil.*
[4]*Department of Animal Science, Texas A&M University, College Station, TX 77843-2471, USA*

Introduction

In ruminants, seasonality and physiological conditions result in fluctuations in feed intake, inducing periods of underfeeding and refeeding throughout the year. These seasonal variations induce fluctuations in live weight in grazing animals (Atti *et al.*, 2000; Mahouachi and Atti, 2005; Kamalzadeh *et al.*, 2009), and consequently in the energy content of the carcass and the whole body. Nonetheless, in hair sheep breeds there is scarce information relative to carcass composition and energy changes in the carcass during periods of weight loss and gain. Moreover, it has been reported that knowledge of body composition of productive animals is of relevance in order to better assess nutrient requirements. The aim of the present work was to evaluate the effect of metabolizable energy intake (MEI) on energy partitioning of muscular and adipose tissues in adult Pelibuey ewes.

Material and methods

Twenty four adult non-pregnant, non-lactating Pelibuey ewes with BW of 37.2±4.0 kg and BCS of 2.5±0.12 in a scale 1 to 5, where BCS 1 represents a thin animal and 5 an obese animal as described by Russell *et al.* (1969), were randomly assigned to four groups of six animals each. One group was slaughtered at the beginning of the experiment for baseline measurements of carcass tissues and energy content. The remaining ewes were randomly allotted to three groups of six animals and were individually housed in metabolic crates, and fed three levels of MEI: low (L), medium (M) and high (H) during 65 days to achieve desirable changes in BW and BCS. MEI was: 0.247, 0.472 and 0.532 MJ/kgBW$^{0.75}$/day for L, M and H levels respectively. At the end of the experiment all ewes were slaughtered. Carcass was split at the dorsal midline in two equal halves, weighed, and chilled at 6 °C for 24 h. After refrigeration, carcass was weighed and the left half of the carcass was completely dissected into carcass fat (subcutaneous and intermuscular), muscle and bone, and each component weighed separately. Dissected tissues of the left carcass were adjusted to the whole carcass. A proportional sample (about 1 kg) was taken of the muscle and adipose tissue of the carcass from each animal, ground and stored at -20 °C in plastic containers. Tissue samples were freeze-dried to determine water content. Dry samples were ground in a hammer mill and analyzed for gross energy. At the end of the experiment, one ewe from treatment L and another from treatment H were removed from the experiment due to illness and their data were not included in the analysis. Data of energy content in tissues of the carcass were analyzed as a completely randomized design using analysis of variance and the Tukey test was used when significant differences among treatments were detected. Statistical tests were carried out with PROC GLM of SAS (SAS, 2002).

Energy contained in muscle and adipose tissues in the carcass showed differences ($P<0.05$) between feeding levels (Figure 1). The total energy content of the carcass in the initial group was 129.05 MJ. The proportions of energy contained as muscle and fat were 76 and 24%, 68 and 32%, and 65 and 35% for L, M and H levels respectively. Energy balances in the carcass (estimated as the energy content in the carcass of initial group minus the final energy content in the carcass of other

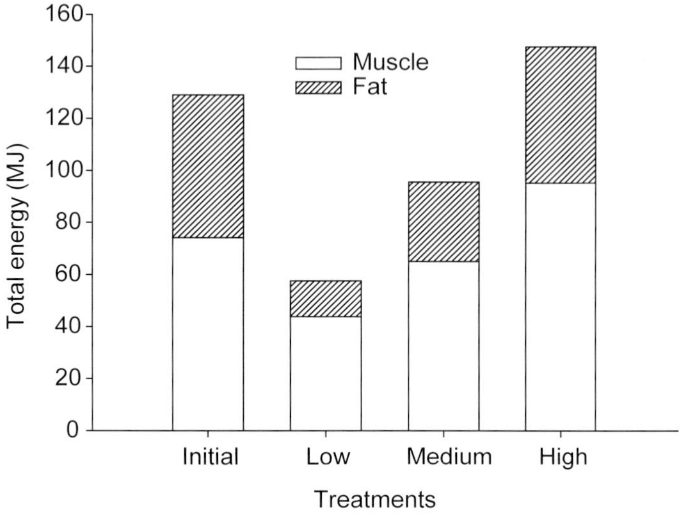

Figure 1. Energy content in the carcass of adult Pelibuey ewes fed three levels of MEI.

groups and divided between the days in experimentation), were -1.099, -0.515 and 0.285 MJ for L, M and H levels respectively. Results indicate that ewes at the L and M levels lost 55% and 26% of their initial energy content in the adipose and muscle tissues of the carcass respectively, while the H level increased its energy content by 14%. Most of the energy stored in the carcass of Pelibuey ewe is contained in muscle tissues.

References

Atti, N., P. Nozière, M. Doreau, C. Kayouli and F. Bocquier. 2000. Effects of underfeeding and refeeding on offals weight in the Barbary ewes. Small Rumin. Res. 38: 37-43.

Kamalzadeh, A., W.J. Koops and A. Kiasat. 2009. Effect of qualitative feed restriction on energy metabolism and nitrogen retention in sheep. South African J. Anim. Sci. 39: 30-39.

Mahouachi, M., Atti, N., 2005. Effects of restricted feeding and re-feeding of Barbarine lams: intake, growth and non-carcass components. Anim. Sci. 81:305-312.

Russell, A. J. F., Doney, J. M., Gunn, R. G., 1969. Subjective assessment of body fat in live sheep. J. Agric. Sci. (Camb.). 72: 451-454.

SAS. 2002. Institute Inc., SAS/STAT. Software, Ver. 9.00, Cary, NC27512-8000. USA.

Mammary gland development in heifers under different metabolizable protein and metabolizable energy ratios

R.L. Albino[1], M.I. Marcondes[1], B.C. Gomes[2], L.G.R. Pereira[2], T.E. da Silva[1] and A.S. Trece[1]
[1]Departamento de Zootecnia, Universidade Federal de Viçosa, Viçosa-MG, Brazil; marcos.marcondes@ufv.br
[2]Empresa Brasileira de Pesquisa Agropecuária, Embrapa Gado de Leite, Juiz de Fora MG, Brazil

Introduction

Mammary development is affected by nutrient intake, which affects the intensity of weight gain. There are several ways to evaluate mammary development, but almost all involve the slaughter of the animals. On other hand, other studies have been conducted with ultrasound to evaluate fat deposition on lean tissue (Wertz *et al.*, 2002). This experiment was designed to evaluate dry matter intake, performance, and mammary gland development in Holstein heifers under different metabolizable protein and metabolizable energy ratios.

Material and methods

The experiment was conducted at the Universidade Federal de Viçosa, Brazil, from December2011 to April 2012. Twenty-five heifers were divided into 5 treatments. The treatments were composed of metabolizable protein and metabolizable energy ratios (MPMER), with 33, 38, 43, 48, 53 gram of metabolizable protein per Mcal of metabolizable energy (ME). The diet with the 43 ratio was based on recommendations of the (NRC, 2001), and two levels above and below this recommendation were also established. All diets were isoenergetic in composition (2.3 Mcal de ME/kg of dry matter), being used corn silage, ground corn, soybean meal and wheat bran to formulate the diet. The dry matter intake (DMI) was controlled daily, and all heifers were weighed in the beginning and final of the trial to calculated daily weight gain (DWG). Ultrasounds were performed in all quarters of each heifer, at the end of each month, to evaluate mammary gland development. The software ImageJ® (NIH, USA) was used to analyze pixels from 6 points within each image (Figure 1). Then, the average pixels were used to analyze density on mammary gland image. The pixels are represented by a scale of 256 grayscale (0=black; 255=white). The DMI, DWG, and feed efficiency (FE) were analyzed in randomized block design. The mammary gland data were analyzed in a split plot design with repeated measures. Statistical analysis was done using SAS (SAS Institute, Cary, NC, USA).

Results and discussion

The MPMER did not affect DMI (P=0.736) or DWG (P=0.090. However, there was a linear decreasing effect of the MPMER in FE (P=0.034), where animals with greater protein level were more efficient for gaining weight (Table 1). It is likely that diets with greater MPMER permit a greater deposit of protein, while diets with lower MPMER had the lipid deposition increased. It is well know that protein deposit is linked to a higher water deposit, which reflects in a greater deposit of lean tissue per kilo of gain. In relation to the mammary gland data, there were significant effects for treatment (P=0.0001), period (P=0.06) and period-treatment interaction (P=0.01). Throughout the experiment, the heifers fed diets with high protein (MPMER 48 and 53) showed a reduction in pixel value. However the heifers fed with low protein (MPMER 33 and 38) showed an increase of pixel value (Figure 2). It is inferred that images generated with high pixel value are associated with a greater fat accumulation in the mammary gland due to greater reflection of fat over the ultrasound waves (Nishimura *et al.*, 2011). Thus, ultrasound images of heifers submitted to high protein treatments might be related with a smaller proportion of fat accumulation in the mammary gland, due the greater reflection of sound waves. On the other hand, heifers that received a low protein diet might have had a greater proportion of fat in the mammary gland.

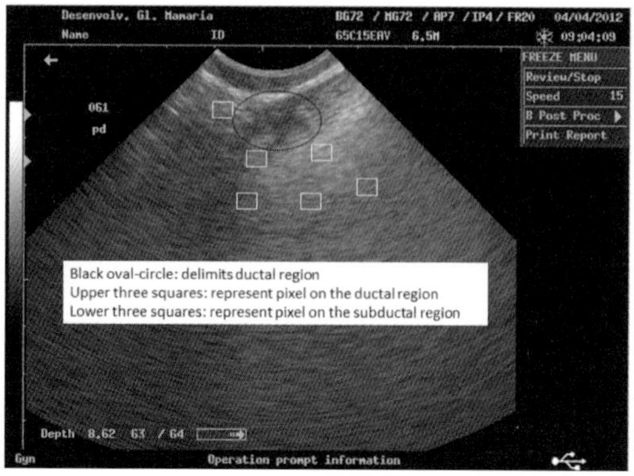

Figure 1. Image representing the method used to analyze mammary gland pictures.

Table 1. Dry matter intake, daily weight gain and feed efficiency.

Itens	MP:ME ratio					*P*-value		CV
	33	38	43	48	53	LINEAR	QD	
DMI (kg/day)	7.09	6.74	6.64	6.84	6.89	0.736	0.327	9.57
DWG (kg/day)	1.01	0.94	0.97	1	1.2	0.091	0.067	18.76
FE	6.92	7.17	6.8	6.82	6.06	0.035	0.128	-

DMI = dry matter intake; DWG = daily weight gain; FE = feed efficiency; QC = quadratic; CV = coefficient of variation.

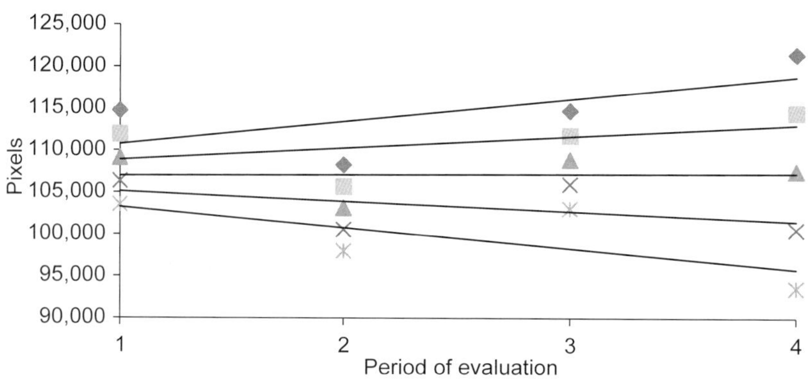

Figure 2. Pixel value variation according to period-treatment interaction. The symbols are representing 33 (◆), 38 (■), 43 (▲), 48 (×), and 53 g of metabolizable protein per Mcal of metabolizable energy (✳).

References

National Research Council (NRC). 2001. Nutrient Requirements of Dairy Cattle. 7th ed. National Academy Press, Washington, DC.

Nishimura, M.;Yoshida, T; EL-khodery, S.; Miyoshi, M.; Furuoka H.; Yasuda, J. and Miyahara, K., 2011. Ultrasound Imaging of Mammary Glands in Dairy Heifers at Different Stages of Growth. *J. Vet. Med. Sci.* 73(1): 19-24.

Wertz, A. E., Berger, L. L., Walker, P. M., Faulkner, D. B., McKeith, F. K. and Rodriguez-Zas, S. L, 2002. Early-weaning and post weaning nutritional management affect feedlot performance, carcass merit, and the relationship of 12th-rib fat, marbling score, and feed efficiency among Angus and Wagyu heifers. J ANIM SCI, 80:28-37.

Effect of starch source and fiber level in mixed diets on lactating Murciano-Granadina goat: Substrate oxidation and milk performance

C. Fernández[1], M.C. López[1] and M. Lachica[2]
[1]Institute for Animal Science and Technology, Universidad Politécnica de Valencia, 46022 Valencia, Spain; cjfernandez@dca.upv.es
[2]Department of Animal Nutrition, Estación Experimental del Zaidín, CSIC, 18100 Armilla, Granada, Spain

Introduction

To achieve the maximum milk production potential, dairy ruminants need to ensure high intake of energy. This might be accomplished by raising the dietary concentration of rapidly degraded non-fiber carbohydrates, such as starch from cereal grains. The increase of starch concentration in diets for dairy cows, however, can lead to undesirable ruminal fermentation, compromising the nutrient supply for milk production and composition. The partial replacement of cereal grain with low starch by-product feeds represents a potential alternative to overcome this limitation. The aim of the present work was to determine the effect of the source of starch (barley and corn) and fiber content, on the substrate oxidation and milk performance of lactating Murciano-Granadina goats.

Material and methods

Twenty Murciano-Granadina goats at mid lactation were used following two crosses over designs. Four diets were used with alfalfa hay as forage source. Diet BS had 41% of barley grain; in diet BF the barley grain was replaced by 41% of fibrous by-products; diet CS had 40% of corn grain; and in diet CF the corn grain was replaced by 40% of fibrous by-products. For diets details see López *et al.* (2013). The goats were allocated on individual metabolism cages. After 10 d of adaptation, feed intake, refusal, total fecal and urine output, and milk yield were recorded daily during 5 d period. Then, heat production from oxidation of nutrients (HPx) was determined during 24 h using a mobile open-circuit respirometry system connected to a head box. Gas exchange measurements for estimation of carbohydrate, fat and protein oxidation were used by Brouwer (1958). He developed equations for calculation of net oxidation of nutrients. Energy associated with the oxidation of protein (OXP), carbohydrate (OXCHO) and fat (OXF) was calculated by the method described by Chwalibog *et al.* (1997) for ruminants. The CO_2/CH_4 ratio of 3 and 1.7 was used for high grain and high forage diets, respectively (Fahey and Berger, 1988). The estimation of oxidation CO_2 (CO_2x) was carried out by reducing CO_2 production with 3 and 1.7 times CH_4 production. The calculations were as follows:

$HPx\ (kJ) = 16.18 \times O_2 + 5.02 \times CO_2x - 5.99 \times Nur$; $OXP = 6.25 \times Nur \times 18.42\ (kJ/g)$; $OXCHO = (-2.968 \times O_2 + 4.174 \times CO_2x - 2.446 \times Nur) \times 17.58\ (kJ/g)$; $OXF = (1.719 \times O_2 - 1.719 \times CO_2x - 1.963 \times Nur) \times 39.76\ (kJ/g)$

Gases were expressed in l/d and urinary nitrogen (Nur) in g/d. The data were analyzed by the MIXED procedure of SAS.

Results and discussion

In order to estimate the CO_2/CH_4 ratio of our diets, the theoretical rumen fermentation balance proposed by Wolin (1960) was followed, and a ratio of 2.6 and 1.7 were found for starch rich and fibrous diets. Slightly different regards the assumed hypothesis. Although no significant effects were observed for HP and k_l, greater milk energy output in BF and CF and higher tissue energy retention in BS and CS were found (López *et al.*, 2013). Average milk yield (Table 1) was 2.24 kg/d, and milk fat content was different ($P<0.05$) among diets (5.49% in higher starch diets and 7.14% in higher

Table 1. Daily milk yields (kg/d), milk composition (%) and oxidation (kJ/kg$^{0.75}$ BW) of protein (OXP), carbohydrate (OXCHO) and fat (OXF) and their contribution (%) to the heat production from substrates oxidation (HPx) of lactating Murciano-Granadina goats (n=20).

	BS	BF	CS	CF	SEM	*P*-value
Milk yield	2.4	2.2	2.2	2.1	0.91	0.2108
Dry matter	14.9^{a}	15.8^{ab}	15.1^{a}	16.7^{b}	0.19	0.0029
Fat	5.5^{a}	6.4^{b}	5.5^{a}	7.9^{b}	0.29	0.0062
Protein	3.9	3.9	4.1	4.0	0.06	0.5716
Lactose	4.7	4.7	4.8	4.7	0.04	0.8730
HPx	754	803	776	760	7.20	0.0626
OXP	100^{a}	101^{a}	97^{a}	138^{b}	4.10	0.0001
OXCHO	512	330	444	334	29.4	0.0704
OXF	142^{a}	373^{c}	235^{ab}	289^{b}	30.1	0.0457
OXP/HPx	13.3^{a}	12.5^{a}	12.4^{a}	18.2^{b}	0.54	0.0001
OXCHO/HPx	67.7^{a}	41.5^{b}	57.3^{a}	43.4^{b}	3.85	0.0461
OXF/HPx	18.9	45.9	30.3	38.4	3.85	0.0741

^{a,b,c} Values within a row with different superscript were significantly different (P<0.05).

NDF diets). This is probably due to the higher NDF content in fibrous than starch diets (35 vs. 47%). The depression in milk fat upon feeding starch rich diets has been explained by a shift from a high availability of fat precursors to glucose and by a shift from lipogenesis to gluconeogenesis. Average HPx was 773 kJ/kg$^{0.75}$ BW/d, and not significant differences were found among treatments. The OXP was greater (P<0.05) for CF vs. others, probably due to lower microbial protein synthesis. A tendency of higher OXCHO for starch rich diets was observed (478 kJ/kg$^{0.75}$ BW/d on average for diets BS and CS, and 332 kJ/kg$^{0.75}$ BW/d for diets BF and CF). Differences (P<0.05) were found for OXF (189 kJ/kg$^{0.75}$ BW/d on average for diets BS and CS, and 331 kJ/kg$^{0.75}$ BW/d for diets BF and CF). Chwalibog *et al.* (1997) pointed out that with increasing the amount of fibrous by-products, more fermentation and less glucose was absorbed directly and oxidized as OXCHO, while more carbohydrate was converted to volatile fatty acids and oxidized as OXF. Therefore, most of the HPx derived from OXF was in fibrous diets (42% for fibrous vs. 25% for starch diets, on average); meanwhile for OXCHO obtained for fibrous and starch diets was on average 43 and 63%, respectively.

Results suggest that inclusion of fibrous by-products in the diet of mid lactation goats reduce OXCHO and increases OXF and fat content without compromising milk yield.

References

Brouwer, E. 1958. On simple formulae for calculating the heat expenditure and the quantities of carbohydrate and fat metabolized in ruminants, from data on gaseous exchange and urine N. Pages 182-194 in Proc. 1th Symposium on Energy Metabolism. EAAP. Publ. 8. Academic Press, London.

Chwalibog, A., A.H. Tauson and G. Thorbek, 1997. Quantitative oxidation of nutrients in growing calves. Z. Ernährungswiss 36, 313-316.

Fahey, G.C. and L.L. Berger, 1988. Carbohydrate nutrition of ruminants. In: Church, D.C. (ed.), The Ruminant Animal. Digestive Nutrition and Physiology. Prentice-Hall, Englewood Cliffs, NJ, pp. 269-297.

López, M.C., C. Fernández and M. Lachica, 2013. Effect of the starch source and fiber level in mixed diets on energy balance of lactating Murciano-Granadinas goats. ISEEP.

Wolin, M.J., 1960. A theoretical rumen fermentation balance. J. Dairy Sci. 43, 1452-1459.

Effect of the starch source and fiber level in mixed diets on the energy balance of lactating Murciano-Granadina goats

M.C. López[1], C. Fernández[1] and M. Lachica[2]
[1]Institute for Animal Science and Technology, Universidad Politécnica de Valencia, 46022 Valencia, Spain; cjfernandez@dca.upv.es
[2]Department of Animal Nutrition, Estación Experimental del Zaidín, CSIC, 18100 Armilla, Granada, Spain

Introduction

The Spanish ruminant production system (FEDNA, 2009) is based on high use of concentrate (40-70%) instead of whole forage rations due to the lack of pasture. Goat livestock occupies the second position in the EU with 30% of the total milk production. By-product feeds have been used extensively in dairy cattle diets as an economical substitute for corn and barley grain. There is an increasing interest in the nutritive value of by-product feeds for dairy ruminant diets.

The aim of this work was to determine the effect of the starch source and fiber content on the energy balance and the efficiency of utilization of metabolizable energy (ME) for milk production (k_l) of lactating Murciano-Granadina goats.

Material and methods

Twenty Murciano-Granadina goats at mid lactation (43±0.5 kg BW) were used following two cross over designs. Four isoproteic and isoenergetic diets [186 g of crude protein (CP)/kg of dry matter (DM) and 19.1 MJ of gross energy (GE)/kg DM, respectively] were used, with alfalfa hay as forage source. The ratio forage:concentrate was 41:59. Diet BS had 41% of barley grain; in diet BF the barley grain was replaced by 41% of fibrous by-products; diet CS had 40% of corn grain; and in diet CF the corn grain was replaced by 40% of fibrous by-products. The fibrous by-product was a blend of 18.5% of soy hulls, 15% of gluten feed and 7.5% of wheat bran. Diet BS and CS had an average starch content of 27%, and BF and CF of 7%. The diet BS and CS had an average NDF content of 35%, and BF and CF of 47%. The goats were allocated to individual metabolism cages. After 10 d of adaptation, feed intake, refusal, total fecal and urine output, and milk yield were recorded daily during 5 d period. Heat production (HP) was determined during 24 h using a mobile open-circuit respirometry system connected to a head box and calculated according to Brouwer (1965). The retained energy was calculated as ME – HP, the retained energy as tissue as ME – HP – milk energy, the RQ as CO_2/O_2, and k_l as corrected milk energy/(ME intake – MEm) assuming MEm=401 kJ/ $kg^{0.75}$ BW/d (Aguilera *et al.*, 1990). The data were analyzed by the MIXED procedure of SAS.

Results and discussion

Greater organic matter (OM) digestibility ($P<0.05$) was observed for diets higher in starch (BS and CS) than fibrous diets (BF and CF). The lowest values in CP and NDF digestibility was for diet CS (Table 1). The significant reduction of OM digestibility was probably due to the increasing NDF content of diets BF and CF combined with the greater value of starch for diets BS and CS. NDF digestibility was numerically greater for fibrous diets probably due to the source of fibrous ingredients (fibrous by-products and alfalfa hay) when compared with starch rich diets, where the only source of fiber was alfalfa hay. No significant differences were found for the intake of ME (1432 kJ/$kg^{0.75}$ BW/d, as average), with higher numerical values for fibrous than starch diets (1441 and 1422 kJ/ $kg^{0.75}$ BW/d on average, respectively). No significant differences among diets were observed in HP with an average value of 788 kJ/$kg^{0.75}$ BW/d. Neither significant difference was found for RQ values (1.0 on average). The retained energy as milk was greater ($P<0.05$) for fibrous diets (549 kJ/

Table 1. Daily intake (kg/d), digestibility (%) and metabolizable energy (ME, kJ/kg$^{0.75}$ BW) intake, heat production, retained energy, efficiency of utilization of metabolizable energy for milk production (k_l) and RQ (CO_2/O_2) of lactating Murciano-Granadina goats (n=20) under four diets based on barley grain (BS), barley grain and fibrous byproducts (BF), corn grain (CS), and corn grain and fibrous byproducts (CF).

	BS	BF	CS	CF	SEM	*P*-value
Dry matter intake	2.02	2.07	2.03	2.09	0.024	0.7733
Digestibility, %						
Organic matter	69.6^a	65.8^b	69.2^a	65.1^b	0.864	0.0280
Neutral detergent fiber	46.3^a	47.6^a	36.6^b	49.5^a	1.893	0.0032
Starch	99.2	99.6	99.1	99.6	0.090	0.7598
Crude protein	78.0^a	75.8^{ab}	69.2^b	70.6^b	0.753	0.0395
ME intake	1414	1439	1430	1443	15.8	0.9260
Heat production	768	819	792	774	7.2	0.0559
Retained energy	646	621	638	669	13.6	0.6593
Milk energy	504^b	544^a	501^b	554^a	13.9	0.0463
Tissue energy	142^a	76^b	137^a	115^{ab}	17.8	0.0451
k_l	0.64	0.60	0.63	0.65	0.007	0.0737
RQ	1.06	0.96	1.03	0.99	0.014	0.0786

^{a,b}Values within a row with different superscript were significantly different ($P<0.05$).

kg$^{0.75}$ BW/d, as average) than starch diets (503 kJ/kg$^{0.75}$ BW/d, as average). The retained energy as tissue was greater ($P<0.05$) for starch diets (140 kJ/kg$^{0.75}$ BW/d, as average) than high fiber diets (96 kJ/kg$^{0.75}$ BW/d, as average). Van Knegsel *et al.* (2007) showed that cows fed a lipogenic diet partitioned more energy to milk than cows fed a glucogenic diet, with a tendency for higher energy mobilization from body fat. No significant differences were observed among diets for k_l (0.63 on average) and the value was similar than the obtained by Aguilera *et al.* (1990) with the same goat breed (k_l=0.67).

In the presents study, the use of fibrous by-products as replacer of barley and corn grain as starch sources, had no effect on ME intake and it increased the retained energy as milk without affect the k_l.

References

Aguilera, J.F., C. Prieto and J. Fonollá, 1990. Protein and energy metabolism of lactating Granadina goats. Br. J. Nutr. 63, 165-175.

Brouwer, E., 1965. Report of sub-committee on constants and factors. In: Blaxter, K.L. (ed.), Proceedings of the Third EAAP Symposium on Energy Metabolism. Publication No. 11. Academic Press, London, pp. 441-443.

FEDNA, 2009. Nutritional Requirements for Dairy Ruminants. ed. Fundación Española para el Desarrollo de la Nutrición Animal, Madrid, Spain.

Van Knegsel, A.T.M., H. van den Brand, J. Dijkstra, W.M. van Straalen, M.J. Heetkamp, S. Tamminga and B. Kemp, 2007. Dietary energy source in dairy cows in early lactation: energy partitioning and milk composition. J. Dairy Sci. 90, 1467-1476.

The development of the gravid uterus of Santa Inês ewes and ewe lambs under two nutritional planes

L.F.L. Cavalcanti[1,2], *I. Borges*[1], *F.A. Souza*[1], *G.L. Macedo Júnior*[3] *and L.O. Tedeschi*[2]
[1]*Departamento de Zootecnia, Universidade Federal de Minas Gerais, Brazil; luigicavalcanti22@gmail.com*
[2]*Department of Animal Science, Texas A&M University, USA*
[3]*Universidade Federal de Uberlândia, Brazil*

Introduction

The Santa Inês is the most common ovine breed in Brazil, and there is a lack of information about the nutrient requirements for this genotype. As a result of the inhospitable environment where this breed was established, the metabolism of these animals evolved different strategies to reserve and utilize nutrients which may influence its requirements. The most significant drain of nutrients of pregnant ewes is the gravid uterus (GU) mainly during the last third of gestation because of the increased fetus growth rate (Rattray *et al.*, 1974). The objective of this study was to measure the weight and composition of the GU of Santa Ines ewes and ewe lambs (primiparous ewes).

Material and methods

The animals (38 ewe lambs and 75 ewes) were mated with rams of known fertility and after 45 days from mating were examined to confirm pregnancy and to split the ewes based on their prolificacy. Ewe lambs with two fetuses were discarded because of the low frequency and difficulty to form groups. The animals were divided into two nutritional planes, *ad libitum* and restricted (15% less protein and energy than the non-restricted group). Ewe lambs were slaughtered at 0, 100, 130, and 140 days of pregnancy, and ewes at 0, 90, 110, 130, and 140 days of pregnancy. The gravid uterus and its content were removed and weighed. Samples were collected and later submitted to dry matter (DM) and crude protein (CP) analysis. An exponential function ($W = W0 \times \exp(k \times t)$), where: W is the weight or CP content (g), W0 is the initial value (g), k is fractional growth rate (1/d), and t is the time in days from mating, was fitted to the data. The lack of fit was assessed based on the F-test and the necessity of additional parameters for each group due to other factors (e.g. nutritional plane, category and number of fetuses) was determined. Additionally, an allometric analysis of CP content versus the GU weight was performed by fitting the function $CPGU = a \times GU^b$, where 'a' is a constant (g), CPGU is the content (g) of CP in the gravid uterus (GU), GU is weight (g) and 'b' is the allometric exponent (dimensionless).

Results and discussion

The GU weight was only affected by prolificacy, being for single gestation (one fetus) described by the equation with W0=222.6±44.11 and k=0.024±.001, regardless of nutritional plane or category. For ewes with two fetuses the parameters obtained were W0=604.3±113.84 and k=0.021±.001. For the CP content, it was impossible to fit one model for both ewes and ewe lambs. The same pattern was observed for the prolificacy. Hence, one equation was developed regardless of the level of nutrition. Three functions were obtained: for all ewe lambs (W0=9.320±3.13, k=0.030±.002); for ewes with single gestation (W0=9.371±4.40, k=0.032±.003) and the last one for ewes with two fetuses (W0=20.595±5.06, k=0.030±.001), as seen at Figure 1. The protein deposition on the gravid uterus in this study was overestimated by the equation of CSIRO (2007) with the same purpose since approximately 60 days of pregnancy, even when compared to the equation for twin gestation. This may be explained by the difference between Santa Ines animals and the traditionally breeds from Australia. Accordingly to Robinson and McDonald (1979) the birth weight is affected by breed and nutritional level, been those factors the main determinants of nutrient deposition rate on gravid uterus.

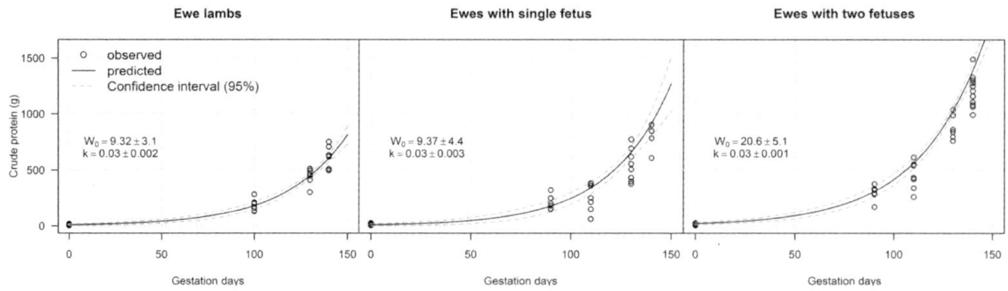

Figure 1. Crude protein content (g) on the gravid uterus during pregnancy for ewe lambs and ewes with single or two fetuses.

Severe restrictions of energy (30% or greater) for long time affect the fetus development (Gao *et al.*, 2009). In this trial the lowest nutritional level was not able to affect the gravid uterus development, what may be explained by the restriction used in this study that probably was not so severe. This result was also observed by (Koong *et al.*, 1975) Koong *et al.* (1975) when they decomposed the k coefficient into three factors: level of nutrition, stage of gestation and number of fetuses. Using a stepwise regression technique they concluded that only the two last factors influenced the growth development, and additionally, that the model for which the number of fetuses was directly related to the initial weight (W0) was more accurate than when related to the growth rate (k), what is quite similar to the results found in the present study for ewes with one or two fetuses. When we used the allometric equation, the exponent 'b' was 1.132±0.04, what means that the CPGU deposition is not isometric to the weight increase of GU. These results suggested that, within the range of the evaluated nutritional planes, the main modulator of GU development and, consequently, of the nutrient requirements for pregnancy for Santa Ines sheep are pregnancy stage (proportion of variance explained, ω^2=0.701), category (ω^2=0.012), and prolificacy (ω^2=0.125). Moreover, the protein is deposited in a faster rate of deposition than the rate of growth of the whole gravid uterus.

References

CSIRO, 2007. Nutrient requirements of domesticated ruminants. Commonwealth scientific and industrial research organisation Collingwood, 270 pp.

Gao F., Y. Liu, and X. Hou, 2009. Effect of maternal undernutrition during late pregnancy on growth and development of ovine fetal visceral organs. Asian Austral. J. Anim. 22,1633-1639.

Koong, L. J., W. N. Garret, and P. V. Rattray, 1975. A description of the dynamics of fetal growth in sheep. J. Anim. Sci. 41,1065-1068

Rattray, P. V., W. N. Garret, N. E. East, and N. Hinman. 1974. Growth, development and composition of the ovine conceptus and mammary gland during pregnancy. J. Anim. Sci. 38, 613-626.

Robinson, J. and I. McDonald, 1979. Ovine prenatal growth, its mathematical description and the effects of maternal nutrition. Ann. Biol. Anim. Bioch. 19, 225-234.

The effect of a limited supply of phenylalanine, threonine, and tryptophan on milk yield and composition

I.H. Iroshan[1], *H. Lapierre*[2] *and L. Doepel*[1]
[1]*University of Calgary Faculty of Veterinary Medicine, Calgary, AB T2N 4N1, Canada; ldoepel@ucalgary.ca*
[2]*Agriculture and Agri-Food Canada, Sherbrooke, QC J1M 0C8, Canada*

Introduction

In dairy cows, milk and milk component yields respond strongly to deletion of all essential amino acids (AA) from an infused AA mixture, but do not respond to deletion of non-essential AA (Doepel and Lapierre, 2010). Lysine and methionine are considered 1st and 2nd limiting AA on corn-based diets (NRC, 2001) while on grass silage-based diets histidine tends to be first-limiting (Vanhatalo *et al.*, 1999). A large proportion of the research in dairy cows has focused on these three AA. Indeed, deletion of these 3 AA from an infusion mixture decreased milk protein yield to the negative control level; however, deletion of the branched-chain AA from the mixture did not alter milk and milk protein yields (Weekes *et al.*, 2006). To date, phenylalanine (Phe), threonine (Thr) and tryptophan (Trp) have received minimal attention, but the importance of these AA should not be overlooked. The digestive flows of these AA are, on average, lower than estimated recommendations (Doepel *et al.*, 2004). Additionally, these AA are usually taken up by the mammary gland in the same amount as that used for milk protein secretion which then raises the question 'will milk protein synthesis decrease if the supply of these AA is limited?' The objective of this study was to determine how milk and milk component yields change when Phe, Thre, and Trp supply is limited.

Material and methods

Five 2nd lactation Holstein cows were used in a 5×5 Latin square with 10-day periods. Cows were fed a diet formulated to supply 100% of the net energy requirement and 70% of the metabolizable protein (MP) requirement (NRC, 2001). The treatments were abomasal infusions of water (Ctl), all AA (TAA) at 925 g/day, all AA excluding Thre (No-Thr), all AA excluding Phe (No-Phe), and all AA excluding Trp (No-Trp). The TAA treatment was estimated to supply, with the diet, 100% of the MP requirement, and AA composition of the TAA was the same as that of casein. Infusates were delivered continuously in 15 l/d of water. Arterial (coccygeal vessel) and venous (mammary vein) blood samples were collected every 2 hours on day 10 between the AM and PM milkings (n=6). Plasma was analysed for urea-N and β-hydroxybutyrate (BHBA). Milk samples from the last 3 days of each period were analyzed for lactose, fat, protein, and milk urea-N (MUN). Data were analyzed using the MIXED procedure of SAS with treatment as the fixed effect, and period and cow as random effects. Treatment differences were determined using pre-planned contrasts, comparing each treatment to TAA.

Results

Dry matter intake tended to be lower for No-Phe vs. TAA (P=0.10). Relative to TAA, milk yield tended to be reduced in CTL, No-Phe and No-Thr cows (0.10<P<0.15; Table 1). Milk protein yield was lower (P<0.01) for CTL and No-Phe than for TAA and tended to be lower (P=0.13) for No-Thr. Milk fat yield (1128 g/d) and lactose yield (1428 g/d) were not affected by treatment (P>0.15). MUN was higher (P<0.01) for TAA than Ctl. Plasma urea-N concentration was increased with TAA compared with CTL, but was lower for TAA than for No-Phe. BHBA arterial concentration and mammary gland extraction rate (36.7%, P=0.82) were not affected by treatment. Deletion of Trp from the infusate did not affect (P>0.15) any measured parameters.

Table 1. Effect of amino acids deletion on dry matter intake (DMI), milk and milk component yield, and metabolite concentrations.[1]

Parameter	Ctl	Treatment				SEM	*P*-value		
		No-Phe	No-Thr	No-Trp	TAA		Ctl TAA	TAA No-Phe	TAA No-Thr
DMI, kg/d	17.8	17.0	17.6	17.2	18.0	0.60	0.69	0.10	0.46
Yield, kg/d									
Milk	29.9	30.1	30.4	32.8	33.4	2.12	0.10	0.12	0.15
Protein	0.78	0.77	0.84	0.88	0.91	0.04	0.01	0.01	0.13
Fat	1.04	1.08	1.16	1.25	1.12	0.07	0.34	0.60	0.71
Lactose	1.36	1.38	1.39	1.51	1.50	0.09	0.15	0.22	0.25
MUN, mg/dl	8.8	13.5	14.6	14.2	14.1	0.94	0.003	0.65	0.70
Plasma metabolites, mM									
Urea-N	6.8	13.2	11.6	11.6	11.2	0.74	0.001	0.03	0.62
BHBA	0.54	0.49	0.46	0.53	0.52	0.04	0.71	0.47	0.22

[1] Pre-planned contrast comparing TAA vs. each treatment; TAA vs. No-Trp not shown: all *P*-values >0.15.

In conclusion, the results of the current study show that a limited supply of Phe and Thr negatively affects milk and milk protein secretion, with Phe limitation on protein yield being stronger than Thr limitation. The elevated urea-N with No-Phe relative to TAA supports that milk protein synthesis was limited by a deficiency of Phe, and thus, AA supply not used for milk protein secretion and therefore in excess was converted to urea. Deficiencies of Phe, Thr and Trp had no effect on milk fat and lactose yield.

References

Doepel, L. and H. Lapierre, 2010. Changes in production and mammary metabolism of dairy cows in response to essential and nonessential AA infusions. J. Dairy Sci. 93, 3264-3274.

Doepel, L., D. Pacheco, J.J. Kennelly, M.D. Hanigan, I.F. López and H. Lapierre, 2004. Milk protein synthesis as a function of AA supply. J. Dairy Sci. 87, 1279-1297.

National Research Council, 2001. Nutritional requirements of dairy cattle. 7th Revised Edn. National Academy Press, Washington D.C.

SAS. 2002. SAS System for Mixed Models. SAS Institute Inc., Cary, NC.

Vanhatalo, A. P. Huhtanen, V. Toivonen and T. Varvikko, 1999. Response of dairy cows fed grass silage diets to abomasal infusions of histidine alone or in combinations with Met and Lys. J. Dairy Sci. 82, 2674-2685.

Weekes, T.L., P.H. Luimes and J.P. Cant, 2006. Responses to amino acid imbalances and deficiencies in lactating dairy cows. J. Dairy Sci. 89, 2177-2187.

Metabolic and hormonal profile of bulls during evaluation of nutritional requirements by respirometric technique under different plane of nutrition

A.L.C.C. Borges, P.A.D. Vivenza, R.R. Silva, E.J. Facury Filho, H.F. Lage, P.H.A. Carvalho, P.R.O. Paes and N.M. Rodriguez
Departament of Animal Science of Veterinary School of Federal University of Minas Gerais, Belo Horizonte, Brazil; analuiza@vet.ufmg.br

Introduction

The knowledge of animal metabolism against different nutritional conditions imposed is of great importance, since the ratio of nutrients absorbed deposited in the body or that are in full use is dependent on metabolic and hormonal status of the animal (Payne and Payne, 1987). The objective of this work was to study the hormonal and metabolic profile of bulls during evaluation of the nutritional requirements of energy, by respirometry technique, in different nutritional plans and undergoing a fasting period.

Material and methods

Fifteen F1-crosbred (*Bos taurus indicus*) × Holstein (*Bos taurus taurus*) bulls, with average weight of 304 kg were used. The animals were randomly assigned into three groups and fed diets to provide mild (maintenance group), medium (medium gain group) and high (*ad libitum* group) weight gains, initially formulated for gains of 0.100 kg, 0.500 kg and 0.900 kg/day. For measurement of fasting heat production, which is the net energy for maintenance (NEm) through respirometric technique, the animals are subjected to 72 h of fasting. The respirometric chamber is performed between 48 and 72 h of fasting. Four times during the fasting were evaluated: the beginning of fasting, 24, 48 and 72 h of fasting. Insulin, glucose, beta-hydroxybutyrate (BHB-β), nonesterified fatty acids (NEFA), urea, creatinine, total protein, albumin, globulins, triglycerides, aspartate aminotransferase (AST) and hematocrit were evaluated. The blood samples were collected by venipuncture/coccygeal artery using a vacuum collection system. The experimental design was a randomized design with a split-plot scheme, since two factors were evaluated nutritional plan (plot factor) and fasting time (subplot factor). The data were analysed by the GLM procedures (SAS, 2001).

Results and discussion

The fasting period of 72 h was characterized by decrease in glucose concentration during the first 48 hours and recovery levels of this component in the maximum time of fasting. This is associated with mobilization of lipid and protein reserves during fasting, which varied between groups, being higher in group *ad libitum* taking these animals the highest values of plasma NEFA and urea, although the β-BHB values were similar in both groups, but increasing with the increased time of fasting. It is suggested that *ad libitum* group (greater body weight) are more susceptible to the effects of fasting, in view of its likely greater energy requirement for maintenance, once the body weight has significant influence on this (Smith and Baldwin, 1974). In situations of food deprivation, they probably have greater energy deficit, signaling the need to mobilize the body in larger proportions body reserves. The table 1 reports the metabolic and hormonal profile of bulls under different nutritional conditions.

There were no differences in hematocrit values and total protein during fasting, indicating that the animals were not dehydrated, so there is no interference of hemoconcentration on these parameters. The response of animals to fasting reveals the efficiency of the maintenance mechanisms of energy within normal in these, ensuring proper physiological condition during the measurement respirometric chamber, because even the maximum period of fasting, plasma glucose concentrations were within the reference values proposed by Kaneko *et al.* (2008), from 2.5 to 4.6 mmol/l.

Table 1. Plasma concentrations of glucose, β-BHB and NEFA, expressed in mmol/l, urea, expressed in mg/dl, and insulin, expressed in uU/ml, during fasting (0, 24, 48 and 72 h), in groups of ad libitum *gain, medium gain and maintenance.*

Component	Group	Time of fasting period (h)[1]				Average
		0	24	48	72	
Glucose (mmol/l)	*Ad libitum*	4.32 a	3.96 a	3.86 a	3.84 a	4.00 a
	Medium gain	4.11 a	3.82 ab	3.27 ab	3.72 a	3.73 b
	Maintenance	3.62 a	2.92 b	2.89 b	3.58 a	3.25 c
β-BHB (mmol/l)	*Ad libitum*	0.391 a	0.419 a	0.681 a	0.915 a	0.602 a
	Medium gain	0.267 a	0.375 a	0.554 a	0.923 a	0.530 a
	Maintenance	0.398 a	0.422 a	0.704 a	0.930 a	0.614 a
NEFA (mmol/l)	*Ad libitum*	0.181 a	0.449 a	0.764 a	1.026 a	0.605 a
	Medium gain	0.121 a	0.288 a	0.764 a	0.942 a	0.529 b
	Maintenance	0.101 a	0.209 a	0.676 a	0.861 a	0.462 b
Urea (mg/dl)	*Ad libitum*	35.66 a	40.60 a	51.07 a	40.43 a	41.94 a
	Medium gain	33.03 a	34.98 a	42.22 ab	32.51 a	35.69 b
	Maintenance	33.37 a	28.68 a	31.15 b	36.63 a	32.46 b
Insulin (uU/ml)	*Ad libitum*	14.88 a	8.78 a	6.78 a	6.31 a	9.19 a
	Medium gain	17.12 a	11.02 a	6.64 a	7.76 a	10.64 a
	Maintenance	13.06 a	4.16 a	4.88 a	6.48 a	7.15 b

[1] a,b,c ($P<0.05$).

Acknowledgements

We would like to thank CNPq, CNPq-INCT, FAPEMIG, CAPES and EPAMIG for their cooperation in carrying out this work.

References

Kaneko, J.J., Harvey, J.W. and M.L. Bruss, 2008. Clinical biochemistry of domestic animals. 6th ed. Academic Press, San Diego. 916p.

Payne, J. M.; Payne, S. The metabolic profile. 1 ed. Oxforf: Oxford University Press, 1987. P.179.

Sas, 2001. Release 8.01. Sas Inst. Inc., Cary, Nc, USA.

Smith, N.E, Baldwin, R.L, 1974. Effects of breed, pregnancy and lactation on weight of organs and tissues in dairy cattle. Journal Of Dairy Science, V.57, N.9, P. 1055-1060.

Methane emission and digestibility of goats subjected to feed restriction

A.R.C. Lima[1], K.T. Resende[1], I.A.M.A. Teixeira[1], T.F.V. Bompadre[1], R.T.S. Frighetto[2] and M.H.M.R. Fernandes[1]
[1]UNESP Univ Estadual Paulista, Department of Animal Science, Via de Acesso Prof Paulo Donato Castellane s/n, 14884-900 Jaboticabal, Sao Paulo, Brazil; mhmrochaf@fcav.unesp.br
[2]EMBRAPA Environment, Brazilian Agricultural Research Corporation, Laboratório de Ecologia Quimica, Rodovia SP-340 Km 127, 13820-000 Jaguariuna, Sao Paulo, Brazil

Introduction

Currently, due to environmental and economic issues, ruminant production system has aimed at maximizing production, concomitant with an increased efficiency. Because of methane is produced in the process of feed energy utilization within the animal, its reduction is usually associated with improved productivity. Therefore, one of the strategies to optimizing energy use in ruminants is to minimize enteric methane (CH_4) emission during the fermentation process, by reducing dry matter intake (Shibata and Terada, 2010). Feed restriction is a common practice on goat production in several parts of world, and can be used to exploiting the compensatory gain, resulting in better efficiency of animal feed energy utilization. Therefore, this study was carried out to evaluate the effect of feed restriction on methane emission and digestibility.

Material and methods

Fifteen dry non-lactating Nubian goats (34.9±2.6) were randomly allocated into 5 groups (blocks) of 3 animals receiving the treatments AL: *ad libitum* intake, MR: moderate restriction (approximately 15% of feed restriction) and SR: severe restriction (approximately 40% of feed restriction). Goats were housed in individual metabolic pens. The experimental diet (15% crude protein, 16.5 MJ GE /kg DM) consisted of 47% dehydrated corn plant and 53% concentrate. The moderate and severe feed restriction was determined daily based on the dry matter (DM) intake of the goats in the ad libitum treatment on the previous day. During a digestion trial, feed intake, feed refusals, feces and urine were collected for 5 d after 20 d adaptation period. Composites of feeds, feed refusals, and feces were dried at 60 to 65 °C for 72 h and ground through a 1 mm screen using a Wiley mill. Composites of urine were passed through a sieve to remove the large particles, and a subsample was taken. Gross energy was determined for feeds, feed refusals, feces, and urine using a bomb calorimeter (Parr Instrument Co.). Measurements of methane emission were performed using the sulphur hexafluoride (SF_6) tracer technique. Data were analyzed in a randomized complete block design, using mixed models of SAS.

Results and discussion

Goats subjected to moderate and severe restriction treatments ate approximately 85% and 60%, respectively, of those fed *ad libitum* (Table 1). Methane emission (P=0.01) was greater in goats fed ad libitum than in goats subjected to moderate and severe feed restriction. However, when methane energy losses were expressed as a proportion of gross energy intake, there were no differences between treatments. Digestible energy (11.5 MJ/kg, P=0.21), metabolizable energy (9.4 MJ/kg, P=0.53) and metabolizability (q_m, P=0.50) was similar for all treatments. It may be a result of using equal diet for all treatments. Moreover, the severity of feed restriction imposed was not enough to cause significant differences in the intake among treatments. In conclusion, feed restriction decreased daily methane emission within the range of intakes considered; however it did not affect efficiency of feed energy utilization.

Table 1. Intake, digestibility and energy utilization of goats subjected to feed restriction.

Variable	Feed restriction				
	NR	MR	SR	*P*-value	SEM
Animals	5	5	5		
BW	45.51	46.33	45.21	0.47	1.02
DMI (g/day)	824.35	705.39	508.58	0.1041	120.87
DMI (g/kg $BW^{0.75}$)	52.86	43.97	31.39	0.0695	5.82
OMI (g/day)	791.97a	677.12b	487.56c	0.013	115.67
Digestibility					
DMD (%)	74.56	69.52	72.80	0.14	2.48
OMD (%)	73.70	68.39	71.97	0.13	2.56
ED (%)	71.25	65.98	70.85	0.18	3.03
Energy utilization					
CH_4(g/day)	22.05	18.87	15.43	0.01	1.89
CH_4(g/100 g OMI)	2.65	2.49	2.77	0.70	0.23
DE (MJ/ kg DM)	11.82	10.99	11.81	0.21	0.495
ME (MJ/ kg DM)	9.75	8.97	9.54	0.53	0.663
q	0.59	0.54	0.57	0.50	0.04
CH_4 losses (%)	8.59	8.00	8.89	0.72	0.78

Acknowledgements

FAPESP project number 2008/58351-5 and 2011/14842-8.

References

Shibata M. and F. Terada, 2010. Factors affecting methane production and mitigation in ruminants. Anim. Sci. J. 81, 2-10.

Efficiency of utilization of dietary nitrogen for milk production by dual-purpose cows fed increasing levels of *Leucaena leucocephala* forage mixed with *Pennisetum purpureum* grass

A. Ruiz-González, A.J. Ayala-Burgos, C.F. Aguilar-Pérez and J.C. Ku-Vera
Department of Animal Nutrition, Faculty of Veterinary Medicine and Animal Science, University of Yucatan, Merida, Yucatan, Mexico; kvera@uady.mx

Introduction

Efficiency of nitrogen utilization (g N in product/g N intake; ENU) by ruminants is low with an average of ~25% (Calsamiglia *et al.*, 2010), but with an ample range of variation (15-40%) suggesting that improvements are possible. It is of paramount importance to increase the efficiency of N utilization in order to improve productive performance (i.e. milk yield) and reduce excretion of nitrogenous products in urine and feces from ruminants to the environment. Silvopastoral systems represent a viable means of increasing levels of animal production in the tropics. Usually, high CP legumes are employed in those systems and the possibility of N loss from the gastrointestinal tract (Poppi and McLennan, 1995) may constraint productive limits.

Material and methods

Five multiparous crossbred (*Bos taurus* × *Bos indicus*) cows with a live-weight of 400±30 kg, were housed in individual metabolic stalls in a 4×4 Latin square arrangement of treatments (Cochran and Cox, 1957). Cows were fed increasing levels (0, 15, 30 and 45%; DM basis); of fresh chopped forage of *Leucaena leucocephala* mixed with chopped *Pennisetum purpureum* grass and cows were supplemented with 2 kg/day of ground corn at the time of milking. Cows were milked manually once daily (6:00 h) and the calf had restricted access to his dam. Milk yield (kg) was recorded daily. A milk sample was taken daily during the last five days of each experimental period to determine N content by the Kjeldahl method (AOAC, 1980; ID 2,062), gross energy was estimated with the equation proposed by Tyrrell and Reid (1965) and the efficiency of energy in milk was estimated with the equation proposed by Yan *et al.*, (1997). Each experimental period lasted for 21 days, 16 days for adaptation and 5 days for measurements of response variables. N intake and N in milk data were analyzed with GLM and orthogonal contrasts performed with the statistical software SAS (2006).

Results and discussion

A correlation was found (R^2=0.92) between N intake and N in milk. Figure 1 predicts N concentration in milk to a known CP intake. Nonetheless, N in milk as a percentage of the total N consumed tended

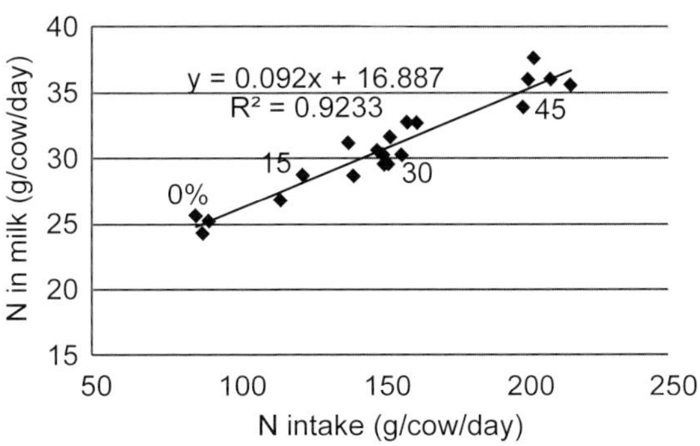

Figure 1. Relationship between nitrogen content in milk and total nitrogen intake (g/cow/day).

to decrease ($P<0.05$) as N intake was increased (Figure 2). Efficiency of N utilization for milk production and the efficiency of ME utilization for milk production decreased ($P<0.05$) as the level of N consumed was increased (Table 1). Daily milk yield was 5.1^c, 5.3^c, 6.4^b and 7.7^a kg (SEM=0.3) for the treatments 0, 15, 30 and 45% Leucaena, respectively. Results agree with Muinga *et al.* (1992) who significantly increased milk yield (7.3, 7.7 and 8.3 kg/d) of crossbred cows fed a basal ration of *Pennisetum purpureum* grass supplemented with increasing levels (0, 1 and 2 kg DM/day) of *Leucaena* forage. There is scope to improve the efficiency of N utilization for milk production in silvopastoral systems in the tropics where high CP legumes are available for incorporation under practical conditions, particularly during the dry season when tropical grasses are deficient in N.

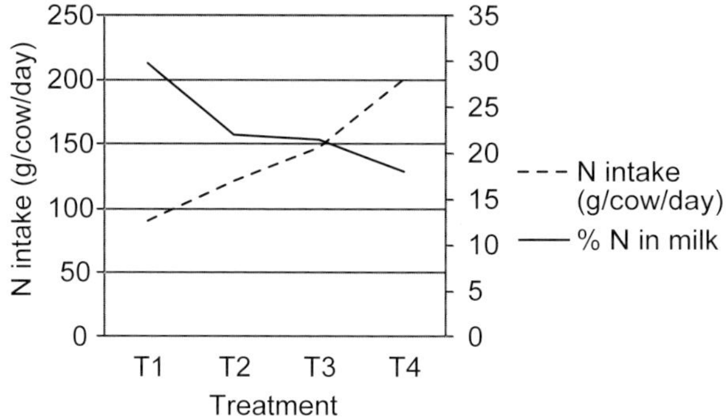

Figure 2. Percentage of consumed nitrogen contained in milk.

Table 1. Milk energy output, efficiency of nitrogen utilization and efficiency of ME utilization for milk production by crossbred cows fed increasing levels of Leucaena *foliage in the ration.*

	% incorporation of foliage of *L. leucocephala* in the ration (DM basis)				SEM	Contrast		
	0	15	30	45		L	C	Q
Milk energy output MJ/d	15.83^c	17.71^{bc}	19.55^{ab}	22.10^a	1.47	*	NS	NS
Efficiency of N utilization in milk	0.29^a	0.22^b	0.21^{bc}	0.18^c	0.04	*	NS	NS
Efficiency of energy in milk	0.62^a	0.61^b	0.60^c	0.60^c	0.10	*	NS	NS

Rows with different literal differ significantly ($P<0.05$).

References

Calsamiglia, S., Ferret, A., Reynolds, C.K., Kristensen, N.B. and van Vuuren, A.M. 2010. Strategies for optimizing nitrogen use by ruminants. Anim. 4:7, 1184-1196.

Cochran, W. G. and Cox, G. M. 1974. Diseños Experimentales. Edit. Trillas. pp. 145-149.

Muinga, R.W., Thorpe, W. and Topps, J.H. 1992. Voluntary food intake, live-weight change and lactation performance of crossbred dairy cows given *ad libitum Pennisetum purpureum* (napier grass var. Bana) supplemented with leucaena forage in the lowlands semi-humid tropics. Anim. Prod. 55, 331-337.

Poppi, D., and McLennan, S. 1995. Protein and energy utilization by ruminants at pasture. J. Anim. Sci. 73: 278-290.

SAS. 2006. Statistical Analysis System Institute. SAS Institute Inc. Versión 9.00. Cary. NC, USA.

Tyrrell, H. F., and J. T. Reid. 1965. Prediction of the energy value of cow's milk. J. Dairy Sci. 48:1215-1223.

Yan, T., Gordon, F.J., Agnew, R.E., Porter, M.G. and Patterson, D.C. 1997. The metabolisable energy requeriment for maintenance and the efficiency of utilization of metabosable energy for lactation by dairy cows offered grass silage-based diets. Lives. Prod. Sci. 51,141-150.

Milk yield and composition, and efficiency of utilization of metabolisable energy for lactation by Pelibuey ewes

J.C. Espinoza-Hernández, A.J. Ayala-Burgos, C.F. Aguilar-Pérez, J.G. Magaña-Monforte and J.C. Ku-Vera
Department of Animal Nutrition, Faculty of Veterinary Medicine and Animal Science, University of Yucatan, C.P. 97300 Merida, Yucatan, Mexico; kvera@uady.mx

Introduction

In the tropical regions of Mexico the main maternal breed used by sheep producers is the Pelibuey. There is little scientific information relative to milk production, energy requirements and efficiency of utilization of metabolisable energy for milk production at different physiological stages by the Pelibuey ewe. Nonetheless, the ewe is the main component of the production of lambs (weaners) for fattening under commercial, practical conditions. This work was carried out to address the above mentioned issues.

Material and methods

The experiment was carried out at the Faculty of Veterinary Medicine and Animal Science, University of Yucatan, located in South Mexico. Twenty-one multiparous (2-3 lambings), lactating Pelibuey ewes (45±6 kg), 2.5-3 years old and rearing two lambs each were used. Ewes were randomly allotted to three treatments of seven ewes each in a completely randomized design, housed in individual stalls and fed three levels of ME intake: Low (L; 0.56 MJ/kg$^{0.75}$/day), Medium (M; 1.12 MJ/kg$^{0.75}$/day) and High (H; 1.67 MJ/kg$^{0.75}$/day), during 49 days as from the fifth day post-lambing. Milk yield measurements were carried out during two non-consecutive days each week according to the technique of the double-weighing of lambs. Lambs were separated from their dams (at 6:00 h), 135 minutes before measurements of milk yield, which were carried out ten times during 24 h. Lambs were allowed to suck their dams for ten minutes to make sure the udder was completely emptied. Milk was sampled every two non-consecutive days each week obtaining a 100 ml sample per ewe during three daily manual milkings (06:00, 12:00 and 17:00 h). For chemical analysis of milk, samples were mixed per treatment per week and a 200 ml aliquot was obtained for fat analysis by the Gerber method by Ling (1963), protein was equivalent to 6.38 times N concentration in milk and was determined by the Kjeldahl method, lactose was assayed with a colorimetric technique by copper reduction. Milk (100 ml) was freeze-dried for DM and GE determination with an adiabatic bomb calorimeter. Efficiency of utilization of ME for lactation (k_l) was determined with the equation described by Yan *et al.* (2006). Data of milk yield, energy balance and k_l were analyzed as a completely randomized design using analysis of variance and the Tukey test was used when significant differences among treatments were detected.

Results and discussion

Mean milk yield showed statistical difference (P<0.05) between treatments M and H from L which may have been due to the effect of ME intake. Fat concentration in milk was not affected (P>0.05) between treatments. Concentration of crude protein showed difference between treatments, treatment M gave the lowest concentration (g/kg) relative to treatments H and L. Lactose and GE concentrations in milk showed statistical differences (P<0.05) between treatments L and M relative to H.

Energy balance was higher (P<0.05) for treatment A (4.088) relative to M (0.041). Efficiency of utilization of ME for lactation was higher (P<0.05) for treatment M (0.631) than for treatment A (0.595). It is concluded that there is a positive relationship between ME intake and milk yield by Pelibuey ewes rearing twins. However, it seems that there is an inverse relationship between ME intake and concentrations of fat, protein, lactose and gross energy in milk. ME intake affects the efficiency of utilization of ME for lactation in the Pelibuey ewe.

Table 1. Means and standard errors (SE) for live weight, ME intake, milk production and composition and efficiency of ME utilization for lactation by Pelibuey ewes fed three levels of ME.

	Treatments[1]			SE		
	Low	Medium	High	Low	Medium	High
Live weight						
Initial (kg)	43.86[a]	48.14[a]	43.14[a]	2.44	2.09	2.89
Weight change (g/d)	-185[b]	-139[b]	58[a]	0.024	0.024	0.017
Intake						
DM intake, ($g/kg^{0.75}/d$)	49.94[c]	99.63[b]	149.18[a]	0.148	0.251	0.757
ME intake (MJ/d)	9.24[c]	19.92[b]	27.30[a]	0.389	0.671	1.352
Means of milk yield and composition						
Milk yield (kg/d)	0.96[b]	1.66[a]	1.74[a]	0.137	0.081	0.048
Milk fat (g/kg)	63.02[a]	56.55[a]	56.55[a]	4.239	0.957	1.317
Milk protein (g/kg)	46.11[a]	42.00[b]	47.00[a]	0.66	0.871	0.849
Milk lactose (g/kg)	43.35[a]	43.72[a]	41.23[b]	0.923	0.528	0.748
Gross energy of milk (MJ/kg)	4.97[a]	4.25[b]	4.13[b]	0.045	0.045	0.06
Efficiency of ME utilization for lactation						
Energy balance (MJ/d)	-	0.041[b]	4.088[a]	-	0.453	0.315
Feed conversion efficiency[2]	-	0.902[a]	0.702[b]	-	0.058	0.054
Gross energetic efficiency[3]	-	0.342[a]	0.257[b]	-	0.017	0.017
k_1 [4]	-	0.631[a]	0.595[b]	-	0.006	0.003

[1] a,b,c,: $P<0.05$ in the same row.

[2] Feed conversion efficiency: milk yield/total DM intake.

[3] Gross energetic efficiency: gross energy content of milk/ME intake.

[4] k_l: efficiency of utilization of ME for lactation.

Reference

Yan, T., C.S. Mayne, T.W.J. Keady and R.E. Agnew, 2006. Effects of dairy cow genotype with two planes of nutrition on energy partitioning between milk and body tissue. J. Dairy Sci. 89, 1031-1042.

Ling ER. 1963. A Textbook of Dairy Chemistry. Vol. 2, 4th ed. Chapman and Hall, London, UK.

Protein turnover and infrared thermography in Nellore bulls classified for residual feed intake

M.L. Chizzotti[1,2], D. Heiderich[2], W.L.B. Aziani[2], M.M. Ladeira[1,2], E.E.L. Valente[2], T. Yanagi Junior[2,3], F.H.M. Chizzotti[1], L. Schiassi[2,3] and D. Lourençoni[2,3]
[1]Instituto Nacional de Ciência e Tecnologia em Ciência Animal, DZO/UFV, 37200-000 Viçosa, MG, Brazil
[2]Universidade Federal de Lavras, Department of Animal Science, DZO/UFLA, 37200-000 Lavras, MG, Brazil; mariochizzotti@dzo.ufla.br
[3]Universidade Federal de Lavras, Department of Agricultural Enginering, DZO/UFLA, 37200-000 Lavras, MG, Brazil

Introduction

Residual feed intake (RFI), the difference between actual and expected intake, is an efficiency trait that is independent of BW and BW gain. The RFI might be related to a lower protein turnover and a lower heat production in efficient animals (Arthur *et al.*, 2004). The aim of this study was to evaluate the relationship of RFI with protein turnover and temperature of eyes and skin.

Material and methods

Seventeen Nellore bulls with average body weight of 378 kg were used. The experimental period was 74 days. The bulls received a 9.7% CP and 2.6 Mcal of DE/kg of DM diet (85% chopped elephant grass and 15% concentrate). Dry matter intake was measured daily. The bulls were classified in low (RFI<-0.5 SD below the mean, n=4), medium (RFI between 0.5 SD the mean, n=9) and high RFI (RFI>0.5 SD above the mean, n=4) according to Koch *et al.* (1963). The protein turnover was evaluated from urinary excretion of 3-methylhistidine (3-MH). A total urine collection was performed during 5-d, using funnels adapted to the animals. The 3-MH in urine was determined by HPLC as well as Total Kjeldahl Nitrogen. On day 36 of the experimental period, determinations of eye and skin temperatures were obtained by infrared thermal image (Fluke Ti 55Ft, Fluke Corporation, USA) and the pictures were analyzed using the software SmartView 3.0. The maximum eye temperature was considered as the maximum value into the eyeball and the average eye temperature was considered as the average temperature in all eyeball area. Similarly, the maximum skin temperature was the maximum temperature in a similar eye area projected under the eyes and the average skin temperature was the average temperature in this area. The data were analyzed using GLM and REG procedures of SAS (9.2) adopting significance level of 0.05.

Results and discussion

Table 1 reports the RFI and urinary excretion of N and 3-MH. It was observed a lower (P=0.045) urinary excretion of 3-MH (in mg/kg of BW/day) in efficient bulls (low RFI). The correlation between RFI value (kg of DM/day) and 3-MH excretion (mg/day) was of 0.45 (P=0.074), indicating that there is a tendency of lower muscular protein degradation in more efficient bulls.

There were no differences (P>0.40) on skin or eyeball temperatures among RFI classes. However, it was found a moderate relation between RFI and maximum eye temperature (r=0.51, P=0.05) (Figure 1). Low correlations were observed between RFI and maximum skin temperature (r=-0.09, P=0.73), average skin temperature (r=0.08, P=0,77) and average eye temperature (r=0.33, P=0.20). Heat increment may contribute to animal inefficiency, thus, a higher heat production should be expected in higher RFI (Montanholi *et al.*, 2009).

Table 1. Residual feed intake (RFI) and daily urinary excretion of 3-methyl-histidine (3-MH) and total N in Nellore bulls classified for RFI.

	RFI group			SEM	*P*-value
	Low	Medium	High		
RFI, kg DM/day	-0.569^{a}	-0.063^{b}	0.725^{c}	0.103	<0.0001
3-MH, mg/day	97.3^{a}	110.2^{b}	147.4^{b}	13.7	0.0506
3-MH,g/kgBW/day	0.201 ^{a}	0.232 ^{b}	0.312^{b}	0.029	0.0452
N, g/day	215.5	191.0	195.8	52.6	0.9283
N,g/kgBW/day	0.432	0.404	0.416	0.115	0.9797

^{a,b,c} superscript letters differ at $P<0.05$.

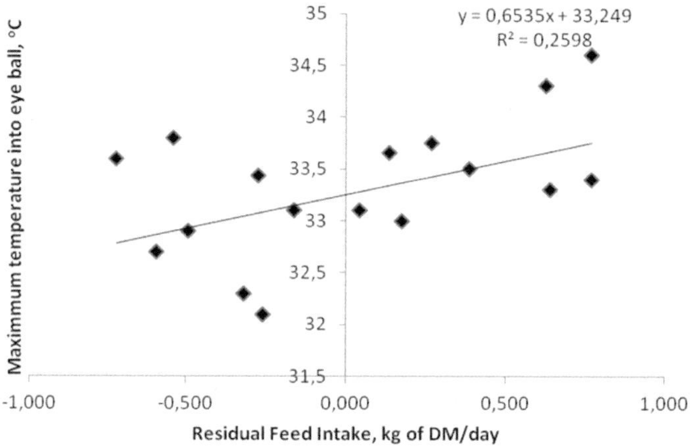

Figure 1. Relationship between the maximum eye temperature and the residual feed intake in Nellore bulls.

In conclusion, low RFI Nellore bulls present a lower excretion of 3-MH excretion (protein turnover). The maximum eye temperature is more related to feed efficiency than the average temperature of eye or skin.

References

Arthur P.F., J.A. Archer and R.M. Herd, 2004. Feed intake and efficiency in beef cattle: overview of recent Australian research and challenges for the future. Austr. J. Exp. Agr. 44, 361-369.

Koch, R. M., L. A. Swiger, D. Chambers, and K. E. Gregory, 1963. Efficiency of feed use in beef cattle. J. Anim. Sci. 22, 486-494.

Montanholi, Y.R.,-K.C. Swanson, F.S. Schenkel, B.W. McBride, T.R. Caldwell, S.P. Miller, 2009. On the determination of residual feed intake and associations of infrared thermography with efficiency and ultrasound traits in beef bulls. Liv. Sci. 125, 22-30.

SAS, 2003. Release 9.2. SAS Inst. Inc., Cary, NC, USA.

Differences in residual feed intake are largely explained by changes in body composition

M.L. Nascimento[1], A.R.D.L. Souza[2], A.S. Chaves[1], S.R. de Medeiros[3], R.R. Tullio[4], M.M. de Alencar[4], A.N. Rosa[3] and D.P.D. Lanna[1]
[1]University of Sao Paulo/ESALQ, Piracicaba, SP, Brazil; lnca@esalq.usp.br
[2]Universidade Federal Rural de Pernambuco, Recife, PE, Brazil
[3]Embrapa Beef Cattle, Campo Grande-MS, Brazil
[4]Embrapa Cattle Southeast, Sao Carlos-SP, Brazil

Introduction

Genetic selection of efficient animals can result in lower costs, reduced environmental impact and increased profitability. The associations between feed efficiency selection and performance are known, however the possible impacts on body composition of Nellore are still unknown. Residual feed intake (RFI), defined as the difference between observed and predicted intake (predicted from metabolic body weight and daily gain) (Koch *et al.*, 1963), has been proposed as an index for genetic selection, however, differences in body composition have been pointed out as a limiting factor in the use of RFI for cattle selection (Herd and Pitchford, 2011). The objective of this work was to examine the relationship between RFI and body composition in Nellore steers.

Material and methods

Dataset from three trials in which the animals had similar nutritional history (grazing systems) were used, totaling 570 steers from 40 bulls (n=341 in trial 1; n=158 in trial 2; n=71 in trial 3). The animals were fed twice daily with 5% of orts. The diets contained corn silage (trial 1, 40% DM) or sorghum silage (trial 2 and 3, 40 and 20% DM, respectively); 13,5 (trial 1), 15,4 (trial 2) and 13,2% CP (trial 3); 2.8 (trial 1), 2.6 (trial 2) and 2.7 Mcal EM per kg DM (trial 3). Individual DMI was measured for at least 70 d, considered the minimum period to assess RFI in beef cattle (Arthur, 2001), and body weight (BW) taken every 14 d.

Fat thickness (FT) and ribeye area (RA) were taken by ultrasound measurements on the 11-12th-rib at the start, middle and final of the 70d period. Final and initial empty BW energy were estimated by an equation for Nellore (EBW Energy = -126,73 + 10.88×RA + 70.01×FT; R^2=0.73), the retained energy was calculated by difference. To estimate the proportion of protein and fat in the gain, the composition of the fat-free dry matter was held constant. Intramuscular ether extract of the *Longissimus* samples were determined.

Average gaily gain (ADG) was estimated by regression between BW and days on feed. RFI was computed by regression of DMI on mid-test $BW^{0.75}$ and ADG using mixed models, where effect of contemporary group (CG) based on feedlot location, year, animal origin and pen type (individual or collective) was considered random. The animals were classified as low, medium and high RFI (mean±0.5 SD). High and low classes were compared by MIXED procedure (SAS, 2008) where RFI class and CG were considered fixed effects, sire as random, and initial age as a covariate.

Results and discussion

As expected, BW and ADG were not related to RFI class (P>0.05) (Table 1). The efficient animals consumed 15.7% less than the inefficient ones (P<0.05). There was no relationship between RFI class and ribeye area (P>0.05). However, efficient animals (low RFI) had lower final fat thickness and intramuscular fat (P<0.05). The greater efficiency of the animals with leaner body composition is related to better utilization of energy intake for muscle deposition compared to adipose tissue. For

Table 1. Performance and body composition of Nellore steers with low and high RFI.

Variable	RFI classes (N)		SE	*P*-value
	Low (174)	High (187)		
RFI, kg/d	-0.636	+0.985	0.032	<0.0001
Initial BW, kg	352.6	349.7	3.42	0.4012
Final BW, kg	439.4	435.9	4.13	0.3966
Gain, kg/d	1.18	1.16	0.025	0.3759
Feed intake, kg/d	7.77	9.22	0.093	<0.0001
F:G Ratio, DMI/gain	7.11	8.33	0.156	<0.0001
Initial ribeye area[1], cm^2	50.8	50.5	0.592	0.6141
Final ribeye area[1], cm^2	60.4	60.4	0.726	0.9287
Ribeye area change[1], cm^2	9.25	9.52	0.560	0.6284
Initial fat thickness[1], mm	2.83	2.91	0.095	0.4015
Final fat thickness[1], mm	4.88	5.29	0.159	0.0089
Fat thickness change[1], mm	2.18	2.63	0.132	0.0007
Retained energy, Mcal/d	3.61	4.00	0.140	0.0054
% fat of the gain	29.0	32.5	1.735	0.0411
% protein of the gain	15.6	14.8	0.382	0.0425
Intramuscular fat, %	2.64	3.10	0.458	0.0008

[1] Measured by ultrasound.

each gram of protein deposited in muscle tissue, three grams of water are also deposited resulting in a greater amount of tissue for the same amount of energy intake with lower energy content (Lofgreen and Garret, 1968).

Using our body composition equation, we estimated that efficient animals retained 0.39 Mcal less energy/day ($P<0.05$). Consequently, empty body gain composition was different, where efficient animals had less fat and more protein ($P<0.05$). Assuming the same estimated efficiency of use of the ME for gain (0.442) according to NRC (1984), the change in estimated body composition, explained 22.2% of the 1.45 kg/d difference in feed intake between high and low RFI classes. This effect is greater than the 5% suggested in the literature based on regression analysis.

References

Arthur, P.F.; Archer, J.A.; Johnston, D.J.; Richardson, E.C.; Parnell, P.F. Genetic and phenotypic variance and covariance components for feed intake, feed efficiency and other post weaning traits in Angus cattle. J. Animal Science, 79, p. 2805-2811, 2001.

Herd, R.M.; Pitchford, W.S. Residual feed intake selection makes cattle leaner and more efficient. Recent Advances in Animal Nutrition, Armidale, v.18, p. 45-58, 2011.

Koch, R.M.; Swinger, L.A.; Chambers, D.; Gregory, K.E. Efficiency of feed use in beef cattle. J. Animal Science, 22, p. 486-494, 1963.

Lofgreen, G.P. and W.N. Garrett, 1968. A system for expressing net energy requirements and feed values for growing and finishing beef cattle. J. Animal Sci. 27, p. 793, 1968.

National Research Council, 1984. Nutrient requirements of beef cattle. 6th ed. Washington: National Academic Press, 1984. 90 p.

SAS, 2008. Release 9.2. SAS Inst. Inc., Cary, NC, USA.

Net protein requirement of pregnancy of goats with single and twin pregnancy

C.J. Härter, D.S. Castagnino, L.D. Lima, A.R. Rivera, H.G.O. Silva, K.T. Resende and I.A.M.A. Teixeira
Faculdade de Ciências Agrárias e Veterinárias, UNESP, Univ Estadual Paulista, Department of Animal Sciences, 14884-900 Jaboticabal, SP, Brazil; harter.carla@gmail.com

Introduction

Pregnancy is an important stage in the animal life, because many transformations occur in the whole body of the pregnant female. However, there is lack of information on nutritional requirements over the whole pregnancy. Most protein recommendations for goats were obtained from data extrapolated from cattle and sheep (Sutton and Alderman, 2000). Although ruminants have similar metabolism, it is known that there are differences between species, especially regarding the efficiency of feed utilization and body composition. Therefore, the objective of this experiment was to determine net protein requirement of Oberhasli and Saanen goats with singleton and twin pregnancy.

Material and methods

A total of 21 Oberhasli (44.01±1.28 kg BW) and 21 Saanen (50.66±1.12 kg BW) were used. Females of each breed were randomly distributed among the treatments according to the number of fetuses (1 or 2) and gestation period (80, 110 and 140 days), in a completely randomized design with a 2×2×3 factorial arrangementof treatments. The animals were fed ad libitum with a diet formulated to meet their requirements, according to NRC (2007). When the female goats reached the specific gestational age they were slaughtered and then there was the withdrawal of the pregnant uterus, mammary gland and all organs and structures were withdrawn. The entire gravid uterus was weighed and then separated into empty uterus (uterus+placenta+placentomes), placental fluid and fetuses, which were weighed separately. The mammary gland was weighed after removing the skin. These components (empty uterus, placental fluid, fetuses and mammary gland) were ground and their samples were freeze-dried for determination of the contents of dry matter, ash, fat according to AOAC (1995) and protein by Dumas combustion method (LECO FP-528 LC). The results of growth of gravid uterus and mammary gland were fitted in the nonlinear Gompertz model:

$$Y = \text{Wbirth} \times e^{-e(\ln(-\ln(\text{Win}/\text{Wbirth})) - b \times \text{days of pregnancy}))} \quad (1)$$

using the NLIN procedure of SAS 9.2 (2009) and the daily requirements were found by the differentiation of the Gompertz :

$$y = Y \times b \times \ln(\text{Wbirth}/Y) \quad (2)$$

Where: Y = protein accretion (g); Wbirth = weight of protein at birth (g); Win = weight of protein at beginning of pregnancy (g); e = 2.718282; b = maximum rate of deposition of protein (g/day); y = protein accretion in g per day.

Results and discussion

The parameters of the equations to predict the protein accretion in the fetus, placental fluid, uterus and mammary gland are presented in Table 1. In both types of pregnancy, Oberhasli goats presented heavier fetuses and greater protein accretion throughout pregnancy than Saanen goats. The total Oberhasli fetuses mass at 140 days of pregnancy was 4.25 kg and 6.27 kg in single and twin pregnancy, respectively, and total Saanen fetuses mass was 3.63 kg and 5.17 kg in single and twin pregnancy, respectively. Protein accretion in Oberhasli was 289.2, 540.8 and 1067.7 g with single pregnancy and 296.9, 644.2 and 1422.42 g with twin pregnancy at 80, 110 and 140 days of pregnancy, respectively. On the other hand, protein accretion in Saanen was 321.0, 492.0 and 867.3 g with single pregnancy and 260.5, 612.2 and 1205.4 g with twin pregnancy at 80, 110 and 140 days of pregnancy. After 80 days of pregnancy net protein requirement of female goats with twin pregnancy (g/day) is higher compared to female goats with single pregnancy. Net protein requirements of Oberhasli goats with single pregnancy were 3.9, 14.8, 16.9 g/day at 80, 110 and 140 days of pregnancy, respectively; and for Oberhasli goats with twin pregnancy were 4.5, 22.1 and 24.0 g/day at 80, 110 and 140 days of pregnancy, respectively. Net protein requirements of Saanen goats with single pregnancy were 2.9, 10.2 and 12.21 g/day and with twin pregnancy 5.6, 18.7 and 17.7 g/day at 80, 110 and 140 days of pregnancy, respectively. The litter size determines the largest fraction of the requirements during pregnancy. In conclusion the Oberhasli goats with twin pregnancy show greater net protein requirement for pregnancy.

Table 1. Parameters generated by nonlinear equation of Gompertz for estimating net protein requirements (g) of pregnant dairy goats.

	Total protein mass at birth (g)[1]	Rate of protein deposition (g/day	Protein mass at beginning of pregnancy (g)	P-value
Oberhasli with singleton pregnancy				
Fetus	1,100	0.0383±0.0062	$3.94\times10^{-57}\pm4.34\times10^{-55}$	<0.0001
Placental fluid	60	0.0428±0.0145	$3.05\times10^{-29}\pm3.29\times10^{-27}$	<0.0001
Uterus	500	0.0118±0.0049	14.48±27.94	<0.0001
Mammary gland	200	0.0196±0.0139	59.18±36.45	<0.0001
Oberhasli with twin pregnancy				
Fetus	1,600	0.0408±0.0057	$5.28\times10^{-72}\pm6.63\times10^{-70}$	<0.0001
Placental fluid	50	0.0314±0.0091	$2.37\times10^{-14}\pm9.06\times10^{-13}$	<0.0001
Uterus	400	0.0146±0.00327	11.30±12.99	<0.0001
Mammary gland	300	0.0145±0.0052	40.27±29.18	<0.0001
Saanen with singleton pregnancy				
Fetus	900	0.0375±0.007	$3.35\times10^{-53}\pm3.88\times10^{-51}$	<0.0001
Placental fluid	30	0.0292±0.0128	$4.17\times10^{-4}\pm0.0060$	<0.0001
Uterus	200	0.0284±0.0173	10.28±40.70	<0.0001
Mammary gland	205	0.0285±0.0286	19.75±78.12	<0.0001
Saanen with twin pregnancy				
Fetus	1000	0.0412±0.003	$5.30\times10^{-62}\pm2.54\times10^{-60}$	<0.0001
Placental fluid	20	0.0411±0.0210	$9.78\times10^{-07}\pm2.7\times10^{-05}$	<0.0001
Uterus	700	0.0117±0.0033	10.49±15.54	<0.0001
Mammary gland	1,200	0.0068±0.0011	19.50±9.691	<0.0001

[1] Value set in the equation to predict the rate of protein deposition and protein mass at beginning of pregnancy.

References

AOAC. 1995. Association of Official Analytical Chemists. Official Methods of Analysis. 16th ed. Washington, D.C.

National Research Council. 2007. Nutrient requirements of small ruminants. Sheep, Goats, Cervids and New World Camelids. National Academy Press. Washington, DC.

Sutton, J. D., Alderman, G. 2000. The energy and protein requirements of pregnant and lactating dairy goats the Agriculture and Food Research Council report. Liv. Prod. Sci. 64, 3-8.

Effect of gender on net energy and protein requirements for growth of goats

A.K. de Almeida, S.P. Silva, D.S. Costa, M.H.M.R. Fernandes, I.A.M.A. Teixeira and K.T. Resende
Department of Animal Science, Universidade Estadual Paulista, Jaboticabal, São Paulo, Brazil; almeida.amelia@gmail.com

Introduction

Currently, the world is facing challenges related to environmental problems, global warming, and demography. In this context, goats may take part on an important socio-economic role with its potential to use marginal land. In addition, these animals adapt easily to various climatic zones, produce under different systems of husbandry, and successfully convert feed into protein sources. However, it is crucially important to understand this species' nutritional requirements, that are determined by several factors such as management, breed, environment, gender, age and others. Nutritional requirements for goats have been based on limited data, and are sometimes calculated by extrapolation of other species needs (ARC, 1980; NRC, 1981). We propose to investigate Saanen goats, from 30 to 45 kg of body weight (BW), the effect of gender on net protein and energy requirements for growth.

Material and methods

The experiment was conducted at the Goat facility of Univ Estadual Paulista/Jaboticabal from June of 2010 to April 2011. We used 48 Saanen goats, among castrated males (n=16), intact males (n=14) and females (n=18), of 30.0±1.09 kg of initial body weight and age of 258±53 days. These animals were randomly allocated into three slaughter groups, conforming body weights, 30, 37.5 and 45 kg. The experimental diet was formulated to meet the species' nutritional requirements according to NRC (2007). We offered a mixed diet consisted of dehydrated corn, cracked corn grain, soybean meal, soybean oil, limestone, and a mineral supplement. The average dry matter intake during the experiment was 1,030±90 g/d, which corresponded to 2.89% of BW. At slaughter, kids were stunned by an electric shock, followed by severing of the jugular vein and carotid artery. All blood was collected and weighed. The digestive tract was weighed before and after it was emptied in order to determine the empty body weight EBW of animals. The EBW was calculated by BW at slaughter minus the weight of the contents of the digestive tract, gallbladder and vesicle. The whole empty body was initially frozen at -6 °C and then cut into small pieces and grind. After grinding and homogenization, samples were freeze dried for DM determination and performed analysis for fat, protein and energy contents. The experimental design was completely randomized in a 3×3 factorial arrangement, where gender and body weight were assumed as fixed effects, and the error was iid $N(0,\sigma^2_{ij})$. To obtain estimates of gain composition, we divided the calculus in phases. The first phase consisted in obtaining the allometric equations to predict protein and energy concentration from EBW, according to ARC, 1980, as follows: Log10 (protein and energy contents) = a + [b × log10 (EBW, kg)]; where component amount is protein and energy contents in the EBW. In the second phase, the equation was differentiated to compute the estimates of gain composition at various EBW: [Component] = $b \times 10^{a} \times EBW^{(b-1)}$, where [Component] is protein and energy contents per unit of empty body weight gain (EWG) (g/kg of gain or kJ/kg of gain). We used the MIXED procedure (SAS Inst. Inc., Cary, NC, USA), to obtain allometric equations, the CONTRAST option to verify difference among gender, and the ESTIMATE option to give us the overall intercept and slope.

Results and Discussion

To our knowledge, our study is the first to determine and compare energy and protein requirements for growth among female, male and castrated male goats of the Saanen breed in the final phase of growth. We observed that as animals grow, the content of energy and fat increases, our data presented in Table

1 demonstrate that females tend to deposit more fat than males. We may also consider that animals were at different proportion of mature weight: 55.5, 48.4 and 41.5 for females, castrate males and intact males respectively. This may explain the huge difference in fat content in the EBW and ADG.

Table 2 shows that the proportion of energy increased as BW increased, however the proportion of protein was nearly constant, suggesting that the fat was the main factor which increased body energy. Although, this difference did not cause distinction among genders for net energy requirements on growth. Since we did not find a difference among genders for protein and energy concentrations, a general allometric equation was proposed.

After differentiation these equations, the retained energy ranged from 16.3 to 19.6 MJ/kg of EWG, and net ranged from 177 to 182 g/kg of EWG as BW ranged from 30 to 45 kg. Therefore, the net energy requirements corresponded with the NRC (2007) recommendations, but we have found lower values for net protein requirements. We have found no differences among gender for Saanen goats ranging from 30 to 45 kg of BW.

Table 1. Effect of gender on performance and body composition at slaughter for Saanen goats.

Variable	Gender			*P*-value
	Female	Castrate male	Intact male	
Water, % EBW	52.05±1.08[c]	57.79±1.02[b]	61.04±1.06[a]	<0.0001
Ash, % EBW	3.93±0.19	3.98±0.18	4.31±0.19	0.32
Protein, % EBW	17.55±0.59	17.08±0.56	16.79±0.58	0.65
Fat, % EBW	28.42±0.73[a]	22.41±0.69[b]	17.14±0.72[c]	<0.0001
ADG, g/d	74.2±7.1[b]	95±6.6[a]	99.6±6.9[a]	0.03

[ab] Distinct letters in the same row, differ at $P<0.05$ by least squares means for gender effect.

Table 2. Allometric equations and relative proportion of protein and energy in the BW at 30, 37.5, and 45 kg Saanen goats.

Variable[1]	Allometric equations	Body weight (kg)			
		30	37	45	RMSE[1]
EBW,kg	EBW, kg = -3.62(±1.02) + [0.92(±0.02) × BW, kg]	24.0	28.6	37.8	0.99
Protein, g/kg of EBW	$Log_{protein}$, g = 2.13(±0.16) + [1.06(±0.11) × LogEBW, kg]	167	169	171	0.03
Energy, MJ/kg of EBW	Log_{energy}, kJ = 3.24(±0.11) + [1.57(±0.07) × LogEBW, kg]	10.4	11.9	13.5	0.04

[1] RMSE is root mean square error. The equations were significant; $P<0.0001$.

Conclusion

We conclude that females represented greater fat accumulation in the carcass, although this difference was not enough to cause distinct energy requirements for growth. Now we can consider that there are no differences among gender for protein and energy requirements for growth of Saanen goats at this phase.

Acknowledgements

FAPESP process number: 2010/02482-4.

References

Agricultural Research Council, 1980. ARC The nutrient requirement of ruminant livestock. London: 351p.
NRC, 2007. National Research Council. Nutrient requirement of small ruminant. Washington, 292 p.

Energy utilization for gain of goat kids

T.F.V. Bompadre, K.T. Resende, M.H.M.R. Fernandes, O. Boaventura Neto, A.N. Mendonca, B. Biagioli and I.A.M.A. Teixeira
UNESP Universidade Estadual Paulista, Department of Animal Science, Via de Acesso Prof Paulo Donato Castellane s/n, 14884-900 Jaboticabal, Sao Paulo, Brazil; mhmrochaf@fcav.unesp.br

Introduction

The total energy body content is based on protein and fat deposition in growing animals, which can be strongly influenced by gender. In this regard, the accretion rate of protein and fat could be helpful for understanding the efficiency of utilization of metabolizable energy (ME) to net energy (NE) for growth (k_g). Previous studies in cattle have found differences in the partial efficiency of use of metabolizable energy (ME) to net energy (NE) for growth between gender, which was 0.54 for bulls, 0.47 for steers and 0.54 for heifers (Chizzotti *et al.*, 2007). However there is a lack of information in this regard with goats. Therefore, this study was conducted to evaluate the effect of the gender on fat and protein deposition in the gain and on energy efficiency for growth in Saanen goat kids.

Material and methods

A total of 48 Saanen goat kids (18 intact males, 12 castrated males and 18 females) with initial body weight (BW) of 4.93±0.1 kg were used. Six non-castrated males and 6 females were slaughtered at beginning of the experiment (baseline animals) to estimate the initial body composition. The reminder was randomly allocated into groups (blocks) of 3 animals of the same gender, subjected to 3 levels of intake: *ad libitum,* 75 and 50% of *ad libitum* intake. The 75 and 50% of *ad libitum* intake levels were determined daily, based on dry matter intake of the kids fed *ad libitum* on the previous day. The entire group was slaughtered when the animal fed *ad libitum* reached 15.2±0.4 kg BW and 101±5 days old. Empty bodies were weighted, ground, mixed and sub sampled for chemical analyses (fat, protein, ash and gross energy). Initial body composition was determined using equations developed from the body composition of the baseline goat kids (Lofgreen and Garret, 1968). Dietary ME, estimated from a metabolism trial, was 7.8 MJ/kg DM. The data were analyzed using mixed models of SAS (SAS Inst. Inc., Cary, NC).

Results and discussion

Table 1 shows body composition of intact male, castrated male and female Saanen goat kids. The equation to predict rate of fat gain on the empty body weight gain (EWG) differed between genders (P=0.02), showing that females had higher fat deposition than entire and castrated males goat kids (Equation 1 and 2). On the other hand, rate of protein gain was not affected by gender (P=0.64) and a general equation was proposed (Equation 3).

$$\text{Fat gain (females)} = -7.26\pm1.79 + 0.25\pm0.02\ \text{EWG, RMSE} = 2.36 \quad (1)$$

$$\text{Fat gain (males and castrated males)} = -3.50\pm1.38 + 0.17\pm0.02\ \text{EWG, RMSE} = 2.36 \quad (2)$$

$$\text{Protein gain (all)} = 1.57\pm0.88 + 0.22\pm0.01\ \text{EWG, RMSE} = 2.38 \quad (3)$$

where fat and protein gain, g/d; and EWG = empty weight gain, g/d.

The equations to predict percentage of fat (FIG) in the gain from retained energy concentration (REc) also differed among genders (P<0.001), suggesting that retained energy as fat had the same pattern for females and castrated males and differed for males (Equation 4 and 5). The percentage of protein (PIG, Equation 6) in the gain from REc was not affected by gender (P=0.10).

$$FIG\ (males) = -25.06 \pm 1.78 + 3.52 \pm 0.17\ REc,\ RMSE = 2.44 \quad (4)$$

$$FIG\ (females\ and\ castrated\ males) = -12.53 \pm 3.81 + 2.52 \pm 0.35\ REc,\ RMSE = 2.44 \quad (5)$$

$$PIG\ (all) = 30.23 \pm 2.66 - 0.48 \pm 0.25\ REc,\ RMSE = 4.56 \quad (6)$$

where FIG = fat in gain as % of EWG, PIG= protein in gain as % of EWG, REc= concentration of retained energy, MJ/kg EWG.

Despite the greater fat gain of females, the efficiency of energy utilization for growth (k_g), calculated as the slope of regression of retained energy (KJ/kg$^{0.75}$ of EBW) on ME intake (KJ/kg$^{0.75}$ of EBW) above maintenance, was not affected by gender (Equation 7).

$$RE = -124.40 \pm 43.18 + 0.47 \pm 0.07\ MEI,\ RMSE = 56.4 \quad (7)$$

where RE= retained energy (KJ/kg$^{0.75}$ of EBW) and MEI= ME intake (KJ/kg$^{0.75}$ of EBW).

Our study indicated that the efficiency of energy utilization for growth does not differ among gender for Saanen goat kids, in early life.

Table 1. Effect of gender on body composition of Saanen kids.

	Gender			P-value
	Male	Castrated	Female	
Water, % EBW	59.77±0.69	60.59±0.81	59.92±0.85	0.74
Ash, % EBW	5.80±0.16	5.81±0.19	5.75±0.20	0.97
Protein, % EBW	25.79±0.47	25.11±0.55	24.39±0.59	0.18
Fat, % EBW	8.31±0.61	8.27±0.72	9.71±0.76	0.29
Energy, MJ/kg EBW	9.05±0.21	8.79±0.25	9.02±0.29	0.70

[1] EBW = empty BW.

Acknowledgements

FAPESP project number 2008/58351-5 and 2012/07177-0.

References

Chizzotti, M.L., S.C. Valadares Filho, L.O. Tedeschi, F.H.M. Chizzotti and G.E. Carstens, 2007. Energy and protein requirements for growth and maintenance of F1 Nellore x Red Angus bulls, steers, and heifers. J. Anim. Sci. 85, 1971-1981.

Lofgreen, G.P. and W.N. Garret, 1968. A system for expressing net energy requirements and feed values for growing and finishing beef cattle. J. Anim. Sci. 27, 793-806.

Protein requirements for growth of male and female Saanen goats kids

F.O.M. Figueiredo, R.F. Leite, M.M. Freire, M.H.M.R. Fernandes, K.T. Resende and I.A.M.A. Teixeira
Department of Animal Science, UNESP Universidade Estadual Paulista, Jaboticabal, SP, Brazil; fernandafigdo@yahoo.com.br

Introduction

Gender plays an important role on deposition rates of different body tissues and due to the variation in the composition of gain, net protein requirements can differ between males and females. Souza *et al.* (2012) reported that bulls had greater protein requirements than castrated males, which were greater than females at the same age because intact males showed greater lean tissue deposition in the body than castrated males and females. In dairy goats, only a few studies dealing with the effect of gender on protein requirements have been published. Thus, the objective of this study was to determine the net protein requirements for growth of intact males, castrated males, and females Saanen goat kids using comparative slaughter technique.

Material and methods

Net protein requirements for growth were obtained using 20 castrated males, 20 intact males and 18 females Saanen goat kids at an average initial body weight (BW) of 15.65 (±0.55) kg in a completely randomized design. The animals were fed *ad libitum* and slaughtered at targeted BW of 15, 23 and 30 kg. Seven castrated males, 8 intact males and 6 females were slaughtered at the beginning of the experiment at an average BW of 16.65 (±0.31) kg, and an initial age of 115.33 (±20.21) days. Another 7 castrated males, 6 intact males and 6 females were slaughtered when they reached 23.10 (±1.00) kg BW at 185.33 (±41.85) days of age. Lastly, another 6 animals of each gender were slaughtered when they reached 31.30 (±1.31) kg BW at 202.33 (±12.66) days of age.

The experimental diet was formulated to meet or exceed the requirements of growing goat kids, according to NRC (2007), and consisted of dehydrated, whole corn plant (Zea mays), cracked corn grain, dehulled, solvent extracted soybean meal, soybean oil, limestone and a mineral supplement fed as a total mixed diet. Dehydrated corn plants consisted of whole corn plants (60 to 70% moisture) chopped when the kernel milk line was approximately two-thirds of the way down the kernel, and dehydrated.

Protein body composition was estimated using direct chemical analyses of the whole empty body. A factorial approach was used to estimate protein requirements for growth.

Simple linear regression analyses were conducted to express empty body weight (EBW) as a function of BW. The protein content of animals was summarized by allometric regressions of the logarithm of observed protein content on the logarithm of empty body weight:

$$Log_{10}Prot = a + b \times Log_{10}EBW \quad (1)$$

where Log_{10}Prot is the log of the total amount of body protein (g), Log_{10}EBW is the log of empty body weight (EBW, in kg), and *a* and *b* are regression parameters.

The first derivative of (Equation 1) with respect EBW yields estimates of the composition of the gain at various EBW:

$$NP_g = b \times 10^a \times EBW^{(b-1)} \quad (2)$$

where NP_g is the net protein requirement for gain (g/ kg EBW gain).

All statistical analyses were conducted using the MIXED procedure of SAS, version 9.2.

Results and discussion

Gender did not significantly affect total body protein ($P<0.05$). Therefore, only results aggregated across genders are presented (Table 1).

The first derivative of the equation for $Log_{10}Prot$ with respect to EBW yields estimates of net protein requirements for growth:

$$NP_g = 184.83 \times EBW^{-0.03} \tag{3}$$

From Equation 3, NP_g decreased from 171.9 to 168.1 g /kg of EBW gain between 15 and 30 kg EBW in Saanen goat kids. Although the decrease in NP_g as EBW increases is small, it is in accord with previous studies that showed a decrease in rate of protein deposition and an increase of rate of fat deposition as body weight increases (Owens *et al.*, 1993). We did not observe an effect of gender on rate of EBW protein accretion between 15 and 30 kg BW in Saanen goats.

Table 1. Equations for estimating protein body composition of female, male and castrated male Saanen goat kids between 15 and 30 kg of body weight.[1]

Traits	Regression equations	BW[2]			
		15	23	30	RMSE[3]
Empty body weight	EBW = -1.140 (±0.316) + [0.832 (±0.013) x BW, kg]	11.3	17.9	23.7	0.37
Total body protein	$Log_{10}Prot$ = 2.283 (±0.055) + [0.976 (±0.044) x Log EBW, kg]	177.2	174.8	173.3	0.01

[1] BW is body weight (kg); EBW is empty BW (kg); $Log_{10}Prot$ is the log of the total amount of body protein (g); $Log_{10}EBW$ is the log of EBW (kg).

[2] Values in the table are the predicted EBW and total body protein accretion rates for the 3 values of BW.

[3] RMSE is root mean square error.

Acknowledgements

FAPESP project number 2008/58351-5 and 2012/07177-0.

References

National Research Council. 2006. Nutrient requirements of small ruminants: Sheep, goats, cervids and new world camelids. Washington, D.C.: National Academy Press.

Owens, F.N.; Dubeski, P.; Hanson, C.F., 1993. Factors that alter growth and development of ruminants. J. Anim. Sci., 71:3138-3150.

Souza, E. J. O., Valadares Filho, S. C., Guim, A., Valadares, R. F. D., Marcondes, M. I., Ferreira, M. A., Prados, L. F., Benedet, P. B., 2012. Protein requirements for females of Nellore, Nellore × Angus and Nellore × Simmental fed on two forage:concentrate ratios. Revista Brasileira de Zootecnia (Braz. J. Anim. Sci.) 41:762-770.

Influence of different lipid supplements and frequencies supply on microbial protein synthesis in beef heifers

M.C.A. Santana[1], T.T. Berchielli[2], G. Fiorentini, R.A. Reis[2], V.C. Modesto[2] and G.T. Pereira[2]
[1]EMATER, Goiânia, GO, Brazil
[2]Department of Animal Science, Universidade Estadual Paulista, UNESP, Jaboticabal, Brazil; mcaspaz@yahoo.com.br

Introduction

Increasing the energy density of the diet by adding lipid sources is a nutritional strategy that is used for finishing cattle, which leads to satisfactory performance. The use of lipids in ruminant diets has been recommended because by reducing methanogenesis the energy efficiency is improved. There are several ways to provide lipids to the diet, which includes grain, oil and calcium salts. According to Allen (2000) the main problem of using lipids rich in unsaturated fatty acids (UFAs) in diets for ruminants is their effect on intake and, consequently, on ruminal digestibility of the fibrous fraction. Also, diets rich in cereal grains cause a pH decrease in the rumen, which inhibits lipolysis and biohydrogenation. To avoid the negative effects just described, the lipid supplement in the form of protected fat has been recommended for ruminants because it is considered an inert ruminal fat source (Harvatine and Allen, 2006). However, the use of fat supplementation is more typical for feedlot cattle than for grazing cattle. In addition, there are only few data on effects of fat supplementation on grazing cattle (Brokaw *et al.*, 2001; Krysl *et al.*, 1991). Likewise, reducing the fat supplementation frequency for grazing cattle, due to great territorial extension in some cases, may enhance logistic management in the farm. Thus, the objective of this study was to evaluate the effects of different lipid sources supplemented daily or thrice weekly on microbial protein synthesis of grazing heifers.

Material and methods

The experiment was carried out during two months. The design used was completely randomized in a 3×2 factorial arrangement analyzed using ANOVA. Significance was set at $P<0.05$ and differences among means were determined using a t-test. The lipid supplement treatments were, soybean grains (SG), soybean oil (SO) and calcium salts (ML) (MEGALAC- E Arm & Hammer, Rio de Janeiro, Brazil) and two supply frequencies; daily (D) or 3 days of week (Monday, Wednesday and Friday) called 'alternate' (A). The supplements contained on average 28.44% of crude protein; 13.46% of ether extract and 27.95% of neutral detergent fiber. With respect to the supplementation frequency, animals fed daily were supplemented at 0.75% of BW, whereas animals fed thrice a week were supplemented at 1.75% of BW. Spot urine samples were collected from six heifers 4 h after feeding, either when heifers urinated spontaneously or following vaginal stimulation. The samples were filtered through 4 layers of cheesecloth and 50 ml was immediately frozen at -15 °C as stock material. One aliquot (10 ml) of urine sample (by each animal) was diluted with 40 ml of 0.018 mM H_2SO_4 and stored at -15 °C for subsequent analysis. Commercial kits were used to analyze urine for creatinine. Allantoin and uric acid in urine samples were measured as described by Chen and Gomes, (1992). The microbial efficiency was obtained using two separate methods; the microbial production (CP g/day) was divided by the TDN according to NRC (2001) calculation, or microbial production was divided by the intake of fermentable organic matter in the rumen (CP/ MOFerm), noting that the lipid fraction was disregarded from the MOFerm.

Results and discussion

Microbial N synthesis estimated by urinary purine derivative excretion was similar between lipid supplements and supply frequencies studied as shown on Table 1. The efficiency of microbial synthesis was not influenced by the form of the provision of lipid supplement and frequencies of

Table 1. Effects of lipids supplements levels and supply frequencies on the microbial N (MN) synthesis and efficiency of MN synthesis (EMS) using urinary purine derivative (P.D) excretion.

	Supplements and frequencies					SEM
	SG	SO	ML	D	A	
Microbial N, g/d	54.07	57.82	54.06	53.78	56.66	1.59
Microbial CP, g/kgW$^{0.75}$	4.91	4.58	4.44	4.53	4.70	0.11
TDN, kg/d	2.66	3.08	2.66	2.76	2.81	0.15
Intake, kg/d	5.5	5.9	5.4	5.6	5.6	0.15
RDTOM, kg/d	4.30	4.36	4.22	4.29	4.42	0.14
EMS, g CP/100 g TDN	13.06	12.80	13.47	12.91	13.65	0.94
EMS, g CP/kg RDTOM	82.63	76.61	80.88	78.91	81.11	4.08

SG = soybean grain; SO = soybean oil; ML = MEGALAC- E; D = daily; A = alternately; RDTOM = ruminally degradable true OM.

supplementation, with mean values of 13.11 MCP g/100 g TDN, which is similar to the NRC (2001) recommendation of 13 g of MCP/100 g of TDN.

The averages observed for the ingestion of dry material were verified at 5.6 kg DM/day and for fermentable organic material, 4.3 kg DM/day, while the apparent digestibility coefficient was observed to be 45.4%. The supplementation frequency can be reduced regardless of lipid source without negatively impact microbial protein synthesis.

References

Allen, M.S, 2000. Effects of diet on short-term regulation of feed intake by lactating dairy cattle. Journal of Dairy Science, v.83, p.1598-1624.

Brokaw, L., B. W. Hess, and D. C. Rule. 2001. Supplemental soybean oil or corn for beef heifers grazing summer pasture: effects on forage intake, ruminal fermentation, and site and extent of digestion. J. Anim. Sci. 79, 2704-2712.

Chen, X. B., M. J. Gomes., 1992. Estimation of microbial protein supply to sheep and cattle based on urinary excretion of purine derivatives an overview of technical details. International Feed Research Unit. Aberdeen, UK: Rowett Research Institute (Occasional publication), 21.

Harvatine, K.J.; Allen, M.S, 2006. Fat supplements affect fractional rates of ruminal fatty acid biohydrogenation and passage in dairy cows. Journal of Nutrition, v.136, p.677-685,

Krysl, L. J., M. B. Judkins, and V. R. Bohman. 1991. Influence of ruminal or duodenal soybean oil infusion on intake, ruminal fermentation, site and extent of digestion, and microbial protein synthesis in beef heifers consuming grass hay. J. Anim. Sci. 69, 2585-2590.

National Research Council – NRC, 2001. Nutrient of requirements of dairy cattle. 7. ed. Washington, D.C.: National Academic Press, 362.

Part 2. Energy and protein interactions, monogastrics

Exploring the biology of energy and protein utilization in non-ruminant animals to improve nutrient utilization efficiencies

C.F.M. de Lange[1], *C.L. Levesque*[2] *and H.R. Martínez-Ramírez*[1]
[1]*Department of Animal and Poultry Science, University of Guelph, Guelph, ON, N1G 2W1, Canada; cdelange@uoguelph.ca*
[2]*South Dakota State University, Brookings, SD 57007, USA*

Abstract

In this review, bioavailability of energy and amino acid, and the biological processes that contribute to nutrient utilization efficiency in non-ruminant animals are discussed. Key limitations of using digestible energy, metabolizable energy, net energy and standardized ileal amino acid digestibility values in diet formulation are highlighted. A modelling system based on nutrient flows is suggested to improve accuracy of representing energy utilization. In nutrient flow models dietary energy sources are characterized based on content and enzymatic digestibility or fermentability of energy yielding nutrients, while the use of nutrients for the various body functions is explicitly represented. In various species animal performance is sensitive to synchronized supply of nutrients; in these cases the dynamics of nutrient absorption should be considered. In regards to amino acid utilization, the effect of microbial fermentation in the upper gut and metabolic inefficiency associated endogenous gut amino acid losses remain to be quantified more accurately. Especially in heat treated ingredients some amino acids may be absorbed in a form that renders them unavailable for metabolism by the animal. Maintenance energy requirements contribute substantially to total energy requirements, and are best expressed as requirements at the cellular level in ATP equivalents. Between animal variability is a key determinant of inefficiency of nutrient use in groups of animals. Various approaches that may be pursued to improve efficiencies of using protein and other energy yielding nutrients are presented.

Introduction

The high feed energy costs, rapidly increasing demand for food animal protein, and concerns about the environmental impact of animal production are important incentives to improve animal production and nutrient utilization efficiencies (NRC, 2010; Moughan, 2012). The utilization of dietary bioavailable energy and protein (i.e. amino acids) for retention in consumable animal proteins involves digestion, absorption, and post-absorptive metabolism. These biological processes are all influenced by factors associated with the animal (species and genotype, gender, physiological state, health status) and the environment (diet composition, thermal and physical environment; e.g. NRC, 1994, 2011, 2012). Energy yielding nutrients, including amino acids, also serve as precursors for bioactive compounds that influence gene expression and endocrine function and may as such also be used to manipulate production efficiencies, including metabolic disorders and gut health (Averous *et al.*, 2003; Jonker *et al.*, 2012; Zijlstra *et al.*, 2012). Optimum dietary energy and protein intake will, therefore, vary between groups of animals, and an understanding of the biology of energy and protein utilization is required for optimizing energy and nutrient utilization for different groups of animals.

In this short review, bioavailability of energy and amino acids, and the biological processes that contribute to nutrient utilization efficiency in non-ruminant animals are discussed. Means to improve efficiencies and areas for further research are identified. The emphasis will be on pigs, primarily because many of the biological processes that contribute to nutrient utilization have been quantified and integrated in mathematical models that allow us to relate nutrient inputs to rates and efficiencies of pork production in a quantitative manner (e.g. Kyriazakis, 1999; Van Milgen *et al.*, 2008; NRC, 2012). Many of the concepts also apply to other species and references to other species are made where appropriate.

One of the biggest challenges in energy and protein nutrition is the prediction and manipulation of voluntary feed intake. This complex topic is beyond this contribution; for reviews on this topic the reader is referred elsewhere (e.g. Kyriazakis, 1999; Torrallardona and Roura, 2009; Richards *et al.*, 2010; Westerterp-Plantenga *et al.*, 2012).

Energy and amino acid bioavailability in feed ingredients

Accurate estimates of bioavailability of energy and nutrients in feed ingredients are required to properly assess dietary requirements of animals, which are independent of dietary feed ingredient composition, and to assign nutritional values to alternative feed ingredients. For effective feed formulation estimates of bioavailability should be additive among feed ingredients. Given the complexities and expenses associates with accurately determining nutrient bioavailability, measures of digestibility or metabolizability are widely used in diet evaluation and feed formulation (Columbus and De Lange, 2012; Moughan *et al.*, 2000; NRC, 1994, 2011, 2012). However, apparent faecal digestibility, which is the simplest measure of nutrient digestibility in mammals and fish, is confounded with enteric fermentation and endogenous gut nutrients, primarily (mucin-)protein, losses from the host. Moreover, the post-absorptive efficiency of nutrient utilization should be considered, especially when assigning bioavailable energy contents to feed ingredients and when evaluating bioavailability of amino acids in feed ingredients that contain (heat) damaged protein.

Digestible energy (DE) and metabolizable energy (ME) systems are still widely used to assess bioavailability of energy in feed ingredients for non-ruminant animals, largely because experimental procedures to establish DE and ME values are relatively simple (Birkett and De Lange, 2001a; Emmans, 1994; Noblet and van Milgen, 2004; NRC, 1994, 2011, 2012). A key limitation of DE and ME systems is the relatively poor prediction of feed efficiency and other aspects of animal performance, especially when feeding diets with extreme fat, fiber or protein contents (e.g. Birkett and De Lange, 2001a; Noblet and van Milgen, 2004). A more mechanistic approach to represent energy utilization, and thus to assign bioavailable energy contents to feed ingredients, is to characterize the content and bioavailability of energy yielding nutrients (Birkett and De Lange, 2001a; Boisen and Verstegen, 1998; Noblet and van Milgen, 2004). This approach will require an extensive data base of enzymatic (ileal) digestibility and fermentability (which may be estimated from fecal minus ileal digestibility) of the dietary energy yielding nutrients (starch, sugars, protein, fat; CVB, 2003), including effects of feed ingredient processing, interactions among nutrients, and the animal's physiological state on these measures of digestibility (Moughan *et al.*, 2000; Noblet *et al.*, 1993; Noblet and van Milgen, 2004).

Standardized ileal digestibility (SID) coefficients are used routinely to estimate the bioavailability of amino acids in feed ingredients for pigs (Stein *et al.*, 2007; NRC, 2012) and have been considered for poultry (Locatelli *et al.*, 2004). The SID values reflect that amino acids that disappear from the hindgut are not available for metabolism by the host – hence the term 'ileal digestibility' – and are corrected for the basal ileal endogenous amino acid losses – hence the term 'standardized'. Basal ileal endogenous amino acid losses represent the difference between amino acids of endogenous origin that are secreted into the upper gut and the amounts reabsorbed prior to the distal ileum of animals that are fed highly digestible protein sources and diets that are devoid of fibers and anti-nutritional factors that induce additional or specific endogenous gut amino acid losses (Stein *et al.*, 2007). The SID values are corrected for the basal ileal amino acid losses only, largely because it is technically too difficult to routinely measure total ileal endogenous amino acid losses (basal losses plus diet specific losses; Stein *et al.*, 2007). Until methodology is in place to routinely measure total ileal endogenous amino acid losses and these losses have been quantified for a wide range of feed ingredients, SID coefficients will provide the basis for estimated bioavailability of amino acids in feed ingredients (Columbus and De Lange, 2012). To improve protein utilization, dietary factors that induce specific endogenous gut amino acid losses should be considered (Tamminga *et al.*, 1995), together with metabolic inefficiencies associated with synthesis and re-absorption of endogenous gut losses (Columbus and De Lange, 2012).

In some instances SID values underestimate bioavailability of amino acids (Columbus and De Lange, 2012). This can occur when amino acids are digested and absorbed in a chemical form that renders them unavailable for metabolism, which applies in particular to lysine that is present in heat treated feed ingredients (Rurtherfurd and Moughan, 2012). Another example is the negative impact of fermentable fiber intake on threonine bioavailability (Libao-Mercado *et al.*, 2006; Zhu *et al.*, 2005), which may be attributed to the impact of fiber-induced enteric fermentative amino acid catabolism and the use of high threonine containing mucus proteins as nitrogen and energy sources for enteric microbes (Libao-Mercado *et al.*, 2006; Fuller, 2012). The latter highlights the potential impact of enteric fermentation in the upper gut of non-ruminants on amino acid utilization. As recently reviewed by Fuller (2012), enteric fermentation also occurs in the upper gut of non-ruminant animals and contributes to *de novo* microbial amino acid synthesis, as well as fermentative catabolism of amino acids. The impact of enteric fermentation in the upper gut of pigs on the net amino acid supply to the host remains to be quantified.

Energy and nutrient availability in feed ingredients, especially high fiber containing co-products from food and biofuel industries, may be improved by additional processing, pre-treatment with enzymes or microbial inoculants, aimed at improving fiber digestibility (Zijlstra *et al.*, 2010; Yáñez *et al.*, 2011) and reducing effects of anti-nutritional factors on nutrient digestibility and endogenous gut nutrient losses (Gilani *et al.*, 2012; Sun *et al.*, 2008; Tamminga *et al.*, 1995).

Post-absorptive energy and nutrient use

In animals absorbed nutrients are required to satisfy requirements for body maintenance functions and production in the form of whole body protein deposition, whole body lipid deposition, growth of products of conception, milk production and so on. Production of nutrients that are retained in the animal's body or secreted with milk represents the balance between synthesis and degradation, both of which require energy and are associated inefficiencies of nutrient use (Gill *et al.*, 1989; Milligan and Summer, 1986).

Largely through genetic selection impressive progress has been made in improving the animals' production potentials (e.g. Kyriazakis, 1999; NRC, 2012). In fact, in many instances the animals' energy intake capacity limits expression of the animals' production potential (NRC, 2012). Recently, we have become more aware of the negative effect of environmental stressors, and especially subclinical levels of disease, on animal production efficiency (e.g. Rakhshandeh and De Lange, 2011; Williams *et al.*, 1997). Therefore, environmental stressors should be minimized and various approaches to reduce the animal's susceptibility to environmental stressors should be pursued, including genetic selection of animals and diet manipulations. The latter is addressed by Dr. Klasing at this symposium.

The classical partitioning of nutrient intake in growing animals (Kielanowski, 1971) requires estimates of maintenance nutrient requirements and nutrient retention of body protein and body lipid deposition, as is presented for ME in Equation 1. In this equation MEm represents maintenance energy requirements, REl and REp energy retention in the form body lipid and body protein, respectively, and kl and kp the corresponding marginal efficiency of using ME intake. This classical approach has been adopted for various animal species and has been expanded to include other aspects of animal production (NRC, 1994, 2011, 2012).

$$\text{ME} = \text{MEm} + (1/\text{kl}) \times \text{REl} + (1/\text{kp}) \times \text{Rep} \quad (1)$$

Clearly for the factorial estimation of energy and nutrient requirements animal production levels should be determined, and increases in production levels will generally increase energetic efficiency and feed efficiency, but these will not be further discussed. In the remainder of this short review nutrient requirements for body maintenance functions and post-absorptive efficiencies of nutrient use are discussed.

Maintenance energy requirements

Accurate estimates of maintenance requirements are critical for energy, simply because of the large contribution of maintenance energy requirements to total energy requirements, especially in non-lactating reproducing animals (NRC, 1994, 2011, 2012).

Even though the concept of maintenance energy requirements is rather convenient for the factorial estimation of dietary energy requirements (Equation 1), its relevance for highly productive animals can be questioned. It may be more appropriate to interpret maintenance energy requirements in producing animals as support costs for production, which means that maintenance energy requirements contribute to the (marginal) inefficiency of using energy intake for production (Labussière *et al.*, 2011). This interpretation has implications for the estimate of 'maintenance' energy requirements that should be independent of the animal's performance level and feeding regime (Birkett and De Lange, 2001a). In the case of growing animals, this is supported by the (statistical) interdependency between estimates of MEm and marginal efficiencies of using ME intake for REl and REp (Birkett and De Lange, 2001a; Noblet *et al.*, 1999), and the empirical observations that growing animals fed at maintenance energy requirements, i.e. at zero body energy balance, will initially gain body protein (and BW) and mobilize body lipid (e.g. Close *et al.*, 1982; Kyriazakis *et al.*, 1993). Moreover, and as discussed later, estimates of MEm will vary with diet nutrient composition. Therefore, it is more appropriate to express maintenance energy requirements as net energy (NEm; i.e. MEm corrected for heat increment of feeding), or ATP equivalents (Birkett and De Lange, 2001a,b; Boisen and Verstegen, 1998; Van Milgen *et al.*, 2001; Figure 1).

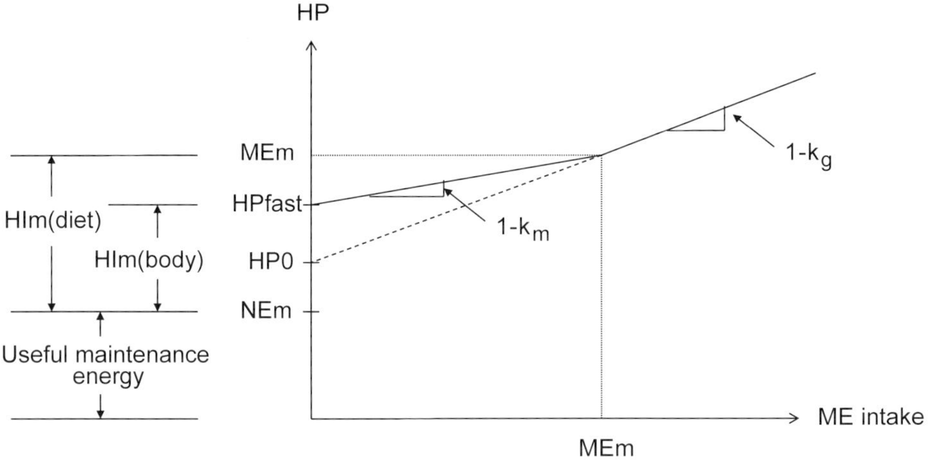

Figure 1. Schematic representation of components of heat production as influenced by metabolizable energy intake. Abbreviations: HIm(body), heat increment of mobilizing body nutrients to provide net energy requirements for maintenance; HIm(diet), heat increment of using metabolizable energy intake to provide net energy requirements for maintenance; HP, heat production; HP0, extrapolated heat production at zero energy intake; HPfast, heat production during fasting; k_g, marginal efficiency of using metabolizable energy intake for body energy retention; k_m, relative energetic efficiency of using body and dietary nutrients for supplying useful energy for body maintenance functions; ME intake, metabolizable energy intake; MEm, metabolizable energy requirements for maintenance; NEm, net energy requirements for maintenance (equivalent to ATP requirements for maintenance functions at the tissue level)(adjusted from Birkett and De Lange, 2001a).

The estimate of NEm will influence directly the absolute NE values and to a lesser extent the relative NE value of feeds and feed ingredients (Noblet *et al.*, 2013; this publication). Unfortunately, NEm cannot be determined directly. Traditionally, basal metabolic rate, representing equilibrium heat production (HP) of a fasting and resting animal has been used to estimate NEm (Figure 1). Basal metabolic rate cannot be measured directly in experiments either, and is generally estimated by measurements of fasting heat production (HPfast), which approaches basal metabolic rate asymptotically in time (Van Milgen *et al.*, 1998). However, the extrapolated plateau HPfast is not a good estimate for NEm either, since it represents both the energy requirements for basic body maintenance functions, which may be interpreted as ATP requirements at the cellular level, as well as heat increment associated with generating ATP from body nutrient stores (Figure 1). Moreover, plateau HPfast is influenced by the previous feeding regime (De Lange *et al.*, 2006). The use of plateau HPfast as an estimate of NEm is not consistent with the definition of NEprod (Birkett and De Lange, 2001a). It is more appropriate to represent NEm as kb × plateau HPfast; where kb is the efficiency of converting energy retained in body nutrients to useful energy contained in ATP (Figure 1). In a similar manner, NEm may be calculated as kd × MEm, where kd is the efficiency of utilizing ME intake for generating ATP. The value km = kd/kb is simply the relative efficiency of using body and dietary nutrients to generate NEm. As an alternative to experimental measurement of (plateau) HPfast, Noblet and van Milgen (2004) estimated HPfast by extrapolating a linear regression of HP against ME intake above MEm to zero ME intake (HP0). The slope in this regression is determined by kg only (Figure 1), where kg is the marginal efficiency of using ME intake for energy retention in BW gain. As such, HP0 has no direct relationship to HPfast and NEm as these are influenced by kb, kd, or km, while there is apparently no relationship between km and kg (Birkett and De Lange, 2001a).

Insufficient quantitative information is deemed available to accurately estimate energy (ATP) requirements for the energy demanding body maintenance functions, such as body protein and body lipid turnover, maintenance of electrolyte gradients across cell membranes (e.g. sodium and potassium pumping), muscle tension, etc. (Baldwin, 1995; Gill *et al.*, 1989; Milligan and Summers, 1986). Therefore, estimates of maintenance energy requirements may be related to the size of metabolically active tissues (Koong *et al.*, 1982; Van Milgen *et al.*, 1998; Yen *et al.*, 1997) or metabolic BW, which may be regarded as a simple predictor of size of metabolically active tissues (Noblet *et al.*, 2013; this publication).

There are differences in maintenance energy requirements between species, especially for growing animals, and these are briefly discussed by Noblet *et al.* (2013; this publication). Within species there is between animal variability in estimated maintenance energy requirements, and this has been the basis for genetic selection based on residual feed intake (Kennedy *et al.*, 1993). The latter is defined as the difference between estimated maintenance energy requirements of individual animals within a population versus the population mean, whereby a negative value implies that an animal is energetically more efficient. Obviously, this approach is sensitive to the model that is used to estimate maintenance energy requirements and environmental conditions. The physiological basis for residual feed intake has been reviewed recently (Young and Dekkers, 2012), and may be attributed to variation in digestive capacity, activity levels, feeding behaviour, immune system stimulation (inflammation and oxidative stress), as well as various aspects of metabolism which will be discussed later.

Maintenance amino acid requirements

Traditionally, maintenance amino acid requirements have been related to the metabolic BW of pigs [i.e. $(BW)^{0.75}$; NRC, 1994, 2011, 2012]. This traditional approach fails to represent the biological processes that contribute to maintenance amino acid requirements and account for factors affecting them. According to Moughan (1999) the main factors that constitute maintenance amino acid requirements are basal endogenous gut amino acid losses, integument amino acid losses, and minimum amino acid catabolism. The latter represents amino acid losses associated with basal whole protein turnover and contributes to the minimum urinary nitrogen excretion. These concepts have been adopted in NRC (2012) for estimating amino acid requirements of pigs. Basal endogenous gut amino acid losses, in both the upper and lower gut, may be related to feed dry matter intake and are the single largest contributor to maintenance amino acid requirements of pigs. Integument losses (e.g. skin, hair) are quantitatively not that important for pigs and may be related to $BW^{0.75}$. According to NRC (2012) insufficient quantitative information is available to generate reasonable estimates of minimum amino acid catabolism. For this reason, NRC (2012) applied the post-absorptive inefficiency (inevitable amino acid catabolism; discussed below) of using intake of SID amino acids for covering endogenous gut and integument amino acid losses to account for minimum amino acid catabolism. Under some conditions the use of some amino acids for the production of non-protein compounds may be quantitatively important. The latter applies, for example, to the use of tryptophan for synthesis of serotonin and kynurenine or cysteine for synthesis of glutathione. The synthesis rates of these compounds remain to be quantified accurately and may be elevated under some conditions, for example during inflammation (Le Floc'h *et al.,* 2009; Rakhshandeh and De Lange, 2011).

Even in animals at moderate levels of production the contribution of maintenance amino acids requirements to total amino acids requirements is relatively small (NRC, 2012). Therefore, there is limited opportunity to improve efficiencies of amino acid utilization by manipulating maintenance requirements. Exceptions are to avoid feeding diets that induce large amounts of specific endogenous gut amino acid losses (Tamminga *et al.,* 2005), or exposing animals to stressors that induce synthesis of non-protein compounds that require amino acids, such as tryptophan (Le Floc'h *et al.,* 2009).

Post-absorptive efficiency of energy utilization

It is well established that energetic efficiencies in animals are influenced by dietary source of energy yielding nutrients (i.e. starch, sugars, protein, fat) and the purpose for which energy is used (i.e. maintenance, body lipid and protein deposition, milk production, etc.) (Black, 1995; De Lange and Birkett, 2001a,b; Van Milgen *et al.,* 2001). For example, direct incorporation of dietary lipids into body lipid has an energetic efficiency of about 90%, while the use of lipid as an energy source for generating ATP is about 66% efficient and de novo endogenous lipid synthesis from digestible starch is about 75% efficient in pigs (Black, 1995). As reviewed by Birkett and De Lange (2001a) the estimated energetic efficiencies in growing pigs of using ME intake for body protein deposition (kp; ranging from 0.44 to 0.60) is lower than that of body lipid deposition (kl; ranging from 0.60 to 0.80); variability of these estimates reflect differences in dietary energy sources across studies, mathematical models used to generate these estimates, assumptions about maintenance energy requirements, and maybe between animal variability in metabolic efficiencies. Energetic inefficiency (i.e. heat increment of feeding) reflects energy costs of feed ingestion and digestion, nutrient absorption, and heat losses associated with metabolic pathways involved in utilization of absorbed energy yielding nutrients for maintenance and production (e.g. Baldwin, 1995; Birkett and De Lange, 2001c; Van Milgen, 2002; Figure 2).

Heat increment of feeding is accounted for in NE systems. In these systems NE is defined as ME intake minus heat increment of feeding and estimated as energy retained in animal products plus NEm. In NE systems for pigs (CVB, 2003; Noblet and Van Milgen, 2004; NRC, 2012) and poultry (Pirgozliev and Rose, 1999) diet NE values are estimated from nutrient compositions. In these empirical NE systems diet effects on energy utilization are represented, while animal effects on the use of energy are not considered or it is assumed that the relative use of energy from various sources for the various purposes is rather constant. A more mechanistic approach to representing diet and animal effects on energy utilization is to mathematically represent nutrient and metabolite flows of energy yielding nutrients (Birkett and De Lange, 2001a,b; Boisen and Verstegem 1998; De Lange and Birkett, 2005; Machiels and Henken, 1986; Pettigrew *et al.,*1992; Van Milgen *et al.*, 2001; Figure 2). Nutrient flow models are more complex than the current empirical NE systems, as they require

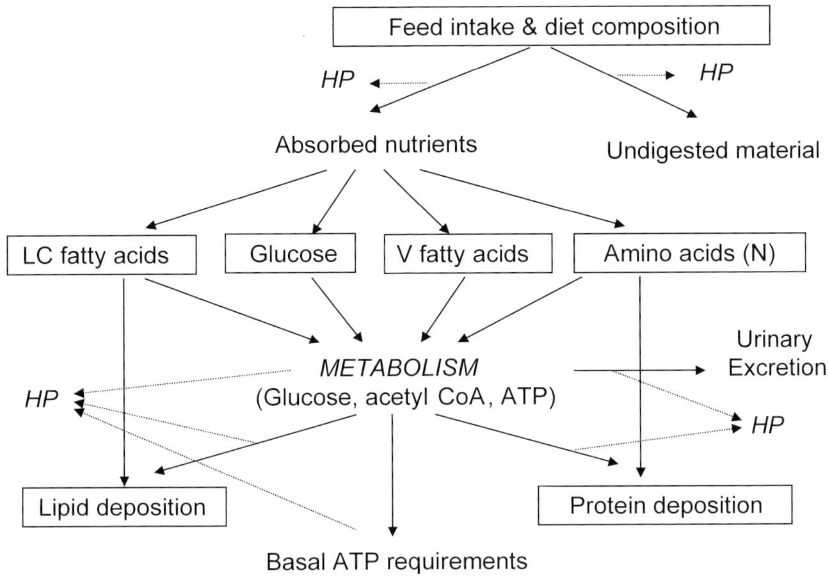

Figure 2. Nutrient and metabolite flows representing utilization of energy yielding nutrients for body protein and body lipid accretion in growing animals (adjusted from Birkett and De Lange 2001b). Abbreviations: ATP = adenosine tri-phosphate; CoA = co-enzyme A; HP = heat production; LC Fatty Acids = long chain fatty acids; V Fatty Acids = volatile fatty acids.

an understanding of nutrient partitioning in animals, and will yield estimates of useful (net) energy for a feed ingredient or diet that varies with the animal's physiological state. The latter makes it difficult to integrate such models with (least-cost) diet formulation. To overcome this difficulty, the concept of 'effective ME' is used in NRC (2012) to integrate the use of empirical NE systems (to represent diet effects on energetic efficiencies) with the use of energy for various purposes (Equation 1), whereby diet 'effective ME' content is calculated from the diet NE content with a typical and fixed conversion (e.g. 0.75 for growing-finishing pigs). This simplification reduces the connection between dietary energy source and use of energy by the animal somewhat.

A benefit from using more mechanistic models of energy utilization is that they allow for a more systematic approach to identify means to improve energetic efficiencies and guide the design of experiments that are targeted towards better understanding of diet and animal effects on key aspects of energy utilization (Figure 2). Nutrient flow representations of energy utilization can be further refined, by making them dynamic and including saturation kinetics for key metabolic pathways. However, in growing animals where the number of alternative pathways for individual nutrient is rather small this may be of limited value. In reproducing animals (e.g. lactating and egg laying) the inclusion of dynamic aspects is likely to be more critical, for representing the potential mobilization of body nutrient stores to support production (e.g. Baldwin, 1995; Pettigrew *et al.*,1992; Birkett and De Lange, 2001a,b). The representation of amino acid utilization can be integrated in nutrient flows models to improve understanding of interactions between energy and protein utilization (e.g. Kyrizakis *et al.*, 1999; Van Milgen *et al.*, 2008; Wies *et al.*, 2004).

Dynamics of nutrient digestion and absorption affect the rate of appearance of nutrients that are available for metabolism and may be influenced to synchronize supply of different nutrients to the animal. The importance of considering the dynamics of nutrient absorption has been shown in broiler chickens (Weurding *et al.*, 2003), veal calves (Van den Borne *et al.*, 2006) and growing pigs (Van den Borne *et al.*, 2007). For example, when feeding different types of starch of similar digestibility to broiler chicks, more slowly digested starch yielded higher efficiencies of amino acid utilization for growth (Weurding *et al.*, 2003). It was hypothesized that rapid digestible starch reduced glucose supply to the lower part of the digestive tract, forcing intestinal tissue to use increased amounts of amino acids as energy source, and reducing amino acid availability to support growth. Alternatively, dynamics of nutrient absorption may influence the animals' endocrinology and thereby the physiological control of nutrient utilization (Zijlstra *et al.*, 2012). Feeding frequency and feed ingredient choice may thus be used to manipulate post-absorptive nutrient utilization.

As mentioned earlier, between-animal variation in energetic efficiencies may in part be attributed to metabolic processes. An important contributor to maintenance energy requirements, or support costs for production, is body protein turnover (Milligan and Summers, 1986; Gill *et al.*, 1989). It may thus be hypothesized that differences in the ratio between protein synthesis and body protein deposition between animals contribute to differences in energetic efficiencies of body protein deposition. However, when comparing growing pigs of two very different genotypes, Landrace and Iberian, under similar conditions this ratio did not differ, while rather large differences were observed in whole body protein deposition (Rivera-Ferre *et al.*, 2006). It is of interest to note that beta adrenergic agonists, such as Ractopamine (Paylean^{TM}) increase protein deposition by increasing protein synthesis and reducing protein degradation (Kim and Sainz, 1992), and thereby reducing the marginal energy cost of body protein deposition. Effects of diet, the environment and the animal on the relationship between protein synthesis and protein deposition remains to be explored further. Another aspect of energy metabolism that is of interest is mitochondrial respiration efficiency, which has been studied in various species, including poultry (Bottje and Carstens, 2009; Ojano-Dirian *et al.*, 2004) and pigs (Lefaucheur *et al.*, 2011). Mitochondria account for about 90% of whole body oxygen consumption and generate ATP through oxidative phosphorylation, which is driven by electron gradients across the inner mitochondrial membrane. Leakage of electrons across the membrane, which is regulated by so-called uncoupling proteins will reduce ATP yield, increase heat production and therefore reduce energetic efficiencies (Ježek *et al.*, 2004). Activity of uncoupling proteins appears related to oxidative stress, which suggests a link with inflammation and therefore the animals' ability to cope with a disease challenge. Initial observations in broiler chickens suggest a relationship between respiratory activity in muscle tissue and feed efficiency (Bottje and Carstens, 2009; Ojano-Dirian *et al.*, 2004), but this needs to be confirmed.

Post-absorptive efficiency of amino acid utilization

The maximum marginal efficiency of utilizing available amino acid intake over and above maintenance amino acid requirements for amino acid retention in body protein deposition is determined largely by inevitable amino acid catabolism (Moughan, 1999). When obtaining estimates of inevitable catabolism, care should be taken to differentiate it from preferential catabolism, which occurs when energy intake limits the expression of the animals body protein deposition capacity, and catabolism of amino acids that are supplied in excess of requirements (De Lange *et al.*, 2001, 2012). The latter is of concern when animals are fed constant dietary levels of amino acids over a wide BW range, because dietary amino acid requirements – expressed as dietary levels – generally decline with increasing BW. Estimates of inevitable catabolism of lysine and threonine in growing pigs have been obtained from observations on individual pigs and in closely controlled serial slaughter studies (NRC, 2012), and are about 25% of available amino acid intake. According to NRC (2012) the rate of inevitable amino acid catabolism is rather constant over a wide range of amino acid intake levels (Figure 3), appears to be largely independent of BW, and increases slightly with improvements in pig performance potential. It should be noted that the efficiency of amino acid utilization for body protein utilization is always lower in groups of pigs than in individual pigs, which can be attributed largely to between-animal variability in amino acid requirements (De Lange *et al.*, 2012). In fact, it can be argued that the main contributor to the diminishing marginal efficiency of amino acid utilization for body protein deposition is between animal variability rather than a marginal increase

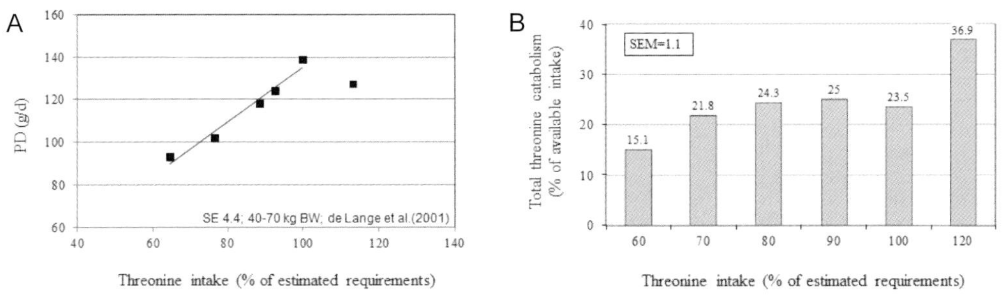

Figure 3. Observed effects of threonine intake on whole body protein deposition (PD; Figure A) and estimated rates of threonine catabolism (Figure B). Results were obtained in a serial slaughter study, using individually housed pigs that were fed casein and cornstarch based diets; dietary protein supply was adjusted weekly to meet the pigs' estimated threonine requirements (adjusted from De Lange et al., 2001, 2012).

in the rate of inevitable amino acid catabolism (Figure 3). Therefore, between animal variability within groups of animals should be considered when establishing the optimum dietary amino acid level for groups of animals (Pomar *et al.*, 2003; Morel *et al.*, 2008), and in NRC (2012) adjustments were made to the inefficiency of utilizing available amino acid intake over and above maintenance amino acid requirements to account for between animal variability.

The rate of inevitable catabolism is influenced by animal and dietary factors. As mentioned earlier, the rate of inevitable amino acid catabolism declines with increasing body protein deposition capacity (NRC, 2012). Observed antagonisms among amino acids that share common catabolic pathways, whereby an excessive supply of one amino acid reduces the utilization efficiency of other amino acids, has in some instances been attributed to increased amino acid catabolism (Langer *et al.*, 2000). This applies in particular to the branch-chained amino acids and has been observed in chickens (Peganova and Eder, 2003), growing pigs (Langer *et al.*, 2000) and sows (Perez-Laspiur *et al.*, 2009). However, in other instances amino acid antagonisms have been attributed to competition for amino acid transporters, as illustrated by reduced uptake of tryptophan in the brain of pigs fed large amounts of neutral amino acids (Henry *et al.*, 1992), and thereby increasing catabolism. These examples illustrate that efficiencies of using amino acids that limit animal production may be reduced, as well as animal productivity, by feeding diets with (severely) imbalanced amino acid profiles (Garlick, 2004).

Conclusions and implications

Key limitations of using digestible energy, metabolizable energy, and standardized ileal amino acid digestibility values in diet formulation are inaccurate predictions of the animals' performance response, especially when feeding diets with extreme nutrient compositions or heat treated ingredients. These limitations are becoming more critical with increased usage of co-products from the food and biofuel industries in animal diets. Energy (and protein) utilization is represented more accurately, improving predictability of animal responses, when it is based on nutrient and metabolite flow models. In nutrient flow models the dietary contents and availability of energy yielding nutrients, and the use of nutrients by the animals for the various body functions can be represented. Nutrient flow representations of energy utilization can be further refined, by making them dynamic and including saturation kinetics for key metabolic pathways. Maintenance energy requirements contribute substantially to total energy requirements, and are best expressed as net energy requirements at the cellular level, in ATP equivalents. Between animal variability is a key determinant of inefficiency of nutrient use in groups of animals, and may be attributed to differences in performance potentials, animal activity, digestive capacity and immune system activity. Various approaches can be used as a means to improve efficiencies of using protein and other energy yielding nutrients, including improving performance potentials, feeding more closely to nutrient requirements for targeted levels of animal performance, reducing the negative effects of dietary fiber and anti-nutritional factors on nutrient utilization, reducing the ratio between protein synthesis and protein deposition, and improving mitochondrial respiration efficiency.

References

Averous, J., A. Bruhat, S. Mordier and P. Fafournoux, 2003. Recent advances in the understanding of amino acid regulation of gene expression. J. Nutr. 133, 2040S-20445S.

Baldwin., R.L. 1995. Modeling Ruminant Digestion and Metabolism. Chapman & Hall, New York.

Birkett, S. and K. de Lange, 2001a. Limitations of conventional models and a conceptual framework for a nutrient flow representation of energy utilization by animals. Brit. J. Nutr. 86, 647-659.

Birkett, S. and K. de Lange, 2001b. Calibration of a nutrient flow model of energy utilization by growing pigs. Brit. J. Nutr. 86, 675-689.

Black, J.L. 1995. Modelling energy metabolism in the pig. Pages 87-102 in P.J. Moughan, M.W.A. Verstegen and M. Visser-Reyneveld (Editors) Modelling growth in the pig. Wageningen Pers, Wageningen, The Netherlands.

Boisen, S. and M.W.A. Verstegen, 1998. Evaluation of feedstuffs and pig diets. Energy or nutrient-based evaluation systems? II. Proposal for a new nutrient-based evaluation system. Acta Agric. Scand. 48, 95-102

Bottje, W.G. and G.E. Carstens, 2009. Association of mitochondrial function and feed efficiency in poultry and livestock species. 87, E48-E63.

Close, W.H., F. Berschauer and R. Heavens, 1982. The influence of protein:energy value of the ration and level of feed intake on the energy and nitrogen balance of the growing pig. Brit. J. Nutr. 49, 255-269.

Columbus, D. and C.F.M. de Lange, 2012. Evidence for validity of ileal digestibility coefficients in monogastrics. Brit J. Nutr. 108, 264-272.

CVB (Centraal Veevoeder Bureau), 2003. Veevoedertabel (Table of Feeding Value of Animal Feed Ingredients). Centraal Veevoeder Bureau, Lelystad, The Netherlands.

De Lange, C.F.M. and S. H. Birkett, 2005. Characterization of the useful energy content in swine and poultry feed ingredients. Can. J. Anim. Sci. 85, 269-280.

De Lange, C.F.M., A.M. Gillis and G.J. Simpson, 2001. Influence of threonine intake on whole-body protein deposition and threonine utilization in growing pigs fed purified diets. J. Anim. Sci. 79, 3087-3095.

De Lange, C.F.M., J. van Milgen, J. Noblet, S. Dubois and S. Birkett, 2006. Previous feeding level influences plateau heat production following a 24 h fast in growing pigs. Brit. J. Nutr. 95, 1082-1087.

De Lange, C.F.M., C. Levesque and B.J. Kerr, 2012. Amino acid nutrition and feed efficiency. Pages 81-90 in J.F. Patience (ed.). Feed efficiency in pigs. Wageningen Academic Press.

Emmans, G.C. 1994. Effective energy: A concept of energy utilization applied across species. Brit. J. Nutr. 71, 801-821.

Fuller, M.F. 2012. Determination of protein and amino acid digestibility in foods including implications of gut microbial amino acid synthesis. Brit. J. Nutr. 108, S238-S246.

Garlick, P.J. 2004. The nature of human hazards associated with excessive intake of amino acids. J. Nutr. 134, 1633S-1639S.

Gilani, G.S., C. W. Xiao and K.A. Cockell, 2012. Impact of antinutritional factors in food proteins on the digestibility of protein and the bioavailability of amino acids and on protein Quality. Brit. J. Nutr. 108, S315-332.

Gill, M., J. France, M. Summers, B. McBride and L. Milligan, 1989. Simulation of the energy costs associated with protein turnover and Na+/K+ transport in growing lambs. J. Nutr. 119, 1287-1299.

Henry, Y., B. Seve, B. Colleaux, P. Ganier, C. Saligaut and P. Jégo, 1992. Interactive effects of dietary levels of tryptophan and protein on voluntary feed intake and growth performance in pigs, in relation to plasma free amino acid and hypothalamic serotonin. J. Anim Sci. 70, 1873-1887.

Ježek, P, M., žáćcková, M. Růžička, E. Škobisová and M. Jabůrek, 2004. Mitochondrial uncoupling proteins – Facts and fantasies Phys. Res. 53, S199-S211.

Jonker, R., M.P K.J. Engelen and N.E.P. Deutz, 2012. Role of specific dietary amino acids in clinical conditions. Brit. J. Nutr. 108, S139-S148.

Kennedy, B.W., J.H. van der Werf and T.H. Meuwissen, 1993. Genetic and statistical properties of residual feed intake. J. Anim. Sci. 71, 3239-3250.

Kielanowski, J. 1971. Energy requirements of the growing pig. In Pig Production, Pages 183-201 in D.J.A. Cole (Editors) Proceedings of the Eighteenth Easter School in Agricultural Science, Nottingham, 1971. Penn State University Press, University Park, PA.

Kim, Y.S. and R.D. Sainz, 1992. Beta adrenergic agonists and hypertrophy of skeletal muscle. Life Sci. 50, 3970407.

Koong, L-J, J. Nienaber, J. Pekas and J-T. Yen, 1982. Effects of plane of nutrition on organ size and fasting heat production in pigs. J. Nutr. 112, 1638-1642.

Kyriazakis, I. (Editor) 1999. Quantitative Biology of the Pig, I. Kyriazakis, ed. Wallingford, UK: CAB International, Wallingford, Oxon, UK.

Kyriazakis, I., D. Dotas and G. Emmans, 1993. The effect of breed on the relationship between feed composition and the efficiency of protein utilization in pigs. Brit. J. Nutr. 71, 849-859.

Labussière E., J. van Milgen J., C. de Lange and J. Noblet, 2011. Maintenance energy requirements of growing pigs and calves are influenced by feeding level. J. Nutr. 141, 1855-1861.

Langer, S., P.W.D. Scislowski, D.S. Brown, P. Dewey and M.F. Fuller, 2000. Interactions among the branched-chain amino acids and their effects on methionine utilization in growing pigs: effects on plasma amino- and keto-acid concentrations and branched-chain keto-acid dehydrogenase activity. Brit. J. Nutr.83, 49-58.

Lefaucheur, L., B. Lebret, P. Ecolan, I. Louveau, M. Damon, A. Prunier, Y. Billon, P. Sellier and H. Gilbert, 2011. Muscle characteristics and meat quality traits are affected by divergent selection on residual feed intake in pigs. J. Animal Sci 89, 996-1010.

Le Floc'h, N., L. Lebellego, J.J. Matte, D. Melchior and B. Sève, 2009. The effect of sanitary status degradation and dietary tryptophan content on growth rate and tryptophan metabolism in weaning pigs. J. Anim. Sci. 87, 1686-1694.

Libao-Mercado, A.J., S. Leeson, S. Langer, B.J. Marty and C.F.M de Lange, 2006. Efficiency of utilizing ileal digestible lysine and threonine for whole body protein deposition in growing pigs is reduced when dietary casein is replaced by wheat shorts. J. Anim. Sci. 84, 1362-1374.

Locatelli, M.L., V. Ravindran and A. Lemme, 2004. Standardized ileal amino acid digestibility in broiler nutrition. J. Anm. Sci. 82 (Suppl. 1), 433-433

Machiels, M.A.M. and A.M. Henken, 1986. A dynamic simulation model for growth of the African Catfish, Clarias gariepinus (Burchell 1822). I. Effect of feeding level on growth and energy metabolism. Aquaculture 56, 29-52.

Milligan, L. and M. Summers. 1986 The biological basis of maintenance and its relevance to assessing responses to nutrients. Proc. Nutr. Soc. 45, 185-193.

Morel, P.C.H., G.R. Wood and D. Sirisatien, 2008. Effect of genotype, population size and genotype variation on optimal diet determination for growing pigs. Acta Horticulturae 802, 287-292.

Moughan, P.J. 1999. Protein metabolism in the growing pig. Pages 299-331 In I. Kyriazakis (Editor) Quantitative Biology of the Pig. CAB International, Wallingford, Oxon, UK.

Moughan, P.J., M.W.A. Verstegen and M. Visser-Reyneveld (Editors), 2000. Feed evaluation – principles and practice. Wageningen Pers, Wageningen, The Netherlands.

Moughan, P.J. (Editor). 2012. Dietary protein for human health. 2012. Br. J. Nutr. 108, S1-S336.

NRC (National Research Council), 1987. Predicting feed intake of food producing animals. The National Academies Press, Washington, DC.

NRC (National Research Council), 1994 Nutrient requirements of poultry. The National Academies Press, Washington, DC.

NRC (National Research Council), 2010. Toward sustainable agricultural systems in the 21st century. The National Academies Press, Washington, DC.

NRC (National Research Council), 2011. Nutrient requirements of fish and shrimp. The National Academies Press, Washington, DC.

NRC (National Research Council), 2012. Nutrient requirements of swine. The National Academies Press, Washington, DC.

Noblet, J., C. Karege, S. Dubois and J van Milgen, 1999. Metabolic utilization of energy and maintenance requirements in growing pigs: effect of sex and genotype. J. Anim. Sci. 77, 1208-1216.

Noblet, J., X.S. Shi and S. Dubois, 1993. Metabolic utilization of dietary energy and nutrients for maintenance energy requirements in sows: basis for a net energy system. Brit. J. Nutr. 70, 407-419.

Noblet J. and J. van Milgen, 2004. Energy value of pig feeds: Effect of pig body weight and energy evaluation system. J. Anim. Sci. 82 (E. Suppl.), E229-E238.

Ojano-Dirain, C. P., M. Iqbal, D. Cawthon, S. Swonger, T. Wing, M. Cooper and W. Bottje, 2004. Determination of mitochondrial function in broilers with low and high feed efficiency. Poultry Sci. 83, 1394-1403.

Peganova, S. and K. Eder. 2003. Interactions of various supplies of isoleucine, valine, leucine and tryptophan on the performance of laying hens. Poultry Sci. 82, 100-105.

Perez-Laspiur, J., J. L. Burton, P. S. D. Weber, J. Moore, R. N. Kirkwood and N. L. Trottier, 2009. Dietary protein intake and stage of lactation differentially modulate amino acid transporter mRNS abundance in porcine mammary tissue. J. Nutr. 139, 1677-1684.

Pettigrew, J.E., M. Gill, J. France and W.H. Close. 1992. A mathematical integration of energy and amino acid metabolism of lactating sows. J. Anim. Sci. 70, 3742-3761

Pirgozliev, V. and S.P. Rose, 1999. Net energy systems for poultry feeds: A quantitative review. World's Poultry Sci. J. 55, 23-36.

Pomar, C., I. Kyriazakis, G.C. Emmans, and P.W. Knap, 2003. Modeling stochasticity: Dealing with populations rather than individual pigs. J. Anim Sci. 81, E178-186E.

Rakhshandeh, A., and C.F.M. de Lange, 2011. Immune system stimulation in the pig. Pages 31-46 in R.J. van Barneveld (Editor) Manipulating pig production XIII. Australasian Pig Science Association Inc. Werribee, Victoria, Australia.

Richard, M.P., R.W. Rosebrough, C.N. Coon and J.P. McMurtry, 2010. Feed intake regulation for the female broiler breeder: In theory and in practice. Poultry Sci. 19, 182-193.

Rivera-Ferre, M. G.; J.F. Aguilera, J.F. and R. Nieto, 2006. Differences in whole-body protein turnover between Iberian and Landrace pigs fed adequate or lysine-deficient diets. J. Anim. Sci. 84, 3346-3355.

Rutherfurd, S.M. and P.J. Moughan, 2012. Available versus digestible dietary amino acids. Brit. J. Nutrition. 108, S306-S314.

Stein, H., M. Fuller, P.J. Moughan, B. Seve and C.F.M. de Lange, 2007. Amino acid availability and digestibility in pig feed ingredients: Terminology and application. J. Anim. Sci. 85, 172-180.

Sun, P, D. Li, B. Dong, S. Qiao, and X. Ma, 2008. Effects of soybean glycinin on performance and immune function in early weaned pigs. Arch. Anim. Nutr. 62:313-21.

Tamminga, S., H. Schulze, J. van Bruchem and J. Huisman, 1995. The nutritional significance of endogenous N-losses along the gastro-intestinal tract of farm animals. Arch. Anim. Nutr. 48, 9-22.

Torrallardona, D. and E. Roura (Editors), 2009. Voluntary feed intake in pigs. Wageningen, The Netherlands: Wageningen Academic Publishers.

Van den Borne, J.J.G.C., J.W. Schrama,.J.W. Heetkamp, M.W.A. Verstegen and W.J.J. Gerrits, 2006. Synchronizing the availability of amino acids and glucose decreases fat retention in heavy pre-ruminant calves. J. Nutr. 136, 2181-2187.

Van den Borne, J.J.G.C., J.W. Schrama, J.W. Heetkamp, M.W.A. Verstegen and W.J.J. Gerrits, 2007. Synchronizing the availability of amino acids and glucose increases protein retention in pigs. Animal 1, 666-674.

Van Milgen, J. 2002. Modelling biochemical aspects of energy metabolism in mammals. J. Nutr. 132, 2154-2162.

Van Milgen, J., Bernier, J.F., Lecozler, Y., Dubois, S. and J. Noblet, 1998. Major determinants of fasting heat production and energetic cost of activity in growing pigs of different body weight and breed/castration combination. Brit. J. Nutr. 79, 1-9.

Van Milgen, J., J. Noblet and S. Dubois, 2001. Energetic efficiency of starch, protein and lipid utilization in growing pigs. J. Nutr. 131, 1309-1318.

Van Milgen, J., J. Noblet, A. Valancogne, S. Dubois and J.Y. Dourmad, 2008. InraPorc: a model and decision support tool for the nutrition of growing pigs. Anim. Feed Sci. and Techn. 143, 387-405.

Westerterp-Plantenga, M.S., S.G. Lemmens and K.R. Westerterp, 2012 Dietary protein – its role in satiety, energetics weight loss and health. Brit. J. Nutr. 108, S-105-112.

Weurding, R. E., H. Enting and M.W.A. Verstegen, 2003. The effect of site of starch digestion on performance of broiler chickens. Anim. Feed Sci. Techn. 110, 175-184.

Weis, R.N., S.H. Birkett, P.C.H. Morel and C.F.M. de Lange, 2004. Independent effects of energy intake and body weight on physical and chemical body composition in growing entire male pigs. J. Anim. Sci. 82, 109-121.

Williams, N.H., T.S. Stahly and D. R. Zimmerman, 1997. Effect of chronic immune system activation on the rate, efficiency, and composition of growth and lysine needs of pigs fed from 6 to 27 kg. J. Anim. Sci. 75, 2463-2471.

Yáñez, J.L., E. Beltranena, M. Cervantes and R.T. Zijlstra, 2011. Effect of phytase and xylanase supplementation or particle size on nutrient digestibility of diets containing distillers dried grains with solubles cofermented from wheat and corn in ileal-cannulated grower pigs. J. Anim. Sci. 89, 113-123.
Yen, T.J. 1997. Oxygen consumption and energy flux of porcine splanchnic tissues. Pages 260-273 in J.P. Laplace, C. Fevrier and A. Barbeau. (Editors). Digestive Physiology in pigs. EAAP Publication no. 88. INRA, France.
Young, J.M. and J.C.M. Dekkers, 2012. The genetic and biological basis of residual feed intake as a measure of feed efficiency. Pages 153-160 in J.F. Patience (Editor) Feed efficiency in Swine. Wageningen Academic Publishers, Wageningen, The Netherlands.
Zhu, C.L., M. Rademacher and C.F.M. de Lange, 2005. Increasing dietary pectin level reduces utilization of digestible threonine intake, but not lysine intake, for body protein deposition in growing pigs. J. Anim. Sci. 83, 1044-1053.
Zijlstra, R.T., R. Jha, A.D. Woodward, J. Fouhse and T.A.T.G. van Kempen, 2012. Starch and fiber properties affect their kinetics of digestion and thereby digestive physiology in pigs. J. Anim. Sci. 2012, 90 (Suppl. 4), 49-58.
Zijlstra, R.T., A. Owusi-Asiedu and P.H. Simmins, 2010. Future of NSP-degrading enzymes to improve nutrient utilization of co-products and gut health in pigs. Livest. Sci. 134, 255-257.

Effects of low birth weight and 3 wk feed restriction on energy metabolism in growing pigs

R. Krüger, S. Görs, K. Martens, M. Derno, B.U. Metzler-Zebeli, C. Nebendahl, H.M. Hammon and C.C. Metges
Institute of Nutritional Physiology 'Oskar Kellner', Leibniz Institute for Farm Animal Biology (FBN), 18196 Dummerstorf, Germany; metges@fbn-dummerstorf.de

Introduction

Intrauterine growth retardation (IUGR) has been shown to affect intestinal growth, resulting in impaired feed utilization, growth performance, and development of skeletal muscle (Quiniou *et al.*, 2002; Wu *et al.*, 2006; Michiels *et al.*, 2013). This leads to altered muscle and fat proportions (Powell and Aberle, 1980; Rehfeldt *et al.*, 2006). Temporary feed restriction (FR) in pigs is associated to physiological changes with respect to energy expenditure (EE) and nutrient oxidation pattern during and after FR compared to adequate dietary conditions (Chwalibog *et al.*, 2004). We explored possible changes in energy metabolism of IUGR pigs before, during and after a 3 wk period of FR, compared to pigs with normal birth weight.

Material and methods

After weaning (d28) female German Landrace pigs with low (L; 0.8-1.2 kg; n=22) or normal birth weight (N; 1.3-1.8 kg; n=24) were fed *ad libitum* until the age of d 79 according to recommendations. The diet consisted of commercial starter feed (46%), oat flakes (34%) and sucrose (20%). Subsequently, half of L and N pigs were continued on *ad libitum* feeding (control C; LC and NC) with grower feed (68%), oat flakes (21%) and sucrose (11%) until d 130. The remaining pigs were fed restrictively (R; LR and NR) from d 80-100 (57% of *ad libitum* intake of C; grower feed) before they were refed *ad libitum* from d 101-130.

Body weight (BW) and feed intake (FI) were monitored. Using indirect calorimetry energy expenditure (EE; kJ/kg $BW^{0.62}$), respiratory quotient (RQ), resting metabolic rate (RMR; kJ/kg $BW^{0.62}$), carbohydrate oxidation (COX; g/kg $BW^{0.62}$, g/kg $BW^{0.62}$/kg FI), and fat oxidation (FOX; g/kg $BW^{0.62}$, g/kg $BW^{0.62}$/kg FI) were calculated on the basis of 24h O_2 consumption and 24h CO_2 production. Energy efficiency (Q) was calculated as the quotient of EE (MJ) and energy intake (MJ) to compare LC and NC pigs. Measurements were performed 4 d before (T1) and 4 d after (T2a) beginning of FR, 4 d before (T2b) and 4 d after (T3a) end of FR, and 25 d after beginning of refeeding (T3b). Statistical analyses were performed with SAS using PROC MIXED procedure and Tukey-Kramer post-hoc test ($P \leq 0.05$).

Results and discussion

From birth to d 131 L pigs were lighter than N pigs (d 131, 65 vs. 70 kg; $P=0.012$), although FI was permanently higher in L pigs (41 vs. 38 g/kg BW on d129; $P=0.04$), but energy efficiency (Q-value) between L and N pigs did not differ.

LC pigs generally exhibited lower FOX (i.e. higher fat synthesis) than NC pigs, irrespective of age (LC vs. NC: -3.5 vs. 0.0 g/kg $BW^{0.62}$; $P=0.02$). Independent of birth weight FR resulted in increased FOX at T2a and T2b to cover energy demands (T2a, R vs. C: 5.7 vs. 0.7, $P<0.001$; T2b, R vs. C: 4.4 vs. -0.1 g/kg $BW^{0.62}$, $P=0.025$), accompanied by decreased COX as a consequence of reduced carbohydrate intake (T2a, R vs. C: 45 vs. 59, $P<0.001$; T2b, R vs. C: 44 vs. 64 g/kg $BW^{0.62}$, $P<0.001$). The RMR (T2b) decreased in order to defend BW (Figure 1). When refed (T3a) EE and RMR of R pigs increased up to the levels of C pigs, whereas COX in R pigs exceeded the level of

C pigs (R vs. C: 87 vs. 72 g/kg $BW^{0.62}$, P=0.001) due to higher FI after FR (41 vs. 43 g feed/kg $BW^{0.62}$, C vs. R; P=0.03). In R pigs COX related to FI was also higher (T3a, R vs. C: 118 vs. 98 g feed/kg $BW^{0.62}$, P<0.001), suggesting glucose utilization to provide NADPH for fat synthesis. At the same time FOX in R pigs decreased below the level of C pigs, indicating replenishment of fat depots (T3a, R vs. C: -10 vs. -4 g/kg $BW^{0.62}$, P=0.001).

While FR acutely affected components of energy metabolism, IUGR pigs had higher COX and lower FOX values (both also in relation to FI) compared to N pigs. The magnitude of FR and refeeding effects in N and L pigs was similar. However, both FR and IUGR induced changes of components of energy metabolism were not persistent after realimentation.

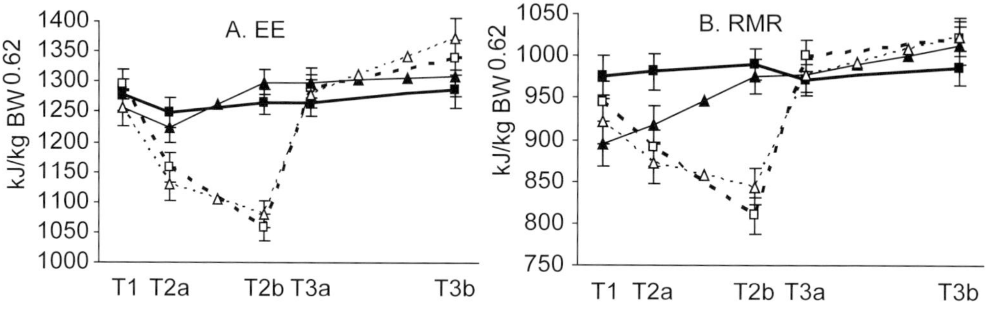

Figure 1. 24h EE (A) and RMR (B) before (T1), during (T2a and T2b) and after (T3a and T3b) FR in pigs with normal or low birth weight compared to ad libitum fed pigs; LSMeans ± SE; n=7-12; NC —■—, NR --□--, LC —▲—, LR ---Δ---

Acknowledgements

The study was supported by the German Ministry of Education and Research within the framework of BioMed-Ernährungsforschung – für ein gesundes Leben, Modul: Biomedizinische Ernährungsforschung‘ (Vision Epifood, FKZ 0315397B).

References

Chwalibog, A., A.H. Tauson, and G. Thorbek, 2004. Energy metabolism and substrate oxidation in pigs during feeding, starvation and refeeding. J. Anim. Physiol. Anim. Nutr. 88, 101-112.

Michiels, J., M. De Vos, J. Missotten, A. Ovyn, S. De Smet, and C. Van Ginneken, 2013. Maturation of digestive function is retarded and plasma antioxidant capacity lowered in fully weaned low birth weight piglets. Br. J. Nutr. 109(1), 65-75.

Powell, S.E. and E.D. Aberle, 1980. Effects of birth weight on growth and carcass composition of swine. J. Anim. Sci. 50(5), 860-868.

Quiniou, N., J. Dagorn, and D. Gaudré, 2002. Variation of piglets’ birth weight and consequences on subsequent performance. Livestock Prod. Sci. 78(1), 63-70.

Rehfeldt, C. and G. Kuhn, 2006. Consequences of birth weight for postnatal growth performance and carcass quality in pigs as related to myogenesis. J. Anim. Sci. 84 (Suppl.), E113-E123.

Wu, G., F.W. Bazer, J.M. Wallace, and T.E. Spencer, 2006. Board invited review: Intrauterine growth retardation: Implications for the animal sciences. J. Anim. Sci. 84, 2316-2337.

Influence of feeding level and energy source on lysine requirements in growing pigs

E.M.A.M. Bruininx[1,2], W.J.J. Gerrits[1], I. Eising[1], P. Vervenne[2], P. Sakkas[1], and J.J.G.C. van den Borne[1]
[1]Animal Nutrition Group, Wageningen University, P.O. Box 338, 6700 AH Wageningen, the Netherlands; joost.vandenborne@wur.nl
[2]Agrifirm Innovation Center, P.O. Box 20018, 7302 HA Apeldoorn, the Netherlands

Introduction

In swine nutrition, minimizing the oxidation of imbalanced amino acids is an important issue in optimizing the use of feed resources. Numerous titration studies have been conducted to determine the optimum lysine-to-energy ratio, increasingly using within-animal titration approaches. Whether or not amino acid requirements depend on the level of feed intake has been subject of debate, with some studies indicating lower lysine requirements with increasing starch intake (Kampman-Van de Hoek *et al.*, 2013), whereas others found no change with feeding level (Bikker *et al.*, 1994).

Starch and sugars are the predominant energy sources in most swine diets, but dietary fat can also be an important energy source. Beneficial effects of high fat levels on the digestion and utilization of protein in pigs have been reported. Protein digestibility increased (Li and Sauer, 1994) and digestible N was retained more efficiently (Berschauer *et al.*, 1983) in response to an iso-energetic exchange of starch by fat. These interactions of energy source with protein metabolism indicate that the optimum lysine-to-energy ratio may depend on the energy source used in swine diets. In addition, it is important to establish whether such relationship would depend on feed intake. The aim of the current study was therefore to determine the effects of energy source (lard vs. starch) and feeding level on lysine requirements in growing pigs using a novel within-animal amino acid titration technique (Kampman-Van de Hoek *et al.*, 2013).

Material and methods

In a 2×2 factorial design, 28 crossbred boars were assigned to one of two feeding levels (FL, 2.2 vs. 2.7 × metabolizable energy requirements for maintenance; ME_m, 458 kJ ME/(kg$^{0.75}$ × d)) and to one of two diets differing in major dietary energy source (ES, starch vs. lard). Starch and fat were iso-energetically exchanged on a digestible energy (DE) basis, i.e. 250 g corn starch was substituted for 104 g lard. A within-animal lysine titration technique was used to assess the responses to changes in lysine-to-energy ratio. After a 14-d adaptation period to housing and feeding conditions, the apparent ileal digestible lysine (AIDLys) supply to pigs was decreased from 1.74 to 0.53 g/MJ DE in 9 equidistant steps of 3 d each with a step size of 15 mg/MJ DE. Urine was quantitatively collected in 24-h intervals and was analyzed for N. Feces were collected over a period of 27 d to calculate apparent total tract digestibility. Retention of digestible N was expressed relative to digestible N intake, indicating the efficiency of digestible N utilization for protein deposition. A linear-plateau model (Koops and Grossman, 1993) was fitted to the N retention data against AIDLys intake (in g/MJ DE) for each pig individually. Data were also analyzed by linear regression procedures. When the non-linear model described the response of N retention to changes in AIDLys intake better ($P<0.1$) than the linear model, lysine requirement could be estimated. Effects of FL, ES and their interaction on parameter estimates were tested by the GLM procedure of SAS.

Results

The linear-plateau model described the response of N retention to changes in AIDLys intake better ($P<0.1$) than the linear model for 25 out of 28 pigs. The estimated AIDLys requirement for these 25 pigs varied from 0.75 to 0.98 g AIDLys/MJ DE but was not affected by FL or ES (Table 1). Other curve fitting parameter estimates were not affected by FL or ES.

Apart from a tendency for starch digestibility there were no interactions between ES and FL. Total tract digestibility of energy was 2.2%-units higher ($P<0.001$; data not shown) for the starch diet than for the fat diet, whereas it tended ($P=0.07$) to be 0.8%-units lower in the 2.7 × ME_m group than in the 2.2 × ME_m group. Total tract protein digestibility tended ($P=0.10$; data not shown) to be 0.9%-units lower in the 2.7 × ME_m than in the 2.2 × ME_m group, but did not differ between the starch and fat diet. The conversion of digestible energy into body weight gain tended ($P=0.07$; data not shown) to be 4% less efficient in pigs fed the fat diet.

Table 1. Effects of feeding level (2.2 vs. 2.7 × ME_m) and energy source (lard vs. starch) on parameter estimates (+ SEM) from a linear-plateau model describing the relation between AIDLys supply (g/MJ DE) and N retention (% of digestible N intake) in growing pigs.

	Feeding level		*P*-value	Energy source		*P*-value
	2.2 × ME_m	2.7 × ME_m		Lard	Starch	
No. of pigs	13	12		13	12	
Intercept, %	11.2±2.0	11.8±2.1	0.821	11.7±2.0	11.3±2.1	0.895
Slope[1]	66.3±3.0	66.2±3.1	0.976	65.5±3.0	67.0±3.1	0.737
Transition point[2]	0.84±0.02	0.86±0.02	0.403	0.85±0.02	0.85±0.02	0.938
Plateau, %	66.7±1.2	68.4±1.3	0.355	67.5±1.2	67.6±1.3	0.949

[1] The slope represents the incremental N retention (% of digestible N intake) per g AIDLys/MJ DE.

[2] The transition point represents the AIDLys requirement in g/MJ DE.

Discussion and conclusions

In line with a study in female pigs (Bikker *et al.*, 1994) our results show that the lysine requirements, expressed as AIDLys relative to DE intake, do not depend on the level of feed intake in growing male pigs. Although other studies (Berschauer *et al.*, 1983; Bruininx *et al.*, 2011) have indicated that post-absorptive effects may contribute to an increased N utilization for growth when feeding high-fat diets, the current study shows that dietary N utilization and lysine requirements in growing pigs are not affected by energy source.

References

Bikker, P., M.W.A. Verstegen, R.G. Campbell, and B. Kemp, 1994. Digestible lysine requirement of gilt with high genetic potential for lean gain, in relation to the level of energy intake. J. Anim. Sci. 72, 1744-1753.

Berschauer, F., U. Ehrensvard, G. Gaus, and K.H. Menke, 1983. Untersuchungen über ernährungphysiologische Wirkungen von Futterfetten in Rationen für wachsende Schweine. Arch. Tierernahr. 33, 683-711.

Bruininx, E.M.A.M., J.J.G.C. van den Borne, E. van Heugten, J. van Milgen, M.W.A. Verstegen, and W.J.J. Gerrits, 2011. Oxidation of dietary stearic, oleic and linoleic acids in growing pigs follows a biphasic pattern. J. Nutr. 141, 1657-1663.

Kampman-van de Hoek, E., W.J.J. Gerrits, C.M.C. van der Peet-Schwering, A.J.M. Jansman, and J.J.G.C. van den Borne, 2013. A simple amino acid dose-response technique to quantify amino acid requirements of individual meal-fed pigs. Submitted to J. Anim. Sci.

Koops, W., and M. Grossman, 1993. Multiphasic allometry. Growth Dev. Aging. 57, 183-192.

Li, S., and W.C. Sauer, 1994. The effect of dietary fat content on amino acid digestibility in young pigs. J. Anim. Sci. 72, 1737-1743.

The use of free amino acids in piglet diets allows the formulation of very low crude protein diets

M. Gloaguen[1,2], N. Le Floc'h[1,2], Y. Primot[3], E. Corrent[3] and J. van Milgen[1,2]
[1]INRA, UMR1348 PEGASE, 35590 Rennes, France; jaap.vanmilgen@rennes.inra.fr
[2]Agrocampus Ouest, UMR1348 PEGASE, 35000 Rennes, France
[3]Ajinomoto Eurolysine s.a.s, 75817 Paris Cedex 17, France

Introduction

Reducing the dietary crude protein (CP) content using free amino acids (AA) enables improving the efficiency of nitrogen utilization while maintaining performance of pigs as long as AA requirements are met. Valine, Ile, His and Leu requirements have been recently estimated (Barea *et al.*, 2009; Van Milgen *et al.*, 2012; Gloaguen *et al.*, 2012), which makes it possible to formulate diets with a very low CP content. The objective of this study was to evaluate to what extent the CP content of diets can be reduced by substituting soybean meal by wheat, barley, and free AA, without affecting performance.

Material and methods

A trial was conducted to measure performance of 84 pigs allotted to six levels of dietary CP. Four diets were based on cereals and soybean meal and provided 17.6, 15.6, 13.5, and 11.8% CP. Two additional low-CP diets without soybean meal were formulated which provided 13.0 and 14.0% CP. For the 13.0% CP diet, L-Arg, L-Pro and L-Gly were added to provide the same level of SID Pro:Lys and (Gly + Ser):Lys as the 15.6% CP diet. L-Glutamate was added as a N source to attain 13.0% CP. In the 14.0% CP diet, L-Glu was added to attain 14.0% CP. All diets were formulated to provide 1.00% standardized ileal digestible (SID) Lys. Different levels of AA supplementation were used to match the requirement of SID Thr:Lys (65%), (Met+Cys):Lys (60%), Trp:Lys (22%), Val:Lys (70%), Ile:Lys (51%), Leu:Lys (100%), His:Lys (32%), Phe:Lys (61%), (Phe + Tyr):Lys (95%), and Arg:Lys (42%). Diets were formulated to provide 9.2 MJ NE/kg, 4.0 g digestible P/kg, 2.9% Ca/digestible P, and an electrolyte balance of 180 mEq/kg.

Six-week-old barrows and female piglets (Pietrain × (Large White × Landrace)) were used. At 5 wk of age, pigs were blocked by sex, BW, and origin (siblings or half-siblings) and housed individually. Piglets within a block (n=6) were allotted to the different dietary treatments. The prestarter diet was gradually replaced by the experimental diets at 10 days post-weaning so that from 12 days post-weaning onwards, pigs were offered the experimental diets only. Pigs had free access to water and feed. The experiment lasted 21 d. Pigs were weighed at the beginning and at the end of the experimental period after an overnight fast for calculation of average daily gain (ADG). Feed intake and feed refusals were measured weekly for calculation of average daily feed intake (ADFI).

Results and discussion

Dietary CP level did not influence ADFI. Lowering the dietary CP content from 17.6 to 13.5% had no effect on ADG (Table 1). However, ADG and feed efficiency were lower for pigs fed the 11.8% CP diet compared to pigs fed the other diets ($P<0.05$). In diets formulated without soybean meal, feed efficiency was lower in pigs offered the 13.0% CP diet than in those offered the 14% CP diet ($P<0.05$). The addition of L-Glu to a very low-CP diet restored feed efficiency to levels observed with higher CP contents. The results of this study show that the CP content of a cereal-soybean meal diet, containing 1.00% SID Lys and supplemented with free Lys, Thr, Met, and Trp, can be decreased by 4 percentage units with the addition of free Val, Ile, Leu, His, and Phe without affecting performance in 10-20 kg pigs. Soybean meal can be totally replaced with cereals and free AA if the CP content of the diet is at least 14%. Therefore, SID levels of 51% Ile:Lys, 100% Leu:Lys, 32%

Table 1. Effect of reducing dietary CP content on performance of pigs.

CP, %	Cereals – SBM – AA				Cereals – AA		Statistics	
	17.6	15.6	13.5	11.8	13.0	14.0	RSD[2]	*P*-value
BW_i, kg	12.7	12.7	12.6	12.6	13.0	12.8	0.6	0.60
BW_f, kg	22.2[a]	22.2[a]	21.9[a]	20.1[b]	21.8[a]	22.3[a]	1.7	<0.01
ADFI, g/d	766	775	779	734	810	782	96	0.55
ADG, g/d	450[a]	454[a]	442[a]	358[b]	420[a]	451[a]	67	<0.01
G:F, g/g	0.59[a]	0.59[a]	0.57[a]	0.49[b]	0.52[b]	0.58[a]	0.06	<0.01

[a-b] Within a row, means without a common superscript differ ($P<0.05$); RSD = residual standard deviation.

His:Lys, 61% Phe:Lys, 95% (Phe+Tyr):Lys, and 42% Arg:Lys appear to be sufficient to maximize growth. The 13.0% CP diet resulted in a reduced feed efficiency. The reduced performance may be due to an N deficiency because the addition of L-Glu to reach 14.0% CP restored feed efficiency. An N deficiency may also explain the reduced feed efficiency observed with the soybean meal based diet with 11.8% CP. The results also indicate that the efficiency of utilizing AA is not lower for free AA compared with protein-bound AA in pigs offered feed ad libitum. In agreement with our results, Le Bellego *et al.* (2001) showed that the N utilization is not affected in low CP diets when at least two meals are fed.

In conclusion, the use of free Val, Leu, Ile, His, and Phe enables a 4 percentage unit reduction in CP content in diets for piglets and soybean meal can be totally replaced with cereals and free AA. However, in these pigs at least 14% CP level is required to ensure the N supply. At very low CP levels, N is a potentially limiting factor for growth in piglets.

References

Barea, R., L. Brossard, N. Le Floc'h, Y. Primot, D. Melchior and J. van Milgen, 2009. The standardized ileal digestible valine-to-lysine requirement ratio is at least seventy percent in postweaned piglets. J. Anim. Sci. 87, 935-947.

Gloaguen M., Le Floc'h N., Primot Y., Corrent E. and van Milgen J., 2012. Response of piglets to the standardized ileal digestible isoleucine, histidine and leucine supply in cereal-soybean meal-based diets. Animal, FirstView, 1-8.

Le Bellego, L., J. van Milgen, S. Dubois, and J. Noblet. 2001. Energy utilization of low-protein diets in growing pigs. J. Anim. Sci. 79, 1259-1271.

Van Milgen, J. *et al.* 2012. Meta-analysis of the response of growing pigs to the isoleucine concentration in the diet. Animal 6, 1601-1608.

Lysine concentration in dietary protein affects performance and pattern of nutrient retention of weaned Iberian piglets

R. Barea, L. Lara, J.F. Aguilera and R. Nieto
Institute of Animal Nutrition, Estación Experimental del Zaidín, Spanish Council for Scientific Research (CSIC), Camino del Jueves s/n, 18100 Armilla, Granada, Spain; rosa.nieto@eez.csic.es

Introduction

The amino acid composition of dietary protein, specifically the essential amino acid composition, is a key factor that affects the efficiency of use of dietary protein and, therefore, growth and protein retention. The balance among essential amino acid for maintenance and production functions has been established in pigs, with some variations according to the physiological state of the animal (BSAS, 2003). Pig genotype is one the factors that might influence such balance, although limited information is available in the literature concerning this issue. The Iberian pig is an obese, slow-growing breed. In previous work we have shown that Iberian pig requirements for total protein differ markedly from those of pigs of conventional genotypes (Nieto *et al.*, 2012), although dietary protein was formulated following the amino acid pattern (g amino acid/kg crude protein) established for conventional pigs. The present work aimed at establishing the dietary lysine (Lys) requirements (g Lys/kg crude protein) for optimum performance of post-weaned Iberian piglets.

Material and methods

Sixty six castrated male Iberian piglets of 10 kg bodyweight (BW) and 40 days of age were used. Piglets were fed one of six experimental diets which differed in their Lys content (55, 60, 65, 70, 75 and 80 g Lys/kg crude protein). L-Lys. HCl was added in increasing concentration to a basal diet containing corn (42%), barley (16%) and soybean meal (30.5%) at the expense of corn starch. Diets were isonitrogenous and isoenergetic (170 g CP and 14.1 MJ ME per kg dry matter). At the start of the trial, six piglets were slaughtered to estimate initial body composition. Ten piglets were allocated into each experimental diet. They were fed twice daily ad libitum and BW was recorded weekly. Piglets were individually housed in an environmentally controlled room (27±1.5 °C) until they reached 25 kg BW when six pigs from each dietary treatment were slaughtered. At slaughter blood, carcass and organs were collected and weighed separately. The carcass was ground, homogenized and freeze-dried for analysis. All analyses have been performed in duplicate. Treatment effects were assessed by analysis of variance using the GLM procedure of SAS. Orthogonal polynomial contrasts were used to determine linear and quadratic effects of Lys addition on performance, carcass nutrient retention and blood metabolites. The experimental protocol was approved by the CSIC Bioethical Committee.

Results and discussion

Average daily feed intake was not affected by Lys supply (ADFI, 890 g dry matter/d; $P>0.05$). Both, daily gain (ADG, g/d) and gain-to-feed ratio (ADG:ADFI) increased linearly with increasing Lys concentration (Table 1; $P<0.001$) and showed no significant quadratic effects. Linear effects of increasing Lys concentration on dietary CP were also detected for ADG expressed per unit of ME intake ($P<0.001$). Carcass gain composition showed that protein increased (linearly, $P<0.001$; quadratically, $P<0.05$) with increasing dietary Lys concentration, reaching maximum values (141 to 146 g protein/kg carcass gain) with diets providing 70 to 80 g Lys/kg CP. Carcass water retention followed a similar pattern as protein, whereas fat concentration in carcass gain decreased linearly ($P<0.01$) and tended to decrease quadratically ($P=0.078$) on increasing dietary Lys, with maximum values for diets providing 55 and 60 g/ Lys kg dietary CP. No effect of dietary Lys was observed for ash carcass gain. Plasma urea concentration tend to decrease with increasing dietary Lys ($P=0.064$ and $P=0.10$ for linear and quadratic effects, respectively). No effect of dietary Lys on the other blood parameters analysed (glucose, creatinine and triglycerides) was detected. The results obtained suggests that optimum growth and carcass protein retention in Iberian piglets can be achieved when dietary protein supply contain at least 70 g Lys/kg CP.

Table 1. Effect of lysine concentration of dietary protein on piglet performance, carcass (CC) nutrient retention and plasma urea concentration.

	g Lys/kg dietary crude protein							*P*-value[1]	
	55	60	65	70	75	80	SEM	Lin	Q
Average daily gain, g	387	433	435	468	469	473	8	0.001	0.249
Gain:feed, g/g	0.450	0.489	0.489	0.512	0.522	0.521	0.005	0.001	0.177
Gain:ME intake, g/MJ	31.8	34.5	34.5	36.2	36.9	36.7	0.4	0.001	0.177
CC protein gain, g/kg	125	132	136	146	141	141	1	0.001	0.044
CC water gain, g/kg	547	580	598	597	590	619	4	0.001	0.200
CC fat gain, g/kg	300	261	239	231	242	214	5	0.001	0.078
CC energy gain, MJ/kg	14.9	13.5	12.7	12.7	13.0	11.9	0.2	0.001	0.128
CC ash gain, g/kg	28.6	26.2	27.7	26.4	26.9	26.4	0.8	0.549	0.772
Plasma urea, mg/100 ml	24.2	22.6	19.1	18.9	20.5	18.7	0.6	0.064	0.100

[1] Contrast: Lin=linear effects; Q=quadratic effects.

Acknowledgements

Funding by the Spanish MINECO (grants ref. AGL2008-0255 and AGL2011-25360) is gratefully acknowledged.

References

BSAS. 2003. Nutrient Requirement Standards for Pigs. Brit. Soc. Anim. Sci., Penicuik, UK, 28 pp.

Nieto, R., L. Lara, R. Barea, R. García-Valverde, M.A. Aguinaga, J.A. Conde-Aguilera and J.F. Aguilera. 2012. Response analysis of the Iberian pig growing from birth to 150 kg body weight to changes in protein and energy supply. J. Anim. Sci. 90, 3809-3820.

Effect of processing of oilseed meals on the apparent ileal protein digestibility and performance in pigs

T.G. Hulshof[1,2], P. Bikker[1] and A.F.B. van der Poel[2]
[1]Wageningen UR Livestock Research, Edelhertweg 15, 8219 PH, Lelystad, the Netherlands; tetske.hulshof@wur.nl
[2]Animal Nutrition Group, Wageningen University, De Elst 1, 6708 WD, Wageningen, the Netherlands

Introduction

Feed ingredients, e.g. oil seed by-products, generally have undergone several processing steps before inclusion in animal diets. Processing of feed ingredients and diets can result in conformational changes of protein structure, the formation of Maillard reaction products and cross-links between amino acids (Bender, 1972). These changes affect protein quality and the latter two reactions especially reduce the amount of the essential amino acid lysine (Mauron, 1990). The effects on protein quality may result in an impaired pig performance presumably related to a decrease in protein, amino acid and lysine digestibility as found for processed diets (González-Vega *et al.*, 2011). The aims of this research were to determine the effects of processing of two oil seed by-products, i.e. soybean meal (SBM) and rapeseed meal (RSM), on apparent ileal crude protein, amino acid and lysine digestibility and performance in growing pigs and to derive criteria to evaluate protein quality in heat treated ingredients. First results of ileal crude protein digestibility are presented here.

Material and methods

Ten barrows (initial BW 24.8±0.28 kg) were suited with a SICV cannula (Mroz *et al.*, 1996) and individually housed in metabolism cages. Four experimental diets (Table 1) were used, consisting of a basal N-free diet supplemented with 35% (un)processed SBM or RSM in combination with 7 or 5% of the sugar rich compound Xylig^{TM}, respectively. SBM and RSM were used as purchased or toasted after mixing with Xylig^{TM}, in order to induce changes in protein quality. The RSM diets were supplemented with synthetic lysine, threonine and tryptophan to levels corresponding to the levels in the SBM diets. In all diets Cr_2O_3 was added as indigestible marker to calculate the apparent ileal crude protein digestibility (AID CP). The experiment consisted of a cross-over design with three periods of 14 days. Feed was provided twice a day at 2.8 times maintenance energy. Ileal chyme was collected during 12 hours on days 9 and 11 of each period and analysed for CP, amino acids and Cr_2O_3. The gain:feed (GF) ratio was calculated as measure for nutrient utilization. The proc mixed procedure in SAS (2008) was used to test for the fixed effects of feed ingredient, processing

Table 1. Nutrient composition (g/kg) of the experimental diets containing either unprocessed or toasted soybean meal (SBM) or rapeseed meal (RSM).

Nutrient	SBM		RSM	
	Unprocessed	Toasted	Unprocessed	Toasted
Dry matter	888	895	888	893
Ash	45	46	39	40
Crude protein	168	170	121	125
Crude fibre	12	11	43	47
Crude fat	30	27	32	35
Starch	423	423	433	436
Sugars	142	136	135	129

treatment, and their interaction on AID CP and GF ratio, with period and animal in the random statement. *P*-values <0.05 were assumed to be significant.

Results and discussion

The results of the AID CP and GF ratio are given in Table 2. The interaction between feed ingredient and processing treatment was not significant for AID CP (P=0.40) and GF ratio (P=0.68) indicating that processing has similar effects on both characteristics for SBM and RSM diets. The animals fed the RSM diets had a significantly lower AID CP than the animals fed the SBM diets. This agrees with other studies indicating that RSM in general has a lower AID CP than SBM (Grala *et al.*, 1998). The type of feed ingredient did not influence the GF ratio despite the lower AID CP content, presumably because of the supplementation of the RSM diets with synthetic lysine, threonine and tryptophan. Processing of the SBM and RSM reduced the AID CP as well as the GF ratio. The lower efficiency of nutrient utilization for tissue deposition as indicated by the reduction in GF ratio might be caused by the reduction in protein digestibility. In conclusion, additional processing of SBM and RSM, i.e. mixing with a sugar rich compound and toasting, reduced the crude protein digestibility and nutrient utilization as indicated by the GF ratio. The effects of individual amino acids will be further examined.

Table 2. Effect of feed ingredient, i.e. soybean meal (SBM) or rapeseed meal (RSM), and processing treatment, i.e. none or toasting, on apparent ileal crude protein digestibility (AID CP) and gain:feed (GF) ratio in pigs.

	SBM		RSM		SEM	P-value		
	Unprocessed	Toasted	Unprocessed	Toasted		Feed ingredient	Processing treatment	Interaction
AID CP	78.4^{c}	66.0^{a}	65.6^{a}	56.1^{b}	1.68	<0.0001	<0.0001	0.40
GF ratio	0.48^{a}	0.43^{ac}	0.45^{a}	0.38^{bc}	0.024	0.10	0.02	0.68

^{a,b,c} different superscripts in the same row indicate significantly (P<0.05) different values.

Acknowledgements

The authors gratefully acknowledge the financial support from the Wageningen UR 'IPOP Customized Nutrition' programme financed by Wageningen UR, the Dutch Ministry of Economic Affairs, Agriculture & Innovation, WIAS, Agrifirm Innovation Center, ORFFA Additives BV, Ajinomoto Eurolysine s.a.s and Stichting VICTAM BV.

References

Bender, A.E., 1972. Processing damage to protein food: a review. J. Food Technol. 7, 239-250.

González-Vega, J.C., B.G. Kim, J.K. Htoo, A. Lemme and H.H. Stein, 2011. Amino acid digestibility in heated soybean meal fed to growing pigs. J. Anim. Sci. 89, 3617-3625.

Grala, W., M. Verstegen, A. Jansman, J. Huisman and P. van Leeusen, 1998. Ileal apparent protein and amino acid digestibilities and endogenous nitrogen losses in pigs fed soybean and rapeseed products. J. Anim. Sci. 76, 557-568.

Mauron, J., 1990. Influence of processing on protein quality. J. Nutr. Sci. Vitaminol. 36, S57-S69.

Mroz, Z., G.C.M. Bakker, A.W. Jongbloed, R.A. Dekker, R. Jongbloed and A. van Beers, 1996. Apparent digestibility of nutrients in diets with different energy density, as estimated by direct and marker methods for pigs with or without ilea-cecal cannulas. J. Anim. Sci. 74, 403-412.

SAS, 2008. Release 9.2. SAS Inst. Inc., Cary, NC, USA.

Determination of the next limiting amino acid in young piglets

A.J.M. Jansman[1], H. van Diepen[1], M. Rovers[2] and E. Corrent[3]
[1]Wageningen UR Livestock Research, P.O. Box 65, 8200 AB Lelystad, the Netherlands; alfons.jansman@wur.nl
[2]Orffa Additives B.V., Vierlinghstraat 51, 4251LC Werkendam, the Netherlands
[3]Ajinomoto Eurolysine s.a.s. 153 Rue de Courcelles, 75817 Paris Cedex 17, France

Introduction

After lysine, methionine/cysteine, threonine, tryptophan, and valine it is not clear which essential amino acid becomes limiting in low protein diets for post weaning piglets. This is related to a scarcity of information on requirement for other essential amino acids in piglets. Isoleucine, leucine and histidine are among the candidates to be the next limiting amino acid (Chung and Baker, 1992; Barea *et al.*, 2009). The aim of the present study was to identify the next limiting amino acid after lysine, methionine/cysteine, threonine, tryptophan, and valine in diets for post weaning piglets in a performance study over a period of 4 weeks in the post-weaning period.

Material and methods

Six dietary treatments were evaluated in eight replicates (pens) with eight piglets (barrows) per pen, weaned at an age of about 28 d. A basal experimental diet with a low crude protein content (140 g/kg) was formulated based on maize, wheat, barley, soybean meal and whey powder. The diet was supplemented with free L-lysine, DL-methionine, L-threonine, L-tryptophan, L-valine and L-phenylalanine to meet the requirement values for these amino acids (10.4 g/kg standardized ileal digestible (SID) Lys; 6.2 g/kg SID methionine + cysteine; 6.7 g/kg SID threonine; 2.3 g/kg SID tryptophan; 7.3 g/kg SID valine and 6.3 g/kg SID phenylalanine; treatment I). The basal diet was supposed to be deficient in isoleucine, histidine and leucine for post-weaning piglets (4.1 g/kg SID isoleucine, 2.6 g/kg SID histidine and 9.3 g/kg SID leucine; 42, 26 and 89% relative to SID Lys, respectively). Treatment II received a diet composed of the basal diet supplemented with 1.2 g/kg L-isoleucine, 0.6 g/kg L-histidine and 1.2 g/kg L-leucine up to the assumed requirement levels of 53, 32 and 101% of the content of SID lysine (NRC, 1998). In three other dietary treatments the supplementation of either L-isoleucine, L-histidine or L-leucine was omitted in the respective diets for treatments III, IV and V resulting in diets containing 79, 81, and 88% of the assumed requirement value for isoleucine, histidine and leucine on a SID basis in the respective diets. A reference treatment (VI) was included fed a diet with a crude protein content of about 169 g/kg, sufficient in all essential amino acids. Except for amino acids, the diets were formulated to be nutritionally complete (CVB, 1996; NRC, 1998). In the week after weaning the diet for treatment VI was given to piglets in all treatments. The experimental diets in pelleted form were fed over a subsequent period of four weeks. Feed intake (FI), body weight gain (BWG) and feed conversion ratio (FCR) were determined over 2-weeks periods.

Results and discussion

Over week 1-2 and 3-4 of the experimental period FI, BWG and FCR were significantly affected by treatment ($P<0.05$) (Table 1). Compared to treatment I, FI and BWG were numerically higher in treatment II in weeks 1-2 and 3-4 ($P>0.05$) and FCR numerically improved in weeks 3-4 ($P>0.05$). BWG and FCR were improved for treatment VI compared to treatments I and II in week 1-2 ($P<0.05$) and compared to treatment I in week 3-4 ($P<0.05$).

Over week 1-2, omitting free L-isoleucine supplementation in treatment III resulted in a lower FI and BWG compared to treatment II ($P<0.05$), while the FCR was not different. Omitting the

Table 1. Feed intake, body weight gain and feed conversion ratio over weeks 1-2 and 3-4 of the experimental period (weeks 0-4).

	FI (kg/d)		BWG (g/d)		FCR	
	wk 1-2	wk 3-4	wk 1-2	wk 3-4	wk 1-2	wk 3-4
I	0.599^{abc}	1.133^{ab}	410^{ab}	684^{ab}	1.466^{b}	1.655^{bc}
II	0.622^{bc}	1.154^{ab}	428^{b}	715^{bc}	1.456^{b}	1.613^{ab}
III	0.576^{a}	1.191^{b}	395^{a}	712^{bc}	1.459^{b}	1.672^{c}
IV	0.604^{abc}	1.107^{a}	416^{ab}	669^{a}	1.455^{b}	1.655^{bc}
V	0.597^{ab}	1.187^{b}	410^{ab}	727^{c}	1.461^{b}	1.632^{abc}
VI	0.629^{c}	1.174^{b}	464^{c}	735^{c}	1.355^{a}	1.599^{a}
P	0.023	0.059	<0.001	0.003	<0.001	0.079
LSD	0.031	0.061	29	35	0.052	0.055

^{a,b,c} Values with a different superscript in the same column within a factor differ at $P<0.05$.

supplementation of L-histidine and L-leucine in week 1-2 did not affect performance compared to treatment II. The reduction in feed intake and body weight gain when omitting the supplementation of L-isoleucine in the first two weeks in treatment III compared to the apparent absence of a response to the omission of supplementation of L-histidine or L-leucine in treatments IV and V suggest that isoleucine was more limiting than histidine and leucine in piglets of this age. Over week 3-4, omitting the supplementation of L-histidine (treatment IV) reduced BWG compared to treatment II ($P<0.05$), while omitting the supplementation of L-isoleucine and L-leucine did not affect performance over this period. The former suggest that, in contrast to the results in week 1-2, the dietary supply of histidine was more limiting in week 3-4 compared to the supply of isoleucine and leucine.

It is concluded that amino acid requirements of piglets change during the post weaning period resulting in isoleucine being the sixth limiting amino acid after lysine, methionine/cysteine, threonine, tryptophan, and valine over the period of weeks 5-7 of age (2-3 weeks after weaning) and histidine being the sixth limiting amino acid during weeks 8-9 of age (4-5 weeks after weaning), when using a diet based on wheat, barley, maize and soybean meal.

References

Barea R., L. Brossard, N. Le Floc'h, Y. Primot, and J. van Milgen, 2009. The standardized ileal digestible isoleucine-to-lysine requirement ratio may be less than fifty percent in eleven- to twenty-three-kilogram piglets. Journal of Animal, Science 87, 4022-4031.

Chung, T.K., D.H. Baker, 1992. Ideal amino acid pattern for 10-kilogram pigs. J. Anim. Sci. 70, 3102-3111.

CVB, 1996. [Requirements for amino acids in piglets and pigs] In Dutch. CVB report 14.

NRC, 1998. Nutrient Requirements of Swine. National Research Council. National Academy Press, Washington, US.

Feeding low protein, amino acid-fortified diets did not affect performance and carcass composition of growing-finishing pigs

J.K. Htoo[1], J. Trautwein[2], J. Gao[1] and G. Dusel[2]
[1]Evonik Industries AG, Rodenbacher Chaussee 4, 63457 Hanau, Germany; john.htoo@evonik.com
[2]University of Applied Sciences Bingen, Berlinstrasse 109, 55411 Bingen am Rhein, Germany

Introduction

Lowering crude protein (CP) level and balancing with supplemental amino acids (AA) in pig diets has environmental and economic benefits. However, reduced performance was observed when CP level was reduced greater than 4%-units in diets for growing pigs although diets were balanced to be adequate in standardized ileal digestible (SID) essential AA (EAA; Guay *et al.*, 2006). The content of nitrogen (N) in low CP diets sometimes may become insufficient for endogenous synthesis of all non-essential AA (NEAA), and this may one of the possible reasons for reduced performance. Thus, the current experiments were conducted to evaluate the effects of balancing low CP diets with supplemental AA and maintaining optimal essential AA:total N ratio on performance, carcass composition and N retention of 24 to 111 kg pigs.

Material and methods

A 83-d trial was conducted with 180 TOPIGS pigs [initial body weight (BW) of 24.4 kg] with 2 pigs per pen and 18 pens per treatment. Pigs were assigned to 5 diets based on corn, barley, wheat, soybean meal (SBM) and rapeseed meal for each of the grower 1 (1.0% SID Lys; d 0-18), grower 2 (0.86% SID Lys; d 19-39), finisher 1 (0.74% SID Lys; d 40-61) and finisher 2 (0.64% SID Lys; d 62-83) phases. The high CP diet (diet 1) and low CP (diet 5; without SBM) were initially produced without inclusion of free AA. Diets 2, 3 and 4 were produced by blending diets 1 and 5 in varying proportions (75, 50 and 25% of diet 1). All diets were analyzed for AA and CP contents, and free AA were added to meet requirements for EAA. All diets contained similar EAA and net energy contents in each phase. Gly, Pro, Arg and Glu were added in diets 4 and 5 of grower 1 and grower 2 diets to balance the EAA:total N ratio to be 50% (Table 1). The step-wise reductions of CP level in the diets are shown in Table 2.

Table 1. Dietary treatments.

1	Conventional grains-SBM based diet (Control)
2	Reduced CP diet (-25% SBM inclusion) + EAA (Lys, Met, Thr)
3	Low CP diet (-50% SBM inclusion) + EAA (Lys, Met, Thr, Trp)
4	Low CP diet (-75% SBM inclusion) + EAA + NEAA (EAA:total N of 50%)
5	Low CP diet (without SBM) + EAA + NEAA + Arg in Grower 1 and 2 (EAA:total N ratio of 50%)

Table 2. Analyzed crude protein content (%)[1] in experimental diets.

Diets	Grower 1	Grower 2	Finisher 1	Finisher 2
1	21.4 (20.4)	21.3 (20.2)	17.0 (17.3)	15.9 (16.1)
2	19.4 (19.8)	19.5 (19.5)	16.0 (16.8)	14.8 (15.5)
3	17.5 (18.9)	17.8 (18.1)	14.9 (15.7)	13.7 (13.9)
4	15.5 (18.2)	16.0 (16.5)	13.9 (14.8)	12.5 (12.7)
5	13.6 (16.9)	14.3 (16.4)	12.9 (13.9)	11.4 (12.4)

[1] Dietary crude protein contents in parentheses are 'with added free amino acids'.

Finisher 2 diets were fed until the pigs reached slaughter BW of approximately 117 kg, and 10 pigs per treatment except diet 3 (9 pigs) were selected for carcass assessment. On d 95, blood was collected from 6 pigs per treatment to determine plasma urea N (PUN) concentration. Additionally, a 5-d N-balance trial was conducted by feeding diets 1 or 5 of finisher 2 phase to 6 pigs (initial BW of 111.0 kg; 3 pigs per treatment) to measure N retention. All data were analyzed by ANOVA using GLM procedure of SAS with pen as the experimental unit.

Results and discussion

During each phase and the overall 83-d period, average daily feed intake (ADFI), average daily gain (ADG) and feed conversion ratio (FCR) as well as final BW were not different among treatments (P>0.05; Table 3). The PUN concentration (d 95) decreased linearly (P<0.001) as dietary CP decreased and was lowest for pigs fed diet 5 which indicates a reduced need for deamination of excess AA (Table 4). The dressing %, lean yield and back fat thickness were not affected by the CP levels. Daily N retention was not affected (P>0.05) but total N excretion was reduced by 32% (P<0.01) by lowering the CP level in finisher 2 diet from 15.9 to 11.4% (Table 5). In conclusion, lowering dietary CP level by replacing SBM with supplemental AA is possible to maintain optimal pig performance when diets are balanced for adequate SID AA and NE, and maintained an EAA:total N ratio of approximately 50%.

Table 3. Effect of dietary protein levels on pig performance (d 0 to 83).

Items	Diet 1	Diet 2	Diet 3	Diet 4	Diet 5	SEM	*P*-value
BW at d 83, kg	100.0	100.4	98.5	98.1	98.0	0.943	0.171
ADFI, g/d	2,218	2,132	2,164	2,096	2,068	39.80	0.085
ADG, g/d	916	915	893	888	886	11.36	0.169
FCR, g/g	2.41	2.35	2.43	2.36	2.33	0.039	0.324

Table 4. Effect of protein levels on plasma urea N level (on d 95) and carcass quality.

Items	Diet 1	Diet 2	Diet 3	Diet 4	Diet 5	SEM	P-value
Plasma urea N, mg/dl	15.41[a]	13.91[ab]	10.82[bc]	9.95[bc]	8.75[c]	0.61	<0.001
BW at slaughter, kg	117.7	117.8	118.3	116.6	117.2	0.553	0.904
Dressing, %	77.2	76.9	77.4	78.0	77.6	0.438	0.425
Lean meat, %	77.2	76.9	77.4	78.0	77.6	0.438	0.425
Back fat, cm	0.87	0.90	1.10	0.99	1.00	0.090	0.412

Table 5. Effect of dietary protein levels on N retention in 111 kg pigs.

Items	Finisher 2 diets		SEM	P-value
	Diet 1 (16.1% CP)	Diet 5 (12.4% CP)		
N intake, g/d	51.3	38.9	0.367	<0.001
N in urine, g/d	32.8	22.2	1.659	0.011
Total N excretion, g/d	40.4	29.3	1.171	0.003
N retained, g/d	10.9	9.6	1.150	0.458

Reference

Guay, F., S.M. Donovan and N.L. Trottier, 2006. Biochemical and morphological developments are partially impaired in intestinal mucosa from growing pigs fed reduced-protein diets supplemented with crystalline amino acids. J. Anim. Sci. 84, 1749-1760.

Changes in protein turnover during pregnancy in pigs at amino acid intake in excess of requirements

S. Moehn[1], M. Rafii[2], P.B. Pencharz[2] and R.O. Ball[1]
[1]Department of Agriculture, Food and Nutritional Science, University of Alberta, Edmonton T6G 2P5, Canada; smoehn@ualberta.ca
[2]Research Institute, Hospital for Sick Children, Toronto M5G 1X8, Canada

Introduction

A series of experiments were completed to determine the amino acid (AA) requirements of pigs for threonine, lysine, tryptophan and isoleucine (Moehn *et al.*, 2011) in early and late pregnancy. The objective of the present analysis was to determine which factors affected protein turnover during pregnancy in pigs when the intake of test AA was above the requirement.

Material and methods

In every experiment, 6 to 7 sows each received 6 diets with graded levels of test AA in both early and late gestation. Protein turnover was determined using a stochastic model based on the ^{13}C enrichment in expired CO_2 and plasma free phenylalanine (Phe) after oral L[1-^{13}C]Phe dosing. Results of all experiments were combined by expressing AA intake as a fraction of the mean requirement determined in each experiment. The observations (n=199 for Phe oxidation [OX] and retention, 128 for Phe flux, body protein breakdown [B] and synthesis [S]) covered AA intakes between 1.003 and 2.468 times the requirement. Other parameters of nutrition, and sow growth and physical characteristics were as described for limiting AA intake in the companion paper. Data was evaluated using the mixed model procedure of SAS with pig nested in experiment as random variable. The start model was the same as used for limiting AA intake, and was reduced iteratively by deleting all non-significant terms until all parameters left were at $P<0.05$.

Results

The resulting models explained a high degree of the variation in the data with $r^2=0.78$ (S) to $r^2=0.86$ (OX). Oxidation and Phe retention reacted oppositely in that OX decreased ($P<0.02$) with increasing gestational age, Phe intake and maternal BW gain, while Phe retention increased ($P<0.03$). Oxidation increased ($P<0.01$) with increasing BW and with an interaction between gestational age and Phe intake, while Phe retention decreased ($P<0.05$). Sow age, alone and in interaction with BW, ME intake and gestational age minimized Phe OX in the 3rd parity ($P<0.03$). Phe retention decreased ($P=0.006$) with sow parity.

Flux and B decreased ($P<0.01$) with increasing gestational age and increased with increasing maternal gain. Flux and B responded quadratically ($P<0.02$) to increasing test AA intake in excess of requirements with increases apparent at test AA intakes over 1.6 times the requirement. An interaction between test AA intake and gestational age increased ($P<0.05$) flux and B, while an interaction between test AA intake and maternal BW gain decreased ($P<0.05$) flux and B. Protein synthesis increased ($P=0.005$) with increasing Phe retention and for sows with greater ME intake ($P=0.08$) while an interaction between ME intake and Phe retention decreased S ($P=0.011$). Subsequent litter size did not affect Phe kinetics.

Discussion

Discussion of protein metabolism in pregnant sows must take into account the competing demands for nutrients of maternal and conceptus growth, especially when conceptus growth accelerates in late pregnancy (NRC, 2012).

The increase in S with increasing Phe retention in sows with greater ME intake indicates that energy intake limited S. Because litter size was not a significant factor for S, the limitation of S occurred for the whole of maternal body and conceptus.

Increasing sow maternal BW gain reduced OX and increased flux and B. Reduction of OX saves AA, which can be utilized for increased Phe retention together with AA derived from increased B towards the end of pregnancy, as indicated by the interaction between test AA intake and gestational age. This agrees with the observed decrease in OX and increase in Phe retention when sows approached parturition, and is indicative of the accelerated fetal growth in late pregnancy (NRC, 2012).

Test AA requirements have been determined over 3-week periods and represent a mean value for these periods. At test AA intakes above requirement, no effect of AA intake on Phe kinetics was expected, based on Salter *et al.* (1990) who found that lysine intake in excess of requirements had no effect on S and B in growing pigs. However, flux and B in pregnant sows increased at test AA intake above requirement, and with an interaction between test AA intake and gestational age. This difference in response of pregnant sows compared to growing pigs may be caused by the exponential increase in fetal growth as sows approach parturition with concomitant increases in AA requirements. Thus, the observed increase of flux and B with increasing test AA intake can be regarded as reflection of increased requirements.

Flux and B decreased when increased AA supply interacted with maternal growth. The reduction of B in the presence of increased test AA indicates that B may be used to supply AA from body tissues when the AA need of conceptus increases. This occurred mainly in late pregnancy when the rapid growth of fetuses increased towards parturition, causing greater demands for AA, as shown by the interaction of test AA intake and gestational age. This increase in flux and B may indicate an increased supply of AA to the conceptus, since the interaction of AA with maternal growth reduced flux and B.

In conclusion, Phe kinetics at AA intake above the requirement were driven by interactions among sow and conceptus growth with test AA intake and stage of gestation. Changes occur predominantly in late pregnancy where the linear increase in conceptus growth increases AA requirements, leading to decreased OX and increased B to satisfy the greater AA need. Thus, Phe kinetics and pregnant sow requirements should be regarded as dynamic processes rather than discrete values.

References

Moehn, S., C. Levesque, R. Samuel and R.O. Ball. 2011. New energy and amino acid requirements for gestating sows. Advances in Pork Production, 22:157-166.

NRC. 2012. Nutrient Requirements of Swine (11th ed.). Nat. Acad. Press. Washington, DC

Salter, D.N., A. Montgomery, A. Hudson, D.B. Quelch and R.J. Elliott. 1990. Lysine requirements and whole-body protein turnover in growing pigs. Br. J. Nutr. 63:503-513.

Effect of reducing dietary energy and protein on growth performance and carcass traits of broilers

C.K. Girish[1], S.V. Rama Rao[2] and R.L. Payne[1]
[1]Health and Nutrition, Evonik (SEA), Pte. Ltd., Singapore; girish.channarayapatna@evonik.com
[2]Project Directorate on Poultry, ICAR, Hyderabad, India

Introduction

Dietary energy in excess of maintenance and production requirements of broilers will lead to increased fat deposition and reduced carcass quality. Low protein diets balanced with supplemental amino acids (AAs) can reduce the need for dietary energy required to break down dietary intact protein and for excretion of the excesses of AAs. Additionally, feeding low protein diets will lead to increased nitrogen utilization while reducing nitrogen excretion. Two experiments were conducted to evaluate the effects of reducing dietary metabolizable energy (ME) and crude protein (CP) at constant levels of standardized ileal digestible (SID) AA contents on growth performance and carcass traits of broiler chickens.

Material and methods

Experiment 1 (Expt 1) was conducted in brooder batteries while Expt 2 was conducted in floor pens. The same dietary treatments were used for both experiments, and the diets were arranged in a 3×3 factorial design with 3 levels (high, medium and low) of ME and CP in a completely randomized design. Each diet was offered to 10 replicates of 5 birds in Expt 1 or 6 replicates of 25 birds in Expt 2. Day-old Cobb 400 commercial broilers were used in both experiments, and the diets were fed *ad libitum* from 0 to 42 d of age in mash form. Diets were formulated to meet Evonik's AA recommendations for starter (0-11 d), grower (12-28 d) and finisher (29-42 d) phases with the exception of the Ile:Lys ratio, which was reduced by 3% to test the current recommendation. The high CP diet was supplemented with DL-Methionine (DL-M), L-Lysine-HCl (L-Lys) and L-Threonine (L-Thr) to reflect current industry practice. Then, the dietary CP level was reduced by 5 and 10% of high CP while maintaining the minimum AA concentrations. As a result, the medium CP diet was supplemented with DL-M, L-Lys, L-Thr and L-Valine (L-Val), and the low CP diet was supplemented with DL-M, L-Lys, L-Thr, L-Val and L-Ile. The concentration of the high, medium, and low ME diets were 2950, 2850 and 2750 kcal/kg, 3050, 2950 and 2850 kcal/kg, and 3150, 3050 and 2950 kcal/kg in the starter, grower and finisher diets, respectively. Similarly the concentrations of CP during the respective phases were 25.2, 23.9 and 22.6%, 21.0, 20.0 and 18.9%, and 19.6, 18.7 and 17.7%.

Body weight (BW) and feed intake (FI) were recorded at 11, 21, 28, 35 and 42 d of age, and body weight gain (BWG) and feed conversion ratio (FCR) were calculated. At the end of experiment, one bird from each pen weighing close to the average body weight (± 5%) of the respective pen was selected for carcass trait measurements. Carcass variables included ready to cook yield (RTC), breast weight (BrW) and abdominal fat (AF). Data were subjected to factorial analyses (Snedecor and Cochran, 1980) to study the effect of interaction between ME and CP and also independent effect of each main factor. Significant difference among different treatment means were compared using Duncan's multiple range test (Duncan, 1955) at $P<0.05$.

Results and discussion

In Expt 1, there were interactions ($P<0.01$) between ME and CP on BWG and FI but not for FCR at d 42 (Table 1). Conversely, in Expt 2, there were interactions ($P<0.05$) on BWG, FI and FCR in birds reared on the floor pens. Birds fed low ME in batteries had higher ($P<0.01$) FCR compared with those fed medium or high ME, but there was no effect of reducing dietary CP on FCR. Birds

Table 1. Dietary effects on growth performance and carcass traits of broilers at d 42.

Diets	Expt 1 (Battery brooder)						Expt 2 (Floor pen)					
	BWG kg	FI kg	FCR kg/kg	RTC	BrW	AF	BWG kg	FI kg	FCR kg/kg	RTC	BrW	AF
				(g/kg live BW)						(g/kg live BW)		
HME.HCP	1.90^{abcd}	3.02^{abc}	1.59	804	221	13.7	2.03^{bc}	3.77^{cd}	1.86^{ab}	788	246	15.0
HME.MCP	1.92^{abc}	2.92^{bc}	1.52	765	213	15.5	2.12^{bc}	3.63^{cde}	1.72^{b}	797	234	16.7
HME.LCP	1.92^{abc}	3.03^{abc}	1.57	769	215	17.3	2.15^{b}	3.89^{bc}	1.81^{ab}	775	233	17.8
MME.HCP	2.02^{a}	3.14^{ab}	1.55	765	206	9.9	2.33^{a}	4.29^{a}	1.85^{ab}	774	225	21.7
MME.MCP	1.79^{cd}	2.80^{c}	1.57	759	217	13.2	2.03^{bc}	3.48^{e}	1.73^{b}	782	246	17.4
MME.LCP	1.86^{bcd}	2.94^{bc}	1.58	761	218	11.8	2.07^{bc}	3.71^{cde}	1.79^{ab}	762	216	12.5
LME.HCP	1.86^{bcd}	2.95^{bc}	1.59	749	214	8.5	2.11^{bc}	3.74^{cde}	1.78^{ab}	786	250	10.2
LME.MCP	1.98^{ab}	3.20^{a}	1.61	750	216	6.6	2.13^{b}	4.07^{ab}	1.91^{a}	766	236	14.5
LME.LCP	1.77^{d}	2.88^{c}	1.62	756	211	12.4	1.94^{c}	3.55^{de}	1.83^{ab}	763	226	14.8
Main effects												
HME	1.92	2.99	1.56^{b}	779^{a}	216	15.5^{a}	2.10	3.76	1.80	787	238	16.5
MME	1.89	2.96	1.57^{b}	762^{b}	214	11.6^{b}	2.14	3.83	1.79	773	229	17.2
LME	1.87	3.01	1.61^{a}	752^{b}	213	9.2^{b}	2.06	3.79	1.84	772	237	13.2
HCP	1.93	3.04	1.58	773	214	10.7^{b}	2.16	3.93^{a}	1.83	782	240^{a}	15.6
MCP	1.90	2.97	1.57	758	215	11.8^{ab}	2.09	3.73^{b}	1.79	782	239^{a}	16.2
LCP	1.85	2.95	1.59	762	215	13.8^{a}	2.05	3.71^{b}	1.81	767	225^{b}	15.0

BWG = body weight gain; FI = feed intake; FCR = feed conversion ratio; RTC = ready to cook yield; BrW = breast weight; AF = abdominal fat; HME = high ME; MME = medium ME; LME = low ME; HCP = high CP; MCP = medium CP; LCP=low CP.

^{a,b,c,d,e} Means having common superscript in a column do not vary significantly.

reared in floor pens had reduced FI when dietary CP was decreased to medium CP level but further reduction to low CP had no effect. Further, there was no effect of reducing dietary ME on FI. There was no interaction (P>0.05) of ME and CP levels on carcass parameters of broilers either in batteries or floor pens. Nevertheless, reducing dietary ME reduced (P<0.01) the relative weights of RTC and AF in broilers raised in battery brooders, but BrW was not affected. Conversely, reducing dietary CP did not affect RTC and BrW but AF was higher (P=0.032) in birds fed low CP compared with those fed high CP. The BrW of broilers fed low CP in floor pens was lower (P=0.041) compared with those fed high or medium CP while there was no effect of reducing dietary ME on carcass traits.

The results suggest that the energy required for protein deposition, as indicated by breast weight, is lower than that for total body weight gain. These experiments also demonstrate that dietary energy can be reduced when the ratio of supplemental AAs to intact protein increases, such as in the case of a low protein diet balanced with supplemental AAs.

References

Duncan, D. B., 1955. Biometrics 11:1-55.

Snedecor, G. W. and W. G. Cochran, 1980. Statistical Methods. Iowa State University Press, Ames.

Metabolic use of a growing diet for red-legged partridge (*Alectoris rufa*) chicks

M. Lachica, J.F. Aguilera, R. Nieto and I. Fernández-Fígares
Department of Animal Nutrition, Estación Experimental del Zaidín, CSIC, Camino del Jueves s/n, 18100 Armilla, Granada, Spain; manuel.lachica@eez.csic.es

Introduction

Red-legged partridge (*Alectoris rufa*) has an enormous ecological and game importance in Spain and is the favorite game species. Intensive breeding is used to restock areas, to re-establish or to maintain the population. More than 4 million birds/yr are released. Paradoxically, there is a lack of information about its protein and energy requirements, which has led to extrapolation of existing knowledge from other partridge species to *A. rufa*. The aim of this study was to determine the metabolic utilization of a specific commercial diet for growing partridge chicks. For this purpose, the feed conversion ratio (FCR), the efficiency of utilization of dietary ME for maintenance (k_m), and the partition of retained energy (RE) as protein (RE_{prot}) and fat (RE_{fat}) were obtained.

Material and methods

Twenty four chicks were allocated by pairs in metabolic cages and fed *ad libitum* with a commercial starter diet (12.0 MJ/kg ME and 280 g/kg CP) for partridge chicks. Feeding, balance and slaughter trials began at 3 wk of age (70 g average BW) and ended at 7 wk. Initial body composition was determined in 4 chicks. Feed intake and BW were recorded weekly. Along the third experimental wk, total excreta was collected to determine energy metabolizability and N balance. Subsequently, a group of 10 chicks was moved to respirometry chambers for 24 h to obtain heat production (HP) from gas exchange measurements. Then, at the end of the fourth wk, 1 chick (210 g average BW) per pair was killed for determining the final body composition. The 10 remaining birds were divided in 2 groups of 5 and HP measured again at a ME intake (MEI) slightly above the theoretical maintenance level, and after 12 h fasting. The RQ (CO_2/O_2) was also determined.

The DM content of feed, excreta and carcass was determined by standard procedures (AOAC, 1990) and total N by the Kjeldahl procedure. The gross energy (GE) was measured in an isoperibolic bomb calorimeter. The MEI was determined as GE content of the ingested feed minus corresponding excreta. Metabolizability was obtained as the ME/GE ratio. The RE and RE_{prot} (retained N $\times$6.25$\times$23.8 kJ/g) were calculated as the difference between the total body energy and protein in the initial and final killed group. The RE_{fat} was the difference between RE and RE_{prot}. The linear equation FR (kJ/kg$^{0.75}$ and d) = $-a + b \times$MEI (kJ/kg$^{0.75}$ and d) was used to obtain the efficiency (b) of utilization of ME for maintenance (k_m). ANOVA-I was used to determine the effect of age on FCR. Tukey multiple range test was used to ascertain the statistical significance of differences (P<0.05).

Results and discussion

The diet ME content was 11.9 MJ/kg and the average metabolizability 0.671±0.0091. Table 1 shows main results. The FCR ranged from 2.55 for the period 4-5 wk to 4.48 at 6-7 wk, being significantly greater (P<0.05) the last one. Under similar experimental conditions, Hermes *et al.* (1984) with *A. graeca* and Ozek *et al.* (2003) with *A. chukar* chicks obtained FCR values that ranged, respectively, 2.94-2.98 and 2.66-2.74 for 0-8 wk. Also Ozek (2004) with *A. chukar* reported FCR values that ranged 2.21-3.40 for 0-2 and 4-6 wk, respectively, and 2.4-4.1 for 0-2 and 6-8 wk, respectively (Ozek, 2006). As the ME level in the present study was similar to the aforementioned studies, the greater FCR could indicate an inadequate CP/ME ratio, particularly in the older chicks. As a precocial bird, *A. rufa* chicks may need to maintain homeothermy faster than other partridges, in an

Table 1. Body weight (BW, g), feed intake (g/d) and feed conversion ratio (FCR) of Alectoris rufa *chicks fed* ad libitum *with a starter commercial diet (mean±standard error).*

	3-4 wk	4-5 wk	5-6 wk	6-7 wk
BW[1]	97.1±4.35	136.9±5.02	178.5±5.64	208.7±5.82
Intake	10.5±0.62	14.3±0.36	17.6±0.55	18.4±0.45
FCR	3.15±0.342[a]	2.55±0.100[a]	2.97±0.074[a]	4.48±0.334[b]

[1] Measured at the end of every wk period; initial BW=71.3±1.98 g; n=20.
[a,b] Values within a row with unlike superscript letter were significantly different ($P<0.05$).

environment that shows a big diurnal temperature gradient, as usually occurs in the Iberian Peninsula. The average RE was 35.0±2.13 kJ/d, with and average RE_{prot} and RE_{fat} of 26.5±1.85 and 8.5±0.33 kJ/d, respectively. The retained protein/CP intake ratio was 0.267. Average MEI was 181.6±6.93 kJ/d and the calculated HP was 146.5 kJ/d (12.3 W/kg BW). Pis (2003) obtained 17.8, 11.3 and 9.46 W/kg BW for *A. chukar* chicks of 1, 4 and 14 wk, respectively. Marjoniemi *et al.* (1995) and Pis (2001) reported values of 21.5 and 17.0 W/kg BW, respectively, for *Perdix perdix* chicks. Our value 12.3 W/kg BW for chicks from 4 to 7 wk may indicate greater energy needs for maintenance than other partridge species of the same age. The gross efficiency for ME utilization was as low as 0.193, as a result of the moderate feed intake with respect to the energy needs for maintenance. The k_m was 0.77±0.046. *Ad libitum*, 0.5×*ad libitum* and fasting RQ values averaged 0.9±0.03, 0.7±0.04 and 0.63±0.08, respectively.

Compared with other partridge species, *Alectoris rufa* shows lower growth rate and poorer FCR. The present study provides evidence that the protein and energy requirements differ from those of other species and are probably slightly higher for maintenance and growth, suggesting a need for adjusting the diet formulation.

Acknowledgements

This research was supported by grant no. AGR 03065 from Junta de Andalucía (Spain).

References

AOAC, 1990. Official Methods of Analysis, 15th ed. Arlington, VA.

Hermes J.C., A.E. Woodard, P. Vohra and R.L. Snyder, 1984. The effect of light intensity, temperature, and diet on growth in Red-Legged partridge. Poult. Sci. 63, 871-874.

Marjoniemi, K., E. Hohtola, A. Putaala and R. Hissa, 1995. Development of temperature regulation in the grey partridge *Perdix perdix*. Wildlife Biol. 1, 39-46.

Ozek, K., 2004. Effect of energy level in the diet on body weight, feed consumption and feed conversion ratio at early growth period in the chukar partridge (*Alectoris chukar chukar*) raised in closed confinement. Revue Méd. Vét. 155, 163-166.

Ozek, K., 2006. The optimum protein content in high-energy starter diet for Chukar Partridge (*Alectoris chukar chukar*). Int. J. Poult. Sci. 5, 522-525.

Ozek, K., O. Yazgan and Y. Bahtiyarca, 2003. Effects of dietary protein and energy concentrations on performance and carcase characteristics of chukar partridge (*Alectoris chukar*) raised in captivity. Brit. Poultry Sci. 44, 419-426.

Pis, T., 2001. Development of thermoregulation in hand-reared grey partridges (*Perdix perdix*). Game Wildl. Sci. 18, 509-520.

Pis, T., 2003. Energy metabolism and thermoregulation in hand-reared chukars (*Alectoris chukar*). Comp. Biochem. Phys. A 136, 757-770.

The amino acid composition of body protein in broilers is affected by the sulphur amino acid supply

J.A. Conde-Aguilera[1,2], C. Cobo-Ortega[1,2], S. Tesseraud[3], M. Lessire[3], Y. Mercier[4] and J. van Milgen[1,2]
[1]INRA, UMR1348 PEGASE, 35590 Saint-Gilles, France; jaap.vanmilgen@rennes.inra.fr
[2]Agrocampus Ouest, UMR1348 PEGASE, 35000 Rennes, France
[3]INRA, UR83, Recherches Avicoles, 37380 Nouzilly, France
[4]Adisseo France SAS, 92160 Antony, France

Introduction

In poultry diets, the sulphur-containing amino acids (AA) Met and Cys are considered the first limiting AA for protein deposition. In the factorial approach, AA requirements are determined using the AA composition of retained protein, the maximum efficiency of AA utilization for growth, and the maintenance AA requirement. It is then assumed that the relative contribution of these traits to the AA requirement is constant, resulting in a constant ideal AA profile. However, there are indications that the AA composition of body protein is affected by genotype, age, gender, and the protein and AA contents of the diet. Broiler studies aiming to assess the response to the total sulphur AA (TSAA) supply mostly focus on performance, carcass yield, and the chemical composition of the whole animal or breast meat. However, as shown in pigs, different tissues may respond differently to a deficient TSAA supply. The objective of this study was to quantify the changes in chemical body composition and meat quality of broilers receiving diets either deficient (TSAA-) or sufficient (TSAA+) in TSAA.

Material and methods

Eighteen birds (Ross PM3 strain) were used in a comparative slaughter study. Six chickens were slaughtered at 7 d of age and the others were divided in 2 homogenous groups of 6 birds each to received either TSAA- or TSAA+ for 35 d and were slaughtered thereafter. Diets were offered slightly below the anticipated *ad libitum* intake, and were formulated so that the nutrient composition (other than TSAA) met or exceeded the NRC recommendations. Diets were changed weekly to account for the change in nutrient requirements during the experiment. The supply of Met and TSAA in TSAA- was about 34% and 22% below those of TSAA+, respectively. At slaughter, birds were divided into blood, feathers, carcass, *Pectoralis major* muscle (PM) and offal for analysis.

Results and discussion

Performance and tissue weight gain were not affected by the TSAA supply (Table 1). In birds receiving diet TSAA-, protein gain was lower in the carcass ($P<0.01$), empty body ($P=0.06$) and PM ($P=0.10$; Table 2). Lipid gain in birds receiving TSAA- was 78% greater in PM ($P<0.001$), 28% greater in abdominal fat ($P<0.05$) and 10% greater in the carcass ($P=0.10$). This increased lipid content in TSAA- birds can be explained through the deamination of AA not used for protein deposition and for which the carbon-chains can be used for lipid deposition. In PM, reducing the TSAA supply resulted in a tendency for an increase in the redness value (a*; $P=0.10$). The AA composition of tissues and tissue gain was affected by the TSAA supply, but the Met and Cys concentrations were changed only in the offal ($P=0.08$). This contrasts with a study with pigs, who responded to a strong Met deficiency by reducing the protein content in the body, and the Met content in body and muscle proteins (Conde-Aguilera *et al.*, 2010). Also, birds receiving diet TSAA- had a greater Ser content in the empty body, carcass, and PM ($P<0.05$) which may reflect an adaptation of methyl metabolism. In contrast, these birds had lower contents of Lys and Glu in the empty body, of Phe, Tyr, Gly, and Glu in PM, and of Ala in the offal ($P<0.05$). Differences in the AA composition of body protein could be the consequence of changing proportions of different types of proteins in the body. Our study indicates that although chickens cope with an AA deficiency predominantly by changing the protein and lipid content in the body, the AA composition is also affected. This questions therefore the use of a constant ideal AA profile in poultry nutrition.

Table 1. Performance and meat quality traits in broilers offered TSAA- or TSAA+ from 7 to 42 d of age.

	TSAA-	TSAA+	RSD	*P*-value
Final BW (g)	2,589	2,632	100	0.48
Average daily gain (g/d)	68.9	70.3	2.1	0.11
Gain-to-feed ratio (g/g)	1.66	1.62	0.06	0.10
Meat quality traits				
pH (24 h)	5.63	5.66	0.03	0.15
Lightness (L*)	51.2	51.4	3.0	0.94
Redness (a*)	0.61	-0.15	0.74	0.10
Yellowness (b*)	13.6	13.2	1.4	0.59
Abdominal fat (g/kg BW)	19.8	15.5	3.1	0.04

Table 2. Composition of the empty body and tissue weight gain in broilers offered TSAA- or TSAA+ from 7 to 42 d of age[1].

	Empty body			Carcass		
	TSAA-	TSAA+	RSD	TSAA-	TSAA+	RSD
Weight gain (g/d)	67.0	68.4	2.6	49.7	51.1	2.3
% of EBW gain				74.3	74.7	1.4
Composition						
N × 6.25 (g/kg)	196†	199	3	185**	191	3
Lipid (g/kg)	141	130	12	134†	121	12
Lys	6.75*	6.91	0.10	7.76	7.87	0.11
Met	2.29	2.30	0.06	2.65	2.60	0.05
Cys	2.01	1.96	0.08	1.08	1.10	0.02
Phe	3.95	3.96	0.06	3.74	3.80	0.06
Tyr	3.10	3.13	0.06	3.12	3.15	0.04
Ser	4.42*	4.20	0.12	3.39**	3.16	0.11
Gly	7.30	7.33	0.20	6.61	6.65	0.30
Ala	6.00	6.00	0.08	6.13	6.04	0.13
Glu	12.61*	12.82	0.16	13.20	13.40	0.22
Pro	6.38	6.31	0.11	5.32	5.20	0.16

Table 2. Continued.

	PM			Offal		
	TSAA-	TSAA+	RSD	TSAA-	TSAA+	RSD
Weight gain (g/d)	11.1	11.5	1.1	12.3	12.5	0.7
% of EBW gain	16.6	16.8	1.2	18.4	18.3	1.1
Composition						
N × 6.25 (g/kg)	230†	237	7	162	157	8
Lipid (g/kg)	16.9***	9.1	1.5	224	212	18
Lys	8.69	8.98	0.36	5.98	6.12	0.42
Met	2.88	2.85	0.13	2.24†	2.42	0.17
Cys	1.19	1.20	0.08	1.20	1.29	0.11
Phe	4.06*	4.25	0.12	3.81	3.78	0.32
Tyr	3.61*	3.81	0.11	3.02	3.02	0.29
Ser	3.36*	3.22	0.09	3.83	3.96	0.32
Gly	4.34*	4.50	0.15	11.13	11.36	0.96
Ala	5.67	5.75	0.22	6.54*	7.06	0.37
Glu	14.00*	14.80	0.59	11.90	12.40	0.93
Pro	3.73	3.72	0.11	7.38†	7.98	0.55

[1]Significant level indicated within a row and tissue component by $*P\leq0.05$, $**P\leq0.01$, $***P\leq0.001$, $^{\dagger}P\leq0.10$. EBW = empty body weight. The AA composition is expressed as grams per 16 grams of N.

References

Conde-Aguilera, J. A., R. Barea, N. Le Floc'h, L. Lefaucheur, and J. van Milgen. 2010. A sulfur amino acid deficiency changes the amino acid composition of body protein in piglets. Animal 4:1349-1358.

Body composition and nutrient partitioning in long term supplementation of betaine and conjugated linoleic acid in mice

J.M. Rodríguez-López, L. González-Valero, M. Lachica and I. Fernández-Fígares
Department of Animal Nutrition, Estación Experimental del Zaidín, CSIC, Camino del Jueves s/n, 18100 Armilla, Granada, Spain; ifigares@eez.csic.es

Introduction

Betaine (N, N, N-trimethylglycine) and conjugated linoleic acid (CLA) have the potential to alter growth and body composition in different animal models (Park *et al.*, 1997; Fernández-Fígares *et al.*, 2008). The effects of these substances are usually evaluated using relatively short studies. Therefore, the present study was conducted to evaluate the effects of long term dietary betaine and CLA supplementation on body composition and nutrients deposition in mice.

Material and methods

Male ICR (CD-1) mice, weighing 14-18 g, were group housed (5 mice per cage, 4 cages per treatment) and fed *ad libitum* Teklad global 14% protein rodent diet containing either no added betaine or CLA (Control), 0.5% betaine, 1% CLA, or 0.5% betaine + 1% CLA for 90 days. Water was freely available. An additional group of 5 mice was slaughtered at the beginning of the experiment to obtain the initial body composition. Feed intake and spillage was recorded every other day and individual weight gain was determined weekly. When mice were slaughtered, carcasses were chemically analysed for water, protein, lean, fat, mineral and energy content. Increases in protein, energy, fat and minerals were calculated as the difference between the final measured composition of experimental mice and the composition assessed from the initial slaughter group. The treatment effect was assessed by ANOVA according to a completely randomized design with mouse as experimental unit using the GLM procedure. Treatment means were compared by Tukey's procedure, and significance was set at $P<0.05$.

Results and discussion

Although studies of CLA effects on body composition are common, CLA effect on nutrient accretion has been studied to a lesser extent. In the present study, long term dietary CLA and betaine + CLA intake markedly affected carcass composition increasing protein, protein:fat ratio, water and minerals and decreasing fat ($P<0.001$) (Table 1). In a similar way, the composition of gain and carcass nutrient deposition of mice fed CLA diet were also affected, so that protein, water and minerals increased whereas fat and energy decreased ($P<0.001$) compared to control and betaine groups. Betaine decreased water content of gain and water deposition compared to Control ($P<0.001$), and showed a synergistic increase in protein:fat ratio when used with CLA ($P<0.001$). Similarly, growing pigs fed with betaine + CLA diets had greater protein deposition than betaine, CLA or control pigs (Fernández-Fígares *et al.*, 2008). The decrease in body fat could be due to increased energy expenditure and energy loss in excreta (Terpstra *et al.*, 2002). CLA decreased body fat and increased lean body mass and ash body content in mice (Park *et al.*, 1997). Nevertheless, mice fed 0.5% CLA high fat diet decreased body fat with no change in protein content (Javadi *et al.*, 2007). Enhanced carcass mineral content and deposition may indicate increased bone formation which may protect against loss of bone mass with aging. In that sense, CLA increased markers of bone formation in mice (Watkins *et al.*, 2001). We have found no reports on the effects of betaine on body composition in laboratory animals. In feed restricted growing pigs, betaine increased protein deposition and decreased carcass fat content (Fernández-Fígares *et al.*, 2002).

Table 1. Effects of dietary betaine (0.5%), CLA (1%) and betaine + CLA on mice carcass composition, composition of gain and the rates of carcass deposition of protein, water, fat, minerals and energy for a period of 90 days.

	Control	Betaine	CLA	Betaine+CLA	SEM
Carcass composition, g/kg					
Protein	194[a]	197[a]	217[b]	214[b]	2.0
Water	659[a]	650[a]	706[b]	721[b]	7.0
Fat	110[a]	115[a]	36[b]	26[b]	8.6
Minerals	373[a]	376[a]	415[b]	404[b]	0.6
Protein:fat	2.3[a]	2.2[a]	7.0[b]	8.8[c]	0.5
Composition of gain, g/kg					
Protein	192[a]	196[a]	228[b]	224[b]	3.3
Water	607[a]	595[b]	680[c]	704[d]	9.9
Fat	163[a]	176[a]	47[b]	31[b]	0.1
Minerals	37[a]	37[a]	44[b]	42[b]	1.0
Deposition, mg/d					
Protein	54[a]	51[a]	63[b]	64[b]	1.7
Water	170[a]	152[b]	187[c]	204[d]	5.8
Fat	48[a]	42[a]	13[b]	9[b]	3.8
Minerals	10[a]	10[a]	12[b]	12[b]	0.4
Total energy, kJ/d	3.2[a]	2.9[a]	2.0[b]	1.9[b]	0.2

[abcd] Within a row, values with different superscripts differ significantly ($P<0.001$).

Overall, long term dietary supplementation of CLA increased protein and mineral and decreased fat deposition in growing mice while betaine had a minor effect.

Acknowledgements

This study was financed by grant AGL2009-08916 from Ministry of Science and Innovation.

References

Fernández-Fígares, I., J.A. Conde-Aguilera, R. Nieto, M. Lachica and J.F. Aguilera, 2008. Synergistic effects of betaine and conjugated linoleic acid on growth and carcass composition of growing Iberian pigs. J. Anim. Sci. 86, 102-111.

Fernández-Fígares, I., D. Wray-Cahen, N.C. Steele, R.G. Campbell, D.D. Hall, E. Virtanen, and T.J. Caperna, 2002. Effect of dietary betaine on nutrient utilization and partitioning in the young growing feed-restricted pig. J. Anim. Sci. 80, 421-428.

Javadi, M., M.J. Geelen, H. Everts, A.G. Lemmens and A.C. Beynen, 2007. Body composition and selected blood parameters in mice fed a combination of fibre and conjugated linoleic acid. J. Anim. Physiol. Anim. Nutr. 91, 492-497.

Park Y., K.J. Albright, W. Liu, J.M. Storkson, M.E. Cook and M.W. Pariza, 1997. Effect of conjugated linoleic acid on body composition in mice. Lipids 32, 853-858.

Terpstra, A.H.M., A.C. Beynen, H. Everts, S. Kocsis, M.B. Katan and P.L. Zock, 2002. The decrease in body fat in mice fed conjugated linoleic acid is due to increases in energy expenditure and energy loss in the excreta. J. Nutr. 132, 940-945.

Watkins, B.A., H.E. Lippman, L. Le Bouteiller, Y. Li and M.F. Seifert, 2001. Bioactive fatty acids: role in bone biology and bone cell function. Prog. Lipid. Res. 40, 125-148.

Effect of immunocastration in combination with addition of fat to diet on quantitative oxidation of nutrients and fat retention in male pigs

N. Batorek[1,2,3,4], J. Noblet[3,4], S. Dubois[3,4], M. Bonneau[3,4], M. Čandek-Potokar[1,2] and E. Labussiere[3,4]
[1]Agricultural Institute of Slovenia, Hacquetova ulica 17, 1000 Ljubljana, Slovenia
[2]University of Maribor, Faculty of Agriculture and Life Sciences, Pivola 10, 2311 Hoče, Slovenia
[3]INRA, UMR 1348 Pegase, 35590 Saint Gilles, France; etienne.labussiere@rennes.inra.fr
[4]Agrocampus Ouest, UMR 1348 Pegase, 35000 Rennes, France

Introduction

Immunocastration (IC) has been studied as one of the possible replacements for surgical castration of male pigs to block reproductive function and eliminate boar taint. It has a positive effect on performance (Batorek *et al.*, 2012) but after IC, increased feed intake (FI) and fat deposition indicate changes at the metabolic level. The objective of the study was to determine the effects of IC and increase of dietary fat on energy (E) metabolism and relationships between digested nutrients and their final outputs for catabolic and metabolic process in fattening pigs.

Material and methods

Digestibility and N and E balance in respiration chambers were determined during 8-9 days in 2 subsequent stages prior and after IC with Improvac® (mean body weight BW: 85 and 135 kg) in 4 replicates (2 groups of 2 littermates each). Pigs were fed at 90% of predicted *ad libitum* ME intake and were allocated to one of 2 groups: Group 1 was fed low fat diet (LF; standard diet with addition of 0.5% of rapeseed oil; 2.5% DM of fat) during stage 1 and 2 and group 2 was fed LF during stage 1 and high fat diet (HF; diet enriched in fat by addition of corn and linseed oils to balance the C18:3:C18:2 fatty acids (FA) ratio between diets; 8.9% DM of fat) during stage 2. Additionally, pigs received a test-meal containing ^{13}C18:2 n-6 FA and $^{13}CO_2$ production was measured (on 0, 1, 5 and 6-7 days after ingestion) in order to calculate the rate of oxidation of this FA. The 7th day was a fasting day with feed deprivation. Heat production (HP) and respiratory quotient (RQ) were calculated according to Brouwer (1965) and components of HP (Van Milgen *et al.* 1997) were used to calculate heat increment (HI) as the sum of HP due to physical activity and thermic effect of feeding. Oxidation of nutrients and their contribution to fat retention (RF) were calculated according to Chwalibog *et al.* (1992). Effect of treatment group, stage and their interaction was tested by the GLM procedure (SAS Inc., Cary, NC, USA).

Results and discussion

To mimic the increase in voluntary FI seen in IM after V2, the ME intake was increased between stages (P<0.01; Table 1). Total RE increased with stage (P<0.01) and was mainly retained as fat (P<0.01) in accordance with increased lipogenesis during stage 2 (P<0.01). RF synthesized from carbohydrates (RF(CHO)) increased with stage (P<0.01; Figure 1), but on HF diet the proportion of RF originating from dietary fat increased at the expense of RF(CHO) (P<0.01). This explains the decreased HP (P<0.01; Table 1) and increased RE and RE as fat (P<0.01) on HF diet, because the utilization of ME originating from lipids instead of carbohydrates is energetically more efficient and results in lower HP (Van Milgen *et al.*, 2001). Nevertheless, the decrease in HI (% ME intake) with HF during stage 2 was only numerical (Table 1). The RQ was always above 1 and it increased with stage (P<0.01; Table 1), but it was lower on HF diet (P<0.01), indicating that there was more oxidation and less synthesis of lipids on HF diet in stage 2. In contrary, production of $^{13}CO_2$ because of oxidation of dietary ^{13}C18:2 n-6 FA increased with stage, but was higher on LF diet from day 2 onwards (Figure 2), indicating that dietary provision of this essential FA was insufficient for pigs on LF diet to satisfy their minimal oxidation rate.

Table 1. Results of energy balance (kJ/kg of $BW^{0.60}$ per day) and respiratory quotient in growing immunocastrated pigs before (stage 1) and after (stage 2) successful immunocastration.

	Stage 1		Stage 2		r.s.d	*P*-value[1]		
Group	1	2	1	2		S	G	S × G
Diet	LF	LF	LF	HF				
Energy balance (kJ/kg of $BW^{0.60}$ per day)								
ME intake	2,270[a]	2,277[a]	2,449[b]	2,492[b]	19	<0.01	0.32	0.10
Heat production	1,346[ab]	1,370[b]	1,371[ab]	1,327[a]	17	0.33	0.77	<0.01
Retained energy as fat	648[a]	600[a]	806[b]	914[c]	30	<0.01	0.48	<0.01
Total retained energy	924[a]	907[a]	1,078[b]	1165[c]	20	<0.01	0.36	<0.01
Lipogenesis	559[a]	513[a]	746[b]	681[b]	38	<0.01	0.26	0.63
Heat increment (% of ME)	23.3	23.8	27.5	24.7	2.9	0.13	0.59	0.30
Respiratory quotient	1.10[a]	1.09[a]	1.14[b]	1.07[a]	0.01	0.14	0.04	<0.01

r.s.d.=residual standard deviation; ME = metabolizable energy; LF = low fat diet; HF = high fat diet.

[1] Tested for effects of stage (S), group (G) and their interaction (S×G).

a-c Least squares means within a row with different superscripts differ ($P<0.05$).

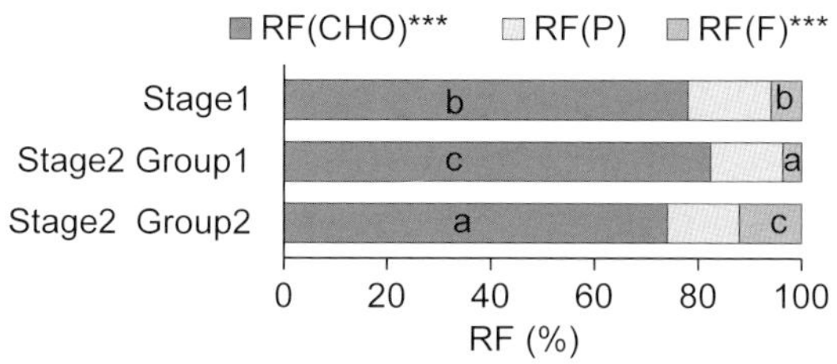

*Figure 1. Percentage of total retained fat (RF) originating from dietary carbohydrates (RF(CHO)), proteins (RF(P)) and fat (RF(F)); *** P<0.01.*

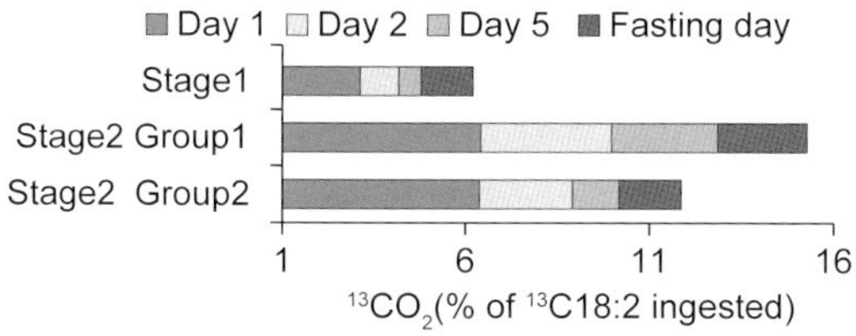

Figure 2. Excreted $^{13}CO_2$ as percentage of consumed $^{13}C18{:}3$ during the 1st, 2nd, 5th and fasting day in respiration chamber.

References

Batorek N., M. Čandek-Potokar, M. Bonneau and J. van Milgen, 2012. Meta-analysis of the effect of immunocastration on production performance, reproductive organs and boar taint compounds in pigs. Animal 6, 1-9.

Brouwer E., 1965. Report of sub-committee on constants and factors. In Energy metabolism. proceedings of the 3rd symposium held at Troon, Scotland, May 1964. K. L. Blaxter. (ed.). London, Academic Press: 441-443.

Chwalibog A., K. Jakobsen, S. Henckel and G. Thorbek, 1992. Estimation of quantitative oxidation and fat retention from carbohydrate, protein and fat in growing pigs. J. Anim. Physiol. Anim. Nutr. 68, 123-135.

Van Milgen J., J. Noblet and S. Dubois, 2001. Energetic efficiency of starch, protein and lipid utilization in growing pigs. J. Nutr. 131, 1309-1318.

SAS. SAS/STAT 9.1 user's guide, 2004. SAS Inst. Inc., Cary, NC, USA.

Net energy content of dry extruded-expelled soybean meal fed to growing pigs using indirect calorimetry

D.E. Velayudhan, J.M. Heo and C.M. Nyachoti
University of Manitoba, Winnipeg, MB R3T 2N2, Canada; deepakvet98@gmail.com

Introduction

Feed is the single most expensive input in commercial pork production and at least 50% of this cost can be attributed in supplying energy to the animal thus making energy financially the most vital component. Swine diets can be formulated on a variety of energy systems such as the digestible energy (DE), the metabolizable energy (ME) and the net energy (NE) systems of which the NE system provides more accurate estimates of the energy available to the animal. Energy values of protein-rich feeds are often overestimated when expressed on a DE or ME system (Noblet *et al.*, 1994). These discrepancies in measurement of available dietary energy have a drastic effect on the economics of pig production and there is, therefore, an ongoing interest in adopting the NE system. The most commonly used protein source in livestock diets is soybean meal (SBM), but it also contains certain antinutritional factors which depress animal growth performance. Studies show that such antinutritional factors are reduced significantly during meal processing (Perilla *et al.*, 1997). One such process is the combination of extrusion with expelling which produces a SBM product called dry extruded-expelled SBM (DESBM). However, published data pertaining to the energy values of DESBM for grower pigs are limited. The aim of this study was to determine the NE content of DESBM in growing pigs using either an indirect calorimetry (IC) or published prediction equations.

Material and methods

Twenty four growing barrows (initial BW = 19.6±0.51 kg) were allotted in a completely randomized design to 4 dietary treatments to give 6 replicates per treatment. Dietary treatments were; a corn soybean meal basal diet (Diet A), a diet containing Diet A and DESBM in a 80:20 ratio with a constant crude protein content compared with the basal diet (Diet B), a diet with 80:20 ratio of basal diet and DESBM with a constant corn:soybean meal ratio (Diet C) and a diet with simple substitution of basal diet with DESBM in 80:20 ratio (Diet D). Pigs were fed in metabolism crates for a period of 16 d at 550 kcal ME/kg $BW^{0.60}$/d to determine DE and ME contents by total collection method. Thereafter, pigs were moved into an indirect calorimeter for a 36 h period where O_2 consumption and CO_2 production were measured to determine heat production (HP) and fasting heat production (FHP). The energy content of DESBM was calculated using the difference method (Woyengo *et al.*, 2010) by subtracting the NE contribution of the basal diet from the NE of the diets containing 20% DESBM.

Mixed procedure of SAS (Software release 9.2; SAS Institute, Cary, NC, USA) was used to analyse the data. The individual pig was considered as the experimental unit.

Results and discussion

Table 1 details the energy and heat production values for the 4 dietary treatments as determined using the IC method. The study demonstrated that the values obtained with the IC method were consistently greater than those obtained with prediction equations. The NE content of DESBM was calculated to be 2,652, 2,548 and 2,540 kcal/kg DM for diets B, C and D, respectively using IC (P<0.0001). Respective values obtained with published equations (Noblet *et al.*, 1994a) were 2,624, 2,530 and 2,436 kcal/kg DM. The discrepancy between the determination technique used was about 1% when diets were formulated with a constant protein content or corn:soybean meal ratio (1.0% and 0.7%, respectively), however, this was 4.1% when diet was formulated with simple substitution technique.

Table 1. Energy and heat production values for diets.

Item	Dietary treatments				SEM	*P*-value
	A	B	C	D		
DE, kcal/kg of DM	3,481	3,462	3,472	3,473	9.40	0.586
ME, kcal/kg of DM	3,392	3,378	3,386	3,386	6.16	0.431
HP, kcal/kg of DM[1]	2,007	1,930	2,038	2,000	53.01	0.637
FHP, kcal/kg of DM	1,547	1,424	1,508	1,468	53.99	0.513
RE, kcal/kg of DM[2]	1,385	1,448	1,348	1,386	52.85	0.686
NE, kcal/kg of DM[3]	2,932	2,872	2,855	2,853	5.66	<0.0001

[1] HP = 3.87 × O_2 + 1.20 × CO_2 -1.43 × urinary N; where HP = heat production (kcal); O_2 = oxygen consumption (L); CO_2 = carbon dioxide production (L).

[2] RE = ME – HP; where RE = retained energy (kcal/d), ME = metabolisable energy (kcal/d).

[3] NE = (RE +FHP)/DMI; where NE= net energy (kcal/kg DM); FHP = fasting HP (kcal/d); DMI = dry matter intake (kg/d).

Conclusion

The results from the present study shows that the NE values of DESBM obtained with the IC method were higher than those obtained with prediction equations for all the three dietary designs; the disparity being least when formulated with a constant corn:soybean meal ratio. So for routine NE determination, diets should be formulated to contain a constant ration of other energy yielding components.

References

Ayoade, D.I., E. Kiarie, M.A. Trinidade Neto and C.M. Nyachoti, 2012. Net energy of diets containing wheat-corn distillers dried grains with solubles as determined by indirect calorimetry, comparative slaughter, and chemical composition methods. J. Anim. Sci. 90, 4373-4379.

Noblet, J., H. Fortune, C. Dupire, and S. Dubois, 1993. Digestable, metabolisable and net energy values for 13 feedstuffs for growing pigs: effect of energy system. Anim. Feed Sci. Technol. 42, 131-149.

Noblet, J., H. Fortune, X. S. Shi, and S. Dubois, 1994. Effect of body weight on net energy value of feeds for growing pigs. J. Anim. Sci. 72, 648-657.

Noblet, J., X.S. Shiand, and S. Dubois, 1993. Metabolic utilization of dietary energy and nutrients for maintenance energy requirements in sows: basis for a net energy system. Br. J. Nutr. 70, 407-419.

Noblet, J., H. Fortune, X. S. Shi, and S. Dubois, 1994. Prediction of net energy value of feeds for growing pigs. J. Anim. Sci. 72, 344-354.

Woyengo, T. A., E. Kiarie and C.M. Nyachoti, 2010. Energy and amino acid utilization in expeller-extracted canola meal fed to growing pigs. J. Anim. Sci. 88, 1433-1441.

Effect of early surgical castration and immune castration on postprandial nutrient profiles in male pigs

N. Le Floc'h[1,2], A. Prunier[1,2], J. van Milgen[1,2], H. Furbeyre[1,2] and I. Louveau[1,2]
[1]INRA, UMR1348 PEGASE, 35590 Saint-Gilles, France; jaap.vanmilgen@rennes.inra.fr
[2]Agrocampus Ouest, UMR1348 PEGASE, 35000 Rennes, France

Introduction

Rearing of entire males (EM) or immunologically castrated male (IC) pigs are two alternatives to surgical castration. Male pigs eat less and exhibit higher growth performance and feed efficiency than early surgically castrated pigs (SC). Late castration by immunization against gonadotrophin-releasing hormone (GnRH) is a relevant strategy to maintain growth performance and prevent the accumulation of molecules responsible for boar taint in meat (Millet *et al.*, 2011). Indeed, IC pigs are similar to EM until immunization is effective. Despite large differences in feed efficiency between EM and SC pigs, the mechanisms involved in these differences have been poorly investigated (Claus *et al.*, 2007). The current study was undertaken to determine whether difference in feed efficiency between SC, IC and EM pigs can be explained by difference in nutrient utilization after a meal by measuring postprandial profiles of plasma nutrient and hormone concentrations.

Material and methods

Male Piétrain × (Large White × Landrace) pigs, 6 SC, 6 EM and 6 IC, were fitted with a jugular catheter under general anaesthesia at 14 weeks of age. Pigs were fed ad libitum a growing diet (CP 16.5%, NE 9.67 MJ/kg, SID Lys 8.4%). Three postprandial plasma nutrient profiles were established on each pig at 16, 18, and 20 weeks of age. These periods were chosen to surround the second injection of Improvac® to IC pigs: the first period being before the decrease in testicular hormones, the other periods being during and after the decrease in testicular hormones in IC pigs. On each test day, blood samples were collected prior to the test meal (400 g), after pigs were fasted overnight, and for 4 hours after the test meal. Plasma concentrations of glucose, urea, amino acids and insulin were determined on each blood samples. Data were analyzed as repeated measured using the MIXED procedure of SAS. The model included the period (age effect), the experimental group (SC, IC and EM), the time and their interactions.

Results and Discussion

The postprandial nutrient profiles were established at a constant age, at an average body weight of 63, 79 and 87 kg for periods 1, 2 and 3, respectively. Plasma insulin concentrations did not differ between the different periods and between IC, SC and EM pigs (data not shown). The average glycaemia (Table 1) did not differ between periods in SC and EM pigs whereas it decreased with age in IC pigs. During period 1, average glycaemia was lower in SC than in the EM and IC pigs but the time-related variations of glycaemia did not differ between the 3 groups (Figure 1). After the second injection of Improvac (periods 2 and 3), average glycaemia was greater in EM pigs than in IC and SC pigs. During period 3, the peak of glucose measured 50 min after the meal distribution was lower in SC and IC pigs compared with EM pigs. Thereafter, glycaemia remained higher than basal values in EM pigs until 180 min whereas it did not differ from basal concentrations in SC and IC pigs.

In both IC and SC pigs, the plasma urea concentrations increased with age whereas those of AA decreased. In EM pigs, both plasma urea and AA concentrations decreased with age. Urea was the highest in SC pigs whereas it did not differ in IC and EM during the three periods.

Table 1. Effects of surgical castration and immune castration on average plasma glucose, urea, and total AA concentrations.

	Period	Experimental groups			Statistics		
		SC	IC	EM	Group	Period	Group × Period
Glucose mg/l	P1	1,132^{x}	1,191^{a,y}	1,193^{y}	0.0009	0.0003	0.21
	P2	1,095^{x}	1,120^{b,x}	1,183^{y}			
	P3	1,092^{x}	1,103^{b,x}	1,178^{y}			
Urea mg/l	P1	253^{a,x}	204^{a,y}	227^{a,x,y}	<0.0001	0.0008	<0.0001
	P2	289^{b,x}	209^{a,y}	231^{a,y}			
	P3	304^{c,x}	239^{b,y}	213^{b,y}			
Total AA µmol/l	P1	5,030^{a}	4,786^{a}	5,006^{a}	0.42	<0.0001	0.17
	P2	4,905^{a}	4,527^{b}	4,774^{b}			
	P3	4,545^{b}	4,557^{b}	4,532^{c}			

^{x,y} $P<0.05$ between experimental groups (on a same line).
^{a,b,c} $P<0.05$ between periods (on a same column).

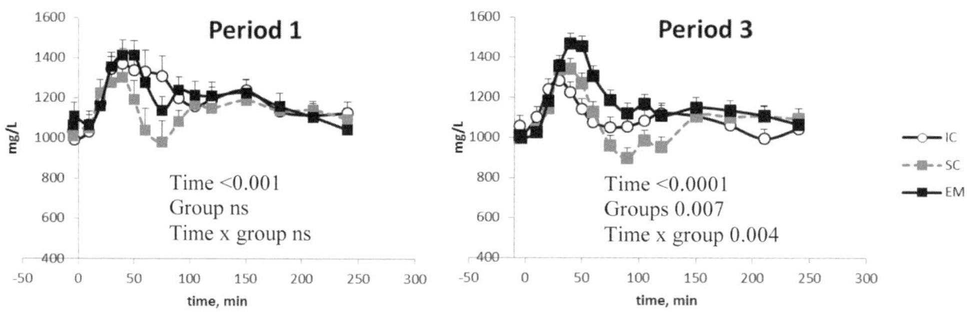

Figure1. Plasma glucose concentrations measured during the 4 hours following the meal distribution.

In summary, plasma glucose profiles were affected by immune castration earlier than urea and AA profiles. The analysis of postprandial glycaemia indicates a greater glucose clearance in both castrated groups, which is not in accordance with data reported in rats (Holmang *et al.*, 1992). Our results suggest that IC pigs would keep the advantages of EM pigs in terms of nitrogen metabolism during the experimental period.

References

Claus, R., M. Lacorn, K. Danowski, M.C. Pearce and A. Bauer, 2007. Short-term endocrine and metabolic reactions before and after second immunization against GnRH in boars. Vaccine 25, 4689-4696.

Holmang, A and P. Bjorntorp, 1992. The effects of testosterone on insulin sensitivity in male rats. Acta Physiol Scand 146:505-510.

Millet S., K. Gielkens, D. De Brabander and G.P.J. Janssens, 2011. Considerations on the performance of immunocastrated male pigs. Animal 5, 1119-1123.

Determination of the valine requirements for growth in pigs from 8 to 18 kg

E. Assadi[1], J.V. Nørgaard[1], N. Canibe[1], B.B. Jensen[1], B. Nielsen[2], K. Blaabjerg[1] and H.D. Poulsen[1]
[1]Department of Animal Science, Aarhus University, 8830 Tjele, Denmark; elhama.soumeh@agrsci.dk
[2]Chr. Hansen A/S, Bøge Allé 10-12, 2970 Hørsholm, Denmark

Introduction

Pig diets are supplemented with crystalline amino acids (AA) to balance the AA profile. This is necessary to avoid poor nitrogen (N) utilization, urinary N excretion, and low growth rates. Valine (Val) is one of the branched-chain AA which cannot be synthesized by the animal and therefore is listed as indispensable AA. In European grower diets, Val is often the fifth limiting AA after lysine (Lys), methionine (Met), threonine (Thr), and tryptophan (Trp) and appears to be a deficient nutrient in pig feedstuffs in regard to the minimum requirement for pigs. The objective of this study was to determine the requirements of Val that supports the maximum animal performance.

Material and methods

To determine the requirements of Val, 120 individually penned pigs with average weight of 9.4±1.4 kg were used in two periods. The experiment started 4-5 days after weaning at 28 days. The pigs were allotted to one of three experimental diets with 0.63:1, 0.67:1, and 0.71:1 of SID Val:Lys ratio. The composition of experimental diets is shown in Table 1.

Each period of the experiment ran for 21 days and average daily gain (ADG), average daily feed intake (ADFI) and feed conversion ratio (FCR) were recorded at the end of each week. Blood samples were taken 3 hours after feeding from the jugular vein at day 18 of the experiment and analyzed for plasma Val, urea and Lys content.

The data were analyzed by the MIXED procedure (SAS, 2011) considering the effect of treatments.

Results and discussion

The performance parameters showed that the ADFI for the 21 day period tended to be higher ($P<0.07$) in the 0.67:1 Val:Lys ratio compared to the lower and upper Val:Lys ratios (0.69 vs. 0.61 and 0.63 kg in the 0.67:1, 0.63:1 and 0.71:1 Val:Lys ratios, respectively). The ADG was not different among the dietary treatments during the 21 day period (0.44, 0.50 and 0.45 kg in the 0.63:1, 0.67:1 and 0.71:1 Val:Lys ratios, respectively). The FCR also was not affected by increasing AA levels in the diets during the experimental period (1.40, 1.40 and 1.43 kg in the 0.63:1, 0.67:1 and 0.71:1 Val:Lys ratios, respectively). The data on blood parameters (Figure 1) showed a significant increase

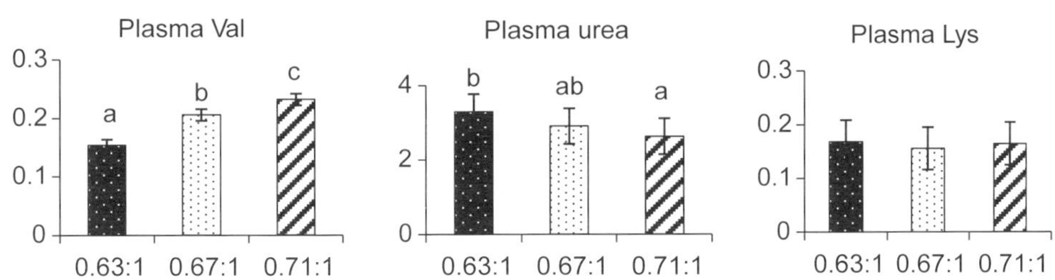

Figure 1. The plasma Val, urea and Lys contents (mmol/l) in the three dietary Val:Lys ratios during the overall period (1-21 day). The higher Val level in the diet increased the plasma Val content (P<0.001) and decreased the plasma urea content (P<0.04) significantly.

Table 1. Composition of three experimental diets with increasing Val:Lys ratios.

Diet composition,%	SID[1] Val:Lys		
	0.63:1	0.67:1	0.71:1
Wheat	64.7	64.7	64.7
Soy protein concentrate	17.5	17.5	17.5
Barley	10.0	10.0	10.0
Fat	2.50	2.50	2.50
Limestone, 38% Ca	1.85	1.85	1.85
Monocalcium phosphate	1.05	1.05	1.05
Salt	0.436	0.436	0.436
Vit-min premix[2]	0.400	0.400	0.400
Microgrits[3]	0.070	0.070	0.070
Phytase[4]	0.020	0.020	0.020
DL-methionine	0.181	0.181	0.181
L-glutamic acid [5]	0.150	0.105	0.059
L-histidine	0.036	0.036	0.036
L-isoleucine	0.036	0.036	0.036
L-leucine	0.147	0.147	0.147
L-lysine HCl	0.531	0.531	0.531
L-phenylalanine	0.039	0.039	0.039
L-threonine	0.236	0.236	0.236
L-tryptophan	0.069	0.069	0.069
L-valine	0.032	0.077	0.123

[1] SID = standardized ileal digestible valine to lysine ratio.

[2] Provided the following per kg of diet: 10,000 IU vitamin A, 2,000 IU vitamin D_3, 94 IU vitamin E, 2.4 mg vitamin K_3, 2.4 mg vitamin B_1, 4.8 mg vitamin B_2, 2.4 mg vitamin B_6, 0.02 mg vitamin B_{12}, 12 mg D-panthotenic acid, 26 mg niacin, 0.2 mg biotin, 200 mg Fe (Fe(II) sulphate), 165 mg Cu (Cu(II) sulphate), 200 mg Zn (Zn(II) oxide), 56 mg Mn (Mn(II) oxide), 0.3 mg KI, 0.3 mg Se (Se-selenite).

[3] Jadis Additiva, Haarlem, the Netherlands.

[4] Natuphos 5000 (100 g/t) (BASF, Ludwigshafen, Germany).

[5] Included to compose isonitrogenic diets.

in the plasma Val ($P<0.001$) and a decrease in the plasma urea content ($P<0.04$) among the dietary treatments. The results of the current study were in agreement with the estimates of the optimum 66% and 67% SID Val:Lys ratio for ADG and ADFI, respectively found by Wiltafsky *et al.* (2009). Gaines *et al.* (2011) also reported that performance increased until a ratio of 65% and the plasma urea content decreased quadratically as the Val:Lys ratio increased in the diet. It can be concluded from the current study that the 0.67:1 SID Val:Lys ratio supports more ADFI and ADG of pigs compared to 0.63:1 and 0.71:1 Val:Lys ratios.

References

Gaines A. M., D. C. Kendall, G. L. Allee, J. L. Usry, and B. J. Kerr. 2011. Estimation of the standardized ileal digestible valine-to-lysine ratio in 13- to 32-kilogram pigs. J. Anim. Sci. 89, 736-742

Wiltafsky, M. K., B. Schmidtlein, and F. X. Roth. 2009. Estimates of the optimum dietary valine to lysine ratio for eight to twenty-five kilograms of body weight pigs. J. Anim. Sci. 87, 2544-2553.

Ideal isoleucine and valine to lysine ratios in low protein diets for growing pigs

C. Wecke, A. Pastor and F. Liebert
Division of Animal Nutrition Physiology, Department of Animal Sciences, Georg-August-University, Kellnerweg 6, 37077 Goettingen, Germany; anja.pastor@agr.uni-goettingen.de

Introduction

Isoleucine (ILE) or valine (VAL) are considered to be potentially limiting the dietary protein quality in protein reduced pig diets supplemented with L-lysine·HCl, DL-methionine, L-threonine and L-tryptophan. Consequently, validated dietary ratios of both amino acids (AA) are necessary to ensure optimal feed protein utilization in fast growing pigs. Therefore, the present study with growing fattening pigs was conducted to derive AA efficiency data of ILE and VAL as related to lysine (LYS) in low protein diets to conclude ideal amino acid ratios (IAAR) of ILE, VAL and LYS, respectively.

Material and methods

N balance studies (n=7) in modern genotype barrows (German Landrace x Piétrain, 55±1 kg average BW) were conducted to measure N utilization and AA efficiency of protein reduced diets (134 g CP/kg) based on barley (78%), field peas (12%), soybean oil (3%), minerals, vitamins, and a mixture of crystalline AAs, respectively. The individual dietary AA ratios related to lysine were above or in line with current recommendations of GfE (2008), except the AA under study. Principles of diet dilution technique were applied to achieve individual limiting position of the AA under study. Additionally, the LYS diluted diet was supplemented with L-glutamic acid (GLU) to examine effects of increased dispensable AA supply on efficiency of lysine as the reference AA. For analyses of N balance data, an exponential N utilization model was used (Liebert and Wecke, 2008). The model parameter application ran according to the procedure as described by Wecke and Liebert (2010). Because the AA efficiency parameter bc^{-1} summarizes the digestion, absorption and post-absorptive utilization processes of the limiting AA and is indirectly related to the physiological requirement per unit protein deposition, the reciprocal relationship between LYS efficiency (as reference) and observed efficiency of individual AA under study was utilized to derive the IAARs according to Samadi and Liebert (2008):

$$IAAR = bc_{Lys}^{-1} : bc_{LAA}^{-1}$$

Statistical analyses ran with software package IBM SPSS Statistics 19.

Results and discussion

Table 1 summarizes the obtained model parameter of protein quality (b) and AA efficiency (bc^{-1}). According to expectation and with reference to the LYS diluted diet with elevated supply of ILE (5.62 g/kg) and VAL (6.55 g/kg), significantly lower ($P<0.05$) data for protein quality were observed due to application of the ILE and VAL diluted diets. Likewise, the individual AA efficiency of the regarding AA diluted diet was significantly improved. Due to the observed significant effects on protein quality and AA efficiency data, the individual position as limiting AA (LAA) was confirmed. In addition, no significant effect of GLU supplementation to the LYS diluted diet on model parameter b and bc^{-1} was observed. In consequence, no deficiency of nonessential AA (NEAA) could be expected. Therefore, the applied dietary EAA:NEAA ratio (47:53) can be stated as adequate.

Table 1. Model parameter of protein quality and AA efficiency and derived IAAR.

Diet[1]	ILE	VAL	LYS	LYS (+GLU)
LAA content (g/kg)	3.75	4.77	9.05	9.05
c_{LAA} (g/16gN)	2.81	3.58	6.71	6.08
Protein quality (*b*)	424^b±8	434^b±16	555^a±12	513^a±10
ILE efficiency (bc^{-1})	**151**^{a}±3	103^c±4	133^b±3	136^b±3
VAL efficiency (bc^{-1})	87^c±2	**121**^{a}±4	114^b±2	116^{ab}±2
LYS efficiency (bc^{-1})	63^b±1	64^b±2	**83**^{a}±2	**84**^{a}±2
IAAR	55	69	100	

[1] Model parameters *b* resp. bc^{-1} in original were multiplied by factor 10^6.

Based on these conditions, the relationship between highest LYS efficiency (as reference) and observed maximal efficiency data for both ILE and VAL was utilized to derive the IAAR. The results are summarized in Table 1.

The derived ratio LYS:ILE:VAL = 100:55:69 is in line with current recommendations of NRC (2012) based on total AA requirement data (100:53:67). BSAS (2006) recommendations are slightly higher for the proportion of standardized ileal digestible ILE (58), but very similar for standardized ileal digestible VAL (70). Generally, the yielded ratios of ILE and VAL to LYS in the present study are higher than current German recommendations (GfE, 2008), assuming an IAAR based on AA composition of the body protein gain (100:49:65) in growing fattening pigs.

References

BSAS (British Society of Animal Science), 2003. Nutrient Requirement Standards for Pigs. (Authors: Whittemore, C. T., M. J. Hazzledine and W. H. Close). BSAS, Penicuik, UK.

GfE (Gesellschaft für Ernährungsphysiologie), 2008. Recommendations for the Supply of Energy and Nutrients to Pigs. Committee for Requirement Standards of the Society of Nutrition Physiology. DLG-Verlags-GmbH, Frankfurt a. M., Germany.

Liebert, F. and C. Wecke, 2008. Models for further developing the evaluation of protein and amino acids as well as for predicting performance from energy and amino acids intake. In Recommendations for the Supply of Energy and Nutrients to Pigs. Committee for Requirement Standards of the Society of Nutrition Physiology. DLG-Verlags-GmbH, Frankfurt a. M., pp. 219-230.

NRC (National Research Council), 2012. Nutrient Requirements of Swine. 11th rev. ed., The National Academies Press, Washington, D.C.

Samadi and F. Liebert, 2008. Modelling the optimal lysine to threonine ratio in growing chickens depending on age and efficiency of dietary amino acid utilisation. Brit. Poult. Sci. 49, 45-54.

Wecke, C. and F. Liebert, 2010. Optimal dietary lysine to threonine ratio in pigs (30-110 kg BW) derived from observed dietary amino acid efficiency. J. Anim. Physiol. a. Anim. Nutr. 94, e277-e285.

Interaction between the valine and tryptophan requirement in young piglets

A.J.M. Jansman[1], H. van Diepen[1], M. Rovers[2] and E. Corrent[3]
[1]Wageningen UR Livestock Research, P.O. Box 65, 8200 AB Lelystad, the Netherlands; alfons.jansman@wur.nl
[2]Orffa Additives B.V., Vierlinghstraat 51, 4251 LC Werkendam, the Netherlands
[3]Ajinomoto Eurolysine s.a.s. 153 rue de Courcelles, 75817 Paris Cedex 17, France

Introduction

After lysine, methionine plus cysteine, threonine and tryptophan, valine is the next limiting amino acid in most common diets for post weaning piglets, in particular in low protein diets. To be able to formulate low protein diets it is essential to have adequate information on the requirement for valine in young piglets and the interaction with the requirements for other amino acids. It is known that an interaction exist between tryptophan and branched-chain amino acids (BCAA; Val, Leu and Ile) as these amino acids share a common transport mechanisms across membranes (Henry *et al.*, 1992; Langer and Fuller, 2000). This suggests that the response towards dietary supplementation of L-valine could be affected by the dietary level of tryptophan. The aim of the present study was to evaluate the interaction between the requirement for valine and tryptophan in young piglets in a performance study in the post weaning period (from about 8 to 24 kg body weight).

Material and methods

Five experimental treatments (I to V) were evaluated each receiving a different experimental diet. A valine and tryptophan deficient, low protein basal diet (169 g CP per kg; 1.8 g SID Trp per kg; 7.0 g SID Val per kg; treatment I) was used supplemented with 1.0 g/kg L-valine (treatment II) or 0.5 g/kg L-tryptophan (treatment III) or both (treatment IV). A reference treatment was included receiving a diet with a crude protein content of 192 g CP/kg and 2.5 g SID Trp per kg and 8.0 g/kg SID Val (treatment V). The diets were based on barley, maize, wheat, soybean meal and peas as main ingredients and contained 10.4 g SID Lys per kg. Each of the five treatments was evaluated in eight replicates (pens with eight male piglets). Feed intake (FI), body weight gain (BWG) and feed conversion ratio (FCR) were measured as the response criteria over an experimental period of four weeks, starting one week post weaning. Except for amino acids, the diets were formulated to be nutritionally complete (CVB, 1996; NRC, 1998). In the first week after weaning the diet for treatment V was given to all piglets.

Results and discussion

Over weeks 1-2, BWG and FCR were significantly affected by the dietary treatments ($P<0.05$). Compared to treatment II (only supplemented with L-valine), BWG was improved (+14%) in treatment IV, receiving the L-valine and L-tryptophan supplemented diet ($P<0.05$). The FCR over this period was improved in treatment IV compared to treatments I, II and III ($P<0.05$). FI over weeks 1-2 did not differ between treatments ($P>0.05$) but tended to be lower in treatment II compared to the other treatments. Over the complete experimental period (0-4 weeks; Table 1), FI was numerically lower in treatment II (only supplemented with L-valine) compared to the other treatments. BWG in treatment II was lower compared to all other treatments ($P<0.05$). BWG was numerically highest in treatment IV (supplemented with both L-valine and L-tryptophan). Over weeks 0-4, FCR was significantly lower in treatment IV compared to all other treatments ($P<0.05$).

Table 1. Feed intake, body weight gain and feed conversion ratio over the experimental period (weeks 0-4).

		FI (kg/d)	% of I	BWG (g/d)	% of I	FCR	% of I
I	CP, 169 g/kg	0.838	100	559 ^{b}	100	1.498^{b}	100
II	I + Val	0.790	94	530 ^{a}	95	1.492^{b}	100
III	I + Trp	0.842	101	565 ^{b}	101	1.493^{b}	100
IV	I + Val + Trp	0.847	101	582 ^{b}	104	1.455^{a}	97
V	CP, 192 g/kg	0.842	100	568 ^{b}	102	1.482^{b}	99
P-value		0.058		0.005		0.042	
LSD[1]		0.042		25.8		0.026	

^{a,b} Values with a different superscript in the same column within a factor differ at $P<0.05$.

[1] LSD = least significant difference ($P<0.05$).

The study revealed an interaction between the dietary supply of tryptophan and valine with regard to the growth performance of post weaning piglets. Supplementation of 1.0 g/kg L-valine to a low tryptophan diet (1.8 g/kg SID Trp) with 169 g/kg CP decreased feed intake and body weight gain, while supplementation of 1.0 g/kg L-valine to a diet with an adequate tryptophan level (2.4 g/kg SID Trp) increased performance and improved feed conversion ratio in piglets over the period of weeks 2 to 5 post weaning. The former emphasizes the value of the ideal protein concept in diet formulation (Chung and Baker, 1996) and thus the need for properly balancing all essential amino acids in practical diets in order to achieve maximum performance in post weaning piglets using low protein diets.

References

Chung, T.K., and D.H. Baker, 1992. Ideal amino acid pattern for 10-kilogram pigs. J. Anim. Sci. 70, 3102-3111.

CVB, 1996. [Requirements for amino acids in piglets and pigs] In Dutch. CVB-documentation report 14.

Henry, Y., B. Sève, Y. Colléaux, P. Ganier, C. Saligaut, and P. Jégo, 1992. Interactive effects of dietary levels of tryptophan and protein on voluntary feed intake and growth performance in pigs, in relation to plasma free amino acids and hypothalamic serotonin. J. Anim. Sci. 70:1873-1887.

Langer, S., and M. F. Fuller, 2000. Interactions among the branched chain amino acids and their effects on methionine utilization in growing pigs: Effects of nitrogen retention and amino acid utilization. Br. J. Nutr. 83:43-48.

NRC, 1998. Nutrient Requirements of Swine. National Research Council. National Academy Press, Washington, US.

Is high protein diet a good nutrition strategy for broiler chickens reared at heat stress condition?

D.M.B. Campos, M.F. Fernandez-Alarcon, F.A. Souza, W.C.L. Nogueira, F.H. Hada, P.R.O. Carneiro and M. Macari
Department of Animal Morphology and Physiology, Sao Paulo State University, Jaboticabal, SP, Brazil; dmbczoo@hotmail.com

Introduction

It is well known that heat stress, in growing broiler chicks, impair the performance. Ain Baziz *et al.* (1996) reported that 53% of the impairment in the performance was due to reduction in feed consumption, but the last 47% was due to the direct effect of ambient temperature, *per se*. One of the approaches that have been proposed to avoid the effect of heat stress in broilers is related to the protein content in the diet, since protein metabolism is deeply involved in the caloric increment, as compared with fat or carbohydrate. According to Gonzalez-Esquerra and Leeson (2006) the Arg:Lys ratio, Met source and time to exposure to heat stress affect protein utilization in hyperthermic birds. Thus, this study was conducted aiming to verify if the increase of protein content in the diet, maintaining the Arg:Lys ratio, affects broiler chicks performance when reared at different environmental conditions.

Material and methods

It was used seven hundred and twenty male chicks from Cobb™ strain. The experiment was performed during the broiler chicks finishing period (21 to 45 days of age) following a 3 protein level (18.91, 21 and 22%) × 3 rearing temperature schedules [thermoneutral *ad libitum* (TN-AL), cyclic heat stress (C-HS, 16h at thermoneutrality and 8 h at 32 °C) and thermoneutral pair-fed (TN-PF) with C-HS], in a factorial design, with four replicates of 20 birds each (total of 9 treatments). From 1-21 days of age, the birds were fed diet containing 21.6% of crude protein (CP) and 3,020 kcal ME/kg. During the experimental period, the diets had the same nutritional and energy levels (3,185 kcal AMEn/kg, 0.7% Ca, 0.33% available P, 0.20% Na, 1.08-1.10% Lys, 0.49-0.54% Met, 0.79% Met+Cys, 0.70-0.75% Thr, 0.20-0.25% Trp, 1.53-1.50% Arg, 1.39 Arg:Lys, 0.45-0.50 Met:Lys), except for crude protein content. Feed intake (FI), feed conversion (FC), daily weight gain (WG), viability (V), productive efficiency index (PEI=(WG×V)/(FC×10), carcass yield (CY), breast (B) and tight+drumstick (T+D) yield were measured. The data were submitted to analysis of variance (General Linear Model, SAS) and significance was compared by Tukey's test at 5% probability.

Results and discussion

The FI was lower and FC was better for 22% CP, in comparison to other levels of CP (Table 1). Some authors reported the same results (Temim *et al.*, 2000; Widyaratne and Drew, 2011), suggesting that the growth efficiency was enhanced by the increased crude protein of the diet in heat stress or thermoneutrality temperature in fast-growing strains.

The treatment C-HS presented worse performance (FI and WG) compared to TN-AL (Table 1). These results could be related to the chicken attempt to reduce the impact of extra heat load due to stress. The reduction of growth efficiency caused by C-HS affected CY, leading to better results in TN-AL treatment. Even under a cyclic heat stress, which allows the chickens to recover, it could be noticed a negative effect in performance and CY, as found by Gonzalez-Esquerra and Leeson (2005) and Temim *et al.* (2000).

There was no effect in V due to the CP treatments or temperature schedules. But, it was observed a significant interaction between the main factors for PEI. The birds fed 22% CP in C-HS condition, presented better performance than those fed 18.91% CP (Table 2). Since PEI takes into account WG, V and FC, slightly changes in these parameters was able to promote this improvement. Based on those results, the increase in the crude protein diet could be a feasible nutritional strategy to reduce the effects of cyclic thermal stress during the broilers' growth period.

Table 1. Performance of broiler chickens from 21 to 45 days old submitted to three dietary protein levels and reared under different ambient temperatures.[1]

CP%	FI (kg)	FC	WG (g)	V (%)	PEI	CY (%)	B (%)	TD (%)
18.91	3.86^{A}	1.57^{A}	102.69	95.00	611.46	72.84	39.78	28.14
21.00	3.84^{A}	1.56^{A}	102.76	95.41	630.86	72.41	39.15	28.68
22.00	3.74^{B}	1.52^{B}	102.89	98.33	668.57	71.97	38.95	28.92
Temp.								
CY-HS	3.62^{C}	1.49^{B}	101.38^{B}	95.42	639.32	71.52^{B}	38.91	29.3^{A}
TN-PF	3.86^{B}	1.59^{A}	100.92^{B}	96.67	612.94	72.18^{AB}	39.58	22.9^{B}
TN-AL, 22 °C	3.96^{A}	1.56^{A}	106.03^{A}	96.67	658.64	73.51^{A}	39.40	28.46^{AB}
P-value								
CP%	0.0001	0.0092	0.9929	0.1153	0.0105	0.3852	0.4662	0.0882
Temp.	0.0128	0.0001	0.0114	0.6943	0.0436	0.0123	0.6113	0.0052
CP%*Temp.	0.1438	0.3211	0.3269	0.4011	0.0120	0.1687	0.8296	0.9072
CV	2.52	2.54	4.13	4.27	6.60	1.06	2.12	1.47

[1] Means followed by same uppercase letters in column do not differ by Tukey test ($P>0.05$).

Table 2. Temperature and CP% interaction on the birds' PEI from 21 to 45 days old.[1]

Temp.	CP%		
	18.91	21	22
C-HS	570.84^{B}	636.23^{AB}	710.89^{A}
TN-PF	585.30^{A}	603.35^{A}	650.17^{A}
TN-AL	678.24^{A}	653.02^{A}	644.67^{A}

[1] Means followed by the same uppercase letters in row do not differ by Tukey test ($P>0.05$).

References

Ain Baziz, H., Geraert, P.A, Padilha, J.C.F.; S. Guillaumin. 1996. Chronic Heat Exposure Enhances Fat Deposition and Modifies Muscle and Fat Partition in Broiler Carcasses. Poult. Sci.75, 505-513.

Gonzalez-Esquerra, R., and S. Leeson.2005. Effects of Acute Versus Chronic Heat Stress on Broiler Response to Dietary Protein. Poult. Sci. 84, 1562-1569.

Gonzalez-Esquerra, R., and S. Leeson.2006. Effect of Arginine:Lysine Ratios and Source of Methionine on Growth and Body Protein Accretion in Acutely and Chronically Heat-Stressed Broilers. Poult. Sci. 85, 1594-1602.

Temim, S., Chagneau, A. M.; Guillaumin, S., Michel, J., Peresson, R., S. Tesseraud. 2000. Does Excess Dietary Protein Improve Growth Performance and Carcass Characteristics in Heat-Exposed Chickens? Poult. Sci. 79, 312-317.

Widyaratne, G. P., and M. D. Drew. 2011. Effects of protein level and digestibility on the growth and carcass characteristics of broiler chickens. Poult. Sci. 90, 595-603.

Evaluation of energy systems in corn and barley based diets and an enzyme complex in broiler chicks

S. Cerrate[1], J. Caldas[3], R. Ekmay[2], J. England[3] and C. Coon[3]
[1]Aviagen Incorporated, Huntsville, AL 35805, USA
[2]Department of Animal Science, Cornell University, Ithaca, NY 14853, USA
[3]Poultry Science Department, University of Arkansas, Fayetteville, AR 72701, USA; ccoon@uark.edu

Introduction

The extensive use of an NE system has not been accepted by the poultry industry because NE feed evaluation is laborious, values have a high variability and the process is expensive. Productive energy (PE), developed by Fraps and Carlyle (1939), was used commercially from 1946-1960 and was unfortunately related to NE. The methodology for the determination of PE is considered the cause for the high variability. De Groote (1968) used one plane of feed intake along with an independent NEm and showed similar variability between ME and NE energy systems. High energy feed costs have caused the poultry industry to dramatically increase the use of exogenous feed enzymes during the past 5 years. The objectives of this study were to compare the ME and NE systems for variability, predicted performance and to determine the extra energy content due to an enzyme complex.

Material and methods

Twenty mash diets from 5 ingredients (corn, soybean meal, pro-plus, barley and poultry fat) and Rovabio® Max at 0.05% (xylanase, B-glucanase, and phytase) were fed to male chicks from 1 to 21 d old. Feed intake was adjusted for each diet to supply similar ME intake per day. There were 3 groups of diets which had excess levels of fat (6, 8, 12, and 13 EE), protein (24, 27.5 and 31% CP), or fiber (14, 16 and 18% neutral detergent fiber) as increasing poultry oil, SBM-Pro-Plus, and barley respectively. Each dietary treatment was fed to 4 replicate floor pens of 25 male and transferred to digestibility cages (6 chicks/cage) and mixed with 0.5% of titanium oxide (TiO_2). At 0 d of age and 21d of age, 5 birds per floor pen were randomly collected to measure protein and fat gain using the Dual Energy X-ray Absorptiometry (DEXA) scanner (Lunar Prodigy from GE®). Dn = $Nutrient_{diet} - Nutrient_{excreta} \times (TiO_{2diet} / TiO_{2excreta})$; Dn = nutrient digestibility. ME, kcal/kg = $GE_{diet} - GE_{excreta} \times (TiO_{2diet} / TiO_{2excreta}) - 8.22 \times (N_{diet} - N_{excreta} \times (TiO_{2diet} / TiO_{2excreta}))$. NE, kcal/kg = (NEm + NEg)/FI; where: NEm = $90 \times 1.15 \times BW_x^{0.75}$, NEg = 9.35 kcal/g × fat gain + 5.66 kcal/g x protein gain. HP = ME – NEg. HI = ME – NE.

Results and discussion

The larger coefficients for NDF in digestibility equations to improve protein (Table 1: Equation 1.1 and 1.2), fat (Equation 2.1 and 2.2), and starch (Equation 3.1 and 3.2) digestibility from chicks fed carbohydrases imply that fiber digestion helps to release the nutrients. There was interaction for ME and enzyme factors, where the extra ME due to carbohydrases were 45 and 213 kcal/kg for corn and barley based diets, respectively (data not shown). On the other hand, the extra NE due to carbohydrases was 113 kcal/kg for all types of diets. There was similar variability between ME (CV=2.6%) and NE systems (CV=2.2%) (P=0.21). Energetic efficiencies (NE/ME) of ME from coefficients of protein, fat, and carbohydrates to predict ME (Equation 4) and NE (Equation 5) were 0.60, 0.90, and 0.69 respectively. The NE intake was a better predictor of BWG rather than ME intake which indicates that NE system is a superior energy system for feed formulation (Figure 1).

In conclusion, the energy evaluation systems (ME vs. NE) affected the hierarchy among digestible nutrients. The NE intake was a better predictor of BWG than the ME intake and both energy systems showed similar variability. The extra ME contents due to carbohydrases were 45 and 213 kcal/as-fed for corn and barley based diets respectively and the extra NE from carbohydrases was 113 kcal/kg as-fed for all types of diets.

Table 1. Regression equations of digestibility coefficient of nutrients (DC N, %DM), ME (kcal/kg as-fed), and NE (kcal/kg as-fed) from dietary and digested nutrients (%DM).

No.	Enzyme	Equations	n	R^2	SEM
1.1	No	DC CP = 0.921(0.204)^a[CP] + 1.658(0.355)^a[EE] -0.806(0.451)^b[NDF]^x + 1.101(0.180)^a[Starch]	45	0.988	7.3
1.2	Yes	DC CP = 0.633(0.109)^a[CP] + 0.930(0.189)^a[EE] + 0.494(0.241)^a[NDF] + 0.948(0.096)^a[Starch]	45	0.988	3.9
2.1	No	DC EE = 2.019(0.186)^a[CP] + 0.524(0.329)^a[EE]^x + 0.994(0.316)^b[NDF]	45	0.993	7.0
2.2	Yes	DC EE = 1.591(0.214)^a[CP] – 0.108(0.378)^a[EE]^x + 2.227(0.364)^a[NDF]	45	0.991	8.1
3.1	No	DC starch = 85.6(3.64)^a – 0.48(0.228)^b[NDF]	45	0.095	3.7
3.2	Yes	DC starch = 75.2(3.90)^a + 0.26(0.244)^{xa}[NDF]	45	0.026	4.0
4	Pooled	ME = 50.8(1.8)[dCP] + 83.0(2.8)[dEE] + 39.1(0.6)[dCHO]	80	0.999	49
5	Pooled	ME = 31.2(4.8)[dCP] + 75.1(7.7)[dEE] + 26.8(1.5)[dCHO]	80	0.996	132

^{a,b} Values within columns of same type of nutrients and with or without carbohydrases having superscript letters differ significantly (P<0.05) according to confidence interval. ^{x}P-value >0.05.

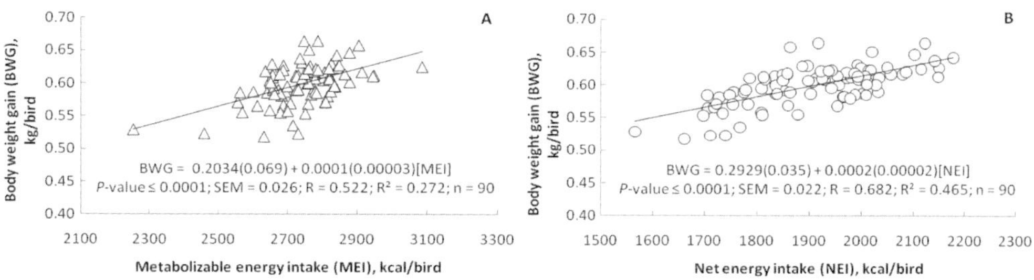

Figure 1. Linear regressions between body weight gain and MEI (A) or NEI (B).

References

De Groote, G., 1968. Energetic evaluation of unstabilized and stabilized fish meals in terms of metabolizable energy and net energy for maintenance and growth. Feedstuffs, Minneap., 40 (51): 26-27;54-55;61.

Fraps, G.S., and E.C. Carlely, 1939. The utilization of the energy of feed by growing chicks. Texas Agricultural Experiment Station Bulletin. 571.

Various fiber fractions as energy supply to exercising horses

C. Brøkner[1], D. Austbø[2], J.A. Næsset[2], D. Blache[3], K.E. Bach Knudsen[4] and A.H. Tauson[1,2]
[1]University of Copenhagen, Department of Veterinary Clinical and Animal Sciences, 1870 Frederiksberg C, Denmark; stinne@sund.ku.dk
[2]Norwegian University of Life Sciences, Department of Animal and Aquacultural Sciences, 1432 Ås, Norway
[3]University of Western Australia, Faculty of Natural and Agricultural Sciences, 6009 Perth, Australia
[4]Aarhus University, Department of Animal Science, 8830 Tjele, Denmark

Introduction

Horses have evolved to eat forages, which primarily are fermented in the hindgut. Results of previous studies indicate that fiber-rich ingredients can be used as only diet for exercising horses (Jansson *et al.*, 2012). However, the knowledge on the concomitant pattern of changes in blood metabolite concentrations due to fiber rich diets horses is limited. This experiment therefore aimed at studying the post-prandial blood plasma response in horses fed diets of different content of soluble and insoluble fiber fractions, while exercising daily.

Material and methods

Four trotter horses with an average body weight of 544 kg were used in a 4×4 Latin square design with a sequence of 24 days adaptation followed by 2 sampling days. The feed rations consisted of only timothy hay (H), hay and molassed sugar beet pulp (Betfor®) combined with either whole oats (OB) or whole barley (BB), hay and a loose chaff based concentrate (Equigard®) (M). The daily rations were divided into three meals: morning at 6 am, afternoon at 3.30 pm after end of sampling, and night at 10 pm. The morning concentrate meals were formulated to contain a maximum of 2 g starch/kg BW. The experimental diets were formulated to be iso-energetic and to fulfill the Danish feeding standards for medium work level (1.5× maintenance). Each horse was subjected to a standard exercise test at the end of each experimental period on an indoor high speed treadmill.

Blood was drawn via jugular vein puncture into heparinized vacutainer tubes at time 0 before the morning meal and again at 3 and 9 h post feeding. Blood plasma was analyzed for metabolites and regulatory hormones. Feed fiber and starch content was analyzed as non-starch polysaccharide (NSP), soluble non-cellulosic polysaccharides (S-NCP), insoluble non-cellulosic polysaccharide (I-NCP) and starch as described by Brøkner *et al.* (2012). Nutrient intake (kg DM) of the diets H, OB, BB, M respectively was as follows: NSP 5.4, 4.1, 3.8, 4.5; S-NCP 0.2, 0.37, 0.39, 0.56; I-NCP 2.5, 1.8, 1.7, 1.8; starch 0.02, 0.95, 1.09, 0.21. Heart beats were recorded by use of portable polar equine training system.

Results were analyzed statistically by use of SAS version 9.2 (SAS Institute Inc, 1999) and results were considered significantly different when P<0.05 and a tendency when P<0.10.

Results and discussion

The diet highest in soluble fiber (M) resulted in the highest plasma glucose and insulin concentrations in the morning after nearly 6 hours fast (Figure 1). This can be ascribed to the gluconeogenic properties of propionic acid through hepatic gluconeogenesis (Jose-Cunilleras and Hinchcliff, 2004). Only the concentration (Table 1) of plasma beta-hydroxybutyrate (BHBA) was significantly affected by diet (P<0.001) and highest on the hay only diet. The hay diet therefore seemed to result in a negative energy balance because the significantly higher BHBA concentration and numerically higher plasma non-esterified fatty acids (NEFA) concentration indicated that the horses were mobilizing fat. This implies that the horses were working harder than medium work level. However, the blood lactate

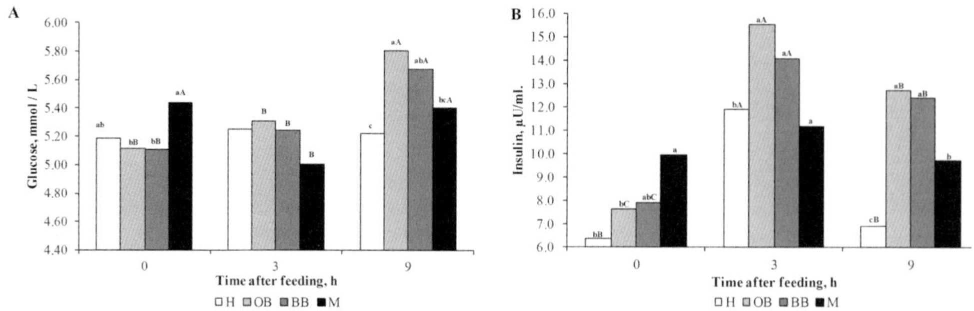

Figure 1. The post-prandial interaction between diet and time on blood plasma glucose (A) and insulin (B) in horses fed different diets (mean of 8 samples). Hay (H), oats and Betfor® (OB), barley and Betfor® (BB), Equigard®, a chaff based concentrate (M). The letters indicate the level of significance (P≤0.05), where the upper case^{A-B} indicate effect of time and the lower case^{a-c} indicate effect of diets.

Table 1. Effect of diet and time on blood plasma BHBA (LSMean, mmol/l) and NEFA (LSMean, µmol/l).[1]

	Diet					Time				*P*-value	
	H	OB	BB	M	SE	0	3	9	SE	Diet	Time
BHBA	0.22^a	0.17^b	0.17^b	0.18^b	<0.01	0.19^a	0.17^b	0.19^a	<0.01	<0.01	<0.01
NEFA	46.6	47.0	34.7	31.5	9.2	45.4^a	21.6^b	52.7^a	8.8	0.24	<0.01
Whole blood after training, mmol/l											
Lactate	2.8	2.9	2.1	2.1	0.8	-	-	-	-	0.8	-

[1] The level of significance was chosen when *P*-value <0.05. The values in rows with different superscripts differ significantly. H = hay; OB = oats and Betfor; BB = barley and Betfor; M = chaff based concentrate; NEFA = non-esterified fatty acids; BHBA = beta-hydroxybutyrate.

immediately after training ranged from 2.1-2.9 mmol/l which shows that the anaerobic threshold had not been reached, and that the horses were working aerobically (Jose-Cunilleras and Hinchcliff, 2004) equivalent to medium work level. The negative energy balance could also be explained by different energy intake even though the experimental diets were formulated to be iso-energetic. It is possible that the energy content in the hay was overestimated.

In conclusion, this results suggested that fiber based diets could fulfill the energy requirements for horses at medium work level. However, there were indications that the horses on the hay only diet were in negative energy balance. Since the horses were not working harder than medium level and the experimental diets were iso-energetic, these results indicate that the feed evaluation system for horses needs further refinement in order to more accurate determine the energy content in diets and in particular diets of high content of insoluble non-starch polysaccharides like hay.

References

Brøkner, C., K.E. Bach Knudsen, I. Karaman, K.L. Eybye and A.H. Tauson. 2012b. Chemical and physicochemical characterization of various horse feed ingredients. Anim. Feed Sci. Tech. 177, 86-97.

Jansson, A., M. Saastamoinen and J. E. Lindberg. 2012. Forage feeding system. In: Saastamoinen, M., M.J. Santos and N. Miraglia, (eds.) Forages and grazing in horse nutrition. EAAP series no. 132. Wageningen Academic Publishers, Wageningen, the Netherlands, pp 289-303.

Jose-Cunilleras, E. and K.W. Hinchcliff. 2004. Carbohydrate metabolism in exercising horses. Equine Comp. Exercise Physio. 1, 23-32.

SAS Institute Inc. 2008: SAS/STAT User's Guide, Version 9.2 Cary, NC, USA.

Changes in protein turnover during pregnancy in pigs when feeding limiting amounts of amino acids.

S. Moehn[1], M. Rafii[2], P.B. Pencharz[2] and R.O. Ball[1]
[1]Department of Agriculture, Food and Nutritional Science, University of Alberta, Edmonton T6G 2P5, Canada; smoehn@ualberta.ca
[2]Research Institute, Hospital for Sick Children, Toronto M5G 1X8, Canada

Introduction

A series of experiments were completed to determine the amino acid (AA) requirements of pigs for threonine (Thr), lysine, tryptophan (Trp) and isoleucine (Moehn *et al.*, 2011) in early and late pregnancy. The objective of the present analysis was to determine which factors caused changes in protein turnover during pregnancy in pigs when the intake of test AA was below requirements.

Material and methods

In every experiment, 6 to 7 sows each received 6 diets with graded levels of test AA in both early and late gestation. Protein turnover was determined using a stochastic model based on the ^{13}C enrichment in expired CO_2 and plasma free phenylalanine (Phe) after oral L[^{13}C]Phe dosing. To combine the results of all experiments, test AA intake was expressed as a fraction of the requirement determined in each experiment, e.g. the Thr intake of 6.35 g/d equaled 0.467 times the requirement of 13.6 g/d, or the Trp intake of 2.02 g/d equaled 0.777 times the requirement of 2.6 g/d. The observations (n=180 for oxidation [OX] and Phe retention, n=129 for flux, protein breakdown [B] and synthesis [S]) were made on 23 sows in their 2nd to 4th parity between day 23 and 110 of pregnancy. Sow body weight (BW) at each study day ranged from 153.8 kg to 294.0 kg. Maternal and total gain in pregnancy was between 6.0 and 60.5 kg and 29.0 and 78.8 kg, respectively. Intakes of test AA ranged from 0.290 to 0.998 of requirement, Phe intake was constant within sows but varied among experiments between 8.52 and 15.40 g/d and metabolizable energy (ME) from 510 to 791 MJ/kg$^{0.75}$. Litter size was between 4 and 20 piglets born alive. Data were evaluated using the mixed model procedure of SAS with pig nested in experiment as a random variable. The start model contained the above parameters, including 2-way and 3-way interactions. The test AA was included in the model without interactions. The starting model was reduced iteratively by deleting all non-significant terms until all parameters left were at $P<0.05$.

Results

The resulting models explained a high degree of the variation in the data with $r^2=0.82$ (S) to $r^2=0.92$ (Ox, g/d). Of the feed-related factors, only AA intake had an effect on Phe kinetics, while ME intake proved non-significant, probably because of the small variability in ME due to the restricted feeding regimen. Increasing AA intake decreased Phe Ox linearly ($P=0.001$), which led to increased Phe retention ($P=0.001$) because S, breakdown (B) and Phe flux were not affected. Increasing BW increased Phe flux, Ox, S and B ($P<0.01$) but decreased Phe retention. Increasing gestational age decreased Ox and B linearly ($P<0.02$) and had quadratic effects on flux and S ($P<0.02$), which reached minima on day 67 and 64 of gestation, respectively. Phenylalanine retention increased linearly throughout pregnancy ($P=0.001$). Phenylalanine retention increased with increasing maternal BW gain ($P=0.001$) and decreased ($P=0.001$) with sow age. For the other parameters of Phe kinetics, sow age interacted with maternal BW (for flux, S and B) or total sow BW gain (for Ox). Phenylalanine OX was less affected by maternal BW gain in older than in younger sows ($P=0.015$). Flux, S and B were greatest in parity 3, and were most affected by maternal BW gain in the 3rd parity. Increasing litter size increased B ($P=0.001$), flux and S ($P<0.01$) but not Ox and Phe retention.

Discussion

The effect of fetal growth on protein metabolism becomes apparent in the increase B, and to a lesser degree of flux and S with increasing litter size. The increase in B would supply additional AA for conceptus growth when AA intake is limiting, while the increase in S represents the greater protein growth with larger litters and maternal growth. The data could not distinguish between maternal and conceptus gain for Phe retention and oxidation. The increase in Phe retention was in agreement with greater total sow BW gain. As expected, increasing the limiting AA intake decreased Ox and increased Phe retention. The lack of impact on S and B, however, was unexpected because growing pigs have shown to increase flux, S and B in response to increasing AA intake (Fuller *et al.*, 1987; Salter *et al.*, 1990). This indicates a qualitatively different response of protein kinetics in pregnancy. Conversely, the decrease in OX and B with progress of pregnancy and concomitant Phe retention can be explained by the increased AA requirement for fetal growth in the latter part of pregnancy. This was aided by the response of flux and S that also increased towards the end of pregnancy after reaching a minimum in mid gestation. In contrast to sows given limiting AA intake, S in non-malnourished women was either not affected by stage of pregnancy (Kalhan *et al.*, 1998) or elevated in mid pregnancy (Willommet *et al.*, 1992).

In keeping with diminishing maternal growth, Phe retention decreased as sows aged. The response of flux, S and B to maternal gain was maximized in the 3rd parity. Although sow BW increased from 2nd to 4th parity, maternal gain was similar in the 2nd and 3rd parity, and only dropped off in the 4th. This may have created a maximum for metabolic demand in the 3rd parity.

In conclusion, beside AA intake, Phe kinetics were affected by the stage of pregnancy and the maternal and conceptus gain. Reduction of OX and B in late pregnancy was the main response to accommodate the rapid fetal growth in late pregnancy when test AA intake was limiting.

References

Fuller, M.F., P. J. Reeds, A. Cadenhead, B. Seve, and T. Preston. 1987. Effects of the amount and quality of dietary protein on nitrogen metabolism and protein turnover of pigs. Br. J. Nutr. 58: 287-300.

Kalhan, S.C., K.Q. Rossi, L.L. Gruca, D.M. Super, and S.M. Savin. 1998. Relation between transamination of branched-chain amino acids and urea synthesis: evidence from human pregnancy. Am. J. Physiol.275 (Endocrinol. Metab. 38: E423-E431.

Moehn, S., C. Levesque, R. Samuel and R.O. Ball. 2011. New energy and amino acid requirements for gestating sows. Advances in Pork Production, 22:157-166.

Willommet, L., Y. Schutz, R. Whitehead, E. Jéquier and E.B. Fern. 1992. Whole body protein metabolism and resting energy expenditure in pregnant Gambian women. Am. J. Physiol. 263 (Endocrinol. Metab.): E624-E631.

Salter, D.N., A. Montgomery, A. Hudson, D.B. Quelch and R.J. Elliott. 1990. Lysine requirements and whole-body protein turnover in growing pigs. Br. J. Nutr. 63:503-513.

Meta-analytical study on the performance and utilization efficiency of different methionine sources by pigs

A. Remus[1], L. Hauschild[1], I. Andretta[2], M. Kipper[2], C.R. Lehnen[2] and R. Isola[1]
[1]Universidade Estadual Paulista, UNESP, Jaboticabal, São Paulo, Brazil; lhauschild@gmail.com
[2]Universidade Federal de Santa Maria, UFSM, Santa Maria, Rio Grande do Sul, Brazil

Introduction

Methionine (Met) is the second limiting amino acid in swine diets. The utilization of a synthetic amino acid for supplying this is a consolidated practice. However, the ideal level of this amino acid to optimize various criteria is still very controversial. Part of the discrepancy between studies may be linked to inter-experimental variability influencing the results. This meta-analysis is presented as a viable alternative by providing techniques that allow control of the variability, uncovering results that would not be noticeable in a smaller population. This study focused on the animal response to ingestion of different levels of methionine.

Material and methods

The database used for the meta-analysis consisted of 20 articles published in journals from 1956 to 2010 (mode: 2006). This study contemplated 6,946 pigs with 3.5 (nursery phase, 53% of studies) to 102.6 kg of body weight (growing-finishing phase, 47% of studies). The metabolic weight ratio was 6.0 kg (mode:13 kg). The following criteria were used for selection of the articles: (1) different levels of methionine in the diets; (2) all other amino acids (AA) were set at 100% of their ideal level; (3) presentation within the article of the nutritional composition of diets. Four treatments were considered, with methionine ratios of: 0.22% basal diet (mode: 0.23%), 0.20% of L-methionine (mode: 0.19%), 0.32% of DL-methionine (mode: 0.30%) and 0.38% (mode: 0.28%) of DL-methionine hydroxy analogue (MHA). The mean composition of diets was: 2,410 kcal of metabolizable energy (mode: 3,152) and 18.57% crude protein. The composition of total AA was: 1.23% of lysine, 0.23% of tryptophan, 0.83% of threonine, 0.76% of methionine plus cysteine and 0.44% of cysteine, 0.88% of calcium and 0.79% of phosphorus. A coding category was established for the methionine source. In addition to this coding category, three other moderating encoding categories were used: (1) general codification; (2) inter codification; (3) intra codification. The criteria utilized for the definition of dependent and independent variables followed the proposals described in the literature (Lovatto *et al.*, 2007; Sauvant *et al.*, 2008). Step 1, the data were analyzed graphically. With this analysis it was possible to generate the correlation hypothesis for the definition of the statistical model. Step 2, the correlation analysis was performed. The statistical model used was covariate adjustment, and all variables were adjusted for metabolic body weight (raised to the 0.60 power). The adjustment was performed to make possible analyzation and to compare pigs with different physiological states (i.e. level of maturity). Step 3, the variance-covariance analysis was carried out using the 'general linear model' procedure. The Stundent-Newman-Keuls multi comparison test using Statistica software was used for comparisons of means.

Results and discussion

The efficiency of different sources of Met was determined by regression of the average daily gain (ADG) as a function of methionine intake. The best efficiency was observed in pigs fed diets supplemented with L-methionine (L-met). These pigs had a gain of 1.44 g/g methionine intake adjusted for $BW^{0.6}$ (L-met = $0.1324 + 1.445\times X$, $R^2 = 90.1\%$). The worst result was observed for piglets supplemented with MHA (MHA= $0.2814 + 0.7184\times X$, R^2= 58.2%). The results on performance and plasmatic urea are presented in Table 1. The methionine intake adjusted for $BW^{0.6}$ was 33% ($P<0.001$) higher when pigs received DL-methionine diets (DL-met), in contrast with basal

Table 1. Performance of pigs receiving different methionine sources adjusted for metabolic body weight (raised to the 0.60 power).

	Basal	L-met	DL-met	HMA	*P*-value
ADG, kg	0.595[a]	0.785[b]	0.693[c]	0.554[a]	<0.001
	0.03*	0.04	0.01	0.02	
ADFI, kg	1.027[a]	1.033[a]	1.119[a]	0.781[b]	<0.001
	0.09	0.13	0.04	0.06	
G:F, kg	0.678[a]	0.626[a]	0.701[a]	0.847[b]	<0.001
	0.03	0.06	0.02	0.03	
Plasma urea N, mg/dl	10.56[a]	7.82[b]	5.49[c]	12.55[a]	<0.001
	1.59	3.44	0.75	1.89	
Methionine intake, g/d	0.398[a]	0.391[a]	0.588[b]	0.392[a]	<0.001
	0.04	0.08	0.02	0.04	

Superscripts: SNK multi comparison test at 95% confidence intervals. * Standard deviation.

diet (BD) ($P<0.001$). The average daily feed intake ($ADFI^{0.6}$) was 24% higher ($P<0.001$) for pigs fed diets supplemented with L-met or DL-met when compared to groups fed with basal or MHA diets. The feed efficiency ratio was 25% better ($P<0.001$) when the pigs were fed diets supplied with L-met compared with MHA. The $ADG^{0.6}$ of piglets supplemented with L-met was 29% higher ($P<0.001$) in contrast with MHA and 12% higher in contrast with DL-met. On the regression of ADG with methionine intake, the best efficiency was observed for pigs fed diets supplemented with L-methionine ($Y = 0.1324 + 1.445 \times X$, $R^2=0.90$).

For the nitrogen (N) balance there was no observed difference ($P>0.05$) among treatments. Plasmatic N was an exception to this; at 56% less in pigs supplemented with DL-met and 38% less in pigs supplemented with L-met in contrast with pigs receiving MHA. Pigs supplemented with MHA had plasmatic concentration of N similar to pigs fed with basal diet.

Conclusion

L-methionine and DL-methionine improved the performance of pigs in contrast with pigs supplemented with DL-methionine hydroxy analogue. It is possible that both sources of methionine can be equal options for supplementation in pigs diets.

References

Lovatto, P.A., Lehnen, C.R., Andretta, I., Carvalho, A.D., and Hauschild, L., 2007, Meta-análise em pesquisas científicas: enfoque em metodologias: Revista Brasileira de Zootecnia, v. 36, p. 285-294.

Sauvant, D., Schmidely, P., Daudin, J.J., and St-Pierre, N.R., 2008, Meta-analyses of experimental data in animal nutrition: animal, v. 2, p. 1203-1214.

Estimating digestible threonine requirements for growing pigs by meta-analysis

R. Isola, L. Hauschild, M.C. Thomaz, N.K. Sakomura and A. Remus
Universidade Estadual Paulista, UNESP, Jaboticabal, São Paulo, Brazil; lhauschild@gmail.com

Introduction

Threonine is the second or third limiting amino acid in pig diets, and can become the first, when supplemented with synthetic lysine. The utilization of a synthetic amino acid for supplying this is a consolidated practice. However, the ideal level of this amino acid to optimize various criteria is still very controversial. Part of the discrepancy between studies may be linked to inter-experimental variability influencing the results. This meta-analysis is presented as a viable alternative by providing techniques that allow control of the variability, uncovering results that would not be noticeable in a smaller population. This study focused on the animal response to ingestion of different levels of threonine.

Material and methods

The database used for the meta-analysis consisted of 14 articles published in journals since 1986 to 2007. This study contemplated 1,187 pigs in the weight range of 5 kg (nursery phase) to 124 kg (finishing phase). To select the articles we used the following criteria: the addition of different levels of dietary digestible threonine (SDThreo) with all other amino acids set at 100% of its ideal level. The average composition of the diets was 3,376 kcal of digestible energy (DE) and 16.22% crude protein, and the total composition of the amino acids lysine 1.04%, 0.21% tryptophan, 0.35% methionine, 0.68% threonine and 0.55% methionine + cysteine.

For the definition of dependent and independent variables the criteria described in the literature was used (Lovatto *et al.*, 2007; Sauvant *et al.*, 2008). Analysis of the database followed three steps. Step 1, the graphic analysis. With this analysis it was possible to generate the correlation hypothesis for the definition of the statistical model. Step 2, the correlation analysis was performed. Finally, in step 3, analysis of variance-covariance procedure by the 'general linear model' was performed, from which the prediction equations were generated. The variables measured by the intake SDThreo were feed intake, weight gain, feed conversion and plasma urea. The statistical model used for covariate adjustment, with the result for plasma urea were adjusted for metabolic body weight (raised to 0.75 power) to enable the analysis and comparison of pigs with different physiological status (at different stages, ages). The data were analyzed by regression analysis and considered only levels below the maximum response for weight gain (BWG). A factorial equation was determined to estimate the demand for SDThreo (SDThreo = $0.049 \times BW^{0.75}$ + BWG) for different body weights. The first component ($0.049 \times BW^{0.75}$) is the maintenance and was extracted from the literature. The second is the requirement SDThreo for growth. The coefficient was estimated by the regression of SDThreo ingested destined for growth (SDThreoG = SDThreo ingested total – SDThreo maintenance) due to the BWG. In regression analysis was used Minitab 16.

Results and discussion

In the regression of weight gain (BWG = -0.01287 + 0.9020 Thr – Thr 0.09671 ** 2, R^2=91.3%), daily feed intake (FI = -0.6099 2885 + Thr – Thr 0.3569 ** 2, R^2=91.5%), feed conversion (FC = 0.6908 – 0.2628 + 0.04996 Thr Thr ** 2, R^2=60.4%) and plasma urea nitrogen (N = 4583-1463 Thr Thr ** 2 + 0.1398, R^2=78.7%) versus SDThreo consumption increased linearly (P<0.05) and quadraticly (P<0.05) for these variables. The regression of SDThreoG according to BWG showed a good fit (SDThreoG = 0.01285×BWG, R^2=0.95). Each gram of weight gain required an intake

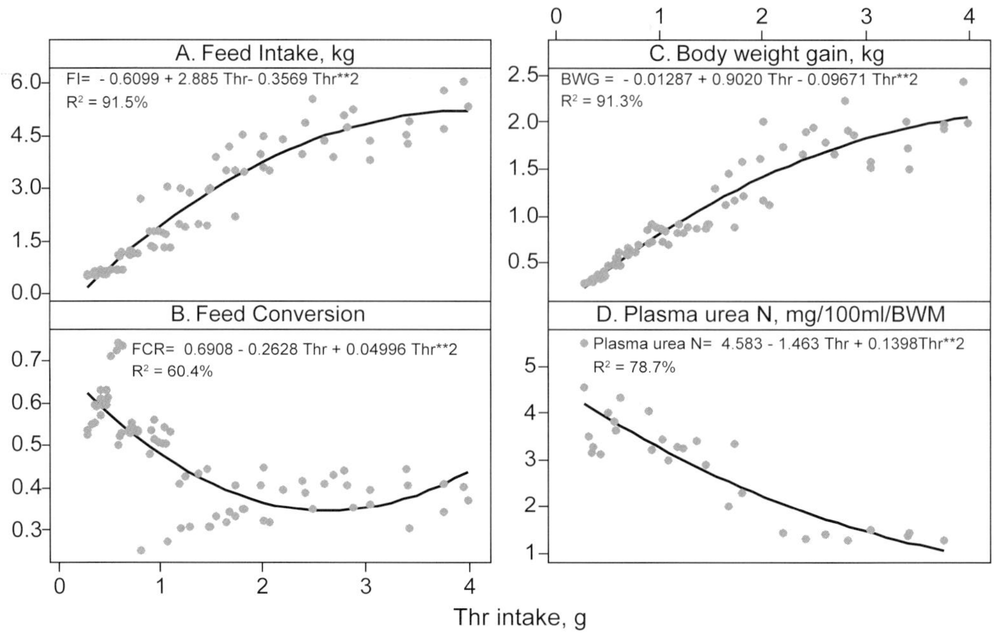

Figure 1. Regression of feed intake (A), feed conversion (B), weight gain (C) and plasma urea (D) due to the intake of threonine.

of 0.0185 g SDThreo to meet the potential growth of the animals. This parameter represents an efficiency of utilization of 77% of SDThreo for weight gain. Based on this equation and the data for consumption and weight gain for a high genetic potential animal was used in the Brazilian Tables of recommended requirements for swine it was estimated requirements of SDThreo for pigs with weights of 22.5, 40, 60, 85, and 110 kg. For these weights the estimated values were 0.729, 0.635, 0.552 and 0.453 SDThreo% of the diet.

Conclusion

The factorial equation structured based on data from the meta-analysis study allowed us to determine the response of pigs at different weights to threonine intake and determine the optimal level of threonine in the diet based on variables available on farm.

References

Lovatto, P.A., Lehnen, C.R., Andretta, I., Carvalho, A.D., and Hauschild, L., 2007, Meta-análise em pesquisas científicas: enfoque em metodologias: Revista Brasileira de Zootecnia, v. 36, p. 285-294.

Sauvant, D., Schmidely, P., Daudin, J.J., and St-Pierre, N.R., 2008, Meta-analyses of experimental data in animal nutrition: animal, v. 2, p. 1203-1214.

Lysine supply in finishing broilers: effect on performances and meat quality

M. Lessire[1], Y. Primot[2], E. Corrent[2], Fraysse Pauline[2], S. Tesseraud[1] and C. Berri[1]
[1]INRA, UR83 Recherches Avicoles, 37380 Nouzilly, France; tesserau@tours.inra.fr
[2]Ajinomoto-Eurolysine s.a.s., 153 rue de Courcelles, 75817 Paris Cedex 17, France

Introduction

In poultry production, there is a constant need for updating birds' requirements especially protein and amino acids, because of economical purposes, genetic improvements and also because parameters of interest are changing from the classical ones i.e. body weight, feed conversion ratio, to more recent ones, i.e. meat yield and processing quality (Berri *et al.*, 2008; Dozier *et al.*, 2010). In chickens, these last characteristics are strongly influenced by the ultimate pH of the meat which depends on muscle glycogen content (Le Bihan-Duval *et al.*, 2008). Recent observations suggest the possibility of modulating these characteristics by varying the lysine intake of broilers (Jlali *et al.*, 2012). Our objective was to determine the digestible lysine requirements of male Ross broilers during the finishing period, taking into account birds growth, feed conversion, edible parts yields, adiposity and meat quality.

Material and methods

Twelve experimental diets were computed using the same raw materials, from the same batches, which were analyzed previously. They presented 4 digestible lysine levels: 7.0, 8.5, 10.0 and 11.5 g/kg, and 3 amino acids profiles: 90, 100 and 110% of the optimum amino acids profile mentioned by Mack *et al.* (1999). Dietary protein content followed lysine content. On the opposite ME value and other nutrients contents were independent of lysine content. As a consequence, the ratio ME/protein decreased with the increase of lysine.

4,600 birds were raised with the same diets till they were 21 days old; then 3,456 with similar weight were retained. The 4×3 dietary treatments were given to these birds divided in 96 floor pens (3 m^2), with 8 pens (36 birds/pen) for each treatment.

At 36 days of age, birds were individually weighted and feed consumption was calculated per pen. In each pen 4 birds were chosen around the mean of the pen and slaughtered. Weight of abdominal fat, breast meat, and tight and drumsticks were measured. Breast meat quality was estimated by measuring pH, drip loss, and color. Lysine requirements were calculated using quadratic broken line models (Pesti *et al.*, 2009).

Results and discussion

Feed consumption decreased with the increase of dietary lysine content. On the opposite, weight gain increased according to a quadratic equation. Without other limiting amino acids digestible lysine requirements can be estimated as 10.5 and 11.3 g/kg for levels 100 and 110% respectively. The corresponding optimum weight gains were 1,580 and 1,592 g (Figure 1). Feed conversion ratio was improved when lysine content increased. The optimum value was estimated at 12.7 g/kg for levels 100 and 110%, with an optimum equal to 1.625. When the other amino acids are limiting (90%), results followed the same trends, but the efficacy of lysine was reduced. Digestible Lysine requirement is globally lower for all the performances criteria than for quality criteria, as already observed by Leclercq (1998). Breast meat yield increased (from 19.3 to 20.2%) and abdominal fat percentage decreased (from 3.2 to 2.2%) with increasing levels of lysine and thus total protein content of feed. Strong effects of amino acids profile and lysine content were observed on the ultimate pH. The lowest values of ultimate pH (5.8 on average) were obtained in the case of a high intake of amino acids relative to lysine (110%). The highest values (6.0 on average) corresponded to the lowest amino acid to lysine ratio (90%). These

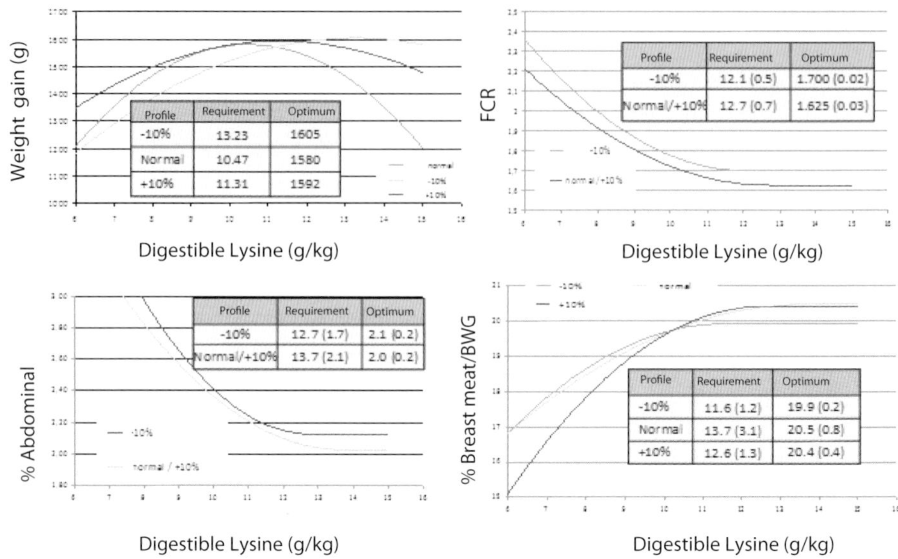

Figure 1. Evolution of performances and yields according to the digestible lysine and other amino acids concentrations.

pH variations had a significant impact on the color and water retention of meat, with the most acid meats having the highest luminance L* and drip loss during storage (L*=50.6 and drip loss=2.7% in average). These results highlight the importance of integrating meat quality criteria in addition to the standard production criteria when defining dietary amino acids requirements of broilers.

Conclusion

Optimizing the lysine content of broiler diets, with an optimum content of the other amino acids appears to be essential in order to improve efficiency and quality of the production. To this, modeling performances in relation with nutrient content seems to be a solution. In this study, digestible lysine requirements, with a same mathematical model, varied from 10.4 to 13.7 g/kg for broilers according to the essential amino acids profiles used and the criteria of response. But this trial shows that performances could be improved (body weight) using amino acids concentrations higher than classical ones, thus indicating a lack of knowledge on finishing broilers requirement or in raw materials digestibility used.

References

Berri, C., J. Besnard, and C. Relandeau, 2008. Increasing dietary Lysine Increases Final pH and Decreases Drip Loss of Broiler Breast Meat. Poult. Sci. 87, 480-484.

Dozier, W.A., A. Corzo, M.T. Kidd, P.B. Tillman, J.P. McMurtry, and S.L. Brandon, 2010. Digestible lysine requirements of male broilers from 28 to 42 days of age. Poult. Sci. 89, 2173-2182.

Jlali, M., V. Gigaud, S. Métayer-Coustard *et al.* 2012. Modulation of glycogen and breast meat processing ability by nutrition in chickens: Effect of crude protein level in 2 chicken genotypes. J. Anim. Sci. 90, 447-455.

Le Bihan-Duval, E., M. Debut, C.M. Berri, *et al.* 2008. Chicken meat quality: genetic variability and relationship with growth and muscle characteristics. BMC Genetics, 9, 6 p.

Leclercq, B., 1998 Specific effects of lysine on broiler production : comparison with threonine and valine. Poult. Sci; 77 : 118-123.

Mack, S., D. Bercovici, G. De Groote *et al.* 1999. Ideal amino acid profile and dietary lysine specification for broiler chickens of 20 to 40 days of age S. Br. Poult. Sci. 40, 257-265.

Pesti, G.M., D. Vedenov, J. A. Cason, and L. Billard, 2009. A comparison of methods to estimate nutritional requirements from experimental data. Br. Poult. Sci. 50, 16-32.

Part 3. Tools and techniques

Proteomic tools help understanding the metabolic adaptation to negative energy balance in dairy cows

B. Kuhla and C.C. Metges
Institute of Nutritional Physiology 'Oskar Kellner', Leibniz Institute for Farm Animal Biology (FBN), 18196 Dummerstorf, Germany; b.kuhla@fbn-dummerstorf.de

Abstract

High-yielding dairy cows have enormous energy and nutrient requirements for milk production which are generally not met by a sufficient feed intake resulting in a negative energy balance (NEB) characterized by mobilisation of body reserves. It is still controversial whether during early lactation the mobilization of body reserves causes insufficient feed intake or insufficient feed intake causes mobilization of body reserves. In order to distinguish between cause and effect, we designed feed-restriction studies modelling NEB as well as follow-up studies on periparturient dairy cows and examined metabolic adaptation processes during NEB. To this end, 2D-gel based proteomic approaches coupled with MALDI-TOF-MS and MALDI-TOF-TOF analyses are often used for the investigation of changes in protein expression, posttranslational modifications (PTMs) and protein identification, while subsequent Western Blots are applied to confirm the existence and expression of individual proteins. Proteomic profiling in tissues obtained from frequent liver and muscle biopsies or from slaughter provides insight into regulatory mechanisms at the translational and posttranslational level. We were able to demonstrate that phosphorylation of the adenosine monophosphate-activated protein kinase (AMPK), a cellular energy key sensor, is increased in hypothalamus and liver but not in skeletal muscle during NEB. Muscle tissue in early lactation showed reduced abundance of muscular cytoskeletal proteins and enzymes involved in glycogen synthesis, fatty acid degradation, and TCA cycling, while the expression of enzymes involved in glycolysis, lactate and ATP production was increased. The functional characterisation of the hepatic oxidative metabolism is of particular interest because of its role to provide substrates for the mammary gland and its involvement in the control of feed intake. While feed restriction down-regulated hepatic proteins associated with fatty acid oxidation, early lactation expression of enzymes participating in fatty and amino acid degradation, TCA cycling, ATP production, and oxidative stress defence was increased. The integration of proteome data with corresponding plasma metabolite and hormone concentrations allowed us to propose an inter-organ crosstalk model in which hepatic and skeletal muscle metabolism in early lactating cows supports gluconeogenesis for milk production while hepatic oxidation of fatty acids interferes with the control of feed intake in the brain.

Introduction

In high-producing dairy cows the transition from late pregnancy to early lactation is characterized by an enormous increase in nutrient and energy demands of the mammary gland in order to express its genetic potential of milk synthesis. These requirements are generally met by an increase of feed intake, mobilization of body fat, protein and glycogen reserves, and the saving of fuels in peripheral organs of lesser priority such as in muscle and liver. The concerted control of body tissue metabolism thus supports the partitioning of macronutrients towards the mammary gland but in parallel also poses a challenge to the metabolism of peripheral organs. Although feed intake normally increases from late pregnancy to early lactation the magnitude of this increase does not match the increase of energy secretion by milk. The gap between insufficient energy intake and increased energy requirements is accompanied by the mobilisation of body reserves and results in a transient NEB in early lactation.

Whether insufficient energy intake during early lactation causes the mobilization of body reserves, or the mobilization of body reserves underlies the insufficient increase of feed intake remains a controversial issue (Ingvartsen and Andersen, 2000; Hammon *et al.*, 2009; Weber *et al.*, 2013). Feed withdrawal, feed restriction or feeding an energy-diluted ration to mid or late lactating dairy cows (Gross *et al.*, 2011; Loor *et al.*, 2007) are frequently used models to study metabolic adaptations to NEB. These experimental models allow to differentiate between cause (insufficient energy intake) and effect (metabolic adaptation), which is generally not possible from association studies by simply monitoring cows during the transit period. A major difference between transiently occurring NEB in early lactation and nutrition-induced NEB is, however, that during early lactation dairy cows are fed *ad libitum* and thus obviously experience satiety which is not the case when feed is withdrawn. Hence, comparative studies involving feed/energy restriction/withdrawal and *ad libitum* feeding of early lactating cows enable the differentiation between metabolic adaptations to NEB during hunger or satiety, respectively. Only recently, Allen *et al.* (2009) summarized experimental evidences suggesting that in early lactation the delay in reaching a proportionately adequate feed intake is due to increased vagus nerve activity driven by an activated hepatic oxidative metabolism.

Even during periods of NEB, the mammary gland is highly supplied with macronutrients, primarily glucose, amino acids and fatty acids serving as important precursors maintaining milk production. While glucose originates predominantly from gluconeogenesis and glycogenolysis of the liver, skeletal muscle protein is broken down to yield amino acids and smaller peptides, and non-esterified fatty acids (NEFA) are mobilized from adipose tissue depots and enter the blood stream. Therefore, changed plasma metabolite concentrations together with an altered intermediary metabolism in liver, skeletal muscle, adipose tissue and mammary gland contribute to a major extent to the metabolic adaptation to NEB. The intermediary metabolism of peripheral organs is not only influenced by substrate availability and product concentration but is also regulated by the endocrine and autonomous nervous system in which the hypothalamus-pituitary axis plays a pivotal role controlling energy balance.

Applications of transcriptomics and the measurements of mRNA abundance identified differentially expressed key enzymes, nuclear and transcription factors in the mammary gland (Bionaz *et al.*, 2012), adipose tissue (Sumner-Thomson *et al.*, 2011), and liver (Loor *et al.*, 2005, 2007) decoding metabolic pathways activated or deactivated in periparturient dairy cows. Although a number of these transcriptional changes do obviously reflect the concentration of the respective protein or even its activity (Li *et al.*, 2012a), signal transduction and distinct metabolic pathways such as adipose tissue lipolysis (Locher *et al.*, 2011) are exclusively triggered through post-translational modifications.

Proteomic profiling is a tool that offers the opportunity to identify changes in protein expression at translational and posttranslational (e.g. phosphorylation, glycosylation, etc.) level and thus enables us to gain further insight into regulatory mechanisms. Proteomic approaches require a combination of stringent separation technologies (2-dimensional gel electrophoresis (2D-GE) or high-performance liquid chromatography (HPLC)), high-resolution mass spectrometry (e.g. MALDI-TOF-MS) and powerful bioinformatic tools. Ultimately, the quantity and quality of the protein identification depends on the quality of the protein databases of farm animals which is consistently improving (Lippolis and Reinhardt, 2008). The proteomic technology allows analysing >1000 different proteins in tissues or body fluids in one single experiment. This provides the researcher with information on cellular processes in response to a certain challenge and allows to identify previously unknown interactions between different proteins (Lippolis and Reinhardt, 2008), First steps in uncovering the proteome of bovine tissues, plasma and red blood cells were made a decade ago (Talamo *et al.*, 2003) in which a few most highly abundant proteins were separated and identified. However, Western blots or other immune-based assays such as RIA or ELISA are still often used to confirm relative expression differences of selected proteins including their posttranslational modifications provided that specific antibodies are available. Thus, the rapidly developing proteome methodology based on progressed mass spectrometry techniques together with established immunoassays and plasma metabolite profiling have contributed a lot to the understanding of metabolic adaptation to NEB in dairy cows.

The aim of this paper is to give an overview on the current state of knowledge in regard to the metabolism of the periparturient dairy cow based on new information gathered by our group and others using proteomic tools.

Hypothalamus and pituitary

The hypothalamus-pituitary axis is the most important control and regulatory system involved in maintaining energy homeostasis or homeorhesis. The hypothalamus receives an array of extracellular humoral and neural information, integrates these input signals to secrete hypothalamic-releasing hormones, which in turn stimulate or inhibit the secretion of pituitary hormones. In turn released pituitary hormones may affect metabolic processes in peripheral tissues, thus forming a feedback control loop.

In a first attempt to dissect mechanism potentially involved in the control of energy homeostasis of dairy cows, we investigated how energy restriction caused by feeding a diet with reduced energy density as compared to *ad libitum* feeding alters the hypothalamic proteome in early-lactating cows (Kuhla *et al.*, 2007). By applying a 2D-GE-based proteomic approach, we found that the glycolytic enzyme glyceraldehyde 3-phosphate dehydrogenase (GAPDH) and the TCA-cycle enzyme aconitase-2 were down-regulated in the restricted group, suggesting an adaptation of brain metabolism in response to diminished substrate availability. On the other hand, ubiquitin carboxy-terminal hydrolase-L1, and heat shock protein 70 kDa-protein-5 were found to be up-regulated pointing to greater proteolysis in the hypothalamus during times of energy deficiency (Kuhla *et al.*, 2007). In feed restricted dairy cows, abundance of [Cu-Zn]-superoxide dismutase was also greater than in *ad libitum* fed counterparts (Kuhla *et al.*, 2007), indicating increased production of reactive oxygen species (ROS), which in turn are increasingly considered as a mitochondrial energy sensor in the brain (Horvath *et al.*, 2009). A cytosolic energy sensor is the adenosine monophosphate-activated protein kinase (AMPK). This intracellular 'fuel gauge' becomes phosphorylated (activated) by AMP when the ATP level is low. In our proteomic approach, we further identified upregulation of 5-aminoimidazole-4-carboxamide ribonucleotide formyltransferase/IMP cyclohydrolase (Kuhla *et al.*, 2007), an enzyme directly involved in AMP production, and Western blot analysis revealed that the ratio of phosphorylated AMPK and total AMPK was elevated in energy restricted dairy cows (Kuhla *et al.*, 2011a). To conclude, we found that the brain adapts to energy restriction not only by adapting its intermediary metabolism but also by activating cell signalling pathways triggering for the maintenance of energy balance. However, we were not able to detect alterations in the expression of hypothalamic-releasing hormones that would suggest an altered secretion of pituitary hormones.

Nonetheless, the pituitary gland accumulates and secretes nine pituitary hormones [e.g growth hormone (GH), prolactin (PRL), luteinizing hormone (LH), thyrotropin (TH), Pro-opiomelanocortin (POMC], which all may carry one or more PTMs or which occur with different peptide lengths. Different PTMs or peptide lengths may exert opposing physiological functions, for example, dimeric but not monomeric growth hormone (GH) and prolactin (PRL) activate various cellular pathways (Langenheim *et al.*, 2006), full-length GH and PRL promote angiogenesis but their proteolytic fragments acquire anti-angiogenic properties (Corbacho *et al.*, 2002), or GH peptides with different lengths possess either insulin-potentiating or anti-insulin properties (Lewis *et al.*, 2000). We asked the question if the expression of the different pituitary hormone variants differ in response to feed restriction. Since pituitary hormone antibodies established for RIAs or ELISAs are generally not suitable to distinguish between different hormone forms, the only method by which this question can be answered is proteomics. In extracts of bovine pituitary gland, the abundance of all pituitary hormone variants identified (GH=44, PRL=16, LH=1, TH=1, POMC=2) did, however, not differ in response to feed restriction despite total plasma PRL but not total plasma LH or GH concentrations were significantly reduced in the feed restricted group (Kuhla *et al.* 2010). Also pituitary total LH and total PRL determined by RIA analysis did not differ between *ad libitum* and feed restricted dairy cows, suggesting that the hormone pool size in the pituitary was sustained after 60-h feed restriction. It rather seemed that the hormone secretion machinery participated in the adaptation to feed restriction because expression of annexin A5 and secretogranin V (neuroendocrine polypeptide 7B2), both stimulating the release of hormones into the blood stream, were increasingly expressed after feed restriction.

Serum, plasma and blood

The sensitivity of 2D-gel-based proteomic approach is generally too low for the detection of proteinous or peptide hormones in plasma or serum. However, by using a proteomic approach based on agarose gel electrophoresis, Piccione *et al.*, (2011) identified several serum markers that were altered during the periparturient period of dairy cows. For example, abundance of α1-globulins including acute phase proteins were increased in early lactation, while β- and γ-globulins were lowest abundant around parturition, indicating increased metabolic stress and a challenged immune system during this time (Piccione *et al.*, 2011). In regard to the latter, Lippolis *et al.* (2006) identified more than 40 proteins in neutrophils that were differentially expressed between day 28 before and 2-3 days after parturition by labelling proteins with isobaric tags for relative and absolute quantitation (iTRAQ). Altered expressions of High Mobility Group Box proteins, histones, myeloperoxidase, proteins involved in arachidonic acid metabolism or with antimicrobial activity were indicative of immunosuppression and reduced neutrophil functionality in effectively combating bacteria in early lactation.

Further efforts have been made in identifying cows affected by metabolic diseases typically occurring during NEB. Subjecting plasma samples from cows with or without milk fever to 2D differential in-gel electrophoresis (2D-DIGE), Xia *et al.*, (2012) recently described upregulation of serpin peptidase inhibitor (angiotensin) and endopin 2B (involved in neural regulation) and downregulation of fibrinogen, albumin, and IgGs in cows affected by milk fever. However, a proteinous serum biomarker specifically predicting the degree of NEB has not been identified so far. It is conceivable that such a biomarker could be found in neutrophils because a large number of metabolic proteins involved in the generation of NADPH and ATP have been identified in a proteomic survey (Lippolis and Reinhardt, 2005), but the experimental evidence is still missing.

Muscle

The skeletal muscle represents the major part of body proteins and some of these proteins may be degraded around parturition. This results in the release of creatinine, 3-methyl histidine, and most proteinogenic amino acids into circulation to be used for milk protein synthesis. However, the degradation of skeletal muscle protein is a relatively slow process that begins around parturition and continuously progresses at least by the 4th week after parturition (Kuhla *et al.*, 2011b). Unexpectedly, the muscular fat content was lowest around parturition, suggesting that fatty acid degradation in muscle tissue is a much earlier event in the adaptation to NEB. Also, muscle glycogen reserves were exhausted within the first two weeks after parturition and were almost replenished after the 4th week of lactation. These observations suggest an early mobilisation of glucose and fatty acids followed by a somewhat delayed mobilisation of amino acids to be either directly used for milk synthesis or for hepatic anabolic processes in early lactation. To study the intermediary metabolic pathways underlying the mobilisation of energy reserves from skeletal muscle, we applied semitendinosus muscle biopsies to 2D-GE-based proteomics. In early lactation, we found two increasingly expressed proteins closely associated with fatty acid degradation namely electron transfer flavoprotein and retinal dehydrogenase 1, in early lactation. This upregulation occurred in parallel with the *post partum* plasma NEFA peak and the nadir of muscle fat content, suggesting a stimulated breakdown of triglycerides and increased fatty acid oxidation in the muscle during early lactation. However, we were not able to identify any enzyme participating in beta-oxidation.

Furthermore, there was increased expression of glycolytic enzymes such as fructose bisphosphate aldolase A, GAPDH, triosephosphate isomerase, enolase, pyruvate kinase and lactate dehydrogenase, but also reduced expression of the glycogenesis enzyme UTP-glucose-1-phosphate uridylyltransferase when compared to the *ante partum* period. The expression of TCA cycle enzymes (aconitase 2, malate dehydrogenase) was also lowered in week 4 *post partum*, whereas the abundance of ATP homeostasis regulating enzymes (creatine kinase and ATP synthase) was upregulated as compared to late pregnancy. Our findings led us to conclude that in early lactation (a) muscle pyruvate is shunted towards lactate and less so towards acetylCoA and (b) that ATP production via the TCA cycle is decreased (Figure 1). The increased skeletal muscle lactate production likely supports hepatic gluconeogenesis to which lactate contributes up to 25% in early lactation (Aschenbach *et al.*, 2010). In Western blot studies we further noticed reduced expression of muscle GLUT4 during early lactation (Kuhla *et al.*, 2011b). Based on these findings we suggested that the decoupled Cori cycle in early lactation supports hepatic gluconeogenesis and thereby glucose partitioning to the mammary gland (Figure 1).

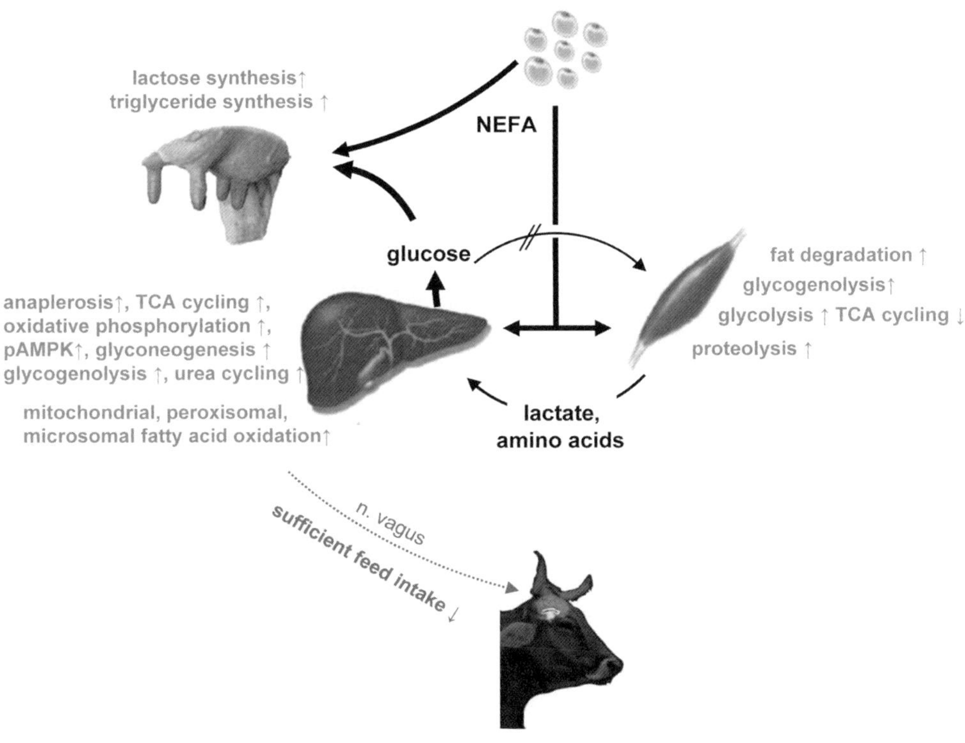

Figure 1. Activation of major metabolic pathways during NEB in early lactation as identified by proteomic approaches. Metabolic and neuronal pathways are in grey while metabolites and macronutrient flows are displayed in black. Amino acids, primarily derived from cytoskeletal proteins, and lactate which originates from muscle glycogenolysis and anaerobic glycolysis, are released from skeletal muscle and utilized by the liver as glucoplastic precursors or by the mammary gland for milk protein synthesis or free AA in milk. Due to prevailing hypoglycaemia in early lactation, glucose is not taken up by the muscle decoupling the Cori cycle. Fatty acids, predominantly released from adipose tissue, serve as substrate for generating energy in muscle and liver in addition to be directly used for milk fat production. The increasingly activated hepatic fatty acid oxidation pathways influence signalling of the vagus nerve and thus prevent sufficient increase of feed intake in early lactation.

There were also numerous indications for an increased proteasomal activity in muscle tissue during early lactation: Reduced expression of heat shock protein beta-1 and alpha-crystallin (both stabilizing myofibrillar proteins), upregulation of bridging integrator protein 1 (which activates caspase-independent processes), reduced abundance of Dj-1 protein (which promotes cell growth) and upregulation of phosphatidylethanolamine-binding protein 1 (a suppressor of cell development and growth) and reduced expression of major cytoskeletal proteins. It appears likely that the observed increased proteasomal activity entails the release of amino acids into the circulation serving either as substrates for milk protein synthesis or in the case of glucoplastic amino acids as precursors for hepatic gluconeogenesis.

Liver

Hepatic gluconeogenesis and glycogenolysis are highly active during early lactation. Because the majority of glucose is used for lactose production in the mammary gland, fatty acids become an important energy source and are intensively mobilized from adipose tissue during this period. However, excessive NEFA concentrations are increasingly oxidized by the liver but only to a certain extent (Grum *et al.*, 1996) because of the reduced availability of oxaloacetate. Consequently, acetyl CoA derived from NEFA oxidation will partially be converted to ketone bodies or re-esterified to triacylglycerides (TG) resulting in the development of fatty liver. The hepatic lipid metabolism has been suggested to be involved in the control of energy balance. For example, increased hepatic fatty acid oxidation leads to increased hepatic ATP production which signals satiety to the brain via vagal afferents and thus prevents sufficient increase of feed intake during early lactation (Allen *et al.*, 2009). Accordingly, the degree of fatty liver is inversely correlated with feed intake and energy balance (Hammon *et al.*, 2009; Weber *et al.*, 2013). To examine how the extent of lipid mobilization influences hepatic oxidative fat metabolism, we retrospectively assigned periparturient Holstein cows either to a group with a high or a low liver fat content *post partum* (Schäff *et al.*, 2012). Feed intake expressed as kg dry matter intake per kg body weight was lower in cows with the higher liver fat content. Liver biopsies taken on day -34, -17, +3, +18, and +30 relative to parturition were applied to 2D-GE-based proteomics and Western blot analysis. In early lactation we found an increased phosphorylation of AMPK and increased expression of β-oxidative enzymes in both groups. For the medium-chain acyl-CoA dehydrogenase, the extent of upregulation was greater in cows developing the higher liver fat content. However, for the further downstream located β-oxidative enzymes, short-chain specific acyl-CoA dehydrogenase, enoyl-CoA hydratase, 3-hydroxyacyl-CoA dehydrogenase 2, and 3-ketoacyl-CoA thiolase as well as 2,4-dienoyl-CoA reductase 1 were lower expressed in cows with the higher liver fat content *post partum* (Schäff *et al.*, 2012). Thus, it seems that cows with the higher fat mobilisation have an increased activation but diminished complete β- oxidation of fatty acids which has also been observed at the level of mRNA (Li *et al.*, 2012b) and enzyme activity (Murondoti *et al.*, 2004) level. Failure of complete mitochondrial β-oxidation seems to result in a proportionally augmented degradation of fatty acids in peroxisoms and microsomes since we found increased expression of the peroxisomal enoyl-CoA hydratase, catalase, electron transfer flavoprotein, and aldehyde dehydrogenase 1 in cows with the higher liver fat content (Schäff *et al.*, 2012).

The lower expression of mitochondrial β-oxidative enzymes in cows developing the higher liver fat content should confine the production of acetyl CoA. An alternative source of acetyl CoA is the breakdown of ketogenic amino acids. Indeed, we found much lower plasma concentrations of leucine particularly during early lactation in cows with the higher liver fat content. Also, the observed increase of serine hydroxymethyltransferase, methylmalonate semialdehyde dehydrogenase, and glycine C-acetyltransferase expression in cows with the higher liver fat content refer to a prevailing catabolism of amino acids whose carbon skeletons enter the TCA cycle via pyruvate. Increased production of pyruvate and phosphoenolpyruvate as indicated by increased expression of L-lactate dehydrogenase and α-enolase refer to a further increased anaplerosis while the higher expression of fumarase points to a greater TCA cycling in cows with the higher liver fat content (Figure 1). The

TCA cycle generates NADH which serves as precursor for oxidative phosphorylation and generation of ATP. Enzymes involved in oxidative phosphorylation like cytochrome c oxidase, ATP-synthase and pyrophosphatase were increased in early lactation and higher expressed in cows with the high as compared to the low liver fat, referring to elevated mitochondrial respiration (at least via complex IV) and ATP production in fatty liver. Intensive mitochondrial oxidation and peroxisomal fatty acid degradation confronts the liver with high amounts of superoxide radicals and H_2O_2 which can be detoxified by Cu/Zn superoxide dismutase and peroxiredoxin 6. The greater expressions of the latter two enzymes in cows with the higher liver fat content indicate an elevated oxidative stress defence. Moreover, the NADPH-dependent glutathione system was negatively affected in cows with fatty liver because enzymes involved in the NADPH-producing pentose phosphate cycle (e.g. transketolase and triposephosphate isomerase) were lower expressed particularly during early lactation. Furthermore, the hepatic glutathione synthesis requires Cys, Glu, and Gly, but plasma Cys concentrations were lower in cows with fatty liver indicating an impaired NADPH-dependent glutathione system and presumably a reduced ability to defend against elevated oxidative stress in these animals (Schäff *et al.*, 2012).

To sum up this part, during early lactation limited mitochondrial but increased peroxisomal β-oxidation and increased ω-oxidation of fatty acids was accompanied by an increased degradation of amino acids. This indicates stimulated anaplerotic reactions and higher TCA cycling followed by increased mitochondrial respiration in fatty liver (Figure 1). The simultaneously increased production of reactive oxygen species as also described in plasma of early lactating dairy cows (Bernabucci *et al.*, 2005), together with a confined availability of anti-oxidative stress enzymes leads to oxidative stress, further affecting health and immune function. Activation of these oxidative metabolic pathways may provide the link for activating the vagus nerve and signalling for disproportionately low feed intake (Figure 1).

In order to distinguish between hepatic oxidative pathways activated during NEB in early lactation and those activated in response to insufficient energy intake, early lactating dairy cows were either deprived from feed for 60 hours or fed *ad libitum* before liver tissue sampling (Kuhla *et al.*, 2009). Feed deprivation increased liver fat content accompanied by augmented AMPK phosphorylation and down regulation of enzymes associated with fatty acid oxidation (acyl-CoA dehydrogenase and thiolase). These results indicate that down regulation of hepatic β-oxidation does not only support the development of fatty liver but is also involved in signalling hunger to the brain (Allen *et al.*, 2009).

Mammary gland

The liver and the mammary gland have complementary metabolic roles during lactation. In a 2D-DIGE approach, in which liver and mammary gland tissue from early lactating cows were differentially labelled, expression differences between tissues (ratio >2) revealed major complementary metabolic pathways in these tissues (Rawson *et al.*, 2012). While – as expected – enzymes involved in gluconeogenesis and β-oxidation of fatty acids were abundantly expressed in liver, enzymes for fatty acid (acetyl CoA carboxylase and fatty acid synthase) and lactose (UDPG-pyrophosphorylase and α-lactalbumin) synthesis were only detected in mammary gland (Figure 1). On the other hand, enzymes involved in the urea cycle or in propionate metabolism were not evident in the mammary gland (Rawson *et al.*, 2012).

Milk fat content dramatically increases from colostrum to mature milk and declines within the first week of lactation. Reinhardt and Lippolis (2008) investigated protein changes in milk fat globule membranes using iTRAQ. Mucin 1 and 15, adipophilin, butyrophilin, and xanthine dehydrogenase and additional proteins associated with various aspects of lipid transport synthesis and secretion such as acyl-CoA synthetase, lanosterol synthase, lysophosphatidic acid acyltransferase, and fatty acid binding protein were upregulated in milk obtained on day 7 compared with colostrum (day 1). In contrast, apolipoproteins A1, C-III, E, and A-IV were lower in milk compared with colostrum.

These data provided new insight into the type of energy and nutrient secretion involved in NEB of early lactation.

In conclusion, we have demonstrated in this paper that the use of proteomics approaches provides a new quality of information on the networking of cellular processes during the transition between pregnancy and lactation in the dairy cow.

References

Allen M.S., B.J. Bradford, and M. Oba, 2009. Board Invited Review: The hepatic oxidation theory of the control of feed intake and its application to ruminants. J. Anim Sci. 87, 3317-3334.

Aschenbach J.R., N.B. Kristensen, S.S. Donkin, H.M. Hammon, and G.B. Penner, 2010. Gluconeogenesis in dairy cows: The secret of making sweet milk from sour dough. IUBMB Life 62, 869-877.

Bernabucci U., B. Ronchi, N.A. Lacetera, and Nardone. 2005. Influence of body condition score on relationships between metabolic status and oxidative stress in periparturient dairy cows. J. Dairy Sci. 88, 2017-2026.

Bionaz M., K. Periasamy, S.L. Rodriguez-Zas, R.E. Everts, H.A. Lewin, W.L. Hurley, and J.J. Loor, 2012. Old and new stories: revelations from functional analysis of the bovine mammary transcriptome during the lactation cycle. PLoS One. 7, e33268.

Corbacho A.M., G. Martinez De La Escalera, and C. Clapp, 2002. Roles of prolactin and related members of the prolactin/ growth hormone/placental lactogen family in angiogenesis. J. Endocrinol. 173, 219-238.

Gross J., H.A. van Dorland, F.J. Schwarz, and R.M. Bruckmaier, 2011. Endocrine changes and liver mRNA abundance of somatotropic axis and insulin system constituents during negative energy balance at different stages of lactation in dairy cows. J. Dairy Sci. 94, 3484-3494.

Grum D.E., J.K. Drackley, R.S. Younker, D.W. LaCount, and J.J. Veenhuizen, 1996. Nutrition during the dry period and hepatic lipid metabolism of periparturient dairy cows. J. Dairy Sci. 79, 1850-1864.

Hammon H.M., G. Stürmer, F. Schneider, A. Tuchscherer, H. Blum, T. Engelhard, A. Genzel, R. Staufenbiel, and W. Kanitz, 2009. Performance and metabolic and endocrine changes with emphasis on glucose metabolism in high-yielding dairy cows with high and low fat content in liver after calving. J. Dairy Sci. 92, 1554-1566.

Horvath T.L., Z.B. Andrews, and S. Diano, 2009. Fuel utilization by hypothalamic neurons: roles for ROS. Trends Endocrinol. Metab. 20, 78-87.

Ingvartsen K.L., and J.B. Andersen, 2000. Integration of metabolism and intake regulation: a review focusing on periparturient animals. J. Dairy Sci. 83, 1573-1597.

Kuhla B., D. Albrecht, S. Kuhla, and C.C. Metges, 2009. Proteome analysis of fatty liver in feed deprived dairy cows reveals interaction of fuel sensing, calcium, fatty acid and glycogen metabolism. Physiol. Genomics. 37, 88-98.

Kuhla B., D. Albrecht, R. Bruckmaier, T. Viergutz, G. Nürnberg, and C.C. Metges, 2010. Proteome and radio immuno assay analyses of pituitary hormones and proteins in response to feed restriction of dairy cows. Proteomics. 10, 4491-4500.

Kuhla B., S. Görs, and C.C. Metges, 2011a. Hypothalamic Orexin A expression and the involvement of AMPK and PPAR-gamma Signaling in Energy Restricted Dairy Cows. Archiv Tierzucht / Archives Animal Breeding. 6, 567-579.

Kuhla B., Kuhla S., P.E. Rudolph, D. Albrecht, and C.C. Metges, 2007. Proteomics analysis of hypothalamic response to energy restriction in dairy cows. Proteomics. 7, 3602-3617.

Kuhla B., G. Nürnberg, D. Albrecht, S. Görs, H.M. Hammon, and C.C. Metges, 2011b. Involvement of skeletal muscle protein, glycogen and fat metabolism in the adaptation on early lactation of dairy cows. J. Proteome Res. 10, 4252-4262.

Langenheim, J.F., D. Tan, A.M. Walker, and W.Y. Chen, 2006. Two wrongs can make a right: dimers of prolactin and growth hormone receptor antagonists behave as agonists. Mol. Endocrinol. 20, 661-674.

Lewis U.J., Y.N. Sinha, and G.P. Lewis, 2000. Structure and properties of members of the hGH family: a review. Endocr. J. 47, S1-8.

Li X., X. Li, G. Bai, H. Chen, Q. Deng, Z. Liu, L. Zhang, G. Liu, and Z. Wang, 2012a. Effects of non-esterified fatty acids on the gluconeogenesis in bovine hepatocytes. Mol. Cell. Biochem. 359, 385-388.

Li P., X.B. Li, S.X. Fu, C.C. Wu, X.X. Wang, G.J. Yu, M. Long, Z. Wang, and G.W. Liu, 2012b. Alterations of fatty acid β-oxidation capability in the liver of ketotic cows. J. Dairy Sci. 95, 1759-1766.

Lippolis J.D., B.D. Peterson-Burch, and T.A. Reinhardt, 2006. Differential expression analysis of proteins from neutrophils in the periparturient period and neutrophils from dexamethasone-treated dairy cows. Vet. Immunol. Immunopathol. 111, 149-64.

Lippolis J.D., and T.A. Reinhardt, 2005. Proteomic survey of bovine neutrophils. Vet. Immunol. Immunopathol. 103, 53-65.

Lippolis J.D., and T.A. Reinhardt, 2008. Centennial paper: Proteomics in animal science. J. Anim. Sci. 86, 2430-41.

Locher L.F., N. Meyer, E.M. Weber, J. Rehage, U. Meyer, S. Dänicke, and K. Huber, 2011. Hormone-sensitive lipase protein expression and extent of phosphorylation in subcutaneous and retroperitoneal adipose tissues in the periparturient dairy cow. J. Dairy Sci. 94, 4514-4523.

Loor J.J., H.M. Dann, R.E. Everts, R. Oliveira, C.A. Green, N.A. Guretzky, S.L. Rodriguez-Zas, H.A. Lewin, and J.K. Drackley, 2005. Temporal gene expression profiling of liver from periparturient dairy cows reveals complex adaptive mechanisms in hepatic function. Physiol. Genomics. 23, 217-226.

Loor J.J., R.E. Everts, M. Bionaz, H.M. Dann, D.E. Morin, R. Oliveira, S.L. Rodriguez-Zas, J.K. Drackley, and H.A. Lewin, Nutrition-induced ketosis alters metabolic and signaling gene networks in liver of periparturient dairy cows. 2007. Physiol. Genomics. 32, 105-116.

Murondoti A., R. Jorritsma, A.C. Beynen, T. Wensing, and M.J. Geelen, 2004. Unrestricted feed intake during the dry period impairs the *postpartum* oxidation and synthesis of fatty acids in the liver of dairy cows. J. Dairy Sci. 87, 672-679.

Schäff C., S. Börner, S. Hacke, U. Kautzsch, D. Albrecht, H.M. Hammon, M. Röntgen, and B. Kuhla, 2012. Increased anaplerosis, TCA cycling, and oxidative phosphorylation in the liver of dairy cows with intensive body fat mobilization during early lactation. J. Proteome Res. 11, 5503-5514.

Sumner-Thomson J.M., J.L. Vierck, and J.P. McNamara, 2011. Differential expression of genes in adipose tissue of first-lactation dairy cattle. J. Dairy Sci. 94, 361-369.

Talamo F., C. D'Ambrosio, S. Arena, P. Del Vecchio, L. Ledda, G. Zehender, L. Ferrara, and A. Scaloni, 2003. Proteins from bovine tissues and biological fluids: defining a reference electrophoresis map for liver, kidney, muscle, plasma and red blood cells. Proteomics. 3, 440-460.

Piccione G., V. Messina, A. Schembari, S. Casella, C. Giannetto, and D. Alberghina, 2011. Pattern of serum protein fractions in dairy cows during different stages of gestation and lactation. J. Dairy Res. 78, 421-425.

Rawson P., C. Stockum, L. Peng, B. Manivannan, K. Lehnert, H.E. Ward, S.D. Berry, S.R. Davis, R.G. Snell, D. McLauchlan, and T.W. Jordan, 2012. Metabolic proteomics of the liver and mammary gland during lactation. J. Proteomics. 75, 4429-4435.

Reinhardt T.A., and J.D. Lippolis, Developmental changes in the milk fat globule membrane proteome during the transition from colostrum to milk. 2008. J. Dairy Sci. 91, 2307-2318.

Sejersen H., M.T. Sørensen, T. Larsen, E. Bendixen, and K.L. Ingvartsen, 2012. Liver protein expression in dairy cows with high liver triglycerides in early lactation. J. Dairy Sci. 95, 2409-2421.

Weber C., C. Hametner, A. Tuchscherer, B. Losand, E. Kanitz, W. Otten, S.P. Singh, R.M. Bruckmaier, F. Becker, W. Kanitz, and H.M. Hammon, 2013. Variation in fat mobilization during early lactation differently affects feed intake, body condition, and lipid and glucose metabolism in high-yielding dairy cows. J. Dairy Sci. 96, 165-180.

Xia C., H.Y. Zhang, L. Wu, C. Xu, J.S. Zheng, Y.J. Yan, L.J. Yang, and S. Shu, 2012. Proteomic analysis of plasma from cows affected with milk fever using two-dimensional differential in-gel electrophoresis and mass spectrometry. Res. Vet. Sci. 93, 857-861.

Quantifying subclinical ruminal drinking using a [^{13}C]-[$^{15}N_2$]-urea based method in veal calves

H. Berends, J.J.G.C. van den Borne and W.J.J. Gerrits
Animal Nutrition Group, Wageningen University, 6700 AH Wageningen, the Netherlands; harma.berends@wur.nl

Introduction

Ruminal drinking (RD) occurs in calves when ingested milk or milk replacer enters the reticulorumen instead of the abomasum, and can be caused by a failure of the reticular groove reflex or by backflow of milk replacer (MR) from the abomasum. In a clinical case, RD often results in chronic maldigestion (Stocker *et al.*, 1999), ruminal acidosis, lack of appetite and recurrent bloat (Van Weeren-Keverling Buisman *et al.*, 1990), and consequently a reduced growth rate. Although clinical cases of RD have been described extensively (Herrli-Gygi *et al.*, 2006; Gentile *et al.*, 2004; Van Weeren-Keverling Buisman *et al.*, 1990), data on subclinical RD are scarce, partly due to methodological issues. Recent studies have shown that subclinical RD can be substantial in veal calves. Between 21 and 35% of an orally supplied dose of Co-EDTA was recovered in the rumen at slaughter (Suárez *et al.*, 2007; Berends *et al.*, 2012). Assuming that nutrients from MR are subject to fermentation in the rumen, subclinical RD may reduce post-absorptive availability of nutrients and result in lower growth rates in calves. Nonetheless, it is unknown if RD is constant over time (i.e. between meals or days), and measuring recovery at slaughter does not allow repetitive measurements within one animal. Among others, solid feed (SF) provision to veal calves may alter the reticular groove reflex and thus RD. In order to study subclinical RD and related factors, there is a need for a method to quantify RD that does not require calves to be sacrificed. In this study such a method was developed and applied to assess the effect of SF intake on RD.

Material and methods

Forty-eight Holstein Friesian male calves, 55±0.3 kg BW and 53±0.7 d of age at start of the experiment, were fed 41 g MR/kg $BW^{0.75}$ per d. Crude protein and fat level of milk replacer were 212 and 192 g/kg DM, respectively. Calves were assigned to 1 of 4 levels of SF intake: 0, 9, 18, or 27 g DM SF/kg $BW^{0.75}$ per d. The SF mixture consisted of 25% chopped wheat straw, 25% chopped corn silage, and 50% non-pelleted concentrate on a DM basis. RD was measured at 164±1.6 kg BW by combining oral administration of [^{13}C]urea and intravenous administration of [^{15}N^{15}N]urea (Figure 1). Incomplete recovery of an oral dose of [^{13}C]urea, provided with MR, in 48-h urine can be

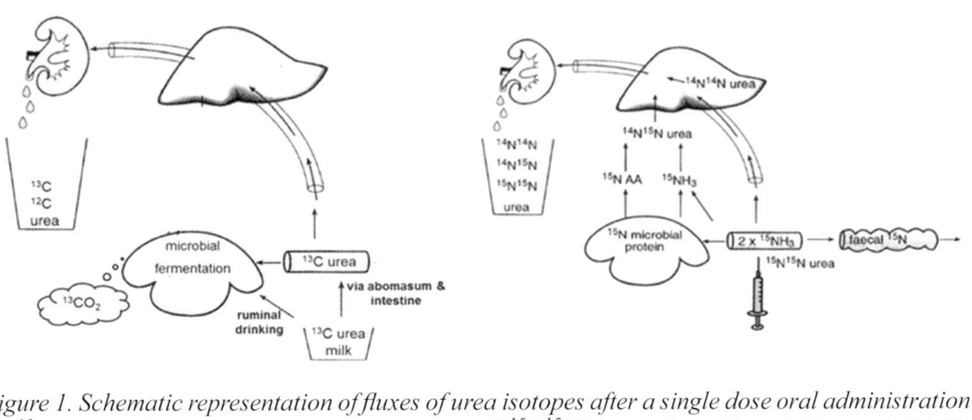

Figure 1. Schematic representation of fluxes of urea isotopes after a single dose oral administration of [^{13}C]urea and 24-h intravenous administration of [^{15}N^{15}N]urea (adapted from Lobley et al.*, 2000).*

explained by RD or by re-entry of absorbed [^{13}C]urea into the gastrointestinal tract, both resulting in fermentation of urea. To correct for the gut entry rate of previously absorbed [^{13}C]urea, we infused [^{15}N^{15}N]urea for 24 h (adapted from Marini and Van Amburgh, 2003) and analysed cumulative urine over 68 h for urea isotopomers and faeces for total ^{15}N enrichment. After monomolecular degradation of urea into N_2, isotopomers (^{14}N^{14}N, ^{15}N^{14}N, ^{15}N^{15}N) were analysed using GC-C-IRMS as described by Marini and Van Amburgh (2003). Infusion of [^{15}N^{15}N]urea was conducted two days after oral administration of [^{13}C]urea.

Results and discussion

Preliminary results show that recovery of [^{13}C]urea, provided with MR, in urine averaged 73±0.3% for calves without SF and decreased (1% per g SF/ kg $BW^{0.75}$ per d; $P<0.05$) with increasing SF intake. The gut entry rate averaged 18% of the urea entry rate for calves without SF and increased with 1.4% per g SF/ kg $BW^{0.75}$ per d ($P<0.05$). When corrected for gut entry, the proportion of [^{13}C] urea absorbed from MR averaged 93±9% and was not affected by SF intake. As a result, subclinical RD was estimated at 0.5±0.6 L, being 7% of the MR provided. In total, 45% of the calves had subclinical RD of 10% of ingested milk or more but this varied substantially between calves (SEM: 9%). Negative estimates for RD (as a % of ingested milk) were obtained for 15 out of 38 calves: 9 calves ranging between -1 and -12%, and 6 calves ranging between -29 and -59%).

It can be concluded that quantification of subclinical ruminal drinking by the dual tracer method (i.e. [^{13}C] and [^{15}N^{15}N]urea) needs refining, and requires comparison to a golden standard, like the Cobalt recovery. The method could be improved when the fate of orally dosed [^{13}C]urea and the gut entry rate are studied simultaneously, to avoid effects of variation between and within days. Ruminal drinking in the current study was generally lower than in previous studies (Suárez *et al.*, 2007; Berends *et al.*, 2012), which may be influenced by the MR level which was lower in the current study. Alternative methods that may provide quantitative and repeatable insight in ruminal drinking may include ultrasonography to measure abomasal volume (Marshall *et al.*, 2005), and/or the acetaminophen absorption test (Sharifi *et al.*, 2009).

References

Berends, H., J.J.G.C. van den Borne, S.J.J. Alferink, C.G. van Reenen, E.A.M. Bokkers, and W.J.J. Gerrits. 2012. Low-protein solid feed improves the utilization of milk replacer for protein gain in veal calves. J. Dairy Sci. 95(11), 6654-6664.

Herrli-Gygi, M., H.M. Hammon, Y. Zbinden, A. Steiner, and J.W. Blum. 2006. Ruminal drinkers: endocrine and metabolic status and effects of suckling from a nipple instead of drinking from a bucket. J. Vet. Med. A53, 215-224.

Lobley, G.E., D.M. Bremner, and G. Zuur. 2000. Effects of diet quality on urea fates in sheep as assessed by refined, non-invasive [^{15}N^{15}N]urea kinetics. Br. J. Nutr. 84, 459-468.

Marini, J.C., and M.E. Van Amburgh. 2003. Nitrogen metabolism and recycling in Holstein heifers. J. Anim. Sci. 81, 545-552.

Marshall., T.S., P.D. Constable, S.S. Crochik, and T. Wittek. 2005. Determination of abomasal emptying rate in suckling calves by use of nuclear scintigraphy and acetominophen absorption. A.J.V.R. 66 (3), 364-374.

Sharifi, K., W. Grűnberg, S. Soroori, M. Mohri, and M.S. Ahrari-Khafi. 2009. Assessment of the acetominophen absorption test as a diagnostic tool for the evaluation of the reticular groove reflex in lambs. Am. J. Vet. Res. 70 (7), 820-825.

Stocker, H., Lutz, H., and P. Rush. 1999. Clinical, haematological and biochemical findings in milk-fed calves with chronic indigestion. Vet. Rec. 145, 307-311.

Suárez, B. J., C.G. Van Reenen, N. Stockhofe, J. Dijkstra, and W.J.J. Gerrits. 2007. Effect of roughage source and roughage to concentrate ratio on animal performance and rumen development in veal calves. J. Dairy Sci. 90, 2390-2403.

Van Weeren-Keverling Buisman, A., J.M.V.M. Mouwen, T. Wensing, and H.J. Breukink. 1990. Intraruminal administration of milk in the calf as a model for ruminal drinking: Morphological and enzymatical changes in the jejunal mucosa. Vet. Res. Commun. 14(2), 129-140.

Estimation of 24-h energy expenditure in dairy cows using the ^{13}C bicarbonate dilution method

A. Münger[1], L.D. Kaufmann[1], S. Görs[2], C.C. Metges[2] and F. Dohme-Meier[1]
[1]Agroscope Liebefeld-Posieux Research Station ALP, P.O. Box 64, 1725 Posieux, Switzerland; frigga.dohme-meier@alp.admin.ch
[2]Leibniz Institute for Farm Animal Biology (FBN), Wilhelm-Stahl-Allee 2, 18196 Dummerstorf, Germany

Introduction

The interest in pasture-based dairy production systems increases because they can be economically attractive and environmentally friendly (Peyraud *et al.*, 2010). However, high yielding dairy cows are not able to cover their nutritional demands just by grazing, and they need additional energy to express their genetic potential for milk production (Kolver and Muller, 1998). So far available data about energy requirements of grazing dairy cows are limited due to the lack of suitable measurement methods. The ^{13}C bicarbonate dilution technique combined with an automatic blood sampling system allows the estimation of energy expenditure (EE) in grazing cows over a 6 h period (Kaufmann *et al.*, 2011). In order to evaluate if EE measured for 6 h can be extrapolated to 24 h an experiment was conducted recording CO_2 production, physical activity and feeding behavior during three 6-h periods per d in dairy cows grazed full-time.

Material and methods

The experiment was carried out with 12 Holstein cows (body weight (BW): 655±41.0 kg; days in milk: 73±12 d; milk yield: 44±4.8 kg/d) and consisted of a 2-wk adaptation period and a 2-wk experimental period. The cows were equally divided between the two experimental weeks so that each cow passed a period of 7-d where data was collected. Cows were on pasture from 07:30 to 14:30 h, 17:00 to 22:30 h and from 23:15 to 04:30 h. Between these grazing periods they were kept in a free-stall barn for preparation of the measurements, feeding of supplement and milking at 05:10 and 16:00 h. The supplement consisted of a cereal-based concentrate in amounts to meet predicted nutrient requirements and was offered in two equal meals at 06:30 and 16:30 h after milking using weighing troughs (Insentec B.V., Marknesse, the Netherlands). For each cow milk yield and milk component contents were determined daily and intake on pasture was quantified by the double alkane technique during the respective experimental week. The CO_2 production of each cow was determined on 1 d from 07:00 to 13:00 h (measurement period (MP) A), from 15:00 to 21:00 (MP B) and from 23:00 to 05:00 h (MP C) using the ^{13}C bicarbonate dilution method (Junghans *et al.*, 2007). After administration of the tracer (0.7 mg $NaH^{13}CO_3$/kg BW) into the jugular vein, blood was sampled with an automatic blood sampling system on pasture. The blood sampling procedure and the calculation of the EE were described by Kaufmann *et al.* (2011). Physical activity of the cows was recorded over 7 d using pedometers and feeding and rumination behavior was investigated over 5 d using a behavior recorder. Differences in EE, physical activity and eating and rumination were analyzed using a linear mixed model with measurement period as fixed effect and cow as random effect. The relationship between EE and physical activity and EE and eating and rumination, respectively, were examined using Spearman correlation. Estimation of the 24 h EE and the comparison which MP was the most reliable to estimate 24-h EE were done by regression analysis.

Results and discussion

During the experimental weeks cows produced daily on average 40.8±4.0 kg of milk with 4.1±0.5% of fat, 3.2±0.2% of protein and 4.7±0.1% of lactose. Total dry matter intake amounted to 19.0±1.5 kg/d composed of 12.1±0.8 kg DM/d herbage and 6.9±1.3 kg DM/d concentrate. Energy expenditure was lower ($P<0.01$) in MP A (270 kJ/kg $BW^{0.75}$) compared to MP B (322) and C (317). Concomitantly, cows spent more time walking ($P<0.001$) in MP C (143 min) than in MP A (72 min) and B (73 min). This was surprising. In general grazing cows are more active at daytime than at night because they eat more and ruminate less during the day (Gibb *et al.*, 1997). This was confirmed by the present results. However, because of the preparation of the measurements cows had to be moved to the barn also during MP C, the night period. Furthermore, in contrast to MP A and B during which cows walked just once from the barn to the pasture, in MP C cows had to cover this distance twice (barn-pasture, pasture-barn). This might explain the difference in EE between MP A and C. However, no correlations ($P>0.05$) existed between EE and physical activity. The clear difference in EE between MP A and B was also astonishing but might be caused by the different activities cows did in these periods. While MP A consisted almost solely of time on pasture, in MP B milking and concentrate feeding were included.

The EE for 24 h was 1204 kJ/kg $BW^{0.75}$ and was estimated based on the data of the three MP with a prediction uncertainty of 12%. The value seems plausible because it is about 20% higher compared to data of grass-fed cows determined in respiration chambers (Bruinenberg *et al.*, 2002). We reported previously that EE of cows grazing full time was about 20% higher compared to cows fed herbage of the same quality in a free-stall barn (Kaufmann *et al.*, 2011).

Further regression analysis showed a low standard error of estimate for each MP when used in conjunction with chosen behavior and activity data to determine EE for 24 h. Differences among periods were minor as well. Therefore, it can be concluded that each of the three 6-h measurement periods can be used to estimate EE of grazing cows for 24 h.

References

Bruinenberg, M.H., Van der Honing, Y., Agnew, R.E., Yan, T., van Vuuren, A.M., Valk, H., 2002. Energy metabolism of dairy cows fed on grass. Livest. Prod. Sci. 75, 117-128.

Gibb, M. J., C. A. Huckle, R. Nuthall, Rook, A. J., 1997. Effect of sward surface height on intake and grazing behaviour by lactating Holstein-Friesian cows. Grass Forage Sci. 52, 309-321.

Junghans, P., Voigt, J., Jentsch, W., Metges, C.C., Derno, M., 2007. The ^{13}C bicarbonate dilution technique to determine energy expenditure in young bulls validated by indirect calorimetry. Livest. Sci. 110, 280-287.

Kaufmann, L.D., Münger, A., Rérat, M., Junghans, P., Görs, S., Metges, C.C., Dohme-Meier, F. 2011. Energy expenditure of grazing cows and cows fed grass indoors as determined by the 13C bicarbonate dilution technique using an automatic blood sampling system. J. Dairy Sci. 94, 1989-2000.

Kolver, E. S., and L. D. Muller. 1998. Performance and nutrient intake of high producing Holstein cows consuming pasture or a total mixed ration. J. Dairy Sci. 81:1403-1411.

Peyraud, J.L., Van den Pol-van Dasselaar, A., Dillon, P., Delaby, L. 2010. Producing milk from grazing to reconcile economic and environmental performances. In: Schnyder, H., Isselstein, J., Taube, F., Auerswald, K., Schellberg, J., Wachendorf, M., Herrmann, A., Gierus, M., Wrage, N. and A. Hopkins (eds.) Grassland in a changing world, Organising Committee of the 23th General Meeting of the EGF and Arbeitsgemeinschaft Grünland und Futterbau der Gesellschaft für Pflanzenbau Vol. 15, 865-879.

The impact of nutritional, animal and farm management factors on variation in milk urea content

A. Bannink[1], *J.W. Spek*[1,2], *J.L. Ellis*[2] *and J. Dijkstra*[2]
[1]*Wageningen UR Livestock Research, Wageningen UR, P.O. Box 65, 8200 AB, Lelystad, the Netherlands; andre.bannink@wur.nl*
[2]*Wageningen University, Wageningen UR, 6708 WD, Wageningen, the Netherlands*

Introduction

Urea excreted in urine is the primary source of undesirable nitrogen (N) emissions from dairy systems, and it accounts for most of the variation in N excretion. Urea is formed by hepatocytes from ammonia generated by catabolism of amino acids or absorbed from the gastrointestinal tract. A minor fraction (<1%) of urea N transported through blood ends up in milk and therefore milk urea N content (MUN; mg N/dl milk) is an indicator of the amount of urea N excreted in urine. Such information may be useful to optimize protein nutrition and reduce N emissions in farming practice. A strong relationship exists between MUN and N excreted in urine, with R^2 values typically ranging between 0.7 and 0.8 (Schröder *et al.*, 2005). However, such high R^2 values apply for the total range of observed MUN and N excretion data. For the narrow range of MUN that is relevant for use as an indicator in practice a much lower R^2 values applies (<0.3). The main part of variation in MUN has then to be attributed to other factors than urea N excreted in urine (UUN, g N/d). The aim of the present work was to explore and obtain a better quantitative understanding of the influence of these other factors on MUN, to let MUN become a more useful indicator of UUN.

Urea dynamics, milk urea and N excretion

The MUN as a concentration (mg N/dl) depends on urea N concentration in blood plasma (PUN; mg N/dl blood plasma). Any nutritional parameter that affects PUN dynamics hence also affects MUN. Therefore, quantifying variation in MUN that is unrelated to variation in UUN requires a quantitative understanding of PUN dynamics, and quantification of the influence of nutritional, management and animal factors on these dynamics.

Nutritional factors influencing PUN dynamics are type of carbohydrates and proteins fed and their site of digestion, affecting the dynamics of ammonia entry in blood plasma from the rumen and amino acid catabolism. Also water dynamics and kidney function affect PUN dynamics. Adding salt to the diet increase urine volume and lowers PUN by 0.06 to 0.10 unit per kg additional urine (Spek *et al.*, 2012a; unpubl.). Water restriction may have an even stronger effect (Burgos *et al.*, 2001; Spek *et al.*, 2013). Renal reabsorption of urea to blood from glomerular filtrate may vary with water dynamics and protein intake (Thornton, 1970; Røjen *et al.*, 2011; Spek *et al.* unpubl.). Finally, PUN may be affected by urea recycling to the gastrointestinal tract. Based on ^{13}C labelled urea infusions, Spek *et al.* (unpubl.) established with an increased N intake a 67% increase of PUN and a 44% increase of urea entry rate in blood plasma but no effect on urea recycling, whereas salt addition and increased urine volume decreased PUN by 17% and decreased urea recycling by 9%.

Next to nutritional factors, management factors such as the moment and frequency of milking and feeding cause variation in MUN. Net urea transfer depends on urea concentration gradient between plasma and milk, and changes in MUN follow those in PUN with a time lag because of the time involved with urea transfer between milk and blood and between milk compartments in the udder. Hence, MUN is not necessarily the integral of changes of PUN in time. Spek *et al.* (2012b) used several infusion protocols with ^{13}C and ^{15}N stable isotope labelled urea to demonstrate urea transfer dynamics. After injection of labelled urea in the udder cistern, enrichment of subsequent fractions of milk collected with milking (from first milk to alveolar milk) varied, depending on the time period between infusion and milking and between various fractions of milk collected. Within half an hour, substantial fractions of labelled urea in milk were transported from cistern to alveoli, and from milk to blood. About a third of isotope injected in the mammary gland was recovered in blood after 2 hours.

Also animal related factors may affect the relationship between PUN and UUN such as body weight. Many animal factors might be covered already by nutritional factors such as feeding behaviour (dominance) and capacity, or milk production capacity. Genetic factors are involved as MUN appears to be a heritable characteristic. In a study of Šebek *et al.* (unpubl.) breeding values for MUN ranged from -2.3 to +2.8 mg/dl for 15,720 measurement weeks of 723 cows in 26 feeding trials. Breeding value of MUN was not related to efficiency of N utilization observed. Hence, unknown, heritable animal factors affect MUN but are unrelated to variation in N utilization and hence UUN at identical N intake, illustrating the existence of factors that should be quantifiable in principle.

Modelling urea dynamics

The dynamics of urea exchange between blood and milk and between udder milk compartments, in combination with the dynamics of urea (ammonia) entry in blood plasma, of milk synthesis and secretion (milking), and of urea reabsorption and urea excretion in urine by the kidneys, may cause MUN to vary independently of UUN. To quantify the urea dynamics in blood plasma, milk, and kidneys, a mechanistic model was constructed consisting of four state variables representing urea pools in blood, kidney (glomerular filtrate) and milk. Flux equations are described by Michaelis-Menten kinetics or mass action forms. Model inputs are the dynamics of the absorption of ammonia, the digestion of protein and cations from the gastrointestinal tract, and of milk synthesis and milking. Model outputs are PUN, MUN, urine volume and UUN. The model is used to improve quantitative understanding of how nutritional and management factors contribute to variation in MUN that is unrelated to UUN.

References

Burgos, M.S., M. Senn, F. Sutter, M. Kreuzer and W. Langhans. 2001. Effect of water restriction on feeding and metabolism in dairy cows. Am. J. Physiol. Regul. Integr. Comp. Physiol. 280, 418-427.

Røjen, B.A., P.K. Theil and N.B. Kristensen. 2011. Effects of nitrogen supply on inter-organ fluxes of urea-N and renal urea-N kinetics in lactating Holstein cows. J. Dairy Sci. 94, 2532-2544.

Schröder, J.J., A. Bannink and R. Kohn, 2005. Improving the efficiency of nutrient use in cattle operations. In: Pfeffer, E. and A.N. Hristov (eds.) Nitrogen and phosphorus nutrition of cattle. CAB International, Wallingford, United Kingdom, pp 255-279.

Spek, J.W., A. Bannink, G. Gort, W.H. Hendriks and J. Dijkstra, 2012a. Effect of sodium chloride intake on urine volume, urinary urea excretion, and milk urea concentration in lactating dairy cattle. J. Dairy Sci. 95, 7288-7298.

Spek, J.W., J. Dijkstra, J.J.G.C. van den Borne and A. Bannink, 2012b. Short communication: Assessing urea transport from milk to blood in dairy cows. J. Dairy Sci. 95, 6536-6541.

Spek, J.W., J. Dijkstra, G. van Duinkerken and A. Bannink, 2013. A review of factors influencing milk urea concentration and its relationship with urinary urea excretion in lactating dairy cattle. J. Agric. Sci. 151, 407-423.

Thornton, R., 1970. Factors affecting the urinary excretion of urea nitrogen in cattle. I. Sodium chloride and water loads. Austr. J. Agric. Res. 21, 131-144.

Metabolizable energy of pure stand alfalfa hay estimated from near infrared spectra

C. Old[1], N. Ohanesian[1], J. Miller[1], R. Hinders[1], W. Vogt[1], J. Oltjen[1] and D. Sapienza[2]
[1]California Chapter, American Registry of Professional Animal Scientists (CaARPAS), Sacramento, CA, USA; anteloperanch31@gmail.com
[2]Sapienza Analytica LLC, Slater, IA, USA

Introduction

Since the 1950s, alfalfa hay sold in California has been priced based on total digestible nutrient (TDN) content estimated from either modified crude fiber (Meyer and Lofgreen, 1959) or acid detergent fiber (ADF). Residues from either procedure can vary in chemical composition for alfalfa hay harvested throughout the California growing season, which begins in February in extreme southern California and ends in October. Since energy based feeding standards are those by which others are judged, CaARPAS undertook this study to determine if metabolizable energy (ME) content of pure stand alfalfa hay could be determined via near infrared (NIR) spectrophotometry.

Material and methods

Approximately 200 samples of pure stand alfalfa hay were collected throughout the alfalfa growing regions of California and western Nevada. Of these, nine were selected as representative of compositional variability in ADF and N content. Samples were cubed and sent to the USDA Dairy Forage Research Center Madison, WI, USA) for evaluation in a metabolism study. Upon receipt, cubes were reground and fed as pellets. Energy losses in feces and urine were determined for each hay fed to wether lambs (n=6) at either 2 (R) or 5% (AL) of body weight. Gaseous energy (GE) losses were estimated utilizing a combined in-silico and in-vitro method. It was recognized that observed urinary energy (UE) losses exceeded expected, as part of a well balanced diet fed to lactating dairy cows. In order to approximate ME under actual conditions GE and UE losses were estimated as 18% of digestible energy (DE) first by the equation ME=0.82 DE (ME1). Methane energy losses were estimated from digestible carbohydrates (in vitro, 0.51 moles CH_4 per 180 g NDF fermented and 0.45 moles CH_4 per 180 g non-structural carbohydrate fermented) plus measured UE, adjusted so that determined (GE + UE) losses averaged 18% of DE (ME2). Samples of all hays were oven dried (62 °C) and ground (1 mm) prior to being placed in a Model 5000 NIR spectrophotometer (Foss NIR Systems, USA). Reflectance values for λ from 950 to 2,500 nm were analyzed using the chemometric utility UNSCRAMBLER (CAMO Software, Norway) to determine NIR prediction models for TDN (AL only), DE and ME (1 and 2) at either level of intake.

Results and discussion

Digestible energy determined at R (mean=2.84, range 2.45 to 3.33 Mcal/kg) was greater ($P<0.001$) than DE determined at AL (mean=2.67, range 2.28 to 3.34 Mcal/kg). Metabolizable energy estimated at R (mean (ME1 and ME2)=2.19, range (ME1) 1.87 to 2.74 Mcal/kg; range (ME2) 1.87 to 2.77 Mcal/kg) was greater ($P<0.001$) than ME estimated at AL. Average estimates for ME (AL) were 2.19 Mcal/kg (ME1 and ME2), the range for ME1 was 2.19 to 2.74 Mcal/kg and for ME2, 2.19 to 2.77 Mcal/kg. TDN (AL only) averaged 60.6 and ranged from 54.2 to 68.8. Response variables (DE, ME and TDN) were predicted from spectral analyses with a minimum R^2 of 0.87 (ME2 R) and a maximum of 0.94 (DE AL and ME1 AL) (Table 1). Meyer and Lofgreen (1959) reported an R^2 of 0.79 for the relationship between TDN and modified crude fiber (crude fiber plus silica), while Bath (1985) reported an identical R^2 for the relationship between TDN and ADF; both studies used wether lambs. In the former study 40 samples of alfalfa hay were evaluated while in the latter nine were evaluated.

Table 1. NIR prediction statistics

Item	R^2	Standard error of calibration
DE (R) Mcal/kg	0.92	0.09
DE (AL) Mcal/kg	0.94	0.07
ME1 (R) Mcal/kg	0.92	0.06
ME1 (AL) Mcal/kg	0.94	0.06
ME2 (R) Mcal/kg	0.87	0.08
ME2 (AL) Mcal/kg	0.92	0.07
TDN (%)	0.90	1.3

Analyses of these data indicates that NIR spectra of pure stand alfalfa hays may be used to predict TDN, DE, and ME content When compared to the system currently used in California to predict alfalfa quality, accuracy is improved, although further testing is required to make the system functional.

References

Bath, D.L. 1985. Accuracy of the modified crude fiber and acid detergent fiber tests to predict TDN of alfalfa hay. Proc. Fifteenth California Alfalfa Symp. 45-49

Meyer, J.H. and G.P. Lofgreen. 1959. Evaluation of alfalfa hay by chemical analyses. J. Animal Sci. 1233-1242

Improving *in sacco* incubation technique to evaluate starch degradability of fresh and fermented corn silage for ruminants

J. Peyrat[1], E. Watremez[1], A. Le Morvan[1], A. Férard[2], G. Cabon[2], P.V. Protin[2], P. Nozière[1] and R. Baumont[1]
[1]INRA, UMR1213 Herbivores, Site de Theix, 63122 Saint-Genès Champanelle, France; julie.peyrat@clermont.inra.fr
[2]ARVALIS-Institut du Végétal, Station expérimentale de la Jaillière, 44370 La Chapelle Saint Sauveur, France

Introduction

Corn silage is the main forage fed to ruminants and represents the most important energetic supply to animals of high level of production (dairy cows and fattening animals). Global methods to predict organic matter digestibility of corn silage were developed in the 1990's (Aufrère *et al.*, 1992). To improve the understanding and the prediction of corn energetic substrates degradation in the rumen, one problem lies in the lack of reference method to evaluate the rate and extent of corn silage starch degradation in the rumen. This issue could lead to errors in the choice of feed supplements in diet formulation. The aim of the present experiment was to determine the most adapted experimental in sacco procedure among current *in sacco* incubation technique to evaluate degradability of fresh corn and corn silage.

Material and methods

Eight corn samples (two genotypes with different types of starch; two stages of maturity, harvested at 27% DM and 42% DM, respectively; two methods of conservation: fresh and fermented) were placed in nylon bags according to the 3 following conditioning methods: dried and 1 mm ground (D1, usual conditioning for *in sacco* technique; Michalet-Doreau *et al.*, 1987), dried and 4 mm ground (D4, a conditioning that is supposed to reduce particle loss through the pores of the bags; Philippeau and Michalet-Doreau, 1997) and fresh-ground (FG, a conditioning which simulates corn arriving in the rumen after chewing). For each of the 24 treatments, the kinetics of starch disappearance was measured in the rumen of three fistulated cows with two replicates per cow and at following incubation times: 2, 4, 8, 16, 24, 48 and 96 h. Then, the effective degradability (ED) was calculated with the step by step method (Kristensen *et al.*, 1982), assuming a particle outflow rate of 0.06 /h. Effects of genotype, maturity, conservation method, sample conditioning and their interactions on degradation parameters were analysed as fixed effects and animal as random effect with the MIXED procedure of SAS (9.1 version, 2002-2003).

Results and discussion

Table 1 reports the effects of conservation method and of sample conditioning on starch degradability. Besides, interaction of method and sample conditioning had a significant effect on starch degradability.

Table 1. Starch degradability (n=144) from corn (fresh and silage) conditioned in different ways (dried and 4mm ground, dried and 1mm ground, fresh ground). Means ± standard errors are presented and statistically compared within each effect (C and S).

	Conservation of corn (C)		Sample conditioning (S)		
	Fresh	Silage	Dried and 4 mm ground (D4)	Dried and 1 mm ground (D1)	Fresh-ground (FG)
2 h degradability (%)	39.2±0.36^a	55.04±0.36^b	32.5±0.42^a	53.8±0.42^b	55.1±0.42^c
Effective degradability (%)	62.4±0.43^a	74.9±0.43^b	58.7±0.44^a	72.7±0.44^b	74.5±0.44^c

^{a,b,c} $P<0.05$.

Effective degradability was significantly higher ($P<0.01$) for corn silage than for fresh corn, regardless of the sample conditioning method. The effective degradability of D1 samples was significantly higher than that of D4 ($P<0.01$) and lower than FG samples ($P<0.01$). This was mainly related to differences in the degradation at 2h that was significantly higher with D1 than with D4 ($P<0.01$) and lower with D1 than with FG samples ($P<0.05$).

Stage of maturity also affected significantly starch degradability. Effective degradability of early stage characterized by 27% DM (74.7%) was significantly higher ($P<0.01$) than that of advanced stage characterized by 42% DM (62.6%). This decrease of effective degradability with stages of maturity were consistent for the two genotypes, the conservation and the sample conditioning methods (significant effect of interaction genotype × stage × conservation × sample conditioning; $P<0.01$).

No significant difference was found between the two genotypes on starch degradability (68.6%; $P=0.26$) although genotype C had a different starch type than genotype B.

This trial shows high differences between starch degradability of fresh and fermented corn. This probably can be explained by partial hydrolysis of starch during fermentation process in silage (Jurjanz and Monteils, 2005). Thus, the use of fresh samples to evaluate corn silage degradability will underestimate starch degradability although it is more convenient for large varieties trials than using fermented samples.

The D1 sample conditioning, frequently used in the *in sacco* studies for evaluation of nitrogen value, leads to high estimates of starch degradability because fine grinding generates high losses of starch particles through the pores of nylon bags (Philippeau and Michalet-Doreau, 1997). Similar high degradability values were obtained with the FG samples. However this method was more difficult in its application than the two others, probably because homogeneous sampling of fresh material is more difficult than of dry samples. The D4 sample conditioning leads to lower estimates of starch degradability, in particular for short incubation times, and the reproducibility was similar for both D4 and D1 conditions.

The expected decrease in starch degradability when the stage of maturity increases (Philippeau and Michalet-Doreau, 1997; Jensen *et al.*, 2005) is clearly confirmed in the present work.

To conclude, our results show a difference in starch degradability between fresh and ensiled corn and quantify the influence of different sample conditioning and confirms the effect of maturity stage. From this study, the most suitable method to evaluate starch degradability of corn silage appears to be the use of silage samples dried and ground at 4 mm.

References

Aufrère, J., D. Graviou, C. Demarquilly, J. Andrieu, J.C. Emile, R. Giovanni and P. Maupetit, 1992. Estimation of organic matter digestibility of whole maize plants by laboratory methods. Anim. Feed. Sci. Technol. 36, 187-204.

Jensen, C., M.R. Weisbjerg, P. Norgaard, T. Hvelplund, 2005. Effect of maize silage maturity on site of starch and NDF digestion in lactating dairy cows. Anim. Feed. Sci. Technol. 118 (3), 279-294.

Jurjanz, S and V. Monteils, 2005. Ruminal degradability of corn forages depending on the processing method employed. Anim. Res. 54, 3-15.

Kristensen, E.S., P.D. Moller and T. Hvelplund, 1982. Estimation of the effective protein degradability in the rumen of cows using the nylon bag technique combined with the outflow rate. Acta Agr Scand. 32, 123-127.

Michalet-Doreau, B., R. Vérité and Chapoutot, P, 1987. Méthodologie de mesure de la dégradabilité in sacco de l'azote des aliments dans le rumen. Bull. Tech. CRZV Theix, 1987.

Philippeau, C. and B. Michalet-Doreau, 1997. Influence of genotype and stage of maturity of maize on rate of ruminal starch degradation. Anim. Feed Sci. Technol. 68, 25-35.

SAS, 2002-2003. Release 9.1. SAS Inst. Inc., Cary, NC, USA.

A titration approach to identify the capacity for starch digestion in milk-fed calves

M.S. Gilbert[1], J.J.G.C. van den Borne[1], A.J. Pantophlet[2] and W.J.J. Gerrits[1]
[1]Animal Nutrition Group, Wageningen University, P.O. Box 338, 6700 AH, Wageningen, the Netherlands; myrthe.gilbert@wur.nl
[2]Centre for Medical Biomics, University Medical Centre Groningen, P.O. Box 30001, 9700 RR Groningen, the Netherlands

Introduction

Calf milk replacers commonly contain 40-50% lactose. For economic reasons, starch is of interest as a lactose replacer. Compared with lactose, starch digestion is generally low in calves. Ileal disappearance of starch was only 60% in calves, whereas lactose disappeared for 97% (Coombe and Smith, 1974). This indicates that the activity of enzymes required for the hydrolysis of starch to glucose limits starch digestion in milk-fed calves. It is however unknown which enzyme system is limiting the rate of starch hydrolysis in the intestinal lumen of calves. In addition, a maximum may exist for the daily quantity of starch that can be hydrolyzed and absorbed. Potentially, enzyme systems may also adapt to the starch fed. Both may be subject to considerable inter-individual variation.

Incomplete intestinal starch hydrolysis and absorption leads to fermentation, subsequently reducing ileal and fecal dry matter (DM) content and pH (Huber *et al.*, 1961; Roy, 1969; Kreikemeier *et al.*, 1991). We used this relation between fermentation and fecal DM content and pH in a titration study where lactose was stepwise exchanged for one of 4 starch products (SPs). These SPs differed in the enzymes required for their complete hydrolysis to glucose. The objectives were (1) to determine which enzyme system limits starch digestion in milk-fed calves and (2) to determine the maximum inclusion level of SPs in the milk replacer.

Material and methods

Forty calves (104±0.5 kg BW) were assigned to either a control diet (with lactose as the only source of carbohydrates) or one of the 4 titration strategies, in which lactose was gradually exchanged for one of 4 SPs. The 4 SPs differed in the enzymes required for their complete hydrolysis to glucose: gelatinized starch (α-amylase and maltase); maltodextrin (α-amylase and maltase); maltodextrin with α-1,6-branching (α-amylase, maltase and isomaltase) and maltose (maltase). Dietary inclusion level of each SP was increased by 3% per week at the expense of lactose. This increase in inclusion level of SPs was stopped when fecal DM content dropped below 10.6%, corresponding to 75% of the average starting fecal DM content, and remained below this value for 3 consecutive weeks. Calves were fed 52 g of milk replacer per kg metabolic BW per day.

Fecal samples were collected from the rectum each week from each calf to determine DM content and pH. The maximum capacity of starch digestion of an individual calf was defined as the inflection point in the relation between inclusion level of the SP and fecal DM content or pH. Differences in treatment averages of these inflection points can be used to identify the rate-limiting enzyme system in starch digestion. The maximum capacity of starch digestion was estimated for each calf using non-linear regression of fecal DM content and pH against the inclusion level of SP.

Results and discussion

For control calves, fecal DM content and pH did not change over time (mean fecal DM content 15.4±0.3%, P=0.481; mean fecal pH 7.5±0.1, P=0.391). Although we hypothesized that inflection points could be estimated to identify the rate-limiting enzyme system in starch digestion, the fecal

DM content and pH already responded to changes in SP inclusion at very low inclusion levels of SPs. Hence, for 27 out of 32 SP-fed calves, linear regression provided a better fit ($P<0.05$) of the data (SP inclusion level vs. fecal DM content or pH) than nonlinear regression. Treatment effects on the slope parameters (fecal DM content or pH against inclusion level) were subsequently analyzed statistically.

Slopes for SP-fed calves were lower (fecal DM content, $P=0.004$; fecal pH, $P<0.001$) than for control calves and did not differ among SPs (Table 1). Fecal DM content dropped by 0.21% (±0.03%) and fecal pH dropped by 0.12 (±0.01) for every % increase in dietary SP. The gradual decrease in fecal DM content and pH with increasing inclusion of SPs indicates that the maximum inclusion level does not mark an inflection point from enzymatic hydrolysis to fermentation but that SPs are already (partly) fermented at lower inclusion levels.

For the current study, SPs were selected on the enzymes required for their hydrolysis. The combination of SPs would lead us to deduce the rate-limiting enzyme for starch digestion in milk-fed calves. As both fecal DM content and pH responded sensitively to increasing inclusion of any of the SPs, the hydrolysis of maltose to glucose by maltase is probably rate-limiting. Therefore, maltose will probably induce intestinal fermentation in starch-fed calves. Surprisingly, the fecal DM content and pH already decreased at low inclusion levels of SPs. Consequently, a maximum capacity of starch digestion could not be identified and fermentation of the SPs was likely substantial.

In conclusion, fecal DM content and pH gradually decreased with incremental inclusion of SPs in calf milk replacer, independent of SP characteristics. This indicates that maltase limits starch digestion in milk-fed calves and that fermentation may contribute substantially to total tract starch disappearance in milk-fed calves.

Table 1. Slopes of the relation between inclusion level of starch products (GS, gelatinized starch; MD, maltodextrin; MDB, maltodextrin with a high degree of branching; MT, maltose, exchanged at the expense of lactose) and fecal DM content or pH in milk-fed calves.

Treatment	Control	GS	MD	MDB	MT	*P*-value
			Mean ± SE			
Number of calves	7	8	8	8	8	
Slopes[1]						
Fecal DM, %	0.04±0.02[a]	-0.19±0.05[ab]	-0.16±0.04[ab]	-0.23±0.03[b]	-0.28±0.09[b]	0.004
Fecal pH	-0.01±0.01[a]	-0.12±0.01[b]	-0.12±0.02[b]	-0.13±0.02[b]	-0.10±0.02[b]	<0.001

[1] Slopes represent the change in fecal DM content or pH per 2.3 days (for control treatment) or per one % increase of a starch product (for starch product treatments).

[ab] Means with different superscripts in the same row differ significantly ($P<0.05$).

References

Coombe, N.B. and R.H. Smith, 1974. Digestion and absorption of starch, maltose and lactose by the preruminant calf. Br. J. Nutr. 31, 227-235.

Huber, J.T., A.D. McGilliard, R.S. Allen and N.L. Jacobson, 1961. Utilization of carbohydrates introduced directly into omaso-abomasal area of stomach of cattle of various ages. J. Dairy Sci. 44, 321-330.

Kreikemeier, K.K., D.L. Harmon, R.T. Brandt, T.B. Avery and D.E. Johnson, 1991. Small intestinal starch digestion in steers – Effect of various levels of abomasal glucose, corn starch and corn dextrin infusion on small intestinal disappearance and net glucose absorption. J. Anim. Sci. 69, 328-338.

Roy, J.H.B., 1969. Diarrhoea of nutritional origin. Proc. Nutr. Soc. 28, 160-170.

Application of washed rumen technique for rapid determination of fasting heat production in steers

D.H. Kim[1], K.R. McLeod[1], J.L. Klotz[2], A.F. Koontz[1], A.P. Foote[1] and D.L. Harmon[1]
[1]Department of Animal and Food Sciences, University of Kentucky, Lexington, KY, USA; dohyungkim77@gmail.com
[2]USDA-ARS, Forage-Animal Production Research Unit, Lexington, KY, USA

Introduction

Traditional measurement of maintenance energy requirements in cattle uses estimates of fasting heat production (HP) made during the third and fourth day of fasting, when the respiratory quotient (RQ) has fallen to approximately 0.7 (Blaxter, 1967). However, results vary with the length (days) of fasting (Blaxter and Wainman, 1966). In addition, this approach has limitations, specifically the actual severity of stress and decline of physical activity induced by the extended fasting period. It was hypothesized that using the washed rumen technique, in conjunction with accurate fasting HP estimates from short-term energy assessment, would more rapidly emulate a fasting state of metabolism, and as a result, provide a more robust measure of fasting HP compared with the traditional methodologies while still excluding most of the energy required for digestion, related tissue deposition, and activity. Therefore, these experiments were conducted to validate the use of the washed rumen technique for a rapid measurement of fasting HP and RQ compared with traditional fasting methodologies.

Material and methods

In Exp. 1, sixteen Holstein steers were divided into two groups of 8 for a comparison of measurements made following feeding (0-24 h for both groups) and fasting (24-48 h; Fasted; 8 steers BW=237±17 kg) or using the washed rumen model (24-48 h; WR; 8 steers BW=322±30 kg). The experiment was conducted after steers were adapted to a cubed alfalfa-based diet at $1.5\times NE_m$ for 10 d. Steers were placed into individual head-boxes and respiratory gas exchange was continuously measured on 2 consecutive d to determine HP and RQ under the fed state (d 11, all steers) and during fasting (d 12, 8 steers) or using the washed rumen technique (d 12, 8 steers). On the day of the washed rumen procedure, the reticulorumen was emptied, washed, and refilled with buffer (NaCl=96; $NaHCO_3$=24; $KHCO_3$=30; K_2HPO_4=2; $CaCl_2$=1.5; $MgCl_2$=1.5 mmol/kg of buffer; aerated with a mixture of 75% N_2 and 25% CO_2) prior to measurement of gas exchange.

In Exp. 2, six Holstein steers (360±22 kg) were used in a replicated 3×3 Latin Square design, with 21 d periods, to determine the effects of prior alimentation on RQ and HP under unwashed (fed state; d 20) and washed rumen (WR; d 21) conditions. Treatments offered were a cubed alfalfa-based diet fed at 1.0, 1.5, and $2.0\times NE_m$. Respiratory gas exchange and washed rumen procedures were conducted as described for Exp. 1. Jugular blood samples were collected prior to the morning feeding and every 4 h for a subsequent 24-h period on d 20 and 21. The HP and RQ for each steer were averaged over day and within each hour, and then analyzed using the MIXED procedures of SAS (SAS Inst. Inc., Cary, NC, USA).

Results and discussion

Exp.1, mean hourly RQ, and daily HP were lower for the WR steers than Fasted steers on average from 8 to 24 h after removal of rumen contents ($P<0.001$ and =0.076, respectively). Fitting RQ data obtained during fasting to a one-phase decay equation showed that plateau was achieved at 0.76±0.01 and 0.72±0.01, and time to plateau was 9 and 8 h, for Fasted and WR steers, respectively. Mean RQ after WR were 0.77, 0.73, and 0.72 (SEM=0.003) for time segments 0 to 8 h, 9 to 16 h, and 17 to 24 h, respectively. Mean fasting HP after WR was 18.75, 16.81, and 16.47 (SEM=0.51 kJ/(h·kg$^{0.75}$)) for time segments 0 to 8 h, 9 to 16 h, and 17 to 24 h, respectively. There were no significant differences in RQ or fasting HP (P=0.225 and P=0.810, respectively) between the time segments of 9 to 16 h and 17 to 24 h post-rumen washing. In contrast, both RQ and HP differed (P=0.090 and 0.081, respectively) across these same time segments for the Fasted group.

Exp. 2, heat production for intakes of 1.0, 1.5 and 2.0 × NE_m were 479.2, 587.1 and 713.4 (SEM=4.0 kJ/(d·kg$^{0.75}$)), respectively. After rumen washing, fasting HP was achieved at 334.4, 356.6 and 409.9 (SEM=2.0 kJ/(d·kg$^{0.75}$)) for 1.0, 1.5 and 2.0 × NE_m prior to fasting, respectively. The plateau RQ were 0.76, 0.83 and 0.87 for intakes of 1.0, 1.5 and 2.0 × NE_m during the fed state, respectively. The RQ were 0.72, 0.71 and 0.71 for WR at 1.0, 1.5 and 2.0 × NE_m intakes prior to fasting, respectively. Mean daily plasma insulin and glucose concentrations were lower for WR than fed state (Table 1, P<0.001), whereas cortisol and non-esterified fatty acid (NEFA) were higher for WR (P<0.001). However, the NEFA was not above threshold levels for a severe energy deficit that can be induced from prolonged fasting.

In conclusion, these studies demonstrate that a fasting state can be rapidly achieved by using the washed rumen technique as opposed to the traditional fasting methodologies without a severe energy deficit. This approach may provide an alternative to the traditional 48 h fasting time. Applying the washed rumen technique may be a more rapid and less stressful means to predict energy required for maintenance in cattle.

Table 1. Comparison of plasma hormones and metabolites[1].

Item	Fed state			WR			SE	P-value[2]		
	1.0 NE_m	1.5 NE_m	2.0 NE_m	1.0 NE_m	1.5 NE_m	2.0 NE_m		W	E	W × E
Cortisol, μg/dl	0.75	0.96	1.10	1.08	1.55	1.84	0.13	<0.001	<0.001	NS
Insulin, μIU/ml	3.76	4.73	6.64	1.79	2.05	2.16	0.39	<0.001	<0.001	<0.001
Glucose, mg/dl	78.31	76.16	74.09	65.01	67.48	67.46	1.78	<0.001	NS	0.086
NEFA, mmol/l	0.17	0.10	0.09	0.55	0.53	0.53	0.03	<0.001	NS	NS
BHBA, mmol/l	0.49	0.50	0.54	0.49	0.52	0.54	0.02	NS	0.09	NS

[1] Data are presented as least squared means of animals were fed alfalfa cubes at levels of 1.0, 1.5 and 2.0 × NE_m based on the BW (Fed state) and incubation of ruminal buffer of 15 kg in the reticulorumen up to 24 h (WR).
[2] W, washing effect; E, energy effect; W × E, their interaction.

References

Blaxter, K. L. 1967. Techniques in energy metabolism studies and their limitations. Proc. Nutr. Soc. 26: 86-96.
Blaxter, K. L. and F. W. Wainman. 1966. The fasting metabolism of cattle. Br. J. Nutr. 20: 103-111.

Challenge models to study the effect of immune system activation on amino acid metabolism in pigs

E. Kampman-van de Hoek[1,2], W.J.J. Gerrits[2], J.J.G.C. van den Borne[2], C.M.C. van der Peet-Schwering[1], H. van Beers[3] and A.J.M. Jansman[1]
[1]Department of Animal Nutrition, Wageningen UR Livestock Research, P.O. Box 65, 8200 AB Lelystad, the Netherlands; esther.vandehoek@wur.nl
[2]Animal Nutrition Group, Wageningen University, P.O. Box 338, 6700 AH Wageningen, the Netherlands
[3]SDa, Autoriteit Diergeneesmiddelen, Yalelaan 114, 3584 CM Utrecht, the Netherlands

Introduction

In response to (non-)pathogenic challenges, the immune system of pigs can be activated. During immune system activation, there is a competition for amino acids (AA) between body protein deposition (growth) and immune function (Sandberg *et al.*, 2007). As a consequence, feed intake and growth are impaired (Williams *et al.*, 1997) and body protein degradation increases to release AA, e.g. for the production of acute phase proteins (APP) (Reeds *et al.*, 1994). Knowledge about the impact of immune system activation on AA metabolism of pigs is limited. The aim of the present study was to compare the use of Complete Freund's Adjuvant (CFA) and turpentine oil (TO) to activate the immune system and to study the effect of immune system activation on AA metabolism in pigs.

Material and methods

Pigs of 30 kg body weight were challenged with either four iv CFA infusions (three on day 0 and one on day 1 (n=3)), a single sc TO injection (n=3), or iv and sc saline as a control (control, n=2). Pigs were fed at a restricted feed intake of 2.7 × the estimated ME requirements for maintenance. Plasma APP concentrations (C-reactive protein (CRP), pigMAP, haptoglobin and albumin) were determined immediately before and at day 1, 2 and 6 after the challenge. White blood cell counts and feed intake were determined daily and a 4-d nitrogen (N)-balance was performed (day 0 to 3). At day 2, a mixture of 7 universally ^{13}C-labelled essential AA (Lys, Met, Trp, Ile, Leu, Val, Phe, Tyr) was infused *iv* as a bolus to study AA dilution kinetics, while feeding the pigs hourly portions. Plasma ^{13}C AA enrichments were measured by isotope ratio mass spectrometry coupled to a gas chromatograph. For each pig and AA a double exponential model was used to fit data of the ^{13}C AA enrichment in plasma:

$$E(t) = a1 \times \exp(b1 \times t) + a2 \times \exp(b2 \times t) \quad (1)$$

where E(t) is the predicted ^{13}C enrichment in plasma AA (APE) at time t (min), and a1, b1, a2, and b2 are parameter estimates from which the irreversible loss rate (ILR, i.e. the sum of disappearance of AA from the plasma pool towards protein synthesis and AA oxidation, in μmol/kg BW·h) was calculated:

$$ILR = d / (a1 / b1 + a2 / b2) \times 60 \quad (2)$$

At 9 d after the start of the challenge, pigs were euthanized after which autopsy was performed with special emphasis to TO injection sites, lung, spleen, liver, and kidney. Data were analyzed by ANOVA with challenge as fixed effect, or by a mixed model with the effect of collection day included as repeated measures.

Results and discussion

One day after the start of the challenge serum haptoglobin concentrations increased 9 fold (P=0.03) in CFA compared to control pigs, but was not increased in TO pigs. Serum CRP concentrations increased three fold in CFA pigs (P=0.06) and in TO-pigs (P=0.03) compared to control pigs. Two days after the start serum haptoglobin increased 7 fold (P=0.02) in both CFA and TO serum compared to control, pigMAP increased 6 fold (P<0.01) only in TO, and CRP increased 4 fold in TO (P=0.01) compared to control. In agreement with our results, haptoglobin and pigMAP concentrations increased up to 6 fold and CRP increased up to 4 fold in response to bacterial and parasitic infections, or inflammation in pigs (Heegaard *et al.*, 2011). CFA increased (P<0.05) eosinophil counts at day 1 and 2. Feed intake was reduced at day 0 and 1 in CFA pigs and at day 1 in TO pigs, but not in control pigs. N retention as % of N intake was similar between groups. The ILR for Lys, Met, Trp, Ile, Leu, Val, Phe and Tyr was numerically lower in CFA pigs compared to control pigs, with the most pronounced numeric difference in ILR for Tyr, followed by Leu, Met, and Phe (Figure 1). In TO pigs, the ILR were almost similar to the control pigs (Figure 1). Autopsy revealed a 1.5 fold greater relative lung weight of CFA pigs than control pigs and infiltration of granulomatous cells in lung (CFA pigs) or at the TO injection sites (TO pigs). In conclusion, the patterns of change in ILR of AA indicate a change in AA utilization in pigs with an activated immune system as induced by CFA administration. TO administration led to less pronounced differences in ILR. Possibly, this change in utilization is due to repartitioning of AA to protein synthesis for the immune system at the expense of synthesis for net body protein deposition (Sandberg *et al.*, 2007). Further studies will be performed to quantify the effect of immune system activation on AA metabolism and N retention.

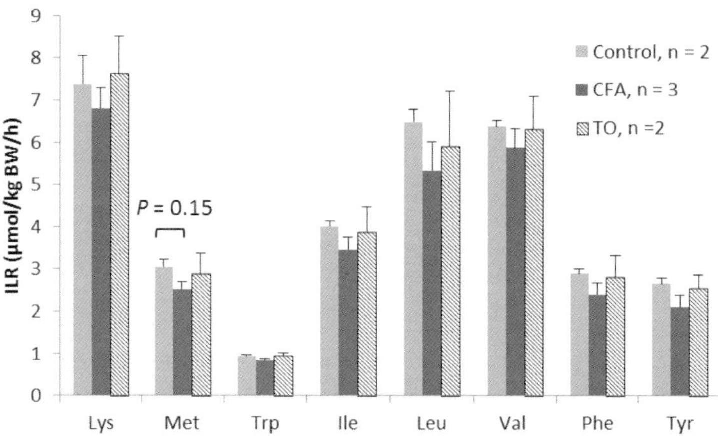

Figure 1. Effect of CFA or TO administration on the irreversible loss rate of plasma amino acids by intravenous bolus infusion of universally labeled ^{13}C-amino acids in growing pigs.

Reference

Heegaard, P., Stockmarr, A., Pineiro, M., Carpintero, R., Lampreave, F., Campbell, F., Eckersall, P., Toussaint, M., Gruys, E., Sorensen, N. 2011. Optimal combinations of acute phase proteins for detecting infectious disease in pigs. Vet. Res. 42: 50.

Reeds, P. J., C. R. Fjeld, and F. Jahoor. 1994. Do the differences between the amino acid compositions of acute-phase and muscle proteins have a bearing on nitrogen loss in traumatic states? J. Nutr. 124:906-910.

Sandberg, F. B., G. C. Emmans, and I. Kyriazakis. 2007. The effects of pathogen challenges on the performance of naïve and immune animals: the problem of prediction. Animal 1: 67-86.

Williams, N. H., T. S. Stahly, and D. R. Zimmerman. 1997. Effect of level of chronic immune system activation on the growth and dietary lysine needs of pigs fed from 6 to 112 kg. J. Anim. Sci. 75: 2481-2496.

Effect of an enzyme complex and dietary nutrients on endogenous losses of amino acids in chicks

S. Cerrate[1], K. Vignale[3], R. Ekmay[2], J. England[3] and C. Coon[3]
[1]Aviagen Incorporated, Huntsville, AL 35805, USA
[2]Department of Animal Science, Cornell University, Ithaca, NY 14853, USA
[3]Poultry Science Department, University of Arkansas, Fayetteville, AR, 72701, USA; ccoon@uark.edu

Introduction

Inconsistencies in expected performance have been noted from feeding exogenous enzymes. Several explanations may exist but it is possible that when feeding diets with exogenous enzymes a metabolic feedback system limits or regulates endogenous digestive enzyme expression, synthesis, and secretion. Labeled amino acids may be an advantage over other methods for measuring endogenous losses because the isotope technique may be used with practical diets. Endogenous amino acid secretions have been shown to be affected by dietary protein, types of fats, and fiber (Siriwan *et al.*, 1989; Danicke *et al.*, 2000) and digestible amino acid requirements should be corrected for the endogenous amino acid losses. The objectives of the present study were to determine endogenous amino acid losses from broilers fed various levels of protein, fat, and fiber with and without added Rovabio Max™.

Material and Methods

Twenty mash diets from 5 ingredients (corn, soybean meal, pro-plus, barley and poultry fat) and Rovabio® Max at 0.05% (xylanase, *B*-glucanase, and phytase) were fed to male chicks from 1 to 21 d old. Dietary treatments had excess levels of fat (6, 8, 12, and 13% EE), protein (24, 27.5 and 31% CP), or fiber (14, 16 and 18% neutral detergent fiber) over a control diet. Each diet was fed to 4 replicate floor pens of 25 chicks and transferred to digestibility cages (6 chicks/cage) and mixed with 0.5% of titanium oxide (TiO_2). One chick per cage was orally infused with ^{15}N-threonine, ^{15}N-cysteine, ^{15}N-methionine, ^{15}N-lysine, and ^{15}N-leucine from 17 to 21 days of age at 2% of dietary amino acids requirements. At 21 d of age, 2 h after the last infusion, ileal digesta of five nonlabeled chicks per cage were sampled and from labeled chicks ileal digesta and wing blood sample were taken.

AIDn (apparent ileal digestibility) = $AA_{diet} - AA_{ileal} \times (TiO_{2diet}/TiO_{2ileal})$

ELA (endogenous losses of amino acids) = $AA_{ileal} \times EFR \times (TiO_{2diet}/TiO_{2ileal})$

EFR (endogenous flow rate) = APE_{ileal}/APE_{plasma}

APE (atom percent excess)$_{ileal\ or\ plasma}$ = $APE_{labeled} - APE_{nonlabeled}$; APE = $^{15}N/(^{15}N+^{14}N)$

Results and discussion

The extra apparent ileal digestible threonine, methionine+cysteine, lysine, and protein from the added enzymes were 0.04, 0.03, 0.04, and 0.89%, respectively (data not shown). The endogenous losses (EL) of ^{15}N-threonine and ^{15}N-cysteine were reduced by Rovabio Max™ (Table 1). The EL of CP using the five labeled amino acids with a profile of amino acid composition of mucin secretion was reduced by Rovabio Max™. The endogenous flow rate (EFR) and EL of ^{15}N-leucine were influenced by the dietary nutrients (Figure 1), where the EL of ^{15}N-leucine was increased with increasing dietary fat, protein or fiber.

$EFR_{(CPm)}=EFR_{Thr}x0.46+EFR_{Cys}x0.28+EFR_{Meth}x0.02+EFR_{Lys}x0.08+EFR_{Leu}x0.16$

$EFR_{(CPp)}=EFR_{Thr}x0.24+EFR_{Cys}x0.08+EFR_{Meth}x0.05+EFR_{Lys}x0.24+EFR_{Leu}x0.39$

Smaller EL of threonine and cysteine from feeding carbohydrase diets suggests that mucin secretions are decreased because threonine and cysteine are the first and third most abundant AA found in mucin. When the extra spared endogenous losses of amino acids due to carbohydrases were compared to their requirements, dietary requirements for threonine, cysteine, and crude protein were reduced by 4.3, 15.2, and 4.2%, respectively. Variation of EL of ^{15}N-leucine due to dietary nutrients indicates that pancreatic secretions are the main regulator of these losses. Leucine is known to be the 3rd most abundant AA in pancreatic protein secretion (Corring and Jung, 1972).

Table 1. Endogenous losses (EL,g/kg DMI, LS Means) of labeled amino acids and crude protein.

Treatments	Thr	Cys	Meth	Lys	Leu	CPm[1]	CPp[1]
Enzymes							
No	0.838[a]	1.775[a]	0.831	3.196	1.153	33.1[a]	33.2
Yes	0.544[b]	1.198[b]	0.730	2.429	1.055	24.7[b]	26.3
SEM	0.068	0.122	0.080	0.417	0.069	2.1	2.6
P-value	0.004	0.002	0.374	0.201	0.323	0.007	0.067
Diets							
SEM	0.151	0.273	0.178	0.933	0.154	4.7	5.9
P-value	0.752	0.633	0.473	0.358	0.0005	0.592	0.239

[1] CPm and CPp were calculated from profile of amino acid composition of mucin and pancreatic secretions, respectively.
[a,b] $P \leq 0.05$.

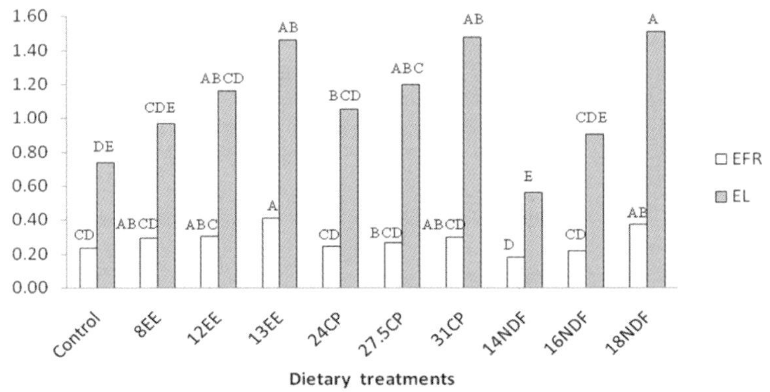

Figure 1. Endogenous flow rate (EFR, APE_{ideal}/APE_{plasma}, LSMEANS) and endogenous losses (EL, g/kg DMI, LSMEANS) of leucine.

References

Corring, T., and J. Jung, 1972. The amino acid composition of pig pancreatic juice. Nutrition Reproduction International 6, 187-190.

Danicke, S., W. Bottcher, H. Jeroch, J. Thielebein, and O. Simon, 2000. Replacement of soybean oil with tallow in rye-based diets without xylanase increases protein synthesis in small intestine of broilers. J. Nutr. 130, 827-834.

Siriwan, P., W. L. Bryden, and E. F. Annison, 1989. Effects of dietary fibre and protein levels on endogenous protein secretions in chickens. Proc. Nutr. Soc. Aust. 14, 143.

Validation of the oral ^{13}C-bicarbonate tracer technique against indirect calorimetry for the estimation of energy expenditure in resting dogs

C. Larsson[1], *R.B. Jensen*[1], *P. Junghans*[2] *and A-H. Tauson*[1]
[1]*Department of Veterinary Clinical and Animal Sciences, University of Copenhagen, Frederiksberg, Denmark; cala@sund.ku.dk*
[2]*Leibniz Institute for Farm Animal Biology, Department of Nutritional Physiology 'Oskar Kellner', Dummerstorf, Germany*

Introduction

To be able to provide guidelines for appropriate nutrition of dogs, a reliable and feasible method for investigations of their energy expenditure (EE) would be very valuable. The ^{13}C-bicarbonate tracer technique (^{13}C-BTT) has been used to assess EE in several species of animals (Junghans *et al.*, 2007; Lachica and Aguilera, 2003) and in humans (Junghans *et al.*, 2008). This stable isotope method is based on the ^{13}C kinetics in breath CO_2 after administration of ^{13}C labeled sodium bicarbonate ($NaH^{13}CO_3$). After tracer administration, the ratio between the ^{13}C and ^{12}C in spot samples of expired air collected over a sufficient period of time can be used to estimate the CO_2 production rate (RCO_2) and thereafter the EE. By using oral administration of the tracer the method can be completely non-invasive, making it an appropriate method for studies in dogs. The aim of this study was to validate the oral ^{13}C-BTT (o^{13}C-BTT) against indirect calorimetry, the 'gold standard' for the estimation of EE. The hypothesis is that the o^{13}C-BTT can be used as a minimal restrictive and non-invasive method to obtain reliable estimates of EE in dogs under near natural conditions.

Material and methods

Eight privately owned dogs of different breeds were included in this experiment. All dogs were of normal body weight (BW) and in the weight range of 15-35 kg. Each dog was measured twice under resting conditions, where the dog was, after an over-night fast, kept in a respiration chamber (temperature = 16 °C; humidity = 68%), for approximately 5 hours. The CO_2 production and O_2 consumption were measured continuously by means of an open-air-circuit respiration unit. Simultaneously, the ratio between ^{13}C and ^{12}C in expired CO_2 was measured online using an infrared isotope analyzer, every 3rd minute after an oral dose of $NaH^{13}CO_3$ (5 mg/kg BW, incorporated in a small piece of liver paté), making comparison between the two methods possible.

The CO_2 production was estimated using the area under the $^{13}CO_2$ enrichment-time curve, the dose of ^{13}C administrated and a recovery factor (RF) for ^{13}C in breath CO_2, as described by Elia (1991). The EE was calculated according to the equation of Brouwer (1965). By using both methods, it was possible to calculate the RF and also the RQ for each measurement. The average values of 0.72 and 0.78 for all measurements, was then used as the estimates for RF and RQ, respectively, in the calculations of RCO_2 and EE when using the ^{13}C-BTT. The values of EE was calculated in kJ/kg $BW^{0.75}$/ h, and then standardized to 24 hours (i.e. kJ/kg $BW^{0.75}$/ d). Data were statistically analysed using the MIXED procedure of SAS®.

Results and discussion

There were significant differences in EE ($P<0.001$) between the dogs with LS-means, calculated from the measurements of both methods, ranging from 304 kJ/kg $BW^{0.75}$/d for dog D, to 511 kJ/kg $BW^{0.75}$/d for dog A. This was also expected due to the large variations between the dogs, i.e. breed, BW and age. However, there were no significant differences ($P>0.05$) between the two methods used for estimation of EE in the present study. The results from this study, shown for each measurement in Figure 1, suggest that the o^{13}C-BTT can be used as a non-invasive method to obtain reliable estimates of EE in different types of dogs during resting conditions, and that the estimates of RF=0.72 and RQ=0.78 can be used when estimating the RCO_2 and EE. However, whether the o^{13}C-BTT gives reliable estimates of EE, and the estimates used for RF and RQ are affected, when used under other circumstances, e.g. stress or physical activity, needs to be further investigated.

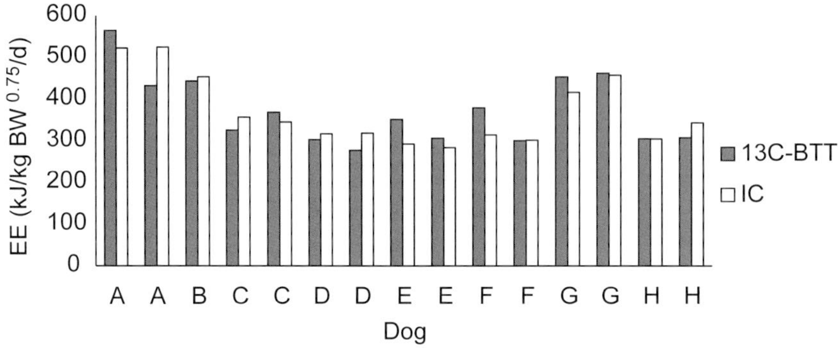

Figure 1. The individual values of energy expenditure (EE; in kJ/kg $BW^{0.75}$/d) estimated from each measurment of eight dogs (A to H) under resting conditions, by using the oral ^{13}C-bicarbonate tracer technique (o^{13}C-BTT) and indirect calorimetry (IC).

References

Brouwer, E., 1965. Report of subcommittee on constants and factors. In: Blaxter, K.L. (editor), Proceedings of the 3rd EAAP Symposium on Energy Metabolism. Academic Press, London. UK, 441-443.

Elia, M., 1991. Estimation of short-term energy expenditure by the labeled bicarbonate method. In: Whitehead R.G. and A. Prentice (editors), New techniques in nutritional research. Academic Press New York, USA, 207-227.

Junghans P., Jentsch W. and M. Derno, 2008. Non-invasive ^{13}C bicarbonate tracer technique for measuring energy expenditure in men – a pilot study. e-SPEN, Eur. e-J. Clin. Nutr. Metab. 3 (2), e46-e51.

Junghans, P., J. Voigt, W. Jentsch, C.C. Metges, and M. Derno, 2007. The ^{13}C bicarbonate dilution technique to determine energy expenditure in young bulls validated by indirect calorimetry. Livest. Sci. 110, 280-287.

Lachica, M. and J.F. Aguilera, 2003. Estimation of energy needs of the free-ranging goat with particular reference to the assessment of its energy expenditure by the ^{13}C-bicarbonate method. Small Ruminant Res. 49, 303-318.

Validation of the ^{13}C-bicarbonate tracer technique against indirect calorimetry for estimation of energy expenditure in resting ponies

R.B. Jensen[1], *C. Larsson*[1], *P. Junghans*[2] *and A.-H. Tauson*[1]
[1]*Department of Veterinary Clinical and Animal Sciences, University of Copenhagen, Frederiksberg, Denmark; ralle@sund.ku.dk*
[2]*Leibniz Institute for Farm Animal Biology, Department of Nutritional Physiology 'Oskar Kellner', Dummerstorf, Germany*

Introduction

Knowledge on the energy requirements of horses is essential for optimal feeding practice. The 'gold standard' for measuring energy expenditure (EE) is indirect calorimetry where the horse has to be confined to a respiration chamber. Alternatively the ^{13}C-bicarbonate tracer technique (^{13}C-BTT) can be used to measure EE (Elia, 1991). After administration of a dose of ^{13}C labeled sodium bicarbonate, the ratio between the ^{13}C and ^{12}C in spot samples of expired air collected over a sufficient period of time can be used to estimate the CO_2 production rate (RCO_2) and EE. Collection of breath samples can be done from the respiration chamber or it can be done under free living conditions with breath bags and a small mask. Combining measurements in the respiration chamber with the ^{13}C-BTT makes it possible to compare the CO_2 production measured with both techniques. Most studies in other species have used intravenous (IV) administration of ^{13}C-bicarbonate (Urschel *et al.*, 2009; Junghans *et al.*, 2007). However, oral administration of ^{13}C-bicarbonate in a single bolus could make the ^{13}C-BTT totally non-invasive. The hypothesis is that the ^{13}C-BTT with IV and oral administration of the tracer in a single bolus can be used for estimation of RCO_2 and EE in sedentary ponies.

Material and methods

Four 3 to 4 years old Shetland ponies weighing 178±48 kg were used in the experiment and they were fed hay (nutritional composition in % of dry matter: dry matter: 85%, ash: 4%, NDF: 72%, ADF: 42%, lignin: 5%, crude protein: 8%, sugars: 12%) twice a day at 08:00 h and 18:00 h (~16 g dry matter/kg body weight daily). Water was available *ad libitum* at all time. The ponies were trained to get used to the respiration chambers, and they were placed in the chambers in the morning 1 h before measurements started. The RCO_2 and EE were estimated at 4 occasions (twice with oral and twice with IV administration of ^{13}C-bicarbonate) with at least one day of rest between measurements. The ponies were kept in the respiration chambers for approximately 11 hours where CO_2 production and O_2 consumption were measured continuously by means of an open-air-circuit respiration unit (temperature = 16 °C; humidity = 70%; volume = 3.5 m^3). Simultaneously, the ratio between ^{13}C and ^{12}C in expired CO_2 was measured online for 11 hours using an infrared isotope analyzer, before (baseline) and every 6 minute after an oral or IV dose of ^{13}C-bicarbonate (2,5 mg/kg body weight; 98 atom % ^{13}C, Sigma Aldrich, St. Louis, USA), making comparison between the two methods and two administration routes possible. The chamber was opened briefly when the oral dose was given, and with the IV dose the pony was removed, injected and replaced in the chamber. The ponies were fed their morning hay immediately after the oral or IV dose of ^{13}C-bicarbonate was given. It was possible to measure $^{13}CO_2$ within 6 min after administration of the isotope, and baseline was reached after approximately 8 hours.

The area under the enrichment-time curve (AUC), the dose of ^{13}C-bicarbonate (D) and a recovery factor (RF) of ^{13}C were used to calculate RCO_2 as described by Elia (1991). Energy expenditure and the respiratory quotient (RQ) were calculated from CO_2 production and O_2 consumption according to Brouwer (1965). The ^{13}C-BTT only estimates CO_2 production, and the mean RQ value was used to estimate EE for this method.

Data was statistically analyzed using the MIXED procedure in SAS®. To study the effect of administration route of ^{13}C-bicarbonate (IV or oral) on the RQ-value and the RF, route was included as fixed effect and pony as random effect in the model. To study the effect of method (IC or ^{13}C-BTT) on CO_2 production and EE, method was included as fixed effect and pony as random effect in the model. Effects were considered significant if $P<0.05$.

Results and discussion

There was no effect ($P>0.05$) of administration route of ^{13}C-bicarbonate on the RQ, however, there was an effect of administration route ($P=0.03$) on the RF (Table 1). The same RQ and the two different RF were then used to calculate the CO_2 production for the ^{13}C-BTT (RQ and RF values are shown in Table 1). The estimated RF found in this experiment is in agreement with results of other experiments (Junghans *et al.* 2007).

There was no significant difference ($P>0.05$) between methods when estimating CO_2 production and EE. The results show that the ^{13}C-BTT can be used to estimate EE with oral administration of ^{13}C-bicarbonate in sedentary ponies. However, the ^{13}C enrichment-time curves were smoother when IV administration of the tracer was used, and that it might be easier to model and calculate AUC when measurements are performed with IV administration under field conditions with spot sampling of breath air instead of continuous sampling of breath samples in the respiration chamber.

Table 1. Effects of administration route (IV or oral) or method (IC or ^{13}C-BTT) for estimation of CO_2 production and energy expenditure in sedentary ponies (LSmean and standard deviation).

	Administration route		SD	*P*-value		
	IV	Oral		Route		
Respiratory quotient	0.80	0.79	0.03	NS		
Recovery factor (%)	0.69	0.76	0.03	0.03		
	Method		SD	*P*-value		
	IC	^{13}C-BTT		Route	Method	R × M
CO_2 production (l/d)	782	793	21	NS	NS	NS
Energy expenditure (kJ/kg$^{0.75}$/d)	433	442	34	NS	NS	NS

References

Brouwer E., 1965. Report of subcommittee on constants and factors. In: Blaxter, K.L. (editor), Proceedings of the 3rd EAAP Symposium on Energy Metabolism. Academic Press, London. UK, 441-443.

Elia M., 1991. Estimation of short-term energy expenditure by the labeled bicarbonate method. In: Whitehead R.G. and A. Prentice (editors), New techniques in nutritional research. Academic Press New York, USA, 207-227.

Junghans P., J. Voigt, W. Jentsch, C.C. Metges and M. Derno, 2007. The ^{13}C bicarbonate dilution technique to determine energy expenditure in young bulls validated by indirect calorimetry. Livest. Sci. 110, 280-287.

Urschel, K.L., T.L. Smith, R.B. Drake, P.A. Harris and R.J. Geor, 2009. Using [^{13}C]Sodium bicarbonate to measure carbon dioxide production in horses at rest. J. Equine Vet. Sci. 29, 375-376.

Biometry and bioelectrical impedance analysis to estimate body composition of fish tambatinga (*Colossoma macropomum* × *Piaractus brachypomus*)

F.T. Andrade[1], M.L.T. Abreu[2], J.B. Lopes[1], E. Lanferdini[2] and A. Saraiva[3]
[1]UFPI, Departamento de Biofísica e Fisiologia, Teresina (PI), 64049-550, Brazil
[2]UFLA, Departamento de Zootecnia, Lavras (MG), 37200-000, Brazil; marvio@dzo.ufla.br
[3]UFV, Departamento de Zootecnia, Viçosa (MG), 36570-000 Brazil

Introduction

Studies of body composition of livestock involve direct methods (proximate analyzes, carcass dissection) and indirect methods (densitometry, dilution, bioelectrical impedance and conductance, x-ray absorptiometry, computed tomography, magnetic resonance, etc.). The bioelectrical impedance analysis (BIA) has seldom been used in fish. The use of BIA to estimate body composition of fish is justified because the technique is rapid, reliable, cheap (Duncan, 2008), nonlethal (Bourdages, 2011), the execution is relatively easy and simple, and it allows field studies (Ellis, 2001). This study was conducted with the objective to assess the use of BIA to estimate body composition and fillet yields of fish tambatinga (*Colossoma macropomum* × *Piaractus brachypomus*).

Material and methods

Two hundred and ten juvenile tambatinga fishes with 136 days of age, initial weight of 37.69±3.32 g, and total length of 12.96±0.37 cm were used in the experiment that ended when fishes reached 318 days, 874.07±71.46 g, and 34.65±1.02 cm total lengths, respectively. Fifteen fishes, randomly chosen, were anesthetized and subjected to fourteen biweekly assessments (BIA and biometric). Electrodes and signal detectors (hypodermic needles Delta 20-5) injected one inch deep in well-defined points on the fishes allowed the measurements of BIA variables: electrical resistance (R) and electric reactance (Xc), in series. Fish biometry data determined were: weight (W), standard length (SL), and volume (V). After euthanasia, fishes' whole carcass with head, skin, and fillet (WC), and fillet with skin (FS) were weighed. The WC samples were ground and homogenized to analyze fat and protein contents. Data of biometry and BIA or just BIA were regressed against the proximate analysis to develop the models. The GLM procedures (SAS Inst. Inc., Cary, NC, USA) were used and Mallow's Cp statistical was calculated to test the adequacy of the number of variables in the models. Data from proximate analysis of fat and protein expressed as percentage of dry matter or grams of whole fish were correlated with the predicted (recovering calculations) using the best models with variables from biometry and BIA or just from BIA.

Results and discussion

The BIA variables (R and Xc) and its parameters (Z, AF and IC) and the biometric variables increased linearly with age ($P<0.001$). Protein decreased ($P<0.05$) and fat increased ($P<0.05$) linearly according to age. Fillet with skin yield exhibited a positive linear effect ($P<0.05$) with age.

The models for prediction of carcass protein and fat on dry matter basis or in grams and skinless fillet yield, according to the data of BIA exhibited determination coefficients (R^2) of 0.650 for fillet with skin yield and 0.943 for fat on dry matter basis.

The best models for carcass protein and fat prediction, in grams or dry-matter basis, and skinless fillet yield are presented in Table 1.

Table 1. Equations (best models) to estimate fat, protein, and fillet with skin in tambatinga.

Component	N	Data of BIA and/or biometry	*P*-value	F-value	R^2
Fat on dry matter basis, %[1]	14	Id, R, CI_{TL}, W and SL	<0.0001	71.05	0.978
Protein on dry matter basis, %[2]	14	Id, R, CI_{TL}, W and SL	<0.0504	3.68	0.697
Fat, g [3]	14	Id, PA, W and SL	<0.0001	740.17	0.997
Protein, g [4]	14	Id, R, CI_{EEI}, W and SL	<0.0001	73.19	0.979
Fillet with skin yield, g[5]	14	Id, PA, W and SL	<0.0415	3.91	0.635

Id = age in days post-hatch; R = electrical resistance; W = weight; SL = standard length; PA = phase angle; CI_{TL} = composition index using total length and reactance; CI_{EEI} = composition index using the separation between electrodes detectors and reactance; n = number of assessments during the experiment.

[1] $y = -8.422926267 - 0.216712691Id + 0.116915731R + 1.901114939CI_{TL} - 0.002513132W + 2.692125898SL$.

[2] $y = 114.3994883 + 0.2419623Id - 0.2438099R - 5.9064319\,CI_{TL} + 0.0318951W - 1.3784901SL$.

[3] $y = 34.76580235 - 0.035504511Id - 0.05527718PA + 0.22293449W - 3.66602166SL$.

[4] $y = -45.24958532 - 0.12274079Id + 0.26813712R - 53.84934130IC_{EEI} + 0.24910634W + 3.13248886SL$.

[5] $y = 5.923697883 + 0.255870176Id + 0.057894834PA - 0.0106793W - 1.490036218SL$.

The correlation among the proximate analysis values and the analogs predicted using the best models with biometric and BIA variables studied also exhibited high coefficients: 0.978 for fat on dry matter basis, 0.839 for protein on dry matter basis, 0.998 for fat in grams, and 0.989 for protein in grams of whole fish.

Identical analysis for fillet with skin yield resulted in lower correlation coefficient (0.796).

In conclusion, for the hybrid tambatinga, bioelectrical impedance analysis allows the estimation of fillet with skin yield and body composition (fat and protein on dry matter basis or in total mass of fish) and the best models include biometric and BIA variables or parameters.

References

Bourdages, C., 2011. Use of bioelectrical impedance analysis (BIA) to predict water and energy content of juvenile Rainbow Trout (Oncorhynchus mykiss). Master of Science dissertation, Faculty of Science of the University of Ontario Institute of Technology, Oshawa, Ontario.

Duncan, M.B., 2008. The Use of BIA for Estimating the Body Composition of Various Fish Species. Master of Science dissertation, Faculty of Virginia Polytechnic Institute and State University, Blacksburg, Virginia.

Ellis, K.J., 2001. Selected body composition methods can be used in field studies. J. Nutr., 131, 1589S-1595S.

Evaluation of the infrared spectroscopy method for the quantification of NANOLIPE marker in feces of dairy cattle

E.O.S. Saliba[1], N.C. Gonçalves[1], G.S.S.C. Barbosa[2], A.L.C.C. Borges[1], N.M. Rodriguez[1], G.R. Moreira[1] and F.A. Silva[1]
[1]Departamento de Zootecnia, Universidade Federal de Minas Gerais, Brazil; saliba@ufmg.br
[2]Campus Florestal, Universidade Federal de Viçosa, Brazil

Introduction

LIPE® is an external marker obtained from lignin isolated from *Eucaliptus grandis* and enriched with phenolic groups that is used to estimate dry matter intake and digestibility (Saliba *et al.*, 2003). Recently, methodological improvements in obtaining LIPE® led to the production of a newer and better version of this marker. Its particles have now reached the nano scale and for this reason it was called NANOLIPE (Saliba *et al.*, 2012). The Fourier Transform Infrared Spectroscopy (FT-IR) is used for the quantification of NANOLIPE in feces. However, one of the key points to ensure the quality of the results is to assess the linearity of the calibration curves used in the determinations (Thompson, 2005). Therefore, the aim of this study was to statistically evaluate the linearity of the FT-IR method in quantifying NANOLIPE in feces of dairy cattle.

Material and methods

A calibration curve was established in fecal matrix. For this purpose, known amounts of the NANOLIPE marker were added in feces of dairy cattle, which have been previously dried at 55 °C for 72 h and ground by a Willey mill type (1 mm particle size). Then, for the calibration and quantification of the marker a Varian® 800 FT-IR Scimitar Series spectrometer was used.

The working range of the method was chosen according to the known average concentration of the marker in routine samples. Thus, four levels of concentration in this range were defined: 0.05, 0.10, 0.15, 0.20 mg of NANOLIPE per g of fecal dry matter. The calibration samples were independently prepared by three different analysts under reproducible conditions. The determination was performed at random read to minimize possible errors at specific levels.

The wave number of 1,035.62 cm^{-1} was chosen because it resulted in the greatest linearity between the response (infrared energy absorption) and the concentration of NANOLIPE. This wave number was the same used for quantification LIPE® in the initial studies of Saliba *et al.* (2003).

Statistical tests were performed to determine three assumptions related to the residuum i.e. it has to follow normal distribution, be independent and homoscedastic. Those tests were: Ryan-Joiner test, Durbin-Watson test and Levene's test, respectively, according to Thompson (2005). Besides, the analysis of variance of the regression equation created allows the evaluation of the FT-IR for the quantification of NANOLIPE marker in feces of dairy cattle.

Results and discussion

The results obtained by the FT-IR spectroscopy in terms of the peak height at 1,035.62 cm^{-1} according to the concentration of NANOLIPE in the feces are shown in Table 1.

The results obtained by the statistical tests previously referred to (Ryan-Joiner, Levene and Durbin-Watson) are shown in Table 2.

Table 1. Average peak height at 1,035.62 cm^{-1} in different fecal concentrations of NANOLIPE measured by FT-IR spectroscopy.

	Fecal concentration of NANOLIPE (mg/g)			
	0.050	0.100	0.150	0.200
Peak height	0.181	0.226	0.271	0.313

Table 2. Statistical evaluation of FT-IR spectroscopy for the quantification of NANOLIPE in feces of dairy cattle (0.05 to 0.20 mg/g).

Parameter	Analytical curve	Results
Normality (R)	0.9750	The residuum follows the normal distribution
Critical value	0.9248	
Homoscedasticity (t_L)	-0.70	The residuum is homoscedastic
Critical value		
Independency (d)	1.62	The residuum is independent
α	>0.10	

R = Ryan-Joiner correlation coefficient; tL = Levene's T statistic; d = Durbin-Watson statistic; α = significance.1

The analysis of variance showed that the ratio between the root mean square (RMS) of the regression and the residue was 4487.63, far exceeding the critical value for $F_{1,6}$ of 5.98 (95% confidence) indicating that regression was statistically significant. The ratio between the RMS of lack of fit and pure error was 2.91 which indicates the goodness of fit compared to the critical value of $F_{2,4}$ which is 6.94 (95% confidence). The correlation coefficient of 0.9998, supported by statistical evaluation is excellent, proving the method to be useful for quantifying the NANOLIPE in feces of dairy cattle.

References

Saliba, E. O. S.; Barbosa, G. S. S. C.; Rodriguez, N. M.; Couto Filho, C. C. C.; Moreira, G.R.; Gonçalves, N. C.; Silva, F. A., 2012. Utilization of nanotechnology to the development of a marker of fecal output in dairy cattle. In: Doreau, M. (editor), Proceedings of the British Society of Animal Science and the Association of Veterinary Teaching and Research Work. Advances in Animal Biosciences, Cambridge, Volume 3: 293.

Saliba, E. O. S.; Rodriguez, N. M.; Pilo-Veloso, D., 2003. Purified lignin extracted from Eucalyptus grandis (PELI), used as an external marker in digestibility trails in various animal species. In: Paim, N., Barcelos, J.O.J, Neto, J.B., Ferreiro, L., Franke, Berbardi, M.L., Dall'Agnol, M. (editors), Proceedings of the 9th World Conference on Animal Production. UFRGS, Porto Alegre: 221-222.

Thompson, M., 2005. Is my calibration linear? In: Thompson, M. (editor). Analytical Methods Committee Technical Brief. The Royal Society of Chemistry, Lodon, AMCTB n. 3: 1-2.

A continuous approach to assess methane production rate in ruminants using respiration chambers

L.F.L. Cavalcanti[1,3], M.A. Fonseca[2,3], J.G.L. Regadas Filho[2,3] and L.O. Tedeschi[3]
[1]Universidade Federal de Minas Gerais, 6627 Av. Antônio Carlos, 31270-901 Belo Horizonte, Brazil; luigicavalcanti22@gmail.com
[2]Universidade Federal de Viçosa, Av. Peter Henry Rolfs, 36570-000 Viçosa, Brazil
[3]Texas A&M University, 230 Kleberg Center, College Station, TX 77843-2471, USA

Introduction

The methane production from ruminants is one of the most discussed topics in animal production since it has been identified as a key problem regarding world climate change. Methane production can be quantified with open-circuit respiration chambers. Usually, the calculation of total methane production is the mean concentration measured along the trial multiplied by the total volume of air exchanged (l/period), resulting in an average value that does not contemplate the production rates variation of methane due to the pattern of ruminal fermentation (Dauncey *et al.*, 1978; Aguilera and Prieto, 1986; Rodriguez *et al.*, 2007). Other authors (Brown *et al.*, 1984) have provided a different approach for gas production calculations in which they included a differential term that is more appropriate for detecting rapid changes in metabolic rate, such as the result of physical activity. However, for continuous modeling processes, such as methane production, that approach might not be suitable. This work evaluates the methane production along the measurement time by modeling the cumulative gas curve to generate gas production rates that can be used to compare factors, such period within a day.

Material and methods

A dataset composed by 10 respirometry trials using Angus steers, with 48 h sampling duration, were grouped and the methane instant concentration over time was multiplied by the instant air flux and summed to obtain the cumulative gas curves. Animals were fed at 2% of their body weight (DM basis), a ration containing 2.76 Mcal/kg of metabolizable energy and 16.6% crude protein, twice daily (7:30 h and 17:30 h). The feeding times were used as a reference to subset each trial resulting in four periods by animal (Figure 1). The logarithm of the cumulative production values in each of

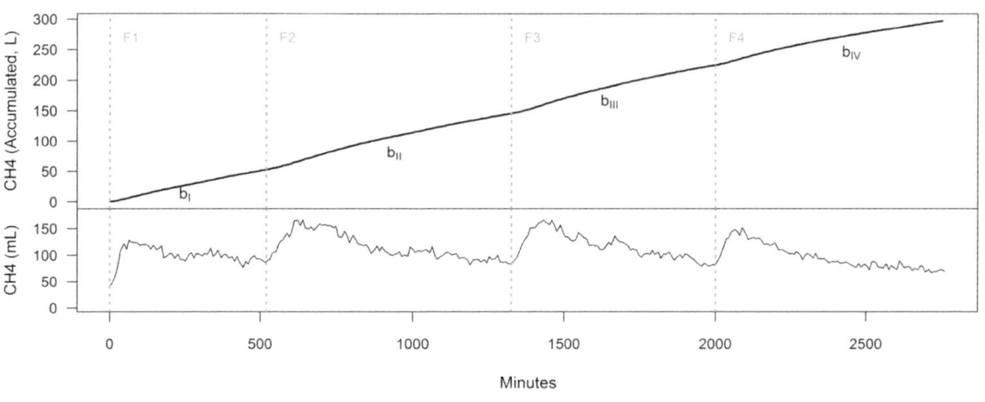

Figure 1. Example of instantaneous and accumulated production of methane in an assay of 48 hours. F1 to F4 represent the four feeding times, F1 and F3 at 07:30 h and F2 and F4 at 17:30 h; b_I to b_{IV} represent the four accumulated curve sections on which linear regression was performed and the slopes were compared.

these periods was used to fit a linear regression. The slopes obtained were summarized and analyzed as a completely randomized design with repeated measures using a mixed model methodology. This step allowed us to model the matrix of covariance assuming the spatial power structure, which was chosen due to the asymmetric time of feeding and the best Akaike's Information Criterion. The effect of period was decomposed into orthogonal polynomials of linear, quadratic, and cubic degrees.

Results and discussion

The cumulative curve obtained for each animal presented a 4-pool behavior in which the moment of acceleration pulses were highly correlated with the feeding times. When we split by period, the data assumed a curvilinear behavior with a higher rate at the very beginning of each curve, as seen in Figure 1. Whereas, when the data was linearized through a logarithmic transformation it had a better linear regression goodness-of-fit, as shown by the greater value of the coefficient of determination (average of 0.9913).

The cubic regression model was adopted ($P<0.05$) to represent the rates of production of each period during the trial time resulting in the following equation: $CH4_r = 1.0466 - 0.0346t + 0.0029t^2 - 0.00006t^3$, where $CH4_r$ is the rate of methane production (l/min) in the time t (hour). The cubic behavior can be explained by the fact that in those trials, the morning meals were 1.2 times bigger than the afternoon's, which increased the intake of fermentable organic matter in the morning period. Therefore, the first and third feeding times resulted in a greater slope for methane production. The methodology proposed in this study is capable of evaluating rates of methane production rather than the total volume as commonly used by other techniques when using respirometry trials. Our approach can be further modified for comparisons of other factors that can affect methane production rates such as different feeds, diets, and breeds. Furthermore, nonlinear functions other than linear regressions can be used to access the production rate and their parameters might be correlated to biological phenomena.

References

Aguilera, J.F., and C. Prieto, 1986. Description and function of an open-circuit respiration plant for pigs and small ruminants and the techniques used to measure energy metabolism. Arch. Anim. Nutr. 36: 1009-1018.

Brown, D., T.J. Cole, M.J. Dauncey, R.W. Marrs and P.R. Murgatroyd, 1984. Analysis of gaseous exchange in open-circuit indirect calorimetry. Med. Biol. Eng. Comput. 22: 333-338.

Dauncey, M.J., P.R. Murgatroyd and T.J. Cole, 1978. A human calorimeter for the direct and indirect measurement of 24 h energy expenditure. Brit. J. Nutr. 39: 557-566.

Rodriguez, N.M., W.E. Campos, M.L. Lachica, I. Borges, L.C. Gonçalves, A.L.C.C. Borges and E.O.S. Saliba, 2007. A calorimetry system for metabolism trials. Arq. Bras. Med. Vet. Zoo. 59: 495-500.

Part 4. Regulation

Mechanisms of regulation of intramuscular fat deposition in porcine muscle by dietary lysine content

M. Katsumata, T. Kyoya, H. Kobayashi, A. Ishida, A. Ashihara and K. Nakashima
NARO Institute of Livestock and Grassland Science, Tsukuba 305-0901, Japan; masaya@affrc.go.jp

Abstract

Feeding a low-lysine diet to pigs promotes the deposition of intramuscular fat (IMF). However, the mechanisms underlying fat accumulation and marbling in porcine muscle remain unclear. As such, the aim of this study was to elucidate the manner in which dietary lysine regulates IMF deposition in finishing pigs. Low dietary lysine levels increased the volume but not the number of intramuscular adipocytes in the *longissimus dorsi* muscle. This observation indicates that lipogenesis in already existing adipocytes plays a pivotal role in IMF accumulation, as opposed to the differentiation of preadipocytes into mature adipocytes. We also found that a low-threonine diet did not affect IMF content in either *longissimus dorsi* or *rhomboideus* muscle. Thus, dietary shortages among the various indispensable amino acids may differentially influence IMF deposition. Furthermore, insulin signaling and triacylglyceride levels were enhanced in cultured adipocytes by cellular exposure to low vs. high concentrations of lysine, suggesting that insulin might similarly contribute to the nutritional modulation of IMF deposition *in vivo*.

Introduction

The production of pork with marbling is particularly important in East Asian countries (e.g. Japan and Korea) due to the eating habits of these regions. For instance, the consumption of fresh pork is two-fold higher than the consumption of processed pork in Japan. For this reason, we recently turned our attention to the nutritional regulation of intramuscular fat (IMF) deposition in porcine muscle.

There are two major approaches for the production of pork with an elevated IMF content: (1) breeding of high IMF pigs, and (2) nutritional regulation of IMF deposition. The first approach yielded two well-known porcine lines that accumulate high amounts of IMF in the *longissimus dorsi* muscle, the synthetic 'TOKYO-X' line and the 'Shimofuri-Red' Duroc line, both developed in Japan (Hyodo 1996; Takasaki *et al.*, 2005; Suzuki *et al.*, 2005). High IMF loin chops derived from these lines consistently receive good reviews in the market, fetching prices that are approximately 60 to 120% more than those of ordinary pork with lower amounts of IMF. Nonetheless, nutritional regulation can be advantageous compared with specialized breeding because one does not need to rear pigs having a specific genetic background.

A number of studies on the promotion of IMF deposition via nutritional regulation were conducted over the past two decades. A typical tactic was to reduce the amount of protein in the diets (Castell *et al.*, 1994; Goerl *et al.*, 1995). Proceedings from a recent ISEP 2007 indicated that elevated activity of stearoyl-CoA-desaturase, a key lipogenic enzyme, led to the enhancement of porcine IMF content in response to reduced dietary protein levels (Doran *et al.*, 2007). Another tactic has been to add a single amino acid to the diet. For example, the addition of leucine or arginine increased the IMF content in porcine muscle (Hyun *et al.*, 2003, 2006; Tan *et al.*, 2009).

In contrast, our present line of attack involves the reduction of lysine levels in the porcine diet. We previously reported that a low dietary lysine levels up-regulated GLUT4 mRNA expression and affected glucose metabolism and substrate oxidation predominantly in the muscle (Katsumata *et al.*, 2001, 2003). Given that the oxidative capacity of muscle affects IMF content (Goto *et al.*, 1994), we went on to show that the shortage of a single amino acid, lysine, augmented IMF deposition in the pig (Katsumata *et al.*, 2005, 2012). However, the mechanisms underlying this process are still unknown. Therefore, the aim of this paper was to investigate the manner in which dietary lysine regulates IMF deposition in finishing pigs.

Effects of low dietary levels of lysine on the volume and number of porcine intramuscular adipocytes

IMF content is mainly dependent on the volume and/or the number of intramuscular adipocytes. The volume and number of intramuscular adipocytes and the IMF content continuously increased in the *trapezius* and *semitendinous* muscles of pigs until the pigs reach 24 weeks of age (Lee and Kaufman, 1974). Thus, this study investigated the effects of low dietary levels of lysine on the volume and number of intramuscular adipocytes in porcine *longissimus dorsi* muscle. The *longissimus dorsi* muscle formed the focus of this work, as this muscle is the most important source of pork in Japan.

Duroc × (Large white × Landrace) crossbred pigs aged 10 weeks were employed in this work. The pigs were assigned to one of two diets: a control diet (n=12) or a low-lysine diet (LL, n=12). The LL diet was designed such that the concentration of lysine was approximately 65% of the dietary requirement given by the NRC (1998). The concentrations of the other indispensable amino acids for the pig met the NRC requirements. The control diet was designed to meet the NRC requirements for all the indispensable amino acids. Half of the pigs assigned to each group were fed the control or the LL diet for 4 weeks, while the remaining six pigs were fed the control or the LL diet for 8 weeks. *Longissimus dorsi* muscle specimens were obtained at 4 and 8 weeks after the beginning of the experiment. In addition, muscle specimens were obtained from 6 pigs at the beginning of the experiment.

The IMF content of the *longissimus dorsi* muscle of the LL group was already higher than that of the control group at 4 weeks ($P<0.05$; 3.8% for the LL group vs. 2.0% for the LL group). At 8 weeks, the magnitude of the difference in the IMF content was further enhanced ($P<0.05$; 6.9% for the LL group vs. 3.3% for the control group). Serial transverse cross sections through the *longissimus dorsi* muscle (10 μm thick) were made by using a cryostat at -20 °C. The sections were stained with Oil-Red-O to detect triacylglyceride. The volume of the intramuscular adipocytes was calculated by using a prediction equation, as described elsewhere (Chikuni *et al.*, 1985). As expected from the results of IMF content, the adipocyte volumes in the LL group were higher at 4 and 8 weeks (Figure1, $P<0.05$), but no differences were observed in adipocytes numbers (Figure 1). These results are inconsistent with those of Lee and Kaufman (1974). Although the exact reasons for this inconsistency are unknown, it can potentially be explained by differences in the breeds of the pigs and/or the types of muscle

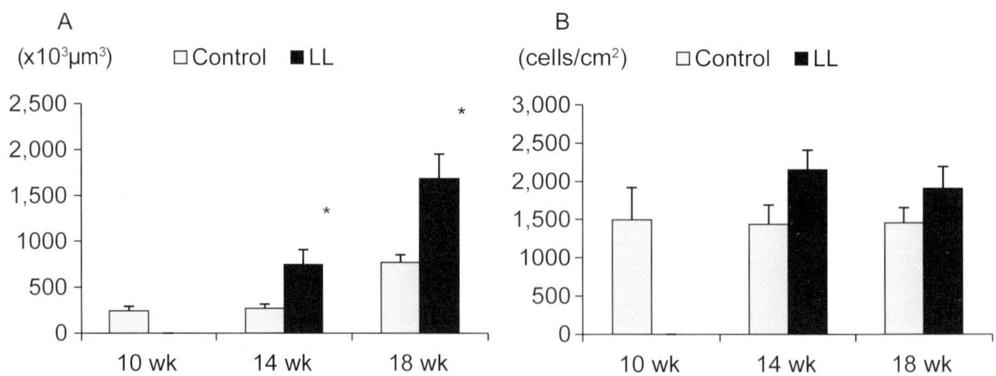

*Figure 1. Effect of low dietary levesl of lysine on the volume and number of intramuscular adipocytes in the porcine longissimus dorsi muscle. (A) Adipocute volume and (B) adipocyte number are shown. Each bar represents mean ± SE (n=6). * Significantly different from the control group (P<0.05).*

employed in the two studies. As far as the *longissimus dorsi* muscle is concerned, we can infer that the accumulation of IMF in the LL group resulted from the hypertrophy of matured adipocytes. This hypothesis is supported by our previous data, which was presented at the ISEP 2010 and showed that PPARγ mRNA levels were higher in the *longissimus dorsi* and *rhomboideus* muscles of pigs fed with a low lysine diet, whereas C/EBPα mRNA levels were not affected (Katsumata *et al.*, 2010). Since C/EBPα is a master regulator of adipocyte differentiation, the result imply that low levels of dietary lysine do not provoke the differentiation of porcine intramuscular preadipocytes into adipocytes.

Effect of low dietary levels of threonine vs. lysine on porcine IMF deposition

The question has been raised as to whether dietary reduction of indispensable amino acids other than lysine would likewise promote IMF deposition in the pig, taking into account the low levels of dietary protein enhanced the fat content in porcine muscle (Castell *et al.*, 1994; Goerl *et al.*, 1995). Therefore, we next focused on threonine, another major limiting amino acid in the pig diet, because feeding a low-lysine diet and a low-threonine diet both enhanced the abundance of GLUT4 mRNA in porcine skeletal muscle (Katsumata *et al.*, 2003, 2004). Hence, porcine muscle might respond similarly to a shortage of either dietary lysine or threonine.

Duroc × (Large white × Landrace) crossbred pigs (age 10 weeks) were assigned to one of four groups (lysine control (LC), low lysine (LL), threonine control (TC), and low threonine (LT)). The LL and LT diets were designed to yield concentrations of lysine and threonine that were approximately 70% of the required concentrations set by the Japanese Feeding Standard for Swine (NARO, 2005). All pigs were fed with the experimental diets for 8 weeks. At the end of the experiment, specimens of serum, *longissimus dorsi* muscle, and *rhomboideus* muscle were collected.

The main findings of this experiment were that although the shortage of lysine in the diet enhanced IMF content in the *longissimus dorsi* muscle and the *rhomboideus* muscle, the shortage of threonine did not (Figure 2). Therefore, our results failed to confirm the hypothesis that a shortage of threonine would also enhance IMF content in porcine muscle, and instead indicate that amino acids have individual, albeit not necessary predictable, effects on the deposition of IMF in the pig.

Lysine has quantifiable anti-obesity actions. For example, recent work showed that dietary lysine reduced the body fat mass in rats by promoting fatty acid oxidation and by inhibiting fatty acid

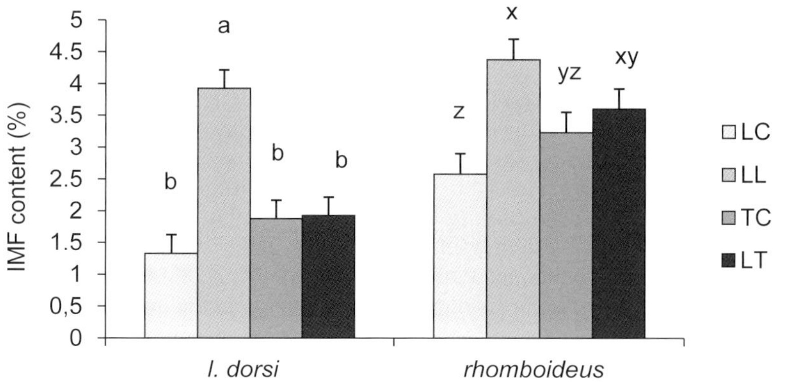

Figure 2. Effect of low dietary levels of lysine and threonine on IMF content in the porcine longissimus dorsi *and* rhomboideus *muscle. Each bar represents the least square mean ± the pooled SE (n=4). a and b (P<0.05); x, y and z (P<0.05).*

synthesis (Kobayashi *et al.*, 2011). Thus, it is conceivable that the shortage of lysine in our study resulted in the reduced oxidation of fatty acids and accelerated deposition of fat in the muscle. Therefore, the differential effects of lysine vs. threonine shortage on IMF accumulation might be explained, at least in part, by contradictory effects of each amino acid on fatty acid metabolism.

The shortage of dietary lysine and threonine both tended to influence serum insulin concentrations (P=0.079, Table 1). The average insulin concentration of the LC and the TC group was 1.52 ng/ml while that of the LL and the LT group was 0.85 ng/ml. In contrast, dietary amino acid levels did not affect serum glucose concentrations (Table 1). These results suggest that the response of peripheral tissues to insulin might be larger in the LL and LT groups than in the control group. Lipogenesis occurs downstream of insulin signaling. Dietary protein deprivation increases the strength of insulin signaling in the liver and muscle of rats (Toyoshima *et al.*, 2004, 2010). Hence, the lysine shortage-promoted deposition of IMF in porcine muscle might be attributable to enhanced insulin signaling. In support of this hypothesis, insulin receptor substrate-1 (IRS-1) mRNA levels were higher for the LL group than for the LC group (data not shown). We are currently investigating the impact of low dietary lysine on the phosphorylation of IRS-1 in porcine muscle.

Some details of the threonine vs. lysine experiment are described elsewhere (Kobayahsi *et al.*, 2012).

Table 1. Effect of low dietary lysine and threonine levels on growth performance and serum insulin and glucose concentrations. Data are expressed as least square means and pooled SE (n=4).

	LC	LL	TC	LT	Pooled SE	AA[1]
Feed intake (g/d)	2,831	2,945	2,736	2,619	101	
Live body weight gain (g/d)	1,018	953	988	823	31	**
Feed efficiency	0.37	0.32	0.36	0.31	0.01	**
Insulin (ng/ml)	1.43	0.62	1.62	1.07	0.34	P=0.079
Glucose (mg/ml)	114	102	114	120	11	

[1] Effects of dietary amino acid levels. **=P<0.01.

Effect of lysine concentration on triacylglyceride accumulation in 3T3-L1 preadipocytes

Feeding a low-lysine diet might result in a reduced supply of lysine to peripheral tissues. In this case, the enhanced deposition of IMF in the pig could perhaps be ascribed to the decreased levels of lysine reaching intramuscular adipocytes. However, living, intact animals do not readily lend themselves to the evaluation of this hypothesis. For instance, low dietary levels of lysine also reduce the concentrations of IGF-1, while concomitantly increasing the concentrations of glucocorticoids (Katsumata *et al.*, 2002; Ishida *et al.*, 2011). These altered hormone concentrations could similarly affect fat accretion in porcine muscle, and thus, we were unable to test whether a decline in the lysine supply directly promoted IMF deposition *in vivo*. To overcome this issue, we conducted *in vitro* experiments to explore the link between lysine concentration and triacylglyceride accumulation in cultured 3T3-L1 preadipocytes. Although the 3T3-L1 preadipocyte is of murine rather than porcine origin, we employed these cells in our work because they represent the most commonly used cell line for studies on adipogenesis.

The concentration of lysine was set at 921 nmol/ml in standard culture medium made with DMEM and FBS (control medium). Low-lysine media were prepared with lysine concentrations set at 461 nmol/ml (×0.5 of the concentration in control medium), 230 nmol/ml (×0.25), and 115 nmol/ml (×0.125). The 461 nmol/ml concentration was similar to the concentration of lysine in the plasma of pigs fed a diet meeting the requirement of lysine, whereas the 115 nmol/ml concentration was similar to the concentration of lysine in the plasma of pigs fed a low-lysine diet. Eight days after the induction of preadipocyte differentiation by exposure to a defined adipogenic cocktail, triacylglyceride levels were measured in the mature adipocytes. As shown in Figure 3, triacylglyceride levels were higher for cells cultured in ×0.125 medium than for cells cultured in control medium. These observations demonstrate that the differentiation of 3T3-L1 preadipocytes in low-lysine media promoted fat deposition in the mature adipocytes.

As mentioned above, insulin signaling might play an important role in the regulation of IMF deposition in porcine muscle. We next measured insulin receptor, IRS-1, and IRS-2 mRNA levels in mature adipocytes (Figure 4). Interestingly, mRNAs levels were higher for adipocytes cultured in ×0.125 medium than for adipocytes cultured in control medium. Furthermore, the degree of IRS-1 tyrosine phosphorylation was elevated for adipocytes cultured in ×0.125 medium (Figure 5), indicating that insulin signaling is involved in the increased deposition of triacylglycerides in 3T3-L1 cell-derived adipocytes cultured in low-lysine media.

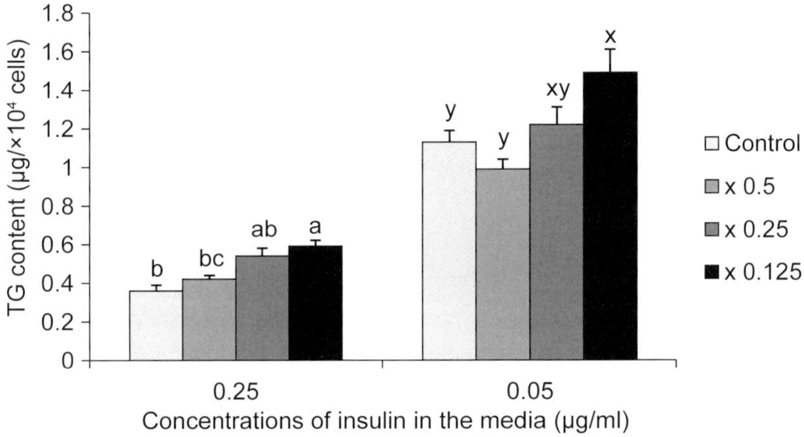

Figure 3. Effect of lysine concentration on triacylglyceride concent in 3T3-L1 adipocytes. Each bar represents the mean ± SE (n=3). a, b, and c (P<0.05); x and y (P<0.05).

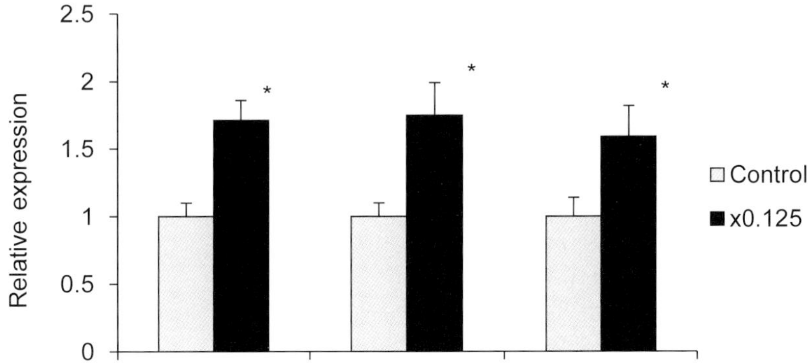

*Figure 4. Effect of lysine concentration on insulin receptor(IR), IRS-1, and IRS-2 mRNA levels in 3T3-L1 adipocytes. Each bars represents the mean±SE (n=6). Means from the control cells are expressed as 1. 18SrRNA levels were used as internal standards. * Significantly different from the control group (P<0.05).*

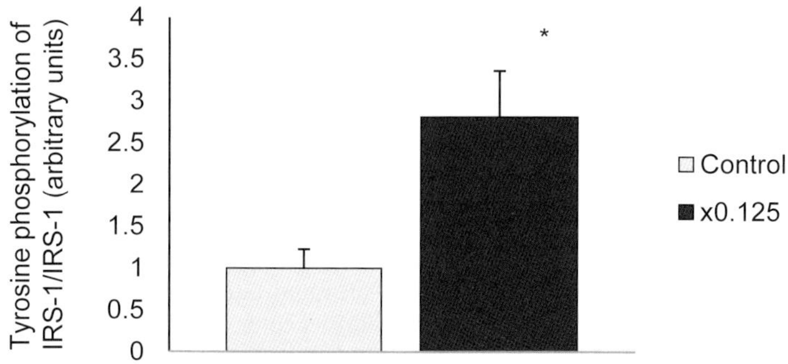

*Figure 5. Effect of lysine concentration on tyrosine phosphorylation of IRS-1 in 3T3-L1 adipocytes. Each bar represents the mean ± SE (n=6). Means from the control cells are expressed as 1. * Significantly different from the control group (*P<0.05*).*

Conclusions

Low level of dietary lysine affected the volume but not the number of intramuscular adipocytes in the porcine *longissimus dorsi* muscle, suggesting that the promotion of lipogenesis in already existing adipocytes contributed to the dietary lysine shortage-enhanced deposition of IMF. Conversely, low dietary threonine level had no impact on IMF accumulation; thus the assorted indispensable amino acids exert differential effects on the buildup of IMF in the pig. The differential effects of lysine vs. threonine shortage might be due to disparities in the ability of each amino acid to modulate fatty acid metabolism. Moreover, the current results suggest that insulin signaling is involved in the regulation of IMF levels by dietary lysine. Further elucidation of the ability of lysine and other indispensable amino acids to influence insulin signaling is extremely important and will form the basis of our future work, given that insulin signaling has an abundance of critical roles in the control of global metabolism.

Acknowledgements

A part of this work was supported by the Program for Promotion of Basic and Applied Research for Innovation in Bio-Oriented Industry.

References

Castell, A.G., R.L. Cliplef, L.M. Poste-Flynn and G. Butler, 1994. Peformance, carcass and pork characteristics of castrates and gilts self-fed diets differing in protein and lysine: energy ratio. Can. J. Anim. Sci. 74, 519-528.

Chikuni, K., M. Jimbu, S. Ozawa, T. Koishikawa, M. Yoshitake and N. Yano, 1985. Relation among back fat thickness, fat cell size and fatty acid composition of swine. Jpn. J. Swine Hus. Res. 22, 181-187.

Doran, O., F.M. Whittington, K.G. Hallet and J.D. Wood, 2007. Effect of a reduced protein diet on expression of lipogenic enzymes in relation to intramuscular fat formation in the pig. In: Ortigues-Marty, I., N. Miraux and W. Brand-Williams (editors), Energy and protein metabolism and nutrition. Wageningen Academic Publishers, Wageningen, the Netherlands, EAAP publication No. 124, 125-126.

Goerl, K.F., S.J. Eilert, R.W. Mandingo, H.Y. Chen and P.S. Miller, 1995. Pork characteristics as affected by two populations of swine and six crude protein levels. J. Anim. Sci. 73, 3621-3626.

Goto, T., H. Iwamoto, Y. Ono, S. Nishimura, K. Matsuo, Y. Nakanishi, R. Umetsu and H. Takahara, 1994. Comparative study on the regional composition of fiber types in M. Longissimus Thoracis with different marbling scores for Japanese Black steers. Anim. Sci. Technol.(Jpn.) 65, 454-463

Hyun, Y., M. Ellis, F.K. McKeith and D.H. Baker, 2003. Effect of dietary leucine level on growth perfoemance, and carcass and meat quality in finishing pigs. Can. J. Anim. Sci. 83, 315-318.

Hyun, Y., J.D. Kim, M. Ellis, B.A. Peterson, D.H. Baker and F.K. McKeith, 2007. Effect of dietary leucine and lysine levels on intramuscular fat content in finishing pigs. Can. J. Anim. Sci. 87, 303-306.

Hyodo, I., 1996. A case study for increasing intramuscular fat content of pork. J. Anim. Breed. 5, 1-9.

Ishida, A., T. Kyoya, K. Nakashima and M. Katsumata, 2011. Muscle protein metabolism during compensatory growth with changing dietary lysine levels from deficient to sufficient in growing rats. J. Nutr. Sci. Vitaminol. 57, 401-408.

Katsumata, M., S. Kawakami, Y. Kaji, R. Takada and M.J. Dauncey, 2001. Low lysine diet selectively up-regulates muscle GLUT4 gene and protein expression during postnatal development. In: Chwalibog, A. and K. Jakobsen (editors), Energy metabolism in animals. Wageningen Pers, Wageningen, the Netherlands, EAAP publication No. 103, 237-239.

Katsumata, M., S. Kawakami, Y. Kaji, R. Takada and M.J. Dauncey, 2002. Differential regulation of porcine hepatic IGF-I mRNA expression and plasma IGF-I concentration by a low lysine diet. J. Nutr. 132, 688-692.

Katsumata, M., M. Matsumoto and Y. Kaji, 2003. Effects of a low lysine diet on glucose metabolism in skeletal muscle of growing pigs. In: Souffrant W.B. and C.C. Metges (editors), Progress in research on energy and protein metabolism. Wageningen Academic Publishers, Wageningen, the Netherlands, EAAP publication No. 109, 187-190.

Katsumata, M., S. Kawakami, Y. Kaji and R. Takada, 2004. Circulating levels of insulin-like growth factor-1 and associated binding proteins in plasma and mRNA expression in tissues of growing pigs on a low threonine diet. Anim. Sci. 79, 85-92.

Katsumata, M., S. Kobayashi, M. Matsumoto, E. Tsuneishi and Y. Kaji, 2005. Reduced intake of dietary lysine promotes accumulation of intramuscular fat in the *Longissimus dorsi* muscles of finishing gilts. Anim. Sci. J. 76, 237-244.

Katsumata, M., A. Ishida, T. Kyoya and K. Nakashima, 2010. Effects of dietary lysine levels on expression of adipogenesis related genes in muscle of growing pigs. In: Crovetto G.M. (editor), Energy and protein metabolism and nutrition. Wageningen Academic Publishers, Wageningen, the Netherlands, EAAP publication No. 127, 681-682.

Katsumata, M., T. Kyoya, A. Ishida, M. Ohtsuka and K. Nakashima, 2012. Dose-dependent response of intramuscular fat accumulation in *longissimus dorsi* muscle of finishing pigs to dietary lysine levels. Livest. Sci. 149, 41-45.

Kobayashi, H., Y. Hirabayashi, T. Ueda and H. Murakami, 2011. Amino acid loses your weights? Effects of dietary lysine on lipid and carbohydrate metabolism. Amino-san Kenkyû 5, 33-37.

Kobayashi, H., A. Ishida, A. Ashihara, K. Nakashima and M. Katsumata, 2012. Effects of dietary low level of threonine and lysine on the accumulation of intramuscular fat in porcine muscle. Biosci. Biotechnol. Biochem. 76, 2347-2350.

Lee, Y.B. and R.G. Kauffmann, 1974. Cellular and enzymatic changes with animal growth in porcine intramuscular adipose tissue. J. Anim. Sci. 38, 532-537.

National Agriculture and Bio-Oriented Research Orgnaization, 2005. Japanese Feeding Standard for Swine, Japan Livestock Industry Association, Tokyo.

National Research Council, 1998. Nutrient Requirements of Swine, 10th edn. National Academy Press, Washington, DC.

Suzuki, K., H. Kadowaki, T. Shibata, H. Uchida and A. Nishida, 2005. Selection for daily gain, loin-eye area, back fat thickness and intramuscular fat based on desired gains over seven generations of Duroc pigs. Livest. Prod. Sci. 97, 193-202.

Takasaki, S., K. Iizuka, A. Suzuki and Y. Itoh, 2005. Chemical, physical and mastication properties of pork from the new 'Tokyo X' breed. J. Cook. Sci. Jpn. 38, 404-409.

Tan, B., Y. Yin, Z. Liu, X. Li, H. Xu, X. Kong, R. Huang, W. Tang, I. Shinzato, S.B. Smith and G. Wu, 2009. Dietary L-arginine supplementation increases muscle gain and reduces body fat mass in growing-finishing pigs. Amino Acids 37, 169-175.

Toyoshima, Y., Y. Ohne, S. Takahashi, T. Noguchi and H. Kato, 2004. Dietary protein deprivation decreases the serine phosphorylation of insulin receptor substrate-1 in rat skeletal muscle. J. Mol. Endocrinol. 32, 519-531.

Toyoshima, Y., R. Tokita, Y. Ohne, F. Hakuno, T. Noguchi, S. Minami, H. Kato and S. Takahashi, 2010. Dietary protein deprivation upregulates insulin signaling and inhibits gluconeogenesis in rat liver. J. Mol. Endocrinol. 45, 329-340.

Impact of dietary l-arginine supply during early gestation on myofiber development in newborn pigs

G. Bee[1], C.E. Pardo[1,2] and M. Kreuzer[2]
[1]Department of Animal Science, Agroscope, 1752 Posieux, Switzerland; giuseppe.bee@alp.admin.ch
[2]ETH Zurich, Institute of Agricultural Sciences, 8092 Zurich, Switzerland

Introduction

During early pregnancy when placental growth is fastest, the level of arginine and its precursor ornithine is elevated in porcine amniotic and allantoic fluid (Wu *et al.*, 2006). This abundance is associated with a high syntheses rate of nitric oxide and polyamine in the porcine placentae (Self *et al.*, 2004). Wu *et al.* (2004) provided evidence showing that both nitric oxide and polyamines play key roles in angiogenesis, which is a critical event during placental growth and fetal development (Town *et al.*, 2005). These observations led to the hypothesis that arginine is important for placental and fetal development. Recently, Bérard and Bee (2010) showed that supplementing the gestation diet with l-arginine, positively affected primary myofiber hyperplasia in semitendinosus muscles (STM) of 75 d old fetuses. Because primary fibers serve as a scaffold for the formation of secondary fibers we hypothesized that offspring from sows fed extra arginine bare the potential for greater fiber hyperplasia and ultimately more efficient postnatal growth. Especially low-birth weight pigs display low myofiber numbers. Thus, the goal of this experiment was to evaluate if arginine supplementation of the dams has a positive effect on muscle development of such piglets.

Material and methods

Five intact (I), 5 unilaterally hysterectomized-ovariectomized (HO; model for crowded intrauterine environment) and 5 unilaterally ovariectomized (OL; model for uncrowded intrauterine environment) sows were subjected to a crossover design in terms of diets (parities 5 and 6). From day 14 to day 28 of gestation they were either offered 25 g/d of l-arginine (Arg) in addition to the standard gestation diet or they received the corresponding control diet (C) which was supplemented with 43 g/d of l-alanine. At farrowing, litter characteristics were assessed. From each litter, 2 female and 2 male piglets with the lowest and medium birth weight were sacrificed. Internal organs and the STM were collected and weighed. Histological analyses of muscles were performed using mATPase staining after pre-incubation at pH 4.5 and 10.2. The data were analyzed with the mixed procedure (SysStat 13), considering the surgical treatment of the sows, the early gestation diet type, gender and the 2- and 3-way interactions as fixed factors.

Results and discussion

As expected, the number of total piglets born and piglets born alive differed ($P<0.05$) among sow groups and was lowest in HO (10.2; 7.8), greatest in I (16.0; 14.3) and intermediate in OL (11.3; 10.0). Similarly, birth weight of offspring born alive differed ($P<0.05$) between the sow groups and was lightest in HO (1.18 kg), heaviest in OL (1.66 kg) and intermediate in I (1.34 kg). The largest variability (standard deviation) in litter birth weight was observed in HO followed by I and OL sows (0.32, 0.26 and 0.18 kg, respectively). Except for birth weight variability, which was lower ($P<0.03$) in offspring from sows fed Arg compared to C (±0.23 vs. ±0.27 kg), litter characteristics were not affected by Arg supply.

Compared to OL, progeny from OH and I sows had lower ($P<0.05$) relative weights of liver (2.90 vs. 2.33; 2.30), kidney (0.84 vs. 0.73; 0.72) and STM (2.30 vs. 2.09; 1.95) and a greater ($P<0.05$) brain-to-liver ratio (0.80 vs. 1.25; 1.46). The latter is also known as brain sparing and is indicative of intrauterine growth retardation. Independent of the sow group, brain-to-liver ratios (1.12 vs. 1.22) tended ($P<0.08$) to be lower in offspring from Arg compared to C sows, which suggest that l-arginine supplementation could alleviate the effect of intrauterine growth retardation.

Supplying Arg to I and OL, but not OH, sows instead of C resulted in greater myofiber hyperplasia and, consequently, a larger muscle area (Figure 1). These differences resulted mainly from a greater myofiber number in the light portion (3.50 vs. 3.24×10^5) of the muscle. The current finding confirms previous assumptions that arginine supply during early gestation positively affects muscle formation and growth potential.

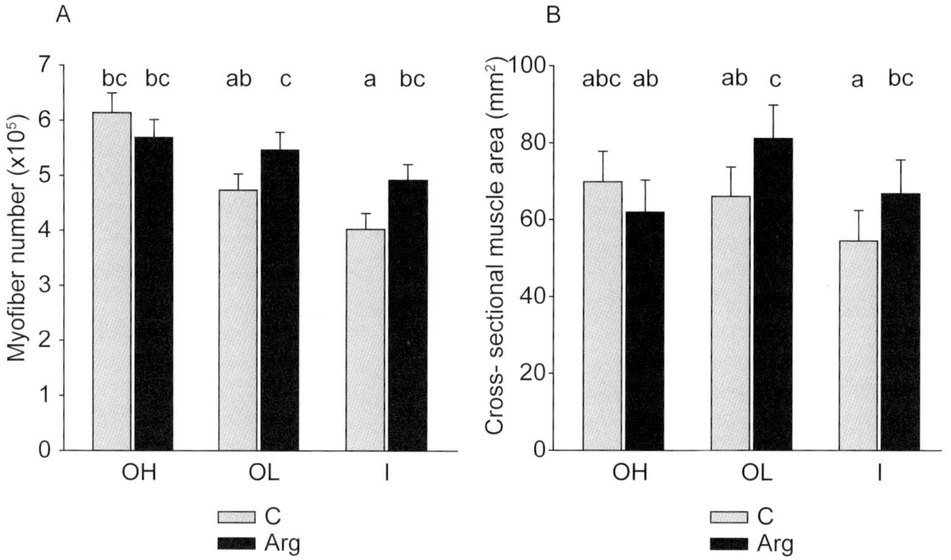

Figure 1. Total myofiber number (panel A) and cross-sectional area (panel B) of the semitendinosus muscle in offspring born from unilaterally hysterectomized-ovariectomized (OH), unilaterally ovariectomized (OL) and intact (I) sows fed either a control diet (C) or a l-arginine (Arg) supplemented diet from day 14 to day 28 of gestation. Least square means within each graph with different superscript differ ($P<0.05$). Error bars represent standard error of least square means.

References

Bérard, J. and G. Bee, 2010. Effects of dietary L-arginine supplementation to gilts during early gestation on foetal survival, growth and myofiber formation. Animal 4, 1680-1687.

Self, J. T., T. E. Spencer, G. A. Johnson, J. Hu, F. W. Bazer, and G. Wu, 2004. Glutamine synthesis in the developing porcine placenta. Biol. Reprod. 70, 1444-1451.

Town, S. C., J. L. Patterson, C. Z. Pereira, G. Gourley, and G. R. Foxcroft, 2005. Embryonic and fetal development in a commercial dam-line genotype. Anim. Reprod. Sci. 85, 301-316.

Wu, G., F. W. Bazer, J. M. Wallace, and T. E. Spencer, 2006. BOARD-INVITED REVIEW: Intrauterine growth retardation: Implications for the animal sciences. J. Anim. Sci. 84, 2316-2337.

Wu, G., F. W. Bazer, T. A. Cudd, C. J. Meininger, and T. E. Spencer, 2004. Maternal Nutrition and Fetal Development. J. Nutr. 134, 2169-2172.

Effect of feed restriction and birth weight on molecular and metabolic response in the liver of pigs

C. Nebendahl[1], R. Krüger[1], S. Görs[1], K. Giggel[1], B.U. Metzler-Zebeli[1], H.M. Hammon[1], E. Albrecht[2], R. Pfuhl[2] and C.C. Metges[1]
[1]Leibniz Institute for Farm Animal Biology, Institutes of Nutritional Physiology, 18196 Dummerstorf, Germany; schmelzer@fbn-dummerstorf.de
[2]Leibniz Institute for Farm Animal Biology, Muscle Biology and Growth, 18196 Dummerstorf, Germany

Introduction

Intrauterine growth restriction (IUGR) in pigs has been shown to increase body fatness and decrease meat quality (Rehfeldt *et al.*, 2006). Moreover, IUGR was reported to lead to altered metabolic and hypothalamic-pituitary-adrenal (HPA) axis function later in life (Poore *et al.*, 2003). Previous experiments indicated that caloric restriction in early postnatal life may affect liver lipid metabolism (Garg *et al.*, 2013). In our study, the following questions were addressed at the molecular and physiological level in pigs: (1) Are there differences in the hepatic transcriptional profile between low and normal birth weight? (2) Are these effects reflected on the tissue composition level? (3) Could the possible birth weight-dependent effects be modified by feed restriction? (4) Are these effects persistent?

Material and methods

Liver tissue of female juvenile pigs (n=42; German Landrace) of low (0.8-1.1 kg, U) and normal (1.4-1.6 kg, N) birth weights was used to study transcriptional and metabolic responses to feed restriction (R, 50% of control diet [C, 13.6 MJ ME/kg], age 78-98 d) followed by subsequent refeeding [14.1 MJ ME/kg] until 131 d of age. Overall, four groups (NC, NR, UC, UR) were included in our study to analyze effects related to birth weight (U vs. N) and/or feed restriction (R vs. C). For whole genome expression studies on porcine-specific Agilent 8x60K multiplex arrays and the analysis of lipid droplets stained with oil red O, liver tissue was taken from animals after overnight fasting (18 h) at ages 75 d (before feed restriction, T1), 98 d (at the end of a 3 weeks feed restriction period, T2) and 131 d (after 5 week of refeeding, T3). Liver studies were performed on three animals per group at each time point. For four-group comparisons, Two-Way Analysis of Variance (ANOVA) was used. When normality test passed ($P \leq 0.05$), Bonferroni-test was used. Not-normally distributed data were log-transformed to obtain normality and to allow ANOVA analysis. After quantile normalization of microarray data, statistical analysis for pairwise comparisons was performed with an unpaired t-test with unequal variance (Welch-test). All results are depicted as means ± SEM, and were considered significantly different when P-values were $P \leq 0.05$. Statistics were done with SigmaPlot 11.0 software.

Results and discussion

At d 75 (T1), hepatic gene expression analysis (Fold Change ≥1.3, $P \leq 0.05$) identified 194 genes (95 up- and 99 down-regulated) involved in lipid metabolism to be differentially expressed in U vs. N animals.

Oil red O-stained liver sections showed 4.6-fold and 3.7-fold increases in the total mean area and number of lipid droplets (LDs) in U versus N, respectively ($P \leq 0.01$). The mean LD size (μm^2) was increased by 24.9% (Figure 1).

However, 3-weeks feed restriction (T2) reduced the total mean area of LDs by 58.3 and 72.7% in U and N animals, respectively ($P \leq 0.01$). Additionally, 451 (311 up- and 140 down-regulated) and

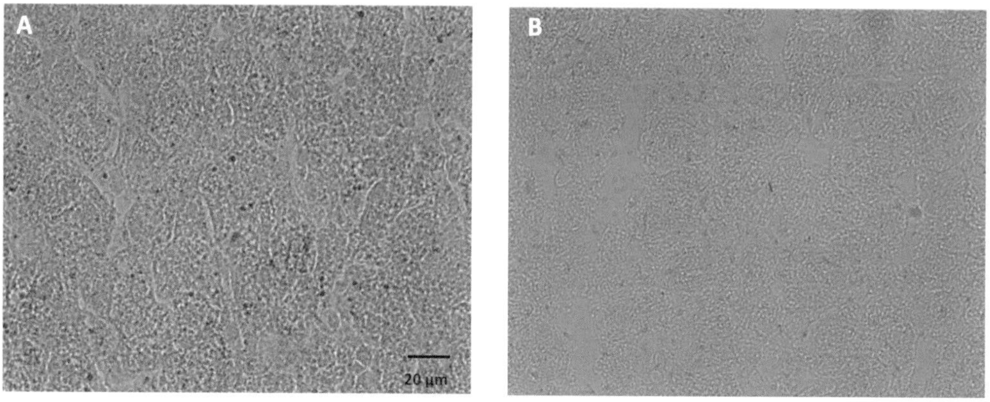

Figure 1. Effects of low (A) and normal (B) birth weight on liver lipid droplet (LD) count and formation in pigs at 75 d of age. Microscopic images show higher total mean area and numbers of lipid droplets, as well as mean LD size in low (A) vs. normal (B) birth weight pigs.

340 (121 up- and 219 down-regulated) genes were found to be differentially regulated in the liver of N and U after feed restriction, respectively. The differentially regulated genes were allocated to functional pathways associated with e.g. protein metabolism. Free arginine concentration was found to be 17% higher in liver samples of UR vs. NR animals ($P \leq 0.05$).

The second objective of our study was to explore whether the observed responses to feed restriction and birth weight persist after 5 weeks of refeeding (T3). In total, 26 long-term regulated genes (6 up- and 20 down-regulated) were identified. In general, refeeding induced the recovery of total mean LD area in U and N animals, respectively. However, in feed-restricted U animals, the mean LD size (μm^2) was still lower by 23.3% as compared to age-matched controls. In contrast, the mean LD size was increased in N (+24.7%, $P \leq 0.01$). Finally, the data suggest that feed restriction programmed juvenile female pigs with low birth weight for an increased rate of hepatic lipolysis in later life.

Acknowledgment

The study was supported by BMBF of Germany in the framework of ‚Ernährungsforschung – für ein gesundes Leben, Modul: Biomedizinische Ernährungsforschung‘ (Vision Epifood, FKZ 0315397B).

References

Garg M., Thamotharan M., Dai Y., Lagishetty V., Matveyenko A.V., Lee W.N. and Devaskar S.U., 2013. Glucose intolerance and lipid metabolic adaptations in response to intrauterine and postnatal calorie restriction in male adult rats. Endocrinol 154, 102-113.

Poore, K.R. and Fowden, A.L., 2003. The effects of birth weight hypothalamo-adrenal axis function in juvenile and adult pigs. J Physiol 547, 107-116.

Rehfeldt, C. and Kuhn, G., 2006. Consequences of birth weight for postnatal growth performance and carcass quality in pigs as related to myogenesis. J Anim Sci 84, 113-123.

Embryo thermal manipulation has long-lasting effects on energy metabolism in chickens

T. Loyau[1], S. Métayer-Coustard[1], C. Praud[1], C. Berri[1], M.J. Duclos[1], S. Tesseraud[1], N. Rideau[1], P. Chartrin[1], C. Hennequet-Antier[1], N. Everaert[2], S. Yahav[3], S. Mignon-Grasteau[1] and A. Collin[1]
[1]INRA, UR83 Recherches Avicoles, 37380 Nouzilly, France; metayer@tours.inra.fr
[2]Department of Biosystems, K.U. Leuven, Kasteelpark Arenberg 30, 3001 Leuven, Belgium
[3]Institute of Animal Science, ARO The Volcani Center, P.O. Box 6, Bet Dagan 50250, Israel

Introduction

Broiler chickens have limited capacities to sustain high temperatures. However, thermal manipulation (TM) during embryogenesis has been shown to lower their body temperature at hatch and to improve thermotolerance until market age (Piestun *et al.*, 2008). This thermotolerance acquisition could partly be due to changes in sensible heat loss, but also in metabolic rate, especially in energy and protein metabolisms of birds. The aim of this study was to evaluate the long-lasting effects of TM during embryogenesis, when coupled or not with a heat challenge at slaughter age (d 34), on plasma metabolites and hormones, cell signaling and the expression of genes involved in muscle metabolism.

Material and methods

Eggs were incubated in control conditions (C; 37.8 °C, 56% Relative Humidity RH) or exposed to thermal manipulation (TM; 12 h/d at 39.5 °C and 65% RH) from day 7 to 16 of embryogenesis. After rearing in standard conditions, half of each group stayed at control ambient temperature (21 °C; C and TM) and the other half was submitted to heat challenge (Ch; 32 °C for 5 h; CCh for control chickens submitted to heat challenge and TMCh for TM chickens submitted to heat challenge) at 34 d of age. At this age, blood was withdrawn from the brachial vein, chickens were slaughtered and the fast-twitch glycolytic *Pectoralis major* (PM) muscle was removed from 9 to 11 male broiler chickens and snap frozen. The expression of genes involved in energy metabolism was analyzed by qRT-PCR and normalized with β-actin, cytochrome B and 18S using Genorm® software. Plasma insulin and thyroid hormone concentrations were measured by RIA. Plasma uric acid, triglyceride and glucose concentrations were measured enzymatically. Cell signaling was explored by western-blot using phospho-specific antibodies directed against kinases involved in protein, carbohydrate and/or energy metabolism as described by Boussaid-Om Ezzine *et al.* (2010). Data were analyzed using the general linear model procedure of SAS, considering incubation and heat challenge within incubation, i.e.Ch(Incubation), as main effects.

Results and discussion

Incubation conditions did not affect plasma uric acid, triglyceride, triiodothyronine, thyroxine and insulin concentrations. However it had significant long-lasting effects on thyroid hormone metabolism by lowering by 40 to 50% the muscle expressions of deiodinases D3 and D2 in TM animals as compared to controls. TM also decreased the expression of the transcription factor PGC1-α involved in mitochondrial biogenesis and metabolic activation (Walter and Seebacher, 2007). Consistently, energy production pathways were modified by incubation conditions at 21 °C, as indicated by decreased muscle citrate synthase and hexokinase 1 expressions in TM compared to control chickens (Figure 1). Moreover, UCP3 was overexpressed in TMCh group as compared to TM and C chickens, possibly related to a limitation of oxidative stress in these animals (Mujahid *et al.*, 2006). This overexpression was associated with a higher glycemia (8.5±0.3 mmol/l) as compared to other groups (from 6.8±0.2 to 7.0±0.3 mmol/l).

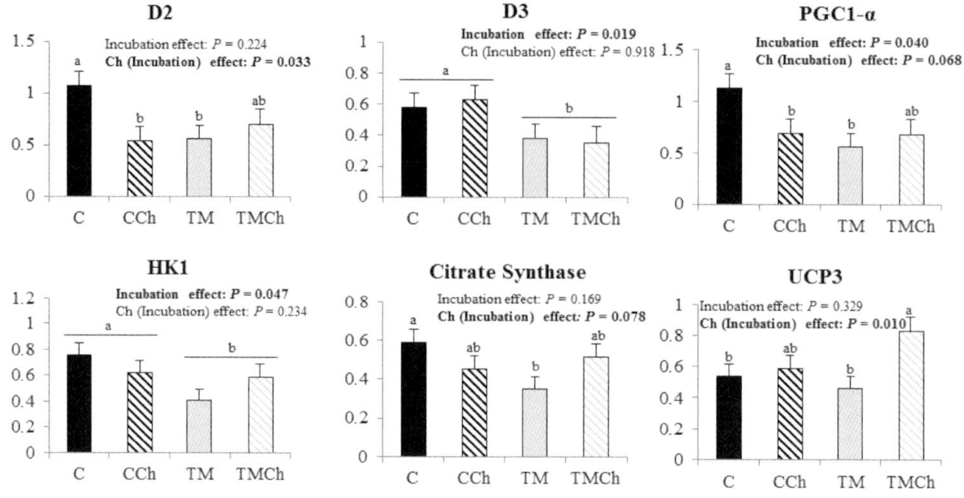

Figure 1. Relative expression of genes involved in muscle energy metabolism. C: control broilers, CCh: control to a heat challenge at d34 (Ch), TM: thermal-manipulated during embryogenesis and TMCh: TM to a heat challenge at d34. Data were analyzed considering incubation and heat challenge within incubation, i.e.Ch(Incubation), as main effects. The different letters indicate significant differences between treatments (a-b: $P<0.1$*). Data are presented as means ± SEM. D2: Deiodinase 2, D3: Deiodinase 3, PGC1-α: PPARγ co-activator 1-α, HK1: Hexokinase 1, CS: Citrate synthase, UCP3: Uncoupling Protein 3.*

The phosphorylation of kinases regulating protein synthesis (S6 ribosomal protein and S6 Kinase 1) was significantly inhibited by heat challenge within incubation conditions, whereas P38 Mitogen-activated protein kinase, involved in stress and inflammatory responses, showed an increased phosphorylation state in TM conditions.

In conclusion, several genes and proteins controlling energy metabolism and stress response in the *Pectoralis major* muscle were affected by embryo thermal manipulation, reflecting a possible long-lasting adaptation, which may modify peripheral metabolism.

Acknowledgements

Funding: Agence Nationale de la Recherche, Project ANR-09-JCJC-0015-01 THERMOCHICK.

References

Boussaid-Om Ezzine, S., N. Everaert, S. Metayer-Coustard, N. Rideau, C. Berri, R. Joubert, S. Temim, A. Collin, and S. Tesseraud, 2010. Effects of heat exposure on Akt/S6K1 signaling and expression of genes related to protein and energy metabolism in chicken (Gallus gallus) pectoralis major muscle. Comp. Biochem. Physiol. B-Biochem. Mol. Biol. 157: 281-287.

Mujahid A., K. Sato, Y. Akiba, and M. Toyomizu, 2006. Acute heat stress stimulates mitochondrial superoxide production in broiler skeletal muscle, possibly via downregulation of uncoupling protein content. Poult. Sci. 85(7), 1259-1265.

Piestun Y., D. Shinder, M. Ruzal, O. Halevy, J. Brake, and S. Yahav, 2008. Thermal manipulations during broiler embryogenesis: effect on the acquisition of thermotolerance. Poult. Sci. 87, 1516-1525.

Walter I. and F. Seebacher, 2007. Molecular mechanisms underlying the development (*Gallus gallus*): a new role of PGC-1a? Am. J. Physiol. Regul. Integr. Comp. Physiol. 293, 2315-2322.

Heat stress-induced overproduction of mitochondrial reactive oxygen species is down-regulated in laying-type chickens

M. Kikusato and M. Toyomizu
Division of Life Sciences, Graduate School of Agricultural Science, Tohoku University, 1-1 Tsutsumidori-Amamiyamachi, Aoba-ku, Sendai 981-8555, Japan; kmotoi@bios.tohoku.ac.jp

Introduction

Heat stress is an environmental factor that gives rise to oxidative stress. We have previously shown that acute heat stress-induced oxidative damage resulted from the overproduction of mitochondrial reactive oxygen species (mitROS) in the skeletal muscle of meat-type chickens. Recently, we further clarified that the overproduction of mitROS resulted from an increase of the mitochondrial membrane potential ($\Delta\Psi$), and that this was accompanied by a decrease in proton leak due to the reduced expression of avian uncoupling protein (avUCP) (Kikusato *et al.*, 2013). Given that avUCP exerts uncoupling activity to dissipate $\Delta\Psi$ and thereby reduce mitROS production, it could be postulated that down-regulation of avUCP expression might be a factor responsible for the overproduction of mitROS under heat stress conditions. On this basis, we hypothesized that the overproduction of mitROS could be down-regulated when avUCP is highly expressed, even under heat stress conditions. We attempt to verify this hypothesis using two different types of chickens (meat-type and laying-type) whose muscle avUCP content is quantitatively different: avUCP expression is higher in the skeletal muscle tissue of laying-type chickens compared with meat-type chickens (Toyomizu *et al.*, 2011).

Material and methods

Meat-type (Ross) and laying-type (WLH) chickens (n=8) at 0 day of age were obtained from commercial hatcheries and fed a standard diet for 3 and 10 wks, respectively. At the end of the respective periods the chickens were transferred to a diet containing 19% crude protein and 2,900 kcal/kg metabolizable energy until body weights reached 1.2 kg. Thereafter, they were divided into two sub-groups for each type, with one of the two groups exposed to 34 °C for 12 h and the other group maintained at 25 °C. Following this treatment the chickens were sacrificed by decapitation, *Pectoralis superficialis* muscles were removed, and mitochondria isolated and incubated at 38 °C in a potassium-based assay medium (0.3% BSA, pH 7.2) containing 50 μM carboxyatractyloside, which inhibits the uncoupling activity of adenine nucleotide translocator. Succinate was added to initiate mitochondrial respiration, and the O_2 consumption rate and membrane potential ($\Delta\Psi$) were simultaneously measured by using electrodes sensitive to O_2 and to the potential-dependent probe $TPMP^+$, respectively. To measure proton leak kinetics, state 4-respiration was titrated with sequential additions of malonate (up to 3.2 mM). In addition, arachidonic acid (150 μM, freshly prepared) was used to activate the uncoupling activity of avUCP. MitROS production was fluorometrically determined by assay with Amplex Red (Ex./Em. = 544/590 nm).

Results and discussion

Body weights of meat- and laying-type chickens decreased in response to the heat exposure, but the extent of the decrease (expressed as a ratio of the weights of heat-stressed versus control chickens) was significantly less in laying-type chickens than in meat-type chickens. Feed intake was also decreased by the heat treatment for both types, but no difference between them was observed. As shown in Figure 1, no difference in basal proton leak between the thermoneutral and heat-stressed groups was observed for meat- or laying-type chickens. In contrast, however, $\Delta\Psi$ at state 4 (furthest right-hand point in the kinetic curve) was significantly increased by heat exposure in meat-type chickens, but was unchanged in laying-type chickens. Mitochondrial proton leak increased markedly for both types of chickens following the addition of arachidonic acid, but the increase was not as significant

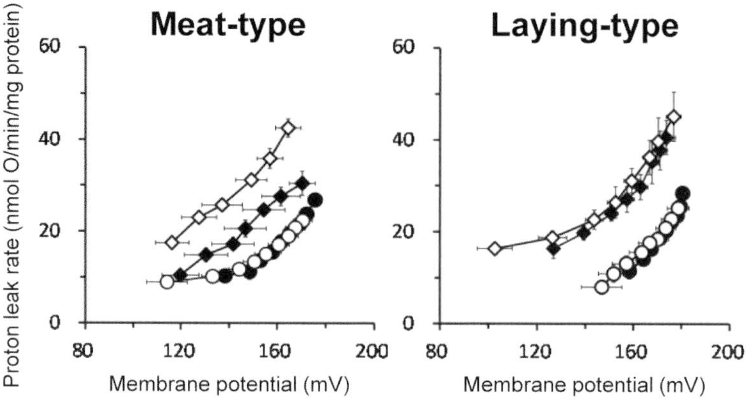

Figure 1. Basal (circles) and arachidonic acid-induced avUCP-activated (diamonds) proton leak in skeletal muscle mitochondria from meat- and laying-type chickens under thermoneutral (open symbols) and heat-stressed (closed symbols) conditions. Values are means±SE, n=4.

for the heat-stressed group compared with the thermoneutral control group in meat-type chickens. No difference in arachidonic acid-induced proton leak was observed between the thermoneutral and heat-stressed groups in laying-type chickens, which suggests that avUCP might be highly expressed even under heat stress conditions.

MitROS production in the presence of arachidonic acid was significantly increased by the heat exposure in meat-type chickens, but was unchanged in laying-type chickens, probably because the avUCP-mediated proton leak was not influenced by the heat stress in the laying-type. In a similar fashion, lipid peroxidation (expressed in terms of malondialdehyde content) in muscle was significantly increased by the heat treatment in meat-type chickens but only marginally in the laying-type. Taken together, these results suggest that avUCP plays an important role in regulating the acute heat stress-induced overproduction of mitochondrial ROS, probably via a decrease in the $\Delta\Psi$ of skeletal muscle mitochondria in birds.

References

Kikusato, M. and M. Toyomizu, 2013. Crucial role of membrane potential in heat stress-induced overproduction of reactive oxygen species in avian skeletal muscle mitochondria. PloS One, in press.

Toyomizu, M., M. Kikusato, Y. Kawabata, M. A. K. Azad, E. Inui and T. Amo, 2011. Meat-type chickens have a higher efficiency of mitochondrial oxidative phosphorylation than laying-type chickens. Comp. Biochem. Physiol. A. 159, 75-81.

Functional roles of aquaporins and urea transporters in urea flux across the ruminal epithelium

M.E. Walpole[1], B.L. Schurmann[1], P. Górka[1,2], G.B. Penner[1], M.E. Loewen[3] and T. Mutsvangwa[1]
[1]Department of Animal and Poultry Science, University of Saskatchewan, Saskatoon, SK, S7N 5A8, Canada; matthew.walpole@usask.ca
[2]Department of Animal Nutrition, University of Agriculture, 31-121 Krakow, Poland
[3]Department of Biomedical Sciences, University of Saskatchewan, Saskatoon, SK, S7N 5A8, Canada

Introduction

In ruminants, urea that is recycled to the rumen is an important source of N for microbial growth. Urea transport (UT-B) proteins facilitate urea movement across the ruminal epithelium, although other mechanisms must be involved as inhibiting UT-B does not completely abolish urea transport (Stewart *et al.*, 2005). Of the aquaporins (AQP), a family of membrane-spanning proteins predominantly involved in water transport, AQP-3, -7, and -10 are also permeable to urea (Litman *et al.*, 2009); however, it is not clear if AQP contribute to urea transport across the ruminal epithelium. Røjen *et al.* (2011) observed that mRNA abundance for AQP-3, -7, and -10 in ruminal epithelium was altered by dietary N content, suggesting that AQP might be involved in trans-epithelial urea transport. Increasing ruminal carbohydrate digestion improves urea transfer to the rumen (Reynolds and Kristensen, 2008); however, the mechanisms responsible for this response remain obscure. Our objectives were to determine: (1) the relative functional roles of AQP and UT-B in ruminal urea transport; and (2) if functional adaptation of AQP and UT-B occurred in response to increased diet digestion.

Material and methods

Twenty five weaned Holstein bull calves (n=5) were assigned to a control diet (CON; 91.5% hay and 8.5% vitamin-mineral premix) or a medium grain diet (MGD; 41.5% barley grain, 50% hay, and 8.5% vitamin-mineral premix) that was fed for 3 (G3), 7 (G7), 14 (G14), or 21 (G21) d. All calves were fed at 2.25% BW at 0800 h. Calves were killed at 1000 h and ruminal epithelium was collected for mounting in Ussing chambers under short-circuit conditions and for analysis of mRNA abundance of UT-B and AQP-3, -7, and -10. To mimic physiologic conditions, the mucosal buffer (pH 6.2) contained no urea, while the serosal buffer (pH 7.4) contained 1 m*M* urea. The serosal-to-mucosal fluxes of ^{14}C-urea ($J_{sm\text{-}urea}$; 26 kBq/10 ml) and ^{3}H-mannitol ($J_{sm\text{-}mannitol}$; 37 kBq/10 ml) were measured, with $J_{sm\text{-}mannitol}$ being used as an indicator of hydrophilic movement. Serosal addition of phloretin (1 m*M*) was used to inhibit UT-B-mediated urea transport, while $NiCl_2$ (1 m*M*) was used to inhibit AQP-mediated urea transport. Gene transcript abundance was measured using real-time qPCR, with GAPDH as a control. Data were analyzed using Proc Mixed as a randomized complete block design, with polynomial contrasts used to test for linear, quadratic, and cubic effects of the duration of adaptation to the MGD. Correlation analysis was performed to examine the relationships between $J_{sm\text{-}urea}$ and mRNA abundance.

Results and discussion

The $J_{sm\text{-}urea}$ tended (P=0.075) to increase linearly as the duration of adaptation to MGD increased (Table 1). Phloretin- (P=0.54) and $NiCl_2$-sensitive (P=0.64) $J_{sm\text{-}urea}$ were not affected by diet. Phloretin-insensitive $J_{sm\text{-}urea}$ tended to increase linearly (P=0.075) as the duration of adaptation to MGD increased; however, $NiCl_2$-insensitive $J_{sm\text{-}urea}$ was unaffected by diet (P=0.11; data not shown). Across treatments, the addition of phloretin or $NiCl_2$ reduced (P<0.001) the $J_{sm\text{-}urea}$ and, when both inhibitors were added simultaneously, $J_{sm\text{-}urea}$ was further reduced (P<0.001; Figure 1). Gene transcript abundance for AQP-3 (P=0.001) and UT-B (P=0.007) increased linearly as the duration of MGD adaptation increased (Table 1). Gene transcript abundance for AQP-3 (P=0.011)

Table 1. Total $J_{sm\text{-}urea}$ (nmol/cm²/h) and mRNA abundance (fold-change relative to control) for UT-B and AQP in bovine ruminal epithelium from animals adapted to grain.

	Days on grain					SEM	Contrasts			
Item	0	3	7	14	21		0 vs. all	Linear	Quad	Cubic
$J_{sm\text{-}urea}$	107.7	112.6	112.8	122.8	144.2	14.5	0.36	0.075	0.60	0.85
Relative mRNA abundance, arbitrary units										
AQP-3	1.00	1.55	2.40	4.47	4.68	0.72	0.014	0.001	0.45	0.47
AQP-7	1.00	0.03	0.04	0.13	0.15	0.29	0.015	0.20	0.12	0.18
AQP-10	1.00	0.13	0.11	-0.02	0.01	0.32	0.019	0.090	0.17	0.38
UT-B	1.00	1.38	1.35	2.44	3.02	0.52	0.096	0.007	0.85	0.78

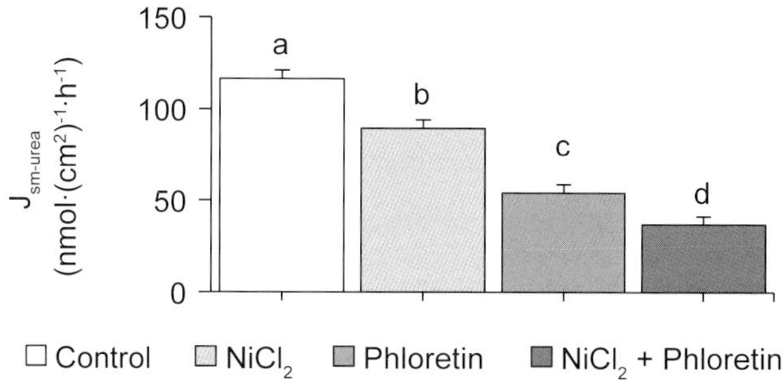

Figure 1. Total, $NiCl_2$-sensitive, and phloretin-sensitive $J_{sm\text{-}urea}$ across bovine ruminal epithelium.

and UT-B (P=0.019) was correlated with $J_{sm\text{-}urea}$. For AQP-7 (P=0.015) and AQP-10 (P=0.019), gene transcript abundance in animals fed MGD was lower relative to those fed CON (Table 1). These results demonstrate that both AQP and UT-B have functional roles in urea transport across the ruminal epithelium, but more work is needed to understand the regulation of these urea transport mechanisms.

References

Litmann, T., R. Søgaard, and T. Zeuthen, 2009. Ammonia and urea permeability of mammalian aquaporins. Handb. Exp. Pharmacol. 190, 327-358.

Reynolds, C. K., and N. B. Kristensen, 2008. Nitrogen recycling through the gut and the nitrogen economy of ruminants: an asynchronous symbiosis. J. Anim. Sci. 86, E293-E305.

Røjen, B., A. S. B. Poulsen, P. K. Theil, R. A. Fenton, and N. B. Kristensen, 2011. Effects of dietary nitrogen concentration on messenger RNA expression and protein abundance of urea transporter-B and aquaporins in ruminal papillae from lactating Holstein cows. J. Dairy Sci. 94, 2587-2591.

Stewart, G. S., C. Graham, S. Cattell, T. P. L. Smith, and C. P. Smith, 2005. UT-B is expressed in bovine rumen: potential role in ruminal urea transport. Am. J. Physiol. Regul. Integr. Comp. Physiol. 289, R605-R612.

Modulation of insulin signaling by n-3 PUFA in chicken liver

S. Tesseraud[1], S. Métayer-Coustard[1], P. Chartrin[1], D. Hermier[2], N. Simon[3], C. Peyronnet[3], M. Lessire[1] and E. Baéza[1]
[1]INRA, UR83 Recherches Avicoles, 37380 Nouzilly, France; sophie.tesseraud@tours.inra.fr
[2]UR914 AgroParisTech-INRA PNCA, 75231 Paris, France
[3]ONIDOL, 75008 Paris, France

Introduction

N-3 polyunsaturated fatty acids (PUFA) are crucial for normal development and organ functioning in vertebrates. Their pleiotropic effects are well documented as regulators of lipid and glucose metabolism, but less known with regard to protein metabolism. Recent studies indicate that the consumption of long-chain PUFA (LC-PUFA) promotes muscle protein anabolism both in farm animals (Gingras *et al.*, 2007) or humans (Smith *et al.* 2011a,b). These studies suggest that n-3 PUFA may act by increasing insulin sensitivity rather than by regulating gene transcription. It should be noted that these studies were performed using n-3 LC-PUFA abundant in fish oils (20 and 22 carbons) and effects of the vegetal precursor α-linolenic acid (ALA, C18:3 n-3) have never been explored. Moreover, the consequences on other tissues than muscle are unknown. Our aim was to investigate the role of n-3 PUFA (ALA or LC-PUFA) on liver protein metabolism in chickens, by focusing on their potential function as co-regulators of the insulin signaling cascade.

Material and methods

Ross male broilers were divided into 3 dietary treatments. Diets were isoproteic (22% CP), isoenergetic (3,000 kcal ME/kg) and had similar lipid supply (6%) with different lipid sources: oleic sunflower oil rich in C18:1 n-9 (control group), fish oil rich in LC-PUFA, and rapeseed and linseed oils providing ALA. At 3 weeks of age, we studied insulin signaling cascade in the liver compared to the *Pectoralis major* (*PM*) muscle in chickens submitted to an *i.v.* injection of insulin or saline. Liver and *PM* muscle lysates were prepared as previously described (Duchene *et al.*, 2008) and subjected to SDS-PAGE gel electrophoresis and western blotting using the appropriate phospho-specific antibody. After washing, membranes were incubated with an Alexa Fluor secondary antibody. Bands were visualized and quantified by the Odyssey® infrared Imaging System. Values are presented as means ±SEM. Data were subjected to ANOVA to detect significant intergroup differences. The means were compared by Fisher's least significant difference test in the case of a significant effect.

Results and discussion

Providing diets enriched in n-3 PUFA, i.e. rich in LC-PUFA as in ALA, for 3 weeks improved chicken body weight (Table 1). Liver weights in the LC-PUFA group were significantly higher than those of the other two groups ($P<0.001$). The *PM* muscle was heavier in chickens fed the ALA diet compared to the other two diets ($P<0.05$). Our data indicated that n-3 PUFA led to different activation patterns of insulin signaling in the liver and the muscle (Figure 1). In the *PM* muscle, ALA-enriched diet may improve insulin sensitivity, with a greater activation of the insulin-induced S6K1/S6 pathway involved in mRNA translation into proteins, thereby potentially increasing muscle protein synthesis and growth. In the liver, conversely to *PM* muscle, insulin-induced phosphorylation levels of S6 were lower in the ALA and LC-PUFA groups compared to the Oleic group. These findings suggest that n-3 PUFA had a tissue-specific effect on insulin signaling. The basic mechanisms sustaining the tissue specific effects of n-3 PUFA (ALA and/or LC-PUFA) remain unclear. One hypothesis is that these differential effects may be due to some differences in membrane fatty acid composition between the liver and muscle, which still requires investigation.

Table 1. Body and tissue weights of 3-wk-old chickens fed Oleic, ALA and LC-PUFA experimental diets and injected intravenously with either insulin or saline.

Diet (D)[1]	Oleic		ALA		LC-PUFA		SEM	P		
Treatment (T)[2]	C	I	C	I	C	I		D	T	D×T
BW, g	986	975	1,010	1,009	1,010	1,006	9.1	<0.01	0.51	0.85
Liver weight, g	22.60	22.68	21.57	22.72	24.84	25.25	0.56	<0.001	0.26	0.65
PM weight, g	62.20	64.08	66.09	68.20	62.55	62.97	1.83	<0.05	0.34	0.89

[1] Oleic, diet containing oleic sunflower oil rich in 18:1; ALA, diet rich in ALA provided by rapeseed and linseed oils; LC-PUFA, diet rich in LC n-3 PUFA provided by fish oil.
[2] Treatment, saline (C) and insulin (I).

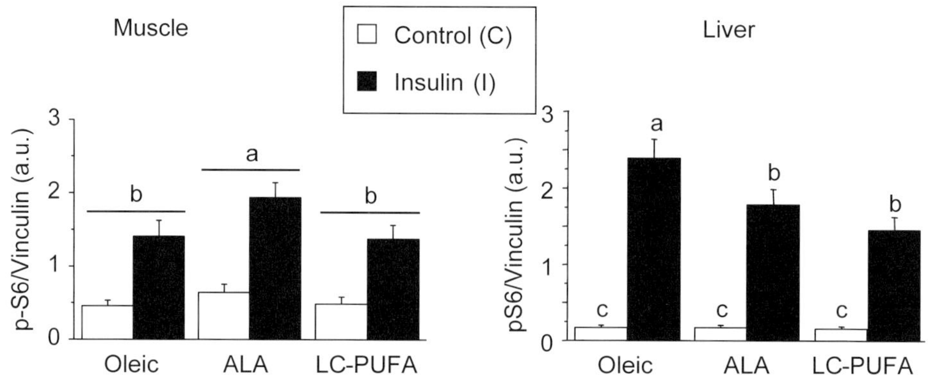

Figure 1. S6 phosphorylation levels in the PM muscle and liver of 3-wk-old chickens fed Oleic, ALA and LC-PUFA diets and injected intravenously with either insulin or saline (control). There was a Diet × Treatment interaction in the liver (letters indicate differences between the 6 groups) but not in the PM muscle (letters indicate differences between diets).

In conclusion, our results suggest that n-3 PUFA can act as co-regulators of insulin anabolic signaling cascade in growing chickens. Further studies more specifically focused on the underlying mechanisms and their consequences on anabolic processes are likely to have new nutritional applications in the future.

References

Duchene, S., S. Métayer, E. Audouin, K. Bigot, J. Dupont, and S. Tesseraud, 2008. Refeeding and insulin activate the AKT/p70S6 kinase pathway without affecting IRSI tyrosine phosphorylation in chicken muscle. Domest. Anim. Endocrinol. 34, 1-13.

Gingras, A.A., P.J. White, P.Y. Chouinard, P. Julien, T.A. Davis, L. Dombrowski, Y. Couture, P. Dubreuil, A. Myre, K. Bergeron, A. Marette, and M.C. Thivierge, 2007. Long-chain omega-3 fatty acids regulate bovine whole-body protein metabolism by promoting muscle insulin signalling to the Akt-mTOR-S6K1 pathway and insulin sensitivity. J. Physiol. 579, 269-284.

Smith, G.I., P. Atherton, D.N. Reeds, B.S. Mohammed, D. Rankin, M.J. Rennie, and B. Mittendorfer, 2011a. Omega-3 polyunsaturated fatty acids augment the muscle protein anabolic response to hyperinsulinaemia-hyperaminoacidaemia in healthy young and middle-aged men and women. Clin. Sci. 121, 267-278.

Smith, G.I., P. Atherton, D.N. Reeds, B.S. Mohammed, D. Rankin, M.J. Rennie, and B. Mittendorfer, 2011b. Dietary omega-3 fatty acid supplementation increases the rate of muscle protein synthesis in older adults: a randomized controlled trial. Am. J. Clin. Nutrition 93, 402-12.

Hormone and metabolite levels differ between cerebrospinal fluid and plasma of periparturient dairy cows

T. Laeger[1], H. Sauerwein[2], C.C. Metges[1] and B. Kuhla[1]
[1]Leibniz Institute for Farm Animal Biology (FBN), Institute of Nutritional Physiology 'Oskar Kellner', 18196 Dummerstorf, Germany; laeger@fbn-dummerstorf.de
[2]University of Bonn, Institute of Animal Science, Physiology & Hygiene Unit, 53115 Bonn, Germany

Introduction

Hormones and metabolites act as satiety signals in the brain and play an important role in control of feed intake. These signals can reach the hypothalamus and brainstem, two major centers of feed intake regulation, via the cerebrospinal fluid (CSF). The contribution of putative anorexic or orexic CSF signals, possibly leading to the insufficient feed intake by high-yielding dairy cows during early lactation has not been studied so far. Therefore, the aim of this study was to elucidate associations existing between both plasma and CSF hormones and metabolites during the periparturient period.

Material and methods

Ten multiparous German Holstein cows were fed *ad libitum* a total mixed ration offered 2-times daily. Energy intake, milk yield, and energy balance (EB) were determined. CSF from the spinal cord was obtained by lumbar puncture after local anesthesia, and blood was withdrawn from the jugular vein before morning feeding on d -20, -10, +1, +10, +20 and +40 relative to calving. Concentrations of β-hydroxybutyric acid (BHBA), glucose, lactate, and non-esterified fatty acids (NEFA) were determined photometrically in blood plasma and CSF. Amino acid concentrations were analyzed by HPLC (C18 column with automated pre-column derivatization and fluorescence detection (Kuhla *et al.*, 2011). The concentration of leptin was measured by a double antibody enzyme immunoassay (Sauerwein *et al.*, 2004). Concentrations of acylated ghrelin were determined by RIA (Börner *et al.*, 2013). All data are presented as means and were analyzed by the residual option in the GLIMMIX procedure of SAS (version 9.3) to construct the block diagonal structure of the residual covariance matrix for each animal. The one-way ANOVA was used to analyze zootechnical variables and two-way ANOVA for the analysis of metabolites and hormones (fixed effects: body fluid and time). A model with the fixed effect body fluid and the covariate day was used to estimate the slopes for plasma and CSF variables for evaluating their parallel course.

Results and discussion

Energy intake declined starting 5 days before parturition (>80 MJ NE_L/d) until calving (51 MJ NE_L/d) and increased thereafter to reach 126 MJ NE_L/d until the 40th day after parturition ($P<0.05$). Energy-corrected milk increased rapidly within the first week after parturition and reached a plateau at around 42 kg/d. All cows were in positive EB prepartum and in negative EB postcalving until the end of the sampling period. While ghrelin did not change during the periparturient period, leptin concentrations decreased after calving and remained low in early lactation in both CSF and plasma (Table 1). Considering the well-known anorexic effect of leptin, diminished leptin concentrations do not explain insufficient feed intake in early lactation. Next to leptin, the metabolites BHBA, lactate, and NEFA showed different responses during the periparturient period, either in plasma or CSF (Table 1). This is presumably due to a different permeability of the blood-brain barrier bordering the periphery from brain and due to an altered brain energy metabolism. In addition, Gln concentrations were increased in CSF on d 1 *postpartum*. Gln may be utilized by the brain for ATP production, the latter signals for satiety (Kola, 2008). Therefore, Gln could act as central anorexigenic signal around parturition. Furthermore, BHBA (Sun *et al.*, 1997) and branch-chained amino acids such as Leu (Morrison *et al.*, 2007) have been shown to exert anorexic responses when administered into the brain. The increased CSF BHBA and Leu concentration in early lactation (Table 1) can be suggested to contribute to the insufficient energy intake after calving.

Table 1. Periparturient plasma (P) and CSF concentrations of hormones and metabolites.

	Fluid[1]	-20	-10	+1	+10	+20	+40
Ghrelin (pg/ml)	P	209	236	127	185	184	125
	CSF	3.6	3.9	4.8	3.5	4.1	3.2
Leptin (ng/ml)[2]	P*	8.5	8.1	5.6	4.3	4.7	5.0
	CSF*	0.6	0.6	0.5	0.4	0.4	0.4
BHBA (mM)[2]	P*	0.3	0.3	0.8	0.9	1.1	1.0
	CSF*	0.1	0.2	0.2	0.2	0.3	0.2
Glucose (mM)	P*	4.3	4.3	4.2	3.6	3.6	3.6
	CSF*	2.7	2.7	2.6	2.0	2.1	2.0
Lactate (mM)[2]	P*	1.5	1.7	1.0	0.7	0.6	0.7
	CSF*	2.2	2.1	2.0	1.8	1.8	1.7
NEFA (μM)[2]	P*	257	402	1,232	1,890	1,681	1,437
	CSF	22.0	14.3	15.1	11.4	8.6	24.1
Gln (μM)	P*	340	355	338	288	262	241
	CSF*	495	505	581	497	450	451
Ile (μM)	P	102	116	68	105	117	112
	CSF*	0.8	1.3	0.9	1.2	1.5	1.6
Leu (μM)	P	134	139	83	121	132	117
	CSF#	2.1	3.3	2.2	2.8	3.3	3.0
Val (μM)	P	244	255	142	212	253	240
	CSF*	2.5	3.8	2.6	3.1	4.4	4.6

[1] * indicates a significant slope ($P \leq 0.05$) of the trend line, whereas # indicates $0.05 < P < 0.1$.

[2] Indicates significantly different slopes of plasma versus CSF trend lines.

References

Börner, S., M. Derno, S. Hacke, U. Kautzsch, C. Schäff, S. Thanthan, H. Kuwayama, H. M. Hammon, M. Röntgen, R. Weikard, C. Kühn, A. Tuchscherer and B. Kuhla, 2013. Plasma ghrelin is positively associated with body fat, liver fat and milk fat content but not with feed intake of dairy cows after parturition. J. Endocrinol. 216, 217-229.

Kola, B. 2008. Role of AMP-activated protein kinase in the control of appetite. J. Neuroendocrinol. 20, 942-951.

Kuhla B., G. Nürnberg, D. Albrecht, S. Görs, H. M. Hammon and C. C. Metges, 2011. Involvement of skeletal muscle protein, glycogen and fat metabolism in the adaptation on early lactation of dairy cows. J. Proteome Res. 10, 4252-4262.

Morrison, C. D., X. Xi, C. L. White, J. Ye and R. J. Martin, 2007. Amino acids inhibit Agrp gene expression via an mTOR-dependent mechanism. Am. J. Physiol. Endocrinol. Metab. 293, E165-E171.

Sauerwein, H., U. Heintges, A. Hennies, T. Selhorst and A. Daxenberger, 2004. Growth hormone induced alterations of leptin serum concentrations in dairy cows as measured by a novel enzyme immunoassay. Livest. Prod. Sci. 87, 189-195.

Sun, M., R. J. Martin and G. L. Edwards, 1997. ICV beta-hydroxybutyrate: effects on food intake, body composition, and body weight in rats. Physiol. Behav. 613, 433-436.

Hepatic α_1- and β_2-adrenergic receptors in dairy cows with different fat mobilization during early lactation

H.M. Hammon, U. Kautzsch, C. Reiko, B. Kuhla, M. Röntgen, M. Derno and C.C. Metges
Nutritional Physiology 'Oskar Kellner', Leibniz Institute for Farm Animal Biology (FBN), Dummerstorf, Germany; hammon@fbn-dummerstorf.de

Introduction

Catecholamines increase around parturition in dairy cows and the hepatic adrenergic system is involved in metabolic adaptation during early lactation, e. g., by stimulation of hepatic glucose production (glycogenolysis and gluconeogenesis) after parturition to cover glucose demands (McDowell, 1983; Nonogaki, 2000; Weber *et al.*, 2013). Effects of adrenaline and noradrenaline are mediated by hepatic α_1- and β_2-adrenergic receptors (AR), which are the dominant subtypes in bovine liver (Carron *et al.*, 2004; Ontsouka *et al.*, 2006). The objective of the present study was to investigate α_1- and β_2-adrenergic receptors (AR) in liver of dairy cows varying in fuel selection during the transition from pregnancy to lactation. Cows differed in their extent of whole body fat mobilization and varied in their respiration quotient (RQ) before and after parturition (Börner *et al.*, 2013). The hypothesis was tested that hepatic α_1- and β_2-AR are involved in regulation of hepatic energy metabolism, that the density of hepatic α_1- and β_2-AR depend on the metabolic type of the cow (high versus low fat mobilization post partum [pp]) and that the number of hepatic AR are related to RQ and carbohydrate (COX) and fat oxidation (FOX) pp in dairy cows.

Material and methods

Nineteen German Holstein cows (>10,000 kg milk/305 d; $\geq 2^{nd}$ lactation) were classified by liver fat concentration (LFC) pp in low (L; <240 mg total fat/g dry matter [DM]; n=9) and high (H; >240 mg total fat/g DM; n=10) fat mobilization around parturition. Cows were studied from dry off up to 40 d pp and were fed total mixed rations *ad libitum*. Milk yield increased pp to 44.1 kg/d by the 5th wk of lactation and did not differ between groups. Liver biopsies were taken on d 3, 18, and 30 pp to measure LFC. During the 2nd wk pp, respiration quotient (RQ) and COX and FOX were determined from gaseous exchange measurements in respiration chambers (Derno *et al.*, 2013). Data for RQ from 16 cows (out of 19 cows) were recently presented by Börner *et al.* (2013). In addition, [U-^{13}C]-glucose was infused to measure endogenous glucose production (eGP) and glucose oxidation (GOx). Cows were slaughtered at d 40 pp and liver samples were snap frozen. For AR measurements, membrane suspensions (2 mg protein/ml) were prepared. Saturation binding assays with increasing concentrations of (^{3}H)-prazosin and (^{3}H)-CGP-12177 were performed for the determination of α_1- and β_2-AR, respectively (Carron *et al.*, 2005; Ontsouka *et al.*, 2006). Maximal binding capacity (B_{max}) and binding affinity (K_D) were calculated using unlabeled phentolamine (α_1-AR) and propranolol (β_2-AR) as competitors. Data are presented as LSMeans ± SE and were analyzed by General Linear Model of SAS with LFC as fixed effect. Correlations were calculated between B_{max} and K_D of AR and eGP, GOx, RQ, COX, and FOX, respectively.

Results and discussion

B_{max} and K_D for hepatic α_1- and β_2-AR were not affected by LFC, respectively. LSMeans for H and L cows were 76.4±9.7 fmol/mg protein (B_{max}) and 0.29±0.06 nM (K_D) for α_1-AR and 73.9±7.6 fmol/mg protein (B_{max}) and 0.61±0.12 nM (K_D) for β_2-AR. In addition, body fat mobilization did not influence eGP and GOx (LSMeans for H and L cows were 1.07±0.04 mmol/[kg BW × h] for eGP and 0.11±0.01 mmol/[kg BW × h] for GOx). However, there was a trend for a positive correlation between B_{max} of α_1-AR and eGP (r=0.61; P=0.1) in H cows, but a strong negative correlation between B_{max} of α_1-AR and eGP (r=-0.93; P<0.001) in L cows. Although hepatic α_1-AR was not influenced by LFC, the relationship between α_1-AR and eGP depended on the degree of fat mobilization after calving. Interestingly, B_{max} and K_D of hepatic β_2-AR were not related to eGP in our cows, but hepatic β_2-AR is involved in stimulation of eGP in human (Rizza *et al.*, 1980). We have recently found a close relationship between hepatic α_1-AR, but not β_2-AR, and eGP in neonatal calves, which may support the present findings that hepatic α_1-AR more than β_2-AR may be involved in regulation of eGP in cattle (Rohrbeck *et al.*, 2012).

Independent of LFC, B_{max} of α_1-AR correlated positively with GOx (r=0.56; $P<0.05$) and COX (r=0.56; $P<0.05$), but negatively with FOX (r=-0.62; $P<0.05$). There were no significant correlations between hepatic β_2-AR (B_{max}) and data for substrate oxidation. These data again point to the importance of hepatic α_1-AR expression for regulation of fuel utilization in dairy cows, whereas adrenergic effects mediated by hepatic β_2-AR may be not regulated at the receptor density level.

In conclusion, our data indicate that the hepatic adrenergic system is involved in regulation of fuel selection in dairy cows during early lactation. Interestingly, α_1-AR, but not β_2-AR, is the dominant adrenergic receptor type that takes part in regulation of glucose supply and substrate oxidation.

Acknowledgements

The research was funded by German Research Foundation (DFG).

References

Börner, S., M. Derno, S. Hacke, U. Kautzsch, C. Schäff, S. ThanThan, H. Kuwayama, H. M. Hammon, M. Röntgen, R. Weikard, C. Kühn, A. Tuchscherer, and B. Kuhla, 2013. Plasma ghrelin is positively associated with body fat, liver fat and milk fat content, but not with feed intake of dairy cows after parturition. J. Endocrinol. 216, 217-229.

Carron, J., C. Morel, H. M. Hammon, and J. W. Blum, 2005. Ontogenetic development of mRNA levels and binding sites of hepatic β-adrenergic receptors in cattle. Domest. Anim. Endocrinol. 28, 320-330.

Derno, M., G. Nürnberg, P. Schön, A. Schwarm, M. Röntgen, H. M. Hammon, C. C. Metges, R. M. Bruckmaier, and B. Kuhla, 2013. Short-term feed intake is regulated by macronutrient oxidation in lactating Holstein cows. J. Dairy Sci. 96, 971-980.

McDowell, G. H., 1983. Hormonal control of glucose homoeostasis in ruminants. Proc. Nutr. Soc. 42, 149-167.

Nonogaki, K., 2000. New insights into sympathetic regulation of glucose and fat metabolism. Diabetologia 43, 533-549.

Ontsouka, E. C., Y. Zbinden, H. M. Hammon, and J. W. Blum, 2006. Ontogenesis of mRNA levels and binding sites of hepatic β-adrenoceptors in young cattle. Domest. Anim. Endocrinol. 30, 170-184.

Rizza, R. A., P. E. Cryer, M. W. Haymond, and J. E. Gerich, 1980. Adrenergic mechanisms of catecholamine action on glucose homeostasis in man. Metabolism 29, 1155-1163.

Rohrbeck, D., J. Steinhoff-Wagner, E. Kanitz, and H. M. Hammon, 2012. Ontogenic changes of hepatic glucocorticoid and $alpha_1$- and $beta_2$-adrenergic receptors in neonatal calves. J Dairy Sci. 95 (Suppl 2), 546-547.

Weber, C., C. Hametner, A. Tuchscherer, B. Losand, E. Kanitz, W. Otten, S. P. Singh, R. M. Bruckmaier, F. Becker, W. Kanitz, and H. M. Hammon, 2013. Variation in fat mobilization during early lactation differently affects feed intake, body condition, and lipid and glucose metabolism in high-yielding dairy cows. J. Dairy Sci. 96, 165-180.

Insulin signaling of glucose uptake in skeletal muscle of lactating dairy cows

S.K. Spachmann, U. Schönhusen, B. Kuhla, M. Röntgen and H.M. Hammon
Institute of Nutritional Physiology 'Oskar Kellner', Leibniz Institute for Farm Animal Biology (FBN), Dummerstorf, Germany; spachmann@fbn-dummerstorf.de

Introduction

Insulin response in skeletal muscles is thought to be impaired in dairy cows during early lactation to favor nutrient supply, especially glucose, towards the mammary gland. However, the molecular mechanisms of insulin action on glucose metabolism in cows and in other ruminants are still not completely understood, particularly during early lactation. Previous studies provide evidence that peripartal insulin resistance in skeletal muscle of ruminants is caused by post-receptor changes of the insulin signaling pathway. The protein content of glucose transporter 4 (GLUT4) is reduced in skeletal muscle of goats and cows at begin of lactation (Balage *et al.*, 1997; Kuhla *et al.*, 2011). On the other hand, lactation does not have an effect on the number, the affinity or the activity of the tyrosine kinase of the insulin receptor (InsR) in sheep (Wilson *et al.*, 1996) and goats (Balage *et al.*, 1992). Furthermore, Duhlmeier *et al.* (2005) showed differences in protein amount of GLUT 1 and 4 in oxidative and glycolytic muscles of lactating cows. We have tested the hypothesis that insulin signaling in skeletal muscle of dairy cows is related to the stage of lactation and differs among various muscle types. Furthermore, we have investigated the influence of different post-calving metabolic status of dairy cows as reflected by divergent fat mobilization and liver fat concentration (LFC) on insulin signaling in muscle tissue.

Material and methods

German Holstein cows (n=19; >10,000 kg milk/305 d; $\geq 2^{nd}$ lactation) were fed twice daily a total mixed ration according to their physiological state. Liver biopsies were taken on d 3, 18, and 30 post partum (pp) to determine LFC. Depending on their mean LFC pp, cows were retrospectively grouped in either low LFC (LLFC) (n=9) or high LFC (HLFC) (n=10). Muscle biopsies of M. semitendinosus (ST) were taken on d 17 ante partum (ap), and d 3 and 30 pp to perform relative quantitative real time RT-PCR using the LightCycler 2.0 (Roche Molecular Biochemicals, Mannheim, Germany) and SYBR Green I as the fluorescent dye (Hammon *et al.*, 2009) to analyze parameters involved in insulin signaling: insulin receptor (InsR), insulin receptor substrate 1 (IRS1), regulatory subunit (p85) and catalytic subunit (p110) of phosphatidylinositol-3-kinase (PI3K), insulin sensitive glucose transporter (GLUT4) and insulin-independent glucose transporter (GLUT1). Hydroxymethylbilane synthase (HMBS) and splicing factor 3 subunit 1 (SF3A1) were used as reference genes (Pérez *et al.*, 2008). Corresponding blood samples for measurement of plasma insulin concentrations were taken before tissue sampling on d 17 ap and d 3 and 30 pp, respectively. Cows were slaughtered on d 40 pp and samples of ST, M. *longissimus dorsi* (LD), and M. masseter (MA) were collected to investigate the insulin signaling pathway on mRNA level in different muscle types. Additionally in all three muscle types the protein abundance of GLUT1 and GLUT4 was determined by Western Blot. Data are presented as LSmeans ± SE and were analyzed by the Mixed Model of SAS 9.2 with group and muscle type or time as fixed effects.

Results and discussion

Plasma insulin concentrations decreased significantly from ap to pp ($P<0.001$) but were not affected by LFC (pooled LSmean ± SE: 0.75±0.1 μg/l, 0.25±0.1 μg/l and 0.35±0.1 μg/l on d 17 ap, and d 3 and d 30 pp). In ST muscle biopsies, mRNA expression of all parameters except for GLUT4 changed with time, with GLUT1 increasing after calving ($P<0.01$), InsR, p85, and p110 being highest around calving ($P<0.01$), and IRS1 tending to decrease after calving ($P<0.06$). Furthermore, InsR showed a group × time interaction ($P<0.05$) and p85 was higher expressed in LLFC than in HLFC ($P<0.01$) during the transition from pregnancy to lactation. These results confirm decreased plasma insulin concentrations pp and the hypothesis that there are insulin signaling post-receptor changes during lactation in bovine muscle tissue, but the gene expression of GLUT4 is not involved in the pp insulin resistance in muscle tissue of cows (Komatsu *et al.*, 2006). However, there might be post-transcriptional regulation of GLUT4 in skeletal muscle, because the protein expression of GLUT4 is reduced with onset of lactation (Kuhla *et al.*, 2011). This assumption is supported by the observed increase of p85 abundance around calving which may lead to a shift from p85-p110 heterodimers to p85 monomers, resulting in a decline of PI3K activity and consequently reduced insulin signaling on GLUT4 translocation (Ueki *et al.*, 2002). Additionally, the tendency of IRS1 to decrease after calving suggests reduced insulin signaling, because IRS1 is the first downstream signal of the intracellular insulin signaling cascade and hence superior for the further signaling steps regulating GLUT4 translocation. The increased GLUT1 mRNA expression after calving in this study supports the idea of an insulin-independent glucose transport for maintaining a basic glucose supply for glycolytic muscle cells during lactation (Duhlmeier *et al.*, 2005). The increased expression of InsR on d 3 pp, mainly in HLFC cows, may indicate an inverse response to the reduced plasma insulin concentrations around parturition (Hammon *et al.*, 2009).

At slaughter on d 40 pp, mRNA expression of all parameters except p110 depend on the muscle type with GLUT4 and InsR being highest and p85 and IRS1 being lowest in MA ($P<0.01$). GLUT1 and GLUT4 were higher ($P<0.05$; $P<0.01$), but IRS1 and p85 were lower ($P<0.01$) expressed in ST than in LD. LFC affected p85 with higher ($P<0.05$) mRNA abundance in LLFC than HLFC. The protein concentrations of GLUT4 corresponded with its mRNA concentrations in all three muscle types, and the protein abundance of GLUT1 were significantly lower ($P<0.01$) in MA than in ST and LD. Our results are in consistence with findings in rodents describing that insulin signaling is related to the type of muscle (Song *et al.*, 1999). Results on GLUT1 and GLUT4 protein abundance were supported by those of Duhlmeier *et al.* (2005).

In conclusion our results indicate that insulin signaling in bovine skeletal muscle depends on skeletal muscle type and stage of lactation, but is barely affected by post-calving fat mobilization.

References

Balage, M., C. Sornet, and J. Grizard. 1992. Insulin receptor binding and kinase activity in liver and skeletal muscles of lactating goats. Am J Physiol. 262(5 Pt 1):E561-568.

Balage, M., J.F. Hoquette, B. Graulet, P. Ferré, and J. Grizard. 1997. Skeletal muscle glucose transporter (GLUT-4) protein is decreased in lactating goats. Anim Sci. 65:257-265.

Duhlmeier, R., A. Hacker, A. Widdel, W. von Engelhardt, and H. P. Sallmann. 2005. Mechanisms of insulin-dependent glucose transport into porcine and bovine skeletal muscle. Am J Physiol Regul Integr Comp Physiol. 289:R187-R197.

Hammon, H. M., G. Stürmer, F. Schneider, A. Tuchscherer, H. Blum, T. Engelhard, A. Genzel, R. Staufenbiel, and W. Kanitz. 2009. Performance and metabolic and endocrine changes with emphasis on glucose metabolism in high-yielding dairy cows with high and low fat content in liver after calving. J. Dairy Sci.92:1554-1566.

Komatsu, T., F. Itoh, S. Kushibiki, and K. Hodate. 2005. Changes in gene expression of glucose transporters in lactating and nonlactating cows. J. Anim. Sci. 83:557-564.

Kuhla B., G. Nürnberg, D. Albrecht, S. Görs, H.M. Hammon, and C.C. Metges, 2011. Involvement of skeletal muscle protein, glycogen and fat metabolism in the adaptation on early lactation of dairy cows. J. Proteome Res. 10, 4252-4262.

Pérez, R., I. Tupac-Yupanqui, and S. Dunner. 2008. Evaluation of suitable reference genes for gene expression studies in bovine muscular tissue. BMC Mol Biol. 9:79.

Song, X.M., J.W. Ryder, Y. Kawano, A.V. Chibalin, A. Krook, and J.R. Zierath. 1999. Muscle fibre type specificity in insulin signal transduction. Am J Physiol. 277(6 Pt 2):R1690-1696.

Ueki, K., D.A. Fruman, S.M. Brachmann, Y. Tseng, L.C. Cantley, and C.R. Kahn. 2002. Molecular balance between the regulatory and catalytic subunits of phosphoinositide 3-kinase regulates cell signaling and survival. *Mol. Cell. Biol. 22(3):965-977.*

Wilson, L.A., S.E. Mills, E. Finley, E. Kilgour, P.J. Buttery, and R.G. Vernon. 1996. Effect of lactation on insulin signal transduction in sheep adipose tissue and skeletal muscle. J Endocrinol. 151(3):469-480.

Influence of *DGAT1* polymorphism on response of dairy cows to ruminal supplementation of linseed oil

A.M. van Vuuren[1], L. Kruijt[1], H.C.A. Widjaja[1], A. Klop[1], A.J. Cnossen[1], D. Anjema[2] and J. van Baal[1,3]
[1]Wageningen UR Livestock Research, P.O. Box 65, 8200 AB Lelystad, the Netherlands; ad.vanvuuren@wur.nl
[2]Central Veterinary Institute, P.O. Box 65, 8200 AB Lelystad, the Netherlands
[3]Animal Nutrition Group, Wageningen University, P.O. Box 338, 6700 AH Wageningen, the Netherlands

Introduction

The *DGAT1* gene encodes for the diacylglycerol acyltransferase enzyme which catalyzes the final step in triacylglycerol synthesis. Various polymorphisms have been reported for DGAT1. One of them, the K232A *DGAT1* polymorphism in which lysine as the 232nd amino acid is replaces by alanine, is associated with increased milk yield, lower fat and protein concentrations (Grisart *et al.*, 2001) and altered milk fatty acid (FA) composition (Schennink *et al.*,2007). Milk FA composition is, on the other hand, also affected by nutritional factors including dietary FA composition (Sterk *et al.*, 2011). Until now, no literature data could be retrieved to establish whether the response in milk FA composition of dairy cows to changes in dietary FA composition is influenced by *DGAT1* polymorphism. Therefore, a study was performed with 8 rumen-cannulated dairy cows in mid-lactation with different *DGAT1* polymorphism, receiving either palm fat or linseed oil.

Material and Methods

Eight rumen-cannulated cows were selected and blocked depending on *DGAT1* polymorphism: 4 cows were genotyped as AA and 4 as KK mutant. Cows were allocated to treatments in a change-over designed experiment with 2 periods. Cows were fed a partial mixed ration (49% grass silage, 33% corn silage, 11% soy bean meal, 5% and rapeseed meal and 2% premix on dry matter basis) ad libitum and 1 kg of concentrates during milking. Dietary treatments were palm fat and linseed oil supplementation at 500 g/d, dosed intra-ruminal in 2 equal portions after milking. Each treatment was given for two weeks. During the last 4 milkings of each period, milk yield was recorded and milk samples were taken to determine fat, protein and lactose content and milk FA composition. Data were statistically analyzed by ANOVA with cow and period as random, and genotype and treatment as fixed effects.

Results and discussion

Cows consumed on average 18.4 kg dry matter/day, including 2.8 kg of crude protein/d, 6.0 kg of NDF/d, and 1.6 kg crude fat/d. Compared to KK genotype, cows with AA genotype produced numerically less milk and had a lower milk fat and protein content (Table 1). The lower milk yield for cows with the AA genotype is in contrast with other observations (Grisart *et al.*, 2001; Schennink *et al.*, 2007), but was attributed to the difference in stage of lactation between cows. Compared to KK genotype, milk fat of cows with AA genotype had higher proportions of oleic and linoleic acids, which was compensated by lower proportions of caproic, caprylic and capric acid (Table 2), the latter in agreement with observations by Schennink *et al.* (2007).

As expected, compared to palm fat, feeding linseed oil resulted in a lower milk fat content. Compared to palm fat, feeding linseed oil resulted in higher proportions of C:18 FA in milk fat, which was compensated by lower proportions of C:14 and C:16 FA. Generally, no genotype × fat source interactions could be observed, except for stearic acid and cetoleic acid (C20:1, n11).

Table 1. Effect of lipid supplementation and DGAT1 *polymorphism on milk yield and milk composition of dairy cows.*

Parameter	Treatment (lipid source)				SED	*P*-value		
	Linseed		Palm					
	DGAT1 genotype					G	L	L×G
	AA	KK	AA	KK				
Milk production								
Milk, kg /d	18.0	21.4	18.8	22.0	0.7	0.50	0.26	0.94
Fat, kg/d	0.59	0.81	0.66	1.14	0.07	0.04	0.01	0.05
Protein, kg/d	0.65	0.89	0.68	0.90	0.03	0.18	0.32	0.85
Milk composition								
Fat, g/kg	33.0	40.7	37.3	51.9	2.4	0.02	0.01	0.10
Protein, g/kg	36.5	42.3	36.6	41.5	0.97	0.03	0.61	0.54

[1] L: lipid source; G: *DGAT1* mutation; L×G: interaction.

Table 2. Effect of lipid supplementation and DGAT1 *polymorphism on milk fatty acid profile (g/100 g FA).*

Fatty acid	Treatment (lipid source)				SED	*P*-value		
	Palm		Linseed					
	DGAT1 genotype					G	L	L×G
	AA	KK	AA	KK				
C4:0	3.9	4.2	4.0	4.2	0.07	0.43	0.18	0.35
C6:0	1.9	2.3	1.9	2.3	0.05	0.03	1.00	0.51
C8:0	0.8	1.1	0.9	1.1	0.04	0.04	0.36	0.36
C10:0	1.7	2.2	1.8	2.2	0.12	0.05	0.43	0.79
C12:0	2.3	2.6	2.4	2.8	0.13	0.16	0.12	0.46
C14:0	8.9	9.2	9.5	9.7	0.26	0.67	0.03	0.95
C16:0	32.3	34.1	22.9	25.5	0.92	0.16	<0.001	0.57
C16:1, n7	2.2	2.2	1.4	1.6	0.08	0.90	<0.001	0.15
C18:0	8.6	9.4	10.0	10.0	0.25	0.59	0.01	0.06
C18:1, n9	24.3	20.7	25.8	21.8	0.45	0.03	0.01	0.56
C18:1, *trans*	2.6	2.6	4.2	4.3	0.49	0.91	0.01	0.84
C 18:1 other	1.3	1.2	2.3	2.1	0.14	0.27	<0.001	0.81
C18:2, n6	1.3	1.2	1.5	1.4	0.04	0.01	0.001	1.00
C18:2, c9,t11	0.8	0.7	1.7	1.6	0.20	0.49	0.01	0.93
C18:2, *trans*	1.2	1.0	3.3	3.0	0.26	0.40	<0.001	0.70
C18:3, n3	0.4	0.4	0.5	0.4	0.03	0.28	0.03	0.56
C 20:1, n11	0.1	0.1	0.2	0.2	0.02	0.67	<0.001	0.03

[1] L: lipid source; G: *DGAT1* mutation; L×G: interaction.

Conclusion

Although K232A *DGAT1* polymorphism altered milk composition and milk FA profile, the response in milk composition and milk FA profile to supplementing linseed oil was similar for AA and KK genotype cows.

References

Grisart, B., Coppieters, W., Farnir, F., Karim, L., Ford, C., Berzi, P., Cambisano, N., Mni, M., Reid, S., Simon, P., Spelman, R., Georges, M., Snel, R., 2001. Positional candidate cloning of a QTL in dairy cattle: Identification of a missense mutation in the bovine DGAT1 gene with major effect on milk yield and composition. Genome Res. 12, 222-231.

Schennink, A., Stoop, W.M., Visker, M.H.P.W., Heck, J.M.L., Bovenhuis, H., van der Poel, J.J., van Valenberg, H.J.F. and van Arendonk J.A.M., 2007. DGAT1 underlies large genetic variation in milk-fat composition. Anim. Genet. 38, 467-473.

Correlations between plasma ghrelin and parameters of fat metabolism in early lactating dairy cows

S. Börner[1], M. Derno[1], H.M. Hammon[1], M. Röntgen[1], S. Thanthan[2], H. Kuwayama[2] and B. Kuhla[1]
[1]Institute of Nutritional Physiology 'Oskar Kellner', Leibniz Institute for Farm Animal Biology (FBN), 18196 Dummerstorf, Germany; b.kuhla@fbn-dummerstorf.de
[2]Agriculture and Veterinary Medicine Inada-cho, Department of Life Science and Agriculture, Obihiro University of Obihiro City, Nishi 2-11, Hokkaido 080-8555, Japan

Introduction

The peptide hormone ghrelin is produced in the ruminal and proximal duodenal wall with a small portion of ghrelin being post-translationally modified by fatty acids at Ser3. Both forms (desacyl and acyl ghrelin) are released into the blood stream and have been initially assigned a role in the control of feed intake. While acyl ghrelin increases feed intake, desacyl ghrelin may exert opposing effects. However, there is accumulating evidence for the involvement of ghrelin in fat allocation and utilization involving liver, muscle and adipose tissue (Barazzoni *et al.*, 2005; Jennings *et al.*, 2011; Rodríguez *et al.*, 2009). Therefore, it seems possible that ghrelin may be involved in body fat metabolism of dairy cows during the transition period. Here we examined the hypothesis that ghrelin could play a role in the extent of body fat mobilization during early lactation in dairy cows.

Material and methods

Sixteen Holstein cows (2nd-4th lactation) were kept in respiratory chambers to determine the respiratory quotient (RQ) during two experimental weeks (wk -6 and +2 relative to parturition). Prior to each experimental period, body weight and back fat thickness (BFT) were measured. On the 1st day of each experimental period animals received ad libitum a total mixed ration (TMR) offered twice daily (07:00 and 15:00 h). Dry matter intake (DMI) was recorded daily and milk yield and milk composition was analyzed for each milking. At the beginning of the 2nd day, animals were feed-deprived for 10 h to study the preprandial ghrelin surge. Afterwards, animals were re-fed *ad libitum* to determine 14 h-compensatory feed intake. On the 1st and 2nd day, serial blood samples were taken in 60 min intervals and analyzed for plasma acyl and total (acyl + desacyl) ghrelin, non-esterified fatty acids (NEFA), and triglycerides. On the 5th day, liver biopsies were taken. Based on the liver fat content (LFC) in week 2 *post partum* (pp), cows were retrospectively assigned to a high (H, n=8, LFC>32% fat/g dry matter (DM)) and low (L, n=8, LFC<31% fat/g DM) fat mobilizing group. Data were evaluated by the MIXED procedure of SAS. Linear correlations were calculated using the CORR procedure of Base SAS. The concentration of the plasma ghrelin peak (after 9 h feed-deprivation) was correlated with the subsequent 14 h-compensatory feed/energy intake or with parameters reflecting fat metabolism.

Results and discussion

Ad libitum and 14-h compensatory feed intake did not differ between groups, neither *ante partum* (ap) nor pp. BFT was higher (P=0.01) but RQ was lower in H cows (P<0.05), both ap and pp. During ad libitum feeding, plasma NEFA concentrations were higher in H cows pp (P<0.01), while group differences were not evident during the feed-deprivation period (P=0.2). Preprandial triglycerides were lower pp (P<0.01), whereas preprandial total ghrelin concentration were lower ap (P=0.03) in H cows. Preprandial acyl ghrelin concentration (P<0.01) and the preprandial acyl:total ghrelin ratio (P<0.02) were higher in H cows, both ap and pp. The pp preprandial acyl ghrelin concentration correlated positively with plasma NEFA concentrations (Table 1). Correlations between pp preprandial total ghrelin and parameters reflecting fat metabolism or feed intake were not evident (Table 1). On the contrary, the preprandial acyl:total ghrelin ratio pp correlated positively with LFC, BFT, and milk fat content, but negatively with RQ, plasma NEFA and TG (Table 1). There were no correlations between acyl ghrelin, total ghrelin or the acyl:total ghrelin ratio with DMI or NE_L intake. These results imply that endogenous ghrelin forms are not major players in controlling feed intake in dairy cows. The acyl:total ghrelin ratio is rather suggested to play a prominent role in determining fat allocation, fatty liver, and fat utilization during the periparturient period.

Table 1. Correlation coefficients between pp preprandial acyl ghrelin, total ghrelin, the acyl:total ghrelin ratio and parameters reflecting fat metabolism or feed intake with $P<0.01$.

	acyl ghrelin	total ghrelin	acyl:total ghrelin ratio
Plasma NEFA	0.56	ns	ns
Plasma TG	ns	ns	ns
Milk fat content	ns	ns	0.68
LFC	ns	ns	0.66
BFT	ns	ns	0.75
RQ	ns	ns	-0.78
DMI	ns	ns	ns
NE_L intake	ns	ns	ns

ns = no significance.

Acknowledgements

Supported by DFG (KU1956/4-1), Germany.

References

Barazzoni R., A. Bosutti, M. Stebel, M.R. Cattin, E. Roder, L. Visintin, L. Cattin, G. Biolo, M. Zanetti, and G. Guarnieri. 2005. Ghrelin regulates mitochondrial-lipid metabolism gene expression and tissue fat distribution in liver and skeletal muscle. Am. J. Physiol. Endocrinol. Metab. 288, E228-235.

Jennings J.S., A.E. Wertz-Lutz, R.H. Pritchard, A.D. Weaver, D.H. Keisler, and K. Bruns 2011. Circulating ghrelin and leptin concentrations and growth hormone secretagogue receptor abundance in liver, muscle, and adipose tissue of beef cattle exhibiting differences in composition of gain. J. Anim. Sci. 89, 3954-3972.

Rodríguez A., J. Gómez-Ambrosi, V. Catalán, M.J. Gil, S. Becerril, N. Sáinz, C. Silva, J. Salvador, I. Colina, and G. Frühbeck,. 2009. Acylated and desacyl ghrelin stimulate lipid accumulation in human visceral adipocytes. Int. J. Obes. (Lond). 33, 541-552.

Effects of realimentation after nutrient restriction during early to mid-gestation on pancreatic digestive enzymes in beef cattle

F.E. Doscher[1], E.A. Nere[1], R.D. Yunusova[1], L.E. Camacho[1], C.O. Lemley[2], L.D. Prezotto[1], K.A. Vonnahme[1], J.S. Caton[1] and K.C. Swanson[1]
[1]Dept. of Animal Science, North Dakota State University, 1300 Albrecht Boulevard, Fargo, ND 58102, USA; faithe.doscher@my.ndsu.edu
[2]Dept. of Animal and Dairy Science, Mississippi State University, 240 Wise Center Dr., Mississippi State, MS 39762, USA

Introduction

Maternal nutrition has been shown to impact fetal pancreatic endocrine development (Dahri *et al.*, 1995). However, less is known about its effects on exocrine function. The objectives were to determine the effect of nutrient restriction and realimentation during gestation on maternal and fetal pancreatic mass and digestive enzymes.

Material and methods

On day 30 of pregnancy, multiparous beef cows (n=46; 581±68 kg) were randomly assigned to dietary treatments: control (C; 100% NRC; n=18) and nutrient restriction (R; 60% NRC; n=34). On day 85 of gestation cows were either slaughtered (C; n=6 and R; n=6), or remained on control (CC; n=12), restricted (RR; n=12) or were realimented to control (RC; n=11). On day 140 cows were either slaughtered (CC; n=6; RR; n=6; RC; n=5) or remained on control (CCC; n=6; RCC; n=5) or realimented to control (RRC; n=6). On day 254 all remaining cows were slaughtered. Maternal and fetal pancreas were weighed and a subsample collected and snap-frozen in liquid nitrogen and stored at -80 °C until analysis for α-amylase and trypsin activities. Data were analyzed as a completely randomized block (breeding group) design within slaughter date using the mixed procedure of SAS. For fetal measurements, sex was also included in the model. Fetal pancreas was not collected at day 85.

Pancreatic tissue (0.25 g) was homogenized in 0.9% NaCl (2.25 ml) using a polytron (Brinkmann Instruments Inc., Westbury, NY, USA). Activity of α-amylase was determined utilizing a kit from Teco Diagnostics (Anaheim, CA, USA). Trypsin activity was assayed using the methods described by Geiger and Fritz (1986) after activation with 100 U/l enterokinase (Glazer and Steer, 1977). One unit (U) of enzyme activity equals 1 μmole product produced per min. Enzyme activity data are expressed as U/g wet tissue, kU/pancreas, and U/kg BW.

Results and discussion

No effects (P>0.05) were observed for maternal pancreas measurements on day 85. Absolute weight of maternal pancreas (g) on day 140 was greater (P<0.04) in CC cows compared to RC and RR cows. Relative weight (% of BW) also varied with CC and RR being greater (P<0.01) than RC. Maternal α-amylase activity on day 140 was not affected (P>0.26) by treatment (Table 1). Moreover, no effects (P>0.33) on fetal pancreatic weight or enzyme activity were observed on day 140 among treatments. Plane of nutrition affects maternal pancreatic weights which could impact pancreatic function.

On day 254, no effects (P>0.10) on maternal pancreatic weight or enzyme activity were observed. Fetal α-amylase activity (U/kg of BW) was greater (P<0.04) in RRC compared to CCC and RCC (Table 2). Nutrient restriction and realimentation during early gestation influenced the activity of α-amylase in fetal pancreas during late gestation suggesting that maternal nutrition may influence the development of the fetal exocrine pancreas.

Table 1. Influence of nutrient restriction and realimentation on pancreatic α-amylase activity in pregnant cows at day 140 of gestation.

Item	Treatment			SEM	*P*-value
	CC[1]	RC[2]	RR[3]		
BW (kg)	607	567	542	32.4	0.22
Pancreas weight					
g	474	354	434	33.2	0.048
% of BW	0.078	0.063	0.079	0.004	0.011
α-Amylase					
U/g	156	90.4	174	52.3	0.44
kU/pancreas	71.5	33.7	71.9	21.84	0.32
U/kg of BW	116	55.6	140	40.34	0.26

[1] Fed at 100% of NRC recommendations from day 30 to 140.
[2] Fed at 60% of NRC recommendations from day 30 to 140.
[3] Fed at 60% of NRC recommendations from day 30 to 85 and 100% of NRC recommendations from day 85 to 140.

Table 2. Influence of realimentation on pancreatic α-amylase activity in ovine fetuses at day 254 of gestation.

Item	Treatment			SEM	*P*-value
	CCC[1]	RCC[2]	RRC[3]		
BW (kg)	36.6	35.4	35.8	2.30	0.77
Pancreas weight					
g	24.4	23.3	26.1	1.63	0.35
% of BW	0.067	0.066	0.073	0.0048	0.24
α-Amylase					
U/g	1.84	5.55	6.04	2.020	0.09
kU/pancreas	0.046	0.131	0.156	0.0505	0.10
U/kg of BW	1.13[a]	3.45[b]	4.23[b]	1.118	0.049

[1] Fed at 100% of NRC recommendations from day 30 to 254.
[2] Fed at 60% of NRC recommendations from day 30 to 140 and 100% of NRC recommendations from day 140 to 254.
[3] Fed at 60% of NRC recommendations from day 30 to 85 and 100% of NRC recommendations from day 85 to 140 to 254.
[ab] Means with uncommon superscripts differ ($P<0.05$).

References

Dahri, S., B. Reusen, C. Remacle, J.J. Hoet, 1995. Nutritional influences on pancreatic development and potential links with non-insulin-dependent diabetes. Proc Nutr Soc. 54(2):345-56.

Geiger, R. and H Fritz, 1986. Trypsin. Pages 119-128 in Methods of Enzymatic Analysis. H. Bergmeyer, ed Acad. Press, New York, NY.

Glazer, G. and M.L. Steer, 1977. Requirements for activation of trypsinogen and chymotrypsinogen in rabbit pancreatic juice. Anal. Biochem. 77:130-140.

Inhibitory effect of ammonia on urea flux across rumen epithelium depends on level of serosal urea

M.E. Walpole, G.B. Penner and T. Mutsvangwa
Department of Animal and Poultry Science, University of Saskatchewan, Saskatoon, SK, S7N 5A8, Canada; matthew.walpole@usask.ca

Introduction

Ruminants perpetually depend on their ability to transfer urea from blood to the rumen in order to maintain a positive N balance. Estimates indicate that 40 to 80% of endogenously-produced urea can be recycled to the rumen where it provides ruminally-available N for microbial growth (Lapierre and Lobley, 2001), so it is important to understand the regulatory mechanisms that control trans-epithelial urea transfer into the rumen. Increasing ruminal ammonia (NH_3-N) concentration decreases trans-epithelial urea transfer from blood into the rumen (Abdoun *et al.*, 2009); however, the mechanisms that are responsible for this response are uncertain. Facilitative urea transporter proteins (UT-B) are expressed in ruminal epithelium and the addition of phloretin, an inhibitor of UT-B function, reduces trans-epithelial flux of urea in isolated ruminal epithelium (Stewart *et al.*, 2005), thus providing proof that these UT-B might have a functional role in ruminal urea transfer. However, the functional role of UT-B in mediating the effects of ruminal NH_3-N on trans-epithelial flux of urea remains obscure. We evaluated the effects of mucosal NH_3 and serosal urea concentrations on total and UT-B-dependent (i.e. phloretin-sensitive) urea flux across the isolated bovine ruminal epithelium.

Material and methods

Six Holstein male calves fed a common diet for ad libitum intake at 0700 h were used. The diet was comprised of 38% alfalfa hay, 31% barley silage, 20% rolled barley and 11% protein-mineral supplement, with a dietary crude protein (CP) content of 14.2% (DM basis). Calves were slaughtered (1000 h) and the ruminal epithelium was collected and mounted in Ussing chambers under short-circuit conditions. To mimic physiological conditions, the serosal buffer (pH 7.4) contained 1.78 (LU) or 7.14 (HU) m*M* urea, whereas the mucosal buffer (pH 6.2) had no urea but contained 2.9 (LA) or 8.8 (HA) m*M* $(NH_4)_2CO_3$. The serosal-to-mucosal flux of ^{14}C-urea ($J_{sm\text{-}urea}$; 42 kBq/15 ml) and ^{3}H-mannitol ($J_{sm\text{-}mannitol}$; 37 kBq/15 ml) were measured, with $J_{sm\text{-}mannitol}$ being used as an indicator of hydrophilic movement. Serosal addition of phloretin (1 m*M*) was used to inhibit UT-B-mediated urea transport. Data were analyzed using the Proc Mixed procedure (SAS, 2001) as a factorial design, with regression analysis used to determine the relationship between $J_{sm\text{-}urea}$ and $J_{sm\text{-}mannitol}$.

Results and discussion

Ruminal NH_3 and blood urea-N concentrations at slaughter averaged 7.34 and 6.32 m*M*, respectively. Flux measurements are presented in Table 1. High NH_3 tended to inhibit total $J_{sm\text{-}urea}$ with HU, but there was no effect of NH_3 on total $J_{sm\text{-}urea}$ with LU (interaction, P=0.055). The phloretin-sensitive $J_{sm\text{-}urea}$ was not affected by serosal urea (P=0.34) or mucosal NH_3 (P=0.36); however, phloretin-insensitive $J_{sm\text{-}urea}$ tended to be higher with LA than HA when incubated with HU but was not different with LU (interaction, P=0.056). The $J_{sm\text{-}mannitol}$ was not affected by urea (P=0.86) or NH_3 (P=0.22) concentration. The $J_{sm\text{-}urea}$ and $J_{sm\text{-}mannitol}$ tended to be positively correlated for HA but not LA in combination with LU (Figure 1). The same pattern was observed with HU treatments where $J_{sm\text{-}urea}$ and $J_{sm\text{-}mannitol}$ tended to be positively correlated for HA but not LA (Figure 1). We conclude that luminal NH_3 has an inhibitory effect on $J_{sm\text{-}urea}$ when serosal urea concentration is high, and that $J_{sm\text{-}urea}$ is mediated via hydrophilic pathways at a low serosal urea concentration but is mediated through non UT-B-dependent pathways when serosal urea concentration is high. Future studies should determine other potential transporters involved in urea movement across the GIT.

Table 1. Urea flux rates [nmol/cm²/h] across ruminal epithelial tissue incubated in Ussing chambers with combinations of high (HU) or low (LU) serosal urea (U), and high (HA) or low (LA) mucosal ammonia (A) with or without phloretin (P).

	Treatment				SE	*P*-value		
	HU		LU			U	A	U × A
Item	LA	HA	LA	HA				
Total $J_{sm\text{-}urea}$	771.8	679.8	168.0	163.7	30.4	<0.001	0.06	0.06
P-insensitive $J_{sm\text{-}urea}$	781.7	717.5	181.7	164.9	34.4	<0.001	0.10	0.06
Total $J_{sm\text{-}mannitol}$	63.6	55.6	51.1	44.9	7.3	0.86	0.22	0.38

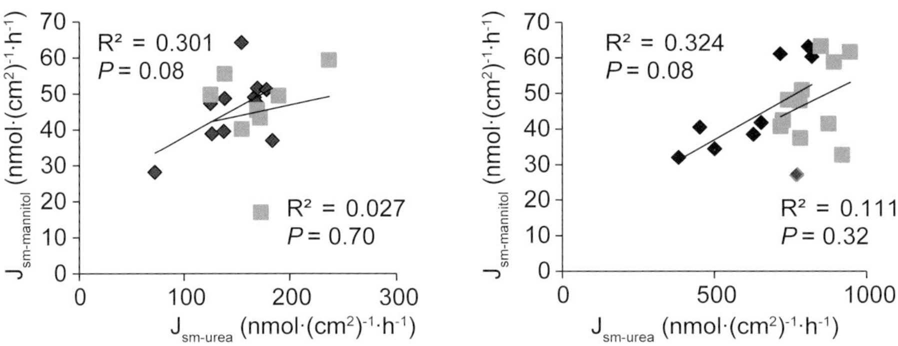

Figure 1. Regression analysis between $J_{sm\text{-}urea}$ and $J_{sm\text{-}mannitol}$ across the ruminal epithelium in Ussing chamber incubations containing 1.78 mM serosal urea with 2.9 (♦) or 8.8 (■) mM ammonia (left panel), or Ussing chamber incubations containing 7.14 mM serosal urea with 2.9 (♦) or 8.8 (■) mM ammonia (right panel).

References

Abdoun, K., F. Stumpff, I. Rabbani, and H. Martens, 2010. Modulation of urea transport across sheep rumen epithelium in vitro by SCFA and CO_2. Am. J. Physiol. Gastrointest. Liver Physiol. 298, G190-G202

Lapierre, H., and G. E. Lobley, 2001. Nitrogen recycling in the ruminant: a review. J. Dairy Sci. 84(E. Suppl.), E223-E236.

Stewart, G. S., C. Graham, S. Cattell, T. P. L. Smith, and C. P. Smith, 2005. UT-B is expressed in bovine rumen: potential role in ruminal urea transport. Am. J. Physiol. Regul. Integr. Comp. Physiol. 289, R605-R612.

Effects of starch derived substrates on pancreatic and mucosal enzyme activities in milk-fed calves

A.J. Pantophlet[1], M.S. Gilbert[2], J.J.G.C. van den Borne[2], R.J. Vonk[1], W.J.J. Gerrits[2], A. Pluschke[3], H.A. Schols[4], M.G. Priebe[1] and J. Roelofsen[1]
[1]Centre for Medical Biomics, University Medical Center Groningen, P.O. Box 30001, 9700 RR Groningen, the Netherlands
[2]Animal Nutrition Group, Wageningen University, P.O. Box 338, 6700 AH Wageningen, the Netherlands; myrthe.gilbert@wur.nl
[3]Centre for Nutrition and Food Sciences, University of Queensland, St. Lucia, 4072, Australia
[4]Laboratory of Food Chemistry, Wageningen University, P.O. Box 8129, 6700 EV Wageningen, the Netherlands

Introduction

α-Amylase and the mucosal enzymes maltase and iso-maltase play an important role in the digestion of starch. Pancreatic α-amylase is responsible for the breakdown of polysaccharides into oligosaccharides such as maltose, maltotriose and dextrin. These oligosaccharides are subsequently converted to monosaccharides by small intestinal mucosal enzymes. Veal calves are fed with milk replacers (MR) which commonly contain 40-50% of lactose, originating from whey. The price of whey, however, is subjected to extremely large fluctuations, and producers of MR are in search of alternatives such as starch-based substrates (SBSs).

The high lactase activity of calves enables rapid digestion of large amounts of lactose. If lactose is replaced by SBSs, other enzymes such as α-amylase, maltase and/or iso-maltase are required for the complete breakdown of these products. However, the capacity and inducibility for enzymatic starch digestion in milk-fed calves is unclear. Therefore, a study was designed to evaluate the effect of SBSs on pancreatic α-amylase and mucosal enzyme activities when gradually introduced in milk-fed claves.

Material and methods

Forty male Holstein Friesian calves (95.3±2.6 kg) at the age of 13 weeks were allotted to 1 of 5 dietary treatments. During the first week, the calves received a milk replacer containing lactose as exclusive source of carbohydrates. Thereafter, one group (n=8) remained on this control diet. In the other groups (n=8 each) lactose was partly replaced by one of four SBSs. Each SBS required different combinations of enzymes for complete breakdown: (1) gelatinized starch (α-amylase and maltase); (2) maltodextrin (α-amylase and maltase); (3) maltodextrin with α-1,6-branching, (α-amylase, maltase and iso-maltase); (4) maltose (maltase).

All diets included Co-EDTA as an indigestible marker. After an adaptation period of 16 weeks, in which the percentage of SBSs was gradually increased to 18% at the expense of lactose, the calves were sacrificed and digesta from the abomasum (AB), the first and second half of the small intestine (SM1 and SM2, respectively) were collected to measure the α-amylase activity (BioVision, Milpitas, USA). Also, mucosal scrapings were collected from three parts of the small intestine (SM1, SM2 and the terminal ileum), to measure lactase, maltase and iso-maltase activity by the Dahlqvist method (Dahlqvist, 1964).

Results and discussion

Preliminary results of the mucosal enzyme activity of 4 calves from the control group and the gelatinized starch group and 2 calves from the other groups are depicted in Figure 1A-C. Lactase activity did not differ between dietary treatments, but differed between segments (P<0.01) and was

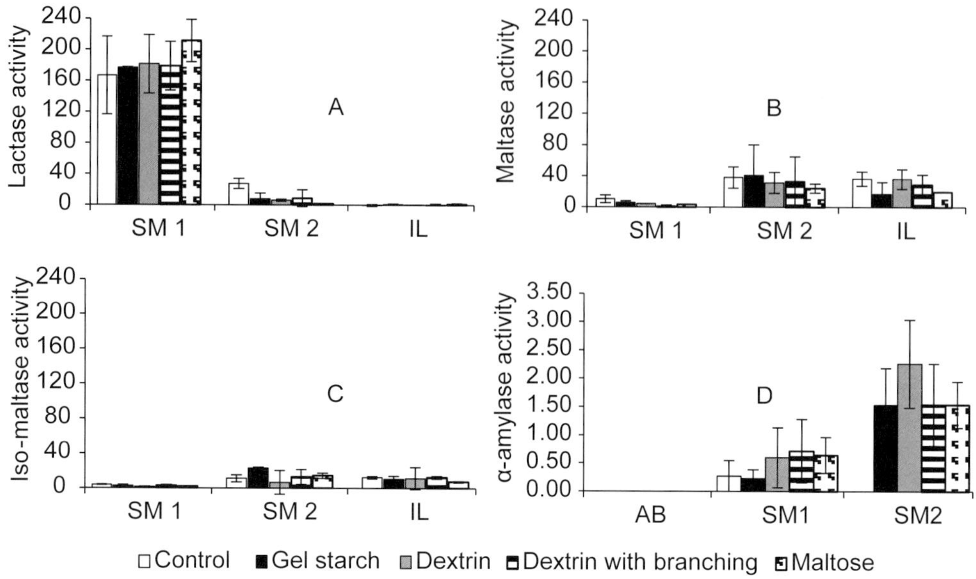

Figure 1. Enzyme activities of: A= Lactase (in U/mg protein), B= Maltase (in U/mg protein), C= Iso-maltase (in U/mg protein), D= α-Amylase (in U/mg Co) for each of the five dietary treatments. SM 1= first half of the small intestine, SM 2= second half of the small intestine and IL= Ileum.

higher in SM1 (177.1±46.6 U/mg protein) than in SM2 (20.614.5 U/mg protein). Lactase activity was almost undetectable in the terminal ileum (0.6±0.7 U/mg protein). Maltase and iso-maltase activities were low in SM1 (4.8±1.1 and 2.5±0.3 U/mg protein, respectively) and higher (P<0.01) in SM2 (41.1±7.3 and 13.8±2.2 U/ mg protein, respectively) and terminal ileum (29.1±3.5 and 10.6±0.8 U/ mg protein respectively). However, there was no significant difference among dietary treatments.

Luminal α-amylase activity was measured in all calves, expressed relative to the indigestible marker (Co), and the results are depicted in Figure 1D. The α-amylase activity in the digesta of the control calves was almost undetectable (0 for AB, 0.3±0.3 and 0 U/g Co for SM1 and SM2, respectively), but increased by feeding of SBSs, particularly in SM2 (1.7±0.3 U/g Co). This was probably due to the induction of microbial α-amylase activity in SM2. However, with the amylase activity kit it was not possible to distinguish between pancreatic and bacterial amylase activity.

Based on these preliminary results we conclude that the activity of mucosal enzymes required for the hydrolysis of starch cannot be induced by feeding SBSs in milk-fed calves. We speculate that a considerable part of the starch disappearance at the terminal ileum in milk-fed calves is caused by degradation by microbial enzymes.

References

Dahlqvist, A., 1964. Method for assay of intestinal disaccharides. Anal. Biochem. 7, 18-25.

Effect of heat stress on intake and metabolism of *Bos taurus* (Angus) and *Bos indicus* (Nellore)

E.E.L. Valente[1], *M.L. Chizzotti*[1], *C.V.R. Oliveira*[1], *M.C. Galvão*[1], *S.S. Domingues*[1], *A.C. Rodrigues*[1] *M.M. Ladeira*[1] *R.A. Ferreira*[1] *and F.H.M. Chizzotti*[2]
[1]*Animal Science Departament, Universidade Federal de Lavras, Lavras, Brazil*
[2]*Animal Science Departament, Universidade Federal de Viçosa, Viçosa, Brazil; mariochizzotti@dzo.ufla.br*

Introduction

During heat stress, ruminants like other homeothermic animals, use pathways to increase heat loss and reduce heat production to maintain euthermia (Bernabucci *et al.*, 2010). Environmentally induced hyperthermia decreases efficiency and production (O'Brien *et al.*, 2010). One of the first noticeable signs of heat stress (HS) during production is reduced nutrient intake, which is presumably an evolutionary strategy to reduce the 'heat increment' of feeding (O'Brien *et al.*, 2010). The aim of this study was to evaluate the effect of different air temperatures on diurnal, nocturnal, and daily feed intake as well as water intake, heart rate, respiratory rate and body temperature.

Material and methods

Sixteen young bulls (8 Angus, *Bos taurus*, and 8 Nellore, *Bos indicus*) with 337±7.4 kg and 16 months of age were used. The experimental period was divided in two 10 days periods, after 7 days of adaptation. Two climate-controlled rooms, with a 12×12 h photoperiod, were used. In period 1 all animals were housed in thermoneutral ambient temperature (TN, 25.4±0.4 °C). In period 2, 8 bulls (4 Angus and 4 Nellore) were subjected to low heat stress (LHS, 28.6±0.2 °C in daytime and ambient temperature at night, 25±0.2 °C) and 8 bulls (4 Angus and 4 Nellore) were subjected to high heat stress (HHS, 31.4±0.4 °C during the daytime and ambient temperature at night, 25±0.2 °C). The diet was composed by 50% of corn silage and 50% of concentrate (16% CP) and offered twice per day at 6 a.m. and 6 p.m. Orts were withdrawn and sampled (at a proportion of 5%) before each feeding.

We measured diurnal, nocturnal and daily feed intake as well as daily water intake. Heart rate was obtained using heart rate monitors (Polar RS800, Finland) on days 9 and 10 of each period, only in Angus bulls at an interval of 1-min for 48 h. Respiratory rate was measured in days 9 and 10 of each period by counting flank movement over a 1-min interval in 4 different times per day. Eye and skin temperatures were measured by thermal imaging (Fluke, USA) on day 10 of each period. Eye temperature was considered as the maximum value in the eyeball and skin temperature was the average of temperature of a similar area under the eyes.

The experiment was conducted under a randomized design using 2×2 factorial arrangement with two genetic groups and two heat stresses. The variables were evaluated according to a completely randomized design with repeated measures over time using the mixed models.

Results and discussion

Diurnal and daily intake of Nellore bulls was not affected (P>0.05) by heat stress, but an increase (P<0.05) was observed in nocturnal intake with high heat stress. However, Angus bulls decreased their diurnal intake by 24% and decreased nocturnal intake by 5% with a total decrease of 15% in daily intake during high heat stress (Table 1). At TN, Angus bulls had higher (P<0.05) daily intake (36.2 g/kg of BW) than Nellore bulls (29.1 g/kg of BW), but during HHS they had similar (P>0.05) intakes (31.6 and 30.2 g/kg of BW, respectively). There was an observed increase (P<0.05) in BR, but eye and skin temperatures and water intake were not affected (P>0.05) with temperature increase.

The HR decreased ($P<0.05$) with heat stress suggesting that occurred reduction in metabolic activity to decrease metabolic heat production. Different anatomical and morphological characteristics partially explain differences in heat tolerance among species and breeds (Gaughan *et al.*, 2010). Thus, Nellore bulls showed more adapted to high temperatures than Angus bulls.

We concluded that Angus feed intake is decreases more than that of Nellore on high air temperature. Angus cattle present lower diurnal intake in high heat stress than on thermoneutral conditions and partially compensates this reduction by increasing the intake in nocturnal period. However, total daily intake of Angus is decreased on high heat stress.

Table 1. Effect of heat stress (means and standard error, SE) on daily intake, heart beat, breath rate and skin temperature according of genetic group.

Genetic group	Treatment[1]			SE	*P*-value
	Thermoneutral	Low stress	High stress		
Daily intake (g/kg of BW)					
Angus	36.2Aa	35.4Aa	31.6b	0.71	0.003
Nellore	29.1B	29.8B	30.2	0.71	0.504
SE	0.67	0.95	0.95		
P-value	<0.001	0.003	0.439		
Heart beat (beats/min)					
Angus	93.0a	89.9ab	87.0b	1.21	0.046
Breath rate (breaths/min)					
Angus	86.3Ab	96.8Aa	104.0Aa	1.83	<0.001
Nellore	29.5Bb	42.0Ba	45.2Ba	1.95	0.003
SE	1.73	2.45	2.65		
P-value	<0.001	<0.001	<0.001		
Skin temperature (°C)					
Angus	34.3b	35.9a	36.2a	0.12	<0.001
Nellore	34.1b	35.9a	35.6a	0.12	<0.001
SE	0.11	0.16	0.16		
P-value	0.278	0.969	0.072		

[1] Subscript lowercase letters within a row and subscript capital letters within a column denote significant difference ($P<0.05$).

Reference

Bernabucci U, Lacetera N, Baumgard LH, Rhoads RP, Ronchi B and Nardone A 2010. Metabolic and hormonal acclimation to heat stress in domesticated ruminants. Animal 4, 1167-1183.

Gaughan JB, Mader TL, Holt SM, Sullivan ML and Hahn GL 2010. Assessing the heat tolerance of 17 beef cattle genotypes. Int J Biometeorol 54, 617-627.

O'Brien, M.D., Rhoads, R.P.,Sanders, S.R., Duff, G.C., Baumgard, L.H. 2010. Metabolic adaptations to heat stress in growing cattle. Domest Anim Endocrin 38, 86-94.

Growth hormone releasing factor and secretion of growth hormone in Iberian and Landrace gilts

J.M. Rodríguez-López, L. González-Valero, M. Lachica and I. Fernández-Fígares
Department of Animal Nutrition, Estación Experimental del Zaidín, CSIC, Camino del Jueves s/n, 18100 Armilla, Granada, Spain; ifigares@eez.csic.es

Introduction

Compared to modern breeds, Iberian pigs have lower growth rate and protein deposition (Nieto *et al.*, 2002). However, factors that limit growth performance of Iberian pigs are still under debate. In a previous study, greater basal serum insulin, leptin and IGF-1 with no difference in growth hormone (GH) compared to Landrace pigs was reported (Fernández-Fígares *et al.*, 2007). The objective of the present work was to explore the possibility that Iberian have lower capacity of GH secretion than Landrace pigs, which could explain their lower growth capacity.

Material and methods

Iberian (n=6) and Landrace (n=6) gilts (29.5±0.36 kg BW) were fitted with permanent carotid artery catheters. Following surgery recovery and after 18 hours fasting pigs were injected intravenously 3.32 μg growth hormone releasing hormone (GHRH)/kg BW and blood samples collected at -15, 0, 15, 30, 45, 60, 75, 90, 120 and 180 min. Plasma samples were frozen (-80 °C) in aliquots until analysis of GH, insulin and amino acids (AA). GH was analysed using a porcine GH RIA kit. Insulin was determined with a porcine insulin RIA kit using human insulin as standard. Plasma free AA were analysed by HPLC using the Waters Pico Tag method which involves pre-column derivatization with phenylisothiocianate and a reverse phase column. Data were evaluated using the mixed procedure of SAS with repeated measures with breed, time of sampling and their interaction in the model statement. Concentration at time zero of the analyte was included as a covariate in the statistical analysis. Significant differences among treatments were assessed using Tukey's multiple-range test.

Results and discussion

Landrace had greater GHRH stimulated GH release than Iberian pigs (50%, $P<0.001$). Previous studies with growing Iberian pigs (15 to 50 kg BW) demonstrated their low genetic potential for lean tissue deposition (Nieto *et al.*, 2002) compared with conventional breeds. Metabolic differences among pigs with distinct genotypes can be expected in blood concentration of metabolites and hormones. We have reported increased serum insulin, leptin and IGF-1, decreased glucose and no change in GH in Iberian compared to Landrace gilts (Fernández-Fígares *et al.*, 2007), although a single point determination was used, which may not be sufficient for determination of pulsatile secretion hormones such as GH. Similarly to the present study, GH clearly differed when extreme breeds were compared and was lower when a reduced growth rate coincides with high daily deposition of fat (Kasser *et al.*, 1981; Wangness *et al.*, 1981).

Increased GH paralleled decreased insulin in the plasma of pigs subjected to a GHRH challenge, so that Iberian showed increased serum insulin compared to Landrace gilts (50%, $P<0.01$). An increase in plasma insulin and glucose after GH injection has been repeatedly reported (Wray-Cahen *et al.*, 1991; Caperna *et al.*, 1993). Although total AA did not change ($P>0.10$), branched chain AA, sulphur AA and aromatic AA increased ($P<0.05$) in Iberian compared to Landrace pigs after a GHRH challenge (6, 124 and 48%, respectively). Administration of GH is associated with rapid changes in protein and energy metabolism. (Dunshea *et al.*, 1992; Caperna *et al.*, 1993). Moreover, a significant decrease in systemic essential, non-essential, branched chain and total AA, concentrations was found in GH treated pigs (Vann *et al.*, 2000; Bush *et al.*, 2002) found. In conclusion, Landrace had greater capacity of GHRH stimulated GH secretion than Iberian pigs, which may explain the lower growth capacity of Iberian pigs.

Table 1. Plasma growth hormone (GH, ng/ml), insulin (μU/ml) and amino acid (AA, mM) concentration in Iberian and Landrace gilts (n=6) after i.v. *injection of porcine growth hormone releasing hormone.*

	Landrace	Iberian
GH	16.4 ± 0.94^{a}	8.2 ± 0.77^{b}
Insulin	5.5 ± 0.59^{a}	8.3 ± 0.54^{b}
Total AA	3.58 ± 0.085^{a}	3.72 ± 0.085^{a}
Branched chain AA	0.53 ± 0.009^{a}	0.56 ± 0.009^{b}
Sulphur AA	0.15 ± 0.014^{a}	0.33 ± 0.014^{b}
Aromatic AA	0.093 ± 0.0040^{a}	0.138 ± 0.0040^{b}

[a,b] Values in the same row with different superscripts are significantly different ($P<0.05$).

References

Bush J.A., G. Wu, A. Suryawan, H.V. Nguyen and T.A. Davis, 2002. Somatotropin-induced amino acid conservation in pigs involves differential regulation of liver and gut urea cycle enzyme activity. J Nutr. 132, 59-67.

Caperna T.J., N.C. Steele, C.M. Evock-Clover, D. Brocht, J.P. McMurtry and R.W. Rosebrough, 1993. Acute effects of administration of porcine growth hormone on circulating levels of hormones and metabolites in 20-, 40-, and 60-kilogram gilts. J. Anim. Sci. 71, 897-905.

Dunshea, F.R., D.E. Bauman, R.D. Boyd, and A.W. Bel, 1992. Temporal response of circulating metabolites and hormones during somatotropin treatment of growing pigs. J. Anim. Sci. 70, 123-131.

Fernández-Fígares, I., M. Lachica, R. Nieto, M.G. Rivera-Ferre and J.F. Aguilera, 2007. Serum profile of metabolites and hormones in obese (Iberian) and lean (Landrace) growing gilts fed balanced or lysine deficient diets. Livest. Sci. 110, 73-81.

Kasser, T.R., R.J. Martin, J.H. Gahagan and P.J. Wangsness, 1981. Fasting plasma hormones and metabolites in feral and domestic newborn pigs. J. Anim. Sci. 53, 420-425.

Nieto, R., A. Miranda, M.A. García and J.F. Aguilera, 2002. The effect of dietary protein content and feeding level on the rate of protein deposition and energy utilization in growing Iberian pigs from 15 to 50 kg body weight. Br. J. Nutr. 88, 39-49.

Vann, R.V., H.V. Nguyen, P.J. Reeds, D.G. Burrin, M.L. Fiorotto, N.C. Steele, D.R. Deaver, and T.A. Davis, 2000. Somatotropin increases protein balance by lowering body protein degradation in fed, growing pigs. Am. J. Physiol. 278, E477-E483.

Wangness, P.J., W.A. Acker, J.H. Burdette, L.F. Krabill and R. Vasilatos, 1981. Effect of fasting on hormones and metabolites in plasma of fast-growing, lean and slow-growing obese pigs. J. Anim. Sci. 52, 69-74.

Wray-Cahen, D., D.A. Ross, D.E. Bauman and R.D. Boyd. 1991. Metabolic effects of porcine somatotropin: Nitrogen and energy balance and characterization of the temporal pattern of blood metabolites and hormones. J. Anim. Sci. 69, 1503-1514.

Effects of a simple or a complex starter microbiota on the gastric transcriptome profile of caesarean derived piglets

D. Priori[1], M. Colombo[1], S.J. Koopmans[2], A.J.M. Jansman[3], G. Schiavo[1], P. Trevisi[1] and P. Bosi[1]
[1]Department of Agricultural and Food Sciences, University of Bologna, 40127 Bologna, Italy; paolo.bosi@unibo.it
[2]Department of Animal Sciences, Adaptation Physiology Group of Wageningen University, Wageningen, the Netherlands
[3]Wageningen UR Livestock Research, Lelystad, the Netherlands

Introduction

Recent research suggests that early exposure of piglets to a diverse microbiota can shape the skills of gut-associated lymphoid tissue to respond to exogenous molecules (Lewis *et al.*, 2012). In pig, a short encounter with a complex microbiota in the early life can be sufficient to influence the intestinal microbiota in the following weeks (Jansman *et al.*, 2012). The priming effect of the complexity of the microbiota that enter the digestive tract, is not fully characterized. The gastric barrier of newborn piglets is not fully active and early contact with microbes can affect functional maturation of the stomach. Thus we aimed at evaluating the effect of complexity of starter microbiota on the gastric transcriptome profile of young pigs.

Material and methods

Twelve caesarean derived piglets were assigned to two treatment groups by litter and body weight on day of birth (d 0), housed in two separate clean laboratory rooms with a hygiene barrier, received serum from sow blood from d 0 to d 2, and were orally dosed a 'simple' microbiota consisting of 10^6-10^7 cfu from each *Lactobacillus amylovorus*, *Clostridium glycolium* and *Parabacteroides* sp from d 1 to d 3. In addition, on d 3 and d 4, the groups received a fecal inoculant obtained from a conventional adult sow (Complex Association, CA), or a placebo inoculant (Simple Association, SA). All piglets were fed a milk replacer diet during day 0 to 5 and a moist diet from day 6 onwards. Pigs were euthanized on d 14 and samples of oxyntic tissue in the stomach were obtained and deep frozen.

On quality-tested mRNA, the analysis of whole transcript expression per each pig was done by Affymetrix©Porcine Gene 1.1 ST array strips. A Robust Multichip Analysis was done on CEL files by Affymetrix Expression Console. Transcripts ID's, in general characterized by several exonic sequences, were associated to 13,406 Human gene names, based on Sus scrofa Ensemble. On gene expression values, exploratory functional analysis was done with Gene Set Enrichment Analysis (C2.BP catalog of gene sets, Molecular Signatures Database v3.0, http://www.broadinstitute.org/gsea/msigdb/Index.jsp), comparing the CA and SA treatment. Normalized enrichment score (NES) was calculated for each gene set, and statistical significance was defined when False Discovery Rate % <25 and *P*-values of NES <0.05.

Results and discussion

Phenotypes SA significantly enriched 138 gene sets compared to CA (Table 1). Most of the 20 pathways with the highest NES were related to cell replication and mitosis, and also to the interplay with xenobiota (Toll like receptors signaling), immune response (chemokine receptors binding chemokines) and host interaction with Human Immunodeficiency Virus factors. Several chemokines characterized by the CXC motif were represented at the top of the core genes characterizing enriched gene sets. Only four sets of genes were significantly enriched in phenotypes CA compared to SA. They relate to the expression of epidermal growth factor (EGF), hormone biosynthesis, signaling of peroxisome proliferator-activated receptors and keratinocyte pathways. Among the core genes within these sets of genes were genes encoding for EGF, deiodinase 1 and adiponectin. The results suggest that early association of newborn pigs with a complex microbiota favorably prevents the activation of pathways related to cell division and inflammatory development in the gastric mucosa at an age of two weeks compared to the association with a simple microbiota.

Table 1. First gene sets enriched in the gastric oxyntic tissue with simplified microbiota association compared to complex microbiota association.

Name	Size	NES	FDR q-val
Mitotic M G1 phases	122	-2.63	0.0000
Mitotic prometaphase	70	-2.55	0.0000
Chemokine receptors bind chemokines	42	-2.45	0.0000
Cell cycle mitotic	236	-2.45	0.0000
DNA replication pre-initation	58	-2.23	0.0005
M G1 transition	48	-2.18	0.0017
Regulation of APC activators between G1 S and early anaphase	57	-2.17	0.0020
CDC20 phospho-APC-mediated degradation of cyclin A	52	-2.14	0.0021
Spliceosome	86	-2.14	0.0020
Toll Like Receptor signaling pathway	79	-2.14	0.0018
CDT1 association with CDC6 ORC origin complex	43	-2.12	0.0024
HIV_infection	140	-2.11	0.0027
SCF β-TRCP mediated degradation of EMI1	40	-2.08	0.0031
Processing of capped intron containing pre-mRNA	93	-2.08	0.0029
S phase	82	-2.05	0.0041
Regulation of ornithine decarboxylase	39	-2.05	0.0039
G1 S transition	78	-2.05	0.0037
G2 M_checkpoints	34	-2.03	0.0052
Cell cycle checkpoints	89	-2.02	0.0063
Host interactions of HIV factors	93	-2.00	0.0067
Influenza life scycle	100	-2.00	0.0066
HIV life cycle	79	-2.00	0.0065
G alpha I signaling events	130	-2.00	0.0063
Synthesis of DNA	71	-1.99	0.0063
Peptide ligand binding receptors	130	-1.97	0.0075

Acknowledgements

Research funded by the European Union (Interplay project, Grant agreement no. 227549).

References

Jansman, A. J. M., J. Zhang, S. J. Koopmans, R. A. Dekker and H. Smidt, 2012. Effects of a simple or a complex starter microbiota on intestinal microbiota composition in caesarean derived piglets. J. Anim. Sci. 90 (Suppl.4): 433-435.

Lewis M. C., C. F. Inman, D. Patel, B. Schmidt., I. Mulder, B. Miller, B. P. Gill, J. Pluske, D. Kelly, C. R. Stokes and M. Bailey, 2012. Direct experimental evidence that early-life farm environment influences regulation of immune responses. Pediatr. Allergy Immunol. 23, 265-269.

Bitter taste receptor genes in pigs: SNP identification by using next-generation semiconductor sequencing

L. Fontanesi[1], F. Bertolini[1], E. Scotti[1], G. Schiavo[1], P. Trevisi[1], M. Colombo[1], P.L. Martelli[2], R. Casadio[2] and P. Bosi[1]
[1]Department of Agricultural and Food Sciences (DISTAL), Division of Animal Production, University of Bologna, Viale Fanin 46, 40127 Bologna, Italy; paolo.trevisi@unibo.it
[2]Biocomputing Group (BiGeA Dept.), University of Bologna, Via San Giacomo, 40126 Bologna, Italy

Introduction

Diet selection in animals is directly affected by their taste perception. It is commonly accepted that the taste chemosensory system in mammals can detect five basic tastes: sweet, salt, sour, umami and bitter (Behrens and Meyerhof, 2009). In particular, bitter sensing evolved as a central warning signal to protect against ingesting potentially toxic bitter-tasting substances. It is sensed by a family of bitter taste receptors (referred as TAS2Rs) that in mammalian genomes are encoded by approximately 10-40 functional TAS2R genes (Bachmanov and Beauchamp, 2007). Variability in these genes could play an important role in feed preferences and feed intake in livestock species. Therefore, it could be also possible to apply information about animals with different variants giving different sensitivity to bitter tastants for diet formulation and feeding practises. Toward this aim, the first step is the identification of polymorphisms at the DNA level analysing livestock genomes.

In this study we resequenced ten *TAS2R* genes in different pig breeds using a next generation sequencing platform (Ion Torrent PGM technology; Rothberg *et al.*, 2011) and identified a quite large number of single nucleotide polymorphisms (SNPs), some of which might potentially have a functional effect.

Material and methods

TAS2R gene sequences were identified mining the Sscrofa10.2 genome version and GenBank database. PCR primers were designed to amplify fragments of 800-1,200 bp covering all coding regions and part of 5'- and 3'-flanking regions including 5'- and 3'-untranslated regions of the selected genes. PCR was carried out on DNA pools constructed using equimolar quantities of genomic DNA of 10-50 pigs of the same breed. Different pools were prepared for pigs of different breeds: 2 pools for Italian Large White, one for Italian Duroc, one for Italian Landrace, one for Casertana and one for Pietrain. The pooling strategy was designed to exploit as much as possible the throughput of the Ion Torrent technology reducing the time and cost of the preparatory works needed to obtain material to be sequenced with this new machine.

Sequencing of the amplified products were sequenced using the Ion Torrent PGM sequencer (Life Technologies). Briefly, PCR products obtained from DNA pools of the same breed were pooled, fragmented for library preparation, barcoded and then pooled again for sequencing reaction on a 316 Ion Torrent chip. Obtained ionograms were automatically selected using barcoding information and compared with reference sequences. Average coverage on the obtained sequences was approximately 20,000×.

Single nucleotide polymorphisms were called using appropriate statistics based on number of animals in the different DNA pools/breeds combination. SNPs identified in the *TAS2R1* gene using the Ion Torrent platform were confirmed by Sanger sequencing of amplified products obtained by genomic DNA from single animals.

Results and discussion

Mining the pig genome and other sequences available in DNA databases (GenBank/EMBL) made it possible to select 10 genes of the TAS2 family with homologous in the human and/or mouse genome. These genes were: *TAS2R1, TAS2R3, TAS2R4, TAS2R7, TAS2R9, TAS2R10, TAS2R16, TAS2R20, TASR38* and *TAS2R39*. Resequencing data obtained with the Ion Torrent platform identified a total of 52 SNPs. The largest number of SNPs was identified in the *TAS2R1* gene (n=12), the lowest was identified in the *TAS2R9* sequences (n=2). Most of the polymorphisms located in coding regions were synonymous substitutions, eight were missense mutations. Four missense mutations were indentified in the *TAS2R1* gene. Difference in allele frequencies between breeds were observed (presence or absence of the two allelic forms in one or another breed) or inferred (based on the number of reads with the alternative alleles) from the sequencing information. All SNPs identified using the Sanger sequencing approach were also identified with the Ion Torrent sequencer indicating that this platform can produce reliable sequence information, improving throughput and speed. To our knowledge, this is the first study that applied this sequencing technology to identify SNPs in a farm animals.

These results represent the first step to investigate potential effects of the variability in taste receptor genes with differences of individual sensitivity to feedstuffs including bitter but not harmful substances. Additional studies are under way to further characterize the targeted genes, including gene expression data (Colombo *et al.*, 2012), and to evaluate association between these polymorphisms and feed intake and related production traits in pigs.

Acknowledgment

Research funded by the European Union (Interplay project, Grant agreement no. 227549) and Italian MiPAAF (Innovagen project).

References

Bachmanov A.A. and G.K. Beauchamp, 2007. Taste receptor genes. Annu. Rev. Nutr. 27, 389-414.

Behrens M. and W. Meyerhof, 2009. Mammalian bitter taste perception. Results Probl. Cell Differ. 47, 203-220.

Colombo M., P. Trevisi, G. Gandolfi and P. Bosi, 2012. Assessment of the presence of chemosensing receptors based on bitter and fat taste in the gastrointestinal tract of young pig. J. Anim. Sci. 90, Suppl 4, 128-130.

Rothberg J.M., W. Hinz, T.M. Rearick, J. Schultz, W. Mileski, M. Davey, J.H. Leamon, K. Johnson, M.J. Milgrew, M. Edwards, J. Hoon, J.F. Simons, D. Marran, J.W. Myers, J.F. Davidson, A. Branting, J.R. Nobile, B.P. Puc, D. Light, T.A. Clark, M. Huber, J.T. Branciforte, I.B. Stoner, S.E. Cawley, M. Lyons, Y. Fu, N. Homer, M. Sedova, X. Miao, B. Reed, J. Sabina, E. Feierstein, M. Schorn, M. Alanjary, E. Dimalanta, D. Dressman, R. Kasinskas, T. Sokolsky, J.A. Fidanza, E. Namsaraev, K.J. McKernan, A. Williams, G.T. Roth and J. Bustillo, 2011. An integrated semiconductor device enabling non-optical genome sequencing. Nature 475, 348-452.

Post weaning growth and metabolism in F_2-offspring of protein restricted mink dams

C.F. Matthiesen and A.-H. Tauson
Department of Veterinary Clinical and Animal Sciences, Faculty of Health and Medical Sciences, University of Copenhagen, Denmark; cmt@sund.ku.dk

Introduction

Malnutrition during fetal life may lead to structural and functional metabolic changes in the offspring, adaptations aiming to maximize the utilization of available nutrients in order to optimize fetal development and survival. Low protein provision during pregnancy in mink has resulted in lower birth weight of F_1- but higher birth weight of F_2-generation offspring (Matthiesen *et al.*, 2010a,b). Furthermore, the relative mRNA abundance of some enzymes in glucose metabolism was affected by fetal life protein provision (Matthiesen *et al.*, 2010a,b). The objectives of the present study were to investigate post-weaning growth and quantitative metabolism in male F_2-generation mink born by dams exposed to either low (L_1) or adequate (A_1) protein provision during fetal life (maternal fetal treatment) and fed either a low (L_2) or adequate (A_2) protein diet during gestation (male kit fetal life treatment), and the male kits' response to a low (L) or an adequate (A) protein diet.

Material and methods

Sixteen F_2 male offspring, four from each maternal and own fetal life treatment (A_1A_2, A_1L_2, L_1A_2, L_1L_2) were used to study post-weaning growth, energy metabolism and substrate oxidation. They were measured in three balance periods including respiration experiments (indirect calorimetry) at 10, 26 and 50 weeks of age. Diets were adequate (A – 31:53:16% of metabolizable energy (ME) from protein; P, fat; F and carbohydrates; CHO) or low (L-19 P:48 F:33 CHO % of ME) in protein and fed during the balance and respiration experiments.

The data were analysed using the mixed procedure in SAS, version 9.2. Fixed effects of maternal-, and kit fetal life protein provision and post weaning diet and their interactions were analysed. Differences were denoted as significant if $P<0.05$.

Results and discussion

The main results are presented in Table 1. Post-weaning growth showed significant ($P<0.001$) interactions between maternal fetal life and own fetal life treatment resulting in A_1A_2 and L_1L_2 having the highest body weights at 50 weeks of age and A_1L_2 and L_1A_2 the lowest. The heat production (HE) was significantly ($P=0.04$) lower in males born by L_1 than A_1 dams in contrast to F_1 males born by L dams which tended to have higher HE than A kits (Matthiesen *et al.*, 2012). The energy retention (RE) was slightly negative in all groups and a significant interaction between maternal fetal life and male kit fetal life treatment ($P=0.04$) resulted in the most negative RE in L_1A_2 males. The only significant effects of post-weaning diet concerned oxidation of protein and carbohydrate which reflected the diet composition. Significant interactions between maternal and kit fetal life treatment resulted in significantly higher oxidation of protein (OXP) and lower oxidation of fat (OXF) for A_1A_2 and L_1L_2 than A_1L_2 and L_1A_2 males, the lower OXP of L_1A_2 males confirming previous findings in F_1 males (Matthiesen *et al.*, 2012).

In conclusion, both maternal and own fetal life protein provision affected post-weaning growth and metabolism of F_2-generation male offspring. The finding that maternal fetal life low protein provision (L_1) resulted in lower HE in their offspring indicates an adaptation towards saving energy. Protein oxidation data were, however, inconclusive due to a lower OXP in L_1A_2 animals indicating a strive

Table 1. The intake of ME, HE, RE, substrate oxidation and body weight of male offspring born by dams exposed to either low (L_1) or adequate (A_1) fetal life protein provision and being exposed to either low (L_2) or adequate (A_2) protein provision during their own fetal life and then fed a low or an adequate protein diet at 10, 26 and 50 weeks of age.

	Treatment				*P*-value				
	A_1A_2	A_1L_2	L_1A_2	L_1L_2	W	D	M	F	M × F
Body weight, g					<0.001	NS	NS	NS	<0.001
10 weeks	1,455	1,335	1,424	1,304					
26 weeks	$3{,}275^{ab}$	$2{,}785^{a}$	$3{,}166^{ab}$	$3{,}479^{b}$					
50 weeks	$3{,}189^{a}$	$2{,}541^{b}$	$2{,}668^{bc}$	$3{,}083^{ac}$					
ME kJ/kg$^{0.75}$	575^{a}	553^{ab}	431^{b}	572^{a}	0.01	NS	NS	NS	0.08
HE kJ/kg$^{0.75}$	633^{a}	627^{a}	620^{ab}	599^{b}	<0.001	NS	0.04	NS	NS
RE kJ/kg$^{0.75}$	-58^{ab}	-73^{ab}	-188^{a}	-10^{b}	<0.001	NS	NS	0.08	0.04
Substrate oxidation in % of HE									
OXP	27^{a}	22^{ab}	20^{b}	27^{a}	NS	<0.001	NS	NS	0.004
OXF	56^{a}	67^{b}	69^{b}	58^{a}	0.01	NS	NS	NS	0.005
OXCHO	17	11	11	15	<0.001	<0.001	NS	NS	0.06

W = week 10, 26 or 50; D = diet – low or adequate; M = maternal fetal life treatment; F = males' own fetal life treatment. Mean values within a row with unlike lower-case superscript letters (a, b, c) were significantly different within treatment combination.

to save protein but not in A_1L_2 and L_1L_2 animals where OXP was similar to controls. Effects of low protein provision to pregnant mink dams may as demonstrated here not only be limited to first generation offspring but also be evident in the following generation.

References

Matthiesen C.F., Blache D., Thomsen P.D., Hansen N.E. and Tauson A.-H. 2010a. Effect of late gestation low protein supply to mink (*Mustela vison*) dams on reproductive performance and metabolism of dam and offspring. Arch. Anim. Nutr. 64: 56-76.

Matthiesen C.F., Blache D., Thomsen P.D. and Tauson A.-H. 2010b. Feeding mink (*Neovison vison*) a protein-restricted diet during pregnancy induces higher birth weight and altered hepatic gene expression in the F_2 offspring. Br. J. Nutr. 104: 544-553.

Matthiesen C.F., Blache D., Thomsen P.D. and Tauson A.-H. 2012. Foetal life protein restriction in male mink (*Neovison vison*) kits lowers post-weaning protein oxidation and the relative abundance of hepatic fructose-1,6-bisphosphatase mRNA. Animal, 6, 50-60.

Investigation of the glucose metabolism of the embryonic and neonatal broiler chicks by injection of insulin

L. Franssens[1], J. Lesuisse[1], A. Koppenol[1,2], Y. Wang[1], E. Willems[1], J. Buyse[1], E. Decuypere[1] and N. Everaert[1]
[1]*Division of Animal-Nutrition-Quality, Lab of Livestock Physiology, KU Leuven, Belgium; lies.franssens@biw.kuleuven.be*
[2]*Division of Animal, ILVO, Belgium*

Introduction

In contrast to most mammals, chickens have blood glucose concentrations that are twice as high as these in mammals. In addition, they develop a resistance to insulin early in life (Mc Cormick *et al.*, 1978; Seki *et al.*, 2001) So far, the response of insulin to elevated glucose levels in embryonic and neonatal chicks is not well documented and the aim is to explore the glucose metabolism by the injection of insulin in the embryonic and neonatal chick.

Material and methods

828 Ross broiler eggs were incubated under standard conditions. Eggs (12 eggs/group) were injected on embryonic day (ED) 16 or ED18 in blood capillaries of the chorioallantoic membrane or neonatal chicks were injected in the blood on postnatal day (D) 2 with 200 μl of insulin (1 μg/g embryo or chick BW) dissolved in a saline solution. An injection of 200 μl saline was used as a sham group. Another group served as a control group and did not receive any treatment. Blood and liver samples were taken before and every two hours after injections (until 23 hours post-injection (PI)). Plasma glucose concentrations were determined. mRNA expression of glucose transporters (GLUT) 1, 2, 3 and 8 in the liver at 7 hours PI was determined, using EF1α1 as a housekeeping gene. Glucose concentrations and mRNA expression levels were analyzed by using the 2-way ANOVA model in SAS 9.2., using treatment and time of injection (glucose concentrations) and treatment and day of injection (mRNA expression levels) with their interaction as variables. When there was a significant difference, a Tukey's test was performed. A 95% confidence interval threshold ($P<0.05$) was set. Data are shown as mean ± SEM. The comparison between the control and saline group was first performed but did not reveal any significant differences.

Results

For every injection day, injection of saline did not affect plasma glucose concentrations compared to non-injected embryos or chicks. On all time points for every injection day, except 21 hours PI on D2, the insulin-injected group had lower plasma glucose concentrations than their sham counterparts. On every injection day, plasma glucose concentrations decreased to a minimum level 7 hours PI. Then, a first significant rise in glucose levels was seen at 15 and 13 hours PI on ED 16 and ED 18, respectively. For embryonic insulin injections, glucose levels did not reach the basal level after 21 hours PI (figure 1A, B). On D2, plasma glucose concentrations first remained low after 7 h PI, then increased significantly at 17 hours PI, reaching control levels at 21 hours post-injection and then decreased again (figure 1C).The mRNA levels of GLUTs in non-injected embryos or chicks were not significantly different from these of the saline group, regardless of day of injection. The insulin injection on ED16, ED18 or D2 did not alter the mRNA levels of GLUT1, 3 and 8 (data not shown). As shown in table 1, the mRNA levels of GLUT2 in the liver of insulin-injected embryos/chicks on ED16 and D2 was significantly lower compared to levels of the saline and control group. On ED18, the mRNA expression levels of this gene did not differ between treatments.

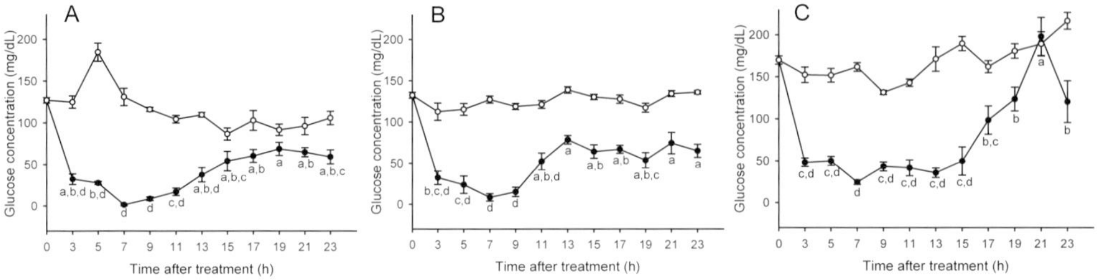

Figure 1. Plasma glucose concentration (mg/dl) of the sham group (white circles) or insulin group (black circles) at different time points (hours, h) before and after insulin injection on ED16 (A), ED18 (B) and D2 (C). Values are means ± SEM; n=12. a,b,c,d,e mean values are significantly different between time points of the insulin-injected group.

Table 1. Fold change (arbitrary units) of GLUT2 at 7 hours PI. Values are means ±SEM; n=5. a and b mean values are significantly different between the groups and the injection days. EF1α1 was used as a reference gene.

Treatment	Injection day		
	ED 16	ED 18	D 2
Control	-7.99E-11[a]±0.40	-3.55E-11[a]±0.29	-4.44E-11[a]±0.40
Sham	0.08[a]±0.32	-0.71[a]±0.21	-0.02[a]±0.38
Insulin	-2.08[b]±0.42	-0.47[a]±0.26	-2.70[b]±0.27

Discussion

On every injection day, the injection of insulin clearly caused a decrease in the plasma glucose concentrations. Interestingly, the recovery pattern back to pre-injection level was different in respect to the injection days. The PI increase in plasma glucose concentrations was more slowly on ED 16 and ED18 when compared with that on D2. Therefore, it seems that embryos are more sensitive to insulin than neonatal chicks. GLUT 2, the major GLUT in liver, is present on the sinusoidal membrane of hepatocytes and allows for the bi-directional transport of glucose, which is adjusted by dietary and hormonal control (Wood and Trayhurn, 2003). In rats, insulin suppresses hepatic glucose production, and this suppression is associated with a decrease in the hepatocyte plasma membrane-bound quantity of the facilitative glucose transport protein GLUT2 (Nathan *et al.*, 2001). Indeed, our results showed that the insulin-injected group had a decrease in GLUT2 mRNA levels on ED16 and D2. In addition, Thorn *et al.* (2012) also demonstrated that GLUT2 mRNA levels were diminished after insulin supplementation in fetal sheep with hypoglycemia. However, reasons of the absence of changes in GLUT2 mRNA level in the insulin injected group on ED18 are unknown and this finding will be further investigated by analyzing additional PI time points.

In conclusion, the sensitivity to exogenous insulin seems to diminish towards the end of the embryonic period and continues during neonatal stage, and a down-regulation of GLUT2 gene expression might be involved.

References

Mc Cormick, K.L., J.A. Widness, J.B. Susa and R. Schwarz, 1978. Effects of chronic hyperinsulinemia on hepatic enzymes involved in lipogenesis and carbohydrate metabolism in the young rat. Biochem. J. 172, 327-331.

Nathan, J.D., P.D. Zdankiewicz, J. Wang, S.A. Spector, G. Aspelund, B.P. Jena, N.E. Seymour, J.P. Geibel and D.K. Andersen, 2001. Impaired hepatocyte glucose transport protein (GLUT2) internalization in chronic pancreatitis. Pancreas 22, 172-178.

Seki, Y., K. Sato, H. Ohtsu and Y. Akiba, 2001. Persistent hypoglycemia is induced by tolbutamide administration in broiler chickens fed a low-carbohydrate diet. Domest. Anim. Endocrinol. 20, 109-122.

Thorn, S.R., S.M. Sekar, J.R. Lavezzi, M.C. O'Meara, L.D. Brown, W.W. Jr. Hay and P.J. Rozance, 2012. A physiological increase in insulin suppresses gluconeogenic gene activation in fetal sheep with sustained hypoglycemia. Am. J. Physiol. Regul. Integr. Comp. Physiol. 303, R861-R869.

Wood, I.S. and P. Trayhurn, 2003. Glucose transporters (GLUT and SGLT): expanded families of sugar transport proteins. Br. J. Nutr. 89, 3-9.

Stage of egg production regulates protein turnover and lysine partitioning for broiler breeders

R.D. Ekmay, K. Vignale, C. Salas, J. England, S. Cerrate and C.N. Coon
University of Arkansas, Poultry Science Department, Center of Excellence for Poultry Science, John Tyson Building, O-211, Fayetteville, AR 72701, USA; ccoon@uark.edu

Introduction

Among the regulators of protein turnover, amino acids play a critical role. Lysine, leucine and isoleucine were found to alter protein synthesis, degradation and its regulation (Yoshizawa, 2004; Nakashima *et al.*, 2005; Tesseraud *et al.*, 2008). Furthermore, Tesseraud *et al.* (2000) reported that differences in the fractional degradation rates (FDR) between fast-growing and slow-growing birds are observed in young birds but not older ones, suggesting a possible age effect.

The current broiler breeder industry practice targets between 440 to 465 kcal ME/d for peak egg production with diets containing a fixed 15-16% dietary protein. Crude protein and lysine intakes often exceed 28 g/d and 1000 mg/day, respectively. A series of studies were conducted to determine if additional protein and energy intake impact protein turnover in broiler breeders during different stages of egg production.

Methods

In the first study, Cobb 700 broiler breeder hens were randomly assigned to one of six dietary treatments in a 2×3 fashion. Two levels of energy (390 and 450 kcal/d) and three levels of protein (22, 24 and 26 g CP/d) were utilized. Thirty-six hens were administered a daily oral dose of 15 mg of 15N-Lys for a period of two weeks. After this enrichment period, hens were allowed a period of two days in which no isotopes were given. After two days, a daily oral dose of 15mg of 2D4-Lys was administered at which point, breast muscle was excised and snap frozen between the 2nd and 4th egg. Muscle and egg were analyzed via GC-MS for the ratio of labeled to unlabeled lysine. Three time periods were assessed: 25, 29 and 45 wk. Data was analyzed as two-way ANOVA and contrasts were performed to determine line shape.

In the second study, ten hens per treatment were given an intravenous flooding-dose of 15N-Phe at 10 ml/kg. After 10 min, birds were slaughtered and breast muscle excised and snap-frozen in liquid nitrogen. Free phenylalanine was quantified via GC-MS and the protein-bound form was quantified via GC-C-IRMS. 3-methylhistidine was utilized to determine protein degradation rates. Protein turnover was determined at wk 22, 26, 31, and 44. Relative expression of calpain 2, proteasome C2 subunit, and F box protein 32 were determined via RT-PCR at wk 20, 25, and 44.

Results and discussion

Results show that over 78% of all labeled lysine was found in breast muscle. In the egg, the portion of total labeled lysine that was deposited varied by age. At wk 25 and 45, the amount of lysine that originated from skeletal muscle and was deposited into yolk was significantly higher than the amount deposited from the diet ($P<0.05$). At wk 29, diet became the major source of lysine deposited into yolk. Diet was the major source of lysine for albumen formation at wk 25 and 29, and an even split with skeletal muscle at wk 45. No difference in the partitioning of lysine was determined by energy or protein intake. Contrasts indicate fractional synthesis rate and FBXO32 increase to a maximum at 25-26 wk and a decrease thereafter. These results suggest that there is an upregulation of protein degradation during the periods of lower egg production. A significant drop ($P<0.05$) in PSMA1 and FBXO32 was observed between 25 and 44 wk and matched the decrease observed in FBR.

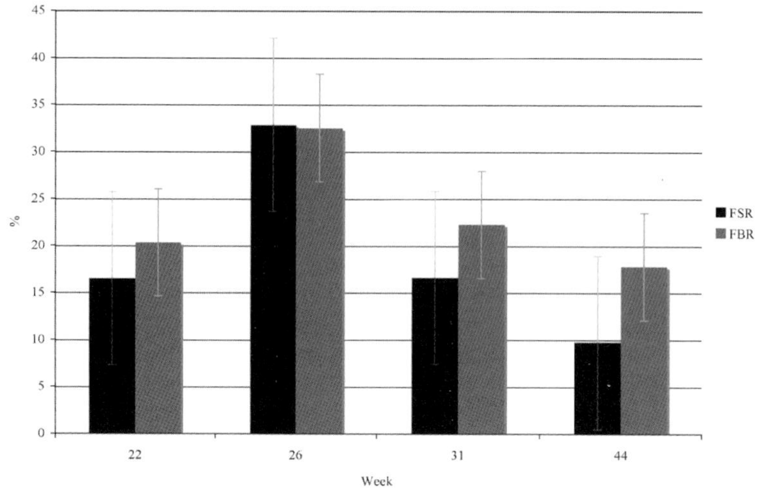

Figure 1. Changes to fractional synthesis (FSR) and breakdown rates (FBR) of Pectoralis *muscle during the transition into sexual maturity and past peak egg production in Cobb 700 broiler breeder hens.*

However, no differences in the expression of calpain were observed. Furthermore, no differences were detected in the levels of fractional synthesis and degradation rates, or the expression of CAPN2, PSMA1, and FBXO32, due to protein or energy intake. In summary, skeletal muscle appears to be a major source of lysine for egg formation during the transition into sexual maturity but not at peak production. Fractional degradation rates are upregulated coming into production but decrease with higher rates of egg production. Results suggest that stage of egg production may play an important role in the regulation of protein turnover. The observed changes in degradation also appear to be mediated by the ubiquitin-proteasome pathway.

References

Nakashima, K., Ishida, A., Yamazaki, M., and A. Hiroyuki. 2005. Leucine suppresses myofibrillar proteolysis by down-regulating ubiquitin-proteasome pathway in chick skeletal muscles. Biochem. Biophys. Res. Commun. 336: 660-666.

Tesseraud, S., Bouvarel, I., Collin, A., Audouin, E., Crochet, S., Seiliez, I. and C. Letterier. 2008. Dily variations in dietary lysine content alter the expression of genes related to proteolysis in chicken pectoralis major muscle. J. Nutr. 139:38-43.

Tesseraud, S., Chagneau, A.M., and J. Grizard. 2000. Muscle protein turnover during early development in chickens divergently selected for growth rate. Poult. Sci. 79: 1465-71.

Yoshizawa F. 2004. Regulation of protein synthesis by branched-chain amino acids in vivo. Biochem Biophys Res Commun. J313: 417-22.

Lipid utilization for egg formation in broiler breeders

C. Salas, R. Ekmay, J. England, S. Cerrate and C.N. Coon
University of Arkansas, Poultry Science Department, Center of Excellence for Poultry Science, Fayetteville, AR 72701, USA; ccoon@uark.edu

Introduction

Energy utilization is a key component of animal production. The study of the mechanisms by which lipids and carbohydrates are mobilized and deposited in the egg, can give us clues about how breeders utilize energy during the production period. Is the dietary energy directly utilized during yolk formation or is there a dynamic metabolic system that consistently provides nutrients to the egg at the expense of body tissue? Are the fatty acids in the yolk derived from direct intestinal absorption, endogenous synthesis, or maternal body stores? Do the systems change as the breeder ages and egg production decreases? The aim of the present studies was to better understand the mechanisms of energy partitioning in broiler breeder hens.

Material and methods

The contribution of dietary fatty acids (FA), tissue FA or newly synthesized FA utilized for the formation of egg yolk during different periods of egg production was determined in broiler breeders using stable isotopes. In Exp 1, 20 broiler breeder pullets (BB) were fed a 25 mg meal of U-13C-linoleic acid (LA) on the first d of oviposition. Two birds were sacrificed each d for the next 10 d and all large yellow follicles were collected from the ovary, identified and stored at -80 °C for analysis. The accumulation of labeled LA acid in the follicles was determined using a GC-MS. The objective was to evaluate the pattern of incorporation of the labeled essential FA into eggs of sexually maturing BB hens. In Exp 2, a group of pullets were fed a daily 15 mg dose of U-13C-glucose (G) 10 d prior to sexual maturity continuing through 10 d following first egg. A 50 mg meal of U-13C-LA was also orally administered 10 d prior to their estimated first egg. On the d of their first egg, each bird received a 25 mg meal of D31-LA. All eggs were collected for the next 10 d. The incorporation of U-13C-LA, D31-LA and labeled palmitic acid (PA) from U-13C-G metabolism in the egg was determined using a GC-MS. The enrichment with three different labeled FA was repeated at peak egg production (30 wk) and 45 wk of age. In Exp 3, broiler breeders were randomly assigned to three different levels of feed allocation to achieve 3 different growth curves during rearing period: 20% heavier (HBW), standard (SBW) and 20% lighter (LBW) than the breeder target. Exp 3 was conducted to determine the effect of body weight and body composition during the egg production cycle on gene expression for lipoprotein lipase (LPL), fatty acid binding protein (FABP), apolipoprotein B (ApoB) and apolipoprotein VLDL-II (ApoVLDL-II). Body composition was determined for each hen using DEXA technology and levels of estradiol in blood were also measured.

Results and discussion

The results of Exp 1 show that first and second d after dosing the birds at first egg, the smaller F5-F7 follicles received a higher relative dose of the labeled LA. This is consistent with other reports, where the peak of incorporation of labeled LA into the yolk was on day 4 or 5 after forced feeding (Burghelle-Mayeur *et al.*, 1990).

The results for Exp 2 indicate that the recovery of the labeled isotopes is altered as the flock ages. At first egg, the deposition of labeled PA is higher when compared to its deposition at 30 wk and 45 wk of age. The deposition of the two LA labels was low for first oviposition and increased at peak production and 45 wk. In older birds, the rate of glucose oxidation has been shown to increase relative to the nonoxidative route of glucose utilization (Buyse *et al.*, 2004). As the hens aged in Exp 2, their incorporation of U-13C-G (as acetate) in the formation of PA is decreased.

The hepatic expression of the selected genes was significantly increased after start of lay and reached a plateau after first oviposition (Figure 1). The expression of LPL significantly decreased throughout the production period for all groups. The levels of expression of all the genes were strongly correlated to the levels of estradiol in each sampling period. The body composition of the birds was significantly different between growth curves at 22 wk and at peak egg production, but these differences did not seem to directly affect the expression of the selected genes. Changes throughout the production period in the mechanisms of lipid formation and mobilization are not only a matter of an increased production of lipoproteins, but also of the increasing inhibition of LPL.

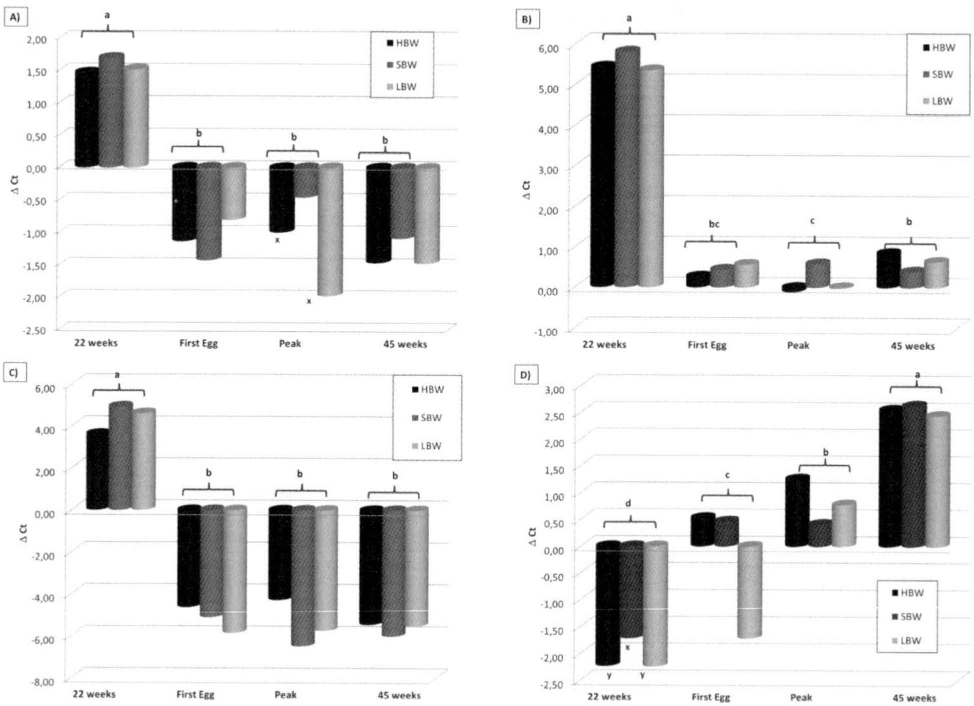

Figure 1. Genetic expression of genes coding for A) ApoB (liver), B) fatty acid binding protein (liver), C) ApoVLDL-II (liver), D) LPL (adipose tissue), determine by RT-PCR for broiler breeders reared in 3 different growth curves. Values are means of ΔCt relative to B-actin.

References

Burghelle-Mayeur C., M. Tixier-Biochard,P. Merat, and Y. Demarne. 1990. Influence of the sex-linked dwarfing gene (dw) on the contribution of dietary lipid to yolk lipid synthesis. Br Poult Sci 31:197-206.

Buyse J., B. Geypens, R.D. Malheiros,V.M. Moraes,Q. Swennen, and Y. Demarne. 2004. Assessment of age-related glucose oxidation rates of broiler chickens by using stable isotopes. Life Sciences 75:2245-55.

The heat-induced production of reactive oxygen species regulates protein content in cultured chick skeletal muscle cells

H. Yoshida, M. Kikusato, K. Souma and M. Toyomizu
Animal Nutrition, Graduate School of Agricultural Science, Tohoku University, 1-1 Tsutsumidori-Amamiyamachi, Aoba-ku, Sendai 981-8555, Japan; toyomizu@bios.tohoku.ac.jp

Introduction

Heat stress is a major factor affecting meat production in the poultry industry, especially in hotter regions of the world. A previous *in vivo* study by our group highlighted that acute heat stress stimulates mitochondrial reactive oxygen species (ROS) production in the skeletal muscle of broiler chickens. Recently, we further clarified that the overproduction of mitochondrial ROS resulted from an increase in the mitochondrial membrane potential, probably due to an increase in substrate oxidation (Kikusato *et al.*, 2010) and a decrease in avian uncoupling protein (avUCP) content (Mujahid *et al.*, 2006). Furthermore, it was reported that heat stress increases protein oxidation and myofibrillar proteolysis in chick cultured skeletal muscle cells (Nakashima *et al.*, 2004). Considering the fact that ROS play a pivotal role in muscle proteolysis in sepsis (Supinski and Callahan, 2007), it could be postulated that heat treatment stimulates mitochondrial ROS production, thereby leading to an increase in skeletal muscle proteolysis. To verify this possibility, we have investigated the effect of heat stress on ROS production, protein content, and the ROS-related expression of genes for avUCP, heme oxygenase-1 (HO-1) and NADPH oxidase 4 (NOX4) in chick skeletal muscle cells in primary culture. We also examine the reducing effects of 4-hydroxy-TEMPO (Tempol), which is a strong antioxidant, on these parameters.

Material and methods

Skeletal muscle cells were isolated from the *pectoralis superficialis* muscles of chicks (Ross strain) at 0 day of age and suspended in basal medium (80% Dulbecco Modified Eagle's Medium and 20% Medium 199) supplemented with 10% fetal bovine serum, 1.5×10^5 U/l penicillin and 0.15 g/l streptomycin. The cell suspension was transferred to uncoated culture dishes to allow fibroblast attachment. The supernatant was collected and myoblasts were counted and seeded onto collagen-coated plates at a density of 4.5×10^4 cells/cm^2. Incubation was carried out at 37 °C in a 5% CO_2-enriched atmosphere of humidified air until sub-confluence. In a subsequent step, the cells were further incubated at normal (37 °C) or high (41 °C) temperature for 6 h in a serum-free basal medium. Where relevant, Tempol (final concentration, 5 mmol/l) was added to cell culture media 1 h before the heat treatment.

Cellular ROS production was determined fluorometrically using CM-H_2DCFDA (λex/λem = 485/535 nm), a cell-permeant molecule that forms a fluorescent adduct upon reacting with ROS. The protein content of muscle cells was measured by the micro-BCA protein assay. Expression levels of genes for avUCP, HO-1 and NOX4 were measured by real-time RT-PCR, with expression referenced to that of 18S ribosomal RNA as an internal control.

Results and discussion

Cellular ROS production in the cultured chick skeletal muscle cells was increased in response to the heat treatment. As expected, the cellular protein content was also decreased by this treatment, confirming that a decrease in protein content occurs concomitantly with an increase in ROS production under heat stress conditions. Also in agreement with our previous *in vivo* study, the expression of avUCP mRNA was significantly decreased by the heat treatment, which suggests that the down-regulation of avUCP might result in the overproduction of mitochondrial ROS in chick muscle cells.

Furthermore, the expression of HO-1 mRNA (HO-1 being a heme-catabolizing enzyme that produces substrates exerting a protective effect against oxidative stress) was significantly decreased by the heat treatment, while the expression of NOX4 mRNA (NOX4 being an enzyme related to superoxide production) was increased by heat treatment. These results suggest that changes in the levels of these enzymes could be partly responsible for the overproduction of ROS in response to heat stress.

The addition of Tempol to culture media attenuated the heat-induced increase in cellular ROS production and the decrease in protein content (Figure 1A), indicating that ROS scavenging by Tempol can suppress the negative effects caused by the heat stress condition; this finding also suggests that the heat-induced ROS production might have led to a decrease in protein content. Also, Tempol reduced the heat-induced decrease in HO-1 expression and the heat-induced increase in NOX4 expression, but did not affect avUCP expression (Figure 1B). These results indicate that Tempol scavenges mitochondrial ROS produced by the down-regulation of avUCP, thereby causing HO-1 and NOX4 gene expression levels to be up- and down-regulated, respectively.

It can be concluded that heat stress induces the overproduction of cellular ROS, which causes a decrease in the protein content of chick muscle cells. Furthermore, this study provides the first evidence that HO-1, NOX4 and avUCP are factors involved in the heat-induced production of ROS.

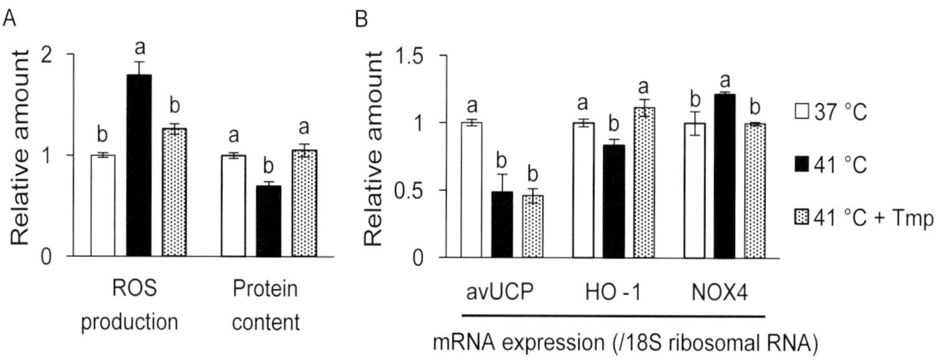

Figure 1. Effect of Tempol on ROS production and protein content (A) and on the expression of avUCP, HO-1 and NOX4 mRNA (B) in cultured chick muscle cells exposed to heat stress. Values are expressed relative to the average at 37 °C, and expressed as the mean ± SE (n=4-5). Statistical analysis was performed with one-way ANOVA followed by the Tukey Kramer multiple comparison test ($^{a,b}P<0.05$).

References

Kikusato, M., A.J.J. Ramsey, T. Amo, and M. Toyomizu, 2010. Application of modular kinetic analysis to mitochondrial oxidative phosphorylation in skeletal muscle of birds exposed to acute heat stress. FEBS Lett. 584, 3143-3148.

Mujahid, A., K. Sato, Y. Akiba and M. Toyomizu, 2006. Acute heat stress stimulates mitochondrial superoxide production in broiler skeletal muscle, possibly via downregulation of uncoupling protein content. Poult Sci 85, 1259-1265.

Nakashima, K., I. Nonaka, S. Masaki, M. Yamazaki, and H. Abe, 2004. Myofibrillar proteolysis in chick muscle cell cultures during heat stress. Animal Science Journal 75, 353-360.

Supinski, G.S., and Callahan, L.A., 2007. Free radical-mediated skeletal muscle dysfunction in inflammatory conditions. J Appl Physiol 102, 2056-2063.

Part 5. Modeling / systems biology

Fasting heat production and metabolic body size in non-ruminant growing farm animals

J. Noblet[1], E. Labussière[1], S. Dubois[1], C.F.M. de Lange[2], R. Barea[2], J. Lasnier[1], V. Rivera[1], M. Warpechowski[3] and J. van Milgen[1]
[1]INRA, 35590 Saint-Gilles, France; etienne.labussiere@rennes.inra.fr
[2]University of Guelph, Canada
[3]Federal University of Parana, Curitiba, Brazil

Introduction

Fasting heat production (FHP) of animals is indicative of their basal metabolic rate and is used to estimate the maintenance energy requirements and for calculating the net energy value of feeds. However, estimates of FHP vary with experimental conditions, as well as with measurement and calculation methods. In addition, FHP is often related to metabolic body weight (MBW) with the idea that FHP per unit of MBW is constant for a given group of animals over a large BW range. For comparing FHP in adult animals of different species, MBW is frequently expressed as $BW^{0.75}$ but validity of the 0.75 exponent has been questioned for growing animals (Noblet *et al.,* 1999; Labussière *et al.*, 2008). The objective of this study is to combine FHP measurements obtained according to a common methodology in four species of growing monogastric farm animals (veal calves, pigs, turkeys and broilers).

Material and methods

In each trial, single housed (calves, pigs) or group-housed (broilers, turkeys) animals were kept for 7 days in a respiration chamber and an energy and protein balance was carried out during the first 6 days. On the 7th day, no feed was provided. The continuous recording of gas exchange in the chamber and of physical activity (with force sensors) of the animals on the fasting day allowed calculating, according to modelling techniques, the asymptotic value of HP at zero activity considered to be FHP (Labussière *et al.*, 2011). The total number of FHP observations was 18 in 25-100 kg BW castrated pigs (fed ad libitum or restricted; de Lange *et al.*, 2006), 55 in 0.6-2.8 kg BW broilers (6 trials; fed ad libitum; unpublished), 19 in 0.4-24.0 kg BW male turkeys (fed *ad libitum*; Rivera-Torres *et al.*, 2010) and 47 in 60-265 kg BW male milk fed calves (fed *ad libitum* or restricted; Labussière *et al.*, 2009). For each species, BW range was divided into 3 intervals to determine the effect of BW on ME intake and FHP (Table 1). The exponent for calculating MBW was estimated from the relationship between FHP and MBW and ME intake during the previous days (log basis) to account for residual heteroscedasticity and effect of feeding level on FHP; the difference with usual coefficients was assessed with the extra sum of square reduction test (Labussière *et al.*, 2008).

Results and discussion

The best exponent of MBW during the growing period was 0.60, 0.70 and 0.85 in growing pigs, broilers and calves, respectively which all differed (P<0.05) from 0.75. In turkeys, it equalled 0.70 and did not differ from 0.75. Differences in the exponent among species can be attributed to differences in the allometric growth of visceral organs relative to the whole body (from 0.7 in pigs to more than 1.0 in calves; Robelin, 1986; Noblet *et al.*, 1999; Rivera-Torres *et al.*, 2011). With increasing BW, FHP decreased in turkeys and calves and tended to decrease in pigs (P=0.11) because of the correlated variations in ME intake (Table 1; Labussière *et al.*, 2011). In close to ad-libitum fed animals, daily FHP averaged 740 kJ/kg $BW^{0.60}$ in pigs, 438 kJ/kg $BW^{0.70}$ in broilers, 413 kJ/kg $BW^{0.70}$ in turkeys and 317 kJ/kg $BW^{0.85}$ in calves over the growing period. The rate of change of FHP per additional ME intake (kJ/kJ) was 0.13 in turkeys, 0.14 in pigs and 0.23 in calves (Table 1). To conclude, specific parameters must be used for MBW of growing non ruminant animals and FHP and maintenance ME requirements must be adjusted for differences in ME intake.

Table 1. Effect of body weight (BW) and previous ME intake on fasting heat production (FHP) of growing pigs, broilers, turkeys and calves.

Species	MBW[1]	BW, kg	ME intake (range), kJ/kg MBW/day	FHP[2], kJ/kg MBW/day	ΔFHP/ΔME[1]
Pigs	$BW^{0.60}$	55.4	2,084[a] (1,578-2,598)	687 (752)	0.14
n=18		70.4	2,082[a] (1,562-2,606)	697 (762)	
		86.7	2,003[b] (1,485-2,528)	631 (707)	
Broilers	$BW^{0.70}$	1.0	1,558[a]	439	-
(6 trials)		1.8	1,472[b]	431	
n=65		2.5	1,398[c]	435	
Turkeys	$BW^{0.70}$	2.6	1,479[a]	458[a] (431)	0.13
n=19		11.6	1,275[b]	453[a] (422)	
		21.1	1,025[c]	391[b] (386)	
Calves	$BW^{0.85}$	72.8	647[a] (525-769)	298[a] (312)	0.23
n=47		150.6	619[b] (533-716)	301[a] (322)	
		236.8	554[c] (467-644)	280[b] (316)	

[abc] LS-means within each species with different superscripts are different ($P<0.05$).

[1] MBW: metabolic BW; ΔFHP/ΔME: slope of the relationship between FHP and ME intake.

[2] In brackets, FHP in close to *ad libitum* ME intake (2580 kJ/kg $BW^{0.60}$/day in pigs, 1260 kJ/kg $BW^{0.70}$/day in turkeys and 710 kJ/kg $BW^{0.85}$/day in calves).

References

De Lange C., J. van Milgen, J. Noblet, S. Dubois and S. Birkett, 2006. Previous feeding level influences plateau heat production following a 24 h fast in growing pigs. Br. J. Nutr. 95, 1082-1087.

Labussière E., S. Dubois, J. van Milgen, G. Bertrand and J. Noblet, 2008. Fasting heat production and energy cost of standing activity in veal calves. Br. J. Nutr. 100, 1315-1324.

Labussière E., G. Maxin, S. Dubois, J. van Milgen, G. Bertrand and J. Noblet, 2009. Effect of feed intake on heat production and protein and fat deposition in milk-fed veal calves. Animal 3, 557-567.

Labussière E., J. van Milgen, C.F.M. de Lange and J. Noblet, 2011. Maintenance energy requirements of growing pigs and calves are influenced by feeding level. J. Nutr. 141, 1855-1861.

Noblet J., C. Karege, S. Dubois and J. van Milgen, 1999. Metabolic utilization of energy and maintenance requirements in growing pigs: effects of sex and genotype. J. Anim. Sci. 77, 1208-1216.

Rivera-Torres V., J. Noblet, S. Dubois and J. van Milgen, 2010. Dynamics of energy utilization in male and female turkeys during growth. Animal 5, 202-210.

Rivera-Torres V., J. Noblet and J. van Milgen, 2011. Changes in chemical composition in male turkeys during growth. Poult. Sci. 90, 68-74.

Robelin J., 1986. Composition corporelle des bovins: évolution au cours du développement et différences entre races. Thesis, Université de Clermont-Ferrand, France.

The evolution of INRA feeding systems for ruminants based on absorbed nutrients and animal responses

P. Nozière[1], D. Sauvant[2], J.L. Peyraud[3] and co-workers[1,2,3,4,5]
[1]UMR Herbivores, Theix, France; pierre.noziere@clermont.inra.fr
[2]UMR Mosar, Paris, France
[3]UMR Pegase, St Gilles, France
[4]UMR Selmet, Montpellier, France
[5]UR Zootechniques, Petit-Bourg; INRA, France

Introduction

The current feed evaluation systems for ruminants still have some limits regarding emerging challenges in ruminant nutrition, particularly for prediction of feed conversion efficiency, product quality, animal health, and emissions to the environment. Further, they do not appropriately cover the case of either low quality or very intensive diets. A working group of INRA's researchers has been constituted for 3 years (the 'Systali' project) to address these issues. Continuously accumulated new published data on digestive, metabolic and productive responses of ruminant to diets have thus been integrated within a 'new generation' of feed unit systems. This is based on flows of nutrients and multiple animal responses to these flows, and allows the updated system to be used on enlarged fields of application.

Material and methods

Large data bases of digestion, metabolic balance and animal performances, representative of the major feeding practices, have been built and interpreted by meta-analysis, allowing dissociation of intra vs. inter experiment variability (Sauvant *et al.*, 2008). To ensure enlargement of fields of application, a particular attention has been devoted to low digestibility (including tropical forages) and high concentrate diets. An integrated mechanistic model of digestion, that ensures consistency across empirical relationships issued from the various meta-analyses, has been built (Sauvant and Nozière, 2013). In contrast with the gut model included in the current INRA systems (INRA, 2007), this new model integrates the effects of the main digestive interactions related to feeding level (FL), proportion of concentrate (PCO), and rumen protein balance (RPB), on transit, substrate degradation, microbial synthesis and ruminal fermentation. There also are improvement in the quantification of endogenous losses and their implication on protein requirements for maintenance. The major equations of this model permit calculation of new nutritive values of feeds and diets, as well as absorbable nutrient flows, and subsequently revisiting the requirements and responses to diets. Multiple marginal responses of animals (i.e. feed efficiency, milk yield, growth rate, milk and carcass composition, risk of acidosis, N and CH4 emissions...) to feeding practices and subsequent nutrient supply are also being provided. These features are currently undergoing extensive validation. A simulation tool, which also integrates models of intake updated at pasture (Delagarde *et al.*, 2011) is being developed, and allows simulation of hundreds of dietary and animal situations, and thereby global checking of the consistency of the values of supply, requirements and responses. Coupled to external data, this allows evaluation and validation of the whole renewed system, across a large range of feeding practices and animal characteristics.

Results and discussion

The global structure of the current INRA (2007) systems has not changed but significant modifications have been adopted. Only the major modifications are presented in the present paper. Dietary protein supply remains a calculation from *in sacco* degradation, however, passage rates are no longer fixed, but modulated according to FL and PCO. The estimation of energy available in the rumen is improved by taking into account digestive interactions and intestinal digestibility of substrates, its efficiency for microbial growth is no longer fixed, but regulated by rumen parameters. The OM digestibility of feeds within a diet is no longer fixed, but regulated by FL, PCO and RPB. Losses of energy through CH4 and urines, that partly compensate digestive interactions, are more clearly dissociated and quantified. A major advance relates to the fact that energy and protein are more clearly interacting, through an iterative effect of the RPB on energy digestion and consequently microbial protein synthesis that itself affects RPB. The protein maintenance requirements, based on fecal endogenous and minimal urinary N losses, have been reconsidered to account for FL. Subsequent recalculation of requirements and responses, and their validation, are still in progress.

Conclusion

The renewed INRA feeding systems for ruminants will be based on a more precise representation of digestive processes and flows of nutrients, and of multiple animal responses to theses flows. They are built from integration of a large amount of data, carefully interpreted, incorporated and validated according to a wide range of feeding practices. While maintaining the functionality and specificity of the current systems, these modifications extend their fields of application, and the spectrum of predicted responses.

References

INRA, 2007. Alimentation des bovins, ovins et caprins – Besoins des animaux – Valeurs des aliments – Tables INRA 2007, Versailles, 307 pp.

Sauvant, D. and P. Nozière, 2013. Integrative model of the digestive tract including the interactions involved in energy and protein digestion in Ruminants. In: J.W. Oltjen, E. Kebreab, H. Lapierre (editors), Energy and protein metabolism and nutrition in sustainable animal production. Wageningen Academic Publishers, Wageningen, the Netherlands, EAAP Series 134:317-318.

Sauvant, D., P. Schmidely, J.J. Daudin and N.R. St-Pierre, 2008. Meta-analyses of experimental data in animal nutrition. Animal 2, 1203-1214.

Delagarde, R., P. Faverdin, C. Baratte and J.L. Peyraud, 2011. GrazeIn: a model of herbage intake and milk production for grazing dairy cows. 2. Prediction of intake under rotational and continuously stocked grazing management. Grass and Forage Science 66, 45-60.

Integrative model of the digestive tract including the interactions involved in energy and protein digestion in ruminants

D. Sauvant[1] *and P. Nozière*[2]
[1]*UMR INRA-AgroParisTech Mosar, Paris, France; sauvant@agroparistech.fr*
[2]*UMR INRA Herbivores, Theix, France*

Introduction

In ruminants, a major limitation of feeding systems in predicting the responses to diets is the insufficient integration of the various interactions involved in energy and protein digestion and metabolism. To bridge this gap, in the new feeding systems built at INRA (Nozière *et al.*, 2013), a focus was made on digestive interactions by combining data issued from meta-analysis and a mechanistic model of digestion, as already proposed (Sauvant and Mertens, 2008).

Material and methods

Three large data bases of results of published digestion studies were built: on cattle digestion (800 experiments-Exp, 2,106 treatments-Tr), on calorimetric studies on cattle, sheep and goats (186 Exp, 1,100 Tr) and on sheep digestion (116 Exp, 384 Tr). These bases have been studied by meta-analysis, using encoding of data allowing careful study of the meta-designs and dissociation of intra vs. inter experiments variability (Sauvant *et al.*, 2008). This phase provided numerous empirical equations quantifying the digestive and some metabolic responses to changes in the diet composition or in feeding practices.

A number of these empirical equations were directly used in the new feeding system to calculate the new versions of net energy (NE) and metabolisable protein (PDI). The other equations related to digestive phenomena that lead to the production of absorbable nutrients, (volatile fatty acids, VFA, glucose and long chain fatty). Equations were also derived that relate to chewing activities, saliva recycling, rumen pH and volume, production of CH4, and transit flow rates, as well as the sizes and adjacent flows of 12 major compartments in the rumen: NDF (degradable-D, or not, in the large or small particles), protein (soluble-S, or D and undegradable), starch (S or D), water, VFA and microbes.

To build the new feed units, three major factors affecting digestive interactions were identified: the dry matter intake per LW (DMI%LW), the proportion of concentrate (PCO) and the rumen protein balance (RPB = ingested CP – duodenal CP = 6.25*[NH3 absorption – UreaN recycling in the rumen]). These 3 factors affect transit of particles and liquids and digestible organic matter (DOM) in the whole tract. They also alter the proportions of gross energy dissipated in CH4 and urine. Feed degradation and by-pass predictions of protein and starch were based on *in sacco* measurements and transits; they were validated on *in vivo* duodenal flows. The calculation of fermentable OM in the rumen (FOM) from DOM was modified to be closer to the actual FOM in the rumen than the previous expressions proposed in the literature (and in the former INRA system). Microbial protein flow at the duodenum was also redefined from FOM, PCO and RPB.

To solve the challenge of insuring global consistency among all these equations, an integrated mechanistic model of gut has been built. Its rumen part, the kernel of the model, is a minimal model based on the 12 above mentioned compartments. In contrast to other mechanistic models of rumen, the current one was built through a top-down procedure using a framework of key structuring relationships issued from the meta-analyses. By this procedure, the most important equations above were used to calibrate the model and to check mutual consistency and for any unexpected interactions. This procedure of optimization was performed with Modelmaker III through virtual experimentations using plausible ranges of values of the most important dietary inputs: DMI%LW, PCO, dietary NDF, CP and starch, effective degradability in sacco of N and starch.

Results and discussion

Major evaluations of the model were performed by comparing the responses of the model with values predicted by the structuring equations. The simulated duodenal flows of protein and starch were very close to the predicted ones and, globally, the equations used to predict the new feed units are mutually consistent. In contrast, predictions of compartments and flows of NDF and H2O were a bit less accurate. For some characteristics, as chewing time, the framework provided more than one equation to calculate that entity according to various factors, which gives fairly different results according to the equation used.

The model has been incorporated within two calculation and simulation tools to ensure its application to small ruminants, evaluate its animal's requirements and responses, integrate with the INRA's fill unit system of intake, and validate the whole renewed system across simulation of hundreds of dietary and animal situations.

Conclusion

A new feeding system model has been constructed from numerous equations extracted from meta-analysis of large datasets of published results ensuring a basis in real digestive phenomena. It includes a better integration of digestive interactions and allows the calculation of not only the new version of the energy and protein units but also the flows of absorbable nutrients and several other key aspects of digestion in ruminants.

References

Nozière P. and D. Sauvant, 2013. The evolution of INRA feeding systems for Ruminants based on absorbed nutrients and animal responses. In: J.W. Oltjen (editor), Energy and protein metabolism and nutrition. Wageningen Academic Publishers, Wageningen, the Netherlands, EAAP Series XXX:xx-xx.

Sauvant, D. and D.R. Mertens, 2008. Use of metaanalysis to build a mechanistic model of responses of rumen digestion to dietary fibre in cattle. Can. J. Anim. Sci. 88, 730.

Sauvant D, P. Schmidely, J.J. Daudin and N.R. St-Pierre, 2008. Meta-analyses of experimental data in animal nutrition. Animal 2, 1203-1214.

Effects of diet composition during the finishing period on protein and lipid deposit in young bulls

J. Agabriel, M. Al-Jammas, I. Ortigues-Marty, C. Villetelle, P. Nozière and F. Garcia-Launay
INRA, UMR1213 Herbivores, 63122 Saint-Genès-Champanelle, France;
jacques.agabriel@clermont.inra.fr

Introduction

The INRA feeding system aims at simulating young bull's growth and body composition in response to dietary changes during the finishing period. Presently, tissue deposition is modified by net energy intake in relation to age and breed, without considering the relative proportion of energy nutrients. But at fixed NE density, rations based on concentrate or maize silage support fatter gains (Garcia *et al.*, 2008), which justifies better characterization of diet composition in prediction models. Our objective is to simulate the absorbed energy nutrient fluxes, their time variation and relative proportions, as well as their effects on lipids and proteins of carcass and visceral depots of finishing bulls. A dynamic model of growing cattle (Hoch and Agabriel, 2004) provides the framework and helps target variables of interest to reveal these effects. The present work aims to quantify diet effects on energy nutrient fluxes of young bulls during fattening and to link them to body, carcass, or visceral lipid content.

Material and methods

We used two complementary approaches: (A1) an analysis of publications chosen from the international literature that allowed estimation of daily nutrient fluxes from published data (n=13), and linking them with fat deposition in carcass and viscera (n=6) and (A2) an analysis of two previous INRA experiments where young bulls, Blond d'Aquitaine – late maturing breed (Micol *et al.*, 2007), and Salers – medium maturing breed (Geay *et al.*, 1996) were individually fed diets (corn silage vs. hay and concentrate; grass silage vs. concentrate). Individual weights, gains, daily intake of each feed, and detailed slaughter data were available.

Diets were estimated from INRA Feed Tables (OMD, Fermentable OM, protein digestible in the intestine PDI, digestible lipids DLi, DE, ME) and from Loncke *et al.* (2009) predictions of net portal appearance (mmol/j/kg LW) of VFA, Glucose, BOH, Lactate and alpha-amino-N. Net portal appearance of each nutrient was expressed on an energy basis. Daily fluxes were calculated during the finishing period for groups in experiments A1 or averaged across each individual in A2. The effects of diets on recalculated nutrients fluxes and body composition were tested. SAS GLM was used for statistical analysis between individual values.

Results and discussion

The calculated patterns between absorbed energy and nitrogen type nutrients differed between diets within experiments and helped to describe the observed differences in fat depots. Using A1 results, we found that total LW gain and carcass weight were greater on groups fed ME rich diets, while the quantity of body lipids at slaughter depended on the composition of ME, especially its content in starch and DLi.

A linear combination of energy from nutrients (R) including DLi and portal amino acids and acetate was significantly associated with the final quantities of carcass lipids (Figure 1). Using individuals (A2), significant differences in portal energy fluxes of VFA and DLi were noticed. Previous linear combination R was individually calculated and compared to carcass lipids (%) or body lipids (%) for the two experiments. A significant effect of diet and breed is observed while individuals' differences remains poorly predicted within group (Table 1).

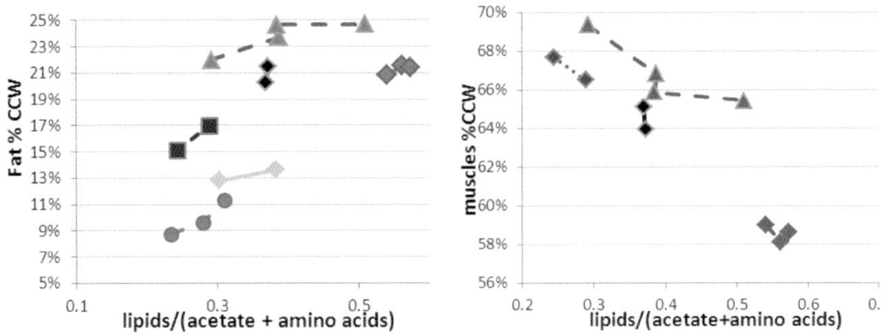

Figure 1. Relationships between a combination of average daily energy from nutrients R (energy of digestible lipids/(energy (portal amino acids and acetate)) and carcass composition (carcass weight =CCW) in chosen experiment (A1).

Table 1. Average values of R and fatness (Lipids carcass%Carcass Weight (CCW), Lipids viscera%CCW) in experiment A2.

Exp	Salers(Geay *et al.*, 1997)				Blond d'Aquitaine(Micol *et al.*, 2007)			
Diets	Hay	Grass silage	Corn silage	*P*-value	Concentrate	Corn silage	Hay	*P*-value
R	0.196 a ±0.002	0.259 b ±0.0006	0.285 c ±0.001	<0.001	0.286 c ±0.007	0.257 b ±0.0003	0.223 a ±0.005	<0.001
LipC%CCW	16.68 a	18.65 b	19.74 b	0.004	9.29 ab	10.87 b	7.99 a	0.002
Lip Visera %CCW	4.58 a	5.58 b	6.03 b	0.006	4.04 ab	4.73 b	3.47 a	0.002

In the future and after validation, evolution of the growth model could incorporate a modulation of the Gompertz function for lipid carcass synthesis by the R combination of energy nutrients.

Acknowledgements

The authors wish to thank the two companies INZO and Limagrain for their support of this research.

References

Garcia F., R.D. Sainz, J. Agabriel, L.G. Barioniand J.W. Oltjen, 2008. Comparative analysis of two dynamic mechanistic models of beef cattle growth. Animal Feed Science and Technology 143, 220-241.

Geay Y., B. Picard, R.D Jailler, R.T. Jailler, A. Listrat, C. Jurie, M.C. Bayle and C. Touraille, 1997. Effets de la nature de la ration sur les performances, les caractéristiques musculaires et la qualité de la viande de taurillons Salers. Renc. Rum. 4, 307-310.

Hoch T. and J. Agabriel, 2004. A mechanistic dynamic model to estimate beef cattle growth and body composition: 1. Model description. Agric. Syst. 81, 1-15.

Loncke C., 2009. Modélisation des relations entre l'alimentation et les flux splanchniques de nutriments énergétiques chez le ruminant. Thèse de Doctorat de l'Institut des Sciences et Industries du Vivant et de l'Environnement : Agro Paris Tech, 456p.

Micol D., H. Dubroeucq, C. Martin, F. Garcia, M.M. Mialon and J. Agabriel, 2007. Utilisation de rations de valeurs contrastées pour la finition de jeunes bovins de race Blond d'Aquitaine. Renc. Rech. Rum. 14, 233-236.

Integrating nutritional and reproductive models to improve reproductive efficiency in dairy cattle

S.L. Shields[1], H. Woelders[2], M. Boer[2,3], C. Stötzel[4], S. Röeblitz[4], J. Plöntzke[4] and J.P. McNamara[1]
[1]Department of Animal Sciences, Washington State University, Pullman, WA, USA; shannon.shields@email.wsu.edu
[2]Animal Breeding and Genomics Centre, Wageningen UR Livestock Research, Lelystad, the Netherlands
[3]Adaptation Physiology Group, Department of Animal Sciences, Wageningen University, Wageningen, the Netherlands
[4]Computational Systems Biology Group, Zuse Institute Berlin, Berlin, Germany

Successful reproduction requires coordination among neural, endocrine and nutritional systems leading to ovulation, insemination and a uterine environment that allows embryonic growth and attachment. These processes are a function of genetics, dietary nutrient composition and intake, and housing and climate. We lack a systems biology approach to study and define the control of reproduction in the dairy cow. Pregnancy rates in the best managed herds only approach 25 to 30%, and may be due to the multifactorial nature of reproductive processes (McNamara, 2010). Lower fertility increases the cost for insemination because of low reproduction performance and remains a primary reason for culling cows in the first three weeks of lactation (Chagas *et al.*, 2007). Additional days the cow is not pregnant beyond the optimal time post-calving are costly to the dairy producer. Low fertility has often been blamed on high rates of milk production, but this is not the only factor affecting reproduction (Sangsritavong *et al.*, 2002). There are three major systems of reproduction that can be affected by genetics, nutrition and management: the hypothalamic-pituitary axis controlling initialization of cycling, the follicular development in the ovary leading to ovulation; and the successful fertilization and growth of an embryo in the uterine environment. Thus, the goal is to improve reproduction efficiency while decreasing any environmental impacts. Better nutritional management, genetic selection for fertility, and more attention to success can improve reproductive efficiency. In order to do all this efficiently, biosystems models can be of great efficacy.

Our hypothesis is that a dynamic, mechanistic, deterministic metabolic systems model of nutrition and reproduction can be used to describe the control of reproductive processes in the dairy cow. Therefore, our objective was to integrate two existing mechanistic, dynamic models of nutritional and reproductive processes in the dairy cow. The objective of this research model is to be suitable for evaluation of data, concepts, and hypotheses regarding underlying genetic, nutritional, and physiological control of reproduction. A model of metabolism (Molly, UC Davis); which describes metabolism of glucose, VFA, and amino acids for fat and protein synthesis and degradation and milk component production, as well as tracking energy transactions (ADP/ATP) (Baldwin, 1995) was integrated with a model of reproductive processes (Stötzel *et al.*, 2011), which describes growth and decay of the follicles and corpus luteum, gonadotropin releasing hormone, follicle stimulating hormone, luteinizing hormone, progesterone, estrogen, oxytocin, and prostaglandin F2α over time. The two models are integrated at specific points: increased body fat and glucose availability decrease the post-partum anovulatory interval; glucose and IGF-I increase rates of follicle stimulating hormone, luteinizing hormone, and follicular growth; and increased degradation of estrogen and progesterone in the liver affect follicular development, maintenance of the corpus luteum and embryonic survival. The energy balance of the cow and resultant flux through the adipose tissue and liver alter the pattern of cyclicity and affects the onset of follicular waves. In addition, glucose flux pre-calving and in early lactation affects IGF-1 concentrations, liver derived IGF-1 regulates final maturation of the dominant follicle (Kawashima *et al.*, 2007). Finally, faster metabolic flux through the liver increases degradation of estrogen and progesterone (Sangsritavong *et al.*, 2002), with a resultant decrease in the signs of visible estrus, reducing the number of inseminations. The reduction in progesterone concentration, prior to insemination and during the luteal phase of the cycle after insemination, is associated with low embryo survival (Diskin *et al.*, 2006).

The model behavior is such that it describes the pattern and direction of response in reproductive processes consistent with literature values. That is: lower energy balance and glucose availability reduce adipose tissue amount and delay the onset of cycling; lower IGF1 (as a function of glucose and energy balance) reduces follicular growth; and increased metabolic rate (conversion of ATP to ADP in the total viscera) as a function of feed intake or milk production increases reduces the peak and average concentrations of estrogen and progesterone. A reduction in body fat of 1 kg will increase DIM at first cycling by 0.15 d (thus, 10 kg less body fat delays first estrus by 1.5 d). Increasing circulating concentration of IGF1 by 20 ng/ml will increase follicular growth by 1 mm in 18 days (versus an average of 12 mm pre-ovulation). Increasing AtAdv (total ATP conversion to ADP) from 372 to 1488 moles/d decreases peak estrogen from 0.342 to 0.281 (arbitrary units; peak $E = 153e^{(0.0004(AtAdv))}$)and degrades the pattern of peaks; progesterone at peak is decreased from 1.023 to 0.913 (peak $P = 274e^{(8X10\text{-}4*AtAdv)}$) and time between peaks is increased.

The model will be tested against other research and field data in order to identify weaknesses in understanding, to use as a basis for further research. The model is to be used in design and testing of specific research hypotheses, integrated over genetics, nutrition and reproduction, to connect the basic biology with animal level function. It can be used in reproductive biology, nutrition and animal management research, and can be used in classes and extension efforts to assist in integrating complex functions toward better management decisions.

References

Baldwin, R. L., 1995. Modelling ruminant digestion and metabolism. Chapman & Hall, New York, USA, pp. 469-518.

Chagas, L. M., J. J. Bass, D. Blache, C. R. Burke, J. K. Kay, D. R. Lindsay, M. C. Lucy, G. B. Martin, S. Meier, F. M. Rhodes, J. R. Roche, W. W. Thatcher, and R. Webb, 2007. Invited Review: New perspectives on the roles of nutrition and metabolic priorities in the subfertility of high-producing dairy cows. J. Dairy Sci. 90, 4022-4032

Diskin, M. G. and D. G. Morris, 2008. Embryonic and early foetal losses in cattle and other ruminants. Reprod. Dom. Anim. 43(Suppl. 2), 260-267.

Kawashima, C., S. Fukihar, M. Maeda, E. Kaneko, C. A. Montoya, M. Matsui, T. Shimizu, N. Matsunaga, K. Kida, Y. I. Miyake, D. Schams, and A. Miyamoto, 2007. Relationship between metabolic homones and ovualtion of dominant follicle during the first follicular wave post-partum in high producing dairy cows. Repro. 133, 155-163.

McNamara, J.P., 2010. Nutrigenomics for improved reproduction, Chapter 19 in Reproductive Genomics in Domestic Animals by Jiang, Z. and T. Ott, eds. 27.

Sangsritavong, S., D. K. Combs, R. Sartori, L. E. Armentano, and M. C. Wiltbank, 2002. High feed intake increases liver blood flow and metabolism of progesterone and Estradiol-17β in dairy cattle. J. Dairy Sci. 85, 2831-2842.

Stötzel, C., J. Plöntzke, and S. Röblitz, 2011. Advances in modeling of the bovine estrous cycle: Administration of PGF2α. ZIB-report 11-17, Zuse Institute Berlin.

Use of thermodynamic equations to improve predictions of volatile fatty acid production by the Molly cow model

S. Ghimire[1], *R.A. Kohn*[2], *P. Gregorini*[3] *and M.D. Hanigan*[1]
[1]*Virginia Polytechnic Institute and State University, Blacksburg, VA 24061, USA; sand72@vt.edu*
[2]*Animal and Avian Sciences, University of Maryland, College Park, MD 20742, USA*
[3]*DairyNZ Ltd., Private Bag 3221, Hamilton, New Zealand*

Introduction

Volatile fatty acids are important products of fermentation in ruminant animals contributing about 72% of the total energy supply (Bergman, 1990). They also affect ruminal hydrogen supply, which dictates methane production. Therefore it is imperative to be able to accurately predict VFA production. The Molly cow model is a dynamic mechanistic model which uses stoichiometry coefficients derived by Murphy *et al.* (1982) to predict VFA production. The coefficients are based on the assumption that substrate supply is a primary determinant of VFA production rates, and interconversions are assumed to either not occur or be constant across diets. Our recent work demonstrated that the model poorly predicted VFA production rates and that accounting for variable interconversion rates may improve prediction accuracy. The objectives of this study were to incorporate VFA interconversion equations and evaluate them in the Molly cow model.

Material and methods

The thermodynamic equations described by Ungerfeld and Kohn (2006) were used to represent interconversions among VFA in the model with ruminal concentrations of VFA and pH the main driving variables. Rate constants for each equation were algebraically calculated from several literature observations, mean estimates were used in the model. Global sensitivity of VFA concentrations and synthesis to the Murphy coefficients and interconversion rate constants was conducted using the Fourier amplitude sensitivity test as described by Saltelli *et al.* (1999) with sampling boundaries of 75 and 125% of initial parameter estimates. For evaluation and parameter estimation purposes, a VFA production dataset containing 8 studies was combined with a data set containing VFA concentration data from 62 studies. The latter was a subset of those used by the NRC committee to formulate the 2001 Nutrient Requirement model. The model was set to simulate for 10 d to ensure steady state, and predictions from the last day were compared to observed values. Parameters that were estimated included the de novo VFA production coefficients for acetate, propionate, and butyrate de novo production from starch, soluble carbohydrate, cellulose, and protein for mixed diets plus the absorption rate constants for each. Residual errors of prediction were used to calculate root means square prediction errors (RMSPE), which were expressed as a percentage of the observed mean, and RMSPE were partitioned into mean bias, slope bias, and dispersion.

Results and discussion

The rederived coefficients were different than the original set. The parameters that were most affected were de novo acetate, propionate, and butyrate synthesis from cellulose, and acetate and butyrate synthesis from hemicelluloses with more than 90% decrease followed by de novo synthesis of propionate from starch and structural carbohydrate (81 and 63% decrease, respectively). Parameters for denovo butyrate synthesis from starch, and acetate and propionate absorption were least affected (7, 10 and 9%, respectively). he results of residual error analyses are presented in Table 1. The revised model predicted net synthesis rates of propionate and acetate substantially better than without the thermodynamic equations primarily due to a reduction in the previously unexplained dispersion error. Predictions of butyrate synthesis were not improved. Ruminal acetate synthesis rates were sensitive to parameters for acetate production from starch, soluble carbohydrate and cellulose and acetate conversion to butyrate (sensitivity coefficients of 0.40, 0.16, 0.07 and 0.23, respectively). Similarly, net propionate synthesis was sensitive to changes in parameters for production from starch, soluble carbohydrate, and butyrate (sensitivity coefficients of 0.58, 0.12 and 0.14, respectively). Net production of butyrate was primarily affected by changes in parameters for interconversion from acetate to butyrate and from butyrate to propionate (0.27 and 0.34, respectively) with little sensitivity to the Murphy coefficients.

This preliminary work indicted that a fully exchanging three pool VFA production model that considers thermodynamic influences better represented observed ruminal production rates for propionate in the Molly cow model. Based on the sensitivity analyses, additional improvements in prediction accuracy, particularly for butyrate, may be possible if interconversion rate constants are parameterized in concert with the denovo VFA synthesis coefficients. The presence of mean bias in predictions of acetate and butyrate concentrations suggests that interconversions with those pools may also need parameterization or that further progress in defining these coefficients may be possible.

Table 1. Residual error analyses for predictions by the Molly cow model with the original Murphy et al. *(1982) coefficients (Initial) and altered to represent thermodynamically driven interconversions among VFA and a newly derived coefficients (Final).*

Variable	N	Observed mean[1]	RMSPE		Mean bias		Slope bias	
			Initial	Final	Initial	Final	Initial	Final
			% of Mean		% of MSPE			
Net synthesis								
Acetate	23	23.1	69	58	4	10	39	13
Propionate	23	10.9	59	38	20	3	1	22
Butyrate	21	6.1	48	46	1	21	54	30
Concentration								
Acetate	193	65.1	27	25	41	13	8	25
Propionate	193	25.3	44	40	45	35	10	11
Butyrate	193	12.6	39	38	41	31	19	21

[1] Values for net synthesis are in mol/d and for concentration are in mmol/l.

Acknowledgements

Research funded by the New Zealand Agricultural Greenhouse Gas Research Centre.

References

Bergman, E.N., 1990. Energy contributions of volatile fatty acids from the gastrointestinal tract in various species. Physiological Reviews 70, 567-590.

Murphy, M.R., Baldwin, R.L., Koong, L.J., 1982. Estimation of stoichiometric parameters for rumen fermentation of roughage and concentrate diets. J. Anim. Sci. 55, 411-421.

Saltelli, A., Tarantola, S., Chan, K.P.S., 1999. A Quantitative Model-Independent Method for Global Sensitivity Analysis of Model Output. Technometrics 41, 39-56.

Ungerfeld, E.M., Kohn, R.A., 2006. The role of thermodynamics in the control of ruminal fermentation. In: Sejrsen, K., Hvelplund, T., Nielsen, M.O. (editors), Ruminant physiology: digestion, metabolism and impact of nutrition on gene expression, immunology and stress. Wageningen Academic Publishers, Wageningen, 55-86.

BR-CORTE 1.0: online software for diet optimization on tropical conditions

S.C. Valadares Filho[1,2], *M.L. Chizzotti*[2,3], *P.A.S. Machado*[1,2], *H.F. Amaral*[1] *and T. Furtado*[1]
[1]*Universidade Federal de Viçosa, Department of Animal Science, Viçosa, MG, 36571, Brazil*
[2]*Instituto Nacional de Ciência e Tecnologia em Ciência Animal, DZO/UFV Viçosa, MG, 37200, Brazil; mariochizzotti@dzo.ufla.br*
[3]*Universidade Federal de Lavras, Department of Animal Science, Lavras, MG, 37200, Brazil*

Precision feeding is essential to improve the economical and environmental efficiency of beef cattle production. BR-CORTE is a free-access, on-line software (www.brcorte.ufv.br) that integrates the nutrient requirement model for Zebu (Valadares Filho *et al.*, 2010a) with the Brazilian feed composition tables (CQBAL 3.0, Valadares Filho *et al.*, 2010b) to optimize diets for beef cattle.

BR-CORTE 2010 (Valadares Filho *et al.*, 2010a) is a mechanistic static model that uses animal and growing system information to estimate the nutritional requirements of *Bos indicus* and their crosses. The model is based on meta-analysis of a dataset of 25 comparative slaughter studies uses as input the physiological stage, breed (Nellore or crosses with Bos taurus), finishing system (pasture or feedlot), sex (bull, steer and heifer), weight and average weight gain to predict the requirements of digestible energy, rumen degradable protein, rumen non-degradable protein and minerals. A summary of the prediction equations for the requirements of energy and protein is reported in Table 1.

The CQBAL 3.0 (www.ufv.br/cqbal) is an on-line feed composition database containing more than 2,600 references from thesis and dissertations reporting feed composition in Brazil (Valadares Filho, *et al.*, 2010b). The software uses feed composition (requires neutral detergent fiber correct for ash and protein, lignin, crude protein, neutral and acid detergent fiber proteins, non-fibrous carbohydrates, ether extract, in situ water soluble fraction and in situ potentially degradable fraction and its rate of degradation) and feeding conditions to predict feed digestible fractions and protein fractions based on equations of Dettman *et al.* (2008) and exports that information for the BR-CORTE 1.0 software.

BR-CORTE 1.0 then uses animal information to predict the requirements for a target weight and feeds selected from the CQBAL 3.0 dataset and their costs per kg as fed, to find the optimal solution of proportion of selected feeds based on the minimum cost of the final diet, using the simplex method and returns the best solution (Murty, 1983). The software allows bounds to limit the proportion of ingredients as well as the proportion of roughage to concentrate. The output reports the proportion of ingredients, in as fed and dry matter basis, and the amount of each nutrient supplied by each ingredient.

The BR-CORTE 1.0 is a useful tool for modeling nutritional requirement of Zebu cattle as well as for teaching and diet formulation.

Table 1. Summary of the BR-CORTE 1.0 prediction equations for the energy and protein requirements of beef cattle.

DMI	Nellore: -2.7878 + ($0.08789 \times SBW^{0.75}$) + ($5.0487 \times ADG$) – ($1.683 \times ADG^2$); Crossbred: -2.6098 + ($0.08844 \times SBW^{0.75}$) + ($4.4672 \times ADG$) – ($1.3579 \times ADG^2$)
EBW	Feedlot: $0.895 \times SBW$; Pasture: $0.863 \times SBW$
EBG	Feedlot Nellore: $0.935 \times ADG$; Feedlot Crossbred: $0.966 \times ADG$; Pasture: $0.955 \times ADG$
EQEBW	Nellore: $(EBW/430) \times 440$; Crossbred:$(EBW/455) \times 440$
NEm	Feedlot:$0.0742 \times EBW^{0.75}$; Pasture $0.0717 \times EBW^{0.75}$
NEg	Bulls:$0.053 \times EQEBW^{0.75} \times EBG^{1.095}$; Steers: $0.064 \times EQEBW^{0.75} \times EBG^{1.095}$; Heifers: $0.072 \times EQEBW^{0.75} \times EBG^{1.095}$; Pasture: $0.052 \times EQEBW^{0.75} \times EBG^{1.062}$
%REp	$1.1404 \times (RE/EBG)^{-1.137}$
Kg	0.327/[0.539 + (%REp/100)]
Km	Nellore: $0.513+(0.173 \times Kg)+(0.100 \times EBG)$; Crossbred: $0.513+(0.173 \times Kg)+(0.073 \times EBG)$
ME	(NEm/km) + (NEg/*Kg*)
TDN	ME/0.82/4.409
NPg	Nellore Bulls: ($238.79 \times EBG$) – ($15.68 \times NEg$); Nellore Steers and Heifers: ($163.73 \times EBG$) – ($4.65 \times NEg$); Crossbred Bulls: ($219.43 \times EBG$) – ($15.01 \times NEg$); Crossbred Steers and Heifers: ($188.71 \times EBG$) – ($7.67 \times NEg$)
K	If SBW $\leq$350 kg: 84.665 – ($0.1179 \times EQEBW$); if SBW >350 kg: 46.9
MPm	Feedlot: $4.0 \times BW^{0.75}$; Pasture: $4.5 \times BW^{0.75}$
MP	MPm + (NPg/*K*)
MCP	$120 \times TDN$
CP	$(1.11 \times MCP)+\{[(MP - (MCP \times 0.64)]/0.80\}$

DMI = DM intake, kg/day; SBW = shrunk BW, kg; ADG = average daily gain, kg/day; EBW = empty BW, kg; EBG = empty body gain, kg/day; EQEBW = equivalent empty BW, kg; NEm = NE for maintenance, Mcal/day; NEg = NE for growth, Mcal/day; %Rep = percentage of energy retained as protein, %; *Kg* = efficiency of utilization of metabolizable energy for growth, %; *K*m = efficiency of utilization of metabolizable energy for maintenance, %; ME = metabolizable energy, Mcal/day; TDN = total digestible nutrients, kg/day; NPg = Net protein for growth, g/day; *K* = efficiency of utilization of metabolizable protein for growth, %; MPm = metabolizable protein for maintenance, g/day; MP = total metabolizable protein, g/day; MCP = microbial crude protein, g/day; CP = total dietary crude protein, g/day.

References

Detmann, E., S. C. Valadares Filho, D. S. Pina, L. Henriques, M. F. Paulino, K. A. Magalhaes, P.A. Silva and M. L. Chizzotti, 2008. Prediction of the energy value of cattle diets based on the chemical composition of the feeds under tropical conditions. Anim. Feed Sci. Tech. 143, 127-147.

Murty, K. G, 1983. Linear programming. John Wiley & Sons Inc.: New York, 482p.

Valadares Filho, S. C., M. I. Marcondes, M. L. Chizzotti and P. V. R. Paulino, 2010a. Nutrient Requirements of Zebu Beef Cattle: BR-CORTE. 2. ed. UFV: Viçosa, 185p.

Valadares Filho, S. C., P. A. Silva, M. L. Chizzotti, H. F. Amaral, K. A. Magalhães, V. R. Rocha Junior and E. R. Capelle, 2010b. Tabelas Brasileiras de composição de alimentos para bovinos. 2. ed. UFV: Viçosa, 502p.

A structural equation model to analyze energy utilization in lactating dairy cows

L.E. Moraes[1], A.B. Strathe[2], E. Kebreab[1], D.P. Casper[3], J. Dijkstra[4], J. France[5] and J.G. Fadel[1]
[1]Department of Animal Science, University of California, Davis, CA 95616, USA; lemoraes@ucdavis.edu
[2]Department of Basic Animal and Veterinary Sciences, University of Copenhagen, Denmark
[3]Dairy Science Department, South Dakota State University, Brookings, SD 57007, USA
[4]Animal Nutrition Group, Wageningen University, 6700 AH Wageningen, the Netherlands
[5]Department of Animal and Poultry Science, University of Guelph, ON, N1G 2W1, Canada

Introduction

Energy balance trials from lactating cows have traditionally been analyzed using the regression approach. However, univariate analysis of mutually interacting animal traits can provide biased parameter estimates if one trait is used as an explanatory variable to model a second trait and the sub-model of the first trait is ignored (Gianola and Sorensen, 2004). Moreover, multivariate methods should be preferred because animal responses are correlated and relationships among responses can be quantified through the use of structural equations. Furthermore, the US current energy evaluation system of dairy cattle (NRC, 2001) relies on energetic parameters from the 1960's but over the past decades genetic improvement of dairy cattle has substantially increased the ability of cows to produce milk. The objective of the present study was to analyze energy utilization in lactating cows using structural equations.

Material and methods

A structural equation model was developed to analyze energy utilization in lactating cows using 1,111 energy balance records from 45 studies conducted from 1963 to 1995. Records represent at least 4 consecutive days of lactating cows in respiration chambers and were collected at the former USDA Energy Metabolism Unit at Beltsville, Maryland. The model was developed under a Bayesian framework, in which minimally informative priors were assigned to all parameters and statistical inference was based on Markov Chain Monte Carlo methods. The response vector, comprised by metabolizable energy (ME) intake, milk energy and tissue energy balance, was modeled through a multivariate normal distribution.

The structural parameters represent the causal relationships among responses and, under our model structure, determined structural parameters can be interpreted as the efficiency of energy utilization. Each response was modeled with a set of exogenous covariates and cross-classified random effects of animal and study. The structural equation model can be represented by: $y_{ijk} = \Lambda_d y_{ijk} + X_{ijk}\beta_d + a_i + s_j + e_{ijk}$, where y_{ijk} is the $(r \times 1)$ response vector of the kth record on animal *i* and study *j*, Λ_d is the $(r \times r)$ matrix of structural parameters describing the causal structure among responses on the dth decade, X_{ijk} is the $(r \times p)$ matrix of exogenous covariates, β_d is the $(p \times 1)$ vector of regression coefficients on the dth decade, a_i and s_j are $(r \times 1)$ vectors of animal and study random effects and e_{ijk} is the $(r \times 1)$ vector of errors. In this notation, *r* represents the number of responses and *p* the number of exogenous covariates. Structural parameters and regression coefficients were allowed to depend on the decade the experiment was conducted (i.e. 1963-1973, 1973-1983 and 1983-1995), estimating decade specific parameters and investigating the changes on these energetic parameters over the years. Maintenance requirements and energetic efficiencies were estimated using the functional form described by Strathe *et al.* (2011). Parameters' 95% credible intervals and Bayesian *P*-values, computed as the proportion of posterior samples which are different from zero, were used to compare the differences between energetic parameters from the second and third decades with estimates from 1963-1973.

Results and discussion

Energetic parameters and their 95% credible intervals are reported in Table 1. The metabolizability (q), estimated as the slope of the regression of ME intake on gross energy (GE) intake, was similar between decades (P=0.94 and P=0.10) as well as the efficiency (k_T) of utilizing tissue energy to produce milk (P=0.81 and P=0.31). However, net energy requirements for maintenance (NE_M) increased substantially over the years (P=0.02 and P<0.001) as well as the efficiency (k_L) of utilizing ME intake for milk production (P=0.06 and P<0.001). Similarly, the efficiency (k_G) of utilizing ME intake for growth increased substantially over the three decades (P=0.005 and P<0.001). Therefore, modern lactating cows have higher maintenance energy requirements than cows from the 1960's and also have a higher efficiency for utilizing ME intake for milk production and tissue gain.

Table 1. Energetic parameters estimates and their 95% credible intervals.

Parameter		Decade	Estimate	95% CI
q	MJ/MJ	1963-1973	0.52	(0.50, 0.53)
q	MJ/MJ	1973-1983	0.52	(0.50, 0.54)
q	MJ/MJ	1983-1995	0.53	(0.51, 0.55)
NE_M	$MJ/kg^{0.75}/day$	1963-1973	0.25	(0.22, 0.29)
NE_M	$MJ/kg^{0.75}/day$	1973-1983	0.32	(0.27, 0.37)
NE_M	$MJ/kg^{0.75}/day$	1983-1995	0.41	(0.36, 0.47)
k_L	MJ/MJ	1963-1973	0.57	(0.55, 0.59)
k_L	MJ/MJ	1973-1983	0.60	(0.57, 0.64)
k_L	MJ/MJ	1983-1995	0.65	(0.62, 0.67)
k_T	MJ/MJ	1963-1973	0.89	(0.83, 0.94)
k_T	MJ/MJ	1973-1983	0.89	(0.83, 0.97)
k_T	MJ/MJ	1983-1995	0.84	(0.74, 0.93)
k_G	MJ/MJ	1963-1973	0.57	(0.53, 0.62)
k_G	MJ/MJ	1973-1983	0.68	(0.61, 0.76)
k_G	MJ/MJ	1983-1995	0.72	(0.66, 0.79)

References

Gianola, D. and D. Sorensen, 2004. Quantitative genetic models for describing simultaneous and recursive relationships between phenotypes. Genetics 167, 1407-1424.

NRC. 2001. Nutrient Requirements of Dairy Cattle. 7th ed. National Academy Press, Washington, DC.

Strathe, A.B., J. Dijkstra, J. France, S. Lopez, T. Yan and E. Kebreab, 2011. A Bayesian approach to analyze energy balance data from lactating dairy cows. J. Dairy Sci. 94, 2520-2531.

Protein deposition potential and modeling of methionine requirements in homozygous (*Na/Na*) and heterozygous (*Na/na*) naked neck meat type chicken

D.R. Khan, C. Wecke and F. Liebert
Division Animal Nutrition Physiology, Department of Animal Sciences, Georg-August-University, Kellnerweg 6, 37077 Goettingen, Germany; dkhan@gwdg.de

Introduction

High ambient temperature (AT) decreases the growth of fast growing chicken because of difficulty in dissipating heat through the feather coverage. Consequently, they become more sensitive to high AT (Yunis and Cahaner, 1999). Introduction of the naked neck gene (*Na*) in modern meat type genotypes could be helpful in tolerating high AT by providing more surface area for heat dissipation compared to their normally feathered (*na/na*) counterparts. The aims of this study were to determine the (1) protein deposition potential and (2) modeling the methionine requirement for these genetics.

Material and methods

Birds of three genotypes (*Na/Na; Na/na; na/na*) were obtained by mating their heterozygous (*Na/na*) parents. In total 144 average weight birds (72 *Na/Na* male and female; 72 *Na/na* male and female) were selected for two N-balance experiments involving both starter and grower periods. Each experimental period was divided into a 5 d adaptation period and two 5d consecutive excreta collection periods. Each balance study utilized 72 birds (36 *Na/Na*; 36 *Na/na*) involving both genders, randomly allotted to five diets with graded dietary protein supply. Due to principles of the diet dilution technique, the dietary amino acid ratio was kept constant. Methionine (Met:Cys = 1:1.01) was identified as the first limiting amino acid in each diet. Methionine requirements depending on age and growth rate (protein deposition) were derived according to Samadi and Liebert (2008).

The daily N maintenance requirement (NMR) was determined using the exponential regression between N intake (NI, mg/$BW_{kg}^{0.67}$) and total daily N excretion (NEX; mg/$BW_{kg}^{0.67}$). The threshold value of the function (Equation 1 and 2) was estimated by application of the Levenberg-Marquardt algorithm within the statistical program package SPSS (Vers. 19).

$$NR = NR_{max}T\ (1 - e^{-b.NI}) \qquad (1)$$

$$ND = NR_{max}T\ (1 - e^{-b.NI}) - NMR \qquad (2)$$

In the equations ND is daily N-deposition, NR is daily N-retention (ND + NMR; mg/$BW_{kg}^{0.67}$); $NR_{max}T$ is theoretical maximum for daily NR (mg/$BW_{kg}^{0.67}$); and b is the slope of the N-retention curve (indicating the feed protein quality independent of N intake).

Results and discussion

Results of both NMR and $ND_{max}T$ estimation are summarized in Table 1. Daily NMR did not reveal any statistical difference between genetics and age periods. However, there was a significant difference between male and female birds. NMR data as observed were applied for further estimation of $ND_{max}T$.

Making use of the yielded model parameters, as an example Methionine requirement data were derived for targeted CP deposition depending on genotype and age (Table 2).

Table 1. Estimates of daily maintenance requirement (NMR) and of theoretical potential for daily N-deposition ($ND_{max}T$).

	Starter period				Grower period			
	Na/Na		*Na/na*		*Na/Na*		*Na/na*	
	NMR	$ND_{max}T$	NMR	$ND_{max}T$	NMR	$ND_{max}T$	NMR	$ND_{max}T$
	$mgN/BW_{kg}^{0.67}/d$							
Male	259	3,849	229	4,053	298	2,674	319	3,094
Female	339	3,613	385	3,890	351	2,489	367	2,560

Table 2. Example for modeling the Met-requirement of male naked neck chicken with dietary Met-efficiency as observed and aimed CP deposition (mean BW starter: 500g; grower: 1500g).

		Na/Na				Na/na			
Item		Starter		Grower		Starter		Grower	
CP deposition (g/d)		9	11	13	15	9	11	13	15
Met-efficiency (bc^{-1})		182	182	252	252	176	176	206	206
Met-requirement ($mg/BW_{kg}^{0.67}/d$)		209	298	327	409	198	274	324	394
Predicted feed intake (g/d)		Met-content needed in the diet (%) depending on feed intake							
Starter	Grower								
60	120	0.34	0.49	0.27	0.34	0.33	0.46	0.27	0.33
70	130	0.30	0.42	0.25	0.31	0.28	0.39	0.25	0.30

Methionine requirements for *Na/Na* were slightly higher as compared to *Na/na* in starter period. However, yielded Met-requirement data were in good agreement with previous studies with normally feathered chicken. In conclusion, for a dietary ratio of Met:Cys near 1:1, reduced feathering did not interfere with Methionine requirements.

References

Samadi and F. Liebert, 2006. Estimation of nitrogen maintenance requirements and potential for nitrogen deposition in fast-growing chickens depending on age and sex. Poult. Sci. 85, 1421-1429.

Samadi and F. Liebert, 2008. Modelling the optimal lysine to threonine ratio in growing chickens depending on age and efficiency of dietary amino acid utilisation. Brit. Poult. Sci. 49, 45-54.

Yunis, R. and A. Cahaner, 1999. The effect of naked neck (Na) and frizzle (F) genes on growth and meat yield of broiler and their interactions with ambient temperatures and potential growth rate. Poult. Sci. 78, 1347-1352.

Linear and non-linear estimates of the efficiency with which metabolizable energy is used for maintenance or gain

C.A. Old[1] *and H.A. Rossow*[2]
[1]*Antelope Ranch, LeGrand, CA 95333, USA; anteloperanch31@gmail.com*
[2]*School of Veterinary Medicine, University of California, Davis, CA 95616, USA*

Introduction

Linear estimates of the efficiency of metabolizable energy (ME) for maintenance (k_m) or gain (k_g), calculated by Lofgreen and Garrett (1968), the basis for the California Net Energy System (CNES), differ from estimates calculated from metabolic pathways; observations in the former are consistent with the statement that k_m is greater than k_g and the converse for the latter. In their analysis these authors describe metabolizable energy intake as both response and predictor variable. A further assumption is that the relationship between retained energy (RE) and ME intake is rectilinear; this assumption is based on a static maintenance requirement. An analysis by Blaxter (1956) and others, showed that maintenance is dynamic. This study was undertaken to evaluate equations representing either linear or nonlinear relationships between ME intake and (RE) to compare estimates of k_m and k_g.

Material and methods

Data from Lofgreen and Garrett (1968) for growing beef steers and heifers were analyzed using either the linear models employed by those investigators, log heat production (HP) = a + b*ME intake (MEI) (1) and MEI = c + RE/k_g (2); a represents log fasting HP (FHP) and c is ME_m (ME intake at RE=0), the point in common. The third equation contained both first order and linear elements and was MEI = $(a_1e^{k1RE} + a_2e^{k2RE})/k_m + RE/k_g$ (3); a_n are parameter estimates, k_n are rate constants. All energy terms used by Lofgreen and Garrett (1968) were scaled by $W^{0.75}$ (W = empty body weight, kg); no such convention was employed in the exponential model. Parameter estimates for the linear analysis are least squares estimators and reliability of parameter estimates was evaluated using a Markov Chain Monte Carlo (MCMC) simulation (1000 simulations) utilizing the method described by Martin *et al.* (2011) through the public domain language R (R, 2011). Parameter estimates for the exponential model were calculated using a non-linear optimization technique (R, 2011). Reliability of parameter estimates k_1, k_2, k_m and k_g was also evaluated using the same technique as for the linear model.

Results and discussion

For the linear model (2) a 95% confidence interval about the intercept ($0.131/W^{0.75}$) ranged from 0.114 to 0.148. Fasting heat production was calculated to be $0.077/W^{0.75}$; k_m is 0.077/0.133, or 0.579. For k_g (0.432), a 95% confidence interval was from 0.360 to 0.539; R^2, while adequate (0.784), suggests that not all the assignable variability in the response variable was captured by variability in the predictor variable. Tests for goodness of fit indicate that equation (2) used by Lofgreen and Garrett (1968) is misspecified (P=0.246) reinforcing the previous statement. Parameter estimates for model 1 were stable; 95% confidence intervals were ±10% about the mean. Model 2 parameter estimates were unstable; MCMC simulation results were largely outside the range of biologically possible range; for model 2 simulation results, 92.9% were biologically infeasible, 65.9% of the simulations indicated either ME_m was less than 0 or k_g greater than 1.

For the second model R^2 was improved (0.864), the estimate of k_m was 0.278±0.001; k_g was 0.662±0.002; these are similar to theoretical maxima for which k_m=0.279 (8 ATP per mole of acetate oxidized) and k_g=0.723. Estimates of k_m for 1000 simulations ranged from 0.250 to 0.305 and for k_g the range was from 0.599 to 0.732; all simulations gave estimates that were not different from those that are biologically possible. Considering that the model is overparametized, this stability is notable. Koong *et al.* (1983) and more recently, Labussierre *et al.* (2011) noted that maintenance requirement is not a constant function of body weight, but rather increases with energy intake. It is possible that the results observed in the second model are simply due to shifting HP from synthesis to maintenance; if this were the case it is expected that k_m and k_g would be correlated. Parameter estimates k_m and k_g from the MCMC simulation were not correlated (r=0.008). It should be noted that for model 3 solutions for k_m and k_g are simultaneous; this is not the case for the linear solutions of k_m and k_g.

This analysis indicates that the parameter estimates regarding efficiencies of ME use for maintenance and gain obtained from linear equations scaled by $W^{0.75}$ are highly unstable and may not reflect the true parameters. While adequate for the intended purposes, that is, prediction of gain in growing and finishing beef cattle, our analysis indicates that these estimates should not be used as the basis for a mechanistic explanation regarding efficiencies of ME utilization. Nonlinear estimates of efficiencies are similar to theoretical estimates obtained from biochemical pathways and more of the variability in ME intake was described by this equation. Incorporation of these concepts into models used to predict input:output relationships for growing and finishing beef cattle may allow for more accurate prediction of performance.

References

Blaxter, K.L. 1956. The nutritive value of feeds as a source of energy: a review. J. Dairy Sci. 39:1396-1424.

Koong, L.J., J.A. Nienaber and H.J. Mersmann. 1983. Effects of plane of nutrition on organ size and fasting heat production in genetically obese and lean pigs. J. Nutr. 113:1626-1631.

Labussierre, E., J. van Milgen, C.F.M. de Lange and J. Noblet. 2011. Maintenance energy requirements of growing pigs and calves are influenced by feeding level. J. Nutr. 141:1855-1861.

Lofgreen, G.P. and W.N. Garrett. 1968. A system for expressing net energy requirements and feed values for growing and finishing beef cattle. J. Anim. Sci. 27:793-806.

Martin, A.D., K.M. Quinn and J.H. Park. 2011. MCMC pack: Markov Chain Monte Carlo in R. J. Stat. Soft. 42: 1-5.

R, D.C.T., 2011. R: A language and environment for statistical computing. R Foundation for Statistical Computing.

Responses of laying hens to methionine and cystine intake

H.C.P. Bendezu, N.K. Sakomura, L. Hauschild, J. Sato, J.C.P. Dorigam and E.P. Silva
Department of Animal Science, University of Agrarian and Veterinary Sciences of UNESP, Jaboticabal, SP 14884900, Brazil; sakomura@fcav.unesp.br

Introduction

The knowledge of the importance of amino acid requirements is related to precise amino acid balance to prevent loss of energy due to an unbalanced diet and maximize economic results. Many studies have been conducted to determine the sum of methionine and cystine (Met+Cys) requirements and the results differ because there are many influential factors such as genetics, management, age, diets, environmental conditions and methodology used. In addition, requirements have been estimated based on the response of the birds over a long period and does not consider the dynamics of body changes in the birds. For this reason, the focus of this experiment was to determine the optimal Met+Cys intake for Dekalb White laying hens in four periods (32 to 48 weeks) according to dilution technique.

Material and methods

Two hundred and eighty-eight Dekalb White laying hens were used during the period from 32 to 48 weeks of age. The hens were distributed in a completely randomized design into eight treatments, six replicates and six hens per experimental unit. Diets were isocaloric (2,850 kcal/kg) with different levels of Met+Cys (0.137; 0.275; 0.414; 0.551; 0.689; 0.792 and 0.895%.) and were obtained by dilution technique (Fisher and Morris, 1970) with the eighth treatment as a control treatment used to verify bird response for Met+Cys limitation. Egg production, mass, feed intake and feed conversion per egg mass were evaluated. Data were analyzed using PROC GLM SAS 9.0. LSMEAN was used to compare treatments means by time. For those that were significant by period, they were subjected to regression analysis according the broken line (BL), quadratic (Q) and BL + Q model for each period (4 weeks) and those that were not significant by period were analyzed by a model for total experiment time (16 weeks).

Results and discussion

The control eighth treatment confirmed that Met+Cys were the limiting amino acids in this experiment. Egg mass was affected by Met+Cys level (P<0.0001) and by the interaction between level and period (P=0.0056). For this reason, the data were analyzed by period. Egg mass increases with age of the hen (Johnston and Gous, 2007). Values of egg mass were adjusted by the BL, Q and BL + Q models for each period. All models had high R^2, however, the BL + Q model was chosen because it has a biological interpretation. The following equations and the results of optimal Met+Cys intake by BL + Q, expressed per day by each period, are numbered sequentially in chronological order:

$$60.27 = -8.91257 + 0.1745 \text{ Met+Cys} - 0.0001064 \text{ Met+Cys}^2; 671 \text{ mg} \quad (1)$$

$$62.43 = -9.76324 + 0.1676\text{Met+Cys} - 0.000094 \text{ Met+Cys}^2; 728 \text{ mg} \quad (2)$$

$$64.56 = -10.516 + 0.16941\text{Met+Cys} - 0.000092 \text{ Met+Cyst}^2; 743 \text{ mg} \quad (3)$$

$$62.30 = -8.43476 + 0.14655 \text{ Met+Cys} - 0.000071 \text{ Met+Cys}^2; 770 \text{ mg} \quad (4)$$

The feed conversion per egg mass (FC) was affected by level (P<0.0001) but there was no significant interaction between level and period (P=0.313). Feed conversion per egg mass covered a wide range, with the value of the lower level treatment much different from the others. For this reason, the Q model did not have any problem in estimating the optimal Met + Cys intake and the result is in agreement others. Equation 5 presented the optimal daily feed intake.

$$FC = 13.17104 - 0.03711\ Met{+}Cys + 0.0000279\ Met{+}Cys^2;\ 665\ mg \qquad (5)$$

References

Fisher C. and T.R. Morris, 1970. The determination of the methionine requirement of laying pullets by a diet dilution technique. British Poultry Science 11, 67-82.

Johnston, S.A. and R.M. Gous, 2007. Modelling the changes in the proportions of the egg components during a laying cycle. British Poultry Science, 48: 347-353.

Assessment of ideal dietary amino acid ratios between branched-chain amino acids for growing chicken

A. Pastor, C. Wecke and F. Liebert
Division Animal Nutrition Physiology, Department of Animal Sciences, Georg-August-University, Kellnerweg 6, 37077 Goettingen, Germany; apastor@gwdg.de

Introduction

Improvement of amino acid (AA) recommendations are a precondition for sustainable conversion of feed to food proteins and have to be based on physiological acceptable and validated procedures to derive both quantitative daily AA requirements and optimal dietary AA ratios. According to earlier applications of an exponential nitrogen (N)-utilization model (Samadi and Liebert, 2006, 2008; Liebert, 2008), two experiments were conducted to obtain new data about ideal amino acid ratio (IAAR) between branched-chain amino acids (BCAA) leucine (Leu), isoleucine (Ile), valine (Val) and lysine (Lys) in meat type chicken diets.

Material and methods

Experiment 1

The N-balance study utilized 72 male chickens (Ross 308, 36/age period) to provide the model parameters (NMR: N-maintenance requirement; $ND_{max}T$: theoretical maximum for daily N-deposition). In total, 8 diets with graded protein supply (6%-34% CP) were fed (n=4-6) over a period of 15 days (5 days adaptation, 2×5 days collection period) for both starter (day 10-20) and grower period (day 25-35). Excreta were collected every 12 hours and stored at -20 °C until further analysis. Samples were analyzed according to the German VDLUFA standards (Naumann and Bassler, 1976-1997). NMR and $ND_{max}T$ were estimated following iteration steps by the Levenberg-Marquardt-Algorithm within the SPSS statistical package (version 19). Observed NMR and $ND_{max}T$ were applied for model parameter b (protein quality parameter, independent on NI).

Experiment 2

The N-balance study examined five diets (n=7), varying in individual AA supply by AA dilution [balanced control diet BD, diets limiting in Lys (80%), Leu (80%), Ile (67%), and Val (64% of the assumed requirement)]. Protein quality parameter (b) was applied to evaluate the individual AA efficiency (bc^{-1}), taking into account the dietary concentration (c) of the limiting AA. According to Samadi and Liebert (2008), observed AA efficiencies were utilized for conclusion of IAAR *[IAAR* $= bc_{Lys}^{-1} : bc_{individual\ BCAA}^{-1}$*]*. However, this proportion is only valid if the AA under study is really in the limiting position in the experimental diet, as indicated by a significant decline of b-value between balanced control diet and the AA diluted diet under study.

Results and discussion

Experiment 1 yielded daily NMR (113mg and 215mg $N/BW_{kg}^{0.67}$ for the starter and grower period, respectively) and daily $ND_{max}T$ (4593mg and 4302mg $N/BW_{kg}^{0.67}$ for starter and grower period, respectively). Observed model parameters were utilized for further model applications. Experiment 2 yielded most efficient N utilization due to the balanced diet BD (Table 1). A significant decline of protein quality (parameter b) was observed following each of the AA diluted diets. However, in the starter period an increased statistical risk had to be accepted to prove significance for further conclusion of Leu requirement data. Consequently, for this age period a further validation of Leu related observations is needed.

Table 1. Results of N balance study (Experiment 2) and derived ideal amino acid ratio for the branched-chain amino acids (Lysine=100). Different superscript letters reveal significant differences among the experimental diets.

	BD[1]	Leu[1]	Ile[1]	Val[1]	Lys[1]
Starter period (10-20d)					
Average BW[3] (g)	$462^{a}\pm155$	$398^{a}\pm150$	$363^{a}\pm142$	$343^{a}\pm142$	$306^{b}\pm92$
NI[4] (mg/d)[7]	$3,555^{a}\pm341$	$3,010^{bc}\pm493$	$3,209^{ab}\pm500$	$2,977^{bc}\pm572$	$2,597^{c}\pm531$
NEX[5] (mg/d)[7]	$1,116^{a}\pm177$	$895^{a}\pm261$	$1,085^{a}\pm265$	$1,114^{a}\pm288$	$887^{a}\pm248$
ND[6] (mg/d)[7]	$2,439^{a}\pm206$	$2,115^{b}\pm275$	$2,124^{b}\pm313$	$1,863^{bc}\pm300$	$1,710^{c}\pm297$
IAAR					
b-value	$221^{a}\pm12$	$215^{b}\pm14$[2]	$199^{b}\pm10$	$185^{c}\pm10$	$191^{bc}\pm9$
$bc_{AA\ experimental\ diet}^{-1}$	42±2	73±4	61±3	40±2	
IAAR (%)[8]		94	55	65	100
Recommendation NRC (1994)		109	73	82	100
Grower period (25-35d)					
AverageBW[3](g)	$1,445^{a}\pm280$	$1,152^{b}\pm207$	$1,141^{b}\pm226$	$1,129^{b}\pm214$	$1,234^{ab}\pm204$
NI[4] (mg/d)[7]	$3,322^{a}\pm341$	$2,299^{c}\pm414$	$2,430^{bc}\pm573$	$2,392^{b}c\pm561$	$2,843^{b}\pm298$
NEX[5] (mg/d)[7]	$947^{ab}\pm150$	$749^{b}\pm214$	$813^{ab}\pm236$	$968^{a}\pm258$	$990^{a}\pm174$
ND[6] (mg/d)[7]	$2,375^{a}\pm231$	$1,550^{c}\pm232$	$1,617^{bc}\pm372$	$1,424^{c}\pm336$	$1,853^{b}\pm161$b
IAAR					
b-value	$258^{a}\pm15$	$218^{b}\pm13$	$217^{b}\pm15$	$191^{c}\pm15$	$216^{b}\pm12$
$bc_{AA\ experimental\ diet}^{-1}$	42±3	80±6	63±5	45±2	
IAAR (%)[8]		106	56	72	100
Recommendation NRC (1994)		109	73	82	100

[1] $P<0.05$.

[2] $P<0.2$ (Duncan test or Games-Howell test, according to Levene test).

[3] BW = body weight.

[4] NI = N intake.

[5] NEX = N excretion.

[6] ND = N deposition.

[7] Related to metabolic BW ($BW_{kg}^{0.67}$).

[8] Based on calculation with sharp (not rounded) figures.

The yielded IAAR for BCAA in meat type chicken was lower than data reported earlier. Breeding progress can be expected as one factor leading to enhanced N utilization and modified requirements. Accordingly, in the starter period we expected a more distinctive limitation of Leu by its dilution down to 80% of the expected requirement. This observation indicates that previously reported Leu recommendations could overestimate the physiological requirement for actual broiler strains. Official AA recommendations (e.g. NRC, 1994) are older than 15 years and could be inadequate for current genotypes. For Leu and Val, a clear differentiation between starter and grower period became obvious. Consequently, we expect different ratios for both of the BCAA between age periods. For Val, we have support for this expectation from ongoing studies.

In conclusion, the derived IAAR for the BCAA was lower than earlier reports, indicating an overestimation of the BCAA with official recommendations. In addition, current results give support for an increased relative importance of Leu and Val with increasing age of the birds.

References

Naumann, K. and R. Bassler, 1976-1997. VDLUFA-Methodenbuch III. Die chemische Untersuchung von Futtermitteln. Mit Ergänzungslieferungen 1983, 1988, 1993, 1997. Darmstadt: VDLUFA-Verlag.

NRC (National Research Council), 1994. Nutrient Requirements of Poultry. 9th Rev. ed., National Academy Press, Washington, D.C.

Liebert, F., 2008. Modelling of protein metabolism yields amino acid requirements dependent on dietary amino acid efficiency, growth response, genotype and age of growing chicken. Av. Biol. Res. 1, 101-110.

Samadi and F. Liebert, 2006. Estimation of nitrogen maintenance requirements and potential for nitrogen deposition in fast-growing chickens depending on age and sex. Poult. Sci. 85, 1421-1429.

Samadi and F. Liebert, 2008. Modelling the optimal lysine to threonine ratio in growing chickens depending on age and efficiency of dietary amino acid utilization. Brit. Poult. Sci. 49, 45-54.

Meta-analysis of the response of growing pigs to valine content of the diet

J. van Milgen[1], M. Gloaguen[1], N. Le Floc'h[1], L. Brossard[1], Y. Primot and E. Corrent[2]
[1]INRA-Agrocampus Ouest, UMR1348 PEGASE, 35590 Saint-Gilles, France; jaap.vanmilgen@rennes.inra.fr
[2]Ajinomoto Eurolysine s.a.s. 75817 Paris Cedex 17, France

Introduction

The use of crystalline amino acids allows reducing the CP in the diet without compromising performance. The extent to which the CP content in the diet can be reduced under practical conditions depends on accurate knowledge of the requirements of amino acids that may become limiting when pigs are offered low-CP diets. Valine is now considered the next-limiting amino acid after Lys, Met (+Cys), Thr, and Trp in cereal-soybean meal based diets. Nevertheless, experimental studies on the response of pigs to the Val supply are scarce. Our objective was to analyze studies for growing pigs fed diets with increasing levels of L-Val.

Data collection and methods of analyses

Twenty-eight dose-response experiments were collected, originating from 20 publications, 9 of which were peer-reviewed. Within an experiment, diets only differed in the supply of L-Val with a minimum of 4 levels of Val. To account for differences in design (e.g. due to differences in body weight or differences in expressing the Val requirement) data were standardized within each experiment. Average daily feed intake (ADFI) and daily gain (ADG) were used as response criteria, and responses were expressed relative to that observed at the highest level of Val-supplementation. The Val content was reported or recalculated from feed ingredients using table values (Sauvant *et al.*, 2004). The standardized ileal digestible (SID) Val content was expressed relative to the Val requirement estimate of the NRC (1998) as a percentage in the diet (using the average body weight during the experiment) or relative to Lys (68% SID Val:Lys), depending on the experimental design. A quadratic regression analysis was carried out for each experiment to assess if the increase in Val content resulted in an increase in ADFI or ADG. Experiments for which there was no indication for a response (P>0.25) or where the magnitude of the response was less than 10% were not considered further. A bent-stick model was then used to analyze the response to the Val supply (Van Milgen *et al.*, 2012). This model is composed of an ascending line and a plateau, connected by quadratic transition phase. The transition accounts for the fact that the Val requirement declines during the experiment while the experimental diets do not change. The model was parameterized to include a Val requirement (midpoint of the transition phase) and plateau for each experiment, whereas the slope of the ascending line was shared across experiments. The Val supply required to maximize ADFI or ADG was also determined.

Results and discussion

Most experiments were carried out in young pigs and the final body weight was less than 35 kg in 25 of the 28 experiments. In 9 experiments, there was no indication for a response to the increasing Val supply. In 4 experiments, the lack of response could be explained because the lowest level of Val was only slightly lower (or even greater) than the NRC requirement estimate. In 3 other experiments, the Val requirement was expressed relative to Lys, but Lys may not have been second-limiting for growth, with Lys levels exceeding the NRC requirement estimate by 20%. For 2 experiments, the Met+Cys and Thr contents of the diet could have been second-limiting for performance (before Lys), thereby limiting the response of the pigs to the increasing Val supply. In 19 experiments (18 in young pigs, 1 in grower pigs), a response to the increasing Val supply was observed (P<0.25). The estimated Val requirements (midpoints of the transition phase) were similar for ADFI or ADG, and both estimates were highly correlated (r=0.97). A very low Val requirement was estimated for an experiment published in the 1950s (53% of the NRC requirement estimate), which was not included in the further analysis. The estimated Val requirements ranged from 86 to 117% of the NRC (1998) requirement estimate for ADFI and from 84 to 114% for ADG (94% on average for both). A Val deficiency resulted in a very strong reduction in performance, with slopes of the ascending lines of 2.34 and 2.69 for ADFI and ADG, respectively. Although the midpoint of the transition phase was considered as 'the requirement', a greater Val supply is required to maximize performance. Between 86 and 120% of the NRC requirement estimate would be required to maximize ADG (average 101%). Expressed relative to Lys, 64% SID Val:Lys would be required to attain the midpoint of the transition phase, while 69% SID Val:Lys would be required to maximize ADG. This increase in Val supply would increase ADG by 5%.

To further analyze the variation in Val requirement estimates, correlations were calculated between the Val requirement estimates and the supply of other essential amino acids (expressed relative to the requirement estimates of the NRC). These correlations were positive for all amino acids, suggesting that the protein supply has an impact on the Val requirement. An increased protein intake may increase protein turnover, resulting in a lower efficiency of Val utilization and an increased Val requirement. The greatest correlation coefficients between the amino acid content and the Val requirement were observed for Phe (0.84), Phe+Tyr (0.79), His (0.75), Thr (0.68), and Lys (0.67). The correlation coefficients with Leu and Ile were lower (0.47 and 0.17, respectively), suggesting that the supply of branched-chain amino acids has no impact on the Val requirement. This contrasts with Ile requirement, where the supply of other branched-chain amino acids or large neutral amino acids other appears to affect the Ile requirement. Although there are indications that an excess supply of Leu aggravates the effect of a Val deficiency (Gloaguen *et al.*, 2011), this effect could not be quantified because response to a Val deficiency was assumed to be similar for all dose-response studies in this meta-analysis.

In conclusion, young pigs respond to a dietary Val deficiency by reducing ADFI which, in turn results in a reduction in ADG. To maximize performance, 69% SID Val:Lys is required. Whether these results, obtained mostly in young pigs, can be extrapolated to grower and finisher pigs remains to be confirmed. The effect of the protein supply on amino acid requirements may require more attention.

References

Gloaguen M., N. Le Floc'h, L. Brossard, R. Barea, Y. Primot, E. Corrent and J. van Milgen, 2011. Response of piglets to the valine content in diet in combination with the supply of other branched-chain amino acids. Animal 5, 1734-1742.

NRC, 1998. Nutrient requirements of swine, 10th revised edition. National Academy Press, 189 pp.

Sauvant D., J.M. Perez and G. Tran, 2004. Tables of composition and nutritional value of feed materials. Pigs, poultry, cattle, sheep, goats, rabbits, horses, fish. INRA, Paris, 304 pp.

Van Milgen J., M. Gloaguen, N. Le Floc'h, L. Brossard, Y. Primot and E. Corrent, 2012. Meta-analysis of the response of growing pigs to the isoleucine concentration in the diet. Animal 6, 1601-1608.

Prediction of dry matter intake in dairy calves

A.L. Silva[1], M.I. Marcondes[1], M.M. Campos[2], F.S. Machado[2], M.M.D. Castro[1] and A.S. Trece[1]
[1]Department of Animal Science, Federal University of Viçosa, Brazil; marcos.marcondes@ufv.br
[2]Brazilian Agricultural Research Corporation, Embrapa Dairy Cattle, Brazil

Introduction

Dry matter intake (DMI) directly affects animal performance; it is the main determinant of nutrients used to meet the requirements for maintenance and production. Therefore, accurate models of dry matter intake in calves is fundamental to formulating diets to meet requirements and the efficient use of the nutrients by the animals (NRC, 2001; Chizzotti *et al.*, 2007). The objective was to establish a model to determine dry matter intake (DMI) and free water intake (WI) for crossbred Holstein-Zebu calves aged between 0 and 60 days.

Material and methods

The experiment was conducted in the Department of Animal Science, Universidade Federal de Viçosa, Brazil. Eighteen male calves, crossbreed Holstein-Zebu, with an initial body weight of 36±5.5 kg, were used. The animals were distributed, according to a completely randomized design, into 3 treatments with 6 replications, and these treatments consisted of three different levels of milk intake, which were 2 (2L), 4 (4L) and 8 (8L) liters per day. The animals had free access to water and concentrate (starter), which was formulated in accordance with the requirements presented in NRC (2001). The animals were fed twice a day at 6:00 and 15:00, and water and starter intake were measured every day at 6:00h. There were performed digestibility assays, at 15 and 45 days of life. During those periods, there were collected samples of feeds offered. Milk samples were dried by lyophilization (Method INCT-CA G-002/1). All samples were analyzed for dry matter, according to the method (INCT-CA G-003/1), describe by Detmann *et al.*, (2012). Environmental variables were also considered: relative humidity (RH), temperature and humidity index (THI) and black globe temperature and humidity index (BGTHI). These estimates were made from daily uptakes of black globe temperature, dry bulb temperature, wet bulb temperature, and maximum and minimum temperatures. Environmental effects, milk intake and age were used in a multiple regression model, considering both linear and quadratic effects, to estimate calves DMI and WI. The test was conducted using the MIXED procedure of the SAS, at the level of significance of 5%.

Results and discussion

The DMI was affected by milk intake ($P<0.001$), THI ($P=0.0453$) and age of the animals ($P<0.001$). However, no significant quadratic effects were observed ($P>0.05$), thus the final regression can be expressed:

$$DMI = 0.4272 + 0.6741 \times M - 0.0059 \times THI + 0.0122 \times day \qquad (1)$$

where DMI= dry matter intake (kg/day), M= milk intake (kg/day of DM), THI= temperature and humidity index (without dimension), Day= Age of animal (days). The DMI observed between treatments varied and increased with increasing the milk intake, as expected. On the other hand, starter intake (SI), followed the opposite behavior, and decreased as the milk intake was increased (Table 1). The reduction in solids intake by calves drinking more milk was due to the fact that these animals have reached satiety by chemical-physiological mechanisms (higher blood glucose) and by physical factors (continuous gut-filling because of curd formation, Khan *et al.*, 2011). The SI can be estimated by the difference between the total DMI and the DMI from milk.

Table 1. Dry matter (kg/day), starter (kg/day DM) and water intakes (l/day,) and prediction equations for dry matter and water intakes.

	L 2[1]	L 4	L 8	Regression
DMI	0.650	0.795	1.114	$DMI = 0.4272 + 0.6741 \times M - 0.0059 \times THI + 0.0122 \times Day$
SI	0.348	0.249	0.123	$SI = DM\ I_{total} - DMI_{milk}$
WI	2.274	1.979	1.273	$WI = -2.9648 + 0.187 \times T_{db} + 0.02756 \times Day + 0.7257 \times SI$

[1] L2: 2 l/day; L4: 4 l/day; L8: 8 l/day; DMI: dry matter intake, SI: starter intake, WI: free water intake.

All the environmental variables (dry bulb temperature, THI, BGHI) affected water intake ($P<0.0001$). However, as the dry bulb temperature had the greater correlation with WI ($r^2=0.185$), the other variables were not included in the model to estimate WI. It was verified significant effects of dry bulb temperature ($P<0.0001$), age in days ($P<0.0001$), and starter intake ($P=0.0028$) on WI. The same way as occurred to DMI, it wasn't observed quadratic effects for any of the variables tested ($P>0.05$) and the multiple linear regression to estimate WI in calves can be expressed:

$$WI = -2.9648 + 0.187 \times T_{db} + 0.02756 \times day + 0.7257 \times SI \quad (2)$$

where WI = free water intake (liters /day), T_{db} = dry bulb temperature (°C), Day= age of animal (days), SI = starter intake (kg/day). The water intake presented a direct relationship with the starter intake (Table 1), and with the percentage of dry matter on the diet, and these were 25.2; 17.7 and 13.9% of the natural matter for the treatments 2L, 4L and 8L, respectively. NRC (2001) mention that among the many factors that can change the free water intake, the percentage of the dry matter on the diet is the main factor. Therefore, higher solid feed intake tends to be followed of the higher water intake, as observed in this study.

It can be concluded that the dry matter intake in calves can be estimated using parameters as milk intake, THI and age of animals. Furthermore, the free water intake can be estimated using parameters as dry bulb temperature, age of animals and starter intake.

References

Chizzotti, M. L., S.C. Valadares Filho, R. F. Diniz Valadares, F. H. M. Chizzotti, M. I. Marcondes, and M. A. Fonseca, 2007. 'Intake, digestibility and nitrogen metabolism in Holstein cows with different milk production levels.' Braz. J. Anim. Sci. 36, 138-146.

Detmann, E., M.A. Souza, and S.C. Valadares Filho, 2012. Métodos Para Análises De Alimentos. ed. E. Detmann, M.A. Souza, and S.C. Valadares Filho. 1st ed. Visconde do Rio Branco: Suprema.

Khan, M A, D. M. Weary, and M. A. G. Von Keyserlingk, 2011. 'Invited Review : Effects of Milk Ration on Solid Feed Intake, Weaning, and Performance in Dairy Heifers.' J. Dai. Sci. 94, 1071-1081.

National Research Council – NRC, 2001. Nutrient Requirements of Dairy Cattle. 7th ed. Washington D.C.: National Academic Press.

Re-evaluation of lysine requirement of broilers based on protein deposition and lysine efficiency

J.C.P. Dorigam, N.K. Sakomura, L. Hauschild and E.P. Silva
Department of Animal Science, University of Agrarian and Veterinary Sciences of UNESP, Jaboticabal, SP 14884900, Brazil; sakomura@fcav.unesp.br

Introduction

The quality of protein is related to concentration and availability of amino acids in feed and differs between geographic regions because of the composition of feedstuffs. Differences in environment, management, genotype and physiological state can also influence the response of birds to amino acid intake. Due to these factors a re-evaluation of amino acid requirements is necessary. The lysine requirement is constantly studied because of its importance as a reference amino acid in the ideal protein concept. Also, the amount of lysine ingested has a direct influence on growth performance because it is used for body protein deposition. Thus a growth model for broilers (Samadi and Liebert, 2007) that takes into account nonlinear functions integrated in a factorial approach considering lysine efficiency, protein deposition, sex and age was used to re-evaluate the optimum dietary lysine level for cobb genotype.

Material and methods

Nitrogen (N) balance studies were conducted for three age periods (6-21; 22-37; and 38-53d) and eighty four broilers of cobb500 genotype were used (42 males and 42 females) per period. The birds were distributed in a completely randomized design with seven treatments (six levels of crude protein (CP): N1=61, N2=124, N3=183, N4=239, N5=295, and N6=364 g/kg in dry matter (DM)) and six replicates for each sex. In the seventh treatment 4 g of L-Lysine HCl (78%) /kg in the diet N1 was added as a control to check if lysine was the first limiting amino acid. The diets were obtained by diluting a high protein diet with a N free diet composed of corn starch to make lysine the first limiting amino acid (4.91g Lys/100 g CP). Each N balance study lasted 15 days (five days of adaptation and two periods of excreta collection of five days). At the end of the collection period the excreta were homogenized and freeze dried for DM and N analysis. The exponential regression between N intake (NI) and N excretion (NEX) allowed estimation of N maintenance requirements (NMR) when NI=0, according to Samadi and Liebert (2007). Based on the exponential function between NI and N retention (NR), the theoretical maximum for NR ($NR_{max}T$) was estimated using the equation: $NR = NR_{max}T\,(1 - e^{-b*NI})$, where b = slope of NR curve. $NR_{max}T$ minus NMR results in the theoretical maximum for N deposition ($ND_{max}T$). For modeling the requirement of limiting amino acid (LAA) an equation was applied: $LAAI = [\ln NR_{max}T - \ln(NR_{max}T - NR)]/16bc^{-1}$; where LAAI is the intake of LAA and bc^{-1} is the slope of LAA concentration (c) on feed protein quality (b).

Results and discussion

The responses of the control treatment were intermediate to those obtained in the treatments N1 and N2, confirming that lysine was the limiting amino acid in diets. The results of the N balance studies are summarized in Table 1. The estimated NMR values in each age period were used to calculate $ND_{max}T$ and are in accordance to results of Samadi and Liebert (2007). The $ND_{max}T$ estimated were 3,746; 3,137 and 2,204 mg/$BW_{kg}^{0.67}$/d for male and 3,620; 3,038 and 2,048 mg/$BW_{kg}^{0.67}$/d for female in the periods I, II, and III, respectively. This continuous decrease with age occurs because the efficiency of amino acid deposition in carcass decreases with age (Sklan and Noy, 2004). Also, sex affects the efficiency of utilization of the amino acid (D'Mello, 2003) leading to different $ND_{max}T$ observed.

Table 1. Applied model parameter depending on age and sex for NMR and $NR_{max}T$.

Sex	Age (days)	Mean BW (g)	NMR (mg/$BW_{kg}^{0.67}$/d)	$NR_{max}T$ (mg/$BW_{kg}^{0.67}$/d)	$ND_{max}T$ (mg/$BW_{kg}^{0.67}$/d)	R^2
Male	6-21	370	220	3,966	3,746	0.98
	22-37	1,438	264	3,401	3,137	0.98
	38-53	3,106	276	2,480	2,204	0.98
Female	6-21	338	225	3,845	3,620	0.98
	22-37	1,362	277	3,315	3,038	0.98
	38-53	2,485	271	2,319	2,048	0.98

The data for modeling lysine requirements are presented in Table 2. Based on efficiency of lysine utilization the required dietary digestible lysine for males and females broilers were 1.07 and 0.9% (6-21d), 1.01 and 0.93% (22-37 d), 0.90 and 0.72% (38-53 d) corresponding to 50, 120 and 160 g of the daily feed intake, respectively. The estimates for age 6-21 d and 22-37 d was lower than actual digestible lysine recommendations (Bernal *et al.*, 2012) of 1.22 for males and 1.24% for females at 10-21 d, and 1.16% for both genders at 22-35 d. However, at 38-53 d the estimative were identical to the requirement of 0.90 and 0.72% for males and females, estimated by Samadi and Liebert (2007) after 50 days with a daily feed intake of 180g. The result indicates that broilers have become more demanding and require more nutrients for maximum performance, primarily at younger ages.

Table 2. Calculation of the lysine requirements for Cobb500 according to efficiency of lysine utilization and CP deposition.

Age	6-21 days		22-37 days		38-53 days	
	Male	Female	Male	Female	Male	Female
CP deposition (g/d)[2]	7.2	6.6	15.0	14.2	17.7	14.1
Lys efficiency (bc^{-1})	58×10^{-6}	62×10^{-6}	66×10^{-6}	71×10^{-6}	96×10^{-6}	104×10^{-6}
Lys requirement (mg/d)	535.5	482.3	1,208.5	1,112.6	1,447.4	1,145.7

[1] Considered 60% of $ND_{max}T$ for simulate practical performance condition.

Acknowledgements

The authors are grateful to FAPESP for the financial support.

References

Bernal L.E.P., F.C. Tavernari, G.R. Lelis, H.S. Rostagno, 2012. Nutritional requirement of digestible lysine for cobb 500 broiler chickens. In: World´s Poultry Congress, 2012, Salvador. Abstract. Salvador: WSPA, 2012. J. Poult. Sci., v. 68, supl. 1, 2012.

D'Mello J. P. F. 2003. Adverse effects of amino acid. In: D'mello JPF Amino acids in animal nutrition. 2ed. Edinburgh: CABI Publishing, pp. 187-202.

Samadi, and F. Liebert, 2007. Lysine requirement of fast growing chickens – effects of age, sex, level of protein deposition and dietary lysine efficiency. J. Poult. Sci. 44, 63-72.

Sklan, D. and Y. Noy, 2004. Catabolism and deposition of amino acids in growing chicks: effect of dietary supply. Poult. Sci. 83, 952-961.

Ideal ratio (relative to lysine) of methionine + cystine and threonine for broiler breeders

A.R. Troni[1], N.K. Sakomura[1], C.F.S. Oliveira[2], E.P. Silva[1], F.G.P. Costa[2] and D.C.Z. Donato[1]
[1]State University of São Paulo, UNESP, Jaboticabal, SP, Brazil; allan_troni@yahoo.com.br
[2]Federal University of Paraíba, UFPB, Areia, PB, Brazil

Introduction

To establish the amino acid requirements for broilers, the ideal ratio of lysine to the amino acids is used. This criterion is based on the small variation in the proportion of amino acids with lysine. Few studies deal with amino acid requirements for broiler breeders. A factorial method for calculating the amino acid requirement was proposed by Fisher *et al.* (1973) for egg production and body weight maintenance, including the standard deviation of requirements and an economic factor. Based on this method, the objective of this study was to determine the efficiency of utilization of methionine and cystine (Met+Cys) and threonine (Thr) and their ideal ratio to lysine (Lys).

Material and methods

Three ten-week (six week adaptation, four week data collection) experiments with 147 broiler breeders housed in individual cages were based on a completely randomized design with seven levels of amino acid and seven replicates. A control to confirm that the studied amino acid was first limiting was included. Diets were based on corn, soybean meal and synthetic amino acids. Egg mass was used in the model: AAI=(a×Emax)+(b×BW)+y, where AAI is amino acid intake, mg/day; a is requirement for egg mass production, mg/g; Emax is egg mass maximum response, g/day; b is body weight maintenance requirement, mg/kg, BW is body weight, kg; and y is the standard deviation (σ) of the requirement: (a × σEmax) + (b×σBW). Coefficients were adjusted by the EFG software module amino acid optimizer. Efficiency of utilization (Effy) was obtained by dividing the deposition of amino acid in the egg (d; mg/g) by the amino acid requirement for egg mass production (a; mg/g), Effy=d/a. The ideal ratio was obtained from the coefficients rate a of Thr (aThr/aLys) and Met+Cys (aMet+Cys/aLys).

Results and discussion

Increasing levels of Lys, Thr and Met+Cys increased broiler breeder egg mass 104.95, 26.72 and 123.88%, respectively (Table 1). The Reading Model partitions amino acid requirements into maintenance and egg production. Based on model parameters, daily requirements of Lys, Thr and Met+Cys for a bird producing an average 45g egg were calculated as 516, 334, and 429 mg, respectively (Table 2). The requirement for maintenance of Lys, Thr and Met+Cys represents 0.66, 0.01 and 7.49% of amino acid intake, respectively. This maintenance requirement is not in accordance to Bonato *et al.* (2011) and Siqueira *et al.* (2011); however, they are similar to those found by Fisher *et al.* (2001). The ratio of lysine to amino acids may vary depending on the maintenance requirement, thus any change in this parameter changes the values (Fisher, 1998). Therefore, we determined the ideal ratio using the parameter a, which represents the egg mass requirement, and does not influence the maintenance requirement. The ideal ratio Thr/Lys was 65%. By considering the maintenance requirement of Lys (37 mg/kg) and Thr (56 mg/kg) found by Nonis and Gous (2008), the ideal ratio is 86% because of the increase of maintenance over the total intake. This result supports the hypothesis of an effect of maintenance on the ideal ratio of amino acids. The technique is a good alternative to find relationships of and efficiency of utilization of amino acids, but the values for maintenance in the literature are discrepant and it is important to rely on the requirements of production of egg mass a to determine the responses.

Table 1. Responses of egg mass (EM, g/day) of broiler breeders to different intakes of amino acids (mg/day).

Treatment	Lysine (220-733)		Threonine (178-594)		Met+Cys (160-638)	
	Intake	EM	Intake	EM	Intake	EM
1	270.3	22.2	237.0	36.3	208.2	20.1
2	384.3	33.9	342.0	40.0	292.7	29.2
3	489.0	39.8	441.0	42.0	486.5	46.0
4	572.1	42.1	528.0	47.9	566.0	44.0
5	658.7	42.2	611.0	50.5	659.8	47.0
6	847.0	42.7	706.7	50.1	750.0	45.0
7	952.5	45.5	886.0	46.0	-	-

Table 2. Parameters of the model, values of efficiency (Effy) and amino acid (AA) rate to lysine (AA/Lys), for broilers breeders.

Amino acid	Emax[1]	BW	σEmax	σBW	a	b	d	Eff	AA/Lys
Lysine	43.8	4.12	10	0.412	11.39	0.80	8.3	73	100
Threonine	46.5	4.30	12	0.645	7.42	0.01	5.7	77	65
Met + Cys	45.5	4.41	5	0.441	8.82	7.37	6.9	78	77

[1] Emax = egg mass maximum response, g/day; BW = body weight; kg; σ = standard deviation; a = requirement for egg mass production, mg/g; b = requirement for BW maintenance, mg/kg; d = amino acid deposition in the egg (mg/g).

Acknowledgements

We thank FAPESP for financial support and CNPq for a scholarship.

References

Bonato, M. A., N.K. Sakomura, J.C. Siqueira, J.B.K. Fernandes and R.M. Gous, 2011. Maintenance requirements for methionine and cysteine, and threonine for poultry. South African Journal of Animal Science 41, 209-222.

Bowmaker, J. E. and R.M. Gous, 1991. The response of broiler breeder hens to dietary lysine and methionine. British Poultry Science 32, 1069-1088.

Fisher, C., 1998. Amino acid requirements of broiler breeders. Poultry Science 77, 124-133.

Fisher, C., R.M. Gous, L. Goddard, 2001. Lysine requirements of broiler breeders. In: Benalve D, (ed.), Australian Poultry Science Symposium v.13. Poultry Research Foundation Sydney, pp. 183-186.

Fisher, C., T.R. Morris and R.C. Jennings, 1973. A model for the description and prediction of the response of laying hens to amino acid intake. British Poultry Science 14, 469-484.

Nonis, M. K. and R.M. Gous, 2008. Threonine and lysine requirements for maintenance in chickens. South African Journal of Animal Science 38, 75-82.

Siqueira, J. C., N.K. Sakomura, H.S. Rostagno, M.A. Bonato, S.R.F. Pinheiro and D.C.N. Nascimento, 2011. Exigência de lisina para mantença determinada com galos de diferentes genótipos. Revista Brasileira de Zootecnia 40, 812-820.

Prediction of net hepatic release of glucose using a 'hybrid' mechanistic model in ruminants applied to positive energy balance

L. Bahloul[1], *I. Ortigues-Marty*[1], *J. Vernet*[1], *H. Lapierre*[3], *P. Nozière*[1] *and D. Sauvant*[2]
[1]*INRA, Unité Mixte de Recherches sur les Herbivores, 63122 St Genès Champanelle, France*
[2]*UMR, Modélisation Systémique Appliquée aux Ruminants, AgroParisTech, 75005 Paris, France; sauvant@agroparistech.fr*
[3]*Agriculture and Agri-Food Canada, Sherbrooke, QC J1M0C8, Canada*

Introduction

Ruminants depend on hepatic gluconeogenesis to meet most of their metabolic demand for glucose which relies on availability of precursors from diet supply and animal requirements (Loncke *et al.*, 2010). Several mechanistic models of the metabolic fate of nutrients across the liver exist that have been parameterized for dairy cows. They cannot be directly used to predict hepatic gluconeogenesis in all types of ruminants in different physiological status. A hybrid mechanistic model of nutrient fluxes across the liver is presently being developed (Bahloul *et al.*, 2012), that is calibrated empirically based on meta-analysis (Sauvant and Mertens, 2008) to be applicable to all types of ruminants in different physiological status and to usual nutritional practices. The objectives of the present work were to test the hybrid liver model in its present state of development to simulate the net hepatic release of glucose when the glucogenic/ketogenic/nitrogenous nutrient profile entering the liver varies. This first application of the model was limited to ruminants in positive calculated energy balance, in which the nutrient fate across the liver is mostly directed by mass action laws.

Material and methods

Net veno-arterial fluxes of nutrients correspond to the inputs (net portal appearance, NPA) and outputs (net splanchnic release, NSR) of the mechanistic model. Nutrients considered are: acetate (C2), propionate (C3), butyrate (C4), glucose, lactate, β-hydroxybutyrate (BHB), total amino acids (TAA), urea, NH_3, O_2, CO_2 and the absorbed fatty acids. Their flows are expressed in mmol of C/h/BW except for NH_3 expressed in mmol of N/h/BW. Ten hepatic compartments are included: C2, C3, C4, BHB, glycogen, protein, glucose, lactate, TAA and CO_2. Conversion fluxes between compartments, including metabolic compartments summarizing the Krebs cycle, represent the main hepatic metabolic pathways. The model based on mass action laws is adjusted on empirical relationships of the net hepatic transfer of nutrients derived by meta-analysis (Loncke, 2009) from FLORA database, allowing estimation of NSR of nutrients. The empirical equation used to predict glucose NSR was based on the sum of the NPA of precursors (Loncke *et al.*, 2010), modified to take into account all AA potentially convertible to glucose. Indeed, if only the NPA of glucogenic AA is considered, a deficit of carbon is calculated to meet glucose output. Sets of experimental data, representative of nutritional practices representing different combinations of ketogenic, glucogenic and nitrogen nutrient inputs, were used to adjust parameters of the conversion fluxes, in cattle and sheep, only in calculated positive energy balance. Nutritional status ranged from 1 to 5 × ME and 1.4 to 6.7 × metabolizable protein maintenance requirements. Parameters were adjusted so that the simulated model outputs converge at best with the values predicted by empirical relationships. The model was evaluated by: (1) calculating the full C balance across the liver; and (2) by comparing outputs (NSR) of glucose simulated by the model with values predicted by the empirical relationships.

Results and discussion

The full C balance across the liver simulated by the model (Figure 1) indicates a tight agreement between C outputs and inputs, with a small deficit averaging 0.0021 mol C/d. Comparison between the model simulated and the empirically predicted NSR of glucose (Figure 2) shows a good (93%)

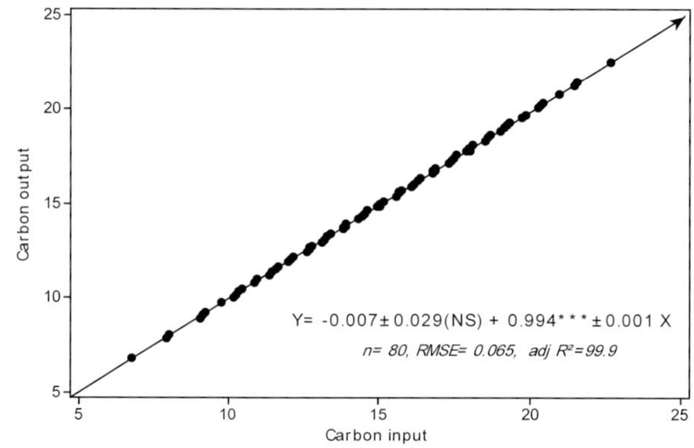

Figure 1. Relationship between the carbon output (Y) simulated by the model and the carbon input (X) in mmol of C/h/BW, ***P<*0.001.*

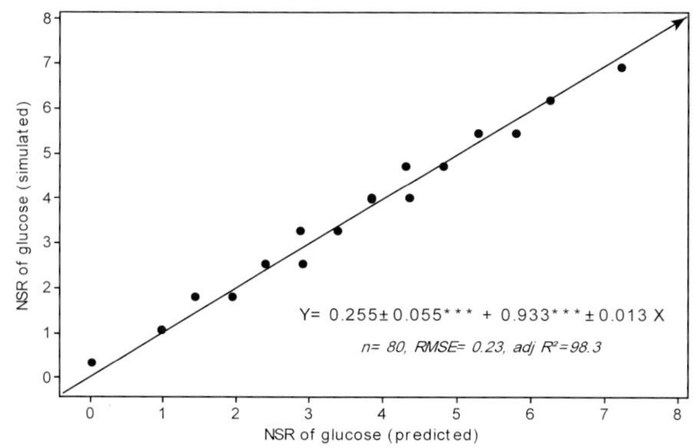

Figure 2. Relationship between the NSR of glucose (Y) simulated by the model and predicted by structural relationship (X) in mmol of C/h/BW, ***P<*0.001.*

simulation. For C3 and TAA, the major glucogenic nutrients, the simulated outputs compare well with the values predicted by the structural relationship. Parameters of mass action laws obtained by model adjustment show that 93%/h and 173%/h of intrahepatic pool of TAA and C3 is metabolized to oxaloacetate (OA) respectively; 124% of intrahepatic pool of OA is converted to glucose. Almost, all TAA up taken by the liver (73%) appeared used for neoglucogenesis. The hepatic uptake of C3 (91%) seems to be converted completely to glucose, only 19%/h and 0.8%/h of intrahepatic pool of C3 are released and converted to pyruvate respectively. A strength of the approach is to identify situations responsible for poor adjustments: an underestimation (0.34 mmol of C/h/BW) of the simulated NSR glucose is observed at elevated TAA NPA (>3 mmol of C/h/BW).

The hybrid model based strictly on mass action laws simulates well the prediction of hepatic release of glucose for ruminants in positive energy balance. The model was found to be stable in term of C balance, with a small negligible deficit in C. The next step consists on the application of the model to ruminants in negative energy balance. If necessary, regulations representing hepatic metabolism will be introduced to take account interaction between nutrients and tissues (nutrient requirements).

Acknowledgements

Financial support from INZO and Limagrain is acknowledged.

References

Bahloul, L., D. Sauvant, J. Vernet, H. Lapierre, P. Nozière and I. Ortigues-Marty, 2012. Mechanistic modeling based on meta-analysis : prediction of the net hepatic fluxes of energetic nutrients in ruminant. Can. J. Anim. Sci. 92, 559-560.

Loncke, C., 2009. Modélisation des relations entre l'alimentation et les flux splanchniques de nutriments énergetiques chez le ruminant. Thèse de doctorat, AgroParisTech. France.

Loncke, C., P. Noziere, S. Lemosquet, J. Vernet, H. Lapierre, D. Sauvant and I. Ortigues-Marty, 2010. Splanchnic release of glucose in relation to dietary supply and animal performances. Renc. Rech. Ruminants 17, 317.

Sauvant, D. and D. Mertens, 2008. Use of meta-analysis to build a mechanistic model of responses of rumen digestion to dietary fibre in cattle. Can. J. Anim. Sci. 88, 730.

Part 6. Products and health

Recent advances in understanding the interactions between nutrients and immunity in farm animals

K.C. Klasing and V.J. Iseri
Department of Animal Science, University of California, Davis, CA, USA; kcklasing@ucdavis.edu

Introduction

Nutrition is an important determinant of the development and efficacy of the immune system. Nutrition is also sometimes used as a management tool to affect changes in the type or magnitude of an immune response. These developments have been the result of difficult research that relies on accepted principles of nutrition to inform the somewhat unknown frontiers of immunology and disease resistance. Historically, plethoric levels of nutrients were thought to boost, improve, or stimulate immunity and these claims were a tool of nutrient marketing because of the intellectual vacuum. These claims illustrate sloppiness in thinking, which fortunately has diminished due to our new appreciation for the role of nutrients in the immune system.

In the case of nutritionally required nutrients, both the absolute amounts and the balance of individual nutrients influence immunity. There is little doubt that severe nutritional deficiencies impair the immune system and increase susceptibility to infectious diseases. However, severe deficiencies that result in classical nutritional pathology are very rare when livestock are fed scientifically formulated commercial diets. Marginal deficiencies that do not markedly impact growth or reproductive output are much more common and the immune system is sensitive to moderate deficiencies of some nutrients. However, the immune system is essential for life and appears to have a very high priority for many nutrients relative to muscle accretion or reproduction and is surprisingly resilient to deficiencies of many nutrients. Understanding which nutrients, when marginally deficient, impair immunity is of great practical importance. Considerable recent progress has been made toward elucidating the mechanisms by which some nutrients are prioritized between leukocytes and other cells types.

Additionally many nutrients, including some that are not essential for growth or reproduction, regulate and modify immune responses. Progress has also been made towards understanding the mechanisms by which nutrients impose their regulatory effects. These developments have permitted the use of surfeit amounts of select nutrients to induce predictable immunoregulatory outcomes for two primary purposes: to decrease the incidence of specific infectious diseases and to minimize the untoward effects of immune responses on growth, reproduction, and the incidence of metabolic diseases (Athanasiadou and Huntley, 2008; Cook, 1991; Ingvartsen and Moyes, 2013; Klasing, 2007; Sordillo *et al.*, 2009; Zebeli and Metzler-Zebeli, 2012). The following review examines recent advances in our understanding of the nutritional needs of the immune system as well as the regulatory actions of nutrients on the immune system. Nutrition also influences the growth of commensal and pathogenic microflora in the gastrointestinal tract; although this area of research is highly relevant and exciting, it is beyond the scope of this review.

The majority of the research on nutrition and immunity has utilized rodents and these studies inform much of what we know about nutrient-leukocyte interactions at a mechanistic level. Chickens have historically been used as a basic research model because of their utility in studying the development of B cells in their bursa. Swine have recently become a favored model for understanding the development of mucosal immunity and are increasingly being used for studies on systemic immunity due to their similarities with humans (Fairbairn *et al.*, 2011). Consequently, basic research using chickens and pigs supplements and extends the lessons we have gained from mouse models.

Meeting the needs for development, maintenance and responses of leukocytes

For most nutrients the most important nutritional strategy for optimizing immunity is meeting the established requirements for maximizing growth, feed efficiency and productivity. For many of the essential nutrients, deficiencies that are sufficiently severe to slow growth or reproduction are also deleterious to the adaptive immune system and markedly increase susceptibility to pathogens (for reviews, see Kidd, 2004; Korver, 2012; Latshaw, 1991). For some nutrients, leukocytes are among the most sensitive to marginal deficiencies of any cell type, while for other nutrients the immune system is largely unaffected by deficiencies. This dichotomy can be explained by several mechanisms. First, with the exception of primary immune tissues of the neonate (thymus, bursa, etc.), most leukocytes in healthy animals are exceptionally inactive and have very low nutrient needs. Second, many lineages of leukocytes have an excellent capacity to compete with other cells when circulating levels of nutrients are low. This high priority for nutrients is greatly enhanced in those cells that respond to an infectious challenge. Third, an immune response is accompanied by the mobilization of nutrients from muscle and other tissues, which supplies adequate amounts of some, but not all, nutrients to leukocytes.

Size and nutrient content of the immune system

There has been considerable conjecture but very little investigation on the size and nutrient content of the immune system. Nutritionists, despite rigorous application of quantitative theory and modeling to nutrient needs for growth or reproduction, have been remiss in applying this fundamental concept to immunity. Iseri and Klasing (2013) recently quantified the amount of lysine that is contained in the cells and protective proteins of the systemic (non-mucosal) immune system of an adult healthy chicken. Lysine was chosen because it is the reference amino acid in the ideal protein system used commonly in non-ruminant nutrition. The cells and proteins of the systemic immune system at maintenance have the same lysine content as 16% of a medium egg or 5% of a pectoralis muscle. During the acute phase an i. v. injection of fixed *E. coli*, the additional lysine accreted into the cells and protective proteins of the immune system is equal to 17% of an egg or 5.4% of a pectoralis muscle. Liver hypertrophies increase markedly during the acute phase of the immune response for the rapid production of acute phase proteins. Hepatic hypertrophy requires the lysine contained in 46% of a medium egg during the first 24 hr post-*E. coli* challenge. Because the liver is recruited to become part of the immune defenses during the acute phase response, it is the most expensive part of the response. The amount of lysine needed for the adaptive phase of the response (antibody production and new lymphocytes) is much less than that needed for the acute phase of the response. In fact with the *E-coli* challenge model system described above, the entire lysine needs for the adaptive response can be provisioned by the decline of the acute phase response and degradation of the acute phase proteins. Interestingly, the additional lysine needs for the synthesis of specific immunoglobulins (3 mg lysine) is very small compared to the other components of the adaptive phase of the response. As a result, the lysine content of an antibody response (specific Ig plus additional cells) is equal to only 1.3% of an egg.

The study by Iseri *et al.* (2013) supports the concept that the cost of an immune response is mostly due to protective processes unrelated to the needs of leukocytes. Even when the hypertrophy of the liver and the massive production of acute phase proteins are included, the amount of nutrients diverted to protective processes accounts for very little of the depression in growth or reproduction that occurs during the response. Clearly other factors such as decreased appetite, increased metabolic rate, and metabolic inefficiencies dominate the nutritional costs of an immune response.

Priority of the immune system for nutrients

The transport of the essential nutrients across the cell membrane has been examined in many cell types, including leukocytes. For example, the essential amino acids lysine and arginine are transported by the cationic amino acid transporter (CAT), of which there are several isoforms and each isoform has different transport kinetics. Chicken muscles express relatively low levels of the high affinity CAT isoform, whereas bursal tissue expresses high levels (Humphrey *et al.*, 2004, 2006). This difference increases dramatically during a lysine deficiency caused by feeding an inadequate diet or by restricting feed intake. The disparity in expression of high-affinity transporters allows the bursa to out-compete muscle for lysine during periods of dietary lysine deficiency and maintain normal numbers of differentiating bursa cells even when muscle growth is severely impaired. Interestingly the thymus has a priority for lysine that is lower than skeletal muscle and this tissue fails to maintain populations of developing thymocytes during periods of lysine restriction. Thus, T-lymphocyte mediate immunity but not antibody responses are impaired by marginal lysine deficiencies.

Changes in the amount and type of glucose transporter expression is coordinated to allow chicken lymphocytes to develop and respond appropriately as neonatal chicks transition from the lipid energy dominant environment of the egg to the carbohydrate energy supplied by the diet (Rudrappa and Humphrey, 2007). The energy consumption of a naïve lymphocyte is relatively small and energy is acquired from the breakdown of glucose, lipids and amino acids (Fox *et al.*, 2005). After antigenic stimulation, glucose consumption increases 20 fold within the first hour, which is facilitated by increases in glucose transporters (Greiner *et al.*, 1994; Humphrey and Rudrappa, 2008). The significant increase in glucose uptake is to not only to provide enough energy for biochemical reactions, but to generate cellular components for the proliferating cell. Glucose is diverted to precursors such as acetyl CoA for fatty acid synthesis, glycolytic intermediates for non-essential amino acids production, and ribose for nucleotides synthesis (Van der Heiden *et al.*, 2009). In chickens, both T and B lymphocytes express high affinity receptors that allow them to compete successfully for glucose when concentrations are low (Humphrey and Rudrappa, 2008; Rudrappa and Humphrey, 2007). This, observation helps to explain why moderate dietary energy restriction does not impinge on immune responses.

Another example of nutrient prioritization occurs in chicken macrophages, which express very high levels of metallothionein when activated. This permits macrophages to sequester high levels of zinc (Laurin *et al.*, 1990) to supply their anabolic needs and endows them with a higher priority for limiting levels of this trace nutrient relative to many other cell types. This is important because the need for zinc markedly increases when cells are activated and become highly anabolic. In summary, it appears that evolution has provisioned some leukocyte lineages with a high priority for many critical nutrients; likely those that were commonly deficient in natural diets as animals evolved. Nutrients that appear to fall into this category include zinc (Mohanna and Nys, 1999), most of the amino acids (Kidd, 2004) except possibly methionine (Cook, 1991; Rama Rao *et al.*, 2003) and energy supplying compounds such as glucose (Humphrey *et al.*, 2004). Conversely, for some nutrients there are not mechanisms that permit leukocytes to establish priority over other cell types and they are affected during deficiencies in a manner similar to all other tissues (e.g. copper; Koh *et al.*, 1996). For some nutrients, the relative priority and requirement for development or response of the immune system has not been examined and compared to the requirement for other processes such as growth or reproduction. This is particularly true for many vitamins and trace minerals and filling this information gap should be a high priority.

Appropriation of nutrients when the immune system responds

The pro-inflammatory cytokines released by phagocytes in response to pathogens induce metabolic changes, including increased protein degradation and insulin resistance, which divert nutrients from skeletal muscle and other tissues. These appropriated nutrients become available for the increased demands of the liver and responding leukocytes (Sirimongkolkasem, 2007). These same cytokines also mediate decreased food consumption. An important question is if the nutrients appropriated during the acute phase response, juxtapositioned with diminished food intake, are sufficient in amount and balance for the immune response. Research in rodent models and humans suggest a mismatch between supply and demand (Obled *et al.*, 2002). It has been suggested that the main contributor to the increased degradation of skeletal muscle is due to the mismatch in amino acid profiles between protective proteins (i.e. acute phase proteins) and skeletal muscle (Reeds and Jahoor, 2001). Furthermore, amino acids are not only used for the production of acute phase proteins, but for increasing liver mass, synthesizing immunoglobulins, and the clonal expansion of leukocytes. Recent work indicate that cysteine is the most limiting amino acid during the acute phase response in chickens (Iseri and Klasing, 2013; Sirimongkolkasem, 2007) and also in rats (Breuille *et al.*, 2006; Breuille and Obled, 2000). This is due to a mismatch between muscle cysteine release and hepatic demand for production of acute phase proteins and glutathione.

Implications for nutritionists

In general, developing T lymphocytes are the most susceptible leukocyte population to nutrient deficiencies due to their inability to compete with other tissues. Severe deficiencies of amino acids like lysine (Humphrey *et al.*, 2006), or branched-chain amino acids (Konashi *et al.*, 2000) that markedly impair growth of chicks cause involution of the thymus but not the bursa. The thymus is more sensitive to deficiencies of the branched-chain amino acids than any other group of amino acids. However, the deficiency must be severe and the thymus and peripheral T cell populations are not sensitive to marginal deficiencies of the branched-chain amino acids (Kidd, 2004; Konashi *et al.*, 2000). With diminished lymphocyte-mediated immunity, exaggerated responses by the innate immune system often occur during infection, which result in robust inflammatory responses and accompanying immunopathology.

Feed restriction is commonly used in the management of animals, especially in breeding females. A wide variety of studies have examined the influence of feed restriction on immunocompetence (Hangalapura *et al.*, 2005; Nir *et al.*, 1996; Praharaj *et al.*, 1996) and the general conclusion is that mild to moderate feed restriction enhances many measures of immunity. Very severe feed restriction impairs cellular immunity but may enhance innate (Hangalapura *et al.*, 2005) or antibody responses (Khajavi *et al.*, 2003). The endocrine and metabolite changes that accompany feed restriction likely mediate many of the effects on immunocompetence in chickens (Nir *et al.*, 1996). In peripartum dairy cows in negative energy balance, derangements in metabolites accompanied by endocrine changes impair immunity and increase susceptibility to infections such as mastitis (Ingvartsen and Moyes, 2013). Preventing these metabolic changes is a key management strategy for optimizing immunity.

Immunoregulatory nutrients

A variety of nutrients modulate the immune system by direct actions on regulatory mechanisms of leukocytes (Ballou *et al.*, 2008; Goff, 2006; Lillehoj and Lee, 2012; Lippolis, 2008; Wallace *et al.*, 2010). Required nutrients with indisputable immunoregulatory actions in rodents and livestock include the long-chain polyunsaturated fatty acids (PUFA) and vitamins A, D, and E. Additionally, some nutrients that are not normally considered as being dietary essential also modulate immunity, including carotenoids, phytonutrients (e.g. conjugated linoleic acids, curcumin, capsicum, genistein, essential oils, etc) and vitamin C. In general those nutrients that are not structural components or co-factors for enzymes are most likely to be immunomodulatory. Unlike increases in nutrients from deficient to sufficient levels, where many indices of immunocompetence are elevated, supplementation of immunomodulatory nutrients causes some components of immunity to be elevated and others to be diminished; in other words the type and intensities of responses have been changed.

Many immunomodulatory nutrients influence the balance of cytokines and eicosanoids released by regulatory cells. For example, PUFAs of the n-3 series enhance the release of less inflammatory eicosanoids compared to n-6 fatty acids (Fritsche, 2006). Several nutrients change the activity of nuclear factor-kappa B (NFκB) and, consequently, modulate cytokine expression. In rodents and chicks, nutrients that decrease NFκB activity and dampen the inflammatory response include a variety of n-3 PUFAs, antioxidants, vitamin A, and lutein. In addition to dampening the inflammatory response by decreasing the release of pro-inflammatory cytokines, these nutrients often shift lymphocyte responses from T-cytotoxic (Th1, cell-mediated) towards antibody responses (Th2). This polarization of adaptive immunity has implications for the susceptibility of experimental animals to authentic infections (Fritsche, 2006; Klasing and Leshchinsky, 1999). Because of their immunomodulatory effects on the Th1:Th2 balance, the benefit of these nutrients depends on the types of pathogens to which animals are exposed. A diet that modulates the immune system towards a Th2 response may be beneficial with some pathogens but be deleterious with others (those protected by Th1 response).

In the few studies that have looked, the specific effects of immunomodulatory nutrients are highly dose dependent. This has been clearly shown for PUFAs, vitamin E and vitamin A, where the immunomodulatory actions follow a bell-shaped curve and high levels may not be as useful as moderate levels (Leshchinsky and Klasing, 2001; Lin and Chang, 2006; Sklan *et al.*, 1994).

Further complicating the picture, the net immunomodulatory influence of a diet is not a simple sum of the actions of individual nutrients because there are robust and sometimes unpredictable interactions between different immunomodulatory nutrients. For example, the anti-inflammatory effect of dietary lutein on chicken macrophages depends on the amount of PUFAs in the diet (Selvaraj and Klasing, 2006; Selvaraj *et al.*, 2011). The converse is also true in that the anti-inflammatory effects of n-3 PUFAs are dependent upon the amount of lutein in the diet. This interaction is mediated by nuclear hormone receptors that respond to these two nutrients (RXR for lutein and PPAR for PUFAs), which in turn affect the expression of NFκB. Similarly, different dietary fatty acids of the n-3 and the n-6 series have separate and interactive effects when supplemented to diets of chicks (Parmentier *et al.*, 2002; Sijben *et al.*, 2001), as does vitamin E and selenium (Singh *et al.*, 2006; Swain *et al.*, 2000).

Implications for nutritionists

Both experimental and clinical results clearly indicate that the specific immunomodulatory actions of a nutrient and its interactions with other dietary nutrients must be understood before application to animal populations because its efficacy is dependent on the milieu of infectious and metabolic diseases that are present. In situations where a specific disease dominates the production losses and where it is clearly known what type of immune response optimally protects against that disease, supplementation of a nutrient that modulates the immune system towards that optimal response is indicated. At this time there are few examples that meet these criteria.

References

Athanasiadou, S. and Huntley, J. F. 2008. Emerging technologies and their applications in interactions between nutrition and immunity to gastrointestinal parasites in sheep. Parasite immunology, 30, 101-111.

Ballou, M. A., Cruz, G. D., Pittroff, W., Keisler, D. H. and DePeters, E. J. 2008. Modifying the acute phase response of Jersey calves by supplementing milk replacer with omega-3 fatty acids from fish oil. J Dairy Sci, 91, 3478-3487.

Breuille, D., Bechereau, F., Buffiere, C., Denis, P., Pouyet, C. and Obled, C. 2006. Beneficial effect of amino acid supplementation, especially cysteine, on body nitrogen economy in septic rats. Clin. Nutr., 25, 634-642.

Breuille, D. and Obled, C. 2000. Cysteine and glutathione in catabolic states. Nestle Nutrition Workshop Series, 3, 173-191.

Cook, M. E. 1991. Nutrition and the immune response of the domestic fowl. Crit. Rev. Poult. Biol., 3, 167-190.

Fairbairn, L., Kapetanovic, R., Sester, D. P. and Hume, D. A. 2011. The mononuclear phagocyte system of the pig as a model for understanding human innate immunity and disease. J Luek Biol, 89, 855-871.

Fox, C. J., Hammerman, P. S. and Thompson, C. B. 2005. Fuel feeds function: energy metabolism and the T-cell response. Nat Rev Immunol, 5, 844-852.

Fritsche, K. 2006. Fatty acids as modulators of the immune response. An. Rev. Nutr., 26, 45-73.

Goff, J. P. 2006. Major advances in our understanding of nutritional influences on bovine health. J. Dairy Sci., 89, 1292-1301.

Greiner, E. F., Guppy, M. and Brand, K. 1994. Glucose is essential for proliferation and the glycolytic enzyme induction that provokes a transition to glycolytic energy production. J Biol Chem, 269, 31484-31490.

Hangalapura, B. N., Nieuwland, M. G., De Vries Reilingh, G., Buyse, J., Van Den Brand, H., Kemp, B. and Parmentier, H. K. 2005. Severe feed restriction enhances innate immunity but suppresses cellular immunity in chicken lines divergently selected for antibody responses. Poult Sci, 84, 1520-1529.

Humphrey, B. D. and Rudrappa, S. G. 2008. Increased glucose availability activates chicken thymocyte metabolism and survival. J. Nutr., 138, 1153-1157.

Humphrey, B. D., Stephensen, C. B., Calvert, C. C. and Klasing, K. C. 2004. Glucose and cationic amino acid transporter expression in growing chickens (Gallus gallus domesticus). Comp Biochem Physiol A Mol Integr Physiol, 138, 515-525.

Humphrey, B. D., Stephensen, C. B., Calvert, C. C. and Klasing, K. C. 2006. Lysine deficiency and feed restriction independently alter cationic amino acid transporter expression in chickens (Gallus gallus domesticus). Comparative biochemistry and physiology. Part A, Molecular & integrative physiology, 143, 218-227.

Ingvartsen, K. L. and Moyes, K. 2013. Nutrition, immune function and health of dairy cattle. Animal, 7 Suppl 1, 112-122.

Iseri, V. J. and Klasing, K. C. 2013. Dynamics of the systemic components of the chicken (Gallus gallus domesticus) immune system following activation by Escherichia coli; implications for the costs of immunity. Dev. Comp. Immuno., In Press.

Khajavi, M., Rahimi, S., Hassan, Z. M., Kamali, M. A. and Mousavi, T. 2003. Effect of feed restriction early in life on humoral and cellular immunity of two commercial broiler strains under heat stress conditions. Br Poult Sci, 44, 490-497.

Kidd, M. T. 2004. Nutritional modulation of immune function in broilers. Poult Sci, 83, 650-657.

Klasing, K. C. 2007. Nutrition and the immune system. Br Poult Sci, 48, 525-537.

Klasing, K. C. and Leshchinsky, T. V. 1999. Interactions between nutrition and immunity: Lessons from animal agriculture. In M. E. Gershwin (ed.), Handbook of Nutrition and Immunology (pp. 363-373). New York: Humana Press.

Koh, T. S., Peng, R. K. and Klasing, K. C. 1996. Dietary copper level affects copper metabolism during lipopolysaccharide-induced immunological stress in chicks. Poult Sci, 75, 867-872.

Konashi, S., Takahashi, K. and Akiba, Y. 2000. Effects of dietary essential amino acid deficiencies on immunological variables in broiler chickens. Br J Nutr, 83, 449-456.

Korver, D. R. 2012. Implications of changing immune function through nutrition in poultry. Anim. Feed Sci. Technol., 173, 54-64.

Latshaw, J. D. 1991. Nutrition--mechanisms of immunosuppression. Vet Immunol Immunopathol, 30, 111-120.

Laurin, D. E., Barnes, D. M. and Klasing, K. C. 1990. Rates of metallothionein synthesis, degradation and accretion in a chicken macrophage cell line. Proc Soc Exp Biol Med, 194, 157-164.

Leshchinsky, T. V. and Klasing, K. C. 2001. Relationship between the level of dietary vitamin E and the immune response of broiler chickens. Poult Sci, 80, 1590-1599.

Lillehoj, H. S. and Lee, K. W. 2012. Immune modulation of innate immunity as alternatives-to-antibiotics strategies to mitigate the use of drugs in poultry production. Poult Sci, 91, 1286-1291.

Lin, Y. F. and Chang, S. J. 2006. Effect of dietary vitamin E on growth performance and immune response of breeder chickens. Asain-Aust. J. Anim. Sci., 19, 884-891.

Lippolis, J. D. 2008. Immunological signaling networks: integrating the body's immune response. J. Anim. Sci., 86, E53-63.

Mohanna, C. and Nys, Y. 1999. Effect of dietary zinc content and sources on the growth, body zinc deposition and retention, zinc excretion and immune response in chickens. Br Poult Sci, 40, 108-114.

Nir, I., Nitsan, Z., Dunnington, E. A. and Siegel, P. B. 1996. Aspects of food intake restriction in young domestic fowl: Metabolic and genetic considerations. World Poult. Sci. J., 52, 251-266.
Obled, C., Papet, I. and Breuille, D. 2002. Metabolic bases of amino acid requirements in acute diseases. Curr Opin Clin Nutr Metab Care, 5, 189-197.
Parmentier, H. K., Awati, A., Nieuwland, M. G., Schrama, J. W. and Sijben, J. W. 2002. Different sources of dietary n-6 polyunsaturated fatty acids and their effects on antibody responses in chickens. Br Poult Sci, 43, 533-544.
Praharaj, N. K., Gross, W. B., Dunnington, E. A., Nir, I. and Siegel, P. B. 1996. Immunoresponsiveness of fast-growing chickens as influenced by feeding regimen. Br Poult Sci, 37, 779-786.
Rama Rao, S. V., Praharaj, N. K., Ramasubba Reddy, V. and Panda, A. K. 2003. Interaction between genotype and dietary concentrations of methionine for immune function in commercial broilers. Br Poult Sci, 44, 104-112.
Reeds, P. J. and Jahoor, F. 2001. The amino acid requirements of disease. Clin. Nutr., Supplement 1, 15-22.
Rudrappa, S. G. and Humphrey, B. D. 2007. Energy metabolism in developing chicken lymphocytes is altered during the embryonic to posthatch transition. J. Nutr., 137, 427-432.
Selvaraj, R. K. and Klasing, K. C. 2006. Lutein and eicosapentaenoic acid interact to modify iNOS mRNA levels through the PPARgamma/RXR pathway in chickens and HD11 cell lines. J Nutr, 136, 1610-1616.
Selvaraj, R. K., Shanmugasundaram, R. and Klasing, K. C. 2011. Effects of dietary lutein and PUFA on PPAR and RXR isomer expression in chickens during an inflammatory response. Comp Biochem Physiol A Mol Integr Physiol, 157, 198-203.
Sijben, J. W., Nieuwland, M. G., Kemp, B., Parmentier, H. K. and Schrama, J. W. 2001. Interactions and antigen dependence of dietary n-3 and n-6 polyunsaturated fatty acids on antibody responsiveness in growing layer hens. Poult Sci, 80, 885-893.
Singh, H., Sodhi, S. and Kaur, R. 2006. Effects of dietary supplements of selenium, vitamin E or combinations of the two on antibody responses of broilers. Br Poult Sci, 47, 714-719.
Sirimongkolkasem, P. 2007. Amino acid partitioning during the acute phase response. Ph.D Thesis, UC Davis, Davis, CA.
Sklan, D., Melamed, D. and Friedman, A. 1994. The effect of varying levels of dietary vitamin A on immune response in the chick. Poult Sci, 73, 843-847.
Sordillo, L. M., Contreras, G. A. and Aitken, S. L. 2009. Metabolic factors affecting the inflammatory response of periparturient dairy cows. Animal health research reviews, 10, 53-63.
Swain, B. K., Johri, T. S. and Majumdar, S. 2000. Effect of supplementation of vitamin E, selenium and their different combinations on the performance and immune response of broilers. Br Poult Sci, 41, 287-292.
Vander Heiden, M. G., Cantley, L. C. and Thompson, C. B. 2009. Understanding the Warburg effect: the metabolic requirements of cell proliferation. Science, 324, 1029-1033.
Wallace, R. J., Oleszek, W., Franze, C., Hahn, I., Baser, K. H. C., Mathe, A. and Reichmann, K. 2010. Dietary plant bioactives for poultry health and productivity. Br Poult Sci, 51, 461-487.
Zebeli, Q. and Metzler-Zebeli, B. U. 2012. Interplay between rumen digestive disorders and diet-induced inflammation in dairy cattle. Res. Vet. Sci., 93, 1099-1108.

Milk fatty acid profile in dairy cows during a negative energy balance in early lactation and feed-restriction in mid-lactation

J.J. Gross[1], *H.A. van Dorland*[1], *R.M. Bruckmaier*[1] *and F.J. Schwarz*[2]
[1] *Veterinary Physiology, Vetsuisse Faculty University of Bern, Bremgartenstrasse 109a, 3001 Bern, Switzerland; josef.gross@vetsuisse.unibe.ch*
[2] *Technische Universität München, Chair of Animal Nutrition, Liesel-Beckmann-Strasse 6, 85350 Freising-Weihenstephan, Germany*

Introduction

Besides diet, mainly lactational stage along with energy balance in dairy cows have an impact on fatty acid (FA) profile in cow`s milk. During the first couple of weeks after parturition, the occurrence of a negative energy balance (NEB) is a common phenomenon observed in dairy cows. The deficiency of nutrients and energy is compensated by mobilization of body reserves, predominantly adipose tissue associated with the release of FA. During insufficient supply and quality of feed, a NEB may also occur later in lactation.

The focus of the present study is set on the interactions between energy balance and milk FA profile in dairy cows. Contrary to earlier studies, the present study investigates effects of both, the NEB in early lactation and a deliberately induced NEB by feed-restriction at around 100 days in milk on milk FA composition.

Material and methods

The animal trial and diets were described previously by Gross *et al.* (2011). According to their energy balance (EB) up to wk 12 p.p. (period 1), 40 multiparous Holstein cows were allocated equably to a control (C, n=20) or a restriction group (R, n=20, -30% of energy requirements) for 3 wks at 100 days in milk (period 2). Thereafter, R-cows were (re)fed similarly as C-cows for 8 wks (period 3). EB of each cow was calculated from daily feed intake, maintenance requirement and milk yield. Blood samples were taken once a week and analyzed for glucose, NEFA and BHBA. Milk samples were analyzed for FA composition by gas chromatography in wk 1, 4, 6 and 12 p.p. of period 1, weekly during period 2 and in wk 1, 2 and 4 of period 3. Changes in milk FA profile over time during lactation and feed-restriction with subsequent realimentation were evaluated by a mixed model in SAS with group, wk and the group x wk interaction as fixed effect. Differences over time and within groups during feed-restriction (period 2) and realimentation (period 3) were detected by the Bonferroni`s t-test ($P<0.05$).

Results and discussion

Lactational stage clearly affected daily milk yield and milk composition. Daily milk yield started at 27.5±0.7 kg/d in wk 1 p.p., peaked in wk 6 p.p. (39.5±0.8 kg/d) and decreased to 32.8±0.8 kg/d in wk 12 p.p. During period 2 restricted cows (27.4 kg/d) had a lower milk yield than control cows (30.5 kg/d; $P<0.05$). Milk yield of restricted cows increased already in wk 1 of realimentation period from 27.6 to 28.6 kg/d and was thereafter even between 0.5 and 1.0 kg/d above the control group (wks 2-8 in phase 3; $P>0.05$).

Energy balance in dairy cows was most negative in wk 1 p.p. and improved with increasing feed intake, but was still negative in wk 6 p.p. In order to compensate for the insufficient energy intake p.p., considerable amounts of body fat were mobilized resulting in elevated plasma concentrations of NEFA and BHBA. During the NEB in early lactation, plasma glucose concentration in the present study showed a nadir of 3.30±0.04 mmol/l in wk 2 p.p., whereas plasma concentrations were highest for NEFA in wk 2 p.p. (0.90±0.06 mmol/l) and for BHBA in wk 3 p.p. (0.98±0.14 mmol/l). After the observed peak of lactation, energy requirements could be met by consumed feed resulting in a positive energy balance. With progressing lactation and improvement of energy balance through increasing feed intake, especially milk FA profile markedly changed. Most changes in milk FA profile took place during the observed NEB from wk 1 to 6 p.p., while FA composition was relatively constant between wk 12 to 21 p.p. FA up to C16:0 showed lowest proportions in wk 1 p.p. that increased to relatively constant proportions from wk 12 p.p. onwards. Saturated FA (SFA), especially C16:0 increased from wk 1 to 12 p.p., while monounsaturated FA (MUFA), mainly represented by C18:1,9c decreased until wk 12 p.p. with improving energy balance. The proportion of polyunsaturated FA (PUFA) was relatively constant from wk 1 up to 21 p.p. Milk FA profile of C-cows was stable during the whole time of period 2 and 3. For R-cows, the proportion of most FA≤C16:0 (e.g. C6:0, C10:0, C10:1, C12:0, C14:0, C16:0) was decreased during the NEB induced by feed-restriction compared to the respective initial values, whereas preformed FA, especially C17:1,9c, C18:0 and C18:1,9c arising from body fat mobilization increased markedly during feed-restriction. These changes occurred rapidly within the first wk of period 2 and disappeared completely within one wk of realimentation (4 d on average). During period 2, SFA decreased, while MUFA (especially C18:1,c9) increased for R-cows compared to C-cows. PUFA were stable during period 2 and 3 in R-cows. Although less in their extent, milk FA clearly showed a similar pattern during the NEB in period 1 and 2.

Short- and medium-chain FA up to C16 increased with the decreasing NEB pp, while long-chain FA, especially C18:1,9c decreased as mobilization of body fat reserves declined. The responses of FA profiles of cows` milk due to a NEB at two lactational stages in the present study – the NEB in early lactation and the deliberately induced NEB by feed-restriction – was similarly directed. Despite the maintenance of a high NEB during the feed-restriction period, changes in milk FA profile were less pronounced compared to changes during the NEB in early lactation and tended to adjust to the initial composition. However, milk FA profile changed within a few days after initiation of the deliberately induced NEB and showed no more differences within the first wk of realimentation compared to control cows. For the dietary composition and feeding regimen in the present study, the close relationship with energy balance makes changes in C18:1,9c as well as in groups of FA (SFA, MUFA, de novo synthesized and preformed FA) suitable indicators of the energy balance in dairy cows.

References

Gross, J., H.A. van Dorland, R.M. Bruckmaier and F.J. Schwarz, 2011. Performance and metabolic profile of dairy cows during a lactational and deliberately induced negative energy balance by feed restriction with subsequent realimentation. J. Dairy Sci., 94, 1820-1830.

The effect of long-term butyrate supplementation to high producing periparturient dairy cows

L. Vandaele[1], *J. Van Eys*[2], *B. D'Heer*[2], *B. Ampe*[1] *and S. De Campeneere*[1]
[1]*ILVO, Animal Sciences Unit, Scheldeweg 68, 9090 Melle, Belgium; leen.vandaele@ilvo.vlaanderen.be*
[2]*Nutriad International NV, Hoogveld 93, 9200 Dendermonde, Belgium*

Introduction

Black *et al.* (1966) indicated that increasing ruminal concentrations of butyrate might be beneficial to dairy cows in early lactation, by preventing the oxidation of pyruvate and by increasing the pyruvate to oxaloacetate conversion. Secondly, extra-mammary tissues can spare glucose by using the increased levels of beta-hydroxybutyrate from butyrate (Holtenius and Holtenius, 1996). Both might lead to better performance in transition cows (DeFrain *et al.*, 2004). This study aimed to explore the effect of daily butyrate supplementation around calving on milk production and composition, and blood glucose level in dairy cattle.

Material and methods

Sixteen Holstein-Friesian multiparous cows were divided in two homogenous groups. Cows in the butyrate group (BUT) received daily 750 g sodium butyrate (3 we pre-calving until 6 we post-calving). Diets of control (CTRL) and BUT were kept iso-nitrogenous and iso-energetic through the inclusion of rumen inert fat into the CTRL diet. Roughage (offered *ad libitum*) was the same for both groups: maize silage and straw during the far off period (the first week of the trial) and pre-wilted grass silage, maize silage and pressed beet pulp during close-up and after calving. Cows were offered a dry cow concentrate before calving and increasing amounts of a protein corrector and concentrates after calving (from 2 kg up to 7.6 kg in 19 days). Daily milk production and weekly bodyweight were recorded. Milk and blood samples were taken weekly. Roughages and concentrates were analyzed to calculate individual energy and protein intake (Dutch protein evaluation system). Data on dietary composition were compared using student's T-test. Post calving data were compared with a longitudinal model with correction for repeated measures within cow (SAS 9.3 for Windows).

Results and discussion

The pre calving CTRL diet contained higher NDF (only far off) and higher NE_L than the BUT diet (far off and close-up), but both diets were similar for DPI and RDPB (Table 1). During the post calving period the dietary composition was similar for CTRL and BUT (Table 1). Daily intake of NE_l (149±5 and 147±5 MJ/d), DPI (2,207±45 and 2,178±55 g/d), RPDB (-128±75 and -139±102 g/d) and DM (20.1±0.3 and 20.4±0.3 kg/day) did not differ between groups post-calving (Table 2).

The mean milk yield of BUT cows was significantly lower than of CTRL cows, but fat and protein content were significantly higher in BUT group (Table 2). The fact that BUT cows seemed to be more sensitive to metabolic problems post calving (two cows with displaced abomasum in BUT and none in CTRL) might partially explain the lower milk yield in the BUT group.

Interestingly, blood glucose levels 2 and 4 h after feeding were significantly lower for BUT then for CTRL. This observation might be related to higher insulin levels as direct response to butyrate as shown by Mineo *et al.* (1990) in sheep. Higher insulin levels might depress glucose levels despite the higher butyric acid concentration.

In conclusion, long term supplementation with butyrate leads to a lower milk yield and lower post calving blood glucose levels. However this did not lead to lower daily fat nor protein production.

Table 1. Dietary composition in CTRL and BUT in g/kg DM (mean±SEM).

	Far off		Close-up		*Postpartum*	
	CTRL	BUT	CTRL	BUT	CTRL	BUT
NDF	375±2.2	366±2.0*	321±3.6	312±4.9	298±3.4	290±4.0
Starch	259±1.8	252±1.9	219±3.6	211±3.9	207±6.6	203±8.7
NEL (MJ/kg)	6.4±0.03	6.3±0.02*	7.2±0.03	7.0±0.01*	7.4±0.04	7.3±0.04
DPI	53±0.4	52±0.3	74±0.7	72±0.9	110±1.4	109±1.5
RDPB	0±2.5	-1±2.6	14±2.7	15±3.2	-6±3.6	-7±5.2

NE_l = net energy lactation; DPI = true protein digested in the small intestine; RDPB = rumen degradable protein balance.
* Results are significantly different between groups within the same period.

Table 2. Least square means (±SEM) of post calving DM intake and performances in both treatment groups, including P*-value.*

	CTRL	BUT	*P*-value
Milk yield (kg/day)	34.9±0.62	32.9±0.70	<0.05
Fat (%)	4.90±0.077	5.27±0.087	<0.01
Protein (%)	3.30±0.029	3.50±0.033	<0.01
Lactose (%)	4.69±0.022	4.61±0.026	<0.05
Fat prod. (kg/day)	1,718.6±36.05	1,695.3±41.03	0.68
Protein prod. (kg/day)	1,146.5±18.51	1,126.7±21.15	0.50
Glucose 2 h (mg/dl)	43±1.5	35±1.7	<0.01
Glucose 4 h (mg/dl)	46±1.4	38±1.6	<0.01

References

Black A.L., J. Luick, F. Moller, R.S. Anand, 1966. Pyruvate and propionate metabolism in lactating cows. Effect of butyrate on pyruvate metabolism. J. Biol. Chem. 241, 5233-5237.

DeFrain J.M., A.R. Hippen, K.F. Kalscheur, D.J. Schingoethe, 2004. Feeding lactose increases ruminal butyrate and plasma beta-hydroxybutyrate in lactating dairy cows. J. Dairy. Sci. 87, 2486-2494.

Holtenius P. and K. Holtenius, 1996. New aspects of ketone bodies in energy metabolism of dairy cows: A review. Zbl. Vet. Med. A. 43, 579-587.

Mineo H., M. Kanai, S. Kato, J.I. Ushijima, 1990. Effects of intravenous injection of butyrate, valerate and their isomers on endocrine pancreatic responses in conscious sheep. Comp. Biochem. Physiol. 95, 411-416.

Peripartal energy balance and peripheral blood mononuclear cell activation in normal and high mobilizing dairy cows

A. Schwarm[1,2], T. Viergutz[3], B. Kuhla[1], H.M. Hammon[1] and M. Schweigel-Röntgen[1]
[1]Institute of Nutritional Physiology „Oskar Kellner', Leibniz Institute for Farm Animal Biology (FBN), Dummerstorf, Germany; roentgen@fbn-dummerstorf.de
[2]Animal Nutrition Group, Institute of Agricultural Sciences, Swiss Federal Institute of Technology (ETH) Zurich, Switzerland
[3]Institute of Reproductive Biology, Leibniz Institute for Farm Animal Biology (FBN), Dummerstorf, Germany

Introduction

In mammals, the time of pregnancy and lactation is highly energy demanding and shifts the priority of energy partitioning from the immune system and body reserves towards the conceptus and the mammary gland (Nelson *et al.*, 2002). The resulting peripartal breakdown of (acquired) immunity has been determined most of all by diminished blood leukocyte DNA synthesis in response to mitogens (e.g. humans: Redwine *et al.*, 2001, cow: reviewed by Mallard *et al.*, 1998). The cell cycle phase preceding the DNA synthetic (S) phase, i.e. immune cell activation (G_1 phase), has received little attention, but the few available data suggest that leukocyte activation may also be impaired by the limited energy/nutrient availability.

Dairy cows with a high liver fat content *postpartum*, i.e. mobilizing body reserves to an above average extent, have a more severe negative energy balance (EB) and a higher risk for metabolic and infectious diseases than cows with a normal liver fat content.

Therefore, the aim of this study was to find out whether the energy-consuming process of peripheral blood mononuclear cells (PBMC) activation – in terms of oxygen consumed over basal levels after *in vitro* stimulation – is differentially altered by EB in normal and high mobilizing cows.

Material and methods

Twelve multiparous high-yielding Holstein cows were retrospectively grouped according to normal (N, 26±3% of dry matter, DM, n=6, mean ± SD) and high (H, 44±6% of DM, n=6) liver fat content (calculated from C, N glycogen and glucose content) at day 18 *postpartum*. Oxygen consumption rate (OCR, as a measure of ATP production) was determined in phytohaemagglutinin (PHA)-stimulated PBMC at week -5, -1, +2 and +5 relative to parturition. OCR of PBMC was assessed fluorometrically 24 h after treatment with PHA (4 μg/ml) or with RPMI 1640 medium (control) (Schwarm *et al.* 2013). The activation index was calculated as the PHA-induced 24 h-increase of OCR above baseline. In addition, we determined cellular lactate production, DNA+RNA synthesis and cell size.

The animals' EB was estimated *antepartum* from metabolizable energy intake and *postpartum* from netto energy intake for lactation and energy-corrected milk yield. Furthermore, we assessed body weight, body condition score, backfat thickness, and selected plasma parameters, e.g. glucose, insulin, non-esterified fatty acid and β-hydroxbutyric acid concentration.

The data were analysed by Student´s *t*-test, Mann-Whitney-U-test, one-way repeated measures analysis of variance (RM ANOVA) and Spearman correlation coefficient (SCC) using Systat 11 (Erkrath, Germany). The significance level was set to α=0.05 and P values between 0.05 and 0.10 were considered as trends.

Results and discussion

H cows started *antepartum* with a higher (week -5, *t*-test, P=0.020) backfat thickness than N cows but had a similar (week -5, U-test, P=0.699) EB. *Postpartum* EB was more negative (*t*-test, P=0.001) in H than N cows at its nadir in the second week in milk, when the energy-corrected milk yield had its maximum.

Basal OCR of PBMC did not change during the studied period of the reproduction cycle [N cows: 1.2±0.4 nmol/min/(10^7 cells), n=6, F=0.35, P=0.791; H cows: 1.1±0.4 nmol/min/(10^7 cells), n=5-6, F=2.03, P=0.159], which is in accordance to previous results (Schwarm *et al.*, 2013). However, peripartal PBMC activation was differentially altered in N and H cows. *Antepartum*, activation indices of PBMC tended to be positively correlated with EB in N cows (n=11, SCC=0.56, P=0.071), but no such relationship was observed in H cows (n=11, SCC=-0.26, P=0.433). It is tempting to speculate, that H cows have a superior nutritional and energetic state *antepartum* because less energy is partitioned to immune function. In contrast, N cows directly channel the surplus energy (of feeding the more energy dense close-up diet) towards immune function. Correspondingly, high levels of plasma glucose coincided in time with high *in vitro* PBMC activation indices in N cows (n=22, SCC=0.46, P=0.031) but not in H cows (n=21, SCC=-0.16, P=0.480). The pattern is reversed *postpartum*: a positive relationship between EB and activation indices of PBMC has been observed only in H cows (n=11, SCC=0.73, P=0.009; N cows: n=12, SCC=0.46, P=0.123). This energy partitioning towards the immune system during early lactation may additionally deteriorate the poor nutritional and energetic state of H cows, without reaching host defense.

We conclude that differences in disease susceptibility between N and H cows may be related to the observed differences in energy partitioning towards immune function. Our results corroborate the assumption that immune modulation during reproduction is a facultative (resource-driven) response rather than an obligatory side effect of going through reproduction per se (French *et al.*, 2007).

References

French, S.S., DeNardo, D.F., Moore, M.C., 2007. Trade-offs between the reproductive and immune systems: facultative responses to resources or obligate responses to reproduction? Am. Nat. 170, 79-89.

Mallard, B.A., Dekkers, J.C., Ireland, M.J., Leslie, K.E., Sharif, S., Vankampen, C.L., Wagter, L., Wilkie, B.N., 1998. Alteration in immune responsiveness during the peripartum period and its ramification on dairy cow and calf health. J. Dairy Sci. 81, 585-595.

Nelson, R., Demas, G., Klein, S., Kriegsfeld, L., 2002. Energetics and immune function. In: Seasonal Patterns of Stress, Immune Function, and Disease. Cambridge University Press, Cambridge, pp. 151-189.

Redwine, L.S., Altemus, M., Leong, Y.M., Carter, C.S., 2001. Lymphocyte responses to stress in *postpartum* women: relationship to vagal tone. Psychoneuroendocrinology 26, 241-251.

Schwarm, A., Viergutz, T., Kuhla, B., Hammon, H. M. and M. Schweigel-Röntgen, 2013. Fuel feeds function: Energy balance and bovine peripheral blood mononuclear cell activation. Comp. Biochem. Physiol. A 164, 101-110.

Impact of CFA and dietary protein supply on acute phase responses and nitrogen retention in pigs

E. Kampman-van de Hoek[1,2], P. Sakkas[2], J.J.G.C. van den Borne[2], W.J.J. Gerrits[2], C.M.C. van der Peet-Schwering[1], H. van Beers[3] and A.J.M. Jansman[1]
[1]Department of Animal Nutrition, Wageningen UR Livestock Research, P.O. Box 65, 8200 AB Lelystad, the Netherlands; alfons.jansman@wur.nl
[2]Animal Nutrition Group, Wageningen University, P.O. Box 338, 6700 AH Wageningen, the Netherlands
[3]SDa, The Netherlands Veterinary Medicines Authority, Yalelaan 114, 3584 CM Utrecht, the Netherlands.

Introduction

During immune system activation, an increased competition occurs between amino acids (AA) for body protein deposition and for immune system functioning (Klasing and Johnstone, 1991; Sandberg *et al.*, 2007). The production of acute phase proteins (APP) has been suggested to increase the demand for especially aromatic AA (Reeds *et al.*, 1994). When muscle protein is mobilized to supply AA for APP production, this leads to an imbalance in AA available for body protein deposition, as the AA composition of APP differs largely from the composition of muscle protein (Reeds *et al.*, 1994). As a consequence, increased AA oxidation and N loss via urine occur. It is hypothesized that the competition between AA increases when dietary protein supply is reduced. In addition, there is increasing evidence that the dietary protein or AA supply can modulate the inflammatory response during immune system activation (Grimble, 2001; Calder and Yaqoob, 2012). The aim of the present study was to quantify the interaction between acute phase protein (APP) responses, induced by immune system activation, and dietary protein supply on nitrogen (N) metabolism.

Material and methods

To quantify the interaction between acute phase protein (APP) responses and dietary protein supply on N metabolism, we challenged 16 pigs (28 kg BW) with Complete Freund's Adjuvant (CFA) at two levels of dietary protein supply. All pigs were surgically fitted with catheters in the jugular vein, housed in metabolic cages and assigned to a diet inducing either an adequate (A) or restricted (R, 70% of A) dietary protein supply, relative to the supply of digestible energy, fed at 2.7× the energy requirements for maintenance. Two consecutive 5 d N-balance measurements were performed using funnels underneath the cages for urine collection under addition of sulphuric acid to prevent NH_3 losses, one before and one 2-7 d after intravenous administration of CFA (0.15-0.20 ml CFA/kg BW, divided over 3 or 4 equal portions in 2 days). Blood samples were collected daily throughout the experimental period up to 8 d after the CFA challenge and analysed for concentrations of acute phase proteins (APP).

Results and discussion

Serum concentrations of haptoglobin, C-Reactive Protein (CRP) and pig Major Acute Phase Protein (pig-MAP) were profoundly increased between d 3 and 6 post CFA challenge ($P<0.01$), and decreased afterwards on d 7 and 8 post challenge. Serum albumin concentrations were reduced post challenge ($P<0.01$). Prior to the CFA challenge, a restricted dietary protein supply reduced the concentrations of CRP ($P<0.05$) and albumin ($P<0.01$), but did not affect the concentrations of haptoglobin and pig-MAP. Post CFA challenge, the serum albumin concentration was significantly reduced by dietary protein restriction ($P<0.01$), while the concentrations of haptoglobin, pig-MAP and CRP were not affected by dietary protein supply. Houdijk *et al.* (2007) also found lower plasma CRP concentrations in pigs with sub-clinical colibacillosis fed a low protein diet, than in pigs fed a high protein diet. In line, plasma albumin concentrations decreased in pigs (Jahoor *et al.*, 1999) and rats (Lunn *et al.*, 1983) in case of a deficient dietary protein supply. Our results indicate that the concentrations of CRP and albumin are affected by the quantitative dietary protein and AA supply. Due to the challenge, N-retention tended to be reduced in A-fed, but not in R-fed pigs (P=0.07). This reduction was mainly caused by increased urinary N losses during the post-challenge period. In line, a reduced N retention due to immune system activation was found by Williams *et al.* (1997) and Daiwen *et al.* (2008). These findings suggest that limitations in dietary protein supply restrict the APP responses to a systemic CFA challenge in growing pigs. Furthermore, our results suggest that the priority of partitioning dietary protein over immune system functioning and body protein deposition depends on the dietary protein supply.

Table 1. Effect of immune system activation induced by CFA administration on N-metabolism (all data expressed in g/ $BW^{0.75}$/d).

Dietary protein	Adequate (A)		Restricted (R)		Pooled	*P*-values		
Challenge	Pre	Post	Pre	Post	SEM	Diet	Chall.	D×C
N intake	2.32	2.32	1.77	1.75	0.061			
Faecal N excretion	0.12	0.12	0.13	0.12	0.003	0.41	0.16	0.34
Urinary N excretion	0.64	0.73	0.42	0.43	0.031	<0.001	0.04	0.14
Total N excretion	0.76	0.85	0.55	0.55	0.031	<0.001	0.09	0.11
N retention	1.57	1.47	1.22	1.20	0.038	<0.001	0.07	0.28

References

Calder, P.C., and P. Yaqoob, 2012. Nutrient regulation of the immune response present knowledge in nutrition. In: Erdman, J.W., Macdonald, I.A., and S.H. Zeisel (editors), Present Knowledge in Nutrition. John Wiley & Sons, Inc. Wiley-Blackwell. Tenth edition: 688-708.

Daiwen, C.,Z. Keying, and W. Chunyan. 2008. Influences of lipopolysaccharide-induced immune challenge on performance and whole-body protein turnover in weanling pigs. Livest. Sci. 113, 291-295.

Grimble, R.F. 2001. Nutritional modulation of immune function. Proc. Nutr. Soc. 6, 389-397.

Houdijk, J.G.M., F.M. Campbell, P.D. Fortomaris, P.D. Eckersall, and I. Kyriazakis. 2007. Effects of sub-clinical post-weaning colibacillosis and dietary protein on acute phase proteins in weaner pigs. Livest. Sci. 108, 182-185.

Jahoor, F., L. Wykes, M. Del Rosario, M. Frazer, and P.J. Reeds. 1999. Chronic protein undernutrition and an acute inflammatory stimulus elicit different protein kinetic responses in plasma but not in muscle of piglets. J. Nutr. 129, 693-699.

Klasing, K.C., and B.J. Johnstone. 1991. Monokines in growth and development. Poult. Sci. 70, 1781-1789.

Lunn, P.G., and S. Austin. 1983. Differences in nitrogen metabolism between protein-deficient and energy-deficient rats with similarly restricted growth rates. Ann. Nutr. Metab. 27, 242-251.

Reeds, P.J., C.R. Fjeld, and F. Jahoor. 1994. Do the differences between the amino acid compositions of acute-phase and muscle proteins have a bearing on nitrogen loss in traumatic states? J. Nutr. 124,906-910.

Sandberg, F.B., G.C. Emmans, and I. Kyriazakis. 2007. The effects of pathogen challenges on the performance of naïve and immune animals: the problem of prediction. Animal 1, 67-86.

Williams, N.H., T.S. Stahly, and D.R. Zimmerman. 1997. Effect of chronic immune system activation on body nitrogen retention, partial efficiency of lysine utilization, and lysine needs of pigs. J. Anim. Sci. 75, 2472-2480.

Nano-nutrition as a method of anticancer therapy

E. Sawosz[1], M. Grodzik[1], S. Jaworski[1], M. Wierzbicki[1], M. Prasek[1] and A. Chwalibog[2]
[1]Warsaw University of Life Science, Division of Biotechnology and Biochemistry of Nutrition, Ciszewskiego 8, 02-786 Warsaw, Poland; ewa_sawosz@sggw.pl
[2]University of Copenhagen, Department of Veterinary Clinical and Animal Sciences, Groennegaardsvej 3, 1870 Frederiksberg, Denmark

Introduction

Nano-nutrition is a method of providing cells with nutritional factors, disintegrated to nanoparticles or transported by nanoparticles. Bioactive substances attached to nanoparticles may penetrate organism, bypass blood system and through the membranes go directly to the cells. Consequently, the transported nutritive substances may be delivered to cells, also to cancer cells switching on the mechanism of programmed cell death, leading to decreased tumour growth.

Graphene is one of carbon allotropes, regarded as the thinnest material in the world (Geim and Novoselov, 2007). It is made of one layer of carbon atoms, forming a hexagonal structure connected by sp^2 type bonds. Moreover, graphene as a carbon structure is not very toxic and used at an optimal level may form drug delivery platform. Graphene may be easy functionalised with amino acids and deliver them to the cells to modify specificity of bonds. Attached to graphene amino acids (arginine and hyroxyproline) may block p53 – MDM2 binding by enriching hydroxyproline or arginine in proline-rich domain of p53 protein, leading to apoptosis of cancer cells.

The objective of the experiments was to use graphene as a platform to deliver arginine and hydroxyproline to glioblastoma multiforme (GM) cancer cells, activate p53 protein and induce apoptosis.

Material and methods

Graphene nanoparticles were obtained from ITE (Warsaw, Poland) and from SkySpring Nanomaterials (Houston, USA). The size of graphene sheets varied around 450 nm.

The experiments were carried out with *in vitro* and *in ovo* methods. Human glioblastoma U87 and U118 cell lines were obtained from the American Type Culture Collection and maintained in DMEM supplemented with 10% fetal bovine serum and 1% penicillin and streptomycin at 37 °C in a humidified atmosphere of 5% CO_2/95% air.

To maintain tumour of GM the fertilized eggs Ross 308 were incubated at 37 °C and 70% humidity. After 7 days the silicone ring with the deposited $3\text{-}4\times10^6$ U87 or U118 glioma cells suspended in 30 µl of culture medium was placed on the chorioallantoic membrane (CAM) in the area of formed blood vessels (Grodzik *et al.*, 2011). The eggs were incubated for 5 days and then experimental solution (control vs. graphene or graphene with attached arginine and hydroxyproline) were injected directly into the tumour. After 5 days the eggs were opened and tumours were resected for further analysis.

Morphology, histology, mortality, expression of protein involved in stress (p-53) as well as and apoptosis and necrosis were measured in vitro with GM cells and using tumour model maintained *in ovo*.

Data were analyzed using one-way analysis of variance using STATGRAPHICS® Plus 4.1.

Results and discussion

Graphene, but also graphene with attached amino acids affected morphology of the GM cells but this effect was different between the U87 and U118 lines. In both cases, graphene showed affinity to GM cells, locating and adhering to the body of cells. However, because of the size (450 μm) the graphene flakes did not enter the celles. The measurements of cell mortality, using the use of Trypan blue dye, demonstrated the dose depended toxic effect of graphene nanoparticles on glioma cells, as previously observed by Chang *et al.*, 2011. Graphene had the highest toxicity at the concentration of 100 mg/kg but was not toxic at the levels less than 20 mg/kg. Toxic effect was observed by evaluating cell membrane functionality and integrity. Analyzing the integrity of membranes between two glioma cell lines significant differences were seen. The membrane integrity in U87 cells was much more disrupted than in U118 even at the low graphene concentrations.

Graphene and also graphene with attached amino acids induced apoptosis in U118 GM and apoptosis and necrosis in U87 GM. Protein p53 is the main activator of apoptosis, but U118 GM showed reduced possibilities to activate apoptosis via p53 induction. It can be supposed that grapheme can activate apoptosis by an alternative way, being in a direct contact with cell membranes and membrane receptors. Amino acids attached to graphene did not significantly affect the observed mechanism.

In experiments with GM tumour graphene injected into the tumour decreased size of tumour and increased number of apoptotic cells. Morphology of the tumour indicated less number of disrupted mitochondria. Expression of p53 and MDM2 protein also changed; however, tendency to increase apoptosis by graphene and graphene attached with amino acids was not correlated with p53, which also may suggest an alternative way of apoptosis.

The results demonstrated that graphene flakes, regardless of attached arginine and hydroxyproline at a level less than 20 mg/kg seem to be not toxic, however, they activate mechanisms of apoptosis in U118 GM cell line and this may indicate anticancer characteristics of graphene.

References

Chang, Y., S. T. Yang, J. H. Liu, E. Dong, Y. Wang, Y. Liu, H. Wang, 2011. In vitro toxicity evaluation of graphene oxide on A549 cells. Toxicology Lett. 200, 201-210.

Geim, A. K. and K. S. Novoselov. 2007. The rise of graphene. Nat. Mater. 6, 183-191.

Grodzik, M., E. Sawosz, M. Wierzbicki,P. Orlowski, A. Hotowy, T. Niemiec, M. Szmidt, K. Miture, A. Chwalibog. 2011. Nanoparticles of carbon allotropes inhibit glioblastoma multiforme angiogenesis in ovo. Int. J. Nanomedicine. 6, 3041-3048.

Differential expression of innate immune system genes in liver of beef cattle with divergent phenotypes for RFI

K.A. Johnson[1], J.J. Michal[1], G.E. Carstens[2], A.N. Hafla[2] and T.D.A. Forbes[2]
[1]Department of Animal Sciences, Washington State University, Pullman, WA 99164, USA; johnsoka@wsu.edu
[2]Department of Animal Science, Texas A&M University, College Station, TX 77843, USA

Introduction

Residual feed intake (RFI) quantifies inter-animal variances in dry matter intake (DMI) independent of variation in body size and productivity (e.g. growth, milk). In beef cattle, Herd and Arthur (2009) estimated that approximately one-third of the biological variation in RFI was due to inter-animal differences in digestion, heat increment, body composition, and activity, with the remaining variation associated with energy expenditure of metabolic processes. Given that RFI is independent of BW and level of production, it is an ideal trait for discovery of genomic regions (Rolf *et al.*, 2011) and functional genes (Chen *et al.*, 2011) associated with efficiency of feed utilization. Some variation in RFI may be caused by individual differences in immune system function. The immune system is an energetically expensive system. A previous study showed that metabolic heat production and oxygen consumption increased 20-30% in mice that were immunized with a relatively benign antigen (Demas *et al.*, 1997). The objective of this work was to examine whether the genes associated with the immune system were differentially expressed by beef cattle with divergent RFI phenotypes.

Material and methods

Transcript abundances in liver from Bonsmara heifers (n=12; Trial 1) and finishing crossbred steers (n=12; Trial 2) with divergent phenotypes for RFI were initially screened with custom-designed NimbleGen bovine gene expression arrays. Several genes involved with innate immune functions and inflammatory responses were identified as biologically significant between the RFI groups. Genes investigated included serpin peptidase inhibitor, clade I (pancipin), member 2 (SERPINI2), tetraspanin 13 (TSPAN13), caspase 3 (CASP3), interleukin 1, beta (IL1B), interleukin 18 (1L18), interferon, alpha-inducible protein 6 (IFI6), interferon-induced transmembrane protein 2 (IFITM2), interferon-induced protein with tetratricopeptide repeats 3 (IFIT3), interleukin 20 receptor, alpha (IL20RA), and folliculin interacting protein 1 (FNIP1).

To further examine the biological significance of these genes, additional liver samples were collected from cattle with divergent RFI. Trial 3 included Santa Gertrudis heifers (n=16) and bulls (n=15), trial 4 comprised crossbred steers (n=16), and trial 5 included Bonsmara heifers (n=19). Cattle in trials 1, 3 and 5 were fed a low-energy diet, whereas, a high-energy diet was fed in trials 2 and 4. Total RNA was extracted from liver, and reverse-transcribed to cDNA. Primers for real-time PCR were designed with QuantPrime and transcript abundances in all 66 cattle were determined using SYBR green-based detection and a BioRad iCycler iQ detection system. Beta-actin was used as a housekeeping gene for data normalization, and transcript abundance expressed relative to a reference sample (pooled cDNA from 4 steers). The PROC MIXED procedure of SAS 9.2 was used to determine the fixed effect of RFI group (high or low) on mRNA expression for each trial separately.

Results and discussion

The expression of immune function genes by heifers and bulls consuming a low energy diet did not vary by RFI phenotype. However, there was a tendency ($P<0.10$) for IFITM2 to be down-regulated and for IL20RA to be up regulated in animals with low RFI phenotypes (more efficient). This is in contrast to efficient steers fed a high grain diet in which IL1B, TSPAN13, and SERPINI2 were significantly down regulated. Others have also reported that SERPINI2 and TSPAN13 were down regulated in liver of cattle with high and low RFI (Chen *et al.*, 2011). Serpins are involved in regulation of development, coagulation, fibrinolysis, and inflammatory processes (Ragg, 2007), while tetraspanins are involved in cell adhesion, fusion, membrane trafficking. Recent evidence suggests that tetraspanins play a key role in the course of pathogenic infection (Monk and Partridge, 2012). Expression of IL1B, TSPAN13 and SERPIN12 were positively correlated (P<0.01) to DMI (0.65, 0.51, 0.75, respectively), feed conversion ratio (0.55, 0.57, 0.56, respectively) and RFI (0.67, 0.58, 0.69, respectively) in steers from trial 2. The other genes examined such as IFIT3 (down regulated), IL20RA (up regulated) tended to be altered in the steers fed a high grain diet in trials 2 & 4. Interleukin 1, beta, IL18, CASP3, IFIT3, IFITM2, IFI6, and IL20RA are genes that are involved in the innate immune response. Examination of the locus in Angus cattle that explained 10% of the variation in DMI revealed that FNIP1 may be important in the regulation of feed intake (US Consortium *et al.*, 2012). In this study, FNIP1 expression was positively correlated (0.46; P<0.05) to DMI in one heifer trial.

The genes involved in the innate immune system may be affected by diet, sex and the RFI phenotype in growing cattle. Pathway analysis of additional genes involved in this system may provide a clearer view of how inflammation and the innate immune system contribute to energy metabolism.

References

Chen, Y., C. Gondro, K. Quinn, R.M. Herd, P.F. Parnell and B. Vanselow, 2011. Global gene expression profiling reveals genes expressed differentially in cattle with high and low residual feed intake. Anim. Genet. 42, 475-490.

Demas, G.E., V. Chefer, M.I. Talan and R.J. Nelson, 1997. Metabolic costs of mounting an antigen-stimulated immune response in adult and aged C57BL/6J mice. Am. J. Physiol. Integr. Comp. Physiol. 273, R1631-R1637.

Herd, R.M., and P.F. Arthur, 2009. Physiological basis for residual feed intake. J. Anim. Sci. 87, E64-E71.

Monk, P.N. and L.J. Partridge, 2012. Tetraspanins – Gateways for infection. Infect. Disord. Drug Targets. 12, 4-17.

Ragg, H, 2007. The role of serpins in the surveillance of the secretory pathway. Cell. Mol. Life Sci. 64, 2763-2770.

Rolf, M.M., J.F. Taylor, R.D. Schnabel, S.D. McKay, M.C. McClure, S.L. Northcutt, M.S. Kerley and R.L. Weaber, 2011. Genome-wide association analysis for feed efficiency in Angus cattle. Anim. Genet. 43, 367-374.

US Consortium for the Genetic Improvement of Feed Efficiency in Beef Cattle.2012. Plant and Animal Genomics Meeting. P0521.

Effect of high-fat by-products pellets in finishing diets for steers

P. Górka[1,2], J.J. McKinnon[1] and G.B. Penner[1]
[1]Department of Animal and Poultry Science, University of Saskatchewan, S7N 5A8, Saskatoon, Saskatchewan, Canada; greg.penner@usask.ca
[2]Department of Animal Nutrition and Feed Management, University of Agriculture in Krakow, 31-120 Krakow, Poland

Introduction

Cereal grain prices have increased in recent years resulting in increased feed costs for feedlot operators. To offset these costs, investigation into alternative energy sources, relative to cereal grains, has been initiated (Marx *et al.*, 2000). Applying single ingredient substitution strategies may limit the use of some byproducts for finishing cattle, such as grain screenings, pea screenings or oat hulls, due to insufficient energy content (Marx *et al.*, 2000) or alternatively may limit the amount of cereal grain that can be replaced. One strategy to eliminate overfeeding of nutrients from high byproduct inclusion rates is to utilize strategic combinations of various byproducts to optimize ruminal and postruminal energy and protein availability (Zenobi *et al.*, 2012). The aim of this study was to determine the effect of including strategically blended high-fat by-product pellets as a partial replacement for barley grain and canola meal in finishing diets for steers.

Material and methods

Two hundred and sixty four crossbreed steers (average BW of 375±0.72 kg) were randomly allocated to one of 12 outdoor pens (22 steers/pen) with each pen randomly allocated to 1 of 3 treatments (4 pens/treatment). The control (CON) diet consisted of 89% concentrate (barley and canola meal), 6% barley silage, and 5% vitamin-mineral pellets (DM basis). High-fat pellets (9.0% crude fat, 29.8% NDF, 5.23 MJ NE_g/kg on DM basis) were included in diets by replacing 30% (HFP30) or 60% (HFP60) of the barley and canola meal relative to CON. The HFP contained (DM basis): 29.5% ground wheat, 26.4% wheat screenings, 14.8% whole canola, 17.4% pea screenings and 11.9% oat hulls. Diets were formulated to be isonitrogenous and isoenergetic with CP and NE_g contents of 13.4% and 5.02 MJ NE_g/kg (DM basis), respectively, and were offered ad libitum once daily. Steers were weighed on 2 consecutive days at the start and end, and monthly weights were taken throughout the study. Feed bunks were cleaned every 2 wk and DMI was calculated based on the amount of feed offered and refused. After completing 155 d on feed, steers were weighed and shipped to commercial slaughter plant and carcasses were evaluated using the Computer Vision Grading System (VBG 2000 e+v Technology GmbH, Oranienburg, Germany). Data were analyzed as a completely randomized design using the mixed model of SAS (Cary, NC, USA) except for carcass grade, which were analyzed using Proc GLIMMIX. Contrasts were used to determine the effect of HFP inclusion and whether the level of inclusion affected the response.

Results and discussion

Initial and final BW, and cumulative ADG were not different among treatments ($P>0.05$; Table 1). Dry matter intake was higher and the gain:feed ratio was lower ($P<0.01$) for HFP30 and HFP60 as compared to CON with no differences observed between HFP30 and HFP60. Hot carcass weight, rib eye area, and marbling score were not affected by treatment ($P>0.05$). Quality grade and carcass yield were not different among treatments ($P>0.05$). Back fat thickness was higher ($P=0.02$) for HFP30 and HFP60 as compared to CON with no differences between HFP30 and HFP60.

Partially replacing barley grain with strategically blended HFP allowed for high inclusion rates of byproducts (18.8 and 37.6% for HFP30 and HFP60, respectively) in finishing diets for beef steers

Table 1. Effect of partially replacing barley grain and canola meal with high-fat pellet (HFP) on BW, ADG, DMI, and carcass yield and carcass grade.

	Treatment			SEM	Contrasts	
	CON	HLP30	HLP60		CON vs. HLP	HLP30 vs. HLP60
Initial BW (kg)	374	374	375	0.2	0.18	0.63
Final BW (kg)	665	663	659	1.8	0.27	0.38
ADG (kg/d)	1.90	1.88	1.86	0.010	0.27	0.38
DMI (kg/d)	10.7	11.3	11.4	0.10	<0.01	0.17
Gain:feed	0.18	0.17	0.16	0.01	<0.01	0.10
Hot carcass (kg)	379	387	382	2.5	0.31	0.39
Rib eye area (cm^2)	94.8	93.9	96.6	0.68	0.78	0.13
Back fat (cm)	0.77	0.93	0.88	0.28	0.02	0.38

and thus the potential to replace a significant amount of high-priced cereal grains. Although HFP inclusion reduced feed efficiency there were no differences in growth performance or carcass yield and quality among treatments suggesting that higher DMI compensated for the potential negative effects feeding diets high in NDF with a small particle size and high in lipid on passage rate and digestibility. Thus, when NE_m and NE_g were calculated based on intake and performance, the NE_m and NE_g (MJ/kg) were 8.33 and 5.61, 7.91 and 5.23, and 7.78 and 5.06 for CON, HFP30 and HFP60, respectively.

The results indicate that up to 60% of barley and canola meal in finishing diets can be replaced with strategically blended HFP without negative effects on ADG of steers and carcass quality. However, gain:feed ratio may be compromised.

References

Marx, T.A., J.J. McKinnon, A.F. Mustafa, D.A. Christensen, and V.J. Racz, 2000. The feeding value of grain screenings for ruminants: Chemical composition and nutrient utilization. Can. J. Anim. Sci. 80: 673-680.

Zenobi, M.G., P. Yu, D.A. Christensen, P.G. Jefferson, H.A. Lardner, and J.J. McKinnon, 2012. Performance of cattle fed diets based on blended by-product pellets varying in rumen avaliale energy and protein content. J. Anim. Sci. 90 (Suppl. 3): 711.

Microbial activity in the large intestine of chickens fed diets containing different sources of inulin-type fructans

M. Taciak, M. Barszcz, A. Tuśnio, E. Święch, Ł. Staśkiewicz and J. Skomiał
Department of Monogastric Nutrition, The Kielanowski Institute of Animal Physiology and Nutrition, Polish Academy of Sciences, 05-110 Jabłonna, Poland; m.taciak@ifzz.pan.pl

Introduction

Inulin-type fructans are water-soluble compounds of plant origin with varying length of polysaccharide chains, with β-2,1 fructosyl-fructose glycosidic bonds. This type of bond makes it resistant to enzymatic hydrolysis by small intestinal digestive enzymes (specific for α-glycosidic bonds). As a result, inulin fructans are indigestible by enzymes of animal origin and are fermented in the large intestine. The inulin fructans can be found in many plants, but by the industry is mainly extracted from chicory and Jerusalem artichoke. Maintaining animal welfare through the modulation of large intestine function may play an important role, therefore the aim of the study was to determine the effect of diets' supplementation with different sources of inulin-type fructans on microbial activity in the large intestine of chickens.

Material and methods

The experiment was performed on 40 male chickens (Ross 308) divided into 5 groups (n=8). Birds were fed from the first day of life cereal diets supplemented with 0.4% of inulin extracted from chicory root with degree of polymerization (DP) ≥10; 0.4% of inulin with DP≥23; 0.8% of dried Jerusalem artichoke and 0.8% of dried chicory, corresponding with 0.4% of pure inulin. Control group was given diet without supplementation. Birds were sacrificed at the age of 14 days and digesta samples were taken from the caecum and colon for analysis of microbial activity indices. Digesta pH was measured using pH-meter, SCFA concentration was analyzed by gas chromatography (Barszcz *et al.*, 2011). Caecal β-glucosidase activity was determined spectrophotometrically (Juśkiewicz *et al.*, 2011). Statistical analysis was performed by one-way ANOVA procedure and *post hoc* Tukey HSD test.

Results and discussion

Neither caecal nor colonic digesta pH was significantly affected by diets (Table 1), however birds fed diet with dried chicory tended to have lower ceacal pH compared to other groups. Bacterial enzyme activity and SCFA concentration in caecal digesta also did not differ significantly between experimental groups. Concentration of SCFA in colonic digesta was affected by experimental diets. Feeding Jerusalem artichoke significantly increased acetic and propionic acid concentrations comparing to chicory and isobutyric acid comparing to chicory and inulin with DP≥10. Increased concentration of isobutyric acid in the colon may be a result of more intensive bacterial catabolism of endogenous or dietary protein (Hughes *et al.*, 2000). Concentration of valeric acid in the caecum and valeric and isovaleric acids in the colon were under detection level.

Contrary to the expectations, microbial activity in the large intestine of chicken is slightly modified by feeding diets with different sources of inulin-type fructans, at the level of 0.4%, since the only difference was found in SCFA concentration in the colon, except for butyric acid concentration. Neither inulin preparations nor dried plants rich in inulin decreased digesta pH and activity of detrimental bacterial enzyme.

Table 1. pH, short chain fatty acids concentrations (µM/g of digesta) and activity of β-glucosidase (U/g of digesta) in the caecum and colon of chicken.

	pH	Acetic	Propionic	Isobutyric	Butyric	Isovaleric	β-glucosidase
Caecum							
Control	6.30	31.91	0.830	0.351	7.92	0.223	273.8
IN_{10}	6.00	27.28	0.875	0.321	10.65	0.211	269.1
IN_{23}	5.90	33.08	0.892	0.327	7.95	0.248	297.9
JA	6.07	31.81	0.963	0.369	7.96	0.259	285.2
CH	5.43	29.14	0.801	0.246	9.14	0.136	263.4
SEM	0.096	0.959	0.039	0.017	0.425	0.023	8.098
P-value	0.053	0.331	0.694	0.139	0.192	0.426	0.696
Colon							
Control	6.71	18.23^{ab}	0.564^{ab}	0.369^{ab}	4.34	bd	na
IN_{10}	6.14	19.31^{ab}	0.516^{ab}	0.287^{a}	4.01	bd	na
IN_{23}	6.09	16.88^{ab}	0.521^{ab}	0.315^{ab}	3.81	bd	na
JA	6.67	28.19^{b}	0.721^{b}	0.520^{b}	3.44	bd	na
CH	6.60	7.75^{a}	0.183^{a}	0.156^{a}	0.97	bd	na
SEM	0.168	1.950	0.053	0.030	0.512	bd	na
P-value	0.663	0.020	0.019	0.002	0.251	bd	na

IN_{10} = inulin extracted from chicory roots with degree of polymerization ≥10; IN_{23} = inulin extracted from chicory roots with degree of polymerization ≥23; JA = dried Jerusalem artichoke; CH = dried chicory; bd = below detection; na = not analyzed.

^{a,b} mean values within a column with different letters differ significantly P≤0.05.

Acknowledgements

Financial support of The National Center for Research and Development. Project No. NR12 0067 10.

References

Barszcz, M., M. Taciak and J. Skomiał, 2011. A dose-response effects of tannic acid and protein on growth performance, caecal fermentation, colon morphology and β-glucuronidase activity of rats. J. Anim. Feed Sci. 20, 613-625.

Hughes, R., E.A.M. Magee and S. Bingham, 2000. Protein degradation in the large intestine: relevance to colorectal cancer. Curr. Issues Intest. Microbiol. 1(2), 51-58.

Juśkiewicz, J., Z. Zduńczyk, E. Żary-Sikorska, B. Król, J. Milala and A. Jurgoński, 2011. Effect of the dietary polyphenolic fraction of chicory root, peel, seed and leaf extracts on caecal fermentation and blood parameters in rats fed diets containing prebiotic fructans. Brit. J. Nutr. 105, 710-720.

Effect of dietary protein and carbohydrates on phenolic compounds formation in the large intestine of pigs

M. Taciak, A. Tuśnio, M. Barszcz, E. Święch, Ł. Staśkiewicz and J. Skomiał
Department of Monogastric Nutrition, The Kielanowski Institute of Animal Physiology and Nutrition, Polish Academy of Sciences, 05-110 Jabłonna, Poland; m.taciak@ifzz.pan.pl

Introduction

Phenolic and indolic compounds, potentially toxic substances, are formed in the gut in bacterial degradation of the aromatic amino acids e.g. phenol, p-cresol and phenylpropionate originate from tyrosine, phenylacetate from phenylalanine and indole, indole propionate and indole acetate from tryptophan (Hughes *et al.*, 2000). This process may be modified by feeding diet containing different carbohydrates and protein sources. The aim of the present study was to assess the influence of protein of animal or plant origin and three types of carbohydrates, used as an energy source for microflora, on the processes of aromatic amino acid degradation in the large intestine of pigs.

Material and methods

A two-factorial experiment was conducted on 36 castrated male pigs of about 15 kg initial body weight. Animals were divided into 6 groups receiving cereal feeds differing in the type of carbohydrates added to diets (cellulose, potato starch, pectin) and the type of protein (potato protein concentrate or casein). After two weeks of experiment pigs were slaughtered and samples of caecal and proximal, middle and distal colon digesta were collected and analyzed for phenol, p-cresol and indole concentration using the gas chromatography method.

Results and discussion

Neither type of carbohydrate nor type of protein in the diet affected concentration of phenol in all segments of the large intestine of pigs. In the caecum, proximal and distal colon concentration of p-cresol and indole was affected by the protein source in the diet. Potato protein concentrate increased concentration of these products, may be as a result of substantial amount of protein passing to the large intestine from ileum because of the apparent resistance of potato protein to enzymatic digestion (Tuśnio *et al.*, 2011). Type of carbohydrate in the diet affected quantity of indole in the caecum and proximal colon, and p-cresol in proximal and distal colon. Both products concentration in the proximal colon, as well as indole in the ceacum were higher in pigs fed diets supplemented with cellulose compared to pectin. Potato starch in the diet resulted in increased concentration of p-cresol in the distal colon also in comparison with pectin. Microbial growth is stimulated by the presence of fermentable carbohydrate. This leads to an increased demand for amino acids by microorganisms, as may by suggested by reduced protein catabolism in the gut. Therefore, the production of potentially harmful compounds, such as phenolic compounds, would be less intensive in the presence of a carbohydrate source (Williams *et al.*, 2001).

Increased concentration of p-cresol and indole, resulting from microbial metabolism of protein, and enhanced by the presence of cellulose, a carbohydrate of low fermentation capability, may be reduced by the addition of highly fermentable pectin.

Table 1. Phenol, cresol and indol concentrations (μM/g of digesta) in the caecum (S), proximal (C25), middle (C50) and distal (C75) colon of pigs.[1]

		PPC			CAS			SEM	*P*-values		
		CEL	PEC	PS	CEL	PEC	PS		Protein	Carbohydrate	P×C
S	phenol	0.66	1.30	0.73	0.70	0.69	0.70	0.28	0.373	0.445	0.435
	cresol	1.30	1.76	1.22	0.90	0.88	0.73	0.28	0.013	0.456	0.654
	indole	0.08	0.02	0.04	0.04	0.01	0.02	0.01	0.007	0.000	0.426
C25	phenol	0.68	0.61	0.68	0.69	0.66	0.73	0.05	0.370	0.364	0.887
	cresol	2.32	1.23	1.95	0.95	0.88	1.11	0.22	0.000	0.022	0.064
	indole	0.09	0.02	0.05	0.03	0.01	0.03	0.01	0.001	0.001	0.042
C50	phenol	0.62	0.63	0.67	0.64	0.59	0.71	0.06	0.823	0.303	0.703
	cresol	2.08	1.85	2.03	1.10	0.89	1.39	0.26	0.000	0.368	0.732
	indole	0.07	0.05	0.04	0.03	0.01	0.02	0.01	0.000	0.062	0.331
C75	phenol	0.58	0.57	0.60	0.56	0.62	0.64	0.04	0.458	0.463	0.623
	cresol	3.20	2.46	4.43	2.94	2.14	3.47	0.62	0.307	0.039	0.821
	indole	0.09	0.06	0.04	0.05	0.04	0.05	0.01	0.223	0.130	0.233

[1] PPC = potato protein concentrate; CAS = casein; CEL = cellulose; PEC = pectin; PS = potato starch.

Acknowledgements

Financial support Project No. N N311 046634.

References

Hughes, R., E.A.M. Magee and S. Bingham, 2000. Protein degradation in the large intestine: relevance to colorectal cancer. Curr. Issues Intest. Microbiol. 1(2), 51-58.

Tuśnio, A., B. Pastuszewska, E. Święch and M. Taciak, 2011. Response of young pigs to feeding potato protein and potato fibre – nutritional, physiological and biochemical parameters. J. Anim. Feed Sci. 20, 361-378.

Williams B.A., M.W.A. Verstegen and S. Tamminga, 2001. Fermentation in the large intestine of single-stomached animals and its relationship to animal health. Nutr. Res. Rev. 14, 207-227.

Microbial activity in the large intestine of piglets fed diets with different sources of inulin

M. Barszcz, M. Taciak, A. Tuśnio, E. Święch, Ł. Staśkiewicz and J. Skomiał
Department of Monogastric Nutrition, The Kielanowski Institute of Animal Physiology and Nutrition, Polish Academy of Sciences, 05-110 Jabłonna, Poland; m.barszcz@ifzz.pan.pl

Introduction

Inulin-type fructans are natural food ingredients present as storage carbohydrates in a number of plants. Currently only two species, both from *Compositae* family, i.e. chicory and Jerusalem artichoke are used by the industry for extraction of these compounds (Kaur and Gupta, 2002). Inulin-type fructans resist digestion by small intestinal enzymes of monogastric animals but are fermented in the large intestine (Roberfroid, 2005). Modulation of large intestine function plays an important role in maintaining animal welfare, therefore the aim of this study was to determine the effect of diets' supplementation with different sources of inulin-type fructans on microbial activity in the large intestine of piglets.

Material and methods

The experiment was performed on 40 castrated male piglets (PIC) divided into 5 groups (n=8). Piglets were fed from the 2nd week of life cereal diets supplemented with 2% of inulin extracted from chicory roots, with degree of polymerization (DP) ≥ 10 (IN_{10}); 2% of inulin with DP ≥23 (IN_{23}); 4% of dried Jerusalem artichoke (JA) and 4% of dried chicory (CH) corresponding to 2% of inulin mentioned before. Control group was given diet without supplementation. Piglets were sacrificed at the age of 50 days and digesta samples were taken from the caecum and proximal, middle and distal colon for analysis of microbial activity indices. Digesta pH was measured using pH-meter, short-chain fatty acids (SCFA) concentration was analyzed by gas chromatography, β-glucuronidase and β-glucosidase activity was determined spectrophotometrically, and statistical analysis was performed by one-way ANOVA and Tukey test.

Results and discussion

Feeding experimental diets did not affect significantly pH of digesta in the piglet large intestine but it affected SCFA concentration (Table 1). Acetic, propionic and butyric acids were the major metabolites in the large intestine, which agrees with previous studies (Loh *et al.*, 2006; Halas *et al.*, 2010). In the caecal digesta, valeric acid concentration was significantly higher in piglets fed diet with JA than in piglets fed diets with IN_{10} and IN_{23} ($P<0.01$). Similarly, in the proximal colon valeric acid concentration in piglets fed JA diet differed significantly from piglets fed IN_{10} diet ($P<0.05$). Valeric acid is considered as an isoacid resulted from protein degradation in the intestine and its concentration increases in piglet fed diets supplemented with inulin, especially that with greater DP (Passlack *et al.*, 2012). There was also a significant effect on acetic acid concentration, which was smaller in piglets fed IN_{10} diet than in piglets fed IN_{23} and CH diet ($P<0.01$). This may indicate that inulin with greater DP affects microbial activity more pronounced than inulin with smaller DP, which was suggested by Passlack *et al.* (2012). There were no differences in SCFA concentration in the middle and distal colon. Bacterial enzymes activity also did not differ significantly between experimental groups, except caecal β-glucosidase activity, which was higher in piglets fed JA diet compared to animals from other groups ($P<0.05$).

Contrary to the expectations, microbial activity in the large intestine of piglets is only slightly modified by feeding diets with different sources of inulin-type fructans. Neither inulin preparations nor dried plants rich in inulin decreased digesta pH and activity of detrimental bacterial enzymes.

They also did not change fermentation profile towards increased production of butyric acid, which is the most important SCFA for colonic epithelium.

Table 1. Short-chain fatty acids concentrations (µM/g of digesta) in the segments of piglet large intestine.

	Acetic	Propionic	Isobutyric	Butyric	Isovaleric	Valeric
Caecum						
Control	35.8	19.0	0.30	10.5	0.19	1.82^{ab}
IN_{10}	33.4	17.7	0.30	9.3	0.22	1.46^{b}
IN_{23}	37.7	17.6	0.35	10.1	0.28	1.61^{b}
JA	33.6	18.2	0.30	9.9	0.27	2.72^{a}
CH	36.4	19.5	0.30	10.3	0.18	1.86^{ab}
SEM	0.98	0.49	0.016	0.41	0.024	0.124
P-value	0.618	0.674	0.878	0.888	0.589	0.010
Proximal colon						
Control	35.3^{ab}	17.9	0.46	10.6	0.42	2.18^{ab}
IN_{10}	30.8^{b}	15.8	0.60	9.1	0.57	1.60^{b}
IN_{23}	39.8^{a}	17.9	0.66	11.8	0.73	2.59^{ab}
JA	34.8^{ab}	18.3	0.47	10.9	0.49	3.17^{a}
CH	39.3^{a}	19.2	0.42	11.5	0.36	2.22^{ab}
SEM	0.92	0.51	0.032	0.41	0.052	0.151
P-value	0.008	0.285	0.094	0.278	0.174	0.015

IN_{10} = inulin extracted from chicory roots with degree of polymerization ≥10; IN_{23} = inulin extracted from chicory roots with degree of polymerization ≥23; JA = dried Jerusalem artichoke; CH = dried chicory

[a,b] means within a column with different letters differ significantly ($P \leq 0.05$).

Acknowledement

Financial support of The National Centre for Research and Development, Project No. NR12 0067 10

References

Halas, D., C.F. Hansen, D.J. Hampson, B.P. Mullan, J.C. Kim, R.H. Wilson and J.R. Pluske, 2010. Dietary supplementation with benzoic acid improves apparent ileal digestibility of total nitrogen and increases villous height and caecal microbial diversity in weaner pigs. Anim. Feed Sci. Tech. 160, 137-147.

Kaur, N. and A.K. Gupta, 2002. Applications of inulin and oligofructose in health and nutrition. J. Biosci. 27, 703-714.

Loh, G., M. Eberhard, R.M. Brunner, U. Hennig, S. Kuhla, B. Kleessen and C.C. Metges, 2006. Inulin alters the intestinal microbiota and short-chain fatty acid concentrations in growing pigs regardless of their basal diet. J. Nutr. 136, 1198-1202.

Passlack, N., M. Al-Samman, W. Vahjen, K. Männer and J. Zentek, 2012. Chain length of inulin affects its degradation and the microbiota in the gastrointestinal tract of weaned piglets after a short-term dietary application. Livest. Sci. 149, 128-136.

Roberfroid, M.B., 2005. Introducing inulin-type fructans. Brit. J. Nutr. 93, Suppl. 1, S13-S25.

Transgenic flax in high-fat diet inhibits inflammatory state development in mice liver

M. Matusiewicz[1], I. Kosieradzka[1], M. Żuk[2] and J. Szopa[2]
[1]Department of Animal Nutrition and Feed Science, Faculty of Animal Science, Warsaw University of Life Science – SGGW, Ciszewskiego 8, 02-786 Warsaw, Poland; magdalena_matusiewicz@sggw.pl
[2]Faculty of Biotechnology, Wrocław University, Przybyszewskiego 63/77, 51-148 Wrocław, Poland

Introduction

Polyphenols are well known as antioxidants very important for health. Polyphenol-rich plants of flax (*Linum usitatissimum* L.) were obtained using two strategies: simultaneous overexpression of three genes coding for enzymes of flavonoid synthesis (chalcone synthase, chalcone isomerase and dihydroflavonol reductase) – line W92, or of gene coding for the enzyme that regulates stability of these compounds – glucose transferase – line GT (Żuk *et al.*, 2011). In comparison with the non-transgenic line, the seed cake extracts of both transgenic flax lines were characterized by an enriched content of flavonoids (kaempferol, quercetin), anthocyanins, lignans and phenolic acids important for antioxidant potential. The objective of the experiment was to determine the effect of diet rich in flaxseed cake of transgenic lines on serum red-ox state as well as on the development of the inflammatory state in mice liver.

Material and methods

Research was conducted on male mice that were derived from outbred stock by crossing four inbred strains: A/St, BN/a, BALB/c and C57BL/6Jn. Five groups of 20 mice were fed (*ad libitum*) for 96 days one of the isoprotein diets: (1) standard diet; (2) high-fat diet; (3) diet supplemented with 30% of seed cake of the non-transgenic flax line (Linola); (4) 30% of seed cake of the transgenic flax line W92; or (5) 30% of seed cake of the transgenic flax line GT (Table 1). The data were examined by one-way analysis of variance, a P-value <0.05 was considered significant.

Table 1. Composition of experimental diets and their nutritional value.

Component	Group diet				
	Standard	High-fat	Linola	W92	GT
Flaxseed cake (%)	-	-	30	30	30
Nutritional value (% DM)					
Total protein	20.0	20.1	19.7	19.7	19.5
Crude fat	2.8	12.1	11.6	11.2	11.9
GE (kcal/100g)	465	515	512	510	514
GE from crude fat (kcal/100g)	27	115	110	106	113

Results and discussion

Mice fed the diet containing seed cake of GT line demonstrated significantly higher level of serum total antioxidant status (TAS), detected by spectrophotometric method, comparing to animals fed high-fat diet and standard diet. Detected spectrophotometrically, concentration of serum thiobarbituric acid reactive substances – TBARS, the products of oxidative degradation of lipids, were significantly lower in all of the flax groups compared to high-fat diet (Figure 1).

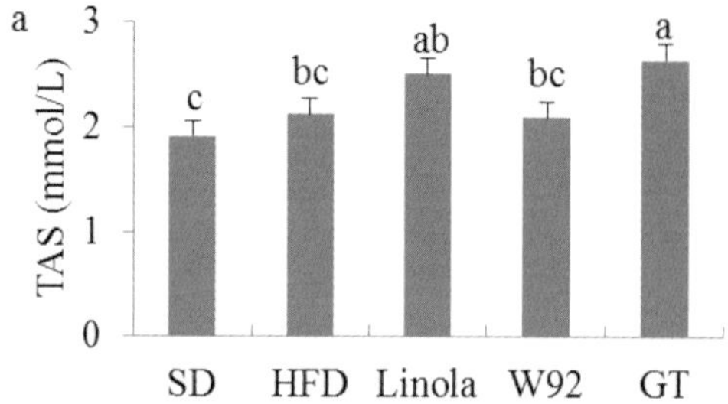

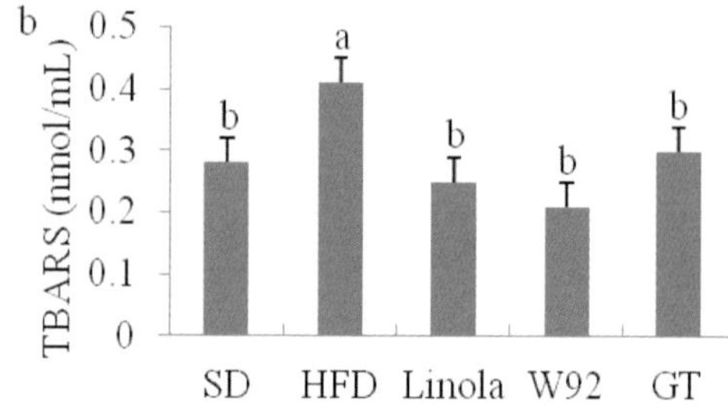

Figure 1. Serum total antioxidant status (TAS) (a) and tiobarbituric acid reactive substances (TBARS) (b) of mice fed different diets. Data are expressed as mean±SEM, $P<0.05$.

Experiments using ELISA tests (Figure 2) showed that the level of proinflammatory cytokine TNF-α was the same in flax groups, the level of this cytokine was decreased in the liver of mice fed line GT and non-transgenic line compared to high-fat diet but these differences were not statistically significant. The concentration of the other proinflammatory cytokine IFN-γ was the same in flax groups, regardless of genetic form. Studies did not show any difference in concentration of IL-6 and adiponectin between the groups. The level of liver C-reactive protein was the same in mice fed seed cake of all of the flax lines and concentration of this protein was significantly smaller in mice fed flaxseed cake of line GT, than mice fed high-fat diet.

The effect of diet supplemented with 30% of seed cake of examined flax transgenic lines on selected parameters of serum red-ox state and inflammatory state in mice liver may result from the changes of biologically-active substances concentration.

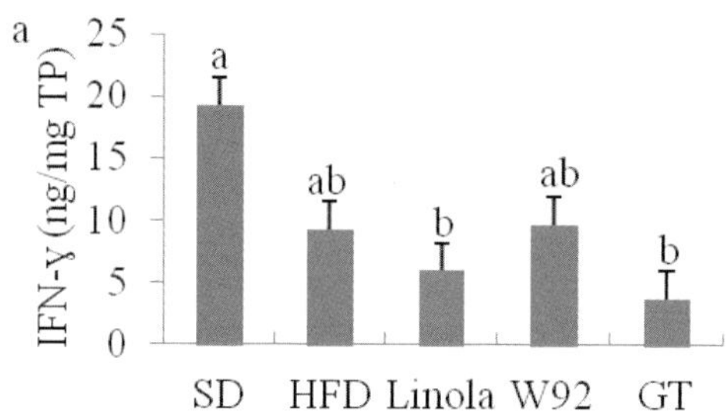

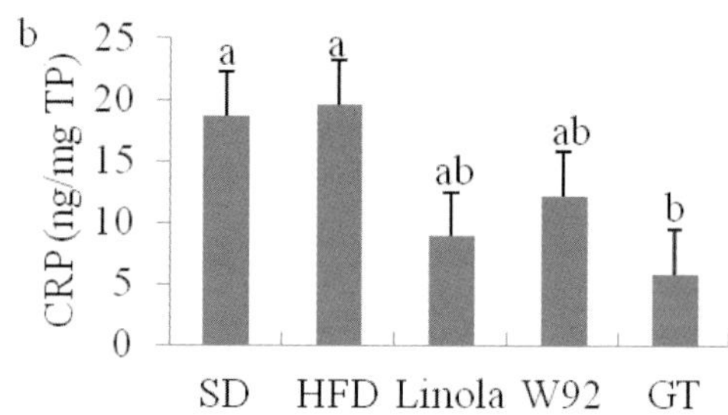

Figure 2. Concentration of inflammatory markers in mice liver, interferon-gamma (IFN-γ) (a), C-reactive protein (CRP) (b). Data are expressed as mean±SEM, $P<0.05$.

References

Żuk, M., A. Kulma, L. Dymińska, K. Szołtysek, A. Prescha, J. Hanuza, J. Szopa, 2011. Flavonoid engineering of flax potentiate its biotechnological application. BMC Biotechnology 2011 11:10.

Part 7. Tissue metabolism

Mechanisms underlying variation in beef cattle feed efficiency: roles of muscle and adipose tissues.

R.A. Hill[1,2], C.M. Welch[1], M. McGee[1], S. Acharya[1], S. Ji[1], C.S. Schneider[1] and J.K. Ahola[2,3]
[1]Department of Animal and Veterinary Science, University of Idaho, ID, USA; rodhill@uidaho.edu
[2]USDA Regional Project W1010 Integrated Approach to Enhance Efficiency of Feed Utilization in Beef Production Systems, USA
[3]Department of Animal Sciences, Colorado State University, CO, USA

Introduction

The potential for the contributions of muscle and adipose tissue to variation in feed efficiency (FE) is large. Both energy storage and energy turnover mechanisms may contribute to variation in efficient use of feed energy inputs (Hill and Ahola, 2012). This paper provides a brief review of some of the mechanisms that contribute to variation in efficiency of energy utilization.

Role of the insulin-like growth factor axis

Due to the relationship between insulin-like growth factor-1 (IGF-I) and linear growth, and the actions of the IGF axis directly in muscle, the IGF axis has been suggested as a source of potential physiological indicators of FE. Gene expression profiling of differentially expressed genes between animals divergent for FE may serve to identify potential candidate genes associated with FE. Red Angus sires divergent for maintenance energy (ME_M) expected progeny difference (EPD) and cross-bred dams were used (AI) to breed progeny across three cohorts (n=222). Progeny plasma IGF-I (at weaning) was negatively correlated (r=-0.22, $P<0.01$) with ME_M EPD (Welch *et al.*, 2011). In a sub-set of animals, (n=37) biopsies of *biceps femoris* muscle were sampled and studied for gene expression using real time PCR. Expression of IGFBP3 was marginally different ($P<0.10$) between the high and low ME_M EPD groups, with gene expression levels in the high ME_M EPD group showing marginally greater IGFBP3 expression (1.5 fold) when compared to the low ME_M EPD group. A difference in IGFBP2 ($P<0.05$) expression was detected between ME_M EPD groups, with gene expression levels in the high ME_M EPD group being greater (approximately 1.8 fold) when compared to the low ME_M EPD group. Skeletal muscle expresses IGFBPs 2, 3, 4, 5, and 6 (Florini *et al.*, 1996). The major functions of these binding proteins are to assist in the transport of IGFs both locally and systemically, to prolong the half-life of IGFs and regulate plasma clearance, and to modulate the interactions of IGFs with their receptors (reviewed in Kokta *et al.* (2004)). These data suggest that modulation of IGF-I at muscle tissue level may affect IGF-1 signaling, as in inefficient animals, it appears that IGF-binding protein levels are higher and this may result in lower availability of local IGF-1. Thus, the IGF axis is implicated in variation in energy utilization in muscle.

Role of redox pathways

In addition, we have studied the potential for variation in oxidative pathways to affect myogenic differentiation and thus the energetic efficiency of muscle growth and development. Reactive oxygen species (ROS) are important regulators of several intracellular signaling pathways. Real time PCR analysis demonstrated that NADPH oxidase-4 (Nox4) isoform is primarily expressed in differentiating myogenic cells and contributes to the generation of ROS during differentiation. Both silencing and over-expression of Nox4 during myoblast differentiation was accompanied by reduced intracellular ROS concentrations and inhibition of differentiation ($P<0.05$). Thus, optimum cellular ROS concentrations appear to be necessary for optimum differentiation and suggests that physiological regulation of ROS concentrations provides a potential mechanism that underpins variation in myogenic energy utilization (Acharya *et al.*, 2013).

Leptin pathway

A polymorphism in the leptin gene (Nkrumah *et al.*, 2004) appears to be related to gain in backfat thickness (P=0.02), ultrasound backfat thickness (P=0.07), and lean meat yield (P=0.007). Further studies by this research group have indicated that polymorphisms in the leptin promoter may have potential to predict intake and possibly FE (Nkrumah *et al.*, 2005). Animals having one of the homozygous genotypes showed significantly higher feed intake (P<0.001), growth rate, metabolic BW (P<0.05), and live weight at slaughter (P<0.10). While at another locus, animals with an homozygous genotype also showed higher feed intake (P=0.001), growth rate (P<0.10), and BW (P<0.01). There were significant differences among the different genotype combinations in ADG (P=0.04) and feed conversion ratio (P=0.019) (Nkrumah *et al.*, 2006). In addition, we have shown that bovine muscle responds to leptin and insulin in a manner that suggests its ability for differential substrate utilization consistent with physiology in which there is low circulating glucose (Strat *et al.*, 2005). Thus, there appear to be mechanisms that link muscle and adipose tissue that have potential to modulate partitioning of energy substrate utilization, and that genetic variation within the leptin axis may affect individual animal physiology, resulting in differences in FE.

Conclusion

These pathways provide targets for further study of efficiency of energy utilization in beef cattle. Evidence presented here suggests that there are interacting genetic and physiological bases that may account for variation in FE, and that these differences manifest in both muscle and adipose and their functional interactions.

References

Acharya, S., A.M. Peters, A.S. Norton, G.K. Murdoch, and R.A. Hill. 2013. Change in Nox4 expression is accompanied by changes in myogenic marker expression in differentiating C2C12 myoblasts. Pflügers Archiv-Europ. J. Physiol. (in press). doi: 10.1007/s00424-013-1241-0.

Florini, J.R., D.Z. Ewton, and S.A. Coolican. 1996. Growth hormone and the insulin-like growth factor system in myogenesis. Endocrin. Rev. 17: 481-517.

Hill, R.A., and J.K. Ahola. 2012. Feed efficiency interactions with other traits: Growth and product quality. In: R. A. Hill (ed.) Feed efficiency in the beef industry. p 145-158. Wiley-Blackwell.

Kokta, T.A., M.V. Dodson, A. Gertler, and R.A. Hill. 2004. Intercellular signaling between adipose tissue and muscle tissue. Domest. Anim. Endocrinol. 27: 303-331.

Nkrumah, J.D., J.A. Basarab, M.A. Price, E.K. Okine, A. Ammoura, S. Guercio, C. Hansen, C. Li, B. Benkel, B. Murdoch, and S.S. Moore. 2004. Different measures of energetic efficiency and their phenotypic relationships with growth, feed intake, and ultrasound and carcass merit in hybrid cattle. J. Anim. Sci. 82: 2451-2459.

Nkrumah, J.D., C. Li, D.H. Keisler, E.L. Sherman, Z. Wang, B.M. Murdoch, and S.S. Moore. 2006. Polymorphisms in the leptin gene and their associations with performance, feed efficiency, and carcass merit of beef cattle. 8th World Congress on Genetics Applied to Livestock Production.: August 13-18, 2006, Belo Horizonte, MG, Brazil.

Nkrumah, J.D., C. Li, J. Yu, C. Hansen, D.H. Keisler, and S.S. Moore. 2005. Polymorphisms in the bovine leptin promoter associated with serum leptin concentration, growth, feed intake, feeding behavior, and measures of carcass merit. J Anim Sci 83: 20-28.

Strat, A.L., T.A. Kokta, M.V. Dodson, A. Gertler, Z. Wu, and R.A. Hill. 2005. Early signaling interactions between the insulin and leptin pathways in bovine myogenic cells. Biochim. Biophys. Acta. 1744: 164-175.

Welch, C. M., G.K. Murdoch, C.S. Schneider, K. Chapalamadugu, K.J. Thornton, J.K. Ahola, J.B. Hall, and R.A. Hill. 2011. Gene expression of Red Angus sired steers and heifers evaluated for residual feed intake. J. Anim. Sci. 89 (E Suppl. 2): 718.

Adipose tissue preferences for acetate and glucose by finishing steers

W.A.D. Nayananjalie[1], T.R. Wiles[1], D.E. Gerrard[2], M.A. McCann[2] and M.D. Hanigan[1]
[1]Dept. of Dairy Science, Virginia Polytechnic Institute and State University, Blacksburg, VA 24061, USA; mhanigan@vt.edu
[2]Dept. of Animal and Poultry Sciences, Virginia Polytechnic Institute and State University, Blacksburg, VA 24061, USA

Introduction

Increased marbling enhances the quality grade of beef, and as such, high starch diets are often used by feedlots to stimulate this type of fat deposition. *In vitro* studies suggest that adipocytes from intramuscular fat (IMF) and subcutaneous fat (SCF) may have different preferences for acetate and glucose (Smith and Crouse, 1984). If such differential precursor preference exists, it should be connected to differential fat synthesis rates from precursors by different depots. The objectives of this study were to assess the acetate and glucose turnover rates, palmitate synthesis rate, and acetate and glucose preference by subcutaneous (SCF), intramuscular (IMF) and visceral adipocytes (VF) in finishing steers.

Material and methods

Experiment was conducted as a complete randomized design. Sixteen, Angus x Simmental steers were fed a corn silage based finishing ration for 120 d. Eight animals (495±14 d of age and 555±18 kg BW) were infused continuously with [$^2H_3$] acetate (1.63 mmol/min), and the remainder (501±12 d of age and 549±20 kg BW) were infused with [U-$^{13}C_6$] glucose (0.07 mmol/min) for 12 h immediately prior to animal harvest. Blood samples were collected before and at the end of infusions and adipose tissue samples from SCF, IMF and VF depots were collected at slaughter. Lipids were extracted from tissue samples and palmitate enrichment in lipids, and acetate and glucose enrichment in blood were determined by GC-MS. Concentrations of DNA in tissue samples were also determined. Plasma enrichments of acetate and glucose, and palmitate enrichment in each depot were used to calculate fractional palmitate synthesis rates (FSRs, %/h). Data were analyzed using the GLIMMIX procedure of SAS and means were separated using Turkey's test. Further, acetate to glucose preference across depots was compared using contrast statements. Data were presented with least square means ± SEM.

Results and discussion

Concentrations of DNA were significantly different by depot where IMF had the greatest DNA concentrations (143±4.9 ng/mg of tissue) and VF the least (79±4.9 ng/mg of tissue). Acetate turnover was significantly greater than that of glucose turnover (0.07±0.01 mmol/min/metabolic body weight (MBW) vs. 0.04±0.01 mmol/min/MBW). Palmitate FSR from acetate were greater than FSR from glucose (0.27±0.04%/h vs. 0.02±0.04; $P<0.05$). However, FSR was not significantly different among the depots for either acetate or glucose ($P>0.05$), and the acetate to glucose preference ratio was not significantly different across depots ($P>0.05$). In conclusion, 13 times more acetate was used for lipid synthesis than glucose, and there were no differences in preference for acetate and glucose among the major fat depots. Thus, diets leading to high glucose supply will not preferentially direct energy storage to intramuscular stores.

References

Smith, S.B. and J.D. Crouse, 1984. Relative contributions of acetate, lactate and glucose to lipogenesis in bovine intramuscular and subcutaneous adipose tissue. J. Nutr. 114: 792-800.

Growth rate in beef cattle affects adipose gene expression and skeletal muscle fiber type

P.A. Lancaster, M.A. Vaughn, J.D. Starkey, E.D. Sharman, C.R. Krehbiel and G.W. Horn
Oklahoma State University, Stillwater, OK 74078, USA; palancaster@ufl.edu
Texas Tech University, Lubbock, TX 79409, USA

Introduction

Growth rate in beef cattle affects fat deposition and skeletal muscle fiber characteristics (Yambayamba and Price, 1991), but individual fat depots appear to be regulated by different mechanisms (Ortiz-Colon *et al.*, 2009). Calkins *et al.* (1981) found that skeletal muscle fiber type composition explained 65% of the variation in marbling score with more oxidative fiber types having a positive correlation with marbling score. Thus, management strategies could be used to influence skeletal muscle characteristics which could enhance intramuscular fat deposition in the finishing phase. The purpose of this study was to evaluate growth rate during the stocker phase on adipose tissue development and skeletal muscle characteristics of growing-finishing beef cattle.

Material and methods

Two experiments were conducted using Angus steers (Exp. 1: n=68; Exp. 2: n=72) assigned to 1 of 4 winter grazing treatments in a completely randomized design: (1) control, grazing dormant native range (DNR) and fed 1.0 kg/d of a 40% CP cottonseed meal-based supplement (CON); (2) grazing DNR and fed a ground corn/soybean meal-based 20% CP supplement at 1% of BW (~2.65 kg/d) (CORN); (3) grazing winter wheat pasture (WP) at a high stocking rate (3.21 steers/ha) to achieve a low rate of BW gain (LGWP); and (4) grazing WP at a low stocking rate (0.99 steers/ha) to achieve a high rate of BW gain (HGWP). At the end of the stocker phase, a subset of steers was slaughtered at similar age (Exp. 1; 3 steers per treatment; 138 d) or hot carcass weight (Exp. 2; 4 steers per treatment; 262, 180, 142, and 74 d for treatments 1-4), and the remaining steers were fed a common finishing diet to an ultrasound predicted rib fat thickness of 1.27 cm. At each harvest, samples of longissimus muscle were collected for fiber typing by immuno-fluorescence microscopy (Exp. 2 only), and samples of intramuscular (IM) and subcutaneous (SC) adipose tissue were collected for mRNA expression of genes involved in adipogenesis (PPARγ, SREBF, C/EBPβ, and Pref1) and lipogenesis (GPDH, FASN, and DGAT2). Growth, carcass, and fiber typing data were analyzed using a general linear model that included treatment. Gene expression data were analyzed using MANOVA with a general linear model that included treatment, adipose tissue and the interaction term when significant. LSmeans were separated using Fisher's protected LSD (α=0.10).

Results and discussion

At the end of the stocker phase in Exp. 1, marbling score increased linearly with ADG (180, 217, 280, 340 ± 18 for CON, CORN, LGWP, and HGWP, respectively), whereas rib fat thickness increased at an increasing rate (0.03, 0.10, 0.17, and 0.85±0.06 cm). There was no treatment x adipose tissue interaction for adipogenic or lipogenic gene expression indicating that adipose tissues responded similarly to the treatments (Figure 1). Adipogenic gene expression was lesser (P<0.05) for CON steers than the other steers and lipogenic gene expression was greater (P<0.05) for LGWP and HGWP steers than CON and CORN steers. Subcutaneous adipose tissue had greater (P<0.05) expression of adipogenic and lipogenic genes than IM. In Exp. 2, LGWP and HGWP steers had greater (P<0.05) marbling score than CON and CORN steers (158, 143, 315, 228 ± 22), but there was no relationship with ADG. In contrast to Exp. 1, there was a treatment x adipose tissue interaction (P<0.05) for both lipogenic and adipogenic gene expression. For adipogenic and lipogenic genes, treatment influenced expression in SC but not in IM. In Exp. 2, the CON steers tended (P<0.15) to have greater percentage of Type 1 (oxidative) muscle fibers at intermediate harvest (33.0, 27.5, 26.0, and 28.0±2.1) compared with the other steers, and the greatest capillary density (66.6, 49.4, 48.2 and 58.6±8.8), but this was not significantly different from other steers. At final harvest, there was no difference (*P*>0.10) in marbling score among steers in Exp. 1 or 2, and there was no effect of treatment on adipogenic or lipogenic gene expression in Exp. 2 (data not shown). In Exp. 2, there was no difference in muscle fiber types among steers at final harvest, but CON steers tended (*P*<0.15) to have greater capillary density than the other steers (67.3, 46.0, 41.3, 50.1±8.3). These data indicate that rate of gain to similar BW affects metabolic pathways in SC and IM differently than rate of gain to similar age, and that low rates of gain may result in skeletal muscle characteristics that are more favorable for IM development.

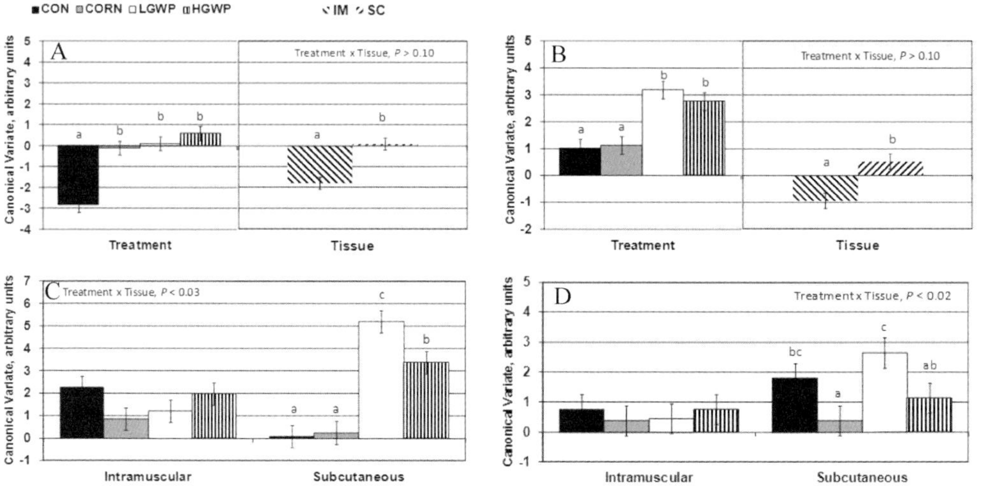

Figure 1. Canonical variate from multivariate analysis of variance of adipogenic (A and C) and lipogenic (B and D) gene expression of intramuscular (IM) and subcutaneous (SC) adipose tissue of steers from intermediate harvest in Exp. 1 (A and B) and 2 (C and D).
[abc] Means within a treatment or tissue without a common superscript differ (P<0.05).

References

Calkins, C.R., T.R. Dutson, G.C. Smith, Z.L. Carpenter and G.W. Davis, 1981. Relationship of fiber type composition to marbling and tenderness of bovine muscle. J. Food Sci. 46:708-710.

Ortiz-Colon, G., A.C. Grant, M.E. Doumi and D.D. Buskirk, 2009. Bovine intramuscular, subcutaneous, and perirenal stromal-vascular cells express similar glucocorticoid receptor isoforms, but exhibit different adipogenic capacity. J. Anim. Sci. 87:1913-1920.

Yambayamba, E. and M.A. Price, 1991. Fiber-type proportions and diameters in longissimus muscle of beef heifers undergoing catch-up (compensatory) growth. Can. J. Anim. Sci. 71:1031-1035.

First evidence of an insulin-sensitive glucose transporter in chicken: GLUT-12

E. Coudert[1], J. Dupont[2], J. Simon[1], E. Cailleau-Audouin[1], S. Crochet[1], M.J. Duclos[1], S. Tesseraud[1] and S. Métayer-Coustard[1]
[1]UR83 Recherches Avicoles, 37380 Nouzilly, France; edouard.coudert@tours.inra.fr
[2]UMR85 PRC, 37380 Nouzilly, France

Introduction

Facilitated transport of glucose into cells is mediated by a family of facilitative-diffusion glucose transporter (GLUT) proteins. In mammals, mostly in adipose and muscle tissues, some GLUTs, called 'insulin-sensitive GLUTs', are recruited at the plasma membrane in response to insulin. Facilitative-diffusion glucose transporter-4 is the best characterized (Bryant *et al.*, 2002). So far, no functional 'insulin-sensitive GLUTs' has been characterized in chicken tissues. This species exhibits some peculiarities for glucose metabolism: a high glycaemia despite the presence of insulin circulating at 'normal' concentrations, and a low sensitivity to exogenous insulin, reminiscent of mammalian type-2 diabetes.

The chicken genome database contains several sequences that are suggested as encoding glucose transporter-like proteins, but none encoding a GLUT-4 homolog. A sequence is predicted as encoding a chicken GLUT-12 homolog (ENSGALG00000013980); it is located on chromosome 3, comprises 4 exons and encodes a 561 amino acid protein exhibiting 73% identity with human GLUT-12. Interestingly, in mammals, GLUT-12 appears to act as an 'insulin-sensitive GLUT' in a way qualitatively similar to GLUT-4 (Stuart *et al.*, 2009).

Material and methods

Tissue distribution of chicken GLUT-12

Ross Male broiler chickens (n=5) were slaughtered at five weeks of age and different tissues were removed and snap frozen into liquid nitrogen.

GLUT-12 mRNA was characterized by reverse transcriptase-polymerase chain reaction (RT-PCR) in various chicken tissues using specific primers for chicken GLUT-12 sequence.

Protein distribution was analyzed by western blot. Tissues lysates were prepared as previously described and subjected to SDS-PAGE gel electrophoresis and western blotting using a commercial GLUT-12 antibody directed against a highly conserved region of the corresponding human protein (87% of identity between human and predicted chicken sequences). After washing, the membranes were incubated with an Alexa Fluor labeled secondary antibody, and the signals were visualized using the Odyssey® infrared Imaging System.

GLUT-12 regulation

In order to evaluate the insulin sensitivity of GLUT-12, we used a previously described model of insulin immuno-neutralization in chickens aged 16 or 17 days (Dupont *et al.*, 2008). Fed chickens were injected with 3 i.v. injections of anti-insulin serum (5 hr-insulin-immunoneutralization) or normal serum (Fed state) at 2 h intervals and compared to chickens fasted for 5 hours and injected with 3 *i.v.* injections of normal serum (Fasted) (n=7 per experimental group). Muscle samples (leg and *Pectoralis major*) were removed and snap frozen. The expression of GLUT-12 was analyzed by qRT-PCR and normalized with β-actin or cytochrome b (cyt b). The data were subjected to ANOVA to detect significant differences, and the means further compared by a Tukey-Kramer test.

In preliminary studies, GLUT-12 translocation was assessed. Membrane proteins were extracted using the Promokine kit (Promocell) according to the manufacturer's recommendations, from muscles of animal fasted, fed and/ or injected with insulin. The enrichment of GLUT-12 in the plasma membrane was measured by western blot.

Results and discussion

GLUT-12 mRNA was expressed in chicken skeletal and heart muscles, i.e. 'insulin sensitive tissues' (Figure 1A, lanes 1-4), where an immunoreactive band, showing the expected size, was also detected (Figure 1B, lanes a-d). No signal was detected in an insulin insensitive tissue, the liver. The immunoreactive GLUT-12 was also detected in purified leg muscle membranes, which increased following insulin administration (data not shown), suggesting translocation of the GLUT-12 protein to the plasma membrane.

GLUT-12 mRNA expression was further characterized *in vivo* in chicken muscle using the insulin immuno-neutralization model. GLUT-12 mRNA expression was significantly lower in fasted and insulin-immunoneutralized conditions compared to the fed condition (Figure 2).

In conclusion, the facilitative-diffusion glucose transporter protein GLUT-12 is present in chicken muscle and could act as an 'insulin-sensitive' transporter in a way qualitatively similar to GLUT-4 in mammals. An extensive characterization of the role of GLUT-12 in chicken is now required.

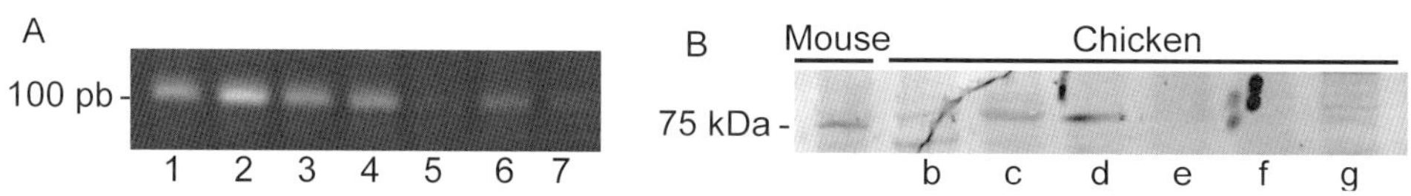

Figure 1. Tissue distribution of GLUT-12. (A) mRNA expression of GLUT-12 in chicken tissues (1, Pectoralis major muscle; 2, Sartorius muscle; 3, Posterior Latissimus Dorsi muscle; 4, heart; 5, duodenum; 6, brain and 7, liver). (B) Protein distribution of GLUT-12 in mouse muscle (a) and in chicken (b, Pectoralis major muscle; c, leg muscle; d, heart; e, brain; f, testis; and g, liver).

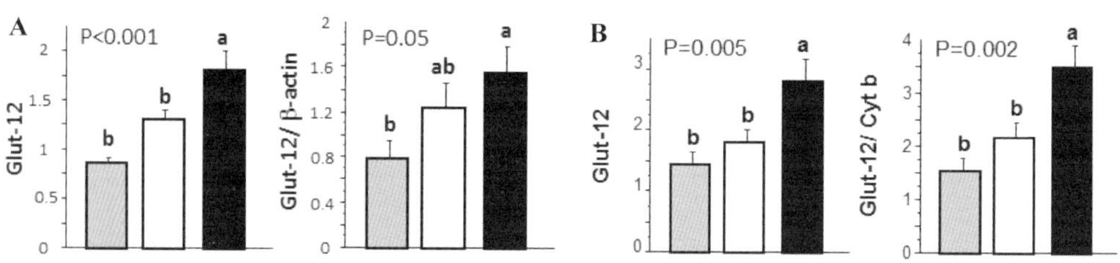

Figure 2. GLUT-12 mRNA expression in leg muscle (A) and in Pectoralis major muscle (B) of chickens in the fed state (in black), fasted (in white) or after 5 hr-insulin-immunoneutralization (in grey). Values are presented as means ± SEM. The means were compared by a Tukey-Kramer test.

References

Bryant, N.J., R. Govers and D.E. James, 2002. Regulated transport of the glucose transporter GLUT4. Nat. Rev. Mol. Cell. Biol. 3, 267-277.

Dupont, J., S. Tesseraud, M. Derouet, A. Collin, N. Rideau, S. Crochet, E. Godet, E. Cailleau-Audouin, S. Métayer-Coustard, M.J. Duclos, C. Gespach, T.E. Porter, L.A. Cogburn and J. Simon, 2008. Insulin immuno-neutralization in chicken: effects on insulin signaling and gene expression in liver and muscle. J Endocrinol. 197(3):531-42.

Stuart C.A., M.A. Howell, Y. Zhang and D. Yin, 2009. Insulin-stimulated translocation of glucose transporter (GLUT) 12 parallels that of GLUT4 in normal muscle. J. Clin. Endocrinol. Metab. 94, 3535-3542.

Influence of mitochondrial function on feed efficiency of broilers with and without growth enhancing levels of minerals supplementation during a coccidiosis challenge

G. Acetoze[1], *R. Kurzbard*[2], *J.J. Ramsey*[1], *K.C. Klasing*[2] *and H.A. Rossow*[1]
[1]*School of Veterinary Medicine, University of California, Davis, CA, USA; gabi_acetoze@yahoo.com.br*
[2]*Department of Animal Science, University of California, Davis, CA, USA*

Introduction

Mitochondrial conversion of energy as NADH and FADH to ATP is an important contributor to energy supply accounting for approximately 20-30% of resting energy requirements (Owen *et al.*, 1978; Zurlo *et al.*, 1990). Therefore changes in mitochondrial efficiency will have large impacts on energetic and feed efficiency (Bottje and Carstens, 2008). Broiler chickens have more efficient muscle mitochondria than laying chickens and this is correlated with their higher feed efficiency and increased growth rates (Bottje *et al.*, 2002). But, it is unknown how diet (mineral levels) affects mitochondrial efficiency in chickens. In 2005, Fariss *et al.* showed protection from oxidative stress by administration of antioxidants such as vitamin E and ubiquinone. Copper and Zinc are known as minerals that have antibiotic and antioxidant properties by reducing the effects of secondary bacterial infection and macromolecular damage by free radicals through the action of superoxide dismutase (Nonn *et al.*, 2003). Understanding the role of mitochondrial efficiency in feed efficiency will aid in feeding decisions and selection of broilers that are more energetically efficient to optimize nutrient use. More energetically efficient broilers will reduce feed intake and costs of production leading to improved feed efficiency and sustainability. This research will assess the influence of feeding increasing levels of antioxidant supplements (copper or zinc) on mitochondrial efficiency and feed efficiency in 21 day old broiler chicks during a coccidiosis challenge.

Material and methods

Sixteen 7 days-old broilers (4 birds/treatment; 2 birds/pen) were randomly assigned to 4 treatments: an infected control diet (Cu 15 mg/kg and Zn 60 mg/kg) + *Eimeria maxima;* 245 mg/kg Cu from Tribasic Copper Chloride + *Eimeria maxima* (TBCC)*;* negative control (Cu 15 mg/kg and Zn 60 mg/kg) – *Eimeria maxima;* and 2,000 mg/kg Zn from ZnO + *Eimeria maxima.* The diet was composed of 49% corn, 40% soybean meal, 6.2% vegetable oil (DM=90.62%, CP=21.37%, Fat=7.7%, ME=2.89 Mcal/day) and was fed for 14 days. Initial and final body weights and dry matter intake were calculated until the culling day. At 21 day of ages, the birds were sacrificed by $C0_2$ and immediately dissected. Approximately 1 gram of liver tissue was used for isolated mitochondria analysis of respiratory control ratio (RCR), proton leak kinetics and membrane damage. Mitochondria were isolated from liver using slight modifications of the procedures of Chappell and Hansford (1972) and Rickwood *et al.* (1987) as previously described (Ramsey *et al.*, 2004). Mitochondrial oxygen consumption was measured using methods described by Harper *et al.* (1998) and Lal *et al.* (2001). All of the mitochondria isolation procedures are well established and have been sited repeatedly in work measuring mitochondrial respiration and/or membrane potential. Integrity of the mitochondrial preparations was determined by measuring RCR, with accepted values greater than 2.0 (State3/State4; Chappell and Hansford, 1972). All measurements were completed in duplicate using mitochondrial protein (1.0 mg/ml) in incubation medium with 5μM rotenone, 0.4 μg nigericin and 8.0 μg of oligomycin. Data were analyzed using R software using ANOVA and a Tukey-Kramer test for treatment mean differences. The statistical model used were $Yij = \mu + Xij\beta + \tau i + \varepsilon ij$, in which Yi is the dependent variable, μ is the overall mean, Xij are the covariates for treatment i and animal j, τi is the fixed effect of treatment, β is the vector of fixed effects and εij are the random errors.

Results and discussion

Average gain/feed values per pen were 1.15 and 1.20 (SEM=0.42) kg feed/kg gain for the treatment containing 245 mg/kg of Cu from TBCC and treatment containing 2,000 mg/kg ZnO, respectively. No statistical significance was found for the effect of treatment on gain/feed efficiency (P=0.83). Oxygen consumption (nmolO_2/mg protein/min) in State 3 and State 4 were 28.2 (SD=4.2) and 9.4 (SD=1.4) for the negative control group, 28.6 (SD=1.6) and 11.0 (SD=0.5) for the positive control group, 23.2 (SD=1.4) and 9.4 (SD=0.7) for 245 mg/kg of Cu from TBCC and 28.9 (SD=3.8) and 10.0 (SD=1.0) for 2,000 mg/kg ZnO. The RCR values were 4.27, 2.40, 2.54 and 2.65 (SEM=0.13) for the negative control group, the positive control group, 245 mg/kg of Cu from TBCC and 2,000 mg/kg ZnO, respectively and were different (P=0.02). Levels of minerals (245 mg/kg Cu as TBCC and 2,000 mg/kg Zn as ZnO) could possibly be decreasing the activity of *Eimeria* maxima by acting as anti-coccidials, protecting the cells from inflammation which is an important process on the production of reactive oxygen species production (ROS) which is related to macromolecular damage, mitochondrial uncoupling, loss of membrane integrity and decrease in oxidative phosphorylation.

References

Bottje, W., Z.X. Tang, M.D. Iqbal, D. Cawthon, R. Okimoto, T. Wang and M. Cooper, 2002. Association of mitochondrial function with feed efficiency within a single genetic line of male broilers. Poult. Sci. 81:546-555.

Bottje, W.G. and G.E. Carstens, 2008. Association of mitochondrial function and feed efficiency in poultry and livestock species. J. Anim. Sci. 87,E48-E63.

Chappell, J.B. and R.G. Hansford, 1972. Preparation of mitochondria from animal tissues and yeasts. In: Subcellular Components: Preparation and Fractionation (ed. Birnie GD). University Park Press, Baltimore. 77-90.

Fariss, M.W., C.B. Chan, M. Patel, B. van Houten and S. Orrenius, 2005. Role of mitochondria in toxic oxidative stress. Mol. Interven. 5, 98-114.

Harper, M.E., S. Monemdgou, J.J. Ramsey and R. Weindruch, 1998. Age-related increase in mitochondrial proton leak and decrease in ATP turnover reactions in mouse hepatocytes. Am. J. Physiol. 275,E197-E206.

Lal, S.B., J.J. Ramsey, S. Monemdjou, R. Weindruch and M.E. Harper, 2001. Evidence that caloric restriction lowers mitochondrial proton leak in skeletal muscle from old rats. J. Gerontol. 56A,B116-B122.

Nonn, L., R.R. Willians, R.P. Erickson and G. Powis, 2003. The absence of mitochondrial thioredoxin 2 causes massive apoptosis, exencephaly, and early embryonic lethality in homozygous mice. Mol. Cell Biol. 23,916-922.

Owen, O.E., G.A. Reichard, G. Boden, M.S. Parel and V.E. Trapp, 1978. Inter-relationships among key tissues in the utrilization of metabolic substrate. Adv. Mod. Nutr. 2,517-550.

Ramsey, J.J., K. Hagopian, T.M. Kenny, E.K. Koomson, L. Bevilacqua, R. Weindruch and M.E. Harper, 2004. Proton leak and hydrogen peroxide production in liver mitochondria from energy restricted rats. Am. J. Physiol. 286,E31-E40.

Rickwood, D., M.T. Wilson and V.M. Darley-Usmar, 1987. Isolation and characteristics of intact mitochondria. In: Mitochondria: A Practical Approach (eds. Darley-Usmar VM, Rickwood D, Wilson MT). IRL Press, Washington, DC. 1-16.

Zurlo, F., K. Larson, C. Bogardus and E. Ravussin, 1990. Skeletal muscle metabolism is a major determinant of resting energy expenditure. J. Clin. Invest. 86,1423-1427.

Expression of amino acid transporter in porcine skeletal muscles during postnatal development

A. Ishida, A. Ashihara, K. Nakashima and M. Katsumata
NILGS, Tsukuba, 305-0901, Japan; aiko15@affrc.go.jp

Introduction

Amino acids are transported across the plasma membrane by the transporters that often overlap in their substrate specificities. Because quality of meat, taste in particular, is dependent on the concentrations and the composition of amino acids in skeletal muscle (Nishimura and Kato, 1988), we aim to change the concentrations or the proportion of amino acids in skeletal muscle by regulating amino acid transporter (AAT) expression in pigs. However, our knowledge on regulation of expression of the AATs in porcine skeletal muscle is limited (Ai-Min *et al.*, 2010, García-Villalobos *et al.*, 2012, Shi-Geng *et al.*, 2009). Skeletal muscle is comprised of myofibre sub-populations which have different metabolic properties. Moreover, properties of porcine myofibres continue to develop postnatally (Davies, 1972). This led us to hypothesize that the expression of the AATs is muscle specifically regulated and its levels change during postnatal development. Therefore, we determined mRNA expression of amino acid transporters in three distinct skeletal muscles at 5 time points during the first 10 weeks postnatally.

Material and methods

A total of 30 male pigs were selected from 6 litters (5 pigs/ litter, birth weight; 1.52±0.17 kg). The pigs were housed under a common practical condition and were fed commercial diets i.e. starting creep feeding on the day 14, weaning on the day 28, and starting grower meal on the day 63. *Longissimus dorsi, rhomboideus*, and *biceps femoris* muscles were taken from six pigs from each litter at 5 slaughter time points: within 24 h after birth (1 day old) and on 12, 26, 45 and 75 days old. The longissimus dorsi muscle has large proportion of a fast-twitch glycolytic muscle. The *rhomboideus* and the *biceps femoris* muscles are mixed slow- and fast-twitch oxido-glycolytic muscles, and the *rhomboideus* muscle has more slow-oxidative muscle than the *biceps femoris* muscle. Therefore, these myofiber type have different characteristics in metabolism due to difference in myofiber type. The levels of mRNA expression of the AAT genes (cationic amino acid transporter (Cat)-1; Cat-2, Cat-3, system y + L-amino acid transporter-1 (y+LAT-1, neutral and cationic amino acid transporter), sodium- and chloride-dependent neutral and basic amino acid transporter B(0+) ($ATB^{0,+}$), B0, + type amino acid transporter ($B^{0,+}AT$)-1, excitatory amino acid carrier 1 (EAAC1, cationic amino acids and cysteine transporter), system ASC neutral amino acid transporter (ASCT) 1, system N2 (SN2, glutamate transporter)) were determined by a real-time RT-PCR method. In order to examine the effects of growth stage, data normalization was carried out using RPL4, HPRT1 and HMBS as the reference genes. The Tukey-Kramer multiple comparisons test was used to determine differences (P<0.05) between the means of the different ages in days.

Results and discussion

Although the mRNA expression of Cat-1 and SN2 on the day 1 was higher than that of the other time points in all the muscle examined (P<0.05), the mRNA expression of Cat-2 and ASCT1 was the lowest on the day 1 and increased with the pigs grew (P<0.05). Although the expression of EAAC 1 was detected in the skeletal muscles, the growth stage did not affect it. The expression of Cat-3, $B^{0,+}AT$ and $ATB^{0,+}$ was not detected or below the limit of accurate quantification in all the muscles examined at any time points. Contrary to our hypothesis, the patterns of the change of mRNA expression of the AATs as the development of the pigs occurred did not differ among the muscles. Consistent with our hypothesis, the expression of mRNAs of the AATs changed as the pigs grew.

Table 1. The expression pattern of porcine AAT genes in different developmental stages of porcine skeletal muscles[1].

Day		1	12	26	45	75	Pooled SEM
BW(kg)		1.5	4.6	9.2	18.8	39.1	
Gene	Muscle						
Cat-1	LD	1.15[a]	0.32[b]	0.30[b]	0.21[b]	0.27[b]	0.06
	RH	1.74[a]	0.85[b]	0.73[b]	0.62[b]	0.46[b]	0.04
	BF	2.36[a]	0.92[b]	0.72[b]	0.59[b]	0.56[b]	0.13
Cat-2	LD	0.75[c]	3.22[b]	3.78[ab]	3.99[ab]	4.63[a]	0.13
	RH	0.35[b]	0.83[a]	0.86[a]	1.05[a]	1.06[a]	0.03
	BF	0.21[d]	0.89[c]	1.02[bc]	1.34[ab]	1.61[a]	0.04
SN2	LD	1.05[a]	0.47[b]	0.37[b]	0.38[b]	0.30[b]	0.04
	RH	1.46[a]	0.86[b]	0.68[b]	0.71[b]	0.54[b]	0.03
	BF	2.23[a]	0.93[b]	0.77[b]	0.76[b]	0.64[b]	0.08
ASCT1	LD	1.55	2.87	2.67	2.40	3.17	0.17
	RH	0.45[b]	0.88[a]	0.85[a]	0.80[ab]	0.86[a]	0.05
	BF	0.53[b]	1.21[a]	1.01[ab]	0.99[ab]	1.03[a]	0.05

[1] Day 1, 12, 26 and 45; n=6, day 75; n=5, LD = *Longissimus dorsi*; RH = *Rhomboideus*; BF = *Biceps femoris*. Data were normalized by 3 stably expressed housekeeping genes as described in material and methods.
[a,b,c,d] Values within a row with different superscripts differ significantly ($P<0.05$).

To our knowledge, this is the first observation of change of the expression of the certain AATs in porcine skeletal muscle during postnatal development. The role of changes in mRNA expression of the AATs in porcine skeletal muscle during postnatal development need to be established.

References

Ai-Min, Z., Z. Xiang-Yan, Z. Jian-Jun, Z. Shi-Geng, H. Zhi-Yi, W. Xiao-Lan, L. TAO and F. Ding-Yuan, 2010. Molecular Cloning, Tissue Distribution and Expression of Porcine y[+]L Amino Acid Transporter-1. Asian-Aust. J. Anim. Sci. 23, 272-278.

Davies, A. S., 1972. Postnatal changes in the histochemical fibre types of procine skeletal muscle. J. Ant. 113, 213-240.

García-Villalobos, H., A. Morales-Trejo, B. A. Araiza-Piña, J. K. Htoo and M. Cervantes-Ramírez, 2012. Effects of dietary protein and amino acid levels on the expression of selected cationic amino acid transporters and serum amino acid concentration in growing pigs. Arch. Anim. Nutr. 66, 257-270.

Nishimura, T. and H. Kato, 1988. Taste of free amino acids and peptides. Food Rev. Int. 4, 175-194.

Shi-Geng, Z., Z. Ai-Nlin, Z. Xiang-Yan, Z. Jian-Jun, Y. ZHANG, H. Zhi-Yi, X. Ping-Wen and F. Ding-Yuan, 2009. Molecular Cloning, Segmental Distribution and Ontogenetic Regulation of Cationic Amino Acid Transporter 2 in Pigs. Asian-Aust. J. Anim. Sci. 22, 712-720.

Rate of rumen epithelial adaptation for sodium and short chain fatty acid absorption

B.L. Schurmann[1], M.E. Walpole[1], P. Górka[1,2] and G.B. Penner[1]
[1]Department of Animal and Poultry Science, University of Saskatchewan, Saskatoon, SK S7N 5A8, Canada; bls144@mail.usask.ca
[2]Department of Animal Nutrition and Feed Management, University of Agriculture in Krakow, 31-120 Krakow, Poland

Introduction

Past studies evaluating ruminal adaptation have largely focused on changes in papillae surface area and epithelial histology as indicators for adaptation. However, abruptly increasing the proportion of dietary concentrate increases the net flux of Na^+ ($J_{net\text{-}Na}$) with marked changes occurring within one week of the dietary change (Etschmann *et al.*, 2009). Given that $J_{net\text{-}Na}$ is driven via an ATP-dependent electrochemical gradient, the increase in $J_{net\text{-}Na}$ indicates that there must also be corresponding increases in energy substrate transport to supply cellular ATP. Presumably, the primary energy substrates would be from apical absorption of short-chain fatty acids (SCFA; Bergman, 1990). The objectives of this study were to establish the timeline for SCFA and Na^+ absorption across the ruminal epithelia following an abrupt increase in diet fermentability.

Material and methods

Twenty-five weaned Holstein bull calves (213±23.0 kg) were randomly assigned to 1 of 5 treatments: the control (CON; 91.5% hay and 8.5% vitamin and mineral supplement) or G3, G7, G14, and G21 which received the same diet (41.5% barley grain, 50% hay and 8.5% vitamin and mineral) but for different durations (3, 7, 14 or 21 d, respectively). All calves were provided feed at 2.25% BW at 0800 h. Reticular pH was recorded every 5 min for 48 h prior to killing at 1000 h. Ruminal fluid was collected to determine the concentration of acetate, butyrate, and propionate, and ruminal tissue from the caudal dorsal blind sac was collected and prepared for mounting in Ussing chambers. Tissues were incubated under short circuit conditions with a mucosal and serosal buffer pH of 6.2 and 7.4, respectively. The mucosal-to-serosal flux of ^{3}H-acetate ($J_{MS\text{-}acetate;}$ 100 kBq/15 ml) and ^{14}C-butyrate ($J_{MS\text{-}butyrate;}$ 72 kBq/15 ml) were measured in parallel. To evaluate the pathway of SCFA absorption (Aschenbach *et al.*, 2009), incubation buffer solutions contained HCO_3^-, were free of HCO_3^-, or were free of HCO_3^- and contained nitrate (40 mM) in order to determine the bicarbonate-dependent, bicarbonate-independent, bicarbonate-independent nitrate-sensitive, and bicarbonate-independent nitrate-insensitive flux. The $J_{net\text{-}Na}$ was measured by difference using the $J_{Na\text{-}MS}$ and $J_{Na\text{-}SM}$ (80 kBq $^{22}Na^+$/15 ml) for tissues paired only when tissue conductance differed by <20%. Data were analyzed as a randomized complete block design using the Mixed Procedure of SAS (version 9.2, Cary, NC, USA) using polynomial contrasts to test whether linear, quadratic, or cubic responses occurred over time.

Results and discussion

Reticular pH decreased (quadratic $P<0.001$) from CON (6.90) to G7 (6.59) then increased to G21 (6.79; SEM=0.028). Total SCFA concentration and the molar proportion of propionate were not affected with means of 84.1 mM (SEM=1.704) and 20.4 mol/100 mol (SEM=0.402), respectively. The molar proportion (mol/100 mol) of acetate decreased (cubic $P=0.008$ and butyrate increased (quadratic $P=0.008$) from CON to G21.

The $J_{MS\text{-}Na}$ and $J_{SM\text{-}Na}$ were not affected, but $J_{Net\text{-}Na}$ increased cubically ($P<0.05$) from 1.16 for CON to 1.66 μmol/(cm^2 × h) for G21, with the highest $J_{Na\text{-}Net}$ (2.41 μmol/(cm^2 × h)) for G7, supportive of previous research (Etschmann *et al.*, 2009; Table 1). Total and bicarbonate-independent $J_{MS\text{-}acetate}$ increased from CON to G7 but decreased thereafter (cubic, $P<0.05$). A quadratic increase ($P<0.05$) in the bicarbonate-independent nitrate-insensitive $J_{MS\text{-}acetate}$ was observed with flux rates of 0.37 for CON, 0.59 for G7 and 0.52 μmol/(cm^2 × h) for G21. Total $J_{MS\text{-}butyrate}$ increased linearly ($P<0.05$) from CON to G21. The bicarbonate-independent nitrate-insensitive $J_{MS\text{-}butyrate}$ increased quadratically ($P<0.05$) from 0.74 for CON to 1.12 μmol/(cm^2 × h) for G21, with maximum uptake at 1.17 for G14.

These results indicate that the ruminal epithelium rapidly adapts to changes in diet fermentability and that increases in SCFA absorption occur in concert with increases in Na^+ absorption. It appears that much of the increase for SCFA absorption is via lipophilic absorption.

Table 1. SCFA concentrations and Na^+, acetate, and butyrate absorption rates.

Item	Treatment[1]					SEM	Contrasts[1]
	CON	G3	G7	G14	G21		
Na uptake, μmol/(cm^2 × h)							
Net[4]	1.16	1.69	2.41	1.58	1.66	0.106	0.042[C]
Mucosal-serosal	1.89	2.81	3.40	2.61	2.90	0.141	NS[A]
Serosal-mucosal	0.73	1.13	0.99	1.03	1.24	0.066	NS
Acetate uptake, μmol/(cm^2 × h)							
Total	1.07	1.12	1.31	1.02	1.14	0.078	0.045[C]
Bicarbonate-dependent	0.61	0.47	0.60	0.48	0.52	0.099	NS
Bicarbonate-independent	0.46	0.64	0.71	0.55	0.62	0.056	0.006[C]
Nitrate sensitive	0.09	0.16	0.12	0.04	0.10	0.063	NS
Nitrate insensitive	0.37	0.49	0.59	0.51	0.52	0.052	0.043[Q]
Butyrate uptake, μmol/(cm^2 × h)							
Total	1.25	1.35	1.53	1.53	1.72	0.106	0.002[L]
Bicarbonate-dependent	0.57	0.47	0.38	0.55	0.70	0.164	NS
Bicarbonate-independent	0.68	0.88	1.15	0.98	1.02	0.133	NS
Nitrate sensitive	-0.06	0.04	0.06	-0.19	0.09	0.132	NS
Nitrate insensitive	0.74	0.84	1.09	1.17	1.12	0.068	0.005[Q]

[1] L = linear; Q = quadratic; C = cubic; NS = all *P*-values were not significant ($P>0.05$).

References

Aschenbach, J.R., S. Bilk, G. Tadesse, F. Stumpff and G. Gabel, 2009. Bicarbonate-dependent and bicarbonate-independent mechanisms contribute to nondiffusive uptake of acetate in the ruminal epithelium of sheep. Am. J. Physiol. Gastrointest. Liver Physiol. 296, G1098-G1107.

Bergman, E.N. 1990. Energy contributions of volatile fatty acids from the gastrointestinal tract in various species. Physiol. Rev. 70, 567-590.

Etschmann, B., A. Suplie and H. Martens, 2009. Change of ruminal sodium transport in sheep during dietary adaptation. Archives of Anim. Nutr. 63, 26-38.

Ruminant-specific molecular and systemic adaptation of renal electrolyte handling to low N intake

S. Starke[1], C. Cox[2], K.-H. Südekum[2] and K. Huber[1]
[1]Department of Physiology, University of Veterinary Medicine Hannover, Bischofsholer Damm 15, 30173 Hannover, Germany; svenja.starke@tiho-hannover.de
[2]Institute of Animal Science, University of Bonn, Endenicher Allee 15, 53115 Bonn, Germany

Introduction

For economical and ecological reasons precise protein rationing is essential in the nutrition of ruminant livestock. This dietary adjustment should consider the endogenous capacity to recycle urea-N for maintaining rumen microbial protein synthesis. However, results of our previous study showed that despite urea recycling, dietary intake of a low N diet affected not only N metabolism but also systemic and renal molecular electrolyte handling in growing male Saanen-type goats (Starke *et al.*, unpublished data). We hypothesized that these changes were also reflected by changes in electrolyte excretion *in vivo*. Therefore, the main objective of the present study was to evaluate the impact of adaptive changes due to low crude protein (LCP) intake on electrolyte handling and excretion *in vivo* in growing goats. To examine whether this adaptation is ruminant-specific, a similar experiment was conducted on growing rats consuming a LCP diet in consideration of the specific differences in nutrition, digestive physiology and metabolism of rats and goats.

Material and methods

Twelve male Saanen-type goats (6 animals per group) and 16 male Wistar rats (8 animals per group) were fed diets either high or low in CP content (as-fed basis: HCP: 20%, LCP: 8%) for 5 weeks. During the 5th week, N, P and Ca balance trials were performed over 6 days in goats and 3 days in rats. Directly at sacrificing blood samples were collected to measure urea, Ca, phosphate (Pi), insulin like growth factor-1 (IGF-1), 1,25-dihydroxyvitamin D_3 (calcitriol), CrossLaps (CTX) and parathyroid hormone (PTH) concentrations. Furthermore, renal expression of the sodium-dependent Pi transporter types IIa and IIc (NaPi IIa and IIc) and the PTH receptor (PTHR) was determined semiquantitatively by Western Blotting. Associations between the measured variables were identified by linear regression analyses. Differences between the feeding groups within one species were tested by unpaired Student's t-test, data comparing the two species were analysed for interaction between species and feeding by Two-Way ANOVA.

Results

Plasma urea concentration (mmol/l; goats: HCP 6.34, LCP 2.38, SEM 0.78, $P<0.001$, rats: HCP 6.25, LCP 4.50, SEM 0.56, $P<0.01$) and urinary N excretion (g/d; goats: HCP 7.17, LCP 1.38, SEM 0.24, $P<0.001$, rats: HCP 0.47, LCP 0.13, SEM 0.02, $P<0.001$) decreased in both species due to LCP intake, whilst daily amounts of retained N were unaffected. In goats, plasma IGF-1 ($P<0.01$) and, by trend, calcitriol ($P<0.1$) concentration decreased (Figure 1A), but Ca and P excretion and retention (g/d) were unchanged. Furthermore, renal PTHR protein expression declined ($P<0.05$). Changes of the modulators of systemic electrolyte handling, IGF-1 and calcitriol, were correlated with changes of other modulators of systemic and also of renal molecular electrolyte handling, like e.g. PTH (calcitriol negatively correlated with PTH; $r^2=0.38$, $P<0.1$) and NaPi IIc (IGF-1 negatively correlated with NaPi IIc; $r^2=0.44$, $P<0.05$). In rats, contrary to goats, plasma IGF-1 and calcitriol concentrations did not change due to LCP intake (Figure 1B). However, daily urine excretion (g/d; HCP 14.41, LCP 9.28, SEM 1.22, $P<0.001$) and daily urinary P excretion (g/d; HCP 0.037, LCP 0.024, SEM 0.003, $P<0.01$) decreased. Furthermore, urinary Ca concentration increased (g/kg; HCP 0.05, LCP 0.08, SEM 0.008, $P<0.001$). Despite these changes, the renal expression of the detected

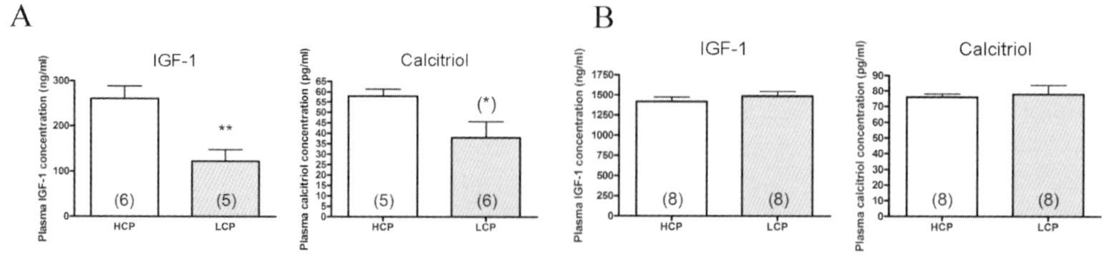

Figure 1. Plasma IGF-1 and calcitriol concentrations in (A) goats and (B) rats during LCP intake compared to the respective HCP group. ** $P<0.01$, (*) $P<0.1$.

proteins remained unchanged. In rats mainly renal molecular modulators of electrolyte homeostasis, like PTHR and NaPi IIc, were correlated with each other.

By direct comparison, in goats with LCP intake plasma urea concentrations decreased stronger than in rats (feeding: $P<0.001$, species: $P<0.05$, interaction: $P<0.05$), accompanied by a greater reduction of urinary and fecal N excretion than in rats, when related to metabolic body size. Furthermore, urinary Ca excretion during LCP intake was more decreased in goats than in rats (feeding: $P<0.05$, species: $P<0.001$, interaction: $P<0.05$; Figure 2).

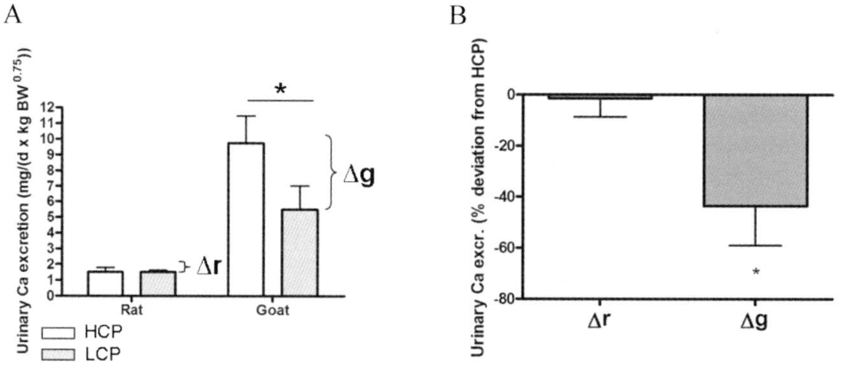

Figure 2. (A) Comparison of urinary Ca excretion related to metabolic body size in rats and goats during LCP intake. (B) Percentage difference of urinary Ca excretion (excr.) of the LCP groups to their respective HCP group in rats and goats. * $P<0.05$.

Summary and conclusion

The main findings of our study were:

1. Both rats and goats exhibited adaptive changes of N metabolism and electrolyte homeostasis due to LCP intake. Particularly the regulation of Ca homeostasis appeared to underlie these adaptive changes.
2. Only in goats, LCP intake decreased concentration of the regulatory hormone IGF-1.
3. Only in goats the observed adaptations resulted in changes in the daily *in vivo* Ca excretion related to metabolic body size.

These results confirm the existence of a goat-specific adaptation of N metabolism and systemic and renal electrolyte handling to low N intake.

Nutrient utilization during inflammation differs between pigs selected for differences in feed efficiency

E. Labussière, E. Merlot, J.-N. Thibault, J. Noblet, N. Le Floc'h and J. van Milgen
INRA, UMR 1348 Pegase, 35590 Saint-Gilles, France; etienne.labussiere@rennes.inra.fr
Agrocampus Ouest, UMR 1348 Pegase, 35000 Rennes, France

Introduction

When based on productive traits, genetic selection does not account for 'non-productive functions' such as the animal's defense systems. Two lines of pigs were divergently selected with a low (RFI-) or a high (RFI+) residual feed intake (RFI), which is calculated as the difference between the actual and the theoretical feed intake required for maintenance and growth (Gilbert *et al.*, 2007). In healthy growing animals, the immune system has low nutritional requirements, which can dramatically increase during inflammation. Whether selection for RFI affects the ability to partition nutrients between growth and immune response is not known. The objective of this study is to compare results obtained in 2 experiments, where the effect of RFI on energy and protein utilization was evaluated in pigs subjected to an inflammatory challenge.

Material and methods

A chronic lung inflammation was induced through an intravenous injection of Complete Freund's Adjuvant (CFA; Melchior *et al.*, 2004). In Exp. 1, the effects of inflammation on rectal temperature, plasma glucose, haptoglobin and IFN-γ concentrations 2 h after the meal were measured in pigs from both lines (n=20; mean body weight – BW: 15-20 kg) from the day before to the 7th day following CFA. On the 8th day following CFA injection, protein synthesis (using a flooding dose of [^{15}N]valine) and proteolytic enzyme activities in the liver and *longissimus dorsi* were quantified and inflammatory cytokines mRNA in the pulmonary lymph nodes were measured. In Exp. 2 (from 3 day before to 3 day after CFA injection), the effects of genotype and inflammation on heat production (HP), respiratory quotient (RQ), N and energy balance and oxidation of dietary [U-^{13}C]glucose (measured as 24 h-$^{13}CO_2$ production on the day before and on the 3rd day after CFA) were measured in pigs from both lines (n=13; BW: 15-30 kg) that were housed individually in a respiration chamber. In both experiments, pigs were offered a commercial starter diet that contained 20.9% CP and 14.9 MJ ME/kg. In Exp. 1, pigs were offered feed *ad libitum* whereas they were restricted at 1.72 MJ ME/kg $BW^{0.60}$/d in Exp. 2 (0.34 on the day of CFA). Data were analyzed (PROC MIXED, SAS, 2004) for the effects of genotype, day and interaction in Exp. 1 and for genotype, period (before or after CFA) and interaction in Exp. 2 according to a mixed linear model. In Exp. 2, RQ was analyzed for the effects of genotype, day as repeated measurement and interaction.

Results and discussion

In both experiments, feed intake did not differ between genetic lines. Plasma haptoglobin concentration was higher (P<0.05) in both lines up to 7 days after CFA but did not differ between RFI lines (Exp. 1). Rectal temperature increased during 3 days following CFA and was higher (P<0.05) on the day of CFA in RFI- than in RFI+ pigs (40.1 vs. 39.8±0.1 °C). At slaughter, IL-1, IL-6, and IL-10 mRNA expression in pulmonary lymph nodes and plasma IFN-γ levels were lower (P<0.10) in RFI- than in RFI+ pigs, suggesting that the inflammation was less severe in RFI- pigs or that they recovered faster, in agreement with the lower proteolytic enzyme activities in the muscle. Differences in utilization of nutrients were observed between the two lines before CFA: HP and oxidation of dietary glucose was lower in RFI- than in RFI+ pigs although glucose was oxidized at a faster rate in RFI- than in RFI+ pigs (Exp.2, Table 1). In contrast to RFI+ pigs, the RQ in RFI- pigs was lower on the day following CFA (1.03) than that measured before or on the 2nd or the 3rd days after CFA injection (1.08 on average). Accordingly, plasma glucose concentration (Exp. 1, Figure 1) and dynamic patterns of glucose oxidation on the 3rd day following CFA injection were not affected by the inflammation in RFI- pigs whereas these parameters were modified in RFI+ pigs (Exp. 2, Table 1). In conclusion, these results indicate that the effects of inflammation on nutrient utilization and metabolism were expressed to a greater extent in RFI- pigs during the 1st day, whereas the effects lasted longer in RFI+ pigs.

Table 1. Effect of genotype (G), period (P) and interaction (P×G) on heat production and kinetics of oxidation of dietary [U-^{13}C]glucose.

Genotype	RFI-		RFI+		Rsd	Significance
Period	Before CFA	After CFA	Before CFA	After CFA		
Heat production (% of ME intake)	65.9	64.0	68.7	66.0	2.1	G*
Respiratory quotient	1.08	1.07	1.06	1.08	0.01	P×G*
^{13}C recovery as $^{13}CO_2$ (% of intake)	48.0	50.2	52.6	49.3	2.4	P×G*
Time to recover 50% of $^{13}CO_2$ (h)	3.6^y	3.7^y	4.8^x	3.9^y	0.2	P**,G*,P×G**

*$P<0.05$; **$P<0.01$.

^{xy} LS-means with different superscript letters within the same row differ $P<0.05$.

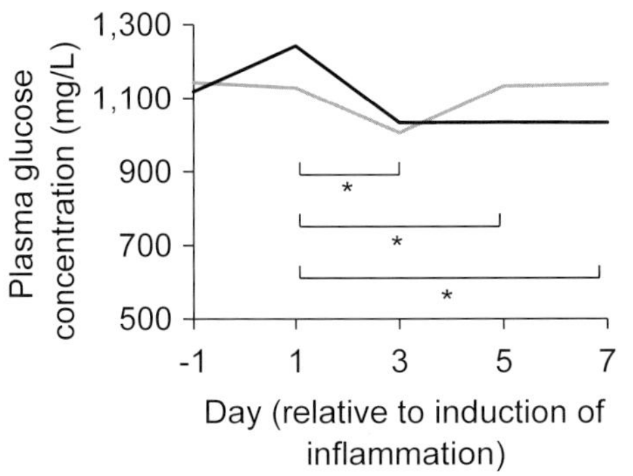

Figure 1. Effect of genetic selection (RFI+ in black and RFI- in grey) on plasma glucose concentration measured two hours after feeding before or after induction of inflammation (Exp. 1; interaction between genotype and day was significant P=*0.05; *P<0.05 for RFI+).*

Acknowledgements

The study was funded by the French National Research Agency (L'Agence Nationale de la Recherche, ANR, ANR-08-GENM038 PIG_FEED Project).

References

Gilbert H., J.P. Bidanel, J. Gruand, J.C. Caritez, Y. Billon, P. Guillouet, H. Lagant, J. Noblet and P. Sellier, 2007. Genetic parameters for residual feed intake in growing pigs, with emphasis on genetic relationships with carcass and meat quality traits. J. Anim. Sci 85, 3182-3188.

Melchior D, B. Seve and N. Le Floc'h, 2004. Chronic lung inflammation affects plasma amino acid concentrations in pigs. J. Anim. Sci 82, 1091-1099.

SAS 2004. SAS/STAT® 9.1 User's Guide.

Differential protein deposition in tissues of growing Iberian and Landrace × Large White pigs under identical nutritional management

R. Nieto, R. Barea, L. Lara, R.A. Márquez and J.F. Aguilera
Institute of Animal Nutrition, Estación Experimental del Zaidín, Spanish Council for Scientific Research (CSIC), Camino del Jueves s/n, 18100 Armilla, Granada, Spain; rosa.nieto@eez.csic.es

Introduction

The present work is part of an experimental program aimed at explaining the biological basis for the lower metabolic efficiency of the Iberian pig, a slow-growing, native obese pig, when compared to conventional lean pig types (Barea *et al.*, 2011).

As a key process for growth, we have focused on protein deposition (PD). Previous work in the literature suggests that pig genotypes with different potential for lean growth might differ not only in the rate of PD but also in the efficiency of use of dietary protein for PD (Fuller *et al.*, 1995), although there are contradictory results regarding this issue (Kyriazakis and Emmans, 1995). Additionally, the distribution of PD among tissues with different metabolic roles may influence the efficiency of whole body (WB) PD. To investigate further these aspects, we performed a comparative study with Iberian and Landrace × Large White (LRW) pigs to assess genotype differences in PD at the WB, carcass and viscera (with clearly different body functions) level during two stages of growth and under identical nutritional management.

Material and methods

The experiment started with 19 Iberian and 19 LRW pigs. Three pigs per genotype were slaughtered at the beginning of the study (15.6±0.1 kg BW) to assess initial body composition. The remaining 32 pigs were fed isoenergetic diets (14.4 MJ ME/kg DM) containing either 130 or 170 g CP/kg DM to match protein requirements for each breed. These CP contents were maintained throughout the experiment for simplicity. Eight pigs per genotype were allocated to each CP level. Pigs were individually housed in an environmentally controlled room. Daily feed allowance was adjusted on a weekly basis according to BW and fixed at 0.8 × *ad libitum* intake of the Iberian pig breed. At 49.6±0.4 kg BW, 16 pigs (4/genotype/CP level) were slaughtered, whereas the rest of pigs (16) continued the study up to slaughter at 115.3±0.6 kg BW. After slaughter, the gastrointestinal tract was emptied and weighed as well as the rest of visceral organs. Four components were obtained for each pig (carcass, head and feet, viscera, and blood) which were kept at -20 °C until analyses. The right half of the carcass and the rest of body components were separately ground, homogenized and sub-samples were freeze-dried and analysed for nutrient composition. The results were subjected to ANOVA for a factorial arrangement of treatments, with included genotype (G), growth stage (P: 15 to 50 and 50 to 115 kg BW), dietary CP content and their interactions using the GLM procedure of SAS. The experimental protocol was approved by the CSIC Bioethical Committee.

Results and discussion

Growth rate was higher for LRW pigs during both growing periods (Table 1; P_G<0.001). The effect of dietary CP is not shown as it was not significant for most of the parameters analysed. On average, WB and carcass PD in LRW pigs was from 0.8 to >2 fold higher than in Iberian pigs (P_G<0.001). No effect of growth stage on WB and carcass PD was detected for Iberian pigs, whereas in LRW both parameters increased ($P_{G\times P}$<0.001) from 19 to 50% in the heavier pigs (*P*<0.05). Average PD in carcass relative to WB was higher in LWR pigs (P_G<0.001) and increased in the heavier pigs of both pig genotypes. On the contrary, PD in visceral tissues relative to total PD was always greater in Iberian pigs (0.117 vs. 0.08 for Iberian vs. LWR pigs, P_G<0.001). The efficiency of PD (WB PD

Table 1. Protein deposition (PD) at whole body (WB), carcass and viscera level of Iberian and Landrace × Large White (LRW) pigs growing from 15 to 50 kg BW (P1) and from 50 to 115 kg BW (P2).

	Iberian		LRW		SEM	*P*-value[1]		
	P1	P2	P1	P2		G	P	G×P
ME intake, MJ/d	21.1	38.0	20.9	37.4	0.3	0.458	0.001	0.744
Protein intake, g/d	235	499	226	500	3	0.491	0.001	0.447
Growth rate, g/d	445	651	594	908	7	0.001	0.001	0.001
WB PD, g/d	53.8	48.2	91.5	118.5	2.2	0.001	0.022	0.001
Carcass PD, g/d	38.3	35.7	71.3	98.3	1.9	0.001	0.004	0.001
Viscera PD, g/d	6.21	5.59	8.30	8.31	0.31	0.001	0.624	0.617
Carcass PD/WB PD	0.711	0.741	0.778	0.829	0.006	0.001	0.002	0.385
Viscera PD/WB PD	0.117	0.117	0.094	0.068	0.003	0.001	0.076	0.073
WB PD/digestible protein intake	0.314	0.127	0.527	0.307	0.008	0.001	0.001	0.329

[1] G = genotype.

to protein ingested or to digestible protein) was considerably lower for Iberian pigs at both stages of growth (P_G<0.001).

Our results show relevant genotype differences in PD pattern during growth that can contribute to explain the lower metabolic efficiency of the Iberian pig when compared to conventional pig types.

Acknowledgements

Funding by the Spanish MINECO (grant AGL2011-25360) is gratefully acknowledged.

References

Barea, R., R. Nieto, F. Vitari, C. Domeneghini and J.F. Aguilera, 2011. Effects of pig genotype (Iberian vs. Landrace×Large White) on nutrient digestibility, relative organ weight and small intestine structure at two stages of growth. Animal 5, 547-557.

Fuller, M.F., M.F. Franklin, R. McWilliam and K. Pennie, 1995. The response of growing pigs, of different sex and genotype, to dietary energy and protein. Anim. Sci. 60, 291-298.

Kyriazakis, I. and G.C. Emmans, 1995. Do breed of pig differ in the efficiency with which they use a limiting protein supply? Br. J. Nut. 74, 183-195.

Intravenous administration of arginine to twin-bearing ewes enhances birth weight and peri-renal fat stores of female offspring in sheep

S. McCoard[1], F. Sales[1], N. Wards[1], Q. Sciascia[1], M. Oliver[2], J. Koolaard[1] and D. van der Linden[1]
[1]AgResearch Limited, Tennent Drive, Private Bag 11008, Palmerston North, New Zealand; sue.mccoard@agresearch.co.nz
[2]The Liggins Institute, University of Auckland, Auckland, New Zealand

Introduction

Advances in genetic selection and breeding have significantly increased the proportion of multiple-bearing pregnancies in sheep. However, competition between littermates in mid to late gestation leads to lower birth weight and increased mortality compared to their singleton counterparts, even when ewes are fed on a high plane of nutrition (McCoard *et al.*, 2000; Wu *et al.*, 2006). Intervention strategies are not currently available to ameliorate the effect of fetal growth restriction *in utero* that results from multiple-bearing pregnancies. Recent research in sheep has indicated that maternal supplementation with L-arginine in mid-late gestation can relieve the restriction on fetal growth induced by under-feeding or increased litter number (quadruplets; Lassala *et al.*, 2011, 2012). The objective of this study was to evaluate the effects of parenteral administration of L-arginine to well-fed twin-bearing ewes from 100 days gestation to birth on lamb birth weight and body composition.

Material and methods

Fifty synchronized Romney ewes were naturally mated to one of two sires to minimize paternal genetic effects on size and weight of fetuses. Twin-bearing ewes were identified following pregnancy diagnosis at 60 days post-mating. Ewes were acclimatized to indoor housing at 80 days of gestation (P80) and fed a lucerne-based pellet diet to meet 100% of NRC feeding recommendations. Between P100 and birth, ewes received an i.v. bolus of either L-arginine-mono-hydrochloride (Arg; 345 μmol/kg BW) or sterile saline solution (control) 3 times a day via a catheter placed in the tarsal vein. At P140, 20 ewes (Arg group n=9; control group n=11) were euthanized and fetal mass and organ mass evaluated. The remaining ewes (Arg group n=13; control group n=12) were allowed to lamb and lamb birth weight was recorded. Blood samples were collected from the ewes and fetuses 1 hour post-bolus administration at P140 and from ewes and lambs 2 hours post-birth. Plasma amino acids were determined by ion-exchange chromatography as previously described (van der Linden *et al.*, 2012). Amino acid data were analysed using the nlme package in R. All other data were analyzed using the MIXED procedure in SAS. Linear models were fitted and included the fixed effects of treatment group (arginine vs. control) for maternal data. The linear model for lamb/fetal data included the fixed effect of treatment group, sex of lamb (female vs. male), treatment by sex interaction and the random effect of ewe to adjust for the twinning effect.

Results and discussion

Ewe live weight and average daily feed intake did not differ (P>0.05) between groups during the treatment period (data not shown). At P140, fetal weight (5.3±0.1 vs. 5.3±0.1 kg) and the weight of the internal organs did not differ (P>0.05) between control and Arg-supplemented groups (data not shown). The notable exception was peri-renal fat stores which were increased by 16% in fetuses from Arg-supplemented ewes relative to controls (8.7±0.4 vs. 7.7±0.4 g, P<0.05). This finding is consistent with maternal Arg supplementation to either under-fed sheep (Satterfield *et al.*, 2011) whereby fetal peri-renal fat stores at 125 days of gestation are increased indicative of Arg stimulation of brown adipose tissue development. At birth, there was a trend for an interaction (P=0.06) between Arg supplementation and sex for birth weight of the lamb. The ewe lambs from Arg-supplemented ewes were 12% heavier at birth compared with controls (5.6±0.1 vs. 5.0±0.2 kg, P=0.03) whereas birth weight of male lambs did not differ between Arg and control groups (5.4±0.2 vs. 5.4±0.1, P=0.80).

At P140, ewes treated with Arg had increased (P<0.05) plasma concentrations (μM; LSM ± SEM) of arginine (992 vs. 109±84), ornithine (394 vs. 137±38), and decreased (P<0.05) plasma concentrations of glycine (516 vs. 795±58), serine (66 vs. 91±8) and methionine (26 vs. 38±2) relative to controls. Fetuses from ewes supplemented with Arg had increased (P<0.05) plasma concentrations of ornithine (367 vs. 213±52) and decreased (P<0.05) plasma concentrations of taurine (47 vs. 84±13), threonine (616 vs. 763± 48), glycine (748 vs. 944±55), methionine (57 vs. 104±7), tyrosine (115 vs. 162±15) and histidine (153 vs. 366±40) relative to control fetuses at P140.

Two hours post-birth, ewes treated with Arg had elevated (P<0.05) concentrations of ornithine (113 vs. 62±12) whereas methionine remained decreased (20 vs. 29±2, P<0.01). Lambs born to ewes treated with Arg had increased (P<0.05) concentrations of isoleucine (181 vs. 80±29) and leucine (404 vs.175± 61) compared to lambs born to control ewes at birth. The differences in the amino acid concentrations of ewes and their offspring at P140 and post-birth are likely due to Arg administration being stopped 2-12 hours prior to birth.

The sex-specific effect of Arg supplementation on lamb birth weight is intriguing. On average, female lambs are 10-15% lighter at birth compared to males, an effect ameliorated in this study. Sex specific effects of nutritional interventions have been reported previously (Jaquiery *et al.*, 2012). However, it is possible that there were insufficient males in each group to detect a treatment effect. Further validation of findings is required. The findings of this study indicate that maternal Arg supplementation ameliorates the effects of fetal growth restriction in female twin lambs. Increased birth weight coupled with peri-renal fat stores may have important implications for the survival of the newborn with importance for both agriculture and medicine.

References

Jaquiery, A.L., M.H. Oliver, M. Honeyfield-Ross, J.E. Harding and F.H. Bloomfield, 2012. Periconceptional undernutrition in sheep affects adult phenotype only in males. J. Nutr. Metab. 123610. doi:10.1155/2012/123610.

Lassala, A., F.W. Bazer, T.A. Cudd, S. Datta, D.H. Keisler, M.C. Satterfield, T.E. Spencer and G., 2010. Parenteral administration of L-arginine prevents fetal growth restriction in undernourished ewes. J. Nutr. 140, 1242-1248.

Lassala, A., F.W. Bazer, T.A. Cudd, S. Datta, D.H. Keisler, M.C. Satterfield, T.E. Spencer and G. Wu, 2011. Parenteral administration of L-arginine enhances fetal survival and growth in sheep carrying multiple fetuses. J. Nutr. 141, 849-855.

McCoard, S.A., W.C. McNabb, S.N. McCutcheon, P.M. Harris and S.W. Peterson, 2000. Muscle growth, cell number, type and morphometry in single and twin fetal lambs during mid to late gestation. Reprod. Fert. Dev. 12, 319-327.

Satterfield, M.C., K.A. Dunlap, D.H. Keisler, F.W. Bazer and G. Wu, 2011. Arginine nutrition and fetal browtn adipose tissue development in nutrient-restricted sheep. Amino Acids. DOI 10.1007/s00726-011-1168-8.

Wu, G., F.W. Bazer, J.M. Wallace, T.E. Spencer, 2006. Board-invited review: intrauterine growth retardation: implications for the animal sciences. J. Anim. Sci. 84, 2316-2337.

Effect of dietary protein concentration and forage type on nitrogen metabolism and nutrient flux across the portal drained viscera and the liver in lactating dairy cows

C.E.S. Barratt, L.A. Crompton, C. Green, D.J. Humphries, R.D. Pilgrim and C.K. Reynolds
School of Agriculture, Policy and Development, University of Reading, P.O. Box 237, Earley Gate, Reading, RG6 6AR, United Kingdom; c.e.barratt@reading.ac.uk

Introduction

With regard to dairy cow nutrition, past efforts have focused on maximising milk output, rather than feed conversion efficiency. More recently, attention has reverted to the efficiency of milk production, and in particular dietary nitrogen (N) utilisation due to increasing concerns regarding nitrogenous emissions from dairy cattle. The contribution of dairy production to environmental pollutants, particularly nitrous oxide and ammonia, is an issue of increasing concern. Current feed prediction schemes concerning metabolisable protein are limited due to an oversimplification of post-absorptive events. There are few recent reliable data on the efficiency of converting dietary N into milk protein and there is still much variation seen between different studies (Hristov and Ropp, 2003; Lee *et al.*, 2009). The present results are from a larger study concerned with describing the absorption and metabolism of nutrients and metabolites by the splanchnic tissues in lactating dairy cows in response to varying amounts of absorbable protein and two forage types.

Material and methods

Six Holstein dairy cows in mid-late lactation, surgically prepared with a rumen fistula and splanchnic blood sampling catheters were fed *ad libitum* in equal meals provided hourly, total mixed rations consisting of a 50:50 mixture (dry matter (DM) basis) of forage:concentrate, with the forage comprised of either 25:75 or 75:25 grass:maize silage (DM basis). Rations were formulated to contain crude protein levels of 12.5, 15 and 17.5% of diet DM giving a 2×3 factorial design. The experiment was conducted as repeated 3×3 Latin squares, with the effect of protein level tested within squares and forage source as the square effect. Within each square treatment periods were 3 weeks long, with blood and digesta sampling taking place in the last week of each treatment period. A total of 8 hourly blood samples sets were obtained and analyzed individually. Net flux of nutrients across the portal drained viscera (PDV) and liver (LIV) including glucose, lactate and β-hydroxybutyrate (BOHB) were calculated as venous-arterial concentration difference times plasma flow rate measured by ρ-aminohippurate dilution. Data were analyzed using mixed models procedures accounting for fixed effects of square, period within square, forage, protein, and forage by protein interaction and random effects of animal.

Results and discussion

Dry matter intake (DMI), N intake (NI), milk yield (MY) and milk N yield increased ($P \leq 0.05$) with increasing diet protein concentration, but were not significantly affected by forage type. Rumen ammonia concentrations, and net PDV release of ammonia (NH_3; Table 1) increased linearly ($P<0.001$) as diet protein concentration increased and were higher when the grass-based diet was fed, as NI was numerically higher for the grass-based diet. Arterial urea concentration and LIV urea production (Table 1) increased ($P<0.001$ and $P<0.06$, respectively) with increasing diet protein concentration and arterial urea concentration was greater ($P=0.06$) when the grass-based diet was fed compared to the maize-based diet. Higher dietary protein levels are associated with the degradation of protein in the rumen, causing increased rumen ammonia concentration (Broderick, 2003) and the capture of ammonia for microbial protein synthesis is energy dependent; therefore the amount and degradability of dietary carbohydrate can influence ammonia absorption. The increment in rumen

Table 1. Dry matter intake, milk yield, N production efficiency and splanchnic nutrient flux.

	Maize			Grass			SEM	*P*-values[1]		
	12.5	15.0	17.5	12.5	15.0	17.5		For	Pro	Int
DMI[2]	20.5	22.3	22.4	19.7	20.3	21.1	0.7	0.185	0.014	0.431
MY[2]	24.8	27.7	29.7	25.2	27.2	31.1	2.5	0.903	0.001	0.582
NI[3]	402	500	583	464	540	638	26	0.128	0.001	0.876
Milk N[3]	137	150	167	139	149	162	13	0.935	0.054	0.938
N Efficiency	0.35	0.30	0.29	0.31	0.28	0.25	0.03	0.332	0.097	0.982
Rumen NH_3[4]	3.13	3.52	4.09	4.19	4.77	5.58	0.38	0.024	0.010	0.750
Arterial urea[4]	4.6	6.2	9.0	6.3	7.4	10.4	0.6	0.058	0.001	0.849
PDV NH_3[5]	301	365	484	367	453	529	23	0.039	0.001	0.580
LIV Urea[5]	567	608	867	403	691	725	101	0.392	0.057	0.422
LIV Glucose[5]	733	747	857	740	742	773	36	0.569	0.037	0.269
LIV BOHB[5]	284	301	312	189	253	411	25	0.609	0.003	0.010

[1] Probability of effects of forage type (For), protein level (Pro) and interaction (Int).
[2] kg/day; [3] g/day; [4] mM; [5] mmol/h.

ammonia was accompanied by an increase in arterial urea concentration. N efficiency (milk N/NI; Table 1) tended to be affected by an increase in protein level (P=0.10), with N capture in milk protein relative to NI decreasing as NI increased with protein level within each diet. Net LIV release of glucose and BOHB both increased linearly (P<0.05) as diet protein concentration, DMI and milk yield increased (Table 1), but the increase in LIV BOHB release was greater for the grass-based diet compared to the maize-based diet (P_{int}=0.01).

These data demonstrate the extent to which forage type (grass versus maize silage) and level of protein supply affect post-absorptive metabolism of N and other metabolites. Results show that there is a strong relationship between NI, net PDV ammonia absorption, net LIV urea production and N use efficiency as protein levels are increased, but this is not affected by forage type.

Acknowledgments

Funded by the Commission of the European Communities (FP7 KBBE-2007-1; REDNEX).

References

Broderick, G.A., 2003. Effects of varying dietary protein and energy levels on the production of lactating dairy cows. J. Dairy Sci. 86, 1370-1381.

Hristov, A.N. and J.K. Ropp, 2003. Effect of dietary carbohydrate composition and availability on utilization of ruminal ammonia nitrogen for milk protein synthesis in dairy cows. J. Dairy Sci. 86, 2416-2427.

Lee, M.R.F. Theobald, V.J. Tweed, J.K.S. Winters, A.L. and N.D. Scollan, 2009. Effect of feeding fresh or conditioned red clover on milk fatty acids and nitrogen utilization in lactating dairy cows. J. Dairy Sci. 92, 1136-1147.

Effect of abomasal amino acid infusion on splanchnic metabolism in *postpartum* transition dairy cows

M. Larsen[1], C. Galindo[2], D.R. Ouellet[3], G. Maxin[3], N.B. Kristensen[1] and H. Lapierre[3]
[1]Department of Animal Science, Aarhus University, Foulum, 8830 Tjele, Denmark; mogens.larsen@agrsci.dk
[2]Département des Sciences Animales, Université Laval, Québec, QC G1V 0A6, Canada
[3]Agriculture and Agri-Food Canada, Sherbrooke, QC J1M 0C8, Canada

Introduction

Postpartum transition cows are in negative energy and protein balance since the high nutrient demand created by initiation of lactation is not met by sufficient feed intake. In comparison to energy, there is a paucity of data on metabolic and production effects of dietary attempts to alleviate the *postpartum* protein deficiency. The aim of the study was to investigate the effect of increasing protein supply on tissue amino acid (AA) metabolism in *postpartum* cows.

Material and methods

Nine multiparous Holstein cows implanted with rumen cannula and permanent indwelling catheters in major splanchnic vessels were used in a generalised randomised incomplete block design with repeated measurements. Cows, blocked according to parity (2nd and 3rd+), were assigned to one of two treatments: continuous abomasal infusion of water (CTRL; n=4) or of a mixture of free AA (AA-CN; casein profile; n=5), initiated at the day of parturition (1st day in milk, DIM). Infusion of AA-CN was administered with half of full dose at 1 DIM, full dose (795±12 g/d) from 2 to 6 DIM, followed by a linear daily decrease (571±2 g/d at 15 DIM and 227±1 g/d at 29 DIM). All cows received the same total mixed ration (163 g CP/kg DM) allocated ad libitum. Six sets of arterial, venous portal, hepatic and mammary venous samples were collected at 45 min intervals at 5, 15, and 29 DIM. Data were analysed with a mixed model including the fixed effects of block, treatment (Trt), DIM, and Trt × DIM; DIM was tested as a repeated measurement using the autoregressive order 1 covariance structure.

Results and discussion

Increasing the early *postpartum* AA supply induced a substantial increase in both milk (7.8±1.3 kg/d), and milk protein yield (Table 1), but milk protein content was unaffected (P=0.31). Arterial concentrations of essential AA (EAA) tended to be greater for AA-CN compared with CTRL (P=0.08; 1,175 vs. 908±84 μM). The increased milk yield was not supported by a greater feed intake as dry matter intake decreased by 1.6±0.6 kg/d with AA-CN. Instead, the increased milk yield seemed to be partially supported by greater mobilisation of body fat, as the arterial concentration of non-esterified fatty acids were higher at 5 DIM for AA-CN as compared with CTRL (723 vs. 382±78 μM), but did not differ by 29 DIM (299±52 μM; interaction: P=0.05). Furthermore, arterial concentrations of urea-N were greater for AA-CN compared with CTRL (P=0.01; 8.5 vs. 5.3±0.6 mM) indicating that the increased AA supply was partially utilised for energetic purposes.

Net portal release of group 1 EAA (His+Met+Phe+Trp+Tyr; Table 1) was greater with AA-CN. Their liver removal, however, increased equivalently resulting in an unaltered net release by splanchnic tissues. Moreover, liver removal relative to total influx of His, Met, Phe, and Trp was greater with AA-CN ($0.01 \leq P \leq 0.08$) indicating an up-regulation of hepatic pathways utilising these EAA. Overall, the deficit between the net splanchnic release and milk protein secretion of group 1 EAA was not reduced in AA-CN cows.

Table 1: Effect of abomasal amino acid (AA) infusion in postpartum *dairy cows on production and tissue metabolism of essential AA (EAA) and non-EAA.*

Item		*Postpartum* Trt		SEM	*P*-values		
		CTRL	AA-CN		Trt	DIM	Trt × DIM
Dry matter intake, kg/d[1]		20.4	18.8	0.4	0.02	<0.01	0.32
Milk, kg/d[1]		38.2	46.0	0.8	<0.01	<0.01	0.29
Milk protein, g/d[2]		1301	1521	43	0.01	0.03	0.50
Group 1 EAA[2]	PDV[3], mmol/h	78	90	4	0.05	0.46	0.44
	Liver, mmol/h	-33	-49	4	0.03	0.53	0.37
	TSP[3], mmol/h	43	42	6	0.89	0.36	0.70
	MG[3], mmol/h	-56	-67	2	0.02	0.10	0.52
	Milk, mmol/h	57	66	2	0.01	0.03	0.46
	PDV:milk ratio	1.39	1.38	0.05	>0.99	0.04	0.51
	TSP:milk ratio	0.75	0.63	0.07	0.15	0.06	0.56
	MG:milk ratio	0.99	1.01	0.01	0.26	0.49	0.95
Group 2 EAA[2]	PDV, mmol/h	121	144	9	0.13	0.53	0.34
	Liver, mmol/h	-4.4	-16	11	0.53	0.65	0.91
	TSP, mmol/h	118	132	12	0.46	0.53	0.27
	MG, mmol/h	-137	-167	5	<0.01	0.26	0.49
	Milk, mmol/h	126	146	4	0.01	0.03	0.46
	PDV:milk ratio	0.98	1.00	0.07	0.88	0.10	0.83
	TSP:milk ratio	0.94	0.91	0.07	0.79	0.13	0.49
	MG:milk ratio	1.09	1.15	0.02	0.02	0.18	0.82
Non-EAA[2]	PDV, mmol/h	217	284	21	0.12	0.25	0.82
	Liver, mmol/h	-140	-242	27	0.06	<0.01	0.64
	TSP, mmol/h	72	50	28	0.62	0.70	0.32
	MG, mmol/h	-150	-174	11	0.22	0.03	0.71
	Milk, mmol/h	244	284	8	0.01	0.03	0.46
	PDV:milk ratio	0.89	1.01	0.08	0.43	0.07	0.78
	TSP:milk ratio	0.29	0.17	0.09	0.47	0.64	0.36
	MG:milk ratio	0.60	0.61	0.04	0.85	0.33	0.63

[1] Average of daily observations from 1 to 29 DIM.

[2] Average of 3 sampling days; Group 1 EAA = His+Met+Phe+Tyr+Trp; Group 2 EAA = Ile+Leu+Val+Lys.

[3] PDV = portal-drained viscera; TSP = total splanchnic tissue; MG = mammary gland.

The net portal release of group 2 EAA (Ile+Leu+Val+Lys) was numerically greater with AA-CN, whereas their small liver removal was unaffected by the increased supply. Consequently, net portal and splanchnic releases of the branched-chain AA and Lys were close to be sufficient to meet milk protein secretion and that was not affected by AA-CN. However, mammary uptake to output ratio increased with AA-CN. The net portal release of non-EAA was numerically greater with AA-CN, but liver removal increased to a greater extent, resulting in an unaltered net splanchnic release. Moreover, as the liver removed a large fraction of non-EAA portal release (except for Glu), there was a substantial deficiency of non-EAA relative to milk secretion with both treatments.

In conclusion, the attempt to alleviate the early *postpartum* protein deficiency by abomasal AA infusion induced a profound increment in milk and milk protein yield. Moreover, the greater AA supply allowed more AA for liver anabolism and catabolism, though the partitioning could not be assessed. Thus, the early *postpartum* protein deficiency persisted, as the cow efficiently utilised the additional AA for anabolic processes intended for the new born calf.

Net portal appearance of amino acids in Iberian compared to Landrace pigs

L. González-Valero, J.M. Rodríguez-López, M. Lachica and I. Fernández-Fígares
Department of Animal Nutrition, Estación Experimental del Zaidín, CSIC, Camino del Jueves s/n, 18100 Armilla, Granada, Spain; ifigares@eez.csic.es

Introduction

Compared to modern breeds, Iberian pigs have lower rates of muscle protein deposition and greater viscera weight. Factors that limit growth performance of Iberian pigs are still unknown. We hypothesized that differences in net portal appearance (NPA) of amino acids (AA) might partially explain the lower growth rate reported in Iberian pigs compared to modern breeds (Nieto *et al.*, 2002).

Material and methods

Net portal absorption of AA was measured in 6 Iberian (Ib) and 6 Landrace (Ld) gilts (28 kg BW) fitted with chronic catheters in portal vein (PV), mesenteric vein (MV) and carotid artery (CA). Blood samples from PV and CA were taken -5 min, and 0.5, 1, 1.5, 2, 2.5, 3, 3.5, 4, 5 and 6 h after feeding 25% of the daily intake of 2 isoenergetic diets (14-14.5 MJ ME/kg DM) with different CP (13 vs. 16%, LCP and HCP, respectively) content in a crossover design, with an adaptation period to the diet of 1 week. Plasma free AA were analysed by HPLC using the Waters Pico Tag method which involves pre-column derivatization with phenylisothiocianate and a reverse phase column. Net portal AA absorption was calculated by multiplying portal-arterial plasma AA concentration difference by PV plasma flow estimated by an indicator-dilution technique infusing para-aminohippuric acid into one MV. Data were subjected to ANOVA using the MIXED procedure. The statistical model applied included the fixed effect of block, diet, breed, sampling time and corresponding interactions. Time of sampling within pig was considered as a repeated measure. Concentration at time zero of the analyte was included as a covariate in the statistical analysis. Significant differences among treatments were tested according to the factorial design.

Results and discussion

Portal and arterial plasma concentration and NPA of AA are presented in Table 1. Feeding increased ($P<0.001$) concentrations of plasma AA in both portal and arterial samples and net portal appearance of AA throughout the postprandial period (data not shown). No interactions of breed or diet with time was found ($P>0.10$). The postprandial increase in portal and arterial concentrations of AA and the subsequent gradual decrease are in agreement with results in the literature (Yen *et al.*, 2004).

Portal and arterial non-essential (NEAA) and total AA were lower ($P<0.05$) in Ib compared to Ld pigs with no difference ($P>0.10$) in portal and arterial essential AA (EAA) concentrations. Furthermore, NPA was lower in Ib compared to Ld for total AA, EAA and NEAA (52, 45 and 50%, respectively; $P<0.001$). Differences in EAA NPA are therefore due to the lower portal blood flow in Ib compared to Ld pigs (Rodríguez-López *et al.*, 2010). Interestingly, no differences in Lys and Met absorption were found between Ib and Ld pigs fed barley-soybean diets of different protein concentration (González-Valero *et al.*, 2012).

Regarding CP content of the diet, portal EAA concentration was greater ($P<0.05$) when pigs were given the HCP, with no differences in portal total AA and NEAA concentrations. Pigs fed the LCP diet had, however, greater arterial total AA and NEAA ($P<0.05$) concentrations although no difference in arterial EAA concentration was found ($P>0.10$). Consequently, NPA of total AA, EAA and NEAA during the 6-h postprandial period was lower ($P<0.01$) when pigs were fed the LCP diet (41, 52 and 35%, respectively) than the HCP diet. Lys and Met NPA were reported to be lower

Table 1. Arterial and portal concentrations ([mM]), and net portal appearance (NPA; mmol/min) of total (T), essential (E) and nonessential (NE) amino acids (AA) in Iberian (Ib) and Landrace (Ld) pigs (n=6 pigs/breed) receiving either 13 (L) or 16% (H) CP diets.

	Ib		Ld		SE	*P*-value		
	LCP	HCP	LCP	HCP		Breed	Diet	Breed×Diet
Portal [TAA]	4.6	4.6	4.8	4.9	0.12	0.008	NS	NS
Arterial [TAA]	4.3	4.0	4.4	4.3	0.10	0.040	0.052	NS
TAA NPA	0.15	0.33	0.40	0.60	0.041	<0.001	<0.001	NS
Portal [EAA]	1.6	1.8	1.6	1.7	0.06	NS	0.031	NS
Arterial [EAA]	1.5	1.5	1.5	1.4	0.05	NS	NS	NS
EAA NPA	0.07	0.15	0.14	0.28	0.018	<0.001	<0.001	NS
Portal [NEAA]	2.7	2.5	3.0	3.0	0.08	<0.001	NS	NS
Arterial [NEAA]	2.6	2.2	2.6	2.7	0.06	<0.001	0.005	<0.001
NEAA NPA	0.07	0.18	0.23	0.28	0.025	<0.001	0.002	NS

when pigs consumed LCP compared to HCP diet (González-Valero *et al.*, 2012). Simoes Nunes *et al.* (1991) reported decreased AA absorption when pigs fed 12% casein were compared to pigs fed 24% casein diets.

Differences in AA NPA absorption may partially explain the disparate growth capacity of Ib pigs compared to modern genotypes. This may be the result of the lower digestibility of protein (Rivera-Ferre *et al.*, 2006; Barea *et al.*, 2011) or higher utilization of AA by the portal-drained viscera in the Ib pigs compare to the Ld pigs or both. Further research on individual AA may modulate the overall conclusion.

References

Barea, R., R. Nieto, F. Vitari, C. Domeneghini and J.F. Aguilera, 2011. Effects of pig genotype (Iberian *vs.* Landrace × Large White) on nutrient digestibility, relative organ weights and small intestine architecture at two stages of growth. Animal 5, 547-557.

González-Valero, L., J.M. Rodríguez-López, M. Lachica and I. Fernández-Fígares, 2012. Differences in portal appearance of lysine and methionine in Iberian and Landrace pigs. J. Anim. Sci. 90, 110-112.

Nieto, R., A. Miranda, M.A. García and J.F. Aguilera, 2002. The effect of dietary protein content and feeding level on the rate of protein deposition and energy utilization in growing Iberian pigs from 15 to 50 kg body weight. Br. J. Nutr. 88, 39-49.

Rivera-Ferre, M.G., J.F. Aguilera and R. Nieto, 2006. Differences in whole-body protein turnover between Iberian and Landrace pigs fed adequate or lysine-deficient diets. J. Anim. Sci. 84, 3346-3355.

Rodríguez-López, J.M., M. Lachica, L. González-Valero and I. Fernández-Fígares, 2010. Energy expenditure of splanchnic tissues in Iberian and Landrace growing gilts. Livest. Sci. 133, 61-63.

Simoes Nunes, C., I. Galibois, A. Rérat, L. Savoie and P. Vaugelade, 1991. Hepatic and portal-drained viscera balances of amino acids, insulin, glucagon and gastrin in the pig after ingestion of casein or rapeseed proteins. Reprod. Nutr. Dev. 31, 217-231.

Yen, J.T., B.J. Kerr, R.E. Easter and A.M. Parkhurst, 2004. Difference in rates of net portal absorption between crystalline and protein-bound lysine and threonine in growing pigs fed once daily. J. Anim. Sci. 82, 1079-1090.

Urea nitrogen absorbed from the hindgut is used efficiently for body protein deposition in pigs fed a diet deficient in non-essential amino acid nitrogen

W. Mansilla[1], D. Columbus[1], J.K. Htoo[2] and C.F.M. de Lange[1]
[1]Department of Animal and Poultry Science, University Guelph, Guelph, ON, N1G 2W1, Canada; wmansill@uoguelph.ca
[2]Evonik Industries AG, Rodenbacher Chaussee 4, 63457, Hanau, Germany

Introduction

The absorption of nitrogen (N) from the hindgut of monogastric animals is thought to be of little value for supporting body protein synthesis. However, N that is absorbed from the lower gut, largely in the form of ammonia, can be used for synthesis of non-essential amino acids (NEAA) or converted to urea. The latter can be excreted in urine or recycled into the upper gut and contribute to microbial-produced amino acids that can be of benefit to the host (Fuller, 2012). The objective of the present experiment was to explore the efficiency of using N absorbed from the hindgut for body protein deposition in growing pigs fed a diet deficient in NEAA-N.

Material and methods

Nine barrow Yorkshire pigs (initial BW 16.9±0.3 kg), were fitted with a simple T-cannula in the caecum. All pigs were fed 3 equal meals daily (8:00, 13:00 and 17:30 h) of a cornstarch, casein and crystalline amino acids-based diet limiting in NEAA-N. Ratios among essential amino acids (EAA) were based on the ideal protein concept (NRC, 1998), while the EAA-N to total N ratio was high (0.75) and above the required ratio (Heger, 1998). Water was supplied with feed in a 3 to 1 ratio.

Pigs were randomly assigned to 1 of 3 treatments, representing a control and 2 different urea-N infusion rates into the caecum (1.5, 3.0 g/d), according to two different (one repeated) 3x3 Latin Square designs and during 3 successive experimental periods. Experimental periods consisted of 5 d adaptation followed by a 4 d N-balance period.

During each N-balance period, urine was collected quantitatively at 24 h intervals in containers with sulphuric acid to keep urine pH below 2. Fecal samples were collected continuously, frozen immediately, pooled per pig and N-balance period, and freeze dried. Urine and feces were analyzed for total N content. Nitrogen retention was calculated as the difference between dietary N intake (corrected for feed wastage) plus infused N and urinary plus fecal N excretion. Data were analyzed using the PROC MIXED procedure of SAS (v. 9.2; SAS Institute Inc., Cary, NC) with pig, period and urea-N infusion rate as sources of variation.

Results and discussion

Due to insufficient feed intake or incomplete urinary N collection, some observations were removed prior to analysis. Whole body N retention differed among treatments (4.86, 6.40 and 7.75 g/d; $P<0.001$) and increased with urea infusion rate (Table 1). Moreover, BW gain increased with urea infusion rate (266, 313 and 360 g/d, $P<0.05$) and there was a difference between the control and the highest urea infusion rate ($P<0.05$). These results are in agreement with Rose and Dekker (1956), who showed that supplying rats fed a diet deficient in NEAA with urea improved growth performance.

Fecal and urinary N excretion did not differ among treatments ($P>0.10$), indicating that all N that was infused into the caecum was absorbed and retained in the body. The marginal efficiency of using

Table 1. Growth performance and aspects on whole body N utilization in growing pigs fed a diet deficient in NEAA-N and infused with varying amounts of urea into the caecum.

	Urea-N infused, g/d			*P*-value
	0 (control) (n=7)	1.5 (n=6)	3.0 (n=8)	
Mean BW, kg	32.1±0.70	31.7±0.70	32.3±1.00	0.41
Whole body N retention, g/d	4.86±0.20[a]	6.40±0.21[b]	7.75±0.19[c]	<0.001
Difference vs. control, g/d	-	+1.54±0.21[a]	+2.89±0.19[b]	<0.001
BW gain, g/d	266±16[a]	313±17[ab]	360±15[b]	<0.05
Difference vs. control, g/d	-	+47±17	+93±15	0.18
Dietary dry matter intake, g/d	922±6.00	906±6.00	914±6.00	0.13
Dietary N intake, g/d	8.79±0.06	8.64±0.07	8.72±0.07	0.13
N infused into caecum, g/d	-	1.49±0.05[a]	3.02±0.05[b]	<0.001
Fecal N excretion, g/d	1.54±0.11	1.57±0.14	1.33±0.10	0.25
Urinary N excretion, g/d	2.38±0.20	2.21±0.22	2.63±0.19	0.39
Estimated ileal N flow, g/d[1]	0.59±0.01	0.59±0.01	0.59±0.01	0.64
Lower gut N disappearance, g/d[2]	-0.96±0.11[a]	0.49±0.11[b]	2.26±0.10[c]	<0.001
Difference vs. control, g/d	-	+1.45±0.11[a]	+3.22±0.10[b]	<0.001
Marginal N retention[3]	-	104.1±9.7	87.4±7.9	0.20

[1] Calculated based on an ileal N digestibility of 93.3% (NRC, 2012).

[2] Calculated as: lower gut N disappearance = (ileal N flow + N infusion) – fecal N flow.

[3] Calculated as: (Increase in N retention vs. control/N infused into lower gut) × 100.

[a,b,c] Values (least square means ± standard error) in the same row followed by different superscripts differ significantly ($P<0.05$).

N infused into the caecum for whole body N retention was not different for the two urea infusion treatments (104.1 and 87.4% for 1.5 and 3.0 g/d of urea-N infused, respectively; $P=0.20$).

In conclusion, these results show that N absorbed from the hindgut can be used efficiently for whole body N retention in pigs fed diets that are deficient in NEAA-N with no excess of EAA, reflecting the use of absorbed non-protein N for endogenous synthesis of NEAA when total N supply is deficient.

References

Fuller, M.F., 2012. Determination of protein and amino acid digestibility in foods including implications of gut microbial amino acid synthesis. Br. J. Nutr. 108, S238-S246.

Heger, J., 1998. Effect of essential: total nitrogen ratio on protein utilization in the growing pig. Br. J. Nutr. 80, 537-544

National Research Council (NRC), 2012. Nutrient requirements of swine. The national academies press, Washington, DC.

Rose, W. and E.E. Dekker, 1956. Urea as a source of nitrogen for the biosynthesis of amino acids. J. Biol. Chem. 223, 107-121.

Supplementation with a leucine pulse during continuous feeding stimulates protein synthesis and suppresses protein degradation pathways in skeletal muscle of neonatal pigs

C. Boutry[1], *S.W. El-Kadi*[1,2], *A. Suryawan*[1], *S.M. Wheatley*[1], *R.A. Orellana*[1], *H.V. Nguyen*[1] *and T.A. Davis*[1]
[1]*USDA/ARS Children's Nutrition Research Center, Department of Pediatrics, Baylor College of Medicine, Houston, TX, USA; claire.regard@gmail.com*
[2]*Department of Animal & Poultry Sciences, Virginia Tech, Blacksburg, VA, USA*

Introduction

Orogastric tube feeding, using either continuous or intermittent bolus delivery, is a common clinical practice in pediatric patients who are unable to feed normally. We have shown, using the neonatal pig as a model for the human neonate, that intermittent bolus feeding (BOL) has a greater stimulatory effect on muscle protein synthesis (PS) than continuous orogastric infusion (CON) (Gazzaneo *et al.*, 2011; El-Kadi *et al.*, 2012). This effect is largely due to the rise in insulin and amino acid (AA) concentrations after BOL feeding which increases the activation of mammalian target of rapamycin (mTOR) and its downstream signaling proteins leading to translation initiation in muscle (Suryawan *et al.*, 2007).

The branched-chain AA, leucine (Leu), acts as a nutrient signal to stimulate PS in skeletal muscle (Anthony *et al.*, 2000). Previously, our laboratory demonstrated that a 1 h parenteral infusion of Leu increases PS in skeletal muscle of neonatal pigs (Escobar *et al.*, 2005). However, the response to Leu was not sustained, likely due to a Leu-induced decrease in the circulating levels of other AA, and particularly the branched-chain AA. Recently, we demonstrated that the Leu-induced stimulation of PS can be sustained up to 24 h if the Leu-induced reduction in essential AA is prevented by infusion (Wilson *et al.*, 2010). Limited evidence suggests that Leu may also suppress muscle protein degradation.

The aim of this study was to determine if administration of a Leu pulse during CON feeding can enhance PS and reduce protein degradation in muscle of the neonate.

Material and methods

Twenty 8-d-old piglets were fed by orogastric tube, for 24 h, equivalent amounts of a milk replacement formula either continuously (CON, n=6) or by bolus, i.e. every 4 h for 15 min (BOL, n=7). A third group (CON+Leu, n=7) was fed continuously and received an additional parenteral Leu infusion (800 μmol/kg/h) for 1 h every 4 h. After 20 h of infusion, blood samples were collected every 15 min for 2 h and every 1 h until 24 h of infusion. After 24 h, a flooding dose of ^{3}H-phenylalanine was injected to determine PS rates. Protein immunoblot analysis by western blot was used to identify the intracellular mechanism involved. Piglets were euthanized and muscle samples were taken at 25.25 h. Data were analyzed using one-way ANOVA and using mixed models for repeated-measures analysis.

Results and discussion

Insulin and glucose concentrations increased 15 min after the bolus meal ($P<0.02$; from 6.4±1.4 to 68.0±7.0 μU/ml and 180.4±4.3 to 216.1±4.3 mg/dl, respectively) and returned to baseline by 2 h; they were unchanged in the CON and CON+Leu groups where they averaged 10.8±3.3 μU/ml and 188.3±4.7 mg/dl, respectively. In the CON+Leu pigs, plasma Leu was higher after the Leu pulse (+206% 1 h after infusion, from 429.9±34.0 to 885.5±34.0 μM, $P<0.0005$). BOL feeding also significantly increased plasma Leu concentration (+286% 1 h after infusion, from 187.1±34.0 to 536.2±34.0 μM) with no change in the CON group (351.9±36.7 μM in average). Leu infusion during CON feeding decreased essential AA concentrations ($P<0.05$: 2572.2±142.5 vs. 1969.6±142.5 μM) compared to CON feeding. Protein synthesis rates in the *longissimus dorsi* muscle were greater ($P<0.0001$) in the CON+Leu (+24%, 14.4±0.8%/d) and BOL (+56%, 18.0±0.9%/d) groups than in the CON group (11.7±0.9%/d). The gastrocnemius ($P<0.003$) and soleus ($P<0.0001$) muscles responded similarly. There was a significant increase of PS after BOL feeding in the small intestine compared to CON and CON+Leu groups ($P<0.005$: 66.0±4.3 vs. 43.6±4.6 and 42.4±4.3%/d, respectively). No difference between groups was found on PS of the heart and liver. Phosphorylation of ribosomal protein S6 kinase 1 and 4E-binding protein 1 and formation of the active eukaryotic initiation factor (eIF) eIF4E•eIF4G complex were also higher in muscle of CON+Leu and BOL than CON suggesting increased translation initiation ($P<0.05$). There was no effect on AMP-activated protein kinase-α, eukaryotic initiation factor 2-α and eukaryotic elongation factor 2 phosphorylation. The ratio of LC3-II to total LC3 in muscle was lower in CON+Leu (0.27±0.1 arbitrary units A.U.) and BOL (0.32±0.1 A.U.) compared to CON pigs (0.70±0.1 A.U.) suggesting reduced autophagy-lysosome system activation ($P<0.05$). There were no differences between groups in indices of the ubiquitin-proteasome pathway, i.e. Atrogin-1 and MURF-1 abundance and FoxO3 phosphorylation.

In conclusion, administration of a leucine pulse during continuous orogastric feeding increases skeletal muscle protein synthesis in neonatal pigs by stimulating translation initiation. The leucine pulse also suppresses the autophagy-lysosome, but not the ubiquitin-proteasome, degradation pathways in skeletal muscle of neonatal pigs.

Acknowledgements

Supported by NIH AR-444474 and USDA/ARS 6250-51000-055

References

Anthony, J.C., Anthony, T.G., Kimball, S.R., Vary, T.C. and Jefferson, L.S. 2000. Orally administered leucine stimulates protein synthesis in skeletal muscle of postabsorptive rats in association with increased eIF4F formation. J. Nutr. 130, 139-145.

El-Kadi, S.W., Suryawan, A., Gazzaneo, M.C., Srivastava, N., Orellana, R.A., Nguyen, H.V., Lobley, G.E. and Davis, T.A. 2012. Anabolic signaling and protein deposition are enhanced by intermittent as compared with continuous feeding in skeletal muscle of neonates. Am. J. Physiol. Endocrinol. Metab. 302,E674-686.

Escobar, J., Frank, J.W., Suryawan, A., Nguyen, H.V., Kimball,S.R., Jefferson, L.S. and Davis, T.A. 2005. Physiological rise in plasma leucine stimulates muscle protein synthesis in neonatal pigs by enhancing translation initiation factor activation. Am. J. Physiol. Endocrinol. Metab. 288, E914-921.

Gazzaneo, M.C., Suryawan, A., Orellana,R.A., Torrazza, R.M., El-Kadi, S.W., Wilson, F.A., Kimball, S.R., Srivastava, N., Nguyen, H.V., Fiorotto, M.L. and Davis, T.A. 2011. Intermittent bolus feeding has a greater stimulatory effect on protein synthesis in skeletal muscle than continuous feeding in neonatal pigs. J. Nutr. 141, 2152-2158.

Suryawan, A., Orellana, R.A., Nguyen, H.V., Jeyapalan, A.S., Fleming, J.R. and Davis, T.A. 2007. Activation by insulin and amino acids of signaling components leading to translation initiation in skeletal muscle of neonatal pigs is developmentally regulated. Am. Physiol. Endocrinol. Metab. 293, E1597-1605.

Wilson, F.A., Suryawan, A., Gazzaneo, M.C., Orellana, R.A., Nguyen, H.V. and Davis. T.A. 2010. Stimulation of muscle protein synthesis by prolonged parenteral infusion of leucine is dependent on amino acid availability in neonatal pigs. J. Nutr. 140, 264-270.

Effects of copper nanoparticles on metabolic rate and development of layer embryos

L. Pineda, E. Sawosz, K.P. Vadalasetty and A. Chwalibog
University of Copenhagen, Department Veterinary Clinical and Animal Sciences, Groennegaardsvej 3, 1870 Frederiksberg, Denmark; ach@sund.ku.dk

Introduction

The poor bioavailability of Copper (Cu) has resulted in high excretion rate of Cu in the faeces of animals. Regardless of inclusion level, the major part of Cu (70-90%) is excreted and its effect on soil microorganisms, plants and aquatic species is today one of the crucial environmental concerns (Gonzales-Eguia, 2009; Zhao *et al.*, 2010).

Copper is an efficient health and growth promoter used as a feed additive for poultry. At present, there are no effective alternatives having similar growth promoting effects as Cu and the withdrawal of Cu from animal diets will cause severe health, performance and economic drawbacks in intensive poultry production.

It was hypothesized that nanoparticles of Cu (CuNano), because of high physical reactivity, may affect O_2 consumption and stimulate growth and development; and thereby can be used as an alternative health and growth promoter for animals. The objective of the study was to investigate the effects of *in ovo* injection of CuNano and timing of injection on metabolic rate (O_2 consumption and heat production, HP) and development of layer hatchlings.

Material and methods

On day 1 of incubation, 192 fertile eggs from 29 week old Lohmann breeder strain were distributed into 4 groups according to the following treatment descriptions: Group 1 – CuNanoD1, Group 2 – CuNanoD10, Group 3 – CuNanoD1+D10 and Group 4 – Control. Colloidal CuNano with 50 mg/kg concentration was introduced via *in ovo* injection. Eggs from Group 1 were injected at day 1 (D1), while eggs in Group 2 at day 10 (D10). In Group 3 eggs were injected twice at D1 and then at D10 of incubation. Eggs in Group 4 did not receive any injections, and served as the non-injected negative control. Gaseous exchange was measured in an open-air-circuit respiration unit, and HP was calculated for 16 and 19-day-old embryos. At 24 hour after hatching, chicks were euthanized and blood samples were collected to evaluate the effects of the *in ovo* CuNano injection on plasma concentration of immunoglobulins (IgG and IgM). Gene expression of nuclear factor kappa-light-chain enhancer of activated B cells (NF-kB) and tumor necrosis factor alpha (TNF-α) on the mRNA level was measured using quantitative polymerase chain reaction method. Yolk free body weight (YFBW) and the relative organ weights were used as a measure of hatchling development.

Data were analysed using GLM procedure (one-way analysis of variance) of SAS (SAS Institute Inc., 2009). The Tukey-Kramer honestly significant difference test was used to test the separation of the means at a significance level of $P<0.05$.

Results and discussion

The O_2 consumption and HP, hence, the metabolic rate was affected by the treatments. *In ovo* injection of CuNano at different days during incubation decreased metabolic rate compared with the control group ($P<0.05$). The residual yolk sac weight in the treated groups was higher than the control group ($P<0.01$), indicating that CuNano injection did not enhance oxidation of fat, which could be associated with the rates of O_2 consumption of embryos in these groups ($P<0.01$). Accordingly, the organ weights relative to YFBW were also lower in embryos injected with CuNano ($P<0.05$). Interestingly, the difference in metabolic rate and organ development between treatments was not reflected in YFBW. The result is not consistent with the reported improvement in growth performance of piglets supplemented with 50 mg CuNano/kg in the diet (Gonzales-Eguia *et al.*, 2009), which can be attributed to the differences in species and to the form of nanoparticles used.

CuNano injection did not influence the plasma concentration of IgM and IgG (both, $P>0.05$; data not shown), suggesting that it did not interact with humoral system. Furthermore, it did not affect mRNA expression of NF-kB and TNF-α (both, $P>0.05$; data not shown), indicating the absence of inflammatory response or immune-suppression properties of CuNano.

The results demonstrated that *in ovo* CuNano injection, regardless of day of injection, altered metabolic rate of embryos and depressed the development of organs, however, did not affect YFBW of hatchling, concentration of immunoglobulins and expression of immune-related genes.

Table 1. Oxygen consumption (O_2), heat production (HP), yolk-free body weight (YFBW), yolk-sac weight (YS), weights relative to YFBW of intestine, liver and heart of layer embryos in ovo injected with copper nanoparticles (CuNano) at different stages of embryonic development.[1]

Parameter	Day of injection of CuNano				SEM	*P*-value
	Control	D1	D10	D1+10		
O_2, ml/h	31.1^a	28.5^b	27.2^b	28.7^b	0.044	0.001
HP, J/h	615.7^a	568.0^b	543.1^b	567.7^b	17.49	0.002
YFBW, g	34.9	34.7	35	36.6	0.08	0.18
YS, g	3.5^a	5.7^b	6.0^b	5.3^b	0.01	<0.0001
Intestine weight, % YFBW	4.6^c	3.9^b	4.0^b	3.6^b	0.004	<0.0001
Liver weight, % YFBW	2.9^a	2.6^b	2.5^b	2.4^b	0.001	0.0002
Heart weight, % YFBW	0.7^a	0.6^b	0.6^b	0.6^b	0.004	0.002

[1] Numbers in a row with a different superscript differ, $P\leq0.05$.

References

Gonzales-Eguia, A., C.-M. Fu, F-Y. Lu and T.F. Lien, 2009. Effects of nanocopper on copper availability and nutrients digestibility, growth performance and serum traits of piglets. Livestock Sci. 126, 122-129.

SAS Institute Inc. 2009. SAS1 Procedure Guide. Version 9.2. Cary (NC): SAS Institute Inc.

Zhao, J., R.B. Shirley, M. Vazquez-Anon, J.J. Dibner, J.D. Richards, P. Fisher, T. Hampton, K.D. Christensen, J.P. Allard and A.F. Giesen, 2010. Effects of chelated trace minerals on growth performance, breast meat yield, and footpad health in commercial meat broilers. J. Applied Poultry Res. 19, 365-372.

Atrogin-1, a muscle-specific ubiquitin ligase, is highly expressed in the smooth muscle of the chicken gizzard

K. Nakashima, A. Ishida and M. Katsumata
NARO Institute of Livestock and Grassland Science, Tsukuba 305-0901, Japan; kaznaka@affrc.go.jp

Introduction

Muscle proteolysis in catabolic conditions is due primarily to activation of the ubiquitin-proteasome proteolytic pathway, whereby the proteins destined to be degraded are linked to a chain of ubiquitin molecules, which targets them of rapid breakdown by the proteasome. Evidence suggests that atrogin-1, an E3 ubiquitin ligase also referred to as MAFbx (muscle atrophy F-box), plays a pivotal role in muscle atrophy (Gomes *et al.*, 2001). Atrogin-1 plays a critical role in development of muscle proteolysis and its gene expression is a reliable index of muscle proteolysis (Ohtsuka *et al.*, 2011).

Atrogin-1 mRNA is expressed in smooth muscle, and its gene expression is increased in the smooth muscle of uterine of the *postpartum* period and in the smooth muscle of intestine of feed deprivation. In chicken, the gizzard is a characteristic avian smooth muscle sac functioning to crush the feed and begin in digestion of the proteins. However, whether atrogin-1 is expressed in the smooth muscle of the chicken gizzard has yet to be investigated.

We previously reported that the expression of atrogin-1 mRNA was increased by fasting, and decreased by refeeding in the skeletal muscle of chicken (Nakashima *et al.*, 2006). However, regulation of atrogin-1 expression in the smooth muscle of the chicken gizzard by food deprivation and nutritional supply has yet to be defined. Thus, in the present study, the effects of fasting and refeeding on the mRNA level of atrogin-1 in the smooth muscle of the chicken gizzard were investigated.

Material and methods

At 11 days of age, four chicks with body weights of 114±1.31 (mean ± SD) were subjected to tissue sampling (skeletal muscle, heart, gizzard, brain and liver) to established tissue distribution of atrogin-1 mRNA.

At 11 days of age, another eighteen chicks of similar body weight were selected and housed in wire-bottomed aluminum cages. They were given free access to a commercial starter diet and water for 3 d. At the start of the experiment, 14-day-old chicks were divided into three groups (n=6/group): fed group, food-deprived group and refed group. Fed chicks were continuously given free access to the diet for 3 days before being killed. Food-deprived chicks subjected to food deprivation for 24 h before they were killed on day 15. Refed chicks subjected to food deprivation for 24 h and then refed for 2 h before being sacrificed on day 16. The tissue samples were frozen with liquid nitrogen and stored at -80 °C until analysis. The mRNA level of atrogin-1 was measured by a real-time RT-PCR method.

Data were analyzed by Paired Student's *t*-test to evaluate tissue distribution of atrogin-1 mRNA or ANOVA and Tukey's multiple comparison test to determine the effect of feeding regime on atrogin-1 mRNA level in smooth muscle. A *P* of value <0.05 was considered statistically significant. Each result is expressed as the mean ± standard deviation (SD). Gene expression levels were estimated on the basis of PCR efficiency and threshold cycle deviation of an unknown sample versus a control. 18S ribosomal RNA was chosen as reference gene.

Results and discussion

As shown in Table 1, the expression of atrogin-1/MAFbx mRNA in the smooth muscle of the gizzard is higher than that of skeletal muscle and heart in chicken.

Although gizzard weight (g) was significantly decreased by fasting (fed, 6.45±0.26 g versus fasted, 5.99±0.37 g, $P<0.05$), refeeding failed to restore these tissues weight to the fed level within 2 h (refed, 5.69±0.28 g). In the smooth muscle of the gizzard, expression of atrogin-1 mRNA was significantly increased by fasting, whereas refeeding significantly suppressed it (Table 2).

In conclusion, we demonstrated here that the smooth muscle of the chicken gizzard expresses high levels of atrogin-1 gene and its expression is associated with the muscle atrophy that occurs during food deprivation.

Table 1. Tissue distribution of chicken atrogin-1 mRNA in 11-day-old chicken.

Tissue	Relative expression
Skeletal muscle	1.00±0.49
Heart	1.22±0.43
Gizzard	13.0±2.91**
Brain	0.36±0.11*
Liver	0.12±0.03**

Results are expressed as ratios to relative to the 18S rRNA levels in skeletal muscle, whose expression level was taken to be equal to 1. Value are means±SD (n=4). Symbols: *$P<0.05$; **$P<0.01$.

Table 2. Effects of fasting and refeeding on the mRNA levels of atrogin-1 in the smooth muscle (gizzard) of chickens.

Treatment	Relative expression
Fed	1.00±0.28^{b}
Fasted	4.75±1.16^{a}
Refed	0.61±0.26^{b}

Results of RNA quantification are expressed as ratios to relative to the 18S rRNA levels in fed chickens, whose expression level was taken to be equal to 1. Values are means±SD (n=6).
^{a,b} $P<0.05$.

References

Gomes M.D., S.H. Lecker, R.T. Jagoe, A. Navon and A.L. Goldberg, 2001. Atrogin-1, a muscle-specific F-box protein highly expressed during muscle atrophy. Proc. Natl. Acad. Sci. USA, 98, 14440-14445.

Nakashima K., Y. Yakabe, M. Yamazaki and H. Abe, 2006. Effects of fasting and refeeding on expression of atrogin-1 and Akt/FOXO signaling pathway in skeletal muscle of chicks. Biosci. Biotechnol. Biochem., 70, 2775-2778.

Ohtsuka A., N. Kawatom, K. Nakashima, T. Araki and K. Hayashi, 2011. Gene expression of muscle-specific ubiquitin ligase, atrogin-1/MAFbx, possitively correlates with skeletal muscle protelysis in food-deprivated broiler chickens. J. Poult. Sci., 48, 92-96.

An *in ovo* ^{13}C-tracer approach to explore liver intermediary metabolism in developing chicken embryos

Q. Hu, U. Agarwal and B.J. Bequette
Animal and Avian Sciences Department, University of Maryland, College Park, MD 20742, USA; bbequett@umd.edu

Introduction

The nutrient components of the egg are finite, thus the chicken embryo must properly coordinate the utilization of these substrates to support developmental requirements. However, the pathways that the macronutrients are allocated towards and the relative contributions the substrates make to the various pathway fluxes at the tissue and whole-body levels of the embryo are yet to be defined. The commercial availability of stable isotope tracers enables researchers to explore the intricacies of individual as well as connecting metabolic pathways networks. The objective of the present study was to employ a constant infusion protocol for the *in ovo* delivery of [$^{13}C_6$]glucose to attain steady-state labeling and high rates of incorporation of ^{13}C into intermediates of gluconeogenesis-glycolysis and the Krebs cycle for accurate measurement of ^{13}C-positional isotopomers for flux analysis.

Material and methods

A sterile solution of [$^{13}C_6$]glucose (7 mg/50 mg saline) was constantly infused (45 mg solution/h) over 8 h into the chorio-allantoic compartment of chicken eggs on embryonic (e) day 14 and 19 (n=4). At the end of infusion, blood was withdrawn from a vitelline vessel, and the whole liver rinsed with ice-cold saline and frozen for analysis. Glucose was extracted from blood with ice-cold acetone and converted to the isopropylidene pentaacetate derivative prior to selected ion monitoring (SIM) by gas chromatography-mass spectrometry (GC-MS) under electron impact mode (Sunny and Bequette, 2010). Liver was homogenized in ice-cold sulfosalicylic acid (10% w/v), amino acids isolated by cation-exchange chromatography and converted to the *t*-butyldimethylsilyl derivitive prior to SIM by GC-MS under electron impact mode. Mass fragmentology analysis of aspartate and glutamate was conducted because these amino acids are in metabolic equilibrium with their corresponding intermediary metabolite oxaloacetate and α-ketoglutarate, respectively. Gluconeogenesis was calculated according to Haymond and Sunehag (2000) with modifications and Krebs cycle and non-essential amino acids (NEAA) fluxes according to Bethold *et al.* (1994). Results were analyzed by using student's *t*-test of SAS 9.2 (SAS Institute, Cary, NC, USA). Probability of $P<0.05$ was considered to be statistically significant.

Results and discussion

Employing the constant infusion approach, Cori cycling, i.e. glucose carbon recycling, tended to be higher in e14 embryos (Table 1) which is similar to earlier work from our lab using a daily injection approach (Sunny and Bequette, 2010). Glucose entry, gluconeogenesis and glycogen turnover were all greater in e14 compared to e19 embryos, which suggests that embryos at later stages of development conserve more glycogen for post-hatch energy requirements.

There was significant incorporation of ^{13}C glucose carbon into many NEAA (aspartate, glutamate, glutamine, serine, glycine and alanine) in the liver indicating the capacity of the embryo to synthesize these amino acids. Importantly, by e11, there is only ~125 mg of preformed carbohydrates remaining in the residual yolk and albumen (Hu *et al.*, 2011), suggesting that non-glucose sources such as glycerol (yolk) and amino acids (albumen) must be significant contributors to NEAA synthesis. Given the high degree of labeling of aspartate and glutamate, it was possible to calculate the proportion of the Krebs cycle oxaloacetate and acetyl-CoA pools in the liver derived from the 3-carbon pool

Table 1. Blood glucose fluxes in e14 and e19 chicken embryos.

		e14	e19	SEM	*P*-value
Cori cycling	(%)	51.1	29.2	6.4	0.09
Fractional gluconeogenesis		0.49	0.32	0.06	NS[1]
Glucose entry	(mg/g BW/d)	21.1	7.2	3.4	<0.05
Gluconeogenesis	(mg/g BW/d)	10.75	2.53	2.15	<0.05
Glycogen turnover	(mg/g BW/d)	10.35	4.70	1.27	<0.01

[1] NS: not significant (P>0.10).

(eg. pyruvate). For the oxaloacetate pool, which is a major anaplerotic reaction to supply carbon skeletons for NEAA synthesis and to maintain Krebs cycle fluxes, the contribution from the 3-carbon pool ranged from 7.5 to 10% (SEM 1.8%) and did not differ between e14 and e19. Thus, substrates that are metabolized via other routes into the Krebs cycle are more significant precursors for NEAA synthesis. For the acetyl-CoA pool, which is the major oxidation route and precursor for fatty acid synthesis, the contribution from the 3-carbon pool ranged from 8.8 to 15% (SEM 3.4%) and did not differ between e14 and e19. Hence, fatty acids are the major oxidative substrates.

In conclusion, this tracer approach employing [$^{13}C_6$]glucose allow for the measurements of Krebs cycle fluxes and activity, gluconeogenesis-glycolysis, glycogen turnover and identification of potential precursors derived from yolk and albumen that contribute to oxidation and NEAA synthesis. The results generated from this study will also lead to a better understanding of the nutritional adequacy provided by the egg contents and their influence on embryo growth and survival.

References

Bethold, H. K., L. J. Wykes, F. Jahoor, P. D. Klein, and P. J. Reeds, 1994. The use of uniformly labeled substrates and mass isotopomer analysis to study intermediary metabolism. Proc. Nutr. Soc. 53, 345-354.

Haymond, M.W. and A. L. Sunehag, 2000. The reciprocal pool model for the measurement of gluconeogenesis by use of [U-13C] glucose. Am. J. physiol. Endocrinol. Metab. 278: E140-E145.

Hu, Q., U. Agarwal, K.R. Somers, K.M. Bailey and B. J. Bequette, 2011. Energy balance regulation and carbohydrate utilization in developing chicken embryos. FASEB J. 25: 774.11.

Sunny, N. E. and B. J. Bequette, 2010. Gluconeogenesis differs in developing chick embryos derived from small compared with typical size broilder breeder eggs. J. Anim. Sci. 88:912-921.

The high-fat diet and flaxseed cake influenced on lipid metabolism in mice selected for body weight

M. Matusiewicz[1], S. Fiedorowicz[1], K. Fiszdon[2], I. Kosieradzka[1] and W. Bielecki[3]
[1]Department of Animal Nutrition and Feed Science, Faculty of Animal Science, Warsaw University of Life Science – SGGW, Ciszewskiego 8, 02-786 Warsaw, Poland; magdalena_matusiewicz@sggw.pl
[2]Department of Genetics and Animal Breeding, Faculty of Animal Science, Warsaw University of Life Science – SGGW, Ciszewskiego 8, 02-786 Warsaw, Poland
[3]Department of Pathology and Veterinary Diagnostics, Faculty of Veterinary Medicine, Warsaw University of Life Science – SGGW, Nowoursynowska 159c, 02-776 Warsaw, Poland

Introduction

It is becoming increasingly evident that consumption of high-fat diet rich in saturated fatty acids results in negative effect on lipid metabolism and red-ox state. Opposite effect was confirmed for flaxseeds that have a number of health benefits result from the content of nutrients including oil rich in polyunsaturated fatty acids, dietary fiber and polyphenolic compounds. Linola cultivar seeds were characterized by reduced content of α-linolenic acid, greater stability and oxidation resistance. The objective of the study was to evaluate the influence of high-fat diet rich in lard and addition of flaxseed cake on serum red-ox state and lipid profile, concentration of liver adiponectin – important cytokine secreted by fat tissue cells as well as effect on the development of the perirenal adipose tissue in mice selected for body weight. The selection had resulted in differences in serum lipid profile and antioxidant state.

Material and methods

Experiment was carried out on male mice that were divergently selected over 130 generations for high (line C) and low (line L) body weight at the postnatal day 21. These lines of animals were derived from outbred stock by crossing four inbred strains: A/St, BN/a, BALB/c and C57BL/6Jn.

The experimental animals were divided into 6 groups with 10 individuals in each, and fed (*ad libitum*) for 96 days one of the isoprotein diets: (1) line C and (2) line L with standard diet, (3) line C and (4) line L with high-fat diet rich in pork lard, (5) line C and (6) line L with diet supplemented with 30% of flaxseed cake (Table 1). Serum total antioxidant status (TAS), lipid peroxides in serum measured as thiobarbituric acid reactive substances (TBARS), total cholesterol (TC) and triglycerides (TG) were detected by the spectrophotometric method. Concentration of adiponectin in liver was determined by enzyme-linked immunosorbent assay.

Results and discussion

Mice fed diet that contains flaxseed cake demonstrated significantly higher level of serum TAS, comparing to animals fed standard diet (Table 2). The TAS was also significantly higher in line L of mice than in line C. Concentration of serum TBARS was significantly lower in flaxseed cake group than in high-fat diet. Feeding flaxseed cake resulted in decreased concentration of serum TC and TG compared to high-fat diet. Line L of mice was characterized by significantly lower TC and TG compared to line C. Two-way ANOVA revealed the interaction in liver adiponectin between diet and mice line. Interaction between diet and mice line demonstrated that feeding SD to line L increased ADP, while feeding HFD or flaxseed cake diet to line L reduced ADP. Visualization of perirenal adipose tissue seemed to appear similar in all groups of mice.

Addition of flaxseed cake to the high-fat diet influenced profitably on serum red-ox state and examined parameters of serum lipid metabolism in mice. Line L was characterized by better values of that parameters than line C. The beneficial effect of diets on liver adiponectin concentration is greater in line L.

Table 1. Composition of experimental diets and their nutritional value (% DM).[1]

Component	Group diet		
	Standard	High-fat	Flaxseed cake
Flaxseed cake (%)	-	-	30
Total protein	20.0	20.1	19.7
Crude fat	2.8	12.1	11.6
GE (kcal/100 g)	465	515	512
GE from crude fat (kcal/100 g)	27	115	110

[1] Perirenal adipose tissue was visualized by scanning electron microscopy (FEI Quanta 200). The results were analyzed by two-way ANOVA with means compared using Tukey's correction, a difference of $P<0.05$ between means was considered to be significant.

Table 2. Two-way ANOVA (diet-mice line) for examined parameters of mice (mean±SEM).[1]

Factor	Red-ox state		Lipid profile		Adiponectin (ng/mg protein)
	TAS (mmol/l)	TBARS (nmol/ml)	TC (mmol/l)	TG (mmol/l)	
Diet	*	*	*	*	*
SD	1.91±0.13[a]	0.28±0.03[ab]	1.54±0.11[a]	1.34±0.18[a]	203.65±47.88
HFD	2.12±0.13[ab]	0.38±0.03[b]	2.64±0.11[b]	1.90±0.18[b]	390.98±47.88
Flaxseed cake	2.51±0.13[b]	0.27±0.03[a]	1.91±0.11[a]	1.29±0.18[a]	323.77±47.88
Mice line	*	NS	*	*	NS
C	2.00±0.10	0.32±0.02	2.29±0.09	1.77±0.15	352.52±39.09
L	2.36±0.10	0.30±0.02	1.78±0.09	1.25±0.15	259.74±39.09
Interaction	NS	NS	NS	NS	*
SD C	1.69±0.18	0.29±0.04	1.77±0.16	1.27±0.25	148.80±67.71
SD L	2.12±0.18	0.27±0.04	1.31±0.16	1.40±0.25	258.50±67.71
HFD C	1.80±0.18	0.38±0.04	2.83±0.16	2.37±0.25	529.60±67.71
HFD L	2.45±0.18	0.39±0.04	2.46±0.16	1.43±0.25	252.37±67.71
Flaxseed cake C	2.50±0.18	0.30±0.04	2.27±0.16	1.68±0.25	379.17±67.71
Flaxseed cake L	2.52±0.18	0.25±0.04	1.56±0.16	0.91±0.25	268.37±67.71

[1] * significant effect ($P<0.05$); NS = non-significant; [a,b,c] values within a factors with different superscripts differ significantly ($P<0.05$).

Branched-chain α-keto acids in plasma of growing chicken: when is the time for blood sampling?

A. Pastor, A. Sünder and F. Liebert
Division Animal Nutrition Physiology, Department of Animal Sciences, Georg-August-University, Kellnerweg 6, 37077 Goettingen, Germany; apastor@gwdg.de

Introduction

Complementary to quantitative requirement studies for the branched-chain amino acids (BCAA) leucine, isoleucine and valine, studies on dynamics of α-keto acids yielded by degradation of the BCAA could provide important additional metabolic insights. Generally, the first step in catabolism of the BCAAs is undertaken by the same enzyme (branched-chain amino acid transferase; Harper *et al.*, 1984) and yields the corresponding branched-chain α-keto acids (BCKA): α-ketoisocaproic acid (KIC), α-ketomethylvaleric acid (KMV) and α-ketoisovaleric acid (KIV), for leucine, isoleucine and valine, respectively. In human beings, a deficiency of this enzyme may create a buildup of the BCAA and their corresponding BCKA, leading to the maple syrup urine disease (Langenbeck *et al.*, 1978). Therefore, determination of the BCKA concentration in blood plasma is a standard procedure in humans. However, for growing chicken, guidelines are not available to indicate the optimal time schedule for sampling. The current study was conducted both to establish a method to detect BCKA in chicken plasma and to yield information about the time course of BCKA in blood from the jugular vein for an improved understanding of the BCAA metabolism in growing chicken.

Material and methods

Eight groups (n=5) of 37-day old male meat type chicken (Ross 308) were fasted overnight (12 hours), subsequently getting access to feed for half an hour. Afterwards, diets were removed and blood samples were taken from the *vena jugularis dextra* at 1, 3, 5, 9, 11, 12 and 23 hours after feed removal. The diet fed met the National Research Council (1994) requirements. Blood samples were centrifuged to yield blood plasma, which was stored at -80 °C until further analysis.

Analysis of BCKA was carried out by high performance liquid chromatography (HPLC), adapted from Kandár *et al.* (2009) for human beings.

Statistical analyses ($P \leq 0.05$) depending on time were carried out using the SPSS statistical software package (version 19.0 for Windows, IBM SPSS Statistics, Inc. Chicago, IL). Data were subjected to a verification of variance homogeneity according to Levene test following Tukey or Games-Howell post-hoc tests.

Results and discussion

The method of Kandár *et al.* (2009) was successfully adapted for determination of BCKA in chicken plasma, making use of several modifications (e.g. concentration of internal standard, injection volume, calibration line). The observed time course of the BCKA related to food deprivation (h) is demonstrated in Figure 1.

Three h after feed removal, the KIC and KIV concentration in blood plasma decreased significantly ($P<0.05$) compared to a feed deprivation of 1 h, but remained relatively stable until 11 h after feed removal. Twelve and 23 h after feed deprivation, only KIC concentration increased significantly compared to 3 h after feed removal, suggesting enhanced catabolism of Leu. For KMV concentration, a significant effect ($P<0.05$) was observed only between 3 h and 12 h of fasting. Accordingly, Leslie and Saunderson (1985) reported the highest BCKA concentration in blood 1 h after feeding.

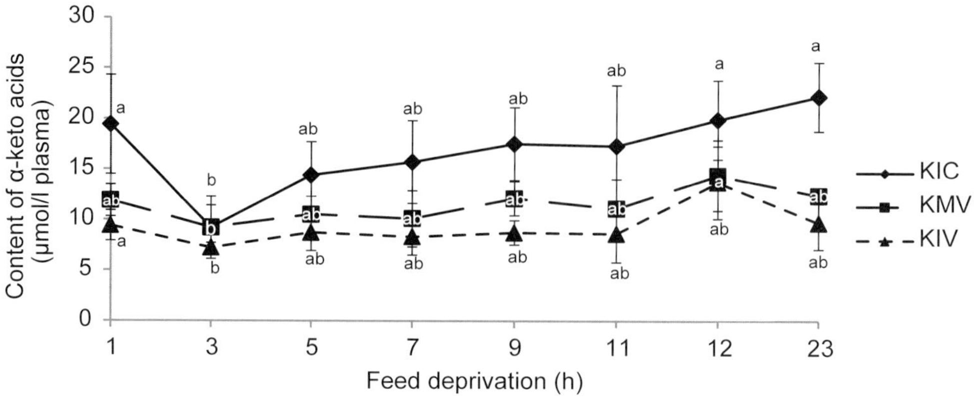

Figure 1. Time course of the branched-chain α-keto acids depending on time of feed deprivation (0 h); values with a different superscript differ within a same line (P<0.05)

It is concluded, that 5 h to 11 h after feeding is the preferable time for blood sampling in meat type chicken for further analysis of BCKA. Within this timeframe, concentration of BCKA demonstrated low variation. Ongoing experiments study effects of varying dietary BCAA supply on BCKA concentration in blood plasma to improve the understanding of BCAA related metabolic processes and interrelationships to the modeling of AA efficiency based requirements in growing chicken.

References

Harper, A., R. Miller and K. Block, 1984. Branched-chain amino acid metabolism. Ann. Rev. Nutr. 4, 409-454

Kandár, R., P. Žáková, J. Jirošová and M. Sladká, 2009. Detemination of branched-chain amino acids, methionine, phenylalanine, tyrosine and α-keto acids in plasma and dried blood samples using HPLC with flourescence detection. Clin. Chem. Lab. Med. 47, 565-572.

Langenbeck, U., U. Wendel, A. Mench-Hoinowski, D. Kuschel, K. Becker, H. Przyrembel and H. J. Bremer, 1978. Correlations between branched-chain amino acids and branched-chain α-keto acids in blood in maple syrup urine disease. Clin. Chim. Acta. 88, 283-291.

Leslie, S. and C.L. Saunderson, 1985. Measurement of branched-chain α-keto acids in the blood plasma pf the domestic fowl and poultry. Comp. Biochem. Physiol. 80B, 99-102.

National Research Council, 1994. Nutrient Requirements of Poultry. 9th Rev. ed. Natl. Acad. Press, Washington, D.C.

Lean accretion and protein turnover are enhanced by intermittent bolus feeding in neonatal pigs

S.W. El-Kadi[1,2], C. Boutry[1], A. Suryawan[1], M.C. Gazzaneo[1], R.A. Orellana[1], N. Srivastava[1], H.V. Nguyen[1], M.L. Fiorotto[1] and T.A. Davis[1]
[1]USDA/ARS Children Nutrition Research Center, Pediatrics, Baylor College of Medicine, Houston, TX, USA; elkadi@vt.edu
[2]Department of Animal and Poultry Sciences, Virginia Tech, Blacksburg, VA, USA

Introduction

Orogastric tube feeding is indicated in neonates with impaired ability to ingest food normally and can be administered by intermittent bolus (INT) or continuous (CON) infusion. Insulin and amino acids play important roles in the regulation of protein synthesis in the neonate. While the sensitivity to insulin is developmentally regulated (Davis *et al.*, 1998) and greatly diminishes with age (Wray-Cahen *et al.*, 1997), amino acids exert a positive effect on muscle protein synthesis throughout life (Denne *et al.*, 1991; Davis *et al.*, 1998; Volpi *et al.*, 1998). In addition to providing amino acids to other organs when dietary supplies are insufficient, high rates of proteolysis are necessary to provide amino acids for ongoing tissue modeling and rapid growth. The two most important proteolytic pathways in skeletal muscle are thought to be the ubiquitin-proteasome and autophagy-lysosome systems (Ventadour and Attaix, 2006). The aim of this study was to determine if these feeding modalities affect growth and lean tissue accretion, and the mechanisms for this response.

Material and methods

Neonatal pigs (n=6, 2-3-d-old, 2.3±0.07 kg BW) were surgically fitted under general anesthesia using sterile techniques with catheters in the carotid artery, jugular vein, and gastrostomy tube. Pigs were assigned to treatment using completely randomized design and fed the same sow milk replacer at 240 ml/kg/d in equivalent amounts either continuously (CON; 10 ml/kg/h) or intermittently (INT; meal given over 15 min at 4 h intervals; 40 ml/kg/bolus) for 21d. Diets supplied 12.8 g/kg/d crude protein and 175 kcal/kg/d. Dual X-ray absorptiometry was used before initiation of the study and 18 d after feeding to determine body composition and linear growth. Feed conversion was calculated at the end of the study based on feed intake and protein and energy retention. Plasma samples, taken over 4 h during the last day of feeding were analyzed for glucose and insulin concentrations as described previously (El-Kadi *et al.*, 2012). Fractional protein synthesis was determined using the flooding dose method (Davis *et al.*, 1999). Pigs were euthanized 1 h following the last meal for the INT group, and at the same time in the CON group. Muscle samples were removed and frozen at -70 °C until analysis. Translation initiation and protein degradation signaling were determined by polyacrylamide gel electrophoresis (El-Kadi *et al.*, 2012). Data were analyzed using the MIXED procedure of SAS (version 8.0). To investigate temporal effects, all samples were considered in a repeated measures analysis. Values presented are means ± SEM.

Results and discussion

We recently demonstrated that 24 h after the initiation of feeding, protein synthesis was enhanced by intermittent bolus feeding compared to continuous feeding with no change in protein degradation (El-Kadi *et al.*, 2012). Whether this acute response is maintained over long-term feeding periods has not been determined.

Weight gain was greater for INT than for CON pigs and resulted in heavier body weights from 9 until 21 d of feeding (5.6 vs. 4.2±0.12 kg) ($P<0.05$). The efficiency of protein and energy retentions were enhanced by 30 and 44% in INT compared to CON fed pigs ($P<0.05$). Lean tissue mass (4.75

vs. 3.53±0.12 kg) and spine length (28 vs. 30±0.4 cm) were greater in INT than CON pigs ($P<0.05$). Glucose and arterial insulin levels measured on the last day of feeding were greater for INT after the meal than for CON pigs ($P<0.05$). Muscle protein synthesis increased by 40% in *longissimus dorsi* (10.6 vs. 7.6±1.41%/d) and 34% in gastrocnemius (9.5 vs. 7.1±0.60) and soleus (11.1 vs. 8.2±0.62) of the INT as compared to the CON group ($P<0.05$).

To establish the mechanism for this response, translation initiation and protein degradation signaling were studied in *longissimus dorsi* muscle. Insulin receptor and insulin receptor substrate-1 expression and protein kinase B phosphorylation in skeletal muscle were unaffected by feeding modality. Formation of the active eukaryotic initiation factor (eIF) 4E-eIF4G complex and phosphorylation of ribosomal protein S6 kinase were higher ($P<0.05$) and phosphorylation of eIF2α was lower ($P<0.05$) in INT compared to CON fed pigs indicating increased translation initiation. Sodium-coupled neutral amino acid transporter 2 expression ($P<0.05$), but not L-type amino acid transporter 1, was higher in INT compared to CON pigs suggesting enhanced glutamine transport. Abundance of muscle-specific ubiquitin ligases, muscle RING finger-1 and F-box protein atrogin-1/MAFbx, was higher for INT compared to CON pigs ($P<0.05$).

These results suggest that intermittent feeding enhances lean tissue accretion as compared to continuous feeding, and that this difference is apparent by 9 d and persists for the duration of feeding. The increased rate of lean tissue accretion in pigs fed intermittently occurs in response to an increased activation of translation initiation. In addition, intermittent feeding enhances lean tissue accretion by increasing amino acid transport and protein turnover.

Acknowledgements

Supported by NIH AR444474 and USDA/ARS 6250-51000-055

References

Davis, T.A., D.G. Burrin, M.L. Fiorotto, P.J. Reeds and F. Jahoor, 1998. Roles of insulin and amino acids in the regulation of protein synthesis in the neonate. J. Nutr. 128, 347S-350S.

Davis, T.A., M.L. Fiorotto, H.V. Nguyen and D.G. Burrin, 1999. Aminoacyl-tRNA and tissue free amino acid pools are equilibrated after a flooding dose of phenylalanine. Am. J. Physiol. Endocrinol. Metab. 277, E103-109.

Denne, S.C., E.M. Rossi and S.C. Kalhan, 1991. Leucine kinetics during feeding in normal newborns. Ped. Res. 30, 23-27.

El-Kadi, S.W., A. Suryawan, M.C. Gazzaneo, N. Srivastava, R.A. Orellana, H.V. Nguyen, G.E. Lobley and T.A. Davis, 2012. Anabolic signaling and protein deposition are enhanced by intermittent compared with continuous feeding in skeletal muscle of neonates. Am. J. Physiol. Endocrinol. Metab. 302, E674-686.

Ventadour, S. and D. Attaix, 2006. Mechanisms of skeletal muscle atrophy. Curr. Opin. Rheumatol. 18, 631-635.

Volpi, E., A.A. Ferrando, C.W. Yeckel, K.D. Tipton and R.R. Wolfe, 1998. Exogenous amino acids stimulate net muscle protein synthesis in the elderly. J. Clin. Invest. 101, 2000-2007.

Wray-Cahen, D., P.R. Beckett, H.V. Nguyen and T.A. Davis, 1997. Insulin-stimulated amino acid utilization during glucose and amino acid clamps decreases with development. Am. J. Physiol. Endocrinol. Metab. 273, E305-E314.

Changes in tissue amino acid composition and protein metabolism in piglets due to a limiting supply of total sulphur amino acids

J.A. Conde-Aguilera[1,2], C. Cobo-Ortega[1,2], N. Le Floc'h[1,2], Y. Mercier[3] and J. van Milgen[1,2]
[1]INRA, UMR1348 PEGASE, 35590 Saint-Gilles, France; jaap.vanmilgen@rennes.inra.fr
[2]Agrocampus Ouest, UMR1348 PEGASE, 35000 Rennes, France
[3]Adisseo France SAS, 92160 Antony, France

Introduction

In growing pigs, amino acid (AA) requirements are usually estimated using body weight gain or N retention as response criteria which imply that the AA composition of total body protein should be constant. However, there are indications that the AA composition of body protein is affected by the protein, energy or AA supply. A deficient total sulphur AA (TSAA; Met+Cys) supply has been shown to modify the AA profile of tissue proteins (Conde-Aguilera *et al.*, 2010). Protein retention results from the rate of protein synthesis exceeding that of protein breakdown. Whether one or both of these processes are affected by a limiting AA supply to preserve the metabolic needs of the animal remains to be explored. The objective of this study was to evaluate the response of protein metabolism to a diet deficient (TSAA-) or sufficient (TSAA+) in TSAA supply by assessing the protein synthesis, proteasome enzyme activity, and the AA composition of different tissues and organs in piglets.

Material and methods

Twelve Piétrain × (Large White × Landrace) piglets were used 14 d after weaning (at 42 d of age) and were fitted with a jugular catheter. They were divided in 2 groups and received either TSAA- or TSAA+ for 10 d. Diets were formulated to meet the nutritional recommendations except for the Met and TSAA supplies in TSAA-, which were 31 and 28% below requirements, respectively. The daily feed allowance was about 25% below the *ad libitum* intake capacity. To measure protein synthesis, piglets were injected with a flooding dose of [1-^{13}C] l-Val and were slaughtered 14 min thereafter. Samples of the liver, kidneys, *longissimus dorsi* (LM), *rhomboideus* (RM) and *semitendinous* (SM) muscles, intestines, kidneys, and skin were collected for analysis.

Results and discussion

The deficient TSAA supply decreased weight gain ($P<0.001$), the protein synthesis rate in LM by 32% ($P<0.01$) and in SM by 17% ($P=0.10$; Table 1). The RNA efficiency was reduced by 40 and 22% for LM and SM, respectively ($P\leq0.03$). Moreover, SM proteasome activity was greater (19%) in piglets receiving diet TSAA- ($P<0.05$). Compared with pigs receiving diet TSAA+, the relative weights of LM and distal jejunum were, respectively lower ($P=0.09$) and higher ($P<0.01$), the Met content of tissue protein decreased in both SM ($P=0.09$) and LM ($P<0.05$), and was greater in the liver ($P<0.05$) in piglets offered diet TSAA- (Table 2). Also, the Cys content was lower in the liver ($P<0.05$) but higher in the distal jejunum and ileum ($P<0.05$) in TSAA- pigs, which may reflect an adaptation of TSAA metabolism to maintain glutathione and tissue redox status (Richie *et al.*, 2004). Regarding the contents of other AA in tissue protein, the responses were minimal for the proximal jejunum, ileum, kidneys and skin, but more important for LM, SM, RM, distal jejunum, and liver. Changes in the AA profile of proteins may be associated with changes in the nature of protein deposited (Conde-Aguilera *et al.*, 2010). In conclusion, the animal shows certain flexibility in response to a limiting TSAA supply, questioning the use of a constant AA profile to establish AA requirements. The functional consequences of these changes remain to be studied.

Table 1. Performance and protein metabolism in piglets offered diet TSAA- or TSAA+.

		TSAA-	TSAA+	RSD	*P*-value
Final BW (kg)		12.3	13.0	1.2	0.24
Weight gain (g/d)		292	375	36	<0.001
Feed conversion (g gain/g feed)		0.78	0.99	0.10	<0.01
Protein synthesis (%/d)	LM	4.33	6.35	0.90	<0.01
	RM	4.20	3.85	0.63	0.39
	SM	4.11	4.95	0.72	0.10
Efficiency of protein synthesis[1]	LM	2.95	4.91	0.46	<0.001
	RM	4.38	4.06	0.62	0.43
	SM	2.98	3.80	0.50	0.03
Proteasome activity[2]	LM	2,170	1,959	384	0.39
	RM	2,741	2,788	513	0.88
	SM	2,284	1,914	265	0.05

[1] Expressed as g of protein synthesized per day per mg of RNA.

[2] Expressed as relative fluorescence units per min per mg of protein.

Table 2. Weight and composition of weight in piglets offered diet TSAA- or TSAA+[1].

Tissue	Diet	Weight (g)	Weight (g/kg BW)	Protein (g/kg BW)	g/16 g N							
					Lys	Met	Cys	Thr	Val	Ile	Leu	His
SM	TSAA-	43.3	3.45	165	8.68	2.78#	1.25	4.57	5.16	4.77	8.18	4.11**
	TSAA+	44.5	3.43	162	8.79	3.02	1.23	4.52	5.21	4.80	8.37	3.45
LM	TSAA-	194	15.4#	176	8.90	2.60*	1.17	4.44	5.00#	4.57	7.86	3.99**
	TSAA+	222	17.1	176	9.86	2.91	1.15	4.58	5.18	4.72	8.00	3.52
RM	TSAA-	5.22	0.42	156	8.03#	2.50	1.26	4.37	4.88*	4.33*	7.75	2.97**
	TSAA+	5.45	0.42	154	8.27	2.60	1.25	4.34	5.07	4.51	7.82	2.82
Liver	TSAA-	308	24.6	174	7.29	2.50*	1.85*	4.55*	5.72**	4.48**	8.61*	2.85*
	TSAA+	313	24.1	169	7.04	2.28	2.00	4.10	5.41	4.24	8.14	2.68
Proximal	TSAA-	148	11.7	135	6.85	2.14	1.44	4.43	4.89	3.93	7.64	2.27
jejunum	TSAA+	146	11.2	131	6.99	2.20	1.49	4.24	4.85	3.93	7.60	2.34
Distal	TSAA-	167	13.2**	126	7.58	2.40	1.58*	4.62#	5.13	4.20	7.95	2.52
jejunum	TSAA+	141	10.8	122	7.45	2.37	1.40	4.41	5.07	4.17	7.85	2.46
Ileum	TSAA-	169	13.5	131#	7.33	2.21	1.40*	4.19	4.73	3.89	7.26	2.19
	TSAA+	162	12.4	119	7.36	2.17	1.35	4.10	4.71	3.92	7.17	2.27
Kidneys	TSAA-	69.2	5.55	136	7.03	2.27	1.60	4.31	5.33	4.11	8.14	2.58
	TSAA+	70.4	5.45	135	6.97	2.21	1.65	4.34	5.38	4.19	8.15	2.68
Skin	TSAA-	35.1	2.67	194	4.45	1.47	1.36	2.71	3.45	2.18	4.54	2.01
	TSAA+	34.1	2.50	186	4.40	1.53	1.19	2.75	3.55	2.13	4.55	2.38

[1] Significant level within a column and tissue by $*P \leq 0.05$, $**P \leq 0.01$, $***P \leq 0.001$, $\#P \leq 0.10$.

References

Conde-Aguilera, J. A., R. Barea, N. Le Floc'h, L. Lefaucheur and J. van Milgen, 2010. A sulfur amino acid deficiency changes the amino acid composition of body protein in piglets. Animal 4, 1349-1358.

Richie, J. P., D. Komninou, Y. Leutzinger, W. Kleinman, N. Orentreich, V. Malloy and J. A. Zimmerman, 2004. Tissue glutathione and cysteine levels in methionine-restricted rats Nutrition 20, 800-805.

Changes in fatty acid composition of intramuscular fat in growing Iberian and Landrace × Large White pigs under identical nutritional management

I. Seiquer, A. Haro, J.F. Aguilera and R. Nieto
Institute of Animal Nutrition, Estación Experimental del Zaidín, Spanish Council for Scientific Research,18100 Armilla, Granada, Spain; rosa.nieto@eez.csic.es

Introduction

Genotype is one of the main factors affecting lipid profile of meat in pigs, although other aspects including diet composition and slaughter body weight (BW) may also have a deep influence on this variable (Wood *et al.*, 2008).

The Iberian pig grows at a slower rate than leaner pig genotypes; it has lower potential for lean tissue deposition, higher whole-body fat content and comparatively lower efficiency for nutrient utilization than pigs from conventional breeds (Nieto *et al.*, 2012). The high quality and consumer acceptability of Iberian meat products are related to its high intramuscular fat content and particular fatty acid (FA) profile compared with conventional pigs. The initial hypothesis of this study is that the pattern of FA unsaturation during growth might contribute to explain the overall low efficiency of energy utilization previously observed in this native obese breed. This study is part of an experimental program aiming at explaining the lower metabolic efficiency of the Iberian pig compared to lean pig types.

Material and methods

The FA composition of intramuscular fat of *longissimum dorsi* (LD) and *biceps femoris* (BF) muscles from Iberian and Landrace × Large White (LRW) pigs at different stages of growth, under identical nutritional management, was investigated. A total of 38 pigs (19/genotype) were involved in the study. Three pigs/genotype were slaughtered at the beginning of the experiment (15.6±0.1 kg BW); the remainder 32 pigs were housed in individual pens and fed isoenergetic diets (14.4 MJ/kg DM) containing either 130 or 170 g CP/kg DM, to match protein requirements of each breed (8 pigs/genotype/CP level). Feeding level was restricted to 0.8 × *ad libitum* intake of the Iberian breed. At 49.6±0.4 kg BW, 16 pigs (4/genotype/CP level) were slaughtered by exsanguination after electrical stunning, whereas the rest of the pigs (16) were slaughtered at 115.3±0.6 kg BW. Within 15 min after slaughter, samples of LD and BF were removed, trimmed of visible intermuscular fat, vacuum packed and stored at -80 °C until analyses. Total intramuscular fat was extracted with chloroform/methanol and FA were transmethylated according to Morrison and Smith (1967). FA profile was obtained by gas chromatography using a Thermo Scientific equipment (Focus GC, Rodano, Italy). FA were expressed as % of total FA methyl esters. Statistical analysis revealed no effect of dietary CP level on the FA analysed; therefore, results for each muscle type were re-analysed by a two-way ANOVA randomized design, including genotype (G) and BW as fixed factors, and their interactions. The experimental protocol was approved by the CSIC Bioethical Committee.

Results and discussion

Significant effects of genotype and BW on the FA composition of intramuscular fat for both muscles was found, except for total saturated FA (SFA), which was not affected by pig genotype (Table 1). Monounsaturated FA (MUFA) increased (P_{BW}<0.001) progressively with BW and were always higher (P_G=0.009) in intramuscular fat from Iberian pig´s muscles, mainly due to the more elevated content of oleic acid (P_G<0.001) compared with LRW pigs. On the contrary, total polyunsaturated FA (PUFA) decreased (P_{BW}<0.001) with increasing BW with percentage values always lower in

Table 1. Major fatty acid composition (% total FA methyl esters) of intramuscular fat in LD and BF muscles from LRW and Iberian pigs slaughtered at different BW.

	LRW (kg)			Iberian (kg)			SEM	*P*-value		
	15	50	115	15	50	115		G[1]	BW	G×BW
LD										
C16:0	25.5	26.9	27.0	25.8	28.1	26.9	0.40	0.606	0.341	0.763
C18:1n9	31.5	35.5	41.3	33.6	40.7	44.6	0.63	0.021	<0.001	0.643
C18:2n6	18.3	10.5	7.4	15.6	6.2	3.9	0.31	<0.001	<0.001	0.765
SFA	40.0	45.3	41.9	40.3	44.8	40.6	0.81	0.795	0.060	0.937
MUFA	37.6	41.4	48.9	40.2	47.5	54.3	0.73	0.009	<0.001	0.742
PUFA	22.4	13.3	9.2	19.5	7.7	5.0	0.43	<0.001	<0.001	0.547
n6:n3	11.9	10.3	12.4	16.3	12.8	12.7	0.21	<0.001	0.002	0.008
BF										
C16:0	23.4	24.5	24.7	25.2	27.2	23.9	0.22	0.015	0.005	0.003
18:1n9	26.6	35.6	39.4	32.4	39.3	45.0	0.42	<0.001	<0.001	0.547
18:2n6	21.4	13.6	10.9	17.3	8.4	7.8	0.34	<0.001	<0.001	0.406
SFA	37.6	40.2	38.7	38.8	42.6	35.8	0.33	0.759	<0.001	0.003
MUFA	32.8	41.7	46.9	39.1	46.2	53.6	0.49	<0.001	<0.001	0.583
PUFA	29.6	18.0	14.4	22.1	11.0	10.6	0.48	<0.001	<0.001	0.245
n6:n3	12.0	10.4	13.3	19.1	13.4	10.7	0.22	<0.001	<0.001	<0.001

[1] G = genotype.

Iberian than in LRW muscles. At the finishing stage, n6:n3 ratio in BF was lower in Iberian than in LRW pigs, and no differences were found for LD ($P_{G \times BW}$=0.008).

The results show relevant differences in the FA composition of intramuscular fat between genotypes, and a clear evolution associated to increasing BW. The composition of the *de novo* synthesized FA would have a deep effect on FA profile of intramuscular fat and might also influence the efficiency of energy utilization.

Acknowledgements

Funding by the Spanish MINECO (grant AGL2011-25360) is gratefully acknowledged.

References

Morrison, W.R. and L.M. Smith, 1964. Preparation of fatty acid methyl esters and dimethyl acetals from lipids with boron fluoride-methanol. J. Lipid Res. 5, 600-608.

Nieto, R., L. Lara, R. Barea, R. García-Valverde, M.A. Aguinaga, J.A. Conde-Aguilera and J.F. Aguilera, 2012. Response analysis of the Iberian pig growing from birth to 150 kg body weight to changes in protein and energy supply. J. Anim. Sci. 90, 3809-3820.

Wood, J.D., M. Enser, A.V. Fisher, G.R. Nute, P.R. Sheard, R.I. Richardson, S.I. Hughes and F.M. Whittington, 2008. Fat deposition, fatty acid composition and meat quality: A review. Meat Sci. 78, 343-358.

Portal-drained viscera heat production in pigs fed betaine and conjugated linoleic acid (CLA) supplemented diets

M.L. Rojas-Cano, M. Lachica, L. Lara, A. Haro and I. Fernández-Fígares
Department of Animal Nutrition, Estación Experimental del Zaidín, CSIC, Camino del Jueves s/n, 18100 Armilla, Granada, Spain; ifigares@eez.csic.es

Introduction

Betaine and CLA have the potential to alter growth and body composition in swine (Fernández-Fígares *et al.*, 2002, 2008) improving feed efficiency, protein accretion rate and reducing body fat. Although their mode of growth promotion is not well understood. We hypothesised that differences in portal-drained viscera (PDV) heat production might partially explain differences in growth in pigs fed betaine and CLA (Fernández-Fígares *et al.*, 2008).

Material and methods

Sixteen barrows (*Sus scrofa mediterraneus*; 19±0.28 kg BW) were individually penned and fed barley-soybean meal-based diets containing no betaine or CLA (Control), 0.5% betaine, 1% CLA, or 0.5% betaine + 1% CLA (Clabet), at 70% of *ad libitum* energy intake. At 30 kg BW three permanent catheters were placed in each pig: in carotid artery and portal vein for blood sampling, and in ileal vein for para-aminohippuric acid (PAH) infusion to measure portal blood flow. Pigs recovered from surgery on metabolic cages and resumed their regular daily intake within three days after surgery. The trial began between 8-10 d after surgery. Forty five min prior blood sampling, a 15 ml pulse dose of PAH (2% w/v) was infused into ileal vein, followed by continuous infusion of 0.8 ml/min until the end of the sampling period. Blood samples were anaerobically taken simultaneously from carotid artery and portal vein every 30 min for 4 h and hourly until 6 h after feeding 1200 g of diet, into an heparinized tube for instantaneous measurement of blood parameters and haematocrit. Plasma was stored at -20 °C until PAH analysis. Whole-blood flow and O_2 consumption rates were based on Fick principle. The assumed energy equivalent for O_2 was 20.4 kJ/l. Data were subjected to ANOVA using the MIXED procedure. The statistical model included the fixed effect of block, diet, sampling time and corresponding interactions. Time of sampling within pig was considered as a repeated measure. Concentration at time zero of the analyte was included as a covariate in the statistical analysis. Significant differences among treatments were assessed using Tukey's multiple-range test.

Results and discussion

Haematocrit and haemoglobin concentration were lower ($P<0.001$) in pigs fed CLA and Clabet diets compared to Control (Table 1). Arterial-portal difference in O_2 concentration was significantly lower ($P<0.001$, 14.3%) in pigs fed Clabet diets compared to Control. Furthermore, O_2 consumption was greater in Control compared to betaine, CLA and Clabet pigs ($P<0.01$, 21-35%). Interestingly, betaine tended towards reduced PDV heat production ($P=0.08$) while CLA and Clabet had decreased PDV heat production compared to Control pigs ($P<0.01$; 27 and 58%, respectively), indicating a potential synergistic effect of betaine and CLA reducing PDV heat production. Visceral tissues have a disproportionate high metabolism with respect to their masses and under certain circumstances may compromise nutrient availability to the periphery (Yen *et al.*, 1989). Splanchnic tissues contribution to total body weight is greater in Iberian than in Landrace pigs (Rivera-Ferre *et al.*, 2005), although heat production (HP) associated to PDV was greater in Landrace than in Iberian pigs (Rodríguez-López *et al.*, 2010). Oxygen concentration difference between arterial and portal venous blood and O_2 consumption by PDV in the present experiment were within the range of published data (Yen and Nienaber, 1992; Rodríguez López *et al.*, 2010). The decreased O_2 consumption and heat production by PDV in pigs fed CLA and Clabet diets would indicate a reduction in energy and nutrient requirement of PDV, with a consequent increase in the portion of absorbed energy and nutrients to be utilized for growth of muscle and other non PDV tissues, which probably accounts for some of the improvement in the rate of weight gain in growing pigs fed Betaine + CLA supplemented diets (Fernández-Fígares *et al.*, 2008).

Table 1. Effects of dietary betaine (0.5%) and CLA (1%) on blood parameters and portal-drained viscera heat production of pigs (n=4/treatment).

	Control	Betaine	CLA	Betaine+CLA	SEM
Haematocrit, %	31.6^{a}	31.7^{a}	30.6^{b}	29.7^{c}	0.29
Haemoglobin, g/l	102^{a}	103^{a}	99^{b}	96^{c}	0.95
Blood O_2 concentration, mmol/l					
Arterial	5.98^{a}	5.95^{a}	5.81^{ab}	5.65^{b}	0.08
Portal	3.98^{ab}	4.12^{a}	3.89^{b}	3.89^{b}	0.06
$C_{CA}\,O_2 - C_{PV}\,O_2$[1], mmol/l	2.05^{a}	1.95^{a}	1.97^{a}	1.75^{b}	0.05
PDV O_2 consumption[2], mmol/min	2.6^{a}	2.1^{b}	2.1^{b}	1.7^{b}	0.14
PDVHP[3], kJ/kg$^{0.75}$/day	123^{a}	102^{ab}	90^{bc}	78^{c}	7.6

[abc] Within a row, values with different superscripts differ significantly ($P<0.001$).

[1] O_2 concentration difference between arterial and portal vein blood.

[2] PDV = portal-drained viscera; [3] PDVHP = portal-drained viscera heat production.

Acknowledgements

This study was financed by grant AGL2009-08916. Ministry of Science and Innovation.

References

Fernández-Fígares, I., J.A. Conde-Aguilera, R. Nieto, M. Lachica and J.F. Aguilera, 2008. Synergistic effects of betaine and conjugated linoleic acid on growth and carcass composition of growing Iberian pigs. J. Anim. Sci. 86, 102-111.

Fernández-Fígares, I., D. Wray-Cahen, N.C. Steele, R.G. Campbell, D.D. Hall, E. Virtanen and T.J. Caperna, 2002. Effect of dietary betaine on energy utilization and partitioning in the young growing feed restricted pig. J. Anim. Sci. 80, 421-428.

Rivera-Ferre, M.G., J.F. Aguilera and R. Nieto, 2005. Muscle fractional protein synthesis is higher in Iberian than in Landrace growing pigs fed adequate or lysine-deficient diets. J. Nutr. 135, 469-478.

Rodríguez-López, J.M., M. Lachica, L. González-Valero and I. Fernández-Fígares, 2010. Energy expenditure of splanchnic tissues in Iberian and Landrace growing gilts. Livest. Sci. 133, 61-63.

Yen, J.T. and J.A. Nienaber, 1992. Influence of carbadox on fasting oxygen consumption by portal vein-drained organs and by the whole animal in growing pigs. J. Anim. Sci. 70, 478-483.

Yen, J.T., D.A.H. Nienaber and W.G. Pond, 1989. Oxygen consumption by portal vein-drained organs and by whole animal in consciuos growing swine. Proc. Soc. Exp. Biol. Med. 190, 2393-2395.

The effect of dietary carbohydrate composition on net portal appearance of nutrients and AA liver uptake in dairy cows fed low protein diets

G. Cantalapiedra-Hijar, J.M. Rodriguez-Lopez, A. Illovies, F. Messad and I. Ortigues-Marty
Institut National de la Recherche Agronomique, UMR 1213, Unité Mixte de Recherches sur les Herbivores, Theix, 63122 St Genès Champanelle, France; gonzalo.cantalapiedra@clermont.inra.fr

Introduction

The efficiency by which ruminants transform the feed N into milk protein has been shown (Huhtanen and Hristov, 2009) to be affected by the dietary carbohydrate composition (CHO; starch vs. fiber). However, no evidences exist on the metabolic adaptations that may be involved. A sparing effect of glucogenic nutrients on amino acids (AA) has been traditionally evoked, but this has not been found at the splanchnic level in a recent study using high protein diets (18-19% CP; Larsen and Kristensen, 2012). The aim of this study was to test whether the absorbed nutrients could impact the liver uptake of essential AA (EAA) when starch replaced fiber as the main dietary carbohydrate in low CP dairy diets.

Material and methods

Five multiparous Jersey cows (365±33 kg of BW) in mid-lactation were used in a 4 x 4 Latin square design, with the fifth cow used as an extra observation. Chronic indwelling catheters were surgically implanted into the mesenteric artery, portal and hepatic veins for blood sampling and mesenteric and ruminal veins for para-aminohippuric acid (PAH) infusions. Four iso-NE_L diets were formulated to provide 2 different CP contents (12.0% and 16.5%) and 2 different CHO (350 g starch and 320 g NDF/kg DM for the starch diets, and 50 g starch and 460 g NDF /kg DM for the fiber diets, respectively). Each experimental period lasted 28 d and feed was offered in equal quantities every hour. On d 27 and 1 hour after the onset of a continuous PAH infusion (38 mmol/h), hourly blood samples were taken over 6 hours from each sampling vessel of each cow. Concentrations of acetate (C2), propionate (C3), butyrate (C4), glucose, lactate, β-hydroxybutyrate (BHBA), ammonia-N, urea-N, EAA and deacetylated PAH were obtained and averaged. Net fluxes of nutrients across the portal drained viscera (NPA) and liver (NHF) were determined. Dry matter intake was measured on the last 3 days of each period. The effects of CP, CHO and CP×CHO on variables were analyzed by GLM procedure while residuals from the regression of EAA-NHF on EAA-NPA were analyzed for the effects of CP and CHO using one-way ANOVA (Minitab14).

Results and discussion

Dry matter intake was similar (P>0.05) across dietary treatments (Table 1). Diets rich in starch vs. fiber decreased the NPA of volatile fatty acids (P=0.01), BHBA (P=0.02) and molar proportion of C2 (P=0.04), while increased the NPA of glucose (P=0.01) and molar proportion of C4 (P<0.01). Low CP diets decreased (P<0.01) the NPA of EAA, ammonia-N and urea-N compared to Normal CP diets. At low protein levels, diets rich in starch promoted a 20% higher EAA-NPA compared to diets rich in fiber, but this effect was not significant (P>0.05). A residuals analysis from the regression of EAA-NHF on EAA-NPA was carried out to evaluate whether the CHO itself could impact the EAA liver uptake (Figure 1). The EAA liver uptake was affected by neither of the studied effects (P>0.05) although a tendency for diets rich in starch (P=0.10) to have higher EAA-NHF (lower uptake) was found. This study confirmed the lack of a significant effect of glucogenic diets on EAA liver uptake (Larsen and Kristensen, 2012). However, the low number of animals used in this type of studies along with some tendencies found in our experiment warrant further research.

Table 1. Dry matter intake and net portal appearance of nutrients promoted by treatments.

	12.0%CP		16.5%CP		SEM	P-values[1]
	Starch	Fiber	Starch	Fiber		
DM intake, kg/d	14.6	14.0	14.7	15.4	0.398	NS
NPA, mmol/min						
VFA	26.0	31.3	24.7	32.9	1.91	CHO*
C2, %	63.0	67.9	62.2	68.9	0.023	CHO*
C3, %	26.7	26.7	27.8	25.7	0.021	NS
C4, %	10.3	5.49	10.0	5.36	0.006	CHO***
Glucose	1.01	-1.14	0.77	-0.91	0.574	CHO*
Lactate	1.79	1.86	1.70	1.76	0.187	NS
BHBA	2.89	3.37	2.80	3.70	0.217	CHO*
Essential AA	3.21	2.68	4.07	4.04	0.153	CP***
Ammonia-N	4.29	4.96	7.49	7.66	0.674	CP**
Urea-N	-1.73	-1.91	-3.93	-3.61	0.416	CP**

[1] NS = non significant; CHO = carbohydrate composition; CP = protein content; * $P<0.05$; ** $P<0.01$ and *** $P<0.001$.

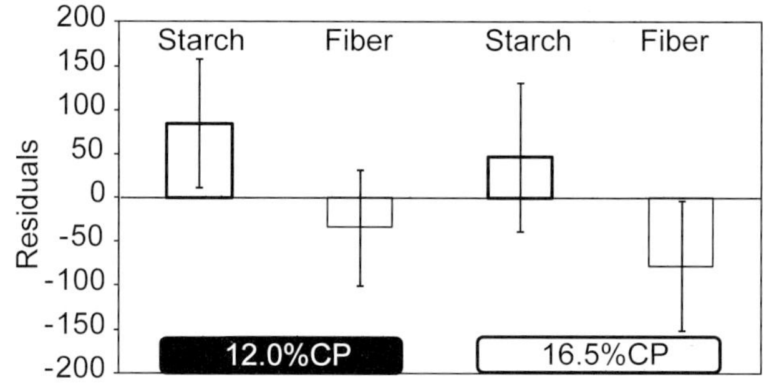

Figure 1. Residuals analysis from the regression of essential AA liver uptake (Y) on essential AA net portal appearance (X) according to the dietary carbohydrate composition (P=0.10) and protein content (P=0.58). $Y = 485(\pm 310;\ P=0.17) - 0.40(\pm 0.09;\ P<0.01)X$ ($r^2 = 0.60$).

Acknowledgment

This study was granted by the Commission of the European Communities; project FP7-BBE-2007-1 'Rednex'. We also thank Adisseo for financial support to conduct this experiment.

References

Huhtanen P., and A.N. Hristov. 2009. A meta-analysis of the effects of dietary protein concentration and degradability on milk protein yield and milk N efficiency in dairy cows. J. Dairy Sci. 92:3222-3232.

Larsen M. and N.B. Kristensen. 2012. Effects of glucogenic and ketogenic feeding strategies on splanchnic glucose and amino acid metabolism in *postpartum* transition Holstein cow. J. Dairy Sci. 95:5946-5960.

Uptake of arterial amino acids by ruminal tissue in Holstein cows under washed rumen conditions

A.C. Storm[1], M. Aguilar[2], M.D. Hanigan[2], N.B. Kristensen[1] and M. Larsen[1]
[1]Department of Animal Science, Aarhus University, Foulum, 8830 Tjele, Denmark; adam.storm@agrsci.dk
[2]Department of Dairy Science, VirginiaTech, Blacksburg, VA, USA

Introduction

Non mesenteric-drained viscera (MDV) metabolism of amino acids (AA) has been proposed to represent as much as 40% of the MDV release of AA (MacRae *et al.*, 1997). There is limited knowledge of the ruminal vein-drained visceral (RDV) use of AA which might be dependent on arterial AA concentration, epithelial metabolism, or both. Ruminal absorption of AA might occur, so ruminal tissue may metabolize AA of both arterial and ruminal origin. The objective of the study was to quantify the effect of dietary CP and ruminal absorption of fermentation products on RDV use of arterial AA when ruminal AA absorption is excluded.

Material and methods

Six Holstein cows catheterized in the rumen vein (RV) and an artery (A) were used in a washed rumen study (Kristensen *et al.*, 2010; Storm *et al.*, 2011). In brief, cows were subjected to two dietary levels of CP (17% and 13% CP, n=3) for 3 wk prior to the washed rumen trial. Further, 3 ruminal buffers (Ammonia, Butyric, and Control) were administrated in a split-plot design with CP as whole plot and buffer as subplot factor arranged in a Latin square. Blood were collected at 9, 20, and 30 min after administration of the buffers. The RV plasma flow was calculated from ruminal clearance and A-RV concentration difference of D_2O. Plasma was analyzed for AA by GC-MS by isotope dilution. Data were analyzed using a model including the fixed effect of CP, buffer, sampling time, period and interactions. Sampling time within cow and buffer was considered as a repeated measure (spatial Gaussian) and cow within CP level as random. Data are presented as LS means ± SEM.

Results and discussion

Overall, net RDV uptake of arterial essential AA (EAA) and total AA was decreased 37 and 33%, respectively (Table 1), when CP was reduced. The A–RV concentration differences of AA were in general not affected by dietary CP ($0.10<P<0.87$), except for Phe, Val, Asp, and Gln, ($P=0.04$, 0.06, 0.08, 0.05, respectively). Nor were the A–RV concentration differences and the RDV uptake of AA affected by buffer ($0.10<P<0.95$), except for A-RV of Asp and Asn ($P<0.01$) and RDV uptake for Asn ($P<0.01$). Thus, low CP induced a decrease in RDV uptake of EAA and total AA that might be related to RV blood flow. The RV blood flow was not increased by CP ($P=0.10$) but by butyrate buffer ($P<0.01$), thereby rejecting a strict RV blood flow related explanation. This indicates that the reduced feed intake and numerically lower RV blood flow with the low CP may interlink in the regulation of RDV AA uptake. However the RDV extraction ratio ([A–RV]/A) were not affected by CP, but ratios of Ala, Asp, Gly, Pro, Ile, Leu, Thr, and Val decreased at low CP and butyrate buffer ($P_{CP \times Buffer}$=0.02, 0.09, 0.002, 0.2, 0.01, 0.003, 0.06, 0.09, respectively). Reduced RDV extraction of AA with low CP and high butyrate absorption indicate that RDV down-regulates branched-chain AA uptake when AA availability is low and metabolism in the ruminal wall is high. Hence, RDV uptake of AA is not only controlled by mass action.

Relating the current RDV uptake of EAA with 17% CP to net PDV release of EAA observed with the same cows at 29 days after caling fed diets at similar intake and CP level (Larsen and Kristensen, 2012) show that RDV uptake of EAA was equal to 5% of net PDV release. Thus, RDV uptake of

Table 1. Ruminal tissue uptake of arterial AA in dairy cows under washed rumen conditions.

Item	Dietary CP		SEM	*P*-value	
	17%	13%		CP	Buffer
Dry matter intake, kg/d	20.5	18.3	0.5	<0.01	-
Ruminal vein blood flow, l/h	280	209	37	0.24	<0.01
Net RDV flux, mmol/h					
EAA	12.3	7.8	1.5	0.10	0.47
His	1	0.9	0.3	0.79	0.29
Ile	1.7	1	0.2	0.04	0.32
Leu	2.8	1.6	0.3	0.05	0.61
Lys	1.7	1.1	0.2	0.09	0.34
Met	0.4	0.2	0.1	0.08	0.36
Phe	1	0.6	0.1	<0.01	0.51
Thr	1.6	1.2	0.2	0.28	0.60
Trp	0.2	0.1	0.1	0.24	0.99
Val	2.1	1.1	0.2	<0.01	0.85
Non-EAA	22.6	15.1	1.5	<0.01	0.27
Ala	4.8	2.9	0.8	0.17	0.87
Asp	0.4	0.1[¥]	0.1	0.08	<0.01
Asn	0.5	0.4	0.1	0.22	0.28
Cys	0.5	0.2	0.1	0.06	0.95
Gln	6	6.6	0.8	0.65	0.55
Glu	0.3[¥]	-0.2[¥]	0.4	0.43	0.83
Gly	7.4	3.8	1.0	0.06	0.18
Pro	2.2	1.3	0.3	0.08	0.36
Ser	-0.3[¥]	-0.4[¥]	0.3	0.86	0.82
Tyr	0.7	0.5	0.1	0.07	0.17
Total AA	34.9	22.9	2.5	0.03	0.43

[¥] Estimate not different from zero (t-test).

EAA could only account for 7.5% of the non-MDV uptake of EAA, assuming that 40% of the MDV release is equals non-MDV uptake. Overall, the data shows that the majority of non-MDV use of arterial AA is related to tissue other than RDV.

References

Kristensen, N.B., A.C. Storm and M. Larsen, 2010. Effect of dietary nitrogen content and intravenous urea infusion on ruminal and portal-drained visceral extraction of arterial urea in lactating Holstein cows. J. Dairy Sci. 93, 2670-2683.

Larsen, M. and N.B. Kristensen, 2012. Effects of glucogenic and ketogenic feeding strategies on splanchnic glucose and amino acid metabolism in *postpartum* transition Holstein cows. J. Dairy Sci. 95, 5946-5960.

MacRae J.C., L.A. Bruce, D.S. Brown, D.A.H. Farningham and M. Franklin, 1997. Absorption of amino acids from the intestine and their net flux across the mesenteric- and portal-drained viscera of lambs. J. Anim. Sci. 75, 3307-3314.

Storm, A.C., M.D. Hanigan and N.B. Kristensen, 2011. Effects of ruminal ammonia and butyrate concentrations on reticuloruminal epithelial blood flow and volatile fatty acid absorption kinetics under washed reticulorumen conditions in lactating dairy cows. J. Dairy Sci. 94, 3980-3994.

Uptake of arterial amino acids by ruminal tissue in periparturient Holstein cows

M. Larsen, A.C. Storm and N.B. Kristensen
Department of Animal Science, Aarhus University, Foulum, 8830 Tjele, Denmark; mogens.larsen@agrsci.dk

Introduction

Ruminal tissue mass increases with increasing feed intake in the periparturient dairy cow requiring amino acids (AA) for protein synthesis. Thus, arterial-ruminal venous (A-V) concentration differences of AA were hypothesised to increase after parturition in response to increasing feed intake. Further, the aim was to assess the ruminal-drained visceral (RDV) uptake of arterial AA in relation to mesenteric-drained visceral (MDV) release of AA.

Material and methods

Nine periparturient Holstein cows catheterised in major splanchnic vessels were used in a complete randomised design with repeated measurements to investigate effects of *postpartum* feeding strategy on splanchnic glucose and AA metabolism. Cows were assigned to 1 of 3 feeding strategies at calving: a glucogenic, a ketogenic, or an alfalfa-glucogenic strategy. Eight sample sets of arterial, ruminal venous, and portal venous blood were collected at -14, +4, +15, and +29 days relative to calving (DRTC). Portal-drained visceral (PDV) plasma flow was measured by down-stream dilution of *para*-aminohippuric acid and RDV plasma flow was assumed to contribute 19.4% to the portal plasma flow (Girard and Desrochers, 2010; Storm *et al.*, 2011). Data were analysed with a model including the fixed effect of treatment, DRTC, and the interaction. DRTC within cow was considered as a repeated measurement (autoregressive order 1). Data is presented as means ± SEM for DRTC as no feeding strategy effects were observed (P>0.11, not Ala), for ruminal A-V differences.

Results and discussion

Ruminal A-V differences were positive for most AA (Table 1), except for Glu and Ser that were negative, and for His, Asn, Asp, Cys, and Gln that did not differ from zero (P<0.05, thus not presented). The general positive ruminal A-V differences indicate an overall RDV uptake of arterial AA, and consequently, no net absorption of AA from the ruminal lumen. Ruminal A-V differences did not change with DRTC either linearly or quadratically (Table 1), except for Ala, Pro, and Ser. However, net RDV uptake of Leu, Lys, Thr, Trp, and Val increased linearly with increasing DRTC when applying the simultaneous increase in RDV plasma flow (Table 1), but did not change for other essential AA (EAA). Loss of AA in sloughed epithelia, ruminal tissue proliferation, and catabolism are the primary fates of the AA taken up in the RDV, however, catabolism would likely be limited during periparturient protein deficiency.

The net PDV release of EAA has been found to comprise between 60 and 80% of the net MDV release of EAA in sheep and cattle, implying that the amount of EAA net utilised in non-MDV tissues is equivalent to the amount of EAA unaccounted for in PDV release. On average in the periparturient period, the non-MDV utilisation of EAA could be estimated to be between 39±1 and 105±3 mmol/h assuming either 80 or 60% PDV recovery of MDV release, respectively. Hence, the RDV uptake of EAA could account for between 11 and 29% of the EAA utilised by the non-MDV tissues. In sheep, Rémond *et al.* (2003) observed the RDV uptake of arterial EAA to account for 30% of non-MDV uptake of EAA. Thus, other non-MDV tissues than the RDV seem to utilise the major part of EAA utilised in non-MDV, likely the abomasum and pancreas synthesising digestive enzymes and hormones.

Table 1. Ruminal tissue uptake of arterial amino acids in periparturient Holstein cows.

Item		Day relative to calving				SEM	*P*-contrasts	
		-14	+4	+15	+29		Linear	Quadratic
Dry matter intake, kg/d		10.0	12.9	16.1	18.6	1.0	<0.01	0.77
Portal plasma flow, l/h		759	980	1312	1406	47	<0.01	0.55
Ruminal A-V differences, μM	Ile	7.8	5.4	5.9	5.3	1.3	0.21	0.54
	Leu	9.7	10.8	10.3	11.1	1.9	0.66	0.94
	Lys	5.5	8.2	7.3	7.7	1.3	0.29	0.41
	Met	1.7	1.6	1.6	1.5	0.4	0.61	0.96
	Phe	3.0	3.5	3.3	2.6	0.6	0.70	0.30
	Thr	6.2	4.3	5.4	5.5	0.9	0.72	0.25
	Trp	0.9	0.8	1.2	1.4	0.2	0.12	0.47
	Val	6.6	7.2	9.2	6.9	1.9	0.75	0.49
	Ala	17.4	16.3	15.5	22.0	1.4	0.06	0.01
	Glu	-2.8	-1.5	-2.5	-1.5	0.8	0.35	0.88
	Gly	24.9	26.8	28.0	29.7	2.3	0.16	0.97
	Pro	4.6	4.6	5.4	7.0	0.9	0.06	0.37
	Ser	-4.7	-8.7	-5.7	-5.6	1.3	0.84	0.05
	Tyr	2.7	2.4	2.4	1.8	0.6	0.36	0.72
Net RDV uptake, mmol/h	EAA	7.8	8.9	10.6	12.8	1.7	0.06	0.56
	Ile	1.2	1.0	1.4	1.6	0.3	0.36	0.49
	Leu	1.5	2.1	2.1	3.2	0.5	0.04	0.52
	Lys	0.8	1.7	1.7	2.1	0.3	0.02	0.54
	Met	0.3	0.3	0.4	0.4	0.1	0.20	0.82
	Phe	0.5	0.7	0.8	0.7	0.1	0.18	0.37
	Thr	1.0	0.8	1.2	1.5	0.2	0.08	0.30
	Trp	0.1	0.2	0.3	0.4	0.1	<0.01	0.12
	Val	1.0	1.5	1.9	2.0	0.3	0.06	0.69
	Ala	2.6	3.3	3.9	6.1	0.4	<0.01	0.05
	Glu	-0.3	-0.3	-0.7	-0.4	0.2	0.53	0.62
	Gly	3.6	5.3	6.2	8.1	0.5	<0.01	0.62
	Pro	0.7	0.9	1.2	2.2	0.2	<0.01	0.08
	Ser	-0.7	-1.9	-1.7	-1.6	0.3	0.04	0.02
	Tyr	0.5	0.5	0.6	0.6	0.1	0.53	>0.99
	Total AA	16	18	24	27	4.4	0.08	0.81
Net PDV release, mmol/h	EAA	60	144	178	245	13	<0.01	0.98
	Total AA	128	303	414	565	26	<0.01	0.48

References

Girard, C.L. and A. Desrochers, 2010. Net flux of nutrients across splanchnic tissues of lactating dairy cows as influenced by dietary supplements of biotin and vitamin B_{12}. J. Dairy Sci. 93, 1644-1654.

Rémond, D., L. Bernard, B. Chaveau, P. Nozière and C. Poncet, 2003. Digestion and nutrient net fluxes across the rumen, and the mesenteric- and portal-drained viscera in sheep fed with fresh forage twice daily: net balance and dynamic aspects. Br. J. Nutr. 89, 649-666.

Storm, A.C., M.D. Hanigan and N.B. Kristensen, 2011. Effects of ruminal ammonia and butyrate concentrations on reticuloruminal epithelial blood flow and volatile fatty acid absorption kinetics under washed reticulorumen conditions in lactating dairy cows. J. Dairy Sci. 94, 3980-3994.

Effects of ethanol on splanchnic nutrient metabolism in sheep at different intake levels

T. Obitsu, K. Nishimura, Y. Udaka, T. Sugino and K. Taniguchi
Graduate School of Biosphere Science, Hiroshima University, 1-4-4 Kagamiyama, Higashihiroshima-shi, 739-8528, Japan; tobitsu@hiroshima-u.ac.jp

Introduction

Ruminants fed silage-based diets are likely to ingest alcohols, due to alcohol production in fermented feeds (Nishino and Shinde, 2007). Although absorbed alcohols are partly detoxified mainly in the liver, this process may affect nutrient utilization by the tissues. In human, absorption and elimination of ingested alcohol are affected by food consumption (Jones *et al.*, 1997). Thus, we investigated the effects of ruminal infusion of ethanol (EtOH) on splanchnic nutrient metabolism in sheep given a diet at different levels.

Material and methods

Six sheep (average body weight: 50±18 kg) fitted with a ruminal cannula and vessel catheters in the splanchnic region were used in a split-plot crossover design. The experiment consisted four 7 day periods. After adaptation period, three sheep were assigned to a low (0.7 times maintenance energy) intake level (L) for 2 weeks (period 1 and 2), then the feeding levels were switched to a high (1.7 times maintenance energy) level (H) for 2 weeks (period 3 and 4). Other 3 sheep were assigned to the reverse sequence of the feeding levels (H to L). At each period, sheep were assigned the treatments with or without administration of EtOH. Sheep were given 4 equal portions of a diet (CP: 12% DM) consisting of 60% concentrate and 40% hay at 6 h intervals. For EtOH treatments, EtOH (1.5 g/kg BW/d, 20% solution) was dosed into the rumen immediately after each feeding for 7 days. Rumen fluid samples were taken at 2 h after feeding on day 5 of each period to measure concentration of EtOH, volatile fatty acids and esters. On day 7, blood samples were collected for 6 h and plasma flow rates were estimated by para-aminohippuric acid dilution. Ruminal characteristics (n=5 or 6), arterial concentrations (n=5 or 6), net portal fluxes (n=5 or 6) and net total splanchnic (TSP) fluxes (n=2 or 3) of plasma metabolites were compared with feeding levels, ethanol infusion and their interactions as main effects, using the MIXED procedure of SAS. For treatments with EtOH, plasma concentration and fluxes of EtOH and clearance rate of arterial EtOH calculated by a two-compartment model were compared between L and H by paired t-test.

Results and discussion

Dry matter intake was higher for H than L intake level (1.21 vs. 0.51 kg/d, SEM=0.065 $P<0.01$), but not affected by EtOH infusion. Ruminal EtOH concentration in sheep infused with EtOH was higher for L compared with H intake level (5.9 vs. 2.9 mmol/l, SEM=0.77, $P<0.10$) at 2 h after feeding. Arterial EtOH concentration peaked (4 mM) at 1.0 (H) and 1.5 h (L) after EtOH dose. The portal recovery of infused EtOH did not differ between L (39±7.0%) and H (45±15.6%). However, clearance rates of arterial EtOH were higher (1.40±0.27 vs. 1.13±0.30 /h, $P=0.07$) for H than L intake. Plasma flows of portal and hepatic vein were higher for H than L intake ($P<0.01$), but not affected by EtOH infusion (Table 1). Arterial acetate concentration increased ($P<0.01$) with EtOH infusion. However, net portal absorption of acetate decreased ($P<0.05$) with EtOH infusion. Interaction of feeding levels and EtOH was observed ($P<0.10$) for net acetate releases by TSP. Arterial concentration and net portal and TSP fluxes of glucose were higher for H than L intake, but not affected by EtOH. Arterial lactate concentration increased ($P<0.05$) with EtOH infusion, but not affected by feeding levels. Net portal releases of lactate decreased ($P<0.05$) but net TSP releases numerically increased with EtOH. Neither arterial concentrations nor net fluxes of 3-hydroxybutyrate and non-esterified

Table 1. Effect of ruminal ethanol (EtOH) infusion on nutrient net flux by splanchnic tissues in sheep at different intake levels.

	High intake[3]		Low intake[3]		SEM	Probability		
	-EtOH	+EtOH	-EtOH	+EtOH		Intake	EtOH	Interaction
Plasma flow rate, l/hr								
Portal[1]	116	110	68	72	7.9	<0.001	0.798	0.413
Hepatic[2]	133	120	86	75	13.3	0.020	0.439	0.941
Arterial concentration[1], mmol/l								
Acetate	0.93	1.56	1.11	1.57	0.19	0.310	<0.001	0.353
Glucose	3.90	3.84	3.36	3.47	0.11	<0.001	0.765	0.315
Lactate	0.69	1.36	1.01	1.46	0.23	0.294	0.012	0.583
Alanine	0.15	0.15	0.21	0.16	0.01	0.018	0.077	0.103
Net portal flux[1], mmol/h								
Acetate	121.7	80.0	53.9	36.8	11.29	0.001	0.030	0.316
Glucose	2.3	3.2	-1.3	-1.8	2.00	0.059	0.912	0.730
Lactate	8.2	3.8	6.3	2.6	1.29	0.238	0.010	0.814
Alanine	3.2	4.7	1.0	1.6	0.48	<0.001	0.041	0.326
Net total splanchnic flux[2], mmol/h								
Acetate	183.9	112.8	67.5	97.7	29.70	0.045	0.447	0.097
Glucose	57.9	35.0	14.9	16.9	10.60	0.012	0.251	0.180
Lactate	13.0	20.3	1.7	3.9	4.41	0.031	0.354	0.615
Alanine	-0.6	-3.6	-1.6	-2.3	1.27	0.819	0.031	0.136

[1] n=5 or 6 for high intake and low intake, respectively.

[2] n=3 except for low intake without EtOH (n=2).

[3] High intake: 1.7 times maintenance energy. Low intake: 0.7 times maintenance energy.

fatty acids were affected by EtOH infusion. Arterial concentration (P=0.155) and net TSP flux (P=0.053) of triglyceride tended to increase with EtOH infusion but not affected by feeding levels. Arterial concentrations and net portal fluxes of most essential amino acids were not affected by EtOH. However, arterial concentration (P=0.08) and net TSP flux (P<0.05) of alanine decreased but net portal alanine absorption increased (P<0.05) with EtOH infusion.

In summary, effects of EtOH load on net splanchnic metabolism of most nutrients, except for acetate, were not modified by feeding levels. Alternation of lactate and alanine fluxes within the splanchnic tissues may reflect the enhanced conversion between these substrates under EtOH load in sheep.

References

Jones A.W., K.A. Jonsson and S. Kechagias, 1997. Effect of high-fat, high-protein, and high-carbohydrate meals on the pharmacokinetics of a small dose of ethanol. Br. J. Clin. Pharmacol. 44,521-526.

Nishino, N. and S. Shinde. 2007. Ethanol and 2,3-butanediol production in whole-crop rice silage. Grassl. Sci. 53, 196-198.

Contribution of amino acids to glucose and lactose synthesis in lactating dairy cows

G. Maxin, D.R. Ouellet and H. Lapierre
Agriculture and Agri-Food Canada, Sherbrooke, Québec J1M 0C8, Canada; gaelle.maxin@clermont.inra.fr

Introduction

Despite a nutritional approach splitting protein and energy inputs in the models used to balance dairy rations, many studies indicated important interactions between energy and protein metabolism. For example, in the dairy cow, increases in milk protein and lactose yields are both induced with increased protein supply. This increase in lactose yield is not always paralleled by an incremental increase in mammary glucose uptake but is associated with an increased uptake of branched-chain amino acids (AA). Although AA are used for gluconeogenesis, Bequette *et al.* (2005) demonstrated *in vitro* that some essential AA could contribute to lactose synthesis by mammary cells supporting earlier observations that arterial glucose would not be the only precursor of milk lactose (Bickerstaffe *et al.*, 1974). Therefore, the objective of this work was to study the interactions between AA and glucose metabolism in dairy cows receiving different AA mixtures and to determine if AA could contribute to lactose synthesis within the mammary gland.

Material and methods

Six dairy cows averaging 708 (SD:70) kg BW and 160 (SD:23) DIM, equipped with rumen fistula and chronic catheters in mesenteric artery, and portal, hepatic and mesenteric veins were used in a replicated 3×3 Latin Square with 14-d periods. Cows were fed a corn-grass silage based diet providing 100 and 67% of their net energy and metabolizable protein (MP) requirements, respectively. Treatments were abomasal infusions of water (CTL, 10 l/d) or a mixture of total AA (1262 g/d; casein profile) estimated to supply with the diet 100% of MP requirements (TAA) or only the essential AA of the TAA mixture (617 g/d; EAA). Milk yield and composition was determined on the last 3 d of each period. On d 14, D-[U-$^{13}C_6$]glucose (8.6±0.1 mmol/h) was infused for 7 h into a jugular vein. Starting at 1 h after the initiation of the infusion, six hourly blood samples were taken simultaneously, from arterial, portal, and hepatic sources to determine the WB and tissue net and total release of glucose. Portal and splanchnic plasma flows were determined by downstream dilution of para-aminohippurate after deacetylation. In addition, cows were milked right after the last blood samples, the milking being performed after oxytocin injection to remove residual milk. Concentrations and isotopic enrichment (IE) of plasma glucose (on all blood samples) and enrichment of galactose and glucose from enzymatically hydrolyzed lactose (from milking at 6 h) were determined by gas-chromatography-mass spectrometry. Data (average for each cow × period) were analyzed according to the replicated 3×3 Latin square using the mixed procedure of SAS (2001) with period and treatment as fixed effect and cow as random effect. Multiple comparisons of means were performed with an adjusted Tukey-Kramer test.

Results and discussion

Dry matter intake was not modified by treatment and averaged 23.2 kg DM/d. Compared to CTL treatment, the infusion of TAA increased milk yield (P<0.01) as previously observed by Doepel and Lapierre (2010). Milk protein yield increased more with TAA treatment compared with EAA treatment (Table 1). The infusion of TAA increased the WB Ra of glucose compared to the other treatments, whereas WB Ra of glucose was not different between CTL and EAA treatments (Table 1). Although both portal and hepatic true glucose release were not affected by treatment, the splanchnic true flux of glucose observed with TAA was numerically higher than for CTL treatment: the increment of WB

Table 1. Effect of different AA mixture supply on milk yield and glucose kinetics in dairy cows.

	Treatments[1]			SEM	*P*-value Trt
	CTL	EAA	TAA		
Milk, kg/d	33.1[b]	35.0[ab]	37.3[a]	2.0	<0.01
Milk protein, kg/d	0.97[c]	1.11[b]	1.21[a]	0.72	<0.01
Glucose kinetics, mmol/h					
Whole body rate of appearance	798[b]	811[b]	902[a]	39	<0.01
Net glucose release					
Portal	43[a]	-51[b]	9[ab]	21	0.03
Hepatic	725	807	780	76	0.53
Splanchnic	768	734	799	76	0.79
Glucose utilization					
Portal	-96[a]	-192[b]	-160[ab]	27	0.06
Hepatic	-17	-7	-31	32	0.83
Splanchnic	-113	-152	-171	33	0.39
True glucose release					
Portal	139	141	169	26	0.67
Hepatic	742	815	811	77	0.38
Splanchnic	882	902	957	81	0.50
Isotopic enrichment (IE) in milk lactose[2]					
Glucose, atom percent excess	1.08	1.11	1.02	0.04	0.19
Galactose, atom percent excess	0.61[a]	0.57[ab]	0.51[b]	0.03	0.03
Galactose IE /Glucose IE	0.56[a]	0.52[ab]	0.50[b]	0.01	<0.01

[1] Abomasal infusion of CTL: water (n=6); TAA: mixture of total AA (n=5); EAA: only essential AA of TAA (n=5).
[2] On sample collected after 6 h of D-[U-$^{13}C_6$]glucose infusion.
[a,b,c] Values with different superscript in the same row are different: $P<0.05$.

Ra of glucose could therefore be mainly related to increased hepatic glucose production, probably by conversion of nonessential AA to glucose.

Six h after the initiation of the D-[U-$^{13}C_6$]glucose infusion, IE of glucose in milk lactose averaged 0.94 (SEM: 0.04) of arterial IE and this ratio was not altered by treatment. This ratio not different from unity ($P>0.10$) suggests that arterial glucose is the major, if not the sole, contributor of carbons to glucose in milk lactose. The IE of galactose in milk lactose was, however, lower than the IE of glucose in milk lactose (Table 1) indicating that other sources of carbon than arterial glucose have been used for galactose synthesis in the mammary gland. Moreover, the ratio of the IE of galactose on IE of glucose in milk lactose decreased with TAA treatment ($P<0.05$) suggesting that the contribution of carbons other than arterial glucose was altered by AA supply.

In conclusion, this study confirms earlier observations that arterial glucose would not be the only precursor of milk lactose, especially as other sources of carbon would contribute to galactose synthesis and suggests that AA could be involved in the milk lactose synthesis.

References

Bequette B.J., S.L. Owens, S.W. El-Kadi, N.E. Sunny and A. Shamay, 2005. Use of ^{13}C-mass isotope distribution analysis (MIDA) to define precursors for lactose and amino acid synthesis by bovine mammary explants. J. Dairy Sci. 88 Supp. 1, 289.

Bickerstaffe R. and E.F. Annison, 1974. The metabolism of glucose, acetate, lipids and amino acids in lactating dairy cows J. Agri. Sci. 82, 71-85.

Doepel, L. and H. Lapierre, 2010. Changes in production and mammary metabolism of dairy cows in response to essential and nonessential amino acid infusions. J. Dairy Sci. 93, 3264-3274.

Contribution of essential amino acids to glucose metabolism and lactose secretion in late lactation dairy cows

H. Lapierre[1], *S. Lemosquet*[2] *and D.R. Ouellet*[1]
[1]*Agriculture and Agri-Food Canada, Sherbrooke, QC J1M 0C8, Canada; helene.lapierre@agr.gc.ca*
[2]*INRA UMR1080 Dairy Production, Saint-Gilles, France*

Introduction

Increased protein supply is known to increase both milk and milk lactose yields as well as whole body rate of appearance (Ra) of glucose (Lapierre *et al.*, 2010). However, despite earlier observations reporting that arterial glucose would not be the only precursor of milk lactose (Bickerstaffe *et al.*, 1974), little attention has been paid to the potential contribution of amino acids (AA) to mammary lactose synthesis. Although, Bequette *et al.* (2006) recently demonstrated *in vitro* that some essential AA could contribute to mammary lactose synthesis, no *in vivo* study has examined the contribution of essential AA to milk lactose secretion and if that contribution would be altered by AA provision. Therefore, the objective of this study was to examine if increased supply of essential AA, in excess of requirement, would affect the contribution of glucose to lactose synthesis in lactating dairy cows.

Material and methods

Four rumen-fistulated dairy cows in late lactation averaging 629 (SD:73) kg were used in cross-over design with two 7-d periods. Throughout the study, cows were fed a fixed amount, restricted at 98% of *ad libitum* intake, of a grass hay-based diet providing 100% of net energy (28.8 Mcal/d) and metabolizable protein (1,907 g/d) requirements (NRC, 2001). Treatments were abomasal infusions of water (CTL) or a mixture of essential AA (EAA_{inf}, 647 g/d; casein profile), estimated to provide 140% of the essential AA requirements (diet + infusion). Milk yield was measured daily and milk composition was determined on the 2 last d of each period. On d 7, D-[6-$^2H_2$]glucose (24.4±0.5 mmol/h) was infused into jugular vein preceded by a priming dose (20.3±1.8 mmol). Blood samples were taken 1:15, 1:45, 2:15, 2:45, 3:15 and 4:15 h after the initiation of the glucose infusion. Furthermore, cows were milked right after the last two blood samples, the milking at 3:15 h being performed after oxytocin injection to remove residual milk. Concentrations and isotopic enrichment (IE) of plasma glucose (on all blood samples) and enrichment of galactose and glucose from enzymatically hydrolyzed lactose (from milking at 4:15 h) were determined by gas-chromatography-mass spectrometry. Whole body glucose Ra was estimated using a non-steady-state model, as proposed by Brockman (1984). Data were analyzed using GLM procedures of Minitab (version 16.1.1.0) with treatment, period and cow as fixed effect.

Results and discussion

Dry matter intake was not affected by treatment (as there was no orts) and averaged 18.0±0.0 kg/d (LSM ± SEM for CTL vs. EAA_{inf}). Milk, protein, fat and lactose yields were not affected (P>0.20) by EAA infusion (Table 1). Milk protein concentration, however, increased (P=0.009) with EAA_{inf} (38.5 vs. 39.9±0.08 g/kg). Plasma glucose concentration was not affected (P=0.77) by treatment (3.77 vs. 3.77±0.004 mM), nor was whole body Ra of glucose (Table 1). The proportion of whole body glucose Ra accounted for by milk lactose secretion, assuming glucose as the sole precursor of lactose, averaged 46% and was not affected by treatment (P=0.94).

In milk lactose, IE of glucose (P=0.02) and galactose (P=0.06) decreased with the EAA_{inf}. In addition, galactose IE was lower (P<0.001) than $glucose_{lactose}$ IE, but EAA_{inf} did not alter this ratio, the IE of galactose being on average 82% that of glucose. After 4:15 h of labeled glucose infusion, the IE

Table 1. Effect of increasing intestinal supply of essential AA (EAA_{inf}) in excess of requirement on milk yield, glucose whole body rate of appearance (WB Ra) and isotopic enrichment (IE) of glucose and galactose during an infusion of labeled glucose.

Parameter	Treatment		SEM	*P*-value
	CTL	EAA_{inf}		
Milk yield, kg/d	22.2	22.8	0.52	0.56
Milk protein yield, g/d	857	905	20.1	0.23
Milk fat yield, g/d	999	1,016	24.0	0.67
Milk lactose yield, g/d	989	996	27.3	0.88
Glucose WB Ra, g/d	2,239	2,286	40.5	0.50
Lactose/glucose WB Ra	0.46	0.46	0.019	0.94
$Glucose_{lactose}$ IE, ape[1]	2.70	2.40	0.030	0.02
$Galactose_{lactose}$ IE, ape[1]	2.23	2.00	0.041	0.06
$Galactose_{lactose}$ IE/$Glucose_{lactose}$ IE[1]	0.82	0.83	0.006	0.39
$Glucose_{lactose}$ IE/$Glucose_{plasma}$ IE[1]	0.59	0.53	0.009	0.06

[1] On sample collected 4:15 h after the initiation of the D-[6-$^2H_2$]glucose infusion.

of $glucose_{lactose}$ was still lower (P<0.01) than the IE of $glucose_{plasma}$, and this ratio was decreased (P=0.06) by EAA_{inf}.

Infusing EAA in cows fed to meet their protein requirement did not affect glucose availability and did not alter milk yield. In contrast, in cows fed a protein-deficient diet, although infusion of EAA did not alter glucose Ra, but it did increase milk yield (Maxin *et al.*, 2013). In the current study, the IE of galactose averaged 82% of $glucose_{lactose}$ IE, suggesting that 18% of galactose did not originate directly from intracellular glucose; this contribution was not affected by EAA_{inf}. The ratio of $glucose_{lactose}$ IE to $glucose_{plasma}$ IE was less than 0.60 after 4:15 h of labelled glucose infusion, compared to the 0.85 observed for the specific radioactivity of lactose on plasma glucose after 5 h of infusion (Bickerstaffe *et al.*, 1974; lactose specific radiaoctivity was not analyzed separately for galactose and glucose) and this porportion was altered by EAA_{inf}, suggesting either there are other contributors to intracellular glucose than arterial glucose uptake or there is exchange of the deuterium between intracellular metabolites or both. This *in vivo* study confirms that glucose is not the sole precursor of mammary lactose synthesis. The effects of AA supply on milk lactose synthesis needs to be directly evaluated through infusion of labelled AA.

References

Bequette, B.J., N.E. Sunny, S.W. El-Kadi and S.L. Owens, 2006. Application of stable isotopes and mass isotopomer distribution analysis to the study of intermediary metabolism of nutrients. J. Anim. Sci. 84, Suppl. E, 50-59.

Bickerstaffe, R., E.F. Annison and J.L. Linzell, 1974. The metabolism of glucose, acetate, lipids and amino acids in lactating dairy cows. J Agric Sci, Cambridge 82, 71-85.

Brockman, R.P., 1984. Validation of an equation for calculation of glucose appearance during nonsteady state in sheep. Can. J. Physiol. Pharmacol. 62, 341-344.

Lapierre, H., C.E. Galindo, S. Lemosquet, I. Ortigues-Marty, L. Doepel and D.R. Ouellet, 2010. Protein supply, glucose kinetics and milk yield in dairy cows. Pages 277-288. *In* EAAP publication No.127. ed. C.M. Crovetto. Wageningen Academic Publishers. The Netherlands.

Maxin. G., D.R. Ouellet and H. Lapierre, 2013. Is there a relation between milk yield and glucose kinetics responses to the supply of different amino acids mixtures? ISEP, Sacramento, CA.

NRC. 2001. Nutrient Requirements of Dairy Cattle. 7th rev. ed. Natl. Acad. Sci., Washington, DC.

Mammary gland from lactating cows responded additively to individual essential amino acids in casein synthesis rate

S.I. Arriola Apelo[1], L.M. Singer[1], X. Lin[3], W.K. Ray[2], R.F. Helm[2] and M.D. Hanigan[1]
[1]Department of Dairy Science, Virginia Polytechnic Institute and State University, VA 24061, USA; sebuy@vt.edu
[2]Department of Biochemistry, Virginia Polytechnic Institute and State University, VA 24061, USA
[3]Animal Science and Technology College, Shandong Agriculture University, Shandong Province 271018, China PR

Introduction

Dairy cattle only capture about 25% of dietary N in milk protein, with the remainder being excreted. By coordinating AA supply with demand, mammary cells reduce recycling of AA to the splanchnic tissues, thereby increasing efficiency. Several essential AA regulate the rate of synthesis of casein by altering the phosphorylation state of intracellular signaling proteins that affect the activity of initiation (eIF) and elongation (eEF) factors (Appuhamy *et al.*, 2012). Therefore, supplementing the specific AA that exert the most influence on pathways controlling eIF and eEF should allow a reduction in CP supply to the animal, which will increase post-absorptive efficiency and reduce N excretion. The objective of this study was to determine the quantitative responses in milk protein synthesis to individual essential AA in mammary tissue.

Material and methods

Mammary tissue was obtained from 5 lactating dairy cows at slaughter. Each quarter was tested for somatic cell count prior to slaughter to ensure the harvested tissue was not from an infected quarter. Animals were milked a maximum of 2 h before slaughter. Mammary tissue slices (0.121±0.016 g) were prepared from the excised tissue and incubated in 5 ml of treatment media at 37 °C while the flasks were oscillated at 60 oscillation/min for 4 h under an atmosphere of 95:5 O_2:CO_2. Preliminary data showed that $^2H_5$ *Phe* enrichment of the pH 4.6 protein precipitate increased linearly from 0.5 to 8 h (0.43±0.04% enrichment increase/h, adjusted R^2=0.93). Following incubation, slices were homogenized in lysis buffer and caseins were precipitated by acidification to pH 4.6. An aliquot of the pellet was trypsinized and $^2H_5$ *Phe* enrichment in the N[34]LLRFFVAPFPE peptide of alpha S1 casein was measured by MALDI TOF-MS-MS. Fractional synthesis rate (FSR, %/h) of alpha S1 casein was calculated as the ratio of area for labeled and unlabeled newly synthesized peptide and the existent one:

$$FSR_{(\%/h)} = \frac{100}{4} * \frac{\frac{m15}{E_{tRNA}{}^3}}{\left(m0 - \frac{m15}{E_{tRNA}{}^3} * (1 - E_{tRNA})^3\right)}$$

where E_{tRNA} represented the % enrichment in the Phe acylated-transfer RNA (Phe-tRNA) pool, m0 represented the area under the curve for peptides with no labeled Phe (m+0), and m15 represented that for peptides with 3 labeled Phe (m+15). Phe-tRNA enrichment was estimated as the ratio of the abundance of the m+15 peptides to m+15 plus a third of m+10 (2 labeled Phe) peptides. The effect of Ile, Leu, Met and Thr on alpha S1-casein FSR was studied with a central composite design consisting of four central runs, 2 axial runs per AA, and a complete 2^4 factorial. The central run was set at 35% of the concentration of each AA in Dulbecco's Modified Eagle Medium (DMEM; 0.147 mmol/l Ile, 0.158 mmol/l Leu, 0.060 mmol/l Met and 0.077 mmol/l Thr). Axial runs were set at 0 and 100% (0.420 mmol/l Ile, 0.450 mmol/l Leu, 0.170 mmol/l Met and 0.220 mmol/l Thr) of DMEM. Factorial runs were set at 20 (0.084 mmol/l Ile, 0.09 mmol/l Leu, 0.034 mmol/l Met and 0.044 mmol/l Thr) and 50% (0.210 mmol/l Ile, 0.225 mmol/l Leu, 0.085 mmol/l Met and 0.110 mmol/l Thr) of DMEM. The experiment was replicated in 5 cows. In each cow, treatments were replicated twice and the average of the two replicates was used for the analysis. Linear, quadratic and one-way linear interaction parameters were estimated with the Glimmix procedure in SAS (Cary, NC). Cow and residual were random effects.

Results and discussion

Phe-tRNA enrichment decreased by the addition of Ile and Met to the media (data not shown). Table 1 shows the linear, quadratic, and one-way interaction parameter values for individual AA effects on casein FSR. The Leu, Met and Thr caused a quadratic response on FSR, and Ile tended to cause the same effect. The quadratic term for Met was greater than for Ile and Leu. However, the first derivative of the term showed that all maximums were within the range of the data. Given this response, it is possible that the tissue responded in a Michaelis-Menten fashion with a plateau starting before the predicted maximum. Media Ile, Leu and Thr concentrations were lower than arterial concentrations observed by Hanigan *et al.* (2004) at the highest level of casein infusion. In that study a quadratic response in milk protein synthesis to increased AA supply was also observed. Therefore, studies at the animal and the tissue level demonstrate that the mammary gland responds in a curvilinear fashion to supply of AA, which is inconsistent with the fixed partial efficiency used in nutrient requirement models. One-way linear amino acid interactions were not significant. Removing non-significant interactions from the model made the four AA to have a significant effect on FSR. Furthermore, 95% confident limits indicated that the linear parameters did not differ among AA. Thus, curvilinear responses in casein FSR to each of the 4 AA studied were similar and additive rather than conditional. Therefore, a cow clearly deficient in Met could be supplemented with Leu, assuming Leu is also marginally deficient, and a response should be observed. Moreover, the response to Leu would be independent of and additive to the response to Met. This observation is in agreement with Hanigan *et al.* (2002)'s prediction using a modeling approach, and it contradicts the single limiting AA theory assumed in nutrient requirement models.

Table 1. Linear and quadratic parameter estimates, standard errors (SE) and P values for the effect of Ile, Leu, Met and Thr on fractional synthesis rates (FSR) of alpha S1 casein.

	Int.[1]	I[2]	L	M	T	I×I	L×L	M×M	T×T	I×L	I×M	I×T	L×M	L×T	M×T
FSR, %/h															
Param.[3]	1.66	-3.99	6.23	6.0	-0.31	-14.9	-23.0	-139	-74.4	7.50	94.2	79.4	9.63	26.3	92.8
SE[4]	1.66	7.52	7.05	18.6	14.0	7.79	6.8	45.6	26.8	25.4	65.2	51.3	60.7	48.3	125
P value	0.4	0.6	0.4	0.7	1.0	0.06	0.001	0.003	0.007	0.5	0.15	0.12	0.9	0.6	0.5

[1] Intercept.

[2] I = isoleucine; L = leucine; M = methionine; T = threonine.

[3] Parameter.

[4] Standard error.

References

Appuhamy, J.A.D.R.N., N.A. Knoebel, W.A.D. Nayananjalie, J. Escobar and M.D. Hanigan, 2012. Isoleucine and leucine independently regulate mtor signaling and protein synthesis in mac-t cells and bovine mammary tissue slices. J. Nutr. 142:484-491.

Hanigan, M.D., L.A. Crompton, B.J. Bequette, J.A. Mills and J. France, 2002. Modelling mammary metabolism in the dairy cow to predict milk constituent yield, with emphasis on amino acid metabolism and milk protein production: Model evaluation. J. Theor. Biol. 217:311-330.

Hanigan, M.D., C.K. Reynolds, D.J. Humphries, B. Lupoli and J.D. Sutton, 2004. A model of net amino acid absorption and utilization by the portal-drained viscera of the lactating dairy cow. J. Dairy Sci. 87:4247-4268.

Effect of amino acid supply on whole body and tissue glucose kinetics in *postpartum* dairy cows

C. Galindo[1], M. Larsen[2], D.R. Ouellet[3], G. Maxin[3], D. Pellerin[1] and H. Lapierre[3]
[1]Département de sciences animales, Université Laval, Québec, QC G1V 0A6, Canada; carloseduardo.galindo@agr.gc.ca
[2]Department of Animal Science, Aarhus University, Foulum, 8830 Tjele, Denmark
[3]Agriculture and Agri-Food Canada, Sherbrooke, QC J1M 0C8, Canada

Introduction

Previous results on dairy cows have shown that an increased amino acid (AA) supply imcreased net hepatic glucose release, whole-body rate of appearance (WB Ra) of glucose, lactose secretion and milk production. However, it is suggested that in the immediate *postpartum* period, the liver glucose release is mainly supported by an important inter-organ transfer of glucogenic carbons such as lactate from peripheral tissues to the liver instead of an increased utilization of AA for gluconeogenesis. The objective of the study was to investigate the effects of supplementing AA on WB Ra of glucose, and tissue metabolism of glucose, lactate and β-OH-butyrate (BHB) in *postpartum* transition dairy cows.

Material and methods

Nine Holstein cows fitted with rumen cannula and indwelling catheters in the portal, hepatic, and two mesenteric veins, and one artery, were used in a generalized randomized complete block design with repeated measures. At calving, cows received the same lactation ration, were blocked according to parity (2nd and 3rd+) and were divided in 2 treatments: abomasal infusion of water (Ctrl; n=4) or abomasal infusion of free AA with casein profile (AA-CN; n=5); AA-CN infusion started with half the maximal dose on day in milk (DIM) 1 and then steadily decreased from 795±12 g/d to 227±1 g/d from DIM 2 to 29 to cover the estimated AA deficit. On DIM 5, 15 and 29, D-[6,6-$^2H_2$] glucose (23.7±0.2 mmol/h) was infused into a jugular vein for 5 h, and 6 blood samples were taken from arterial, portal, hepatic, and mammary sources at 45 min-intervals, starting one hour after the initiation of the glucose infusion. Trans-organ flux were calculated as veno-arterial differences times plasma flow (splanchnic: downstream dilution of deacetylated *para*-aminohippurate; mammary: Fick principle using Phe+Tyr). Data were analysed with a mixed model including the fixed effects of block, treatment (Trt), DIM, and Trt × DIM and random effect of cow; DIM was tested as a repeated measurement using the appropriate covariance structure.

Results and discussion

Milk production on the 3 days around sampling days increased numerically (P_{Trt}=0.15) with AA infusion (Table 1) and as DIM progressed, dry mater intake and milk production increased (P_{DIM}<0.01). Glucose and lactate arterial concentrations were not altered by AA infusion, but glucose arterial concentration increased (P_{DIM}<0.01) and lactate arterial concentration decreased numerically (P_{DIM}=0.15) as lactation progressed. Arterial concentrations for both BHB and non-esterified fatty acids were higher in AA-CN than in Ctrl cows at DIM 5 (1.26 vs. 0.66; 0.87 vs. 0.58 mM, respectively) but not at DIM 29 (0.88 vs. 0.86, SEM: 0.06; 0.25 vs. 0.34, SEM: 0.07 mM respectively; $P_{Trt \times DIM}$<0.05).

Despite increased AA supply in AA-CN cows and increased demand for milk production, WB Ra of glucose was not affected by AA-CN (P_{Trt}=0.23); similarly splanchnic true flux was unaltered by treatments and is numerically equivalent to WB Ra. Splanchnic utilization of glucose was unaffected by AA-CN and represented on average 16% of WB Ra. Mammary glucose utilization increased with AA-CN infusion (P_{Trt}=0.03), but the proportion relative to WB Ra was unaltered by treatment, averaging 78%, and increased gradually as lactation advanced (P_{DIM}=0.01). Lactate

Table 1. Effect of amino acid infusion (AA-CN) on milk yield, glucose kinetics and glucose, lactate and β-OH-butyrate tissue metabolism in postpartum *dairy cows.*[1]

Item	Treatment		SEM	*P*-values		
	Ctrl	AA-CN		Trt	DIM	Trt × DIM
Dry matter intake, kg/d	20.6	19.4	1.0	0.42	<0.01	0.16
Milk, kg/d	41.9	47.4	2.5	0.15	<0.01	0.68
Body weight change, kg	-63	-105	17	0.14	0.02	0.50
WB Ra[2] of glucose, mmol/h	698	770	41	0.23	0.41	0.79
Arterial concentration, mM						
Glucose	3.13	3.07	0.13	0.74	<0.01	0.52
Lactate	0.28	0.32	0.02	0.31	0.15	0.75
β-OH-butyrate	0.83	1.14	0.06	<0.01	0.02	0.03
Glucose flux[3], mmol/h						
Portal net flux	-37	6	24	0.23	0.29	0.35
Hepatic net flux	661	627	81	0.77	0.21	0.39
Splanhnic net flux	626	630	66	0.96	0.07	0.47
Mammary net flux	-549	-620	28	0.09	0.01	0.28
Portal utilization[4]	-123	-117	23	0.84	0.73	0.85
Hepatic utilization	17	-6	21	0.45	0.93	0.36
Splanchnic utilization	-105	-123	29	0.66	0.74	0.45
Mammary utilization	-529	-620	26	0.03	0.01	0.20
Portal true flux[5]	86	120	11	0.03	0.02	0.13
Hepatic true flux	643	633	78	0.92	0.23	0.46
Splanchnic true flux	731	752	8	0.86	0.09	0.28
Lactate flux, mmol/h						
Portal net flux	208	185	19	0.40	<0.01	0.36
Hepatic net flux	-286	-322	36	0.49	0.21	0.88
Splanchnic net flux	-78	-139	27	0.14	0.13	0.58
Mammary net flux	9	-14	21	0.44	0.29	0.54
β-OH-butyrate flux, mmol/h						
Portal net flux	193	222	37	0.58	0.03	0.85
Hepatic net flux	330	408	45	0.24	0.12	0.23
Splanchnic net flux	524	636	53	0.16	0.23	0.60
Mammary net flux	-231	-271	22	0.22	0.02	0.79

[1] Mean values of the 3 sampling days. [2] Whole body rate of appearance. [3] A positive flux indicates release and a negative flux indicates uptake by the tissue. [4] Utilization from plasma supply. [5] True flux = net flux – utilization.

and BHB net fluxes were not affected by AA infusion, but splanchnic tissue tended to take up more lactate (P_{Trt}=0.14) and to release more BHB (P_{Trt}=0.16) whereas the mammary uptake of BHB was numerically higher for AA-CN cows.

Conclusion

The increased supply of AA in *postpartum* dairy cows did not increase WB Ra of glucose but increased glucose mammary glucose uptake in line with the increased milk yield. These results indicate that in early lactation the AA have priority for metabolic pathways other than gluconeogenesis suggesting that their contribution to hepatic gluconeogenesis may increase as lactation progresses. Consequently, other glucogenic substrates such as lactate and the mobilization of body fat would play an important role in this interval transition to support the energy demands.

Effects of metabolizable protein supply on N efficiency: plasma amino acid concentrations in dairy cows

D.R. Ouellet, D. Valkeners and H. Lapierre
Agriculture and Agri-Food Canada, 2000 College, Sherbrooke, QC, J1M 0C8, Canada; daniel.ouellet@agr.gc.ca

Introduction

Efficiency of N utilization improves with low N intake but this practice remains a viable option only if milk production is not compromised. With this strategy in mind, we studied in a broader perspective the effect of reducing metabolizable protein (MP) supply on urea kinetics, endogenous N secretions, milk yield, N efficiency and plasma amino acid (AA) concentrations. This manuscript reports the three latter.

Material and methods

Four lactating cows were used in a replicated incomplete 3×3 Latin square design (28-day experimental period). Three concentrates (50% of diet; corn and soybean meal-based) were formulated and fed in a total mixed ration with 45% grass silage and 5% hay, to supply NE_L meeting requirement (126 MJ/d), but incremental amounts of MP: 1,430 (Low), 1,920 (Med) and 2,160 (High) g MP/d, corresponding to 72, 98 and 111% of estimated MP requirements (NRC, 2001). Cows were fed every other hour, and blood samples were collected on day 27 at 8:00, 10:00, 12:00 and 14:00 h and on day 28 at 9:00, 11:00, 13:00 and 15:00 h. Concentrations of AA were determined on individual sample using isotopic dilution technique (Calder *et al.*, 1999). Dry matter intake (DMI), milk production, and urinary purine derivatives were measured during the total collection of feces and urine (reported in Valkeners *et al.*, 2007) on days 13-20 of each period. Data were analyzed using the MIXED procedure of SAS/STAT® using the mean of each cow × period. Linear and quadratic orthogonal polynomial contrasts were used to delineate MP supply effect.

Results and discussion

The DMI did not differ between treatments (P=0.14) whereas, milk production decreased from High to Low (P_{linear}=0.06; Table 1). Milk protein concentration was not affected (P=0.49) by MP supply, but milk protein yield decreased (P_{linear}=0.03) from High to Low. Proportion of non-protein N in milk decreased (P_{linear}<0.01) with decreased MP supply. Efficiency of N utilization increased (P_{linear}<0.01) with decreased MP supply. The increased efficiency of N utilization at low MP was related to greater urea recycling to the gut, increasing from 66 to 92% of urea entry rate (Valkeners *et al.*, 2007).

Decreasing MP supply decreased concentration of Lys (P_{linear}<0.08) and most branched-chain AA (Val and Ile: P_{linear}<0.05; Leu: $P_{quadratic}$=0.11) whereas amongst Group 1 AA (His, Met, Phe+Tyr and Trp), His (P_{linear}<0.01) and Trp (P_{linear}<0.07) decreased. This low His concentration at Low supply might be related to the high proportion of bacteria-N, 79% (SEM 1.5%) of the total duodenal N flow, estimated from urinary purine derivatives, compared with 71 and 66% for Med and High, respectively. Across the non-essential AA, only Ala (P_{linear}=0.08) and Glu (P_{linear}<0.001) concentrations were altered by MP supply and they increased with decreased protein supply. These 2 non-essential AA are known to be active contributors in inter-organ movement of N to maintain N homeostasis especially during periods of low N supply.

In conclusion, peripheral concentrations of Group 1 AA, extracted in high proportion of hepatic inflow by the liver, were reduced to some degree for Trp but largely for His by lowered MP supply.

This emphasizes the importance of monitoring His supply at low protein supply, as microbial protein has a low proportion of His. On the other hand, the branched-chain AA which are poorly extracted by the liver, responded to protein supply and could be related to a stimulation of protein synthesis in peripheral tissues including the mammary gland (Lei *et al.*, 2012).

Table 1. Dry matter intake, milk yield and composition and plasma amino acid (AA) concentration of cows receiving incremental metabolizable protein (MP) supply.

Item	MP level			SEM	*P* value, contrast	
	Low	Medium	High		Linear	Quadratic
Dry matter intake, kg/d	16.5	17.2	16.8	0.4	0.16	0.11
Milk yield, kg/d	27.6	29.7	29.1	0.87	0.06	0.15
Milk N, %	3.13	3.15	3.22	0.08	0.34	0.52
Milk true protein yield, kg/d	0.83	0.89	0.89	0.02	0.03	0.27
Milk N: N intake	0.38	0.32	0.28	0.01	0.01	0.97
AA concentrations, μM						
Histidine	18.2	42.2	42.2	4.2	0.009	0.17
Isoleucine	91.2	104.5	116.2	4.8	0.02	0.59
Leucine	132.1	127.0	151.9	7.1	0.20	0.11
Lysine	49.9	55.1	62.4	4.7	0.08	0.53
Methionine	15.8	17.7	16.8	0.7	0.24	0.23
Phenylalanine	37.8	39.4	40.6	1.8	0.39	0.34
Threonine	86.0	90.3	82.5	8.7	0.83	0.44
Tryptophan	44.6	49.4	48.1	1.5	0.07	0.19
Valine	150.1	188.3	216.2	13.4	0.02	0.70
Alanine	223.9	211.3	199.5	11.0	0.08	0.71
Aspartate	18.8	17.4	15.8	2.8	0.43	0.86
Asparagine	38.2	42.7	37.3	1.9	0.89	0.08
Glutamate	69.5	60.8	58.5	2.4	0.001	0.15
Glutamine	264.4	277.9	269.6	15.4	0.67	0.53
Glycine	226.8	223.9	197.0	10.6	0.16	0.27
Proline	70.1	75.9	75.4	3.0	0.24	0.58
Serine	72.9	75.4	72.3	4.0	0.99	0.45
Tyrosine	48.0	43.7	44.4	2.2	0.50	0.60

References

Calder, A.G., K.E. Garden, S.E. Anderson and G.E. Lobley, 1999. Quantitation of blood and plasma amino acids using isotope dilution electron impact gas chromatography/mass spectrometry with U-^{13}C amino acids as internal standards. Rapid Commun. Mass Spectrom. 13, 2080-2083.

Lei, J., D. Feng, Y. Zhang, F.-Q. Zhao, Z. Wu, A. San Gabriel, Y. Fujishima, H. Uneyama and G. Wu, 2012. Nutritional and regulatory role of branched-chain amino acids in lactation. Frontiers in Bioscience 17, 2725-2739.

National Research Council, 2001. Nutrient requirements of dairy cattle 7th rev. ed. Natl. Acad. Press, Washington, D.C.

Valkeners, D., H. Lapierre, J. Marini, and D.R. Ouellet, 2007. Effects of metabolizable protein supply on nitrogen metabolism and recycling in lactating dairy cows. In Energy and protein metabolism and nutrition. EAAP publication No. 124, Pages 417-418.

Small intestinal, stomach complex, and total gastrointestinal tract masses are decreased relative to body weight in high efficiency steers

A.M. Meyer, K.W. Christensen, W.J. Means, S.L. Lake and S.I. Paisley
Department of Animal Science, University of Wyoming, Laramie, WY 82071, USA; ameyer6@uwyo.edu

Introduction

Although both scientific and industry interest in feed efficiency of ruminant animals has increased greatly in recent years, physiological mechanisms underlying differences in individual feed efficiency remain largely unknown. The gastrointestinal (GI) tract and liver are not only essential for nutrient acquisition and utilization, but also use 40 to 50% of whole body energy expenditure (Ferrell, 1988). Despite recent research in the area (Basarab *et al.*, 2003; Mader *et al.*, 2009; Meyer *et al.*, 2012), the role of visceral organs, especially that of the GI tract, in feed efficiency is unclear. We hypothesize that a portion of the individual differences observed for feed efficiency can be attributed specifically to GI tract size and function through its energy use and ability to digest and absorb nutrients. The objective of this study was to investigate GI tract and other visceral organ masses in market weight steers classified as high and low efficiency based on finishing period residual feed intake (RFI).

Material and methods

Hereford-Angus crossbred steers (n=59; from one contemporary group) were fed a common finishing diet (11.4% CP, 2.0 Mcal NE_m/kg, 1.35 Mcal NE_g/kg; DM basis) for 57 d using the GrowSafe Feed Intake System (model 4000E, GrowSafe Systems Ltd., Airdrie, AB, Canada). Twelfth rib fat thickness was determined using ultrasound at the conclusion of the feed intake test. For all steers with at least 1.02 cm 12th rib fat thickness (n=40), RFI was calculated as the expected feed intake subtracted from the actual feed intake. Expected feed intake was determined by regressing average daily gain and metabolic midweight on actual feed intake. The 20% most efficient (low RFI, n=8) and 20% least efficient (high RFI, n=8) steers were slaughtered 5 or 7 d after the end of the feed intake test. At slaughter, visceral organs were dissected, trimmed of fat, stripped of digesta, and weighed. Data were analyzed with PROC MIXED of SAS 9.2 using RFI class (high vs. low efficiency) as a fixed effect in the model.

Results and discussion

Omasal mass (kg) was less (P=0.02) for high efficiency than low efficiency steers, although other visceral organ masses (kg), body weight (BW), and hot carcass weight (HCW) were unaffected (P>0.11) by RFI class. When expressed relative to BW, total GI tract mass (g/kg BW) was less (P=0.02) for high efficiency compared with low efficiency steers. In addition, small intestinal and omasal masses (g/kg BW) were less ($P \leq 0.04$) and stomach complex mass (g/kg BW) tended (P=0.10) to be less in high efficiency steers than low efficiency steers. Conversely, high efficiency steers tended (P=0.10) to have greater pancreatic mass (g/kg BW) compared with low efficiency steers. Similar results were observed when organ masses were expressed relative to HCW. Total GI tract, small intestinal, and omasal masses (g/kg HCW) were less ($P \leq 0.03$) and stomach complex mass (g/kg HCW) tended (P=0.06) to be less for high efficiency steers than low efficiency steers. Despite this, other visceral organ (large intestine, liver, spleen, kidney, omental and mestenteric fat, heart, and lung) relative masses (g/kg BW or HCW) were unaffected (P>0.10) by RFI class.

Results of this study suggest that GI tract masses relative to BW, especially those of the small intestine and stomach complex, are decreased in high efficiency steers. Moreover, differences in stomach complex mass appear to be driven by the omasum. Decreased tissue mass may result in reduced energy and nutrient use, explaining some variation observed in efficiency of nutrient utilization in feedlot steers. Additional data from our laboratory suggest that more efficient cattle have less small intestinal mass, but more dense intestinal mucosa (Meyer *et al.*, 2012), which may enable these animals to maintain digestive and absorptive functions with less tissue mass. In conclusion, growth and function of GI tract tissues are potential targets for development of strategies to improve feed efficiency in beef cattle. Further research is planned to investigate growth, vascularity, and gene expression of these tissues to better understand differences in nutrient use and function that may affect metabolic efficiency.

Table 1. Visceral organ masses and relative masses of high and low efficiency steers.

Item[1]	High efficiency	Low efficiency	SEM[2]	*P*-value
Total GI tract, kg	23.1	25.0	1.0	0.20
g/kg BW	42.7	46.0	0.9	0.02
g/kg HCW	67.6	73.9	1.6	0.01
Stomach complex, kg	17.2	18.4	0.8	0.31
g/kg BW	31.8	33.8	0.8	0.10
g/kg HCW	50.3	54.3	1.4	0.06
Omasum, kg	4.38	5.49	0.31	0.02
g/kg BW	8.10	10.11	0.47	0.01
g/kg HCW	12.8	16.2	0.8	0.009
Small intestine, kg	4.13	4.60	0.20	0.12
g/kg BW	7.63	8.50	0.28	0.04
g/kg HCW	12.1	13.7	0.5	0.03
Large intestine, kg	1.80	2.01	0.12	0.24
g/kg BW	3.32	3.72	0.20	0.18
g/kg HCW	5.27	5.97	0.33	0.15
Liver, kg	6.51	6.79	0.31	0.53
g/kg BW	12.0	12.5	0.3	0.26
g/kg HCW	19.0	20.1	0.4	0.11
Pancreas, kg	0.600	0.515	0.047	0.21
g/kg BW	1.10	0.94	0.07	0.10
g/kg HCW	1.75	1.50	0.10	0.13

[1] Total GI tract = stomach complex + small intestine + large intestine; BW = body weight; HCW = hot carcass weight.
[2] SEM for n=8 per RFI class.

References

Basarab, J.A., M.A. Price, J.L. Aalhus, E.K. Okine, W.M. Snelling and K.L. Lyle, 2003. Residual feed intake and body composition in young growing cattle. Can. J. Anim. Sci. 83, 189-204.

Ferrell, C.L, 1988. Contribution of visceral organs to animal energy expenditures. J. Anim. Sci. 66, 23-34.

Mader, C.J., Y.R. Montanholi, Y.J. Wang, S.P. Miller, I.B. Mandell, B.W. McBride and K.C. Swanson, 2009. Relationships among measures of growth performance and efficiency with carcass traits, visceral organ mass, and pancreatic digestive enzymes in feedlot cattle. J. Anim. Sci. 87, 1548-1557.

Meyer, A.M., K.M. Cammack, S.I. Paisley, P. Moriel, W.J. Means, M. Du, J.S. Caton and B.W. Hess, 2012. Correlation of feed efficiency and small intestinal growth in finishing cattle born to dams fed varying levels of nutrients during early to mid-gestation. J. Anim. Sci. 90, (Suppl. 2): 70.

Whole body oxidative metabolism in dairy cows with a different liver fat content in early lactation

M. Derno, S. Börner, H.M. Hammon, M. Röntgen and B. Kuhla
Institute of Nutritional Physiology 'Oskar Kellner', Leibniz Institute for Farm Animal Biology (FBN), 18196 Dummerstorf, Germany; b.kuhla@fbn-dummerstorf.de

Introduction

During the transition from late pregnancy to early lactation, high-yielding dairy cows mobilize large amounts of adipose tissue resulting in increased plasma concentrations of non-esterified fatty acids (NEFA). NEFA are mostly oxidized to CO_2 whereas at excessive concentrations, NEFA are re-esterified to form triacylglycerides leading to the development of fatty liver. Thus, the capacity for fatty acid oxidation – among others – determines the fat load of the liver. We hypothesized that whole-body fat oxidation adapts to the extent of fat mobilization during early lactation.

Material and methods

Fifteen German Holstein cows (2nd to 4th lactation, >10,000 kg/305 d in at least one previous lactation) were fed *ad libitum* a total mixed ration (dry period: 5.8 MJ NE_L/kg DM, 13% utilizable protein, lactation period: 7.1 MJ NE_L/kg DM, 16% utilizable protein) from week 3 *antepartum* (*ap*) until week 3 *postpartum* (*pp*). In week 3 *ap* and week 2 *pp*, body weight (BW) and ultrasonically measured back fat thickness (BFT) were determined and animals were kept in respiratory chambers (Derno *et al.*, 2009) to which they were well adapted before. Therein, animals were fed *ad libitum* for 24 h to determine dry matter intake (DMI) and milk yield. On the subsequent day, cows were feed-deprived for 10 h and a blood sample was taken in an EDTA-containing tube via extension tubing from the outside of the chamber to analyze plasma NEFA concentration. Concentrations of O_2, CO_2 and CH_4 (measured continuously in 6 min intervals) were analyzed to calculate basal heat production (HP) and whole body fat oxidation (FOX) according to HP (kJ) = 16.18 O_2 (l) + 5.02 CO_2 (l) – 2.17 CH_4 (l) – 5.99 N (g) and FOX (g) = 1.69 O_2 (l) – 1.69 $CO_{2\ (metabolic)}$ (l) using the last measuring interval after 10 h feed deprivation as described previously (Derno *et al.*, 2013). At the end of the 3rd week *pp*, a liver biopsy was taken and based on liver fat content (LFC) as a measure of body fat mobilization cows were allocated to high (HLFC, n=7) and low (LLFC, n=8) LFC groups (G). Data were evaluated by the two-way repeated measure ANOVA procedure of SigmaPlot 11.0 including the effect of G, period (P), and the interaction of G × P followed by post-hoc comparison using the Tukey-Kramer procedure. Linear correlations were calculated using SigmaPlot software and the Pearson correlation coefficient is provided.

Results and discussion

Before and after parturition, BFT was greater ($P_G \leq 0.01$) in HLFC than LLFC cows and differences were larger between G after than before parturition ($P_{G\times P<}$0.01), but DMI and energy corrected milk yield were comparable between groups (Table 1). Furthermore, DMI, FOX, HP, and plasma NEFA concentrations were lower ($P_P \leq 0.001$) during the dry period than in early lactation in all animals (Table 1). Before parturition, NEFA concentrations were comparable between cow groups but *pp*, HLFC cows developed higher ($P<0.05$) plasma NEFA concentrations as compared to LLFC cows ($P_{G\times P}$=0.07), which parallels the fat content of the liver. Furthermore, *ap* HP was not different between groups (P_G=0.33) but in early lactation, HP was higher ($P<0.05$) in HLFC than in LLFC cows ($P_{G\times P}$=0.065), Considering the same DMI before the 10 h-feed deprivation period in HLFC and LFFC cows, differences in *pp* HP are not attributable to dietary HP but should be of catabolic origin.

Table 1. Zootechnical parameters, heat production, whole-body fat oxidation and plasma NEFA concentrations in cows developing a high (HLFC) and low (LLFC) liver fat content postpartum.

Item	*antepartum*		*postpartum*		P_{group}	P_{period}	$P_{group \times period}$
	LLFC	HLFC	LLFC	HLFC			
BFT[1]	12.1^{a}	21.1^{b}	13.0^{c}	23.5^{b}	0.01	<0.01	<0.01
Milk yield			42.8	40.4	0.4		
DMI, kg/d	10.9^{a}	10.2^{a}	16.4^{b}	15.9^{b}	0.641	<0.001	0.86
NEFA, μM[2]	459^{a}	394^{a}	$1,432^{b}$	$1,800^{c}$	0.123	<0.001	0.07
FOX, g/6 min[2]	1.0^{a}	5.8^{b}	4.0^{c}	12.0^{d}	0.001	<0.001	0.1
HP, $kJ/kg^{0.75}d$[2]	586^{a}	623^{a}	748^{b}	876^{c}	0.016	<0.001	0.065

[1] Note: Back fat thickness still increased until parturition above pp values presented.

[2] Determined after 10 hours feed deprivation.

[a-d] Values with different superscripts differ significantly ($P<0.05$, Tukey-Kramer test).

FOX increased from the *ap* to *pp* period and strong correlations between *ap* FOX and *pp* FOX ($r=0.735$; $P<0.01$) were found. These results indicate that increased FOX determined 10 h after feed deprivation between *ap* and *pp* is primarily due to increased body fat mobilization and subsequent fatty acid oxidation and less due to the higher *pp* DMI ingested before the 10 h-feed deprivation period.

Moreover, FOX was higher in HLFC than in LLFC cows not only *pp* but already *ap* despite there were no indications on body fat mobilization in HLFC cows during week 3 *ap* ($P_{GxP}=0.1$). Based on the higher FOX in HLFC compared to LLFC cows already before parturition, we suggest that a high *ap* BFT is accompanied with a high basal FOX, both subscribing the risk for overwhelming fat mobilization and development of fatty liver *pp*.

Acknowledgements

Supported by DFG, Germany.

References

Derno, M., H.G. Elsner, E.A. Paetow, H. Scholze and M. Schweigel, 2009. Technical note: a new facility for continuous respiration measurements in lactating cows. J. Dairy Sci. 92, 2804-2808.

Derno, M., G. Nürnberg, P. Schön, A. Schwarm, M. Röntgen, H.M. Hammon, C.C. Metges, R.M. Bruckmaier and B. Kuhla, 2013. Short-term feed intake is regulated by macronutrient oxidation in lactating Holstein cows. J. Dairy Sci. 96, 971-980.

Lipolytic capacity of visceral adipose tissue in the dairy cow

Á. Kenéz[1], *L. Locher*[1], *A. Rizk*[2], *S. Dänicke*[3], *J. Rehage*[4] *and K. Huber*[1]
[1]*Department of Physiology, University of Veterinary Medicine Hannover, Bischofsholer Damm 15, 30173 Hannover, Germany; akos.kenez@tiho-hannover.de*
[2]*Department of Surgery, Anesthesiology and Radiology, Faculty of Veterinary Medicine, Mansoura University, Elgomheryia St. 15-17, 35516 Mansoura, Egypt*
[3]*Institute of Animal Nutrition, Federal Research Institute for Animal Health, Bundesallee 50, 38116 Braunschweig, Germany*
[4]*Clinic for Cattle, University of Veterinary Medicine Hannover, Bischofsholer Damm 15, 30173 Hannover, Germany*

Introduction

Catecholamine-induced lipolysis in adipose tissue (AT) depots is considered to be the key metabolic pathway for providing endogenous energy in times of high energy demand in the peripartal dairy cow (McNamara, 1991; Koltes and Spurlock, 2011). It has been described in humans and rodents that subcutaneous and visceral AT differ in their lipolytic activity (Tavernier *et al.*, 1995; Reynisdottir *et al.*, 1997). However, in dairy cows, only the subcutaneous adipose tissue (SCAT) has been examined regarding its lipolytic capacity (McNamara and Hillers, 1986), and little is known about the bovine visceral adipose tissues. Locher *et al.* (2011) have revealed that intrinsic expression of hormone-sensitive lipase (HSL) in retroperitoneal adipose tissue (RPAT) is greater than in SCAT in dairy cows antepartum; thus, it is hypothesized that the lipolytic response of RPAT is enhanced, too. Therefore, this *ex vivo* study aimed to determine the rate of lipolysis in catecholamine-stimulated AT explants.

Material and methods

The SCAT and RPAT biopsy samples were collected from 38 dry Holstein-Friesian cows 6 weeks before parturition. After 20 min of pre-incubation, triplicates of SCAT and RPAT slices were incubated in Dulbecco's Modified Eagle Medium containing 2% fatty-acid-free BSA for 90 min at 37 °C, in the absence (basal) or presence of 10^{-6} M isoproterenol (non-selective agonist of beta-adrenoceptors). Free glycerol and free non-esterified fatty acid (NEFA) content of the incubation media was measured, and values were corrected for tissue wet weight. The NEFA:glycerol ratio was calculated. Data were analyzed by two-way ANOVA for factor *Tissue*, factor *Treatment* and respective interactions.

Results and discussion

Release rates of glycerol and NEFA are shown in Figure 1 and Figure 2, respectively. Mean basal release of glycerol and NEFA did not differ between RPAT and SCAT. In both AT, release of glycerol and NEFA increased after isoproterenol-stimulation, but this increase was higher in RPAT than in SCAT.

Our results showed for the first time that RPAT and SCAT of dairy cows did not respond equally after challenging them with the same catecholamine stimulus. In particular, RPAT had a more pronounced lipolytic responsiveness to catecholamine-stimulation than SCAT (Figure 1 and 2). This is most likely due to the difference in the amount of HSL protein in RPAT and SCAT, with significantly higher HSL expression in RPAT in dairy cows antepartum (Locher *et al.*, 2011). The higher lipolytic response of RPAT indicated that RPAT was already able at day 42 antepartum to provide higher lipolytic capacity and therefore was prepared to release more NEFA at times of peak energy demand *postpartum*. Accordingly, being capable of releasing more NEFA, RPAT may contribute to a greater extent to the plasma NEFA peak commonly seen in high yielding dairy cows *postpartum*.

Mean basal NEFA:glycerol release ratio did not differ between the two AT (Figure 3). After isoproterenol-stimulation, the ratio increased in RPAT, but not in SCAT.

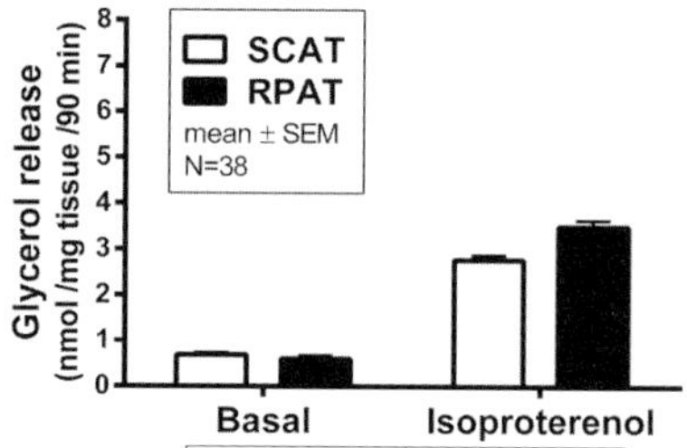

Tissue	*P*=0.001
Treatment	*P*<0.001
Interaction	*P*<0.001

Figure 1. Glycerol release in adipose tissue explants.

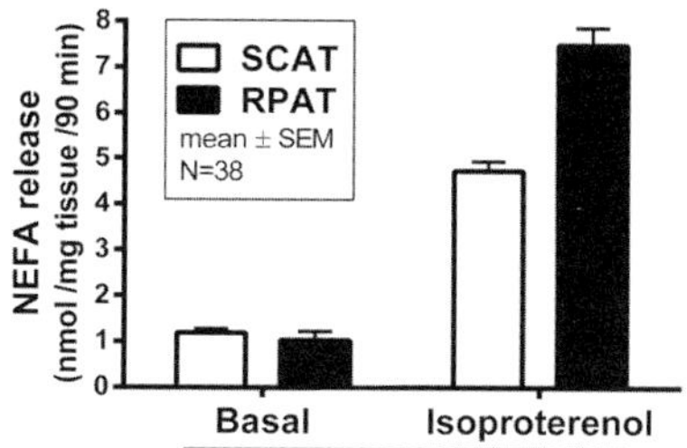

Tissue	*P*<0.001
Treatment	*P*<0.001
Interaction	*P*<0.001

Figure 2. NEFA release in adipose tissue explants.

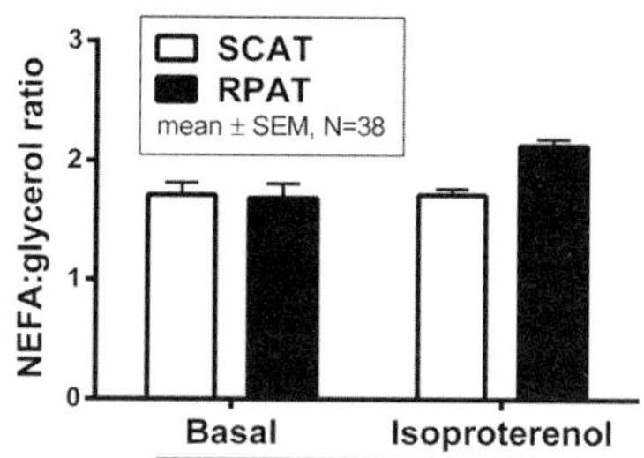

Tissue	*P*=0.029
Treatment	*P*=0.013
Interaction	*P*=0.012

Figure 3. NEFA:glycerol release ratio.

Our results showed that the ratio of NEFA release to glycerol release was constant under basal conditions in both tissues and in SCAT after isoproterenol stimulation as well, but was significantly greater in isoproterenol-stimulated RPAT (Figure 3). Again, this suggests a high impact of RPAT on the development of a plasma NEFA peak *postpartum*. The total hydrolysis and total release of triglycerides would result in a NEFA:glycerol ratio of 3:1. The actual value of release ratio is considered to be an index of NEFA re-esterification within the adipocytes in human studies (Zierath *et al.*, 1998). However there is no evidence concerning the role and significance of re-esterification and membrane transport mechanisms in altering NEFA and glycerol release in dairy cow AT. Therefore, the exact causal factors which led to higher NEFA release rates in isoproterenol-stimulated RPAT still have to be elucidated.

References

Koltes, D.A. and D.M. Spurlock, 2011. Coordination of lipid droplet-associated proteins during the transition period of Holstein dairy cows. J. Dairy Sci. 94, 1839-1848.

Locher, L.F., N. Meyer, E.M. Weber, J. Rehage, U. Meyer, S. Dänicke and K. Huber, 2011. Hormone-sensitive lipase protein expression and extent of phosphorylation in subcutaneous and retroperitoneal adipose tissues in the periparturient dairy cow. J. Dairy Sci. 94, 4514-4523.

McNamara, J.P., 1991. Regulation of adipose tissue metabolism in support of lactation. J. Dairy Sci. 74, 706-719.

McNamara, J.P. and J.K. Hillers, 1986. Regulation of bovine adipose tissue metabolism during lactation. 2. Lipolysis response to milk production and energy intake. J. Dairy Sci. 69, 3042-3050.

Reynisdottir S., M. Dauzats, A. Thörne and D. Langin, 1997. Comparison of hormone-sensitive lipase activity in visceral and subcutaneous human adipose tissue. J. Clin. Endocrinol. Metab. 82, 4162-4166.

Tavernier G., J. Galitzky, P. Valet, A. Remaury, A. Bouloumie, M. Lafontan and D. Langin, 1995. Molecular mechanisms underlying regional variations of catecholamine-induced lipolysis in rat adipocytes. Am. J. Physiol. 268, E1135-1142.

Zierath, J.R., J.N. Livingston, A. Thörne, J. Bolinder, S. Reynisdottir, F. Lönnqvist and P. Arner, 1998. Regional difference in insulin inhibition of non-esterified fatty acid release from human adipocytes: relation to insulin receptor phosphorylation and intracellular signalling through the insulin receptor substrate-1 pathway. Diabetologia 41, 1343-1354.

Effects of nutrient restriction on liver and small intestine energy use in pregnant beef cows

L.D. Prezotto[1], L.E. Camacho[1], C.O. Lemley[2], F.E. Doscher[1], J.S. Caton[1], K.A. Vonnahme[1] and K.C. Swanson[1]
[1]Department of Animal Science, North Dakota State University, Fargo, ND 58102, USA; ligia.prezotto@ndsu.edu
[2]Department of Animal & Dairy Sciences, Mississippi State University, Mississippi State, MS 39762, USA

Introduction

The extensive use of grazing systems for beef cattle and the high variation in forage quality throughout the year has an important impact on production. Changes in forage quality and availability alter the nutritional and physiological status of cows during gestation. Nutrient restriction during this crucial time plays a large role on cow weight change and on fetal growth and development. The objectives were to determine how nutrient restriction and realimentation during gestation influences energy utilization and oxygen consumption of intestinal and liver tissues of the cow and the developing fetus.

Material and methods

Forty six cross-bred multiparous gestating and non-lactating beef cows were individually-fed a common hay diet using Calan gates to meet or exceed NRC recommendations until d 30 of gestation. Pregnant cows were grouped by d of insemination and body weight, and randomly assigned to one of three treatment groups: 1) 100% NRC recommendations throughout the experimental period (CCC; n=18); 2) 60% NRC recommendations from d 30 until 85 of pregnancy, then realimented to 100% of recommendations (RCC; n=16); or 3) 60% of NRC recommendations from d 30 until 140, then realimented to 100% of recommendations (RRC; n=12). Cows were weighed and slaughtered at d 85 (n=6 and 6 for R and C), 140 (n=6, 5 and 6 for RR, RC and CC), and 254 (n=6, 5 and 6 for RRC, RCC and CCC) of gestation. Fetuses were immediately removed and weighed. Maternal and fetal liver and small intestinal mass was recorded and tissues were sampled for immediate analysis. For *in vitro* O_2 consumption analysis, duplicate tissue subsamples (200±10 mg) were placed into test chambers containing 3 ml of buffer and a Clarke polariographic electrode and O_2 consumption measured for 5 min. Maternal and fetal observations are expressed as absolute values and values relative to maternal or fetal body weight (BW), respectively. Data were analyzed using a completely randomized block (breeding group) design using PROC MIXED of SAS, with group used as a random variable and treatment as the main effect. When greater than 2 means were tested and the overall $P<0.05$, means were compared using the pdiff statement in SAS. Means followed by different lower case letter (a, b) are different ($P\leq0.05$).

Results and discussion

At d 85, maternal small intestinal mass increased ($P=0.04$) in R as compared to C cows (R=4,616; C=4,127±148 g). Fetal small intestinal mass relative to BW decreased ($P=0.02$) in R as compared to C cows (R=16; C=19±0.7 g/kg BW). Moreover, fetal small intestinal O_2 consumption relative to BW decreased ($P=0.05$) in R as compared to C cows (R=0.13; C=0.22±0.03 mol/min/kg BW). At d 140 of gestation, maternal liver weight decreased ($P=0.01$) in RR as compared to CC cows (RR=3,723^{a}; RC=4,143^{ab}; CC=4,368^{b}±198 g). Total O_2 consumption in the maternal liver decreased ($P=0.01$) in RR as compared to CC cows (RR=28^{a}; RC=43^{a}; CC=61^{b}±6.3 mol/min). Total O_2 consumption in the maternal small intestine relative to BW also decreased ($P\leq0.03$) in RR cows as compared to CC cows (RR=0.05^{a}; RC=0.08^{ab}; CC=0.09^{b}±0.01 mol/min/kg BW). Fetal liver weight increased ($P=0.03$) in RC and RR cows when compared to CC cows (RR=76^{a}; RC=86^{a}; CC=67^{b}±4.9 g).

Fetal small intestinal O_2 consumption relative to BW increased (P=0.03) in the RC cows when compared to RR and CC cows (RR=0.17^b; RC=0.39^a; CC=0.20^b±0.05 mol/min/kg BW). At d 254 of pregnancy, maternal jejunal weight increased in RCC but decreased in RRC and CCC (P=0.04; RRC=903^b; RCC=$1{,}343^a$; CCC=$1{,}005^b$±130 g). All the other parameters measured did not differ between the treatment groups (P>0.10) are not presented herein.

Tissue mass and metabolic activity influence energy use by tissues. In response to maternal nutrient restriction we observed an increase in small intestinal mass in cows and decrease in fetuses at d 85. This may indicate that cows are adapting to nutrient restriction by increasing intestinal mass in an attempt to increase the digestion and utilization of feed. Moreover, realimentation appears to be able to reverse such effects.

Oxygen consumption (and thus energy expenditure) in the gut accounts for a large proportion of whole body oxygen consumption and it has been observed to vary depending on level of intake (Webster, 1980), physiological status (Reynolds and Huntington, 1988), and animal age (Vatnick *et al.*, 1989). Our results show that intestinal and hepatic O_2 consumption in maternal and fetal tissues decreased in response to nutrient restriction. The decrease in O_2 consumption with no change or an increase in mass may indicate an improvement in energetic efficiency of these tissues during nutrient restriction. However, this change was reversed during the realimentation period.

In conclusion, nutrient restriction in pregnant cows during early and midgestation affects tissue mass and energy utilization within the small intestine and liver. This may indicate that cows and fetuses are adapting to restriction by increasing or decreasing mass in an attempt to increase the digestion and utilization of feed to generate energy. However, maternal realimentation has the potential to reverse these effects.

References

Reynolds, C.K. and G.B. Huntington, 1988. Partition of portal-drained visceral net flux in beef steers. 1. Blood flow and net flux of oxygen, glucose and nitrogenous compounds across stomach and post-stomach tissues. Br. J. Nutr. 60, 539-551.

Vatnick, I., A.W. Bell, J.M. Kelly and B.W. McBride, 1989. Gestational changes in hepatic oxygen consumptioin of the ovine fetus. *In* Energy metabolism of farm animals. Y. Van Der Honing and W.H. Close, comilers. Pudoc, Wageningen, the Netherlands, 159-162.

Webster, A.J.F., 1980. Energy cost of digestion and metabolism in the gut. *In* Digestive physiology and metabolism in ruminants. Y. Ruckebusch and P. Thivend, eds. MTP Press, Lancaster, PA, 469-484.

The effect of pregnancy on weight change, visceral organ mass and circulating serum metabolites in mature beef cows

K.M. Wood[1], C.J. Fitzsimmons[2], S.P. Miller[1], B.W. McBride[1] and K.C. Swanson[3]
[1]Dept. of Animal and Poultry Science, University of Guelph. Guelph, ON, Canada; kwood@uoguelph.ca
[2]Agriculture and Agri-Food Canada / University of Alberta, Edmonton, AB, Canada
[3]Dept. of Animal Sciences, North Dakota State University, Fargo, ND, USA

Introduction

During the last trimester of gestation, fetal growth dramatically increases and results in increased nutrient demand in order to support growth (NRC, 1996). Freetly *et al.* (2008) suggested that cows may be able to reduce maintenance energy costs in order to support the energetic demands of the conceptus. Although visceral organs account for approximately 10% of bodyweight (BW) they represent about 50% of total energy costs (Reynolds *et al.*, 1991) and can be attributed to maintenance energy costs and energy expenditures (McBride and Kelly, 1990). Little is known about the effect of pregnancy in mature beef cows on visceral organ mass and metabolism. By better understanding the metabolic role of pregnancy, there may be opportunities to better understand maintenance energy costs and improve overall feed efficiency.

Material and methods

Eighteen non-lactating, mature Simmental/Angus crossbred cows, pregnant (PREG; n=9) and non-pregnant (OPEN; n=9) were used in a replicated randomize complete block design. Cows were blocked (n=3) by date of parturition such that each block was slaughtered 4-5 wk prior to expected parturition (approximately 250 d of gestation). Each block of the three blocks contained two cows per treatment, with the exception of block three in the second replicate, which contained one cow on each treatment. The second replicate contained only blocks 1 and 3; the first replicate contained 12 cows and the second replicate contained 6 cows. The first block was slaughtered at d 89 of the trial and each block every 7 d thereafter.

Cows were individually fed a ration containing high grass haylage and 30% wheat straw for ad libitum intake (CP, %DM=10.3; NDF, %DM=62.0; NEm=1.10 Mcal/kg DM, NRC,1996). Individual feed intake was measured using Calan gates. Cows were weighed, ultrasounded for rib (over the 12th and 13th rib) and rump fat, and a serum sample obtained three to five d prior to slaughter. At slaughter, organs were removed, trimmed of fat and weighed. Serum was analyzed for beta-hydroxybutyrate (BHBA), non-esterified fatty acids (NEFA), glucose, urea, total cholesterol and triiodothyronine (T3).

Data were analyzed using Proc Mixed (SAS, 2008) as a replicated randomized complete block with pregnancy, cow age, and block nested within replicate as fixed effects and significance determined at $P \leq 0.05$.

Results and discussion

Liver (actual weights, relative to BW, and relative to HCW) were heavier ($P \leq 0.02$; Table 1) in OPEN than in PREG cows. Rumen and kidney fat weight were also greater ($P \leq 0.04$) in OPEN than in PREG, but only when expressed relative to BW. As expected, uterus weight was greater ($P \leq 0.001$) in PREG cows. Kidney, lung, heart, pancreas, spleen, lower gut, and total visceral fat weights (actual, relative to BW or to HCW) were not different ($P \geq 0.06$) between PREG and OPEN cows. Pre-slaughter PREG cows had greater BHBA, NEFA and urea concentrations ($P \leq 0.04$) and lower (P=0.04) total cholesterol concentration. These differences in liver mass and metabolites suggest that pregnant cows alter their metabolism to meet the energetic demands of the growing fetus. However,

Table 1. Visceral organ and internal fat mass in pregnant (PREG) and non-pregnant (OPEN) mature beef cows.

Item	OPEN	PREG	SEM	*P*-value
Body weight pre-slaughter, kg	784	845	26.9	0.12
Liver, g	7,347	6,576	215.9	0.02
Liver/body weight, g/kg	9.42	7.82	0.201	<0.001
Liver/hot carcass weight, g/kg	18.53	15.97	0.457	0.002
Rumen, kg	15.5	13.9	0.62	0.08
Rumen/body weight, kg/kg	0.020	0.017	0.0009	0.01
Rumen/hot carcass weight, kg/kg	0.039	0.034	0.0019	0.06
Kidney fat, g	13,167	10,812	1,077.4	0.13
Kidney fat/body weight, g/kg	16.62	12.73	1.231	0.04
Kidney fat/hot carcass weight, g/kg	32.47	25.99	2.316	0.06

the following measurements did not differ between PREG and OPEN: pre-slaughter BW (P=0.12); average daily gain (P=0.2); pre-slaughter ultrasound measures of rib or rump fat ($P \geq 0.3$); and hot carcass weight or measured back fat ($P \geq 0.4$). Dry matter intake did not differ (P=0.20) and averaged 12.8 kg DM/d for OPEN and 13.4 kg DM/d for PREG (SEM= 0.36).

These data indicate that PREG cows may metabolize energy reserves and alter their metabolism in order to meet the energetic demands of the growing fetus, without affecting their own growth. More research is needed to determine the underlying metabolic processes involved in these differences in circulating metabolites and liver, rumen and kidney fat mass and if these changes may lead to underlying differences in feed efficiency.

References

Freetly, H.C., J.A. Nienaber and T. Brown-Brandl, 2008. Partitioning of energy in pregnant beef cows during nutritionally induced body weight fluctuation. J. Anim. Sci. 86, 370-377.

McBride, B.W. and J.M. Kelly, 1990. Energy cost of absorption and metabolism in the ruminant gastrointestinal tract and liver: a review. J. Anim. Sci. 68, 2997-3010.

National Research Council. 1996. Nutrient Requirements of beef cattle. National Academy

Press, Washington, DC.

Reynolds, C.K., H.F. Tyrrell and P.J. Reynolds, 1991. Effects of diet forage-to-concentrate ratio and intake on energy metabolism in growing beef heifers: whole body energy and nitrogen balance and visceral heat production. J. Nutr. 121, 944-1003.

SAS, 2008. Release 9.1.3. SAS Inst. Inc. Cary, NC, USA.

Skeletal muscle fatty acid oxidation in lactating dairy cows during early lactation

C. Schäff, H.M. Hammon, M. Röntgen and B. Kuhla
Institute of Nutritional Physiology 'Oskar Kellner', Leibniz Institute for Farm Animal Biology (FBN), 18196 Dummerstorf, Germany; b.kuhla@fbn-dummerstorf.de

Introduction

The mobilization of adipose tissue of dairy cows during early lactation is reflected by increased fatty acid (FA) plasma concentrations. The capacity of the liver for complete oxidation of non-esterified FA (NEFA) is limited leading to an increased formation of ketone bodies, re-esterification, and accumulation of triacylglycerides in the liver. This is accompanied by a lower feed intake and a high incidence of metabolic disorders (Drackley *et al.*, 2005). The skeletal muscle may also oxidize FA and thus relieve the liver from FA load. Here we hypothesized that skeletal muscle FA degradation adapts to the extent of fat mobilization during early lactation. .

Material and methods

German Holstein cows (2nd to 4th lactation, >10,000 kg/305 d in a previous lactation, n=19) were fed twice daily a total mixed ration according to their physiological state. From 3 weeks before until 5 weeks after parturition weekly blood samples were taken for the photometric analysis of total plasma NEFA and glucose concentrations (Abx Pentra 400; Horiba), and the determination of the plasma FA profile and 3-methylhistidine (3-MH) concentration by HPLC (GC-2010 and GCMS-QP2010; Shimadzu). As the extent of fat mobilization is reflected by the liver fat content (LFC), liver biopsies were obtained on days 3, 18, 30 post-partum (PP). Based on the mean LFC, cows were grouped into the ten highest (H) and nine lowest (L) mobilizing cows. Muscle biopsies were obtained by shoot biopsy on days -17, 3, 30 relative to calving and used to quantify the phosphorylation of AMP-activated protein kinase (pAMPK) by ELISA (ELISA DuoSet IC; R&D Systems) and the mRNA expressions by quantitative real-time reverse transcriptase PCR (qRT-PCR). Total RNA was extracted with Trizol Reagent. Integrity and quality were proofed by gel electrophoresis and on a NanoPhotometer. RNA was reverse-transcribed and the obtained cDNA applied to qRT-PCR carried out on a LightCycler 2.0 (Roche) for the following genes: peroxisome proliferator-activated receptor family (PPAR α, γ, δ, and PPARγ-coactivator 1α (PGC1α)), carnitine palmitoyltransferase 1α and β (CPT1α, β), the β-oxidative enzymes 3-hydroxyacyl-CoA dehydrogenase (HADH), very long-chain acyl-CoA dehydrogenase (ACADVL), 3-ketoacyl-CoA thiolase (ACAA2), and uncoupling protein 3 (UCP3). mRNA expressions was quantified according to the $2^{-\Delta\Delta CT}$ method using splicing factor 3 subunit 1 (SF3A1) as housekeeping gene. Back fat thickness (BFT) and *m. longissimus dorsi* thickness (MLDT) were determined once a week. All data were analyzed by repeated measures ANOVA using the mixed procedure of SAS (SAS/STAT software in the SAS System for Windows, release 9.2.) with group, time, and group × time as fixed effects. Least square means were calculated and presented with their standard error.

Results and discussion

BFT and MLDT were higher in H cows (P_{group}<0.05) and decreased PP as expected (P_{time}<0.001). Total NEFA concentration peaked in the third week PP (P_{time}<0.001) and was higher in H than in L cows (P_{group}<0.001). The ratio between unsaturated and saturated FAs was higher in H than in L cows (P_{group}<0.01) and continuously increased over time (P_{time}<0.001). Also 3-MH, considered as a marker of muscle breakdown, peaked around calving (P_{time}<0.001). Group × time interactions were not observed for the variables analyzed. These results indicate that the mobilization of adipose and muscle tissue PP is more pronounced in H as compared to L cows.

The activation (phosphorylation) of the energy sensor AMPK in muscle tissue was not different between groups and did also not change over time. This finding indicates that in early lactation the muscle is sufficiently supplied with ATP energy, despite reduced plasma glucose concentrations (P_{time}<0.001) and muscle tissue degradation. To be able to maintain cellular energy homeostasis, muscle tissue seems to switch from glucose to FA utilization in early lactation.

Long-chain FAs are transported into the mitochondria where they are oxidized. Several key regulators are involved in these processes and were analyzed for their mRNA expression. CPT1α and CPT1β, which facilitate the entrance of FAs from the cytosol into the mitochondria, were upregulated in early lactation with CPT1β being highest on d 3 PP and CPT1α on d 30 PP (P_{time}<0.001). The expression of PPARs, which comprise important nuclear FA receptors and transcription factors regulating FA oxidation, changed during the periparturient period (P_{time}<0.05). In detail, we found increased expression of PPARγ in early lactation (P_{time}<0.05) being higher in H than in L cows (P_{group}<0.05). In contrast, expression of PPARα continuously decreased with time (P_{time}<0.001). PPARδ (the predominant form in muscle tissue) was highest expressed on day 3 PP and revealed an interaction ($P_{time \times group}$<0.05). UCP3, which signals for the transport of FAs into and out of mitochondria and so promotes fat oxidation in cattle (Brennan *et al.*, 2009) was also increasingly expressed at day 3 PP and showed an interaction ($P_{time \times group}$<0.05). In contrast, PGC1α, known to be a coactivator of PPARδ and for transcription of UCP3 and CPT1β (De Lange *et al.*, 2007), dropped markedly after calving (P_{time}<0.001). On the other hand, enzymes involved in β-oxidation, namely HADH (P_{time}<0.07), ACAA2, and ACADVL (each P_{time}<0.05) were highest expressed on day 3 PP.

Our results indicate that around calving transport and oxidation of FAs into muscle cells are highly activated which may occur as adaptation to the onset of fat mobilization in early lactation. PPARδ's promoting effect on FA oxidation is likely not dependent on the interaction with PGC1α. Within the first 4 weeks of lactation, however, muscular FA oxidation decreases to the ante partum level, although plasma NEFA concentrations are still elevated. Hence it seems likely that muscular FA oxidation is generally more dependent on the stage of lactation than on the extent of fat mobilization. The latter assumption is supported by the finding that only minor mRNA expression differences were observed between groups. Consequently, FA oxidation in skeletal muscle may relieve the liver from too high fat load already in the beginning of lactation but becomes less important in week 5 of lactation.

Acknowledgements

Supported by DFG, Germany

References

Brennan, K.M., J.J. Michal, J.J. Ramsey and K.A. Johnson, 2009. Body weight loss in beef cows: I. The effect of increased beta-oxidation on messenger ribonucleic acid levels of uncoupling proteins two and three and peroxisome proliferator-activated receptor in skeletal muscle. J. Anim. Sci. 87, 2860-2866.

Drackley, J.K., H.M. Dann, G.N. Douglas, N.A. Janovick-Guretzky, N.B. Litherland, J.P. Underwood and J.J. Loor, 2005. Physiological and pathological adaptations in dairy cows that may increase susceptibility to periparturient diseases and disorders. Ital. J. Anim. Sci. 4, 323-344.

De Lange, P., M. Moreno, E. Silvestri, A. Lombardi, F. Goglia and A. Lanni, 2007. Fuel economy in food-deprived skeletal muscle: signaling pathways and regulatory mechanisms. FASEB J. 21, 3431-3441.

Transcriptome profile comparison between beef and dairy adipose pooled mRNA reveals differences

J.M. Thomson[1,2], P. Stothard[2] and J.P. McNamara[3]
[1]Montana State University, Animal and Range Sciences, P.O. Box 172900, Bozeman, MT 59717, USA; jennifer.thomson@montana.edu
[2]University of Alberta, Department of Agriculture, Food, ad Nutritional Sciences, 1400 College Plaza, Edmonton, AB T6G 2P5, Canada
[3]Washington State University, Department of Animal Sciences, 116 Clark Hall, Pullman, WA 99164, USA

Introduction

Recent research has found evidence of selection in *Bos taurus* when compared to wild relatives from the Bovina subtribe. The Bovina subtribe is thought to have diverged from *B. taurus* between one and two million years ago (MacEachern *et al.*, 2009). This research was also able to show large deviations from neutrality in two distinct breeds of cattle: the Holstein which has been subjected to artificial selection primarily on the basis of milk production; and Angus, a commercial beef breed which has had broader selection on traits related to growth rate, meat quality, and suitability for a more extensive production system (MacEachern *et al.*, 2009; Hayes *et al.*, 2008).

The uptake of next generation sequencing (NGS) technologies has expanded our ability to examine the gene expression and transcriptome differences between these breeds. This technology allows a broader look at gene expression than its predecessor, microarrays, as the number and specificity of the probes on the array do not limit it. It provides a powerful tool to examine tissue characteristics and improve our annotation and understanding of the relationship of gene expression and tissue function (Harhay *et al.*, 2010). Bioinformatic approaches have been developed that allow for comprehensive functional analysis of the large gene lists generated by RNA-seq methods including gene set enrichment analysis (GSEA) and modular enrichments analysis (MEA) (Huang *et al.*, 2008). The aim of the present experiment was to compare the functional transcriptome of beef and dairy cattle adipose tissue in order to better understand the role of the tissue in the divergently selected breeds.

Material and methods

The RNA was isolated from adipose tissue samples from 8 North American Holstein-Friesian animals; four mature lactating animals and 4 growing heifers and 10 crossbred beef animals. The crossbred beef animals were primarily of Hereford and Angus genetics. The RNA was pooled within breed if it had an RNA integrity number greater than 7.5. This was done to produce a single representative sample and to reduce the background variation due to animal-to-animal genetic difference. The pooled RNA was enriched for mRNA using poly-T oligo-attached magnetic beads and then cDNA was synthesized. The cDNA was fragmented and sequencing adapters were added. The cDNA libraries were sequenced according to manufacturer recommendations. The cDNA libraries from the crossbred beef animals were sequenced on a Genome Anlyzer II (Illumina) and the adipose from the Holstein animals was sequenced on a SOLiD 4 platform (Applied Biosystems). Even though two NGS platforms were used we were confident in the comparison because of the high repeatability of RNA-seq experiments across platforms (Jiang *et al.*, 2011; Nookaew *et al.*, 2012) when a minimum of 20 million mapped reads is used. We generated 60 million mapped reads from dairy adipose and 40 million mapped reads from beef adipose. The reads were assembled and normalized expression values calculated using the Genesifter NGS pipeline (Geospiza). Both Genesifter (Geospiza) and DAVID bioinformatics software version 6.7 were used to perform gene ontology and pathway enrichment analyses. We used a Fisher's Exact Test and a Bonferroni adjustment for multiple comparisons to determine differential expression. We then filtered our data based on an adjusted P-value of <0.01 and at lease a 2 fold difference between breeds.

Results and discussion

We found 16,214 transcripts that were differentially expressed in the adipose tissue of the two breeds. Some of the most highly expressed genes in the Holstein adipose tissue included fatty acid binding protein 4 (FABP4), cytochrome c oxidase 2 (COX2), and ATP synthase subunit 6 (ATP6). Fatty acid binding protein 4 (FABP4) was also high in beef cattle but other highly expressed genes were not shared including osteonectin (SPARC), and vimentin (VIM) and matrix Gla protein (MGP) which is a growth inhibitor. In general beef cattle had enriched biosynthetic pathways when compared to dairy and increased expression of genes related to chemokine signaling. Dairy cattle showed enrichment of pathways related to oxidative phosphorylation and cellular remodeling when compared to beef cattle adipose tissue.

This provides evidence that genetic selection can alter gene expression, tissue metabolism and function, and may provide insight into how we can improve genetic selection efficiency.

References

Harhay, G. P., T.P. Smith, L.J. Alexander, C.D. Haudeenschild, J.W. Keele, L.K. Matukumalli, S.G. Schroeder, C.P. Van Tassell, C.R. Gresham, S.M. Bridges, S.C. Burgess and T.S. Sonstegard, 2010. An atlas of bovine gene expression reveals novel distinctive tissue characteristics and evidence for improving genome annotation. Genome Biology 11:R102.

Hayes, B. J., A.J. Chamberlain, S. MacEachern, K. Savin, H. McPartlan, I. MacLeod, L. Sethuraman and M. Goddard, 2008. A genome map of divergent artificial selection between *Bos Taurus* dairy cattle and *Bos Taurus* beef cattle. Animal Genetics 40:176-184.

Huang, D.W., B.T. Sherman and R.A. Lempicki, 2008. Bioinformatics enrichment tools: paths toward the comprehensive functional analysis of large gene lists. Nuc Acids Res 37:1-13.

Jiang, L., F. Schlesinger, C.A. Davis, Y. Zhang, R. Li, M. Dalit, T.R. Gingeras and B. Oliver. 2011. Synthetic spike-in standards for RNA-seq experiments. Genome Res 21:1543-1551.

MacEachern, S., B. Hayes, J. McEwan and M. Goddard, 2009. An examination of positive selection and changing effective population size in Angus and Holstein cattle populations using a high density SNP genotyping platform and the contribution of ancient cattle polymorphism to genomic diversity in Domestic cattle. BMC Genomics10:181.

Nookaew, I., M. Papini, N. Pornputtapong, G. Scalcinati, L. Fagerberg, M. Uhlen and J. Neilsen, 2012. A comprehensive comparison of RNA-seq based transcriptome analysis from reads to differential gene expression and cross-comparison with microarrays. Nuc Acid Res 40:10084-10097.

Effects of omitting the dry period on plasma progesterone and prolactin during lactogenesis and on colostrum IgG content in dairy cows during the periparturient period

H.A. van Dorland[1,2], R.S. Zbinden[1], G. Remmelin[3], B. Kemp[4], A.T.M. van Knegsel[4] and R.M. Bruckmaier[1]
[1]Veterinary Physiology, Vetsuisse Faculty University of Bern, Bern, Switzerland; anette.vandorland@bfh.ch
[2]Present address: School of Agricultural, Forest and Food Sciences, Bern University of Applied Sciences, Zollikofen, Switzerland
[3]Livestock Research, Wageningen University and Research Centre, Lelystad, the Netherlands
[4]Adaptation Physiology Group, Wageningen University, Wageningen, the Netherlands

Introduction

Omitting the dry period represents a strategy that may reduce the metabolic stress in early lactating cows (e.g. Andersen *et al.*, 2005). A drawback of continuous milking with no dry period might compromise colostrum quality in goats (Caja *et al.*, 2006): however, insufficient studies have been carried out to confirm this in dairy cows.

The objective of this study was to evaluate the effects of omitting the dry period on key hormone patterns during lactogenesis and the IgG content of colostrum in periparturient dairy cows.

Material and methods

Twenty Holstein-Friesian dairy cows were randomly assigned to two experimental groups of which one group was managed without a dry period (0 d dry period; DRY0), and the other group was managed with a dry period of 60 days (DRY60). Prepartum, dry cows received a dry cow ration, lactating cows received a lactation ration supporting 25 kg of milk. *Postpartum*, all cows received a lactation ration. Dry cow ration consisted of grass silage, corn silage, wheat straw, rapeseed meal and concentrate in a ratio of 37:23:23:11:6 (DM basis). Lactating ration consisted of grass silage, corn silage, rapeseed straw, soybean meal and concentrate in a ratio of 48:32:2:12:6. Dry matter intake was recorded daily between day -7 and day 3 relative to calving date. Milk yields were recorded daily, and milk samples were collected daily for analysis of IgG concentration from day 7 pre- until day 3 *postpartum* for DRY0, and from day 0 until day 3 *postpartum* for DRY60 cows. Furthermore, from day 5 pre- until day 3 *postpartum*, blood was sampled daily from the coccygeal vein and plasma concentrations of progesterone and prolactin were determined.

Data were analyzed with the MIXED procedure of SAS (SAS Institute, 2001), including treatment (dry period length), sampling time-point and their interaction as fixed effects. The sampling time points were treated as repeated factor within subjects (animals).

Results and discussion

Dry matter intake did not differ ($P_{treatment}$>0.05) between DRY0 and DRY60 cows (14.2±0.5 vs. 13.9±0.5 kg, resp.). Daily milk production between day 0 and day 3 post-calving averaged 22.3±1.9 kg in DRY60 cows, and 13.5±1.4 kg in DRY0 cows ($P_{treatment}$<0.05).

Plasma prolactin concentration did not differ between DRY0 and DRY60 cows and followed a similar pattern ($P_{treatment \times day}$>0.05), starting to increase significantly from one day before calving in both groups (P_{day}<0.05; Figure 1A). Progesterone dropped prepartum (P_{day}<0.05) and followed a similar pattern in both groups, but was significantly lower on one day before calving in DRY60 compared to DRY0 (Figure 1B).

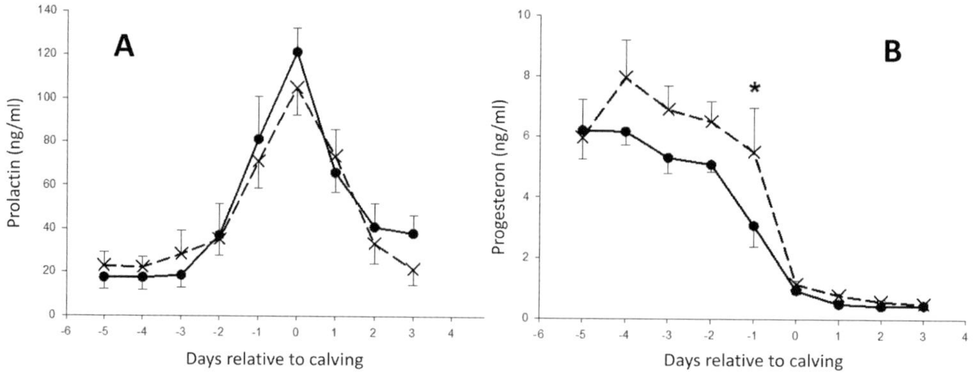

*Figure 1. Concentrations of prolactin (A) and progesterone (B) in cows without dry period (+; DRY0) and in cows with a dry period of 60 days (●; DRY60). A * represents a difference between DRY0 and DRY60 (P<0.05).*

These observations show that the endocrine profiles of hormones supporting lactogenesis remained largely unaffected by the omission of the dry period. The only significant difference between DRY0 and DRY60 for progesterone one day before calving may point out to a stronger decrease in progesterone concentration for DRY0 and could also imply a faster calving process after the progesterone drop.

With regard to IgG concentration, IgG level decreased in both groups from calving up to day 3 *postpartum* and no difference was observed between the DRY60 and DRY0 cows. In contrast, calculated total IgG mass in first colostrum was higher ($P_{treatment}$<0.05) for DRY60 compared to DRY0 cows (1,006±149 g vs. 274±103 g).

DRY0 cows compared to DRY60 cows had a lower milk yield ($P_{treatment}$<0.05) at the first *postpartum* milking (13.5±2.0 kg vs. 21.8±1.7 kg). The continuous milking prevented accumulation of IgG mass in the mammary gland in DRY0 cows, which was compensated by less dilution through less milk synthesis. Therefore, the supply of colostrum to the newborn calf was not compromised in DRY0 compared to DRY60 cows, as the newborn calf would ingest only a small amount of the total first colostrum (generally, requirement is 4 kg of first colostrum) and IgG concentration in milk on day 1 was similar in both groups.

References

Andersen, J.B., T.G. Madsen, T. Larsen, K.L. Ingvartsen and M.O. Nielsen, 2005. The effects of dry period versus continuous lactation on metabolic status and performance in periparturient cows. J. Dairy Sci. 88, 3530-3541.

Caja, G., A.A.K. Salama and X. Such, 2006. Omitting the dry-off period negatively affects colostrum and milk yield in dairy goats. J. Dairy Sci. 89, 4220-4228.

SAS Institute. 2001. SAS User's Guide: Statistics. Version 8 ed. SAS Inst. Inc., Cary, NC.

Part 8. Environmental sustainability

The contribution of animal production to agricultural sustainability

M. Doreau[1], *H.P.S. Makkar*[2] *and P. Lecomte*[3]
[1]*INRA, UMR Herbivores, 63122 Saint-Genès Champanelle, France; michel.doreau@clermont.inra.fr*
[2]*Animal Production and Health Division, FAO, Rome, Italy*
[3]*CIRAD, UMR SELMET, 34398 Montpellier, France*

Abstract

Due to the increasing demand for animal products for a growing world population, use of a large amount of feed ingredients in competition with human food coupled with the urgent need for mitigating environmental effects, livestock farming has to transform. New models of sustainability, that are acceptable to all stakeholders, must be explored. Reducing negative environmental impacts of various agricultural practices is a major global challenge. However, to avoid making inappropriate decisions, sustainability should be understood and addressed by combining indicators that are relevant at farm, country and world levels and not solely based on the emission of greenhouse gases. Sustainability has to go beyond productive efficiency, embracing eco-efficiency concepts and including social equity, and ethical dimensions of development. There is a large diversity of systems, from low-productive or extensive to high-productive or intensive, and countries are at different stages of development, from developing, fast developing to developed. Hence solutions are likely to be very different. Co-existence of different livestock production systems and addressing their sustainability using multiple indicators is a key to success.

Introduction

Livestock farming enters a paradoxical situation. In a changing context, marked by escalating costs of inputs, decreasing natural resources and ongoing climate change, livestock production systems have to transform in the short- to medium-term. For the expected 9 billion people by 2050 the world will need to produce much more food with the same agricultural area, or with a smaller area, if the ongoing soil degradation is not abated and/or biofuel production keeps growing. Increases in income and urbanization in emerging countries and resulting changes in food habits lead to inclusion of a higher amount of animal-source foods (meat, milk, eggs) in the diet of humans (FAO, 2009). Transformation of biomass into edible animal foods has a lower efficiency than direct vegetal food production. The former requires more natural resources per MJ of energy or kg of protein yield, and thus leads to higher emissions. The desire to protect environment and biodiversity, and to improve animal well-being are raising new questions regarding the manner by which animal foods are being produced or will be produced. For example, how can the increase of the current production be considered while reducing environmental effects and maintaining equity in the recognition and sharing of economical and social impacts of the goods and services produced through livestock? How can animal agriculture be made a driver for agricultural sustainability? In the future, countries and societies will prioritize objectives of producing livestock differently, depending on factors such as income levels, role of smallholders relative to large scale producers, importance of and prospects for import or export, degree of pressure on, and degradation of, natural resources, and social and ethical concerns. The objectives will be prioritized differently according to the country's stage of economic development. Addressing these divergent goals is a challenge for the next decades.

Livestock systems, which are very diverse throughout the world (Robinson *et al.*, 2011), should take better into account the efficiency with which natural resources, carbon, nitrogen and energy, as well as human and animal resources are used to produce goods and provide multiple functions in diverse landscapes. There is currently a major societal debate on both positive and negative impacts of livestock production and their future capability to provide ecosystem services. The differentiated role of intensive, mixed crop-livestock and extensive or low-input systems, their productive performances in context to their natural, renewable or non-renewable resource use, social contributions and environmental impacts, are often controversial and challenging subjects. Across the world, the diverse agro-ecological contexts present a wide spectrum of assets and constraints, resulting in geographical diversity of livestock farming systems performing multiple functions in the territories where they developed. Sustainability indicators for such diverse systems should be broadened and balanced, to go beyond traditional assessments of productive performance.

Sustainability in agriculture – a concept difficult to arrive at a consensus

The classical definition of sustainability demands balance amongst three pillars corresponding to environmental, economic and social issues. Criteria for their assessment are numerous, and there is no consensus on their applicability. Environmental issues include the use of natural resources and its impacts on the environment. An obvious natural resource capital is land, the use of which for agriculture is increasingly becoming limiting, despite a recent and extensive deforestation of primary forest and the presence of still large uncultivated areas, for example in Siberia. In addition, on arable land there is an increasing competition between the production of human food, animal feeds and biofuels. Other natural resources used by agriculture are either non-renewable (e.g. fossil energy or phosphorus) or renewable but vulnerable (e.g. freshwater). In the latter case, a shortage has consequences on global ecosystems and human activities. Environmental impacts may be global, e.g. greenhouse gases (GHG) are emitted in the world's atmosphere, wherever they are produced; however, other impacts are restricted to a small territory or a watershed, e.g. soil and water pollutants. Social issues do not have the same meaning in developed and developing countries, in rural and urban societies. Lebacq *et al.* (2013) distinguished social issues based on farm-targeted criteria (working conditions, education, way of life) and society-targeted criteria (multifunctionality, acceptability of agricultural practices, quality of products). A common point to all social issues is the difficulty in finding an agreement amongst stakeholders for the criteria and then giving them relative weights. Economic evaluation is generally done in terms of profitability, a short-term criterion, whereas it should also consider macro-economic changes and future agricultural policies, as well as resilience to climatic and market hazards. In addition, the functional expression of income is extremely variable, and can be calculated per unit of product, or per hectare on-farm, or on-farm plus off-farm, etc. with different conclusions for profitability according to the unit chosen.

The mandatory need to feed 9 billion people requires addressing trade-offs between sustainability at farm level and at global level. The efficiency of production is not given due consideration in some production systems. For example, it is unclear whether low-input diets with a low animal productivity, which may provide a good income to the farmer (e.g. organic farming), are sustainable at a large scale for providing food to fast developing countries. It is also necessary to integrate natural changes, such as the effect of climate change on agriculture (change in crop yield, in crop species) and the adaptive capacity of animals which may help to cope with these changes. Breeds/species which were the most adapted may be less adapted to different natural conditions in future.

Addressing the environmental issues in many developing and transition countries is important but it does not divert focus from the main objective which is to increase the national production for meeting the increasing food needs. The social importance of family households (farmers) becomes low at the expense of urban society in many developed countries, whereas in the least developed countries smallholders who represent a major proportion of the population play a central role. Most societal issues related to environment, security and welfare are of higher importance for countries with consumers having a high standard of living. Attaining well-being of the agricultural community is aimed throughout the world but with large differences among countries due to differences in their developmental stage. There is a need to go beyond the sustainability assessment of existing systems, and to redesign systems for the long term by anticipating economic, social and environmental trade-offs and shocks that will come from future development of the global production. Table 1 proposes

Table 1. Criteria for agricultural sustainability: their nature and importance differ according to the level where they are considered (only a selection of criteria has been mentioned).

	Environment	Social	Economic
Farm level	Use of organic manure	Human well-being Increasing livelihood Cultural role of animals	Profit Resilience
Country or territory level	National GHG inventory Soil and water pollution Use of water	Employment Animal welfare Food security	Import-export balance Food sovereignty
World level	Use of natural resources Air pollution Land use, deforestation	Reduction in poverty	Achieving production target

a set of sustainability criteria to address issues at different levels, from local level (the farmer) to world level (planet and mankind).

Animal production and the environment

A significant part of the environmental impact and intensive use of non-renewable resources are related to food production. The role of livestock in greenhouse gases (GHG) emissions, water and soil pollution, loss of natural resources and of biodiversity, among others, has been stressed by FAO (Steinfeld *et al.*, 2006). As farm animals consume plants, animal-source foods have a higher environmental impact than vegetal food per kg of edible product. In addition, a significant part of GHG emissions comes from enteric methane produced by ruminants. Comparing impacts based on GHG emission per kg of product might not be appropriate, because of differences in dry matter, energy and protein contents of products. So making comparison based on energy or protein yield is expected to give a better insight. Nevertheless, Vieux *et al.* (2012) surprisingly showed that replacing meat with fruit and vegetables on an iso-caloric basis did not decrease total GHG emissions. A comparison on an energy basis does not account for a major specificity of animal foods, which is protein supply. The expression per kg of protein is especially useful to compare milk, eggs and meat of different origin. It has thus been shown that all impacts (GHG, eutrophication, energy and land use) per kg protein are higher for beef than for other animal products (Gill *et al.*, 2010, De Vries and de Boer, 2010). What is often questioned about livestock production is that the products derived from 35% of arable land are consumed by animals largely as feeds. This land could have been used for producing food for direct consumption by humans. It is a simplistic view because a large part of these products are co-products and by-products which are human non-edible. Moreover livestock and their feeds also provide a buffer in national and international markets, which can be drawn upon in case of food shortages. In the previous world food crises of the last 4 decades 1974/75 and 1981/82, overall grain supplies to animal sector fell significantly at world level (FAO, 2009). In addition, many geographical areas, such as mountains and rangelands, are covered by grass because they are not suitable for crop production.

High water demand of livestock is often cited as a major environmental issue. A frequently cited figure is 15,000 liters of water needed to produce one kg of beef. This calculation includes 'green' water, i.e. water need for producing plants which are consumed by animals. Plant water need corresponds to evapotranspiration, which is positively related with crop/forage yield and rainfall. As a consequence, extensive systems require the largest amounts of water, because they need large land areas. Such a calculation is made independent of water scarcity: when including green water, the water requirement for world annual beef production is higher than the world's freshwater reserves

(Doreau *et al.*, 2012). In order to determine the effective impact of livestock on water resources, recently improved methodologies have been developed. They use a life cycle assessment (LCA) approach, and consider only blue water, i.e. water consumed as liquid by the animals and on-farm water use for servicing and irrigation to produce feeds. Some of the authors propose a weighted value accounting for the risk of water reserve depletion (Kounina *et al.*, 2013).

Animal production has also positive effects on the environment. In developed countries, it often adds to the quality of landscape for urban people. In developing countries, an adequate manure management restores soil fertility in crop-livestock systems, and in pastoral areas management of grasslands is a unique way to use soil and to protect it against erosion (Blanfort *et al.*, 2011). These environmental services are often difficult to take into account because the value of the quality of landscapes with natural grasslands compared to cereal fields, for example, is difficult to quantify, e.g. in monetary terms. Recently, in a LCA approach, two groups of authors (Nguyen *et al.*, 2012b, for French beef; Ripoll-Bosch *et al.*, 2013, for Spanish lamb) made an economic allocation between income from selling meat and from subsidies targeted on the use of low-fertilized grasslands from this production, and on the multi-functionality of livestock. These initial attempts, though commendable, need to be improved.

As a partial solution to the livestock-related environmental issues, rationalisation of consumption of animal products is an option. This would consist of a decrease in animal product consumption by people in developed countries and by high- and middle-income classes in urban areas in the fast developing economies and an increase in consumption by those living in developing countries particularly in Africa where the animal product consumption is very low. However, this will not solve the problem, as consumption is expected to increase in developing countries, resulting in a substantial increase in GHG emissions from the livestock sector in these countries, although the GHG emission will stabilize in developed countries (FAO, 2009). The need for improving production efficiency, for example by improving animal breeding, nutrition and health, could help to limit the environmental impact of the livestock sector (Thornton, 2010).

Challenges in reducing environmental impacts from animal production

During the last decade many studies aimed at finding ways to reduce environmental impacts, using different approaches, among them LCA being the most used, have been conducted. These studies were either on partial systems (fattening bulls or pigs, animals on pasture) or complete systems (beef herd in a whole year, total pig production for a whole year). Results were very diverse, because of differences in the system boundaries used in the LCA methodology, in the origin of the data, in the emission allocation methods and in the choice of emission/pollution factors. Large difference among systems or countries exist (FAO, 2010 for milk; Basset-Mens and van der Werf, 2005 for pig), but within a system application of an array of options requiring important changes in the system are necessary to considerably decrease impacts (Nguyen *et al.*, 2013 for beef). However, between-farm comparisons show large differences for the same system (Doreau and Dollé, 2011 for GHG in dairy farms): acquisition of good environmental practices by farmers is a first step. There are a lot of public recommendations for decreasing emissions especially GHGs (UNFCCC, 2008) but they are often unstructured and do not account for interactions between options and trade-offs between impacts. A valuable tool for policy makers is the establishment of marginal abatement cost curves which combine the efficiency of options and their cost on a long-term perspective (Moran *et al.*, 2008). However, finding the best available options may be hampered by the lack of accurate knowledge of technical efficiency, and the difficulty to predict economic trends and prices in the future. Once options are identified, policy makers have two major instruments at their disposal: changes in taxes and subsidies structure, differing in nature and efficacy (Gerber *et al.*, 2010).

For a few years, many countries focused their efforts on reducing GHG emissions, which emerged as an enormous challenge due to the on-going global warming and need for an immediate answer. The use of non-renewable energy and of the phosphorus, or localized degradation of water quality also required urgent actions. An intervention that reduces one impact may increase another one in production systems; for example cattle fattening diets are ranked differently for GHG emissions, eutrophication and energy use (Nguyen *et al.*, 2012a). Organic farming leads to lower changes than conventional farming in GHG emissions per kg milk, because the higher methane emission due to lower cow productivity is compensated for by lower nitrous oxide and carbon dioxide emissions due to lower inputs. Other impacts such as eutrophication and acidification potential or energy use are generally lower for organic farming in ruminants (De Boer, 2003); differences are less marked for pig meat (Basset-Mens and van der Werf, 2005) because crop yields and pig growth rates are lower in organic farming.

When impacts are expressed per ha instead of per kg of animal product, comparison between farming systems give widely different results. Impacts per ha are lower for grass-based extensive systems including organic farming. This unit is useful for a territorial approach, when the objective is for example to reduce pollution at a watershed level, or to reduce GHG in a national inventory. However, these systems result in an increase in land use. At a country level, increasing land use by livestock may be positive for local management, for example in mountain areas; but at world level it may result in deforestation of primary forest due to the pressure on land for agricultural activities if the global increase in food production is aimed at.

A largely debated issue is the extent of carbon sequestration by soils, especially pastures, which contributes to reducing net emissions of GHG. Permanent grasslands and no-tillage practices are efficient techniques. According to IPCC (2007) the increase in carbon sequestration by soils may contribute to 89% of GHG mitigation from agriculture by 2030. Although this estimate appears excessive, integrating C sequestration in GHG balances should be more frequent; to our knowledge it has been done only for 2 years (e.g. Doreau *et al.*, 2011) but the amount of carbon storage per ha of grasslands that has been measured in these studies is extremely variable (from 0.12 to 2.5 T C/ha per year); moreover when crops or leys replace permanent grasslands there is a C release. In order to account for both existence of C sequestration and criticisms relative to uncertainty, it may be recommended to present first GHG raw emissions then GHG balance due to land use and land use change, which comprises the effect of C storage or release, and the effect of deforestation of primary forest when soybean is used.

Most agricultural LCA are established from the cradle to the farm gate; environmental cost of product processing, distribution, domestic use and waste recycling are seldom taken into account. These processes increase environmental costs, especially energy use and GHG emissions; post-farm gate GHG are 17-22% and 1-10% of the total in industrialized and developing countries, respectively (FAO, 2010). Besides the calculation of impact per kg of produced foods, it is necessary to assess post-production losses. Food waste may go beyond 30% of production for dairy products and meat, especially in developed countries. Post-harvest grain losses have long been mentioned in developing countries; losses in animal-source foods have been less intensively studied but may be lower due to the importance of family or local consumption. For example in India the losses in animal-source foods are less than 5% of the production (Nanda *et al.*, 2012). A recent paper stresses the risk of increase in waste in fast developing countries due to the rapid increase in food production (Parfitt *et al.*, 2010). In developing countries, the main reason for waste is the lack of infrastructure and the low organizational level for trade and transport; while in developed countries, the marketing push of the supermarkets leading to excessive purchases by customers, the inability to sell foods before the expiry date by supermarkets, the level of demand by the consumer and non-consumption within the shelf life of the products are major sources of retail and consumer food losses (Hodges *et al.*, 2011).

Sustainable animal production in the future

Many studies show there is still underutilized potential to reduce the burden of the sector on the environment, and strengthen the positive role that the livestock sector may have in mitigating climate change, nutrient recycling, protection of biodiversity and the provision of other environment services. Significant progress can be made through the development of regulatory frameworks and incentives for environmental services. For extensive systems, the payment of environmental services such as carbon sequestration and biodiversity protection (for example through silvopastoral systems; see Murgueitio *et al.*, 2011) are good examples. It has potential for application in many regions of the world e.g. Latin America and Africa. In intensive systems, more efficient use of inputs (water, nutrients, energy) through innovations in technologies and adaptation and improvement in the existing practices and/or restoration of strong linkages between livestock and crops can lead to obvious environmental gains. In economic terms, in a finite space where there is strong competition between agricultural and non-agricultural uses of land as well as among the crops themselves for use as human food, animal feed or industrial transformation processes, improving animal performance remains central to productive efficiency. The improvement in skills, knowledge, techniques and tools in the areas of animal genetics, feed and feeding, animal health and management and their use will help achieve the productive efficiency required to maintain the competitiveness of products and industries and to meet multiple expectations from livestock production.

In fast developing countries, aside from the traditional agricultural sector, the mobilization of public and especially private funds for land acquisition and realization of investments in structures of very large size, may help to develop new farming systems and contribute in these countries to increase production capacity and efficiency. These systems are different from the model that prevailed during the twentieth century in developed countries (EU, US, Canada, Japan, etc.) wherein the farm modernization and the growth in the quantity produced was made through specialization of family size farms. These development models will certainly be capable of competing, and will also be complementary to the dynamics that are created across the regions and the quality niches that they each address. However, it is likely that due to their size, which concentrates effluents in a same place, and sometimes to their financial objectives, these industry-like enterprises often raise environmental questioning.

Beyond the economic, environmental and societal assessments of the situations and of the stakes that will dictate the future, transforming knowledge and skills into practice, and using them to improve the efficiency of resource and land use for the sector growth is a major challenge. To address the complexity of issues raised for making changes in the operating practices in the livestock sector, it is necessary that all actors (from the primary resource producer to the consumer) get better connected. The different sources of knowledge to which they have access should lead to the adoption of new technologies for production as well as of new modes of consumption of animal products. This will require an even greater flow of knowledge and innovations among all stakeholders whether these are in developed or developing countries.

Sustainability of low-productive livestock systems

For low productive livestock systems, a major criticism is the environmental impact that they produce, especially when expressed as GHG emission per kg of product, milk or meat. The FAO (2010) compared GHG emissions for milk production in different countries. When annual milk yield is lower than 2,000 kg, GHG emissions per kg increase rapidly. Since the full genetic potential of animals in the field is not realized (Cunningham, 2005), increase in productivity by improving animal nutrition, management and health practices will decrease GHG emission per unit of animal product. Also rearing a high genetic potential cow with 8,000 kg milk production per year in the areas that now produce 2,000 kg or less per year is also not a realistic option due to lack of feed resources and unsuitable climatic factors. Creating artificial conditions required to express full genetic potential of such high producing animals is not expected to be sustainable and increase the GHG emission per kg milk when GHG emission in the processes required to create the artificial conditions are taken into account. It may also be noted that low-productivity systems could also be economically viable due to low input costs. For smallholder livestock farmers in developing countries, a major pillar of sustainability is the social one. Indeed, livestock have a cultural role. Some societies as Peul or Masai cannot be dissociated from cattle herds. Beyond these specific cases, livestock farming (or crop-livestock farming) is essential for the livelihood of many rural communities in Africa, Asia or Latin America. Herds are a capital which may buffer cash flow problems. Animal gift and lending are a mechanism of solidarity for very poor people, and/or they contribute to a social network favoring interdependency between communities (Alary *et al.*, 2012). In addition, livestock farming by smallholders also contributes to food security at country level, in addition to providing valuable minerals and vitamins to pregnant women and growing children. Owing to the lack of infrastructures for food transport, meeting locally the physiological needs and maintaining social equilibrium are of utmost importance, which animal husbandry offers to farmers. There is a need to better value these social dimensions the livestock play for such communities and bring them into the holistic equation of sustainability rather than labeling such systems unsustainable based on only GHG emission per kg of animal product.

Many farmers have very low income, either because of poor management of their farm (for example excessive inputs with regard to output) or because of unfavorable natural conditions, for example mountains or poor soils which lead to low crop and/or forage yields. The question is why to maintain livestock in such places where the production is less efficient compared with that in better natural conditions. From a social point of view, livestock maintains human presence, which would not occur in case of uncultivated land or forest. Moreover these areas do not compete with crops for use as human food or for biofuel production due to low productivity. The challenge is to improve economic profit, when farms are vulnerable to both climatic and economic shocks, such as drought and when there is drop in product price. For ruminants, the first solution is to take advantage of the high adaptive capacity of the animals, by alternating undernutrition and refeeding periods to decrease global feed cost (Blanc *et al.*, 2006). Bio-economical modeling allows finding the most appropriate strategies; for examples for beef farms increasing purchased feeds and grass surface for haymaking are proposed for a better herd resilience (Mosnier *et al.*, 2010). Darnhofer *et al.* (2011) recommend diversification of farm activities in order to improve resilience of the whole farm. Specific systems related to a territory sometimes may give an opportunity for producing niche products that consumers are ready to pay more, owing to a higher sensory quality or a unique geographical origin. Appropriately managed grazing land and supportive institutional and policy frameworks can enhance productivity and livelihoods in addition to providing large benefits in the form of carbon sequestration, protection of biodiversity and water services.

Increasing sustainability in intensive systems

Faced with the need to increase meat, milk and egg production, the development of intensive systems is often recommended (Steinfeld *et al.*, 2006). The intensive model used for several decades aimed to increase production per animal or per unit of surface. However, it has become necessary to improve the environmental performance (sometimes called eco-efficiency) by decreasing inputs while maintaining or only slightly reducing productivity. For milk production, this is possible, for example, by reducing grass N fertilization and stocking rate (Basset-Mens *et al.*, 2009), by improving energy use efficiency through means of nutrition or genetic (Reynolds *et al.*, 2011) or by drastically decreasing protein supply to animals (Fanchone *et al.*, 2013). This is more difficult for pig and especially poultry production, but the use of synthetic amino acids decreases soybean use and thus environmental costs and land use change (Mosnier *et al.*, 2011). There are other examples showing that technology or biotechnologies may help. However, sometimes, efficient biotechnological solutions are not acceptable from an ethical point of view or customer perception. This is the case for instance for phytase-producing transgenic pigs which help to reduce phosphorus losses, whereas phytase from fungal or bacterial origin is widely used. Another way to reduce inputs is to improve animal fertility and health. Unfortunately, animals with higher productivity have a higher sensitivity to various diseases; and in cows, there is a negative relation between animal productivity and fertility. These undesirable effects are due to the past genetic selection which aimed at only increasing yield. Genetic selection using other criteria, for example higher disease resistance, reproductive efficiency, and longer productive life, that help to increase efficiency of production should be targeted (Scollan *et al.*, 2010). Also precision agriculture i.e. on-farm development of monitoring techniques (e.g. physical and chemical sensors, image and sound recordings, etc.) and provision of all inputs to the places and at times (using computerized tools) required by the animal, thereby maximizing resource use efficiency.

The realization of profitability of well-managed intensive systems leads to extreme situations: farms having a huge number of animals, and exploitation of animal potential by maximizing productivity. These farms are widespread in North America but are increasingly found in developing countries as well. In developing countries they compete with smallholders who do not have the same access to infrastructure and resources, and hence become extinct with time. Such farms limit imports of animal products but profit generally does not benefit the local population. In addition to raising these major social issues in developing countries, these huge intensive farms concentrate manure at one place with negative environmental consequences such as soil and water pollution if they are not managed properly. According to Scollan *et al.* (2010) good management of manure and use of biogas are attractive options to overcome these environmental problems.

It is worth noting that high productive farms may look sustainable based on GHG emission per kg animal product but might not be sustainable in the future due to increasing cost of cereals and fossil energy. Equally important is to consider animal welfare and ethical dimensions of feeding high grain diets to animals. Also, such systems disrupt the nitrogen cycle through transport of high amounts of soybean e.g. from Southern America to other parts of the world. In addition any disruption in trade or volatility of the cost of these feed inputs can be catastrophic for such farms. In order to address these issues, a recent approach promotes agro-ecology, a way to improve or to maintain efficiency by means of ecological solutions. Application of the principles of agro-ecology to livestock has been extensively described by Dumont *et al.* (2013). In addition to a decrease in inputs, agro-ecology advocates an increase in animal and vegetal biodiversity, the optimization of metabolic functioning of farming systems and an improvement in management for maintaining animal health. These techniques generally result in better sustainability and resilience to shocks (Thornton, 2010). A classical improvement of eco-efficiency based on technology and optimization of animal functions is not exclusive of an agro-ecological approach. It is applied in different systems. Some options are presented in Table 2.

The dependence on cereals in many countries of the South will grow and their use for animals will make the systems less resilient and more prone to food-feed-fuel competition. An improved use of fibrous materials, forages or industrial by-products (Bocquier and Gonzalez-Garcia, 2012; FAO, 2012) could reduce the dependence on cereals. The dynamics of concentration of livestock farming in peri-urban situations produces nutrient surplus and associated latent pollution in these areas. This is in addition to the other challenge of nutrient depletion in the rural areas. In the cropping systems, the loss of carbon as well as problems of fragility and fertility of tropical soils remains a major issue. These issues together with the rising cost of fertilizers and energy, GHG emissions associated with manufacturing and transportation of various inputs, and the scarcity and competition for various resources (e.g. phosphorus, water) suggest re-designing the system to have closer integration between livestock and crops, especially in developed countries.

Table 2. Some examples of solutions to improve eco-efficiency by different approaches.

	Technical and technological approach	Agro-ecological approach
Genetics	Increasing genetic merit (productivity, productive efficiency, longevity, fertility)	Choosing breeds adapted to the ecosystem and to the climate, resilient multi-purpose (e.g. milk and meat) breeds
Nutrition	Optimizing digestive functions, improving metabolic efficiency for milk secretion and muscle accretion, feeding balanced rations that meet nutrient requirements, using additives, adapting feeding to reproduction cycle	Using more feeds produced on farm, diversifying feeds, use of co-products, choosing reproduction periods according to resources, providing adequate clean water
Animal health	Conceiving healthy diets, monitoring animal activities, using quick and efficient disease diagnostic, monitoring and control methods	Using natural products against parasites and diseases and for enhancing immune response by animals, improving animal and farm hygiene, taking care of animal comfort
Use of resources	Preferring feeds with low environmental impacts including low water and land use, managing herd to limit non-productive animals	Shifting from mineral to organic fertilization, optimizing manure application, favoring plant biodiversity
Manure management	Improving slurry storage (type, duration) for lower losses, producing biogas	Better integrating crops and livestock

Conclusion

Defining sustainability for livestock farming is a challenge, because it requires balancing multiple and changing objectives of the three pillars of sustainability: profit, planet and people including ethics in an array of dimensions from local to global, which is not easy to achieve. There is a need and room for coexistence of very diverse systems, each of them being adapted to a set of environmental and socio-economical conditions in different parts of the world. In this paper, the diversity of these systems in terms of farm structure and natural resource use has been outlined, and the means through which their sustainability could be enhanced have been discussed. In any system, key to improving sustainability lies in improving the multiple criteria of efficiency of the animal and the herd. The challenge for policy makers is to elaborate with stakeholders the roadmaps for realization of livestock production systems that efficiently utilize natural resources while respecting ethical and socio-cultural dimensions of people, which may differ from region to region.

References

Alary, V., G. Duteurtre and B. Faye, 2011. Elevage et sociétés: les rôles multiples de l'élevage dans les pays tropicaux. In: Elevage en régions chaudes (Coulon J.B., Lecomte P., Boval M., Perez J.M., eds). INRA Prod. Anim. 24, 145-156.

Basset-Mens, C. and H.M.G. van der Werf, 2005. Scenario-based environmental assessment of farming systems: the case of pig production in France. Agric. Ecosyst. Environ. 105, 127-144

Basset-Mens, C., S. Ledgard and M. Boyes, 2009. Eco-efficiency of intensification scenarios for milk production in New Zealand. Ecol. Econ., 68, 1615-1625.

Blanc, F., F. Bocquier, J. Agabriel, P. D'Hour and T. Chilliard, 2006. Adaptive abilities of the females and sustainability of ruminant livestock systems. A review. Anim. Res. 55, 489-510.

Blanfort, V., M. Doreau, J. Huguenin, J. Lazard, V. Porphyre, J.F. Soussana and B. Toutain, 2011. Impacts et services environnementaux de l'élevage en régions chaudes. In : Elevage en régions chaudes (Coulon J.B., Lecomte P., Boval M., Perez J.M., eds). INRA Prod. Anim. 24, 89-112.

Bocquier, F. and E. González-Garcia, 2010. Sustainability of ruminant agriculture in the new context: feeding strategies and features of animal adaptability into the necessary holistic approach. Animal 4, 1258-1273.

Cunningham, E. P., 2005. The genetic dimension. In: Knowledge agriculture: perspectives towards a new model of milk production. R. Keenan & Co., Co. Carlow, Ireland, 9-11.

Darnhofer, I., S. Bellon, B. Dedieu and R. Milestad, 2010. Adaptiveness to enhance the sustainability of farming systems. A review. Agron. Sustain. Dev. 30, 545-555.

De Boer, I.J.M., 2003. Review. Environmental impact assessment of conventional and organic milk production. Livest. Prod. Sci. 80, 69-77.

De Vries, M. and I.J.M. de Boer, 2010. Comparing environmental impacts for livestock products: A review of life cycle assessments. Livest. Sci. 128, 1-11.

Doreau, M. and J.B. Dollé, 2011. Strategies for reducing greenhouse gas emissions in dairy production. A European perspective. Proc. 47th Eastern Nutrition Conference, Montréal, Canada, 55-77.

Doreau, M., H.M.G. van der Werf, D. Micol, H. Dubroeucq, J. Agabriel, Y. Rochette and C. Martin, 2011. Enteric methane production and greenhouse gases balance of diets differing in concentrate in the fattening phase of a beef production system. J. Anim. Sci. 89, 2518-2528.

Doreau, M., M.S. Corson and S.G. Wiedemann, 2012. Water use by livestock: a global perspective for a regional issue? Anim. Front., 2 (2), 9-16.

Dumont, B., L. Fortun-Lamothe, M. Jouven, M. Thomas and M. Tichit, 2013. Prospects from agroecology and industrial ecology for animal production in the 21st century. Animal, 7, 1028-1043.

Fanchone, A., P. Nozière, J. Portelli, B. Duriot, V. Largeau, M. Doreau, 2013. Effects of nitrogen underfeeding and energy source on digestion, nitrogen partition, and milk production in dairy cow. J. Anim. Sci. 91, 895-906.

FAO, 2009. The state of food and agriculture. Livestock in the balance. FAO, Rome, Italy, 166 p.

FAO, 2010. Greenhouse gas emissions from the dairy sector: A Life Cycle assessment. FAO, Rome, Italy, 94 pp.

FAO, 2012. Biofuel co-products as livestock feed. Opportunities and challenges (H.P.S. Makkar, ed). FAO, Rome, Italy, 534 p.

Gerber, P., N. Key, F. Portet and H. Steinfeld, 2010. Policy options in addressing livestock's contribution to climate change. Animal 4, 393-406.

Gill, M., P. Smith and J.M. Wilkinson, 2010. Mitigating climate change: the role of domestic livestock. Animal 4, 323-333.

Hodges, R.J., J.C. Buzby and B. Bennett, 2011. Postharvest losses and waste in developed and less developed countries : opportunities to improve resource use. J. Agric. Sci. 149 (S1), 37-45.

IPCC, 2007. Agriculture. In Climate Change 2007: Mitigation. Contribution of Working Group III to the Fourth Assessment Report of the IPCC on Climate Change (B. Metz, O.R. Davidson, P.R. Bosch, R. Dave, L.A. Meyer, eds), Cambridge University Press, UK.

Kounina, A., M. Margni, J.B. Bayart, A.M. Boulay, M. Berger, C. Bulle, R. Frischknecht, A. Koehler, L. Milà i Canals, M. Motoshita, M. Núñez, G. Peters, S. Pfister, B. Ridoutt, R. van Zelm, F. Verones and S. Humbert, 2013. Review of methods addressing freshwater use in life cycle inventory and impact assessment. Int. J. Life Cycle Assess. 18, 707-721.

Lebacq, T., P.V. Baret and D. Stilmant, 2013. Sustainability indicators for livestock farming. A review. Agron. Sustain. Dev. 33, 311-327.

Moran, D., M. MacLeod, E. Wall, V. Eory, G. Pajot, R. Matthews, A. McVittie, A. Barnes, B. Rees, A. Moxey, A. Williams, P. Smith, 2008. UK marginal abatement cost curves for the agriculture and land use, land-use change and forestry sectors out to 2022, with qualitative analysis of options to 2050. Report to the Committee of Climate Change, SAC, Edinburgh, UK.

Mosnier, C., J. Agabriel, M. Lherm and A. Reynaud, 2010. A dynamic bio-economic model to simulate optimal adjustments of suckler cow farm management to production and market shocks in France. Agric. Syst. 102, 77-88.

Mosnier, E., H.M.G. van der Werf, J. Boissy and J.Y. Dourmad, 2011. Evaluation of the environmental implications of the incorporation of feed-use amino acids in the manufacturing of pig and broiler feeds using Life Cycle Assessment. Animal 5, 1972-1983.

Murgueitio, E., Z. Calle, F. Uribe, A. Calle and B. Solorio, 2011. Native trees and shrubs for the productive rehabilitation of tropical cattle. ranching lands. Forest Ecol. Manag. 261, 1654-1663.

Nanda, S.K., R. K. Vishwakarma, H.V.L. Bathla, A. Rai and P. Chandra, 2012. Harvest and post harvest losses of major crops and livestock produce in India. All India Coordinated Research Project on Post Harvest Technology (ICAR), Ludhiana. pp. 137, September 2012.

Nguyen, T.T.H., H.M.G. van der Werf and M. Doreau, 2012a. Life cycle assessment of three bull-fattening systems: effect of impact category on ranking. J. Agric. Sci., 150, 755-763.

Nguyen, T.T.H., M. Doreau, M. Eugène, M.S. Corson, F. Garcia-Launay, G. Chesneau and H.M.G. van der Werf, 2012b. Effect of type of ration and allocation methods on the environmental impacts of beef-production systems. Livest. Sci. 145, 239-251.

Nguyen, T.T.H., M. Doreau, M. Eugène, M.S. Corson, F. Garcia-Launay, G. Chesneau and H.M.G. van der Werf, 2013. Effect of farming practices for greenhouse gas mitigation and subsequent alternative land-use on environmental impacts of beef-cattle production systems. Animal, 7, 860-869.

Parfitt, J., M. Barthel and S. Macnaughton, 2010. Food waste within food supply chains : quantification and potential for change to 2050. Phil. Trans. R. Soc. B. 365, 3065-3081.

Reynolds, C.K., L.A. Crompton and J.A.N. Mills, 2011. Improving the efficiency of energy utilisation in cattle. Anim. Prod. Sci 51, 6-12.

Ripoll-Bosch, R., I.J.M. de Boer, A. Bernués and T. Vellinga, 2013. Accounting for multi-functionality of sheep farming in the carbon footprint of lamb: A comparison of three contrasting Mediterranean systems. Agric. Syst. 116, 60-68.

Robinson, T.P., P.K. Thornton, G. Franceschini, R.L. Kruska, F. Chiozza, A. Notenbaert, G. Cecchi, M. Herrero, M. Epprecht, S. Fritz, L. You, G. Conchedda and L. See, 2011. Global livestock production systems. FAO, Rome, Italy, and ILRI, 152 pp.

Scollan, N.D., D. Moran, E.J. Kim, and C. Thomas, 2010. The environmental impact of meat production systems. Report to the International Meat Secretariat, 65 p., http://www.meat-ims.org/old-site/IMSReview-final-20710.pdf

Steinfeld, H., P. Gerber, T. Wassenaar, V. Castel, M. Rosales and C. de Haan, 2006. Livestock's long shadow: environmental issues and options. FAO, Rome, Italy, 390 p.

Thornton, P.K., 2010. Livestock production: recent trends, future prospects. Phil. Trans. R. Soc. B. 365, 2853-2867.

UNFCCC, 2008. Challenges and opportunities for mitigation in the agricultural sector. Technical paper FCCC/TP/2008/8, 101 p.

Vieux, F., N. Darmon, D. Touazi and L.G. Soler, 2012. Greenhouse gas emissions of self-selected individual diets in France: Changing the diet structure or consuming less? Ecol. Econ. 75, 91-101.

Environmental, social, and economic footprints of current and past beef production systems

K.R. Stackhouse-Lawson[1], J.O. Reagan[1], B.J. Isenberg[2], E.J. Pollak[3], T. Battagliese[4], B. Ulhman[4], C. Barcan[4], I. Schulze[5], J. Silva[5] and C.A. Rotz[2]
[1]National Cattlemen's Beef Association, Centennial, CO 80112, USA; kstackhouse@beef.org
[2]USDA-ARS, Pasture Systems and Watershed Management Research Unit, University Park, PA 16802, USA
[3]USDA-ARS-NPA, Roman L. Hruska U.S. Meat Animal Research Center, Clay Center, NE 68933, USA
[4]BASF Corporation, Nutrition and Health, Florham Park, NJ 07932, USA
[5]BASF Corporation, Fundação Espaço ECO, 09844-900 Sao Bernardo do Campo, Brazil

Introduction

The beef industry has defined sustainability as meeting the growing demand for beef by balancing environmental responsibility, economic opportunity, and social diligence. Measuring sustainability is challenging, as the beef supply chain is one of the most complex food systems in the world. As the first and largest research project of this kind, this study represents an innovative approach toward creating a more sustainable beef product. Our objective is to establish a sustainability baseline (including environmental, economic, and social footprints) for the US beef industry by quantifying life cycle inputs and outputs for beef production over time.

Material and methods

To determine the sustainability of beef production, a combination of models were used. The USDA-ARS Integrated Farm System Model (IFSM) was used to simulate environmental and economic footprints from cradle to farm-gate. The socio-eco-efficiency tool (SEEBALANCE®) extends this analysis by determining the environmental, economic, and social impacts of beef from cradle to grave providing a comprehensive assessment of sustainability.

The IFSM is a process-level farm model that simulates crop growth, feed production and use, animal growth, and returning manure nutrients to the land to predict the environmental impacts and economics of agriculture production systems (Rotz *et al.*, 2005). For the current study, relevant information for the US Meat Animal Research Center (USMARC) beef operation was gathered and used to establish model parameters. The USMARC farm, cow-calf and feedlot operations were simulated to evaluate performance, environmental impact and economics.

The environmental impacts and economics of beef production at the USMARC were combined with primary data from the packer, case ready, retail, and consumer segments of the beef value chain for 2005 and 2011 using SEEBALANCE®. The SEEBALANCE® analysis includes environmental, social, and economic considerations as determined by method of life cycle analysis (Kölsh *et al.*, 2008). This approach quantified US beef sustainability considering economic, social and ecological impacts along all segments of the beef value chain.

Results and discussion

Integrated farm system model: USMARC

A 25-year simulation of the USMARC's current production system gave a carbon footprint of 11 kg of CO_2e per kg of live weight sold, which is consistent with other experiments (Johnson *et al.*, 2003; Capper, 2011; Stackhouse-Lawson *et al.*, 2012). The energy required to produce that beef (energy footprint) was 25.9 MJ/kg. The total water required (water footprint) was 21,300 l/kg of live weight sold, and the water footprint excluding that obtained through precipitation was 2,800 l/kg. The simulated total cost of producing their beef was about $2.20/kg of live weight sold, which agreed with USMARC production records.

SEEBALANCE®

Table 1 quantifies the environmental, social and economic considerations of the beef supply chain expressed in 0.45 kg of minimally processed boneless edible consumed beef (UB). Overall, the sustainability of the US beef industry, given the present assumptions, has improved by 7% in 6 yr.

Table 1. Environmental, social and economic sustainability indicators for the beef supply chain.

Sustainability indicators	2005	2011	% change
Economic (expressed per UB[1])			
Consumer price ($)	5.24	5.55	6
Environmental (expressed per UB)			
Energy use (MJ)	521	511.0	-2
Resource consumption (mg $silver_e$)	5.05	4.96	-2
Water consumption (L)	2,418	2,336	-3
Solid waste (kg municipal $waste_e$)	0.19	0.18	-7
Greenhouse gases (kg CO_2e)	23.7	23.6	-1
Photochemical ozone creation potential (g C_2H_4e)	0.026	0.026	0
Acidification potential (g SO_2e)	336	327	-3
Ozone depletion potential (g CFCe)	0.013	0.013	0
Water emissions (grey water (l diluted $water_e$)	4,981	4,487	-10
Land use (m^2a)	21.4	20.5	-4
Social (normalized and weighted[2])			
Occupation illnesses and accidents	0.90	0.60	-32
Toxicity potential	1.00	0.84	-16

[1] User benefit (UB) 0.45 kg of minimally processed boneless edible consumed beef.

[2] Social indicators are normalized and weighted based on severity of incident or chemical.

References

Capper, J.L, 2011. Replacing rose-tinted spectacles with a high-powered microscope: The historical vs. modern carbon footprint of animal agriculture. Anim. Front. 1, 26-32.

Johnson, D. E., H. W. Phetteplace, A. F. Seidl, U. A. Schneider, and B. A. McCarl, 2003. Management variations for us beef production systems: Effects of greenhouse gas emissions and profitability. Proc. 3rd Int. Methane and Nitrous Oxide Mitigation Conf., Bejing China. 953-961.

Kölsh, D., P Saling, A. Kircherer, A. Grosse-Sommer, and I. Schmidt, 2008. How to measure social impacts? A socio-eco-efficiency analysis by the SEEBALANCE® method. Int. J. Sustainable Development. 11, 1-23.

Rotz, C.A., D.R, Buckmaster, and J.W. Comerford, 2005. A beef herd model for simulation of feed intake, animal performance, and manure excretion in farm systems. J. Anim. Sci. 38, 231-242.

Stackhouse-Lawson, K.R., C.A. Rotz, J.W. Oltjen, and F.M. Mitloehner, 2012. Carbon footprint and ammonia emission of California beef production systems. J. Anim. Sci. 90, 4641-4655.

Effect of fat supplementation and stage of lactation on methane emission in dairy cows

L. Alstrup, M.R. Weisbjerg and P. Lund
Department of Animal Science, AU Foulum, Aarhus University, Denmark; lene.alstrup@agrsci.dk

Introduction

Methane is produced in ruminants as a consequence of fermentation of organic matter in the rumen. The methane produced represents an energy loss to the animal and will vary with feed composition and intake; under extreme circumstances 2-12% of gross energy intake is converted into methane, but in an intensive dairy production values between 3-7% are more realistic (Martin *et al.*, 2008). It has been shown that supplementation of fatty acids (FA) to the feed decreases methane emission on a short term (Brask *et al.*, 2012). Although supplementation with fat to the diet is the most promising dietary strategy to reduce enteric methane emission (Grainger and Beauchemin, 2011), research in long term effects of supplementing fat, and of days in milk (DIM) on methane production is lacking. The aim of the present experiment was to study the effect of FA supplementation, effect of stage of lactation, and their interaction on methane emission.

Material and methods

Six Holstein Friesian cows were blocked in three groups according to DIM and randomly assigned to a ration with either no fat supplementation (NO FAT) or fat supplementation (FAT) in the form of rolled rapeseed. Rations consisted of app. 55% forage (50% maize silage and 50% grass clover silage) and 45% concentrate on dry matter (DM) basis. Content of fatty acids in rations were 18 and 47 g/kg DM for no fat and fat supplementation, respectively. Methane production was measured for 2×48 h in 17 m^3 open-circuit respiration chambers. Feed intake and milk yield was recorded in the chamber period. The animals were housed individually and chambers were only opened twice daily for milking and subsequent feeding. Methane was measured as the accumulated amount in l over 24 hours. Corrections were made for the measurements during the openings of the chambers for milking and feeding. Methane production was measured at approximately 50, 125, 170 and 220 DIM, and at 50 DIM all cows received the NO FAT ration. Experimental data were analysed with the MIXED procedure of SAS, using recordings at 50 DIM as covariate and ration, cow and DIM as class variables. Cow was set as random effect to account for repeated measurements.

Results and discussion

Methane emission in l/day increased significantly (P=0.03) from time point 125 DIM to 220 DIM; from 619 l/day to 721 l/day for cows receiving NO FAT and from 561 l/day to 585 l/day for cows receiving FAT (Figure 1). Methane as l/kg DM intake (DMI) also increased significantly (P=0.002) from time point 125 DIM to 220 DIM; from 29.0 to 31.8 and from 28.0 to 30.5 for NO FAT and FAT respectively, and tended to be reduced for FAT (P=0.08).

In the same period DMI (kg/24 h) increased for cows receiving NO FAT from 21.9 kg/day to 23.1 kg/day, whereas cows receiving FAT had a decrease in feed intake from 19.6 kg/day to 18.8 kg/day (Figure 2). Milk yield, measured as kg energy corrected milk (ECM), decreased significantly (P=0.009) throughout lactation. From time point 125 DIM to 220 DIM milk yield decreased from 32.9 kg/day to 25.9 kg/day and from 33.9 kg/day to 30.2 kg/day for NO FAT and FAT respectively.

The increase in daily methane emission throughout lactation was not only related to increased DMI, as L methane/kg DMI also increased throughout lactation. According to Garnsworthy *et al.* (2012) an increased methane emission despite an increase in DMI with stage of lactation may be due to

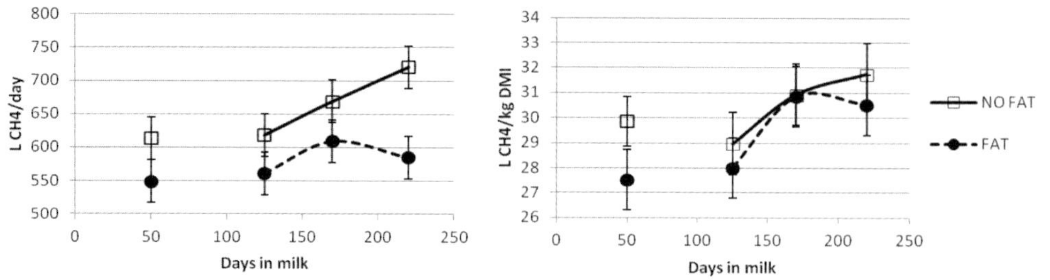

Figure 1. Methane emission per day and per kg dry matter intake (DMI) throughout lactation.

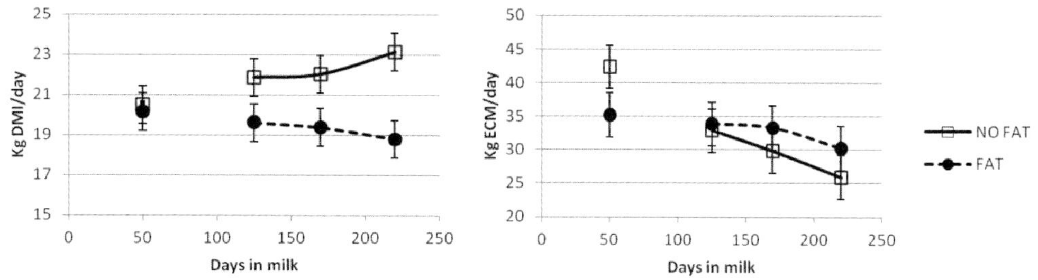

Figure 2. Kg dry matter intake (DMI) and kg energy corrected milk (ECM) per day throughout lactation.

an increased proportion of forage in diet, leading to higher methane emission per kg DMI. In the current study the proportion of forage was kept constant, so the increased methane emission cannot be explained by changed forage/concentrate ratio.

In conclusion, this study indicates that methane production increases with days in milk, though season could be confounded with DIM. Further it is suggested that supplementation with rapeseed to the ration could reduce this increase in methane production, both as absolute methane production in liters and as methane production related to DMI. Further, rapeseed supplementation could increase milk yield.

References

Brask, M., P. Lund, M.R. Weisbjerg, A.L.F. Hellwing and T. Hvelplund. 2012. Methane Production and Digestion in Dairy Cows fed Different Physical Forms of Rapeseed as Fat Supplement. J. Dairy Sci. doi:10.3168/jds.2011-5239.

Garnsworthy, P.C, J. Craigon, J.H. Hernandez-Medrano and N. Saunders. 2012. Variation among individual dairy cows in methane measurements made on farm during milking. J. Dairy Sci. 95, 3181-3189

Grainger, C. and K.A. Beauchemin. 2011. Can enteric methane emissions from ruminants be lowered without lowering their production? Anim. Feed Sci. Technol. 166-167, 308-320.

Martin, C., J. Rouel, J.P. Jouany, M. Doreau and Y. Chilliard. 2008. Methane output and diet digestibility in response to feeding dairy cows crude linseed, extruded linseed, or linseed oil. J. Anim. Sci. 86, 2642-2650.

Methane emission and protein precipitating ability of condensed tannins from warm-season perennial legumes

H.D. Naumann[1,4], *L.O. Tedeschi*[1], *A.E. Hagerman*[2], *B.D. Lambert*[3,4] *and J.P. Muir*[4]
[1]*Texas A&M University, College Station, TX, USA; hnaumann@tamu.edu*
[2]*Miami University, Oxford, OH, USA*
[3]*Tarleton State University, Stephenville, TX, USA*
[4]*Texas A&M AgriLife Research, Stephenville, TX, USA*

Introduction

Enteric methane (CH_4) emissions by ruminants represent a decrease in energy availability to the animal. Forages containing condensed tannins (CT) may suppress enteric CH_4 emissions and bind to proteins, allowing protein to escape microbial degradation in the rumen resulting in ruminal bypass protein. Objectives of this study were to evaluate the ability of CT from warm-season perennial legumes commonly consumed by ruminants to suppress CH_4 emissions and bind to protein, and to assess the potential use of these forages in sustainable ruminant production systems.

Material and methods

Nine species of warm-season perennial legumes were evaluated (Table 1). Two separate extractions were performed on oven-dried plant tissue from each species evaluated. Condensed tannins were determined by a butanol-HCl method (Terrill *et al.*, 1992). Protein-precipitating ability (PB) was determined by reacting crude plant extracts with bovine serum albumin (Hagerman and Butler, 1978) and analyzing protein-phenolic residues for N. Methane production was determined using an *in vitro* gas production technique (Tedeschi *et al.*, 2009). Replications consisted of two fermentation events, 12/7/2011 and 1/23/2012, where each plant species was fermented in each of two fermentation chambers. Fermentation chamber was considered a random variable, whereas fermentation flasks within each fermentation chamber were considered random factors. Two ruminally-cannulated steers un-adapted to forage containing CT were used for rumen fluid collection. Forages were individually fermented anaerobically in rumen fluid for 48h. Methane measurements were made by gas chromatography following fermentation. Data were analyzed using the GLIMMIX procedure of SAS (SAS Inst. Inc., Cary, NC, USA).

Table 1. Total condensed tannin (CT) concentrations (TCT; % DM), methane production (CH_4; g/kg DM) and amount of protein bound (PB; g/kg DM) by CT from warm-season perennial legumes in vitro.

Plant	TCT	CH_4	PB
Acacia angustissima var *hirta*	8.9^{b}	0.6^{e}	65.6^{d}
Arachis glabrata	0.0^{d}	38.2^{a}	0.0^{e}
Desmanthus illinoensis	8.1^{b}	24.9^{b}	75.3^{bc}
Desmodium paniculatum	12.5^{a}	7.9^{de}	64.7^{d}
Lespedeza cuneata	8.3^{b}	15.1^{cd}	78.1^{ab}
Lespedeza stuevei	11.7^{a}	4.9^{e}	72.3^{c}
Leucaena retusa	3.2^{c}	40.7^{a}	4.4^{e}
Mimosa strigillosa	11.7^{a}	7.6^{de}	70.9^{c}
Neptunia lutea	8.3^{b}	19.7^{bc}	80.0^{a}

a-e Within a column, LS means without a common superscript differ ($P<0.05$).

Results and discussion

Total CT (TCT), PB and CH_4 are shown in Table 1. Total CT was greatest for *D. paniculatum*. *L. stuevei* and *M. strigillosa* ($P<0.05$). Other than *A. glabrata* (CT negative control), *L. retusa* had the least TCT ($P<0.05$). Condensed tannins from *N. lutea* and *L. cuneata* demonstrated the greatest PB ($P<0.05$), whereas CT from *L. retusa* demonstrated almost no PB at all. Condensed tannins from *L. retusa* did not suppress CH_4 emissions *in vitro*. Fermentations containing *A. angustissima* suppressed *in vitro* CH_4 emissions to the greatest degree ($P<0.05$). Regressions of CH_4 on TCT and PB on TCT (Figure 1a-b) indicated that TCT could explain up to 44% of variation in CH_4 production and up to 69% of variation associated with protein precipitating ability. The negative correlation between TCT and CH_4 indicated that for every unit increase in TCT, CH_4 decreased by 3.27 g/kg DM. Total CT and PB were positively correlated such that for every unit increase in TCT, PB increased by 6.7 g/kg DM. Protein-precipitating ability of CT (Figure 1c) explained only 37% of the variation associated with CH_4 production.

The biological activity of CT relative to suppression of CH_4 emissions and PB differed depending on source of CT and were affected by TCT concentration. Results suggest that enteric CH_4 suppression is predominately affected by factors other than protein precipitation by CT. Of the species evaluated, CT from *L. cuneata* demonstrated the greatest combined ability to bind to protein and suppress CH_4 production *in vitro*.

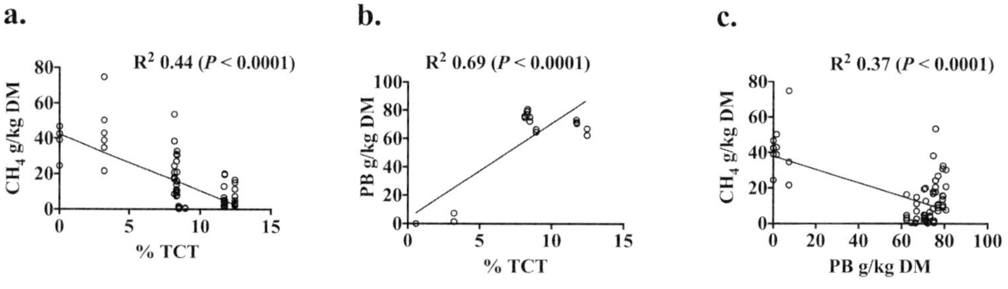

Figure 1. The effects of (a.) total condensed tannins (% TCT) on methane (CH_4 g/kg DM) production, (b.) total condensed tannins (% TCT) on the amount of protein bound (PB g/kg DM) by condensed tannins and (c.) the amount of protein bound (PB g/kg DM) by CT on methane (CH_4 g/kg DM) production in vitro.

References

Hagerman, A.E. and L.G. Butler, 1978. Protein precipitation method for the quantitative determination of tannins. J. Agric. Food Chem. 26, 809-812.

Tedeschi, L. O., P. J. Kononoff, K. Karges, and M. L. Gibson, 2009. Effects of chemical composition variation on the dynamics of ruminal fermentation and biological value of corn milling (co) products. J. Dairy Sci. 92, 401-413.

Terrill, T. H., A. M. Rowan, G. B. Douglas, and T. N. Barry, 1992. Determination of extractable and bound condensed tannin concentrations in forage plants, protein concentrate meals and cereal grains. J. Sci. Food Agric. 58, 321-329.

Short-term dose effects of feeding monensin on methane emissions from lactating Holstein dairy cattle

S.E. Place[1], Y. Pan[2], Y. Zhao[2] and F.M. Mitloehner[2]
[1]Department of Animal Science, Oklahoma State University, Stillwater, OK 74078, USA; splace86@gmail.com
[2]Department of Animal Science, University of California Davis, CA 95616, USA

Introduction

Monensin is a feed additive used in dairy cattle diets to improve feed efficiency that may reduce methane(CH_4) emissions; however, past results have been variable, which could be due to the dose of monensin included in the diet. The objective of this study was to test the short-term dose effects of monensin on eructated CH_4 emissions from lactating dairy cows.

Material and methods

Twenty multiparous lactating Holstein cows were stratified by days in milk and randomly assigned to one of four treatments provided in a pelleted supplemental top dress feed (CON, LOW, MED, HIGH containing 0, 175, 368, and 518 mg/cow/day of monensin, respectively). All cows were fed the same basal total mixed ration (TMR) throughout the experiment and CON top dress for 19 d (PRE period), then their respective treatment top dress for 21 d (MON period), and then returned to the CON top dress for 21 d (POST period). Cows were milked and fed twice daily at 04:30 and 16:30 h, with feed offered ad libitum. Eructated and respired gas emissions were sampled for each cow individually on the last day of each period via a ventilated hood system described in detail by Place *et al.* (2011).

All statistical analysis was conducted using Proc Mixed procedures in SAS version 9.3 (SAS Institute Inc., Cary, NC, USA). Dry matter intake (DMI) was included as a covariate in the CH_4 emissions models.

Results and discussion

Methane emissions, milk production, DMI, and milk composition were similar across treatments in the MON period (Table 1).

The change in CH_4 emissions from the PRE to MON period across treatments varied (6.4, 6.1, 1.6 and 3.9 g/cow/h for the CON, LOW, MED, and HIGH treatments, respectively), with MED having a lower change in CH_4 emissions per cow and per kg of milk compared to CON (Figure 1). Changes in CH_4 emissions per cow and per kg of milk from the PRE to POST period were not different across treatments, suggesting there were no carryover effects of monensin in the POST period.

Two of the previous studies feeding lactating dairy cows TMR have shown reductions in CH_4 emissions with monensin fed at a rate of 24 mg/kg DM (Sauer *et al.*, 1998 doses ranged from 348 to 427.2 mg/cow/d, Odongo *et al.*, 2007 ranged from 307 to 708 mg/cow/d), while another found no effect of monensin on CH_4 emissions with monensin fed at 600 mg/cow/d (approximately 21-22 mg/kg DM) (Hamilton *et al.*, 2010). The lower increase in CH_4 emissions per cow and per unit of milk for the 368 mg/cow/d MED treatment is within the lower ranges fed by Sauer *et al.* (1998) and Odongo *et al.* (2007). There may be an effective dose of saturation for monensin, over which all rumen microorganisms are affected equally; however, further investigation is required. In conclusion, during a short-term feeding period, monensin had over time dose effects on CH_4 emissions, but the effects do not seem to be linear in nature.

Table 1. Treatment least-squares means (n=5) per cow for CH_4 emissions and animal performance by period.

Item	Period	Treatment				SEM
		CON	LOW	MED	HIGH	
CH_4, g/h	PRE	16.2	20.2	23.1	19.8	0.337
	MON	22.6	26.3	24.7	23.7	0.417
	POST	19.8^{a}	23.0^{ab}	28.7^{b}	24.0^{ab}	0.544
CH_4: Milk yield, g/h: kg/d	PRE	0.351	0.469	0.554	0.438	0.008
	MON	0.512	0.599	0.581	0.534	0.011
	POST	0.473^{a}	0.542^{ab}	0.701^{b}	0.557^{ab}	0.014
DMI, kg/d	PRE	25.8^{a}	28.6^{ab}	29.0^{b}	26.6^{ab}	0.159
	MON	27.4	29.2	29.2	27.5	0.135
	POST	27.7	30.0	29.2	28.5	0.122
Milk yield, kg/d	PRE	40.0	43.5	43.0	41.2	0.300
	MON	43.4	46.3	44.3	43.6	0.343
	POST	43.3	47.3	43.0	44.8	0.335
Milk fat %	PRE	3.55	3.45	3.59	3.56	0.105
	MON	3.39	3.85	3.55	3.50	0.118
	POST	3.41	3.51	3.67	3.57	0.088
Milk protein %	PRE	2.96	3.12	3.16	3.01	0.041
	MON	2.94	3.08	3.12	2.98	0.036
	POST	3.14	3.32	3.33	3.16	0.048

[a,b] Within row least squares means without common subscript letters differ ($P<0.05$).

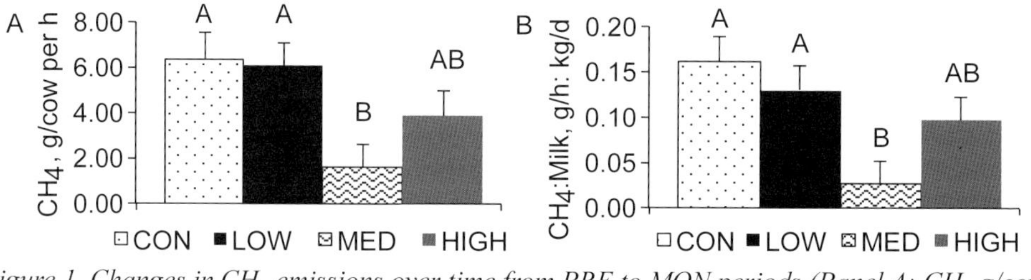

Figure 1. Changes in CH_4 emissions over time from PRE to MON periods (Panel A: CH_4 g/cow/h, Panel B: CH_4 g/h: milk yield kg/d). A, B indicate significant differences ($P<0.05$*).*

References

Hamilton, S.W., E.J. DePeters, J.A. McGarvey, J. Lathrop, and F.M. Mitloehner, 2009. Gas emissions, animal performance, and bacterial population structure responses to dietary monensin in lactating dairy cows. J. Environ. Qual. 39, 106-114.

Odongo, N. E., R. Bagg, G. Vessie, P. Dick, M. M. Or-Rashid, S. E. Hook, J. T. Gray, E. Kebreab, J. France, and B. W. McBride, 2007. Long-term effects of feeding monensin on methane production in lactating dairy cows. J. Dairy Sci. 90, 1781-1788.

Place, S. E., Y. Pan, Y. Zhao, and F. M. Mitloehner, 2011. Construction and operation of a ventilated hood system for measuring greenhouse gas and volatile organic compound emissions from cattle. Animals. 1, 433-446.

Sauer, F. D., V. Fellner, R. Kinsman, J. K. Kramer, H. A. Jackson, A. J. Lee, and S. Chen, 1998. Methane output and lactation response in Holstein cattle with monensin or unsaturated fat added to the diet. J. Anim. Sci. 76, 906-914.

Methane emission from sheep is related to concentrations of rumen volatile fatty acids

C.S. Pinares-Patiño[1], H. Kjestrup[1], S. MacLean[1], E. Sandoval[1], G. Molano[1], R. Harland[1], S. Hickey[2], E. Young[2], K. Dodds[2], K. Knowler[2], N. Pickering[2] and J. McEwan[2]
[1]AgResearch Grasslands, Private Bag 11008, Palmerston North, New Zealand; cesar.pinares@agresearch.co.nz
[2]AgResearch Invermay, Private Bag 50-034, Mosgiel 9053, New Zealand

Introduction

Globally, ruminants are the most important source of emission of methane (CH_4). Animal-to-animal variation in CH_4 emission has genetic basis (Pinares-Patiño *et al.*, 2011), hence offering a potential mitigation avenue through animal breeding. However, for this approach to progress to practical application a rapid and reliable method of ranking animals for their CH_4 emissions is required. Microbial fermentation of feed in the rumen produces volatile fatty acids (VFA), hydrogen (H_2), carbon dioxide (CO_2), ammonia and heat. A last step in the process is the reduction of CO_2 to CH_4 by Archaea using H_2 as a source of energy. Formation of both acetic and butyric acids is accompanied by the production of H_2 and CO_2, whereas propionic production involves a net uptake of H_2, hence VFA profiles may be used to predict CH_4 emission rates (Benchaar *et al.*, 2001). This controlled study conducted with sheep explored the relationship between rumen VFA and CH_4 emission.

Material and methods

Over three years 1,081 lambs (including 96 males) of progeny test programs were phenotyped for their CH_4 emissions using respiration chambers (Pinares-Patiño *et al.*, 2012) while fed on pelleted lucerne (19% crude protein, 43% neutral detergent fibre and 10 MJ ME/kg DM). Feeding was at 2.1 times maintenance requirements, with equal size meals delivered at 08:30 and 16:00 h. There were 24 respiration chambers available. Lambs were 6-10 months-old and 30-45 kg liveweight. They were managed in lots of 96 animals. Acclimatisation in pens (21 d) was followed by two measurement periods (P1 and P2, 10-15 d interval). In each period, individuals' feed dry matter intake (DMI) were measured in metabolic crates (4 d) and then in respiration chambers (2 d). Both in P1 and P2, individuals were randomly allocated to groups (4, each of 24 animals) and respiration chambers. At P1 and P2, soon after exit from the chambers (pre-feeding stage), rumen contents (20-50 ml) were sampled by stomach-tubing for VFA analysis (Sun *et al.*, 2012).

Methane yield (g/kg DMI) was calculated from daily CH_4 emission and DMI. Data for VFAs were analysed as their $\log_e (x + 1)$, where x was concentration (mM) or molar %. Heritability (h^2) and repeatability of both CH_4 yield and VFA were calculated. Fixed effects for CH_4 yield were birth flock and contemporary group (cg): recording year+lot+group+period, and for VFAs the cg: birth year+birth flock+sex. Random effects were animal and the permanent environmental effects period and recording year were fitted to estimate repeatability within year (VFA and CH_4 yield) and across year (CH_4 yield), respectively.

Results and discussion

The mean CH_4 emission was 24.3 g/d, whereas the CH_4 yield was 15.8 g/kg DMI. Heritability (±s.e.) for CH_4 yield was 0.16±0.03, with repeatability within and across years of 0.41±0.04 and 0.25±0.04, respectively. The mean of total concentration of VFA was 51.1 mM, whereas the concentrations of acetic, propionic and butyric acids were 34.9, 8.8 and 4.0 mM. The mean molar % of acetic, propionic and butyric acids were 68.4, 17.0 and 7.7, respectively, with acetate/propionate ratio of 4.1.

Concentrations (as $\log_e$) of total and major VFA had higher estimates of h^2, repeatability and genetic correlation (r_g) with CH_4 yield than when expressed on molar % (as $\log_e$) basis (Table 1). This finding contrasts with those of Robinson *et al.* (2010), who from a study involving 12 rumen-cannulated sheep concluded that VFA concentrations had poor relationships with both daily CH_4 emission and CH_4 yield. However, experimental conditions between the later study and this study were largely different (e.g. feeding conditions, number animals, etc.). There is no agreement in the literature on the validity and representativeness of sample of rumen contents collected via stomach tube, but representativeness of stomach tube sample seems be related to feeding time and depth of insertion (Shen *et al.*, 2012). In the present study, sampling took place at fasting and the operation was completed within one minute.

Results from this study in highly controlled conditions suggest that concentrations of VFA in rumen samples obtained by stomach tubing in a fasting state are heritable and potentially useful to estimate CH_4 yield. However, this finding needs to be corroborated.

Table 1. Rumen VFA ($\log_e$) heritability (h^2), repeatability and correlation (r_g) with CH_4 yield.

	mM			molar %		
	h^2	rep	r_g	h^2	rep	r_g
VFA	0.10±0.04	0.33±0.03	0.92±0.10			
Ace	0.09±0.04	0.34±0.03	0.95±0.10	0.04±0.03	0.08±0.03	-0.01±0.28
Pro	0.10±0.04	0.31±0.03	0.78±0.15	0.09±0.04	0.15±0.03	-0.18±0.17
But	0.09±0.04	0.28±0.03	0.86±0.13	0.04±0.04	0.18±0.03	0.36±0.34
Ace/Pro				0.10±0.04	0.11±0.03	0.08±0.18

Acknowledgements

This study was funded by the New Zealand Pastoral Greenhouse Gas Research Consortium

References

Benchaar, C., C. Pomar and J. Chiquette, 2001. Evaluation of dietary strategies to reduce methane production in ruminants: a modelling approach. Can. J. Anim. Sci. 81, 563-574.

Pinares-Patiño, C.S., J.C. McEwan, K.G. Dodds, E.A. Cárdenas, R.S. Hegarty, J.P. Koolaard and H. Clark, 2011. Repeatability of methane emissions from sheep. Anim. Feed Sci. and Technol. 166-167, 210-218.

Pinares-Patiño C.S. and G.C. Waghorn, 2012. Manual on respiration chambers designs. GRA, New Zealand (http://www.livestockemissions.net/reports,listing,73,technical-manual-on-respiration-chamber-designs.html)

Robinson, D.L., J. Goopy and R.S. Hegarty, 2010. Can rumen methane production be predicted from volatile fatty acid concentrations? Anim. Prod. Sci. 50, 630-636.

Shen, J.S., Z. Chai, L.J. Song, J.X. Liu and Y.M. Wu, 2012. Insertion depth of oral stomach tubes may affect the fermentation parameters of ruminal fluid collected in dairy cows. J. Dairy Sci. 95, 5978-5984.

Sun, X.Z., D. Pacheco, S. Hoskin and D.W. Luo, 2012. Rumen fermentation characteristics are influenced by feeding frequency in sheep fed forage chicory and perennial ryegrass at two feeding levels. Proc. N.Z. Soc. Anim. Prod. 72, 111-116.

Energy efficiency and methane emission by sheep fed sorghum silages at different maturation stage

F.S. Machado[1], N.M. Rodríguez[2], M.N. Ribas[2], F.P. Pôssas[2], L.C. Gonçalves[2] and L.G.R. Pereira[1]
[1]Brazilian Agricultural Research Corporation, Embrapa Dairy Cattle, Brazil; fernanda@cnpgl.embrapa.br
[2]Department of Animal Science, Federal University of Minas Gerais, Brazil

Introduction

The importance of sorghum as a forage crop is growing in many regions of the world due to its high productivity and ability to utilize efficiently water even under drought conditions (Sanchez *et al.*, 2002). The introduction of calorimetric studies in tropical conditions is important for conceptual advances in roughage evaluation, enabling the best way of utilization, optimizing livestock performance (Rodriguez *et al.*, 2007). The objectives of this study were to examine the effects of sorghum genetics and stages of maturity at harvest on efficiency of energy use and methane production by sheep.

Material and methods

Forty five mature wether sheep were housed individually in metabolic cages and fed at 60-80 g DM/kg $BW^{0.75}$ per day. The treatments were silages of the sorghum hybrids BRS 610, BR 700 and BRS 655, harvested at three maturation stages (milk, soft dough and floury). The sorghum hybrids were developed by Embrapa Maize and Sorghum (Sete Lagoas, Minas Gerais, Brazil). Diets were fed for a 20 day adaptation followed by a 5 day collection period. Oxygen consumption, carbon dioxide production, and methane emissions were monitored for a period of 24 h on individual sheep one after the other, in an open-circuit respiration chamber for small ruminants, after adaptation period of two days in another similar respiration chamber. The chambers were made of transparent acrylic resin plates with external dimensions of 1.2 m (wide) × 2.0 m (height) × 2.1 m (length). The equipment and procedure utilized for the respiration study were described by Rodriguez *et al.* (2007). Heat production and fasting heat production were calculated as per Brouwer's equation (Brouwer, 1965).

The experimental design utilized was completely randomized in a 3×3 factorial arrangement. Analysis of variance (ANOVA) was used to analyse data using the General Linear Model Procedure (SAS, 2001). Main effects and interactions of hybrid and maturation stage were evaluated. Treatments means were differentiated using SNK test (SAS, 2001).

Results and discussion

There were no differences among the treatments for the apparent digestibility of gross energy and metabolizability ($P>0.05$). A significant ($P<0.005$) interaction between sorghum hybrid and maturation stage was observed for the efficiency of ME utilization for maintenance (K_m). Silage harvested at the soft dough stage had higher net energy to gross energy ratio (NE/GE) than that harvested at floury stage ($P<0.05$).

No differences were observed among the methane output, as liter per day, by sheep that received silage of the hybrid BRS 610 (without tannin) and those sheep fed silages of the hybrids BRS 700 and BR 655 (with tannin). Likewise, Oliveira *et al.* (2006) found no effect of diets containing sorghum silage with low and high tannin content on methane emission by beef cattle. The methane emission expressed as g per kg of digestible dry matter intake (g/kg DDM) and g per kg of digestible neutral detergent fiber intake (g/kg DNDF) differed among maturation stages, with soft dough stage having lower emission than floury stage ($P<0.05$).

The lack of differences in total daily methane output indicates that sorghum genetics and maturity at harvest should not be a strategy to reduce enteric methane emissions from ruminants.

Table 1. Energy efficiency and methane emission by sheep fed silages of the sorghum hybrids BRS 610, BR 700 and BRS 655 in three maturation stages (milk, soft dough and floury).

Hybrid	MS	ADGE	q_m	K_m	NE (%GE)	CH_4 (l/d)	CH_4 (g/kg DDM)	CH_4 (g/kg NDFD)
BRS 610	M	50.80	0.44	0.69_{Aa}	30.71	20.93	29.85	59.39
	SD	57.97	0.52	0.76_{Aa}	40.12	19.15	24.06	46.69
	F	51.54	0.46	0.68_{Aa}	31.46	22.44	26.99	73.02
BR 700	M	55.77	0.50	0.68_{Aa}	34.51	18.01	23.88	49.42
	SD	48.00	0.43	0.72_{Aa}	31.38	16.71	24.23	51.57
	F	52.07	0.45	0.71_{Aa}	32.49	24.46	34.36	67.46
BRS 655	M	54.33	0.47	0.61_{Ab}	28.98	22.28	33.75	61.98
	SD	51.68	0.46	0.78_{Aa}	35.99	13.64	24.26	46.14
	F	49.16	0.42	0.53_{Bc}	22.58	20.32	32.10	64.91
SEM		2.86	0.03	0.03	2.72	3.36	3.43	7.59
Main effects								
Hybrids								
BRS 610		53.44	0.47	0.71	34.10	20.84	26.97	59.70
BR 700		51.95	0.46	0.70	32.79	19.73	27.49	56.15
BRS 655		51.72	0.45	0.64	29.18	18.75	30.04	57.68
Maturation stage								
M		53.63	0.47	0.66	31.40_{ab}	20.41	29.16_{ab}	56.93_{ab}
SD		52.55	0.47	0.75	35.83_{a}	16.50	24.19_{b}	48.13_{b}
F		50.92	0.44	0.64	28.84_{b}	22.41	31.15_{a}	68.46_{a}
Probabilities, P≤								
H		0.7284	0.5182	0.0020	0.0871	0.7486	0.5091	0.8482
MS		0.5122	0.3569	0.0000	0.0116	0.1049	0.0489	0.0088
H x MS		0.1246	0.1113	0.0015	0.0583	0.6703	0.2836	0.7173

MS = maturation stage; M = milk; SD = soft dough; F = floury; ADGE = apparent digestibility of gross energy; qm = metabolizability; Km = efficiency of ME utilization for maintenance; NE/(%GE) = net energy to gross energy ratio (%); DDM = digestible dry mater; NDFD = digestible neutral detergent fiber; SEM = standard error of mean; H = significance for the hybrid effect; MS = significance for maturation stage effect; H × MS = significance for hybrid × maturation stage interaction.

Means of the hybrids followed by the same capital letters in columns and means of the maturations stages followed by the same lowercase letters in columns do not differ significantly by SNK test ($P<0.05$).

References

Brouwer, E. 1965. Report of subcommittee on constants and factors. Proc. 3rd Symposium on Energy Metabolism. Eur. Assoc. Anim. Prod. Pub. No. 1, 441-443.

Oliveira, S. G., T. T. Berchielli, M. S. Pedreira, O. Primavesi, R. Frighetto, A. A. Lima. 2007. Effect of tannin levels in sorghum silage and concentrate supplementation on apparent digestibility and methane emission in beef cattle. Anim. Feed Sci. Technol. 135, 236-248.

Rodríguez, N. M., W. E. Campos, M. L. Lachica, I. Borges and L. C. Gonçalves. 2007. A calorimetry system for metabolism trials. Arq. Bras. Med. Vet. Zootec. 59, 495-500.

Sanchez, A. C.,P. K. Subudhi, D. T. Rosenow, H. T. Nguyen. 2002. Mapping QTLs associated with drought resistance in sorghum (*Sorghum bicolor* L. Moench). J. Plant Mol. Biol. 48, 713-726.

SAS, 2001. Release 8.01. SAS Inst. Inc., Cary, NC, USA.

Methane emission from lactating cows fed diets with different forage base

S. Colombini, L. Rapetti, G. Galassi, L. Malagutti, M. Pirondini and G.M. Crovetto
University of Milan, Department of Agricultural and Environmental Sciences, via Celoria 2, 20133 Milan, Italy; matteo.crovetto@unimi.it

Introduction

Enteric methane (CH_4) emissions from ruminant livestock are major contributors to anthropogenic emission of greenhouse gases (GHG) and diet manipulation is the most direct mean of lowering CH_4 emissions from ruminants (Beauchemin *et al.*, 2008). Nutrition and feeding management is a very broad area, with many opportunities for mitigating CH_4 emissions (Knapp *et al.*, 2011). A typical diet fed to dairy cattle in Northern Italy is based on corn silage as main forage; however, corn requires several agronomical inputs which also contribute to GHG emissions. Sorghum is growing in popularity as a silage crop, primarily because it requires less water and agronomical inputs than corn.

Colombini *et al.* (2012) showed that whole plant grain sorghum silage can replace corn silage in dairy cow TMR without negative effects on milk production. In the same study, the forage sorghum, probably chopped too long, negatively affected dry matter intake (DMI) and milk production. The aim of the present research was to study the effect of the TMR evaluated in the previous study (Colombini *et al.*, 2012), and with a different forage base: corn (CS), whole plant grain sorghum (WPGS) or forage sorghum (FS) silages, on enteric CH_4 emissions from dairy cows.

Material and methods

Six lactating primiparous Italian Friesian cows were used. Three diets with a different silage base were tested: CS, WPGS, or FS. Diets were balanced to have a content (% DM) of 11.0, 36.0, and 26.0 of metabolizable protein, NDF, and starch, respectively. Because of the different fiber and starch contents, the three forages were included in the experimental diets in different proportions, and corn meal was needed at different proportions to compensate for the lower starch content of the WPGS and FS silages. The cows were fed the experimental diets in a replicated 3×3 Latin square design with 28-d period. In each period cows were placed in individual respiration chamber to register CH_4 production for seven days continuously and to collect the daily amount of feces produced. Data were statistically analyzed by the Mixed procedure of SAS (SAS Institute, 2001).

Results and discussion

Dry matter intake, milk yield and CH_4 emission of cows fed the different experimental diets are given in Table 1. A linear trend was observed for an effect of the kind of silage on DMI (P=0.07). Particularly, the FS diet resulted in a lower DMI, probably due to the greater particle size of FS diet. Despite a trend for a difference in DMI, total methane production (L/d) was not affected by diet. The passage rates of the iNDF fraction (%/h), predicted according the equation of Krizsan *et al.* (2010), were: 2.19, 2.19 and 2.12 for diets CS, WPGS and FS, respectively. Consequently, a higher retention time in the rumen for FS diet can be hypothesized which in turn increases the extent and rate of ruminal dietary fermentation as confirmed by total tract fiber digestibility which was higher (P=0.01) for SF diet (54.1%) than CS (51.4%) and WPGS (48.6%), as reported in a previous paper (Colombini *et al.*, 2012). As a consequence, there were differences among diets when CH_4 production (l) was expressed per kg of DMI (P=0.09) or per kg of NDF intake (P=0.02). Particularly, cows fed FS diet had the highest methane production for kg of NDF intake (71.3 vs. 63.4 and 67.6 l/kg NDF intake, for FS, CS and WPGS, respectively). Methane production per milk yield (l/kg) was: 18.7, 19.7 and 20.4 for CS, WPGS and FS, respectively (P=0.12). Methane energy (% of the gross intake energy) was higher for SF diet than CS and WPGS diets (P=0.04). Significant linear regressions were found between

CH_4 (l/d) and starch and NDF intakes (kg/d): CH_4=388+17.3×starch intake (R^2=0.18, P=0.08) and CH_4=276+28×NDF intake (R^2=0.48, P=0.001). Significant linear and quadratic regressions were found between the percentage of CH_4 gas energy loss on total gross energy (MGE) and the percentage of the dietary particles retained on a 19 mm sieve (DP19): MGE=4.88+0.039×DP19 (P=0.002; R^2=0.50); MGE=4.69+0.099×DP19–0.002×$DP19^2$ (P=0.002; R^2=0.59).

Conclusions

Sorghum forage is an interesting crop for low agronomic inputs. However, in this study, despite a lack of significant difference in absolute CH_4 emission, the SF diet increased the percentage of CH_4 energy loss and CH_4 emission per kg of NDF intake.

Table 1. Dry matter intake, milk yield and CH_4 emission of the cows on experiment.

		CS	WPGS	FS	SE	P-value
DMI	kg/d	20.0	20.0	18.2	0.53	0.07
Milk yield	kg/d	25.4[a]	24.6[ab]	23.6[b]	0.43	0.05
CH_4	l/d	467	483	478	16.1	0.74
CH_4/DMI	l/kg DMI	23.8	24.3	26.3	0.94	0.09
CH_4/NDF	l/kg NDF intake	63.4[b]	67.6[ab]	71.3[a]	2.20	0.02
CH_4/MILK	l/kg milk	18.7	19.7	20.4	0.75	0.12
CH_4 energy	% gross energy	5.1[b]	5.2[b]	5.8[a]	0.21	0.04

a,b Means in the same rows with different superscript are different for P<0.05.

References

Beauchemin K.A, M. Kreuzer, F. O'Mara and T. A. McAllister, 2008. Nutritional management for enteric methane abatement: a review. Aust. J. Exp. Agr. 48:21-27.

Colombini S., G. Galassi, G.M. Crovetto and L. Rapetti, 2012. Milk production, nitrogen balance, and fiber digestibility prediction of corn, whole plant grain sorghum, and forage sorghum silages in the dairy cow. J. Dairy Sci. 95:4457-4467.

Knapp J.R., J.L. Firkins, J.M. Aldrich, R.A. Cady, A.N. Hristov, W.P. Weiss, A.D.G. Wright and M.D. Welch, 2011. Cow of the future research priorities for mitigating enteric methane emissions from dairy. Working draft. Innovation center for U.S. dairy. http://www.usdairy.com/Public%20Communication%20Tools/CowoftheFutureWhitePaper_7-25-11.pdf

Krizsan S.J., S. Ahvenjärvi and P. Huhtanen, 2010. A meta-analysis of passage rate estimated by rumen evacuation with cattle and evaluation of passage rate prediction models. J. Dairy Sci. 93:5890-5901.

SAS Institute. 2001. User's Guide: Statistics. Release 8.01. SAS Inst. Inc., Cary, NC.

Effect of condensed tannins on methane emission and ruminal microbial populations

M. Rira[1,2], C. Marie-Magdeleine[2], H. Archimède[2], D.P. Morgavi[1] and M. Doreau[1]
[1]INRA, UMR1213 Herbivores, 63122 Saint-Genès-Champanelle and Clermont Université, VetAgro Sup, UMR Herbivores, BP 10448, 63000, Clermont-Ferrand, France; michel.doreau@clermont.inra.fr
[2]Unité de Recherches Zootechniques, INRA, 97170 Prise d'Eau Petit-Bourg, Guadeloupe, France

Introduction

Enteric methane (CH_4) produced by domestic ruminants represents approximately 15% of the global emissions of this potent greenhouse gas. For reducing rumen CH_4 emission various compounds have been tested as feed additives. Among these compounds, tannins are considered a promising group of natural additives. A meta-analysis by Jayanegara *et al.* (2012) showed that condensed and hydrolysable tannins might reduce CH_4 production. However, it is still unclear (1) whether tannin supplementation reduces rumen CH_4 in every situation and (2) to which extent this is associated with adverse effects on digestibility and their potential toxicity to some rumen micro-organisms (Goel *et al.*, 2005). In this experiment we investigated the effect of tanniniferous tropical plants on enteric CH_4 production and on numbers of methanogens, protozoa, and total and main cellulolytic bacteria.

Material and methods

Two sheep breeds were used, Texel (T, n=4) of temperate origin and Blackbelly (B, n=4) of tropical origin in two 4×4 Latin square designs. Diets, given ad libitum twice daily, consisted in tropical natural grassland based on *Dichanthium spp.* fed alone (C) or in association with 3 different tanniniferous forages given as pellets at 44% of the daily ration on average. The tanniniferous forages were leaves of *Leucaena leucocephala* (L), *Glyricidia sepium* (G) or *Manihot esculenta* (M). Total contents in condensed tannins measured by the vanillin-H_2SO_4 method were 75, 39 and 92 g/kg dry matter (DM) for L, G and M, respectively. Intake, total tract digestibility and CH_4 production using the SF_6 method were determined. For microbial parameters, rumen contents were sampled before the morning feeding. Microbial groups (total bacteria, *Fibrobacter succinogenes, Ruminococcus albus, Ruminococcus flavefaciens* and total methanogens) were enumerated by quantitative PCR (qPCR) using group-specific primers targeting the *rrs* gene for bacteria and the *mcrA* gene for methanogens. Protozoa were counted by microscopy. Statistical analyses were performed using the mixed procedure of SAS with period, diet, breed, and the diet x breed interaction as fixed effects and animal as random effect. Statistical differences were declared significant at $P \leq 0.05$. Orthogonal contrasts between control diet and all tanniniferous forages were also determined.

Results and discussion

The addition of tannin-rich plants given as pellets increased DM intake probably due to the physical presentation (Table 1). Within tannin-rich plants, it was higher for M than for G diet. Intake per kg body weight was higher for Blackbelly than for Texel. Organic matter digestibility did not differ among diets and breeds, although contrast analysis showed a higher digestibility for C than for other diets. Daily CH_4 production did not vary among diets and breeds, but CH_4 production per kg DM intake was higher with C diet compared with tannin-rich diets. Within these latter diets, CH_4 production was higher for G than for M and L diets. Concentration of total bacteria and *R. flavefaciens* was higher for C and L diets than for G and M diets; concentration of *R. albus* was lowest for C diet. The methanogens population was higher for Texel than for Blackbelly. In contrast the addition of condensed tannins did not influence the population of protozoa and *F. succinogenes.*

Table 1. Intake, digestibility, methane emission and ruminal microbial populations in sheep fed tropical grassland hay (C) alone or associated with tannins-containing plants Leucaena leucocephala *(L),* Glyricidia sepium *(G) and* Manihot esculenta *(M).*

Breed	Texel				Blackbelly				SEM	*P* value[1]
Diet	C	L	G	M	C	L	G	M		
DM intake, g/kg body weight/d	16.28	24.45	20.72	27.68	16.62	25.96	24.49	31.54	1.877	B 0.04 D<0.01
Organic matter digestibility, %	69.66	64.94	62.69	62.36	70.12	65.74	62.78	66.97	3.639	NS
CH4, g/kg DM intake	32.06	24.23	28.83	24.28	29.98	15.89	24.98	15.26	2.147	B 0.03 D<0.01
Protozoa, $\log_{10}$ cells/ml	4.97	5.01	5.04	5.03	4.96	5.01	4.97	5.06	0.046	NS
Total bacteria[2]	11.91	11.90	11.80	11.84	11.89	11.96	11.85	11.81	0.041	D 0.04
F. succinogenes[2]	9.71	9.79	9.61	9.72	9.67	9.85	9.64	9.68	0.079	NS
R. Albus[2]	8.00	8.79	8.33	8.49	7.60	8.42	8.34	8.16	0.261	D 0.04
R. flavefaciens[2]	8.96	8.71	8.53	8.65	8.94	9.03	8.67	8.85	0.154	D 0.05
Methanogens[3]	10.00	9.85	9.72	9.84	9.76	9.71	9.70	9.74	0.082	B 0.04

[1] B = breed; D = diet. Breed × diet interaction was never significant.

[2] *rrs* copy number /g DM ($\log_{10}$)

[3] mcrA copy number /g DM ($\log_{10}$)

Our results confirm that tannin-rich plants can limit CH_4 production per kg DM intake. The low effect of *Glyricidia sepium* on reduction of CH_4 production could be explained by a low tannin concentration or by the intrinsic characteristics of *Glyricidia sepium* tannins. The presentation of tannin-rich plants as pellets probably decreased ruminal DM retention time which resulted in an increase in DM intake and can partially explain the reduction in CH_4 production. As methanogens and protozoal numbers were not changed, further research is necessary to elucidate the relations between methane production and microbial activity in tannin-rich diets.

Acknowledgements

This research is a part of the AnimalChange project funded by the European Community's FP7 Programme. It was also funded by other EU funds (FEDER, FEADER) and by the Guadeloupe region. The first author received a fellowship from the Algerian Ministry of Higher Education and Research.

References

Goel, G., A.K. Puniya, C.N. Aguilar and K. Singh, 2005. Interaction of gut microflora with tannins in feeds. Naturwissenschaften 92, 497-503.

Jayanegara, A., F. Leiber, and M. Kreuzer, 2012. Meta-analysis of the relationship between dietary tannin level and methane formation in ruminants from in vivo and in vitro experiments. J. Anim. Physiol. Anim. Nutr. 96, 365-375.

Methane emission by cattle supplemented with additives in Brazil

A.L.C.C. Borges[1], M.P. Fonseca[1], R.R. Silva[1], H.F. Lage[1], J.A.S. Rodrigues[2], L.C. Gonçalves[1], I. Borges[1] and E.O.S. Saliba[1]
[1]Department of Animal Science, Federal University of Minas Gerais, UFMG, Brazil; analuiza@vet.ufmg.br
[2]Brazilian Agricultural Research Corporation, Brazil

Introduction

Energy is a limiting factor to life and productive functions of animals. The knowledge of gas production by animals according to the diet demonstrates the partial efficiency of energy utilization as an important indicator of feed utilization. Virginiamycin is a non-ionophore antibiotic, with potential effect on improvement of ruminal fermentation and features indicative of an increase in dry matter intake. This antibibiotic acts by penetrating the bacterial cell wall, binding to the 50 S ribosomal subunit and blocking protein synthesis through the inhibition of peptide bond formation (Cocito, 1979). Monensin, an ionophore antibiotic, has antimicrobial activity against Gram positive bacteria and subsequent alterations in ruminal fermentation products, namely an increase in propionate at the expense of acetate and methane (Nagaraja *et al.*, 1997). This study aimed to evaluate the inclusion of additives monensin, virginiamycin and its combination on methane emission determined by respirometric chamber in F1 Crossbred Holstein x Gir bulls.

Material and methods

Twenty F1-crossbred (*Bos taurus indicus*) × Holstein (*Bos taurus taurus*) bulls (initial average body weight 150 kg, final average body weight 275 kg) were randomly assigned to four groups of five animals each and fed four diets: a control diet (C, with 50% sorghum silage with Tanzania grass (*Panicum maximum Jacq vr. Tanzânia l.*), 9.89% soybean meal, corn meal 37.23%, 0.72% urea, in dry matter basis), a diet with inclusion of monensin (M), a diet with inclusion of virginiamycin (VM) and a diet with inclusion of monensin and virginiamycin (M+VM). Monensin and virginiamycin were included in the formulation of concentrates at doses of 22 and 30 mg/kg dry matter, respectively. Roughage and concentrate were supplied at a ratio of 50:50. The diets were fed twice daily as a total mixture ration. Dry matter intake was monitored during all the experiment. The animals were fed *ad libitum*. The production of methane was obtained individually by respirometric chamber through the open circuit system available at the Animal Science Department of Veterinary School of UFMG and described by Rodriguez *et al.* (2007). The experiment was conducted in a completely randomized design. Data were analyzed by SAEG, described by Ribeiro (2001), and means were compared by Tukey test at 5% probability ($P<0.05$).

Results and discussion

Table 1 reports the production of methane by the supplemented bulls. The methane emission was lower ($P<0.05$) for animals supplemented with both additives, when expressed in liters by day (l/d), liters by kilogram of dry matter (l/kg DM) and liters by kilogram of digestible dry matter (l/kg DIG DM). The reduction in methane production with combined use of monensin and virginiamycin is probably more related the reduction of the precursors of methane, H_2 and formate. A direct effect might have resulted from a change in ruminal pH (Blaxter & Clapperton, 1965), but, since no reduction was observed in feed intake and digestibility of the fiber, and diets had 38,6% neutral detergent fiber (NDF), this does not seem likely. Rather changes in the gastrointestinal microbiota may have been responsible. Further studies have to show whether the known greater efficiency in converting to metabolizable energy and improving the utilization of metabolizable energy is also realized under tropical conditions.

Table 1. Mean values, probability (P-value) and coefficient of variation (CV) of production of methane by supplemented bulls.

Variables[1]	Treatment				*P*-value	CV (%)
	C	M	VM	M+VM		
CH_4 (l/day)	185.62[a]	153.78[ab]	171.57[ab]	137.70[b]	0.0298	14.66
l CH_4/kg DM	28.86[a]	24.97[ab]	27.94[ab]	23.27[b]	0.0342	11.48
l CH_4/kg DIG DM	48.21[a]	41.63[ab]	45.15[ab]	38.84[b]	0.0484	11.61
l CH_4/kg NDF	68.81[a]	58.01[a]	63.75[a]	57.27[a]	0.2926	16.78
l CH_4/kg DIG NDF	163.78[a]	140.56[a]	135.10[a]	147.75[a]	0.2853	28.81
l CH_4/kg OM	31.36[a]	25.74[a]	30.89[a]	26.94[a]	0.1520	15.40
LCH_4/kg DIG OM	51.70[a]	42.62[a]	49.93[a]	45.07[a]	0.2909	17.06
l/kg $BW^{0,75}$	3.19[a]	2.53[a]	2.95[a]	2.42[a]	0.0548	16.36
l CH_4/kg gain	120.66[a]	107.25[a]	127.18[a]	108.24[a]	0.2016	14.27

[a,b] $P<0.05$.

[1] Methane production expressed in liters per day (l/day), liters per kilogram of dry matter (l/g DM) liters per kilogram of digestible dry matter (l/kg DIG DM) liters per kilogram of insoluble fiber neutral detergent (l/kg NDF) liters per kilogram of neutral detergent fiber digestible (l/kg DIG NDF) liters per kilogram of organic matter (l/kg OM) liters per kilogram of digestible organic matter (l/kg DIG OM), liters per kilogram of metabolic weight (l/kg $BW^{0.75}$) and liters per kilogram of gain (l/kg gain).

Conclusion

The supplementation with 30 mg/kg DM of virginiamycin associated with 22 mg/kg DM of monensin likely promotes energy efficiency improvement of diet, due to the reduction in methane production without compromising nutrient intake.

Acknowledgment

The authors thank CNPq, CNPq-INCT, FAPEMIG, CAPES and EMBRAPA CORN AND SORGHUN, for their cooperation in carrying out this work.

References

Blaxter, K.L., J.L. Clapperton, 1965. Prediction of the amount of methane produced by ruminants. Br. J. Nutr. 19, 511-522.

Cocito, C, 1979. Antibiotics of the virginiamycin family, inhibitors which contain synergistic components. Microb. Rev. 43, 2, 145-198.

Hobson, P.N. and C.S. Stewart, 1997. Rumen Microbial Ecosystem. In: Nagaraja, T.G., C.J. Newbold, C.J. Van Nevel and D.I. Demeyer, Manipulation of rumenal fermentation, Blackie Academic and Professional, London, 523-632.

Ribeiro Jr. J.I. SAEG (A system for statistical analysis), 2001. Federal University of Viçosa, Viçosa, MG, Brazil.

Rodríguez, N.M., Campos, W.E., Lachica, M.L., Borges, I., Gonçalves, L.C., Borges, A.L.C. C. Saliba, E.O.S, 2007. A calorimetry system for metabolism trials. Arq. Bras. Med. Vet. Zoot. 59, 2, 495-500.

Effect of paddy rice diets on performance in chickens under thermoneutral and heat stress conditions

F. Nanto, C. Ito, M. Kikusato, S. Ohwada and M. Toyomizu
Graduate School of Agricultural Science, Tohoku University, Aoba-ku, Sendai, Japan; f_nanto@bios.tohoku.ac.jp

Introduction

The global demand for corn to be used in the production of feed and fuel is increasing at a rapid rate. Many types of grain have been proposed as substitutes for corn in broiler chicken diets, thereby serving as alternative sources of dietary carbohydrates. We have already demonstrated that paddy rice show some potential for use as a substitute for corn in poultry feed under thermoneutral conditions (Nanto *et al.*, 2012). To the authors' knowledge, virtually no information is available to confirm a direct effect of the rice feeding on performance on chickens exposed to heat stress. Heat stress is a major issue for the poultry industry because of the growth retardation and high mortality. We have previously found that heat stress induces oxidative damage, resulting in growth performance in birds (Azad *et al.*, 2010). Furthermore, it has been reported that intestinal morphology was altered by heat stress (Quinteiro-Filho *et al.*, 2010). From these findings, we could hypothesize that oxidative stress and intestinal morphology might be factor responsible for growth retardation under heat stress conditions. Therefore, the present study was conducted to not only determine the effects of whole-grain paddy rice-based diets on growth performance in birds under heat stress conditions, but also to clarify the involvement of intestinal morphological alterations and oxidative stress in the performance under chronic heat stress conditions.

Material and methods

Forty-eight chicks (ROSS) at 0 day old were divided into 3 groups that were fed one of the following 3 experimental diets ad libitum for 28 day: one of corn-based diets (CP: 20%, ME:3.1 kcal/g, fat content:5.5%), another two ME levels of whole-grain paddy rice-based diets: standard ME rice diet (CP: 20%, ME: 3.1 kcal/g, fat content: 11.5%) and low ME rice diet (CP: 20%, ME: 2.8 kcal/g, fat content: 5.5%). At 21 d of age, birds in the each groups were randomly divided into two sub-groups (n=8), and then one of the two groups were exposed to heat stress (33 °C for 7 days) while the remaining group was maintained at 24 °C. At the end of the experiment, all chickens were sacrificed by decapitation. Tissues were immediately frozen, powdered in liquid nitrogen and stored at -80 °C until analyzed.

A one cm length of intestinal tissue from each chicken was collected from the midpoint of the duodenum and stored in 10% formalin neutral buffer solution prior to morphological analysis, which was performed on 3-4 μm-thick sections cut on a microtome. Sections were stained using the hematoxylin-eosin method. Villus height and crypt depth were measured with the aid of a microscope from 10 randomly selected villi and associated crypts on two sections per chicken. Plasma endotoxin concentrations were determined by a chromogenic Limulus amoebocyte lysate (LAL) end-point assay (QCL-1000, Lonza Group Ltd., Basel, Switzerland). Lipid peroxidation levels in tissues were determined colorimetrically in terms of the production of TBARS. Chemiluminescence (CL) intensity of the *pectoralis* major muscle was measured as free radical reaction and lipid peroxidation after incubation for 10 minutes at 100 °C with N_2 flow. Plasma ceruloplasmin (Cer) concentration that is an indicator of inflammatory response was measured with the *p*-phenylenediamine colorimetric method.

Data were analyzed using the statistical analysis system. Data were first analyzed by a general linear model analysis of variance procedure and the means were compared using Duncan's least significance multiple-range test. Correlation analysis was assessed using the Pearson correlation coefficient.

Results and discussion

Under thermoneutral condition, body weight gain (BWG) of birds fed both rice-based diets was significantly decreased ($P<0.05$) compared with corn-based diet, with the standard-ME rice diet-fed chickens showing a reduced tendency to lose weight compared with low-ME diet-fed animals (Figure 1A). On exposure to chronic heat stress, BWG of birds fed corn-based diet were significantly decreased. Heat-stressed birds fed both standard- and low-ME rice diets showed a decrease in the BWG compared with control diet-fed heat-stressed birds, but birds fed low-ME rice diet showing a tendency of improved growth compared with standard-ME diet-fed birds. On exposure to heat stress, the villus height: crypt depth ratio was significantly decreased in birds fed a control diet (Figure 1B). Heat-stressed birds fed both standard- and low-ME rice diets showed slightly decrease in the villus height: crypt depth ratio compared with control diet-fed heat-stressed birds. Meanwhile, plasma endotoxin concentration (which is one of indexes of intestinal barrier dysfunction) was significantly increased by heat stress in birds fed either corn- or rice-diet, and the degrees of the increase due to heat stress was enhanced in both standard- and low-ME rice diet groups compared to control diet-fed heat-stressed ones (Figure 1C). We further investigated the relationship between the intestinal morphology and plasma endotoxin concentration of birds under thermoneutral and heat-stressed conditions. As a result, a negative correlation between plasma endotoxin concentration and the villus height: crypt depth ratio was observed (r=-0.534, $P<0.0001$; Figure 2), suggesting that intestinal morphology damage might trigger influx of endotoxin from intestine into blood under heat stress conditions. Moreover, negative correlation between the plasma endotoxin concentration and CL intensity of muscle (r=0.4286, P=0.0029) or TBARS content of the liver (r=0.3367, P=0.020) implied that the plasma endotoxin might induce oxidative damage. There is no significant correlation between the plasma endotoxin concentration and plasma Cer concentration among all the groups.

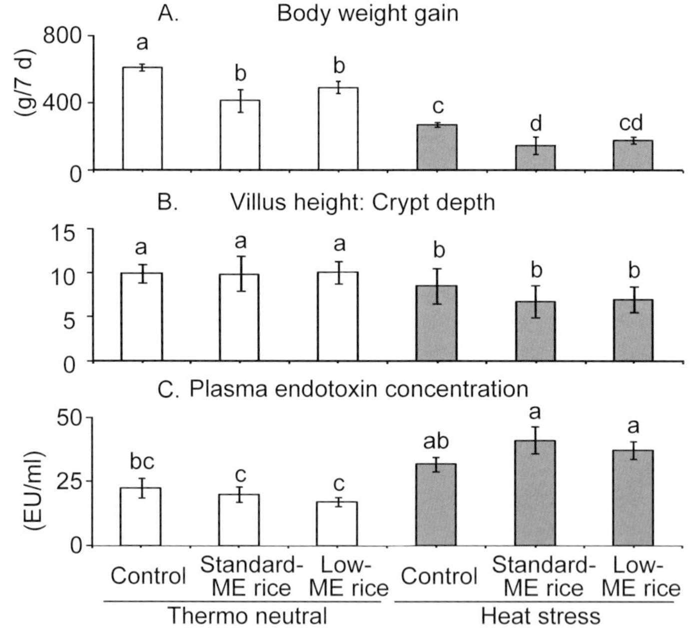

Figure 1. Effects of corn-based diet and two ME levels of whole-grain paddy rice-based diets (standard and low) on body weight gain (A), villus height: crypt depth (B) and plasma endotoxin concentration (C) of broiler chickens exposed to thermo neutral and chronic heat stress (33 degrees, 7 d). Values are means ± SE, n=8 per group.
[a,b,c] P<0.05, *for each treatment, values with different letters are statistically different.*

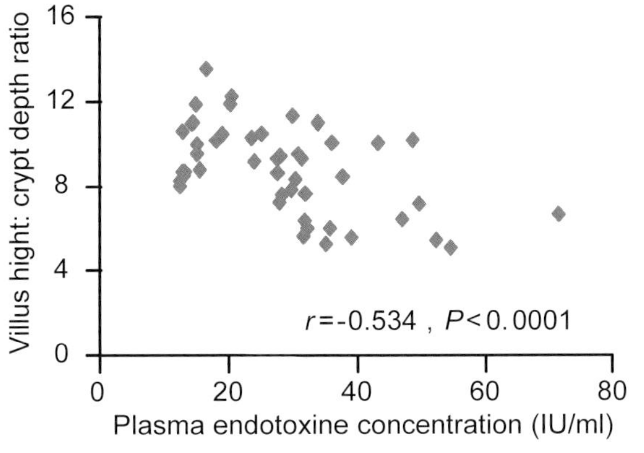

Figure 2. The relationship between the villus height and plasma endotoxin concentration.

Nevertheless, heat-exposed birds fed standard ME diet showed remarkable increases of plasma endotoxin and Cer compared with control diet-fed heat-stressed birds.

Taken together, the present study demonstrated that the growth performance due to chronic heat stress in the standard-ME rice diet-fed chickens was further decreased compared to that of the corn-fed chickens, but this decrease was ameliorated in the low ME rice diet-fed chickens. Furthermore, we suggest that heat stress-induced intestinal morphology damage might be partly responsible for the increases in oxidative damage and inflammatory response, probably via an endotoxin influx into blood.

References

Azad, M.A.K., Kikusato, M., Maekawa, T., Shirakawa, H. and Toyomizu M, 2010. Metabolic characteristics and oxidative damage to skeletal muscle in broiler chickens exposed to chronic heat stress. Comp. Biochem. Physiol. A, 155: 401-406.

Nanto, F., Kikusato, M., Ito, C., Sudo, S. and Toyomizu M, 2012. Effects of Dehulled, Crushed and Untreated Whole-Grain Paddy Rice on Growth Performance in Broiler Chickens. The Journal of Poultry Science, 49: 291-299.

Quinteiro-Filho, W.M., Ribeiro, A., Ferraz-de-Paula, V., Pinheiro, M.L., Sakai, M., Sá. L.R., Ferreira, A.J. and Palermo-Neto J, 2010. Heat stress impairs performance parameters, induces intestinal injury, and decreases macrophage activity in broiler chickens. Poultry Science, 89: 1905-14.

Effect of invertebrates on growth performance and feeding behavior of red-legged partridge (*Alectoris rufa*) chicks

M. Lachica and I. Fernández-Fígares
Department of Animal Nutrition, Estación Experimental del Zaidín, CSIC, Camino del Jueves s/n, 18100 Armilla, Granada, Spain; manuel.lachica@eez.csic.es

Introduction

Red-legged partridge (*Alectoris rufa*) has a big ecological importance in Spain and it is under huge hunting pressure. So it is bred intensively in game farms for restocking areas where the number of birds is low or even it has been exterminated. About 4 mill birds/year are released. Their natural diet includes plants, mainly Gramineae, and also a high proportion of invertebrates in the first wk of life, which is linked with the bird survival index (Green, 1984). The sudden diet change from the farm to the field seems to be responsible of a low survival index (5-10% in the first year). In this phase, studies about nutrition and appetency for animal *vs*. vegetal feed are scarce or non-existent. The aim of this work was to determine the impact of the inclusion of invertebrates in a commercial diet for partridge chicks over its metabolic utilization by obtaining the feed conversion ratio (FCR), the total protein and gross energy (GE) intake, and the voluntary intake of animal *vs*. vegetal food during the first 6 wk of life.

Material and methods

Twenty-eight 1-2 d old (11.8±0.2 g body weight (BW)) chicks hatched in our lab were allocated at random by pairs in metabolic cages and fed *ad libitum*. A control (C; n=14) group was fed with a commercial starter diet (908 g/kg dry matter (DM), 279 g/kg crude protein (CP) and 18.0 MJ/kg GE), and a larvae (L; n=14) group could choose between the starter diet and alive *Calliphora sp.* larvae (324 g/kg DM, 175 g/kg CP and 8.6 MJ/kg GE). Intake and BW were recorded every 3 d for 45 d for determining FCR (g DM feed/g BW gain). Analysis of variance was used to determine the diet and age effect on FCR, the diet effect on total protein and GE intake, and the proportion of larvae intake in the group L. Bonferroni multiple range test was used to ascertain the statistical significance of differences.

Results and discussion

Table 1 shows the results obtained. BW, similarly as observed by Liukkonen-Anttila *et al.* (2002) with *Perdix perdix*, and FCR were improved in the L (2.05 vs. 2.31). FCR increased from day 30 until the end when larvae intake decreased, indicating an inadequate protein/energy ratio. A significant diet×age interaction was found. For the C group the FCR ranged from 1.48-3.65. Ozek (2006) reported FCR ranging 2.2-3.1 in *Alectoris chukar* fed diets of similar CP content and growth period. For the L group, FCR ranged from 0.85-3.37. No significant differences were observed between C and L groups in protein (3.65 vs. 3.55 g/d) and GE intakes (236 vs. 211 kJ/d) indicating a better protein quality for the L diet since Met and Cys level is higher in larvae than in plants (Anonymous, 1970). The larvae/commercial feed intake ratio was 0.69, 0.71 and 0.58 for the 1st, 2nd and 3rd wk, respectively. Rueda *et al.* (1992) obtained ratios of 0.81, 0.69 and 0.51 for the same periods in *Alectoris rufa* in the wilderness. The ratio decreased and reached a steady ratio (0.35) from the 4th wk until the end of the experiment indicating that birds would always eat invertebrates if they are available in the habitat.

The inclusion of invertebrates in the diet of hand-reared partridge chicks during their first wk of life improved their growth may be attributable to the intake of protein of better quality. This could be of importance for the survival of the red-legged partridge in restocking areas.

Table 1. Effect of diet (= group), sampling period (SP[1]; = age) and its interaction on the feed conversion ratio (FCR, g DM feed/g BW gain) of Alectoris rufa *chicks fed* ad libitum *with a commercial starter diet (Control-C; n=14) or based on the starter and alive* Calliphora sp. *larvae (Larvae-L; n=14). Body weight (BW, g), intake (g) and percentage of larvae intake (%) in every SP.*

	FCR	BW[2]	Intake[2]	FCR		Larvae intake
Diet				C	L	
C	2.31^{a}					
L	2.05^{b}					
SP[1]						
1	1.37^{ab}	14.5	3.6	1.88^{ab}	0.85^{a}	65.5^{def}
2	1.54^{ab}	18.4	10.8	1.74^{a}	1.33^{ab}	72.0^{f}
3	1.37^{a}	27.5	18.1	1.48^{a}	1.26^{ab}	70.1^{ef}
4	1.67^{abc}	37.7	27.7	1.93^{ab}	1.40^{abc}	71.1^{f}
5	1.48^{ab}	51.7	29.9	1.64^{a}	1.31^{ab}	64.9^{def}
6	1.61^{abc}	68.2	36.7	1.66^{a}	1.57^{abc}	51.2^{bcdef}
7	1.67^{abc}	86.5	40.9	1.85^{a}	1.50^{abc}	47.8^{abcde}
8	1.98^{abc}	105.2	50.5	2.06^{ab}	1.90^{bcd}	53.8^{cdef}
9	2.18^{cd}	124.6	51.5	2.22^{abc}	2.14^{cd}	30.6^{ab}
10	3.11^{ef}	141.0	71.0	3.20^{de}	3.01^{ef}	30.9^{abc}
11	2.04^{bc}	160.7	47.9	2.04^{ab}	2.04^{bcd}	40.5^{abcd}
12	2.71^{de}	179.0	52.1	2.93^{cde}	2.49^{de}	27.8^{a}
13	3.03^{ef}	195.3	60.1	2.70^{bcd}	3.36^{f}	38.9^{abc}
14	3.41^{f}	208.5	62.2	3.65^{e}	3.16^{ef}	38.4^{abc}
15	3.51^{f}	227.4	68.8	3.64^{e}	3.37^{f}	38.0^{abc}
Diet×SP	0.0073					
SEM[3]	0.029			0.043	0.040	1.16

[1] Sampling period every 3 d.

[2] Measured at the end of every 3 d period.

[3] Standard error of mean; ^{a,b,c,d,e,f} within a row, values with different superscripts differ significantly (P<0.05).

Acknowledgements

This research was supported by grant no. AGR 03065 from Junta de Andalucía (Spain).

References

Anonymous, 1970. Amino-acid content of foods and biological data on proteins. In: F.A.O. Nutritional Studies No. 24. Rome, 285 pp.

Green, R.E., 1984: The feeding ecology and survival of partridge chicks (*Alectoris rufa* and *Perdix perdix*) on arable farmland in East Anglia. J. Appl. Ecol. 21, 817-830.

Liukkonen-Anttila, T., A. Putaala and R. Hissa, 2002. Feeding of hand-reared grey partridge *Perdix perdix* chicks-importance of invertebrates. Wildl. Biol. 8, 11-19.

Ozek, K., 2006. The optimum protein content in high-energy starter diet for chukar partridge (*Alectoris chukar chukar*). Int. J. Poult. Sci. 5, 522-525.

Rueda, M.J, J.R. Baragaño and A. Notario, 1992. Alimentación natural de la perdiz roja (*Alectoris rufa L.*). In: La perdiz roja, gestión del hábitat, 1ª ed. Fundación 'La Caixa', Barcelona, 27-39.

Approach to determine the amino acid composition of the natural diet of red-legged partridge (*Alectoris rufa*)

I. Fernández-Fígares and M. Lachica
Department of Animal Nutrition, Estación Experimental del Zaidín, CSIC, Camino del Jueves s/n, 18100 Armilla, Granada, Spain; manuel.lachica@eez.csic.es

Introduction

Red-legged partridge (*Alectoris rufa*) plays a very important role in Spanish ecosystems being part of the diet of 32 predators (3 reptiles, 9 birds and 20 mammals; Yanes *et al.*, 1998). Also, it has a big economic importance and, as the favorite game species, it is subjected to an enormous hunting pressure. As a result, it is bred intensively in game farms to satisfy the hunters' demand and restock areas where its number is too low. About 4 millions of birds/yr are released. Hunting season is performed during the reproductive rest when partridges are in maintenance conditions. NRC (1994) reported the nutrients requirements for game birds without specific data for partridges and in many cases the reader is addressed to requirements of turkeys. Nevertheless, the requirements of free-living birds potentially differ significantly from those of domestic species that have been inbred for improvement of meat and egg production (Murphy and King, 1982). Therefore, *Alectoris rufa* requirements still remain essentially unknown with no information about the amino acids (AA) composition of the natural diet or differences related to gender. The objective of this work was to establish a first approach to determine the AA composition of the natural diet of adult red-legged partridge.

Material and methods

The content of crops and gizzards of 17 birds (7 females and 10 males) hunted during the same hunting season (October-February; nonbreeding season and postbreeding moult passed) in 'Sierra de Baza' Natural Park (Granada, Spain) was analyzed for dry matter, organic matter (OM) and AA. After protein hydrolysis, AA were determined by HPLC according to the Waters Pico Tag method. AA concentration was expressed as mg/g OM to avoid interference by inorganic compounds. To determine the gender effect on AA profile including essential (EAA), non-essential AA (NEAA) and total AA (TAA), ANOVA-I was used. Bonferroni multiple range test was used to ascertain the statistical significance of differences.

Results and discussion

Table 1 shows the results obtained. Diet content of females had a higher ($P<0.05$) concentration of individual EAA than males (Arg, Ile, Leu, Lys and Phe) and a tendency for higher ($P<0.06$) Met, His, Thr and Val; the same occurred with individual NEAA ($P<0.09$). Likewise females presented greater values than males for EAA ($P<0.05$), NEAA ($P<0.07$) and TAA ($P<0.05$). Except for Arg, dietary AA concentration of females was more than twice the one of males.

Readiness of feather loss is an escape behavior for captured birds which is significantly related to susceptibility to predation (Møller *et al.*, 2006). It differs significantly between males and females. The role of females in parental care is more important than males in dichromatic species, which may promote their escape behavior (Moller *et al.*, 2011). On the other hand, the metabolizable energy efficiency for feather synthesis (around 5%) is lower than the one estimated for body protein (40-60%) in growing homeotherms. Moreover, dietary AA play an important role in the development of feathers (Murphy and King, 1984a,b). Females depend entirely on exogenous sources of protein to satisfy their AA requirements for molting (Hipes and Hepp, 1995). There is a reduction of S retention during molt by a loss of bone chondroitin sulfate producing a cyclic osteoporosis (Meister,

Table 1. Concentration (mg/g OM) of individual, essential (EAA), nonessential (NEAA) and total amino acids (TAA) in females (n=7) and males (n=10) of Alectoris rufa.

	EAA	His	Arg	Thr	Val	Met	Ile	Leu	Phe	Lys	
Female°	25.8^a	1.68	5.08	1.66	2.11	0.60	1.57^a	3.24^a	1.98^a	2.86^a	
Male	11.6^b	0.57	2.62	0.67	0.94	0.30	0.67^b	1.30^b	0.78^b	1.09^b	
SEM	3.16	0.269	0.532	0.239	0.286	0.076	0.206	0.438	0.270	0.369	
	NEAA	**Ala**	**Asp**	**Cys**	**Glu**	**Gly**	**Pro**	**Ser**	**Tyr**		**TAA**
Female	14.1	2.45	3.36	0.39	5.22	1.85	2.24	1.94	1.68		39.9^a
Male	4.9	1.17	1.23	0.18	1.83	0.90	0.80	0.70	0.73		16.5^b
SEM	2.34	0.330	0.517	0.052	0.802	0.237	0.387	0.300	0.255		5.41

^{ab} Values within a column with different superscript letters were significantly different (P<0.05).

1951). Thus, the protein for feather replacement relies on food protein sparing body proteins from depletion (Murphy and King, 1984b).

There were differences in the concentration of AA in the natural diet of females and males. It seems important for the females to have a greater intake of AA than males to maintain a pool of available AA to replace feather protein loss even out of the specific molt season. It may be advisable to use diets differing in the proportion of individual AA in the game farms during the maintenance phase for breeding birds.

Acknowledgements

This research was supported by grant no. AGR 03065 from Junta de Andalucía (Spain).

References

Hipes D.L. and G.R. Hepp, 1995. Nutrient-reserve dynamics of breeding male Wood Ducks. Condor 97, 451-460.

Hohman W.L. and R.D. Crawford, 1995. Molt in the annual cycle of Ring-necked Ducks. Condor 97, 473-483.

Meister W., 1951. Changes in histological structure of the long bones of birds during molt. Anat. Rec. 111, 1-21.

Møller A.P., J.T. Nielsen and J. Erritzøe, 2006. Losing the last feather: feather loss as an anti-predator adaptation in birds. Behav. Ecol. 17, 1046-1056.

Møller A.P., S.S. Christiansen and T.A. Mousseau, 2011. Sexual signals, risk of predation and escape behavior. Behav. Ecol. 22, 800-807.

Murphy M.E. and J.R. King, 1982. Semi-synthetic diets as a tool for nutritional ecology. Auk 99, 165-167.

Murphy M.E. and J.R. King, 1984a. Sulfur amino acid nutrition during molt in the white-crowned sparrow. 1. Does dietary sulfur amino acid concentration affect the energetics of molt as assayed by metabolized energy? Condor 86, 314-323.

Murphy M.E. and J.R. King, 1984b. Sulfur amino acid nutrition during molt in the white-crowned sparrow. 2. Nitrogen and sulfur balance in birds fed graded levels of the sulfur-containing amino acids. Condor 86, 324-332.

NRC, 1994. Nutrient requirement of poultry. National Academy Press, Washington, DC.

Yanes, M., J. Herranz, J. de la Puente and F. Suárez, 1998. La perdiz roja: Identidad de los depredadores e intensidad de la depredación. In: I Curso sobre la perdiz roja. FEDENCA-GRUPO Editorial V, 135-147.

Part 9. Baldwin symposium

DGLC = -GLCUSE
GLCUSE = Fmax (1-Echarge)
1 - KGLC
GLC
PF = Y*

The life and legacy of Dr. Ransom Leland ('Lee') Baldwin V

September 21, 1935 – November 30, 2007

R.L. Baldwin, VI[1] *and C.C. Calvert*[2]
[1]*Bovine Functional Genomics Laboratory, USDA ARS, Beltsville, MD, USA; ransom.baldwin@ars.usda.gov*
[2]*Department of Animal Sciences, University of California, Davis, CA, USA*

Professor Ransom Leland ('Lee') Baldwin was born on September 21, 1935, in Meriden, Connecticut. Lee was the eldest of three children raised on the family dairy farm. He grew up milking dairy cows and this background led to his lifelong commitment to the dairy industry. Lee attended the University of Connecticut and earned a B.S. in animal industries.

Subsequently, he attended Michigan State University and earned a M.S. in dairy nutrition before his Ph.D. in biochemistry and nutrition in 1963. He was a National Science Foundation fellow from 1957 to 1961. Lee joined the faculty at the University of California at Davis in 1963, attaining the rank of professor in 1970. From 1992 to 2000 he served as Sesnon Professor of Animal Science. Lee finally retired in April 2001, following a 38-year career. He met his wife, Mary Ellen, upon graduation from High School and they were married on June 1, 1957, following completion of his bachelor's degree from the University of Connecticut. Together they raised their three children Cheryl Choate, Randy Baldwin, and Robert Baldwin and a foster child, Angel Starr, in Davis, California and had six grandchildren.

Lee was truly a Professor and intertwined his research and teaching endeavors during his long career at UC Davis. He was a passionate educator with an exceptional gift for challenging his students to integrate knowledge from different disciplines. His primary undergraduate course was lactational biology where he enthusiastically challenged his students and taught critical-thinking skills. His lactation course was specifically designed to impart an integrated understanding of biochemical, genetic, nutritional, physiological, and structural factors relating to mammary gland development, the initiation and maintenance of lactation, the composition of milk and limits to productivity. A strong emphasis was placed on using knowledge from basic mathematics, chemistry, and biology to solve problems in animal production. Lee also taught the graduate level nutritional energetics course, which began during a time of historic change. The nutritional energetics course evolved from the classical approach of '*The Fire of Life*' (Kleiber, 1961) to focusing on nutritional energetics, from animal-level input:output relationships, to one that emphasized the specific biochemical and physiological bases for energy expenditures. While the graduate students were learning metabolism and energetics, Lee was teaching them how to think. His teaching developed his research, just as his research added to his teaching. These teaching and research efforts became the foundation for the development of integrated systems approach to research and application in animal nutrition, an approach that has been adopted in most courses in nutritional energetics across the world.

An example of Lee's early commitment to the science of integrative biology and to training future generations of students was his paper published rather early in his career entitled: 'Estimation of Theoretical Calorific Relationships as a Teaching Technique; a Review' (1968). This paper is one of the first to focus on the actual biochemical processes and control elements that make up the practical animal-level outcomes in growth and lactation.

Lee maintained an extremely productive research and graduate training program spanning five decades. From his earliest efforts as an assistant professor of animal sciences at UC Davis to his retirement as Sesnon professor (above rank) and member of the National Academy of Sciences (in 1993), Lee never lost sight of the importance of a rationally thought-out and followed research plan, integrated seamlessly with a similar graduate training program. Today dozens of graduate students mentored by Lee make up a significant core of productive animal scientists in education and industry around the world. He wrote in the introduction to 'Modeling Ruminant Digestion and Metabolism' (1995) that his research program is best characterized as one which couples experimental reduction and analysis with the use of mathematical modeling to achieve synthesis, integration and effective utilization of knowledge of underlying function in the solution of problems in ruminant animal production.

He started his career as a basic ruminant microbiologist, subsequently moving his focus to whole animal and tissue level pathway energetics, then onto tissue growth and development, and finally, to practical animal feeding strategies using integrative approaches. The results of his studies of ruminal microbiology and tissue metabolism are the material upon which microbial and digestive and metabolic elements of all lactating cow-feeding systems worldwide are based, including the National Research Council, the Cornell Net Carbohydrate and Protein System, CPM Dairy, and the various systems in use in Europe and Australasia. An innovative feature of his research on tissue metabolic functions has been the use of experimental protocols that reveal kinetic properties of tissues and enable quantitative and time-dynamic evaluations of factors that determine patterns of nutrient utilization. Lee was one of the first to champion the combination of *in vivo* and *in vitro* experiments to extract the greatest possible knowledge and quantitative descriptions from every experiment. Clearly he is generally considered among the founding fathers of modeling in animal science, for introducing process-based simulation modeling using differential equations. Starting in the 1960s through the 21st century, his focus remained on the development of biochemical, mechanistic, dynamic, computer-assisted models of ruminal function and tissue metabolism, primarily in lactating cattle. Lee's summation of this work led to the development of 'Molly', a computer model of rumen and tissue metabolism in the lactating cow, which he named after a particularly patient cow of his childhood. The uniqueness of Lee's approach is that it accommodates effects of current feeding practices in practical use upon subsequent performance that influence milk yields in full lactations, including resultant economic benefits and that it traces metabolism of individual nutrients rather than aggregate entities such as metabolizable energy. Because different nutrients are used for different purposes at differing efficiencies, this model predicts effects of diets of greatly differing composition upon lactational performance in a superior fashion. Today Lee's model and, more importantly, his integrated approach continues to define ruminant food production in Europe, the United States, Australia, Canada, and New Zealand.

Through his pioneering animal research and teaching activities Baldwin created an entire generation of nutritional scientists, and altered the basic philosophy and application of agricultural research and education funding and conduct worldwide. In addition to the scientific and educational impact, it is a matter of historical record that application of his research and findings has led to a major increase in efficiency of animal production and an improvement in the nutritional health of millions of people.

References

Baldwin R. L., 1995. Modeling Ruminant Digestion and Metabolism. London: Chapman & Hall.
Klieber M., 1961. The fire of life: an introduction to animal energetics. John Wiley and Son, New York, USA.

Application of mathematical modelling in animal nutrition, physiology and energy balance

J. France
Centre for Nutrition Modelling, Department of Animal and Poultry Science, University of Guelph, Guelph N1G 2W1, ON, Canada; jfrance@uoguelph.ca

Lee Baldwin was a pioneer of mathematical modelling in animal science, and is widely regarded within the subject as the founding father of process-based simulation modelling using differential equations. And rightly so. He was an integrationist and a visionary. Lee was the first to champion the integration of in vivo and in vitro experimentation and mathematical modelling to extract as much knowledge as possible from an experimental programme. Indeed his mantra was *in vivo, in vitro, in silico*. He saw beyond statistics and biometrics into the wider realms of mathematics. He believed that mathematics as employed in the physical and engineering sciences had much to offer the natural sciences, and that the methods of operations research and applied mathematics (mathematical physics) in particular should be actively embraced in applied biology. He was, for example, among the first to apply linear programming to basic ruminant digestion studies (Reichl and Baldwin, 1975), and a great admirer of Sir Kenneth Blaxter's work, not only on energy metabolism, but on his use of differential equations to interpret digesta flow measurements (e.g. Blaxter *et al.*, 1956). Indeed Lee tended to subscribe to that well-worn cliché of differential equations being Sir Isaac Newton's key to the universe.

This summary gives a synoptic account of Lee's evolution as a mathematical modeller, much of it in his own words. Lee undertook graduate studies at the University of Michigan in the early 1960s and was profoundly influenced by his two principal advisors:

> Professor Roy S. Emery, major professor for my M.S. degree and co-mentor during my Ph.D: his training encouraged me to develop as a quantitative scientist – 'Do your arithmetic and take a course in differential equations.' Professor W. A. Wood wondered in my interview with him as a prospective student why I used statistical techniques in my M.S. thesis – 'What was wrong with your data?' – and taught me biochemistry, biochemical methods and theology, and most importantly perhaps, to take students for what they are and help them develop. His advice in this regard is the reason that I have helped train many successful Ph.D. students, who, in turn, have contributed so much to our research program over the years.

Thus Lee's credo – quantification and integration – were firmly established by the very outset of his career.

His development as a mathematical modeller was marked by his belief in taking sabbatical study periods. His first sabbatical was in the late 1960s with David Garfinkel at the University of Pennsylvania in Philadelphia to learn-in-depth mathematical descriptions of biochemical equations, particularly those of carbohydrate metabolism. He was highly impressed with and influenced by Garfinkel's work, especially his work on modelling brain metabolism (Garfinkel, 1966), which truly was the cornerstone of his devotion to mathematical aspects of biology and the genesis of Lee's efforts to model the rumen.

In the 1970s he took sabbatical periods 'down under' with Marc Ulyatt at DSIR in Palmerston North, New Zealand, and with John Black at CSIRO in Blacktown near Sydney, Australia. In Palmerston North he worked on modelling rumen function (Baldwin *et al.*, 1977) and in Sydney focused on modelling the nutritional and physiological status of different organs and tissues (Baldwin and Black, 1979). His work in Australia laid the foundations for the growth modelling of Oltjen *et al.* (1986) and the current California beef model (Garcia *et al.*, 2008). By the end of the 1970s, Lee (along with

John Black) was generally recognised as the world's leading animal modeller, and his precepts for modelling rumen function and post-absorptive metabolism were firmly established.

In September 1983, Lee took his final sabbatical at the Grassland Research Institute, in Hurley near Henley-on-Thames, England. The Hurley group (John Thornley, David Beever, Maggie Gill and myself), in collaboration with John Black, were actively involved in modelling ruminant digestion and metabolism at the time. Lee's primary goal was to fuse the Davis (Baldwin *et al.*, 1977) and Hurley (France *et al.*, 1982) rumen models, and the Davis (Baldwin *et al.*, 1980) and Hurley (Gill *et al.*, 1984) metabolism models, to produce new rumen and metabolism models for subsequent integration into a whole cow model. The purpose of the rumen model was to transform feed inputs into nutrient supply, and was parameterised for North American diets; it was not originally intended for grass-based diets or to account for microbial interactions. The objective of the metabolism model was to provide a detailed description of partition of nutrients and energy balance within the cow. His other goal was to learn the computer simulation language ASCL, recently developed at Cambridge University by George Mitchell with the capacity to handle non-linear kinetics easily, which we were using. By April 1984, the first in-depth mathematical model of the lactating dairy cow was completed, and a series of papers written – one on the metabolism model, one on the rumen model, one on the whole cow model, and one on solving stiff equations (Baldwin *et al.*, 1987a,b,c; France *et al.*, 1992). The last paper, which illustrates Lee's life-long interest in mathematics, arose because the cow model originally exhibited stiffness (due to the rate parameters having different orders of magnitude), a problem not easily handled back then, and this issue had to be resolved in order to run it speedily. It is interesting to note the gap between the papers being written and their publication. Mainstream nutrition journals were highly reluctant to publish non-experimental papers at that time.

The first working version of the whole cow model was unimaginatively called cow1.csl before eventually becoming Molly. Discussing the first outputs from cow1 and possible changes to the coding, I remember Lee becoming a little agitated:

> 'We can't call it cow1, she must have a name'. OK Lee, what shall we call it? 'We'll name her after one of the cows on my father's farm; let's call her Molly.' No Lee, that would be rather imprudent around here at the moment. 'I guess so; what about Daisy then?' No, that's the acronym Reading University use for their computerised dairy management information system and it will cause confusion. 'Myrtle then?' OK Lee, Myrtle it is!

So Molly started life as Myrtle. Two years later, back in California, Lee made several alterations to the computer code and Myrtle became Daisy. Subsequently, and in Lee's words:

> Over the 6 year period (1986-1992) during which DAISY reigned, many piecemeal changes were introduced and the flow of biological logic became disjointed, e.g. DAISY got old; after all, most cows are culled before they complete six lactation cycles. Therefore the program was re-organized, corrected, and formatted to form MOLLY, named after the very docile, patient cow to which my father assigned the task of teaching me to milk by hand when I was 8 or 9 years old. Perhaps Molly will provide me and associates with a continuing opportunity to learn.

Lee's capstone achievement in modelling came with the publication of his seminal book Modelling Ruminant Digestion and Metabolism, in which a detailed account of Molly was presented (Baldwin, 1995). Since his retirement, improvements to the cow model have continued and the current version is more user-friendly. Molly's adoption is flourishing and she is currently popular not only in North America but also in Australasia and Europe. But much more importantly, the integrated approach and vision that Lee championed have become accepted practice throughout animal science globally. His legacy is therefore an outstanding one.

A fuller appreciation of Lee's life and achievements can be found in the memoir by Baldwin VI *et al.* (2010).

References

Baldwin R.L., 1995. Modeling Ruminant Digestion and Metabolism. London: Chapman & Hall.

Baldwin R.L., J. France and M. Gill, 1987a, Metabolism of the lactating cow: 1. Animal elements of a mechanistic model. J. Dairy Res. 54:77-105.

Baldwin R.L., J.H.M. Thornley and D.E. Beever, 1987b. Metabolism of the lactating cow: 2. Digestive elements of a mechanistic model. J. Dairy Res. 54:107-131.

Baldwin R.L., J. France, D.E. Beever, M. Gill and J.H.M. Thornley, 1987c. Metabolism of the lactating cow: 3. Properties of mechanistic models suitable for evaluation of energetic relationships and factors involved in the partition of nutrients. Journal of Dairy Research 54:133-145.

Baldwin R.L. and J.L. Black, 1979. Simulation of the effects of nutritional and physiological status on the growth of mammalian tissues: description and evaluation of a computer program. CSIRO Aust. Anim. Res. Lab. Tech. Pap. No. 6, 1-35.

Baldwin R.L., L.J. Koong and M.J. Ulyatt, 1977. A dynamic model of ruminant digestion for evaluation of factors affecting nutritive value. Agric. Systems 2:255-288.

Baldwin, R.L., N.E. Smith, J. Taylor and M. Sharp, 1980. Manipulating metabolic parameters to improve growth rate and milk secretion. J. Anim. Sci. 51:1416-1428.

Baldwin R.L. VI, C.C. Calvert, J.G. Fadel, J. France and J.P. McNamara. 2010. Ransom Leland ('Lee') Baldwin V 1935-2007: A biographical memoir. National Academy of Sciences, Washington, D C.

Blaxter K.L., N. McC. Graham and F.W. Wainman 1956. Some observations on the digestibility of food by sheep, and on related problems. Br. J. Nutr. 10:69-91.

France J., J.H.M. Thornley and D.E. Beever, 1982. A mathematical model of the rumen. J. agric. Sci., Camb: 99:343-353.

France J., J.H.M. Thornley and K.A. Crist, 1992. On solving stiff equations with reference to simulating ruminant metabolism. J. Theoret. Biol. 156:525-539.

Garcia, F., R.D. Sainz, J. Agabriel, L.G. Barioni and J.W. Oltjen, 2008. Comparative analysis of two dynamic mechanistic models of beef cattle growth. Anim. Feed Sci. Tech. 143:220-241.

Garfinkel, D., 1966. A simulation study of the metabolism and compartmentation in brain of glutamate, aspartate, the Krebs cycle, and related metabolites. J. Biol. Chem. 241:3918-3929.

Gill M., J.H.M. Thornley, J.L. Black, J.D. Oldham and D.E. Beever, 1984. Simulation of the metabolism of absorbed energy-yielding nutrients in young sheep. Br. J. Nutr. 52:621-649.

Oltjen J.W., A.C. Bywater, R.L. Baldwin and W.N. Garrett, 1986. Development of a dynamic model of beef cattle growth and composition. J Anim. Sci. 62:86-97.

Reichl J.R. and R.L. Baldwin, 1975. Rumen modelling: rumen input-output balance models. J. Dairy Sci. 58:879-890.

Contributions of Lee Baldwin to lactation biology

J.P. Cant
Centre for Nutrition Modelling, Department of Animal and Poultry Science, University of Guelph, Canada; jcant@uoguelph.ca

Lee Baldwin began his studies of lactation in the early 1960's by characterizing metabolic changes that occurred in the mammary glands before and after parturition. He considered the mammary glands a rare and fascinating experimental model of organ development and differentiation that could be induced to occur repeatedly in the adult animal. His ambitious undertaking was to catalog the expression levels of literally dozens of mammary enzymes and their associated intermediary metabolites at various stages of pregnancy and lactation in the rat, cow, and guinea pig (Baldwin, 1966; Milligan and Baldwin, 1966). He was conducting proteomics and metabolomics decades before the terms were invented.

In the quest to understand regulation of milk synthesis, Baldwin developed an experimental approach that was to remain a cornerstone of his style. The approach was to apply perturbations to the animal, measure lactational performance, collect tissues at slaughter and measure an array of metabolite and enzyme levels, and incubate tissues in vitro with a range of radiolabelled substrates to assess treatment effects on fluxes through metabolic pathways. Because of the large number of measurements and the complexity of the system, contradictions and anomalies would often arise in the interpretation of data. Baldwin saw these discrepancies as opportunities to advance knowledge because obviously something was missing in the explanation. He sought to improve the explanations with the help of mathematical modelling.

Arguably, Lee Baldwin's greatest contribution to the discipline of lactation biology was the school of thought he espoused. His approach was one of perturbation, analysis, and synthesis, where analysis and synthesis refer, respectively, to the breaking of a whole into parts, and the reassembly of a whole from its parts. Baldwin was an admirer and a scholar of Antoine Lavoisier, who wrote in 1790, 'Chemistry affords two general methods of determining the constituent principles of bodies, the method of analysis, and that of synthesis… In general it ought to be considered as a principle in chemical science, never to rest satisfied without both these species of proofs.' When I read these words, I cannot help but think of Baldwin's constant reminders of the need to practice synthesis after analysis.

At the time Baldwin started his career, it was just becoming known which hormones were responsible for the onset and maintenance of lactation, while the intracellular second messengers of these lactogenic hormones and their mechanisms of action were deep in the fog. The accepted metabolic pathways by which lactose, short- and long-chain fatty acids, and triacylglycerol were synthesized in mammary epithelial cells contained large gaps. By now, most of the metabolic pathways by which milk components are synthesized have been worked out and our horizons have expanded with the use of the modern tools of molecular biology that capture a wider panorama of the mammary landscape than in the past. Some of the key findings from analysis of recently collected data are that: nutritive effects on milk synthesis are due in large part to changes in secretory cell number through proliferative or apoptotic mechanisms (Capuco *et al.*, 2001); prolactin affects mammary gene expression through a JAK-STAT signalling pathway (Rui *et al.*, 1994); activation of the ser/thr kinase Akt by lifting of the repressive effects of progesterone at the end of gestation is central to the initiation of copious milk secretion (Rudolph *et al.*, 2003); growth hormone influences insulin sensitivity and fat mobilization from adipose tissues via synthesis of an inhibitory subunit of the PI3-kinase (Del Rincon *et al.*, 2007); the decline in milk production post-peak is due to a limited capacity of the secretory cell to perform any function other than milk synthesis (Lemay *et al.*, 2007).

Many of these hypotheses were developed from an analysis of large volumes of data generated from modern transcriptomics or proteomics studies. There is widespread optimism that the ability to measure expression levels of thousands of genes at once is going to transform our understanding of biology, and yet just under the surface lies a vein of skepticism that anything meaningful can be extracted from the reams of data that are so easily obtained. Baldwin, as an early practitioner in the field, set out some important tenets as to how to proceed through the tangle. He taught that (1) nutrient fluxes or other such changes over time must be recorded in association with the panoramic snapshot of the gene expression levels, and (2) hypotheses developed analytically from the expression data should be written in mathematical form so that they can generate predictions of the observed fluxes and complete the analysis/synthesis proof. It is interesting to see that systems biology, in which biological hypotheses are expressed mathematically for synthetic purposes, is experiencing a rebirth under the umbrella of bioinformatics in an attempt to extract understanding from high-density biological data.

It is exactly as Baldwin taught.

References

Baldwin, R.L., 1966. Enzymatic activities in mammary glands of several species. J. Dairy Sci. 49, 1533-1542.

Baldwin, R.L. and L.P. Milligan, 1966. Enzymatic changes associated with the initiation and maintenance of lactation in the rat. J. Biol. Chem. 241, 2058-2066.

Capuco, A.V., D.L. Wood, R. Baldwin, K. McLeod and M.J. Paape, 2001. Mammary cell number, proliferation and apoptosis during a bovine lactation: relation to milk production and effect of bST. J. Dairy Sci. 84, 2177-2187.

Del Rincon, J.-P. K. Iida, B.D. Gaylinn, C.E. McCurdy, J.W. Leitner, L.A. Barbour, J.J. Kopchick, J.E. Friedman, B. Draznin and M.O. Thorner, 2007. Growth hormone regulation of p85 expression and phosphoinositide 3-kinase activity in adipose tissue: mechanism for growth hormone-mediated insulin resistance. Diabetes. 56, 1638-1646.

Lavoisier, A.L. 1790. Elements of chemistry, in a new systematic order, containing all the modern discoveries. transl. Kerr, William Creech, Edinburgh.

Lemay, D.G., M.C. Neville, M.C. Rudolph, K.S. Pollard and J.B. German, 2007. Gene regulatory networks in lactation: identification of global principles using bioinformatics. BMC Syst. Biol. 1, 56.

Rudolph, M.C., J.L. McManaman, L. Hunter, T. Phang and M.C. Neville, 2003. Functional development of the mammary gland: use of expression profiling and trajectory clustering to reveal changes in gene expression during pregnancy, lactation, and involution. J. Mam Gl. Biol. Neoplas. 8, 287-307.

Rui, H., R.A. Kirken and W.L. Farrar, 1994. Activation of receptor-associated tyrosine kinase JAK2 by prolactin. J. Biol. Chem. 269, 5364-5368.

Modeling animal growth with Lee Baldwin

Roberto D. Sainz
Department of Animal Science, University of California, Davis, CA 95616, USA; rdsainz@ucdavis.edu

Lee Baldwin's research career focused on elucidating the physiological functions underlying animal performance. Early on, he realized that the number of metabolic variables and the interactions among them were too great to be fully understood and manipulated without the use of quantitative modeling techniques. At first, this entailed the construction and use of static analytical models that evolved into more complex dynamic forms. Always, the emphasis was on improving our understanding, or as he put it, 'advancing the science' rather than simply achieving the best fit to data. Baldwin never considered himself a modeler, rather a scientist who used modeling as part of the overall research program. For Baldwin, models were most useful for:

1. integration of existing concepts and data regarding animal function into a consistent mathematical framework;
2. evaluation of those same concepts and data for adequacy in describing growth and body composition under different feeding regimes, growth trajectories, etc.;
3. identification of critical experiments and measurements;
4. evaluation of alternative hypotheses for probable adequacy in explaining the effects of genotype, nutrition, and other growth manipulations.

Nutritional models for growing animals should enable accurate predictions of nutrient requirements, calculation of responses of defined animals to defined feeds and calculation of optimal nutritional strategies. Baldwin recognized that the traditional approach applied in development of feeding systems, based upon empirical fits to extensive data sets, would not be able to extend our knowledge of nutritional biology. He quoted Sir Kenneth Blaxter's view that 'we must advance through utilization of 'First Principles''. This led him to adopt the approach espoused by Robinson (1971) that 'quantitative representation of the growth process and effects of endogenous and exogenous agents thereupon must accommodate the contributions of hyperplasia, hypertrophy and accretion to the growth process' (Baldwin, 1995). Underlying that notion is the principle that understanding of governing biological mechanisms coupled with appropriate mathematical modeling tools should result in greater accuracy and wider application than can be achieved with the traditional, empirical approach.

Baldwin's first effort at modeling animal growth occurred during a sabbatical leave in Australia, when he was able to collaborate with John Black at CSIRO. The resulting model (Baldwin and Black, 1979) of growth of tissues and organ systems was based on the realization that relative organ weights contribute to differences in fasting heat production and that nutritional, physiological and environmental interactions determine patterns of nutrient utilization, growth rates and body composition. Initially, allometric equations were used to estimate tissue and organ weights in growing animals, based on published data. No consistent relationship was found in the values of ***a*** and ***b*** among and within species. This lack of interspecific applicability of numerical inputs to the allometric equation was considered unacceptable.

Robinson (1971) developed the concept first proposed by Enesco and Leblond (1962) that growth of tissues, organ systems and whole animals occurs in three phases: hyperplasia, hypertrophy and accretion. Hypertrophic growth was defined by Robinson (1971) as, essentially, increases in the amount of protein per unit of DNA to some genetically defined maximum. These concepts led to three premises that apparently determine growth in mammals:

1. The primary genetic determinant of organ size is the final number of nuclei, represented by the amount of deoxyribonucleic acid (DNA).

2. Each unit of DNA specifies, on a genetically defined basis for each tissue and each species, information required for the ultimate formation of a specific amount of cell material. Whether information specified by a unit of DNA leads to the formation of cell material is dependent upon the nutritional and physiological status of the animal.
3. The specific activities of enzymes responsible for tissue metabolism and growth vary as an exponential function of organ size. Also, the kinetic properties of enzymes are reasonably constant across species.

Using a model based upon these three premises, Baldwin and Black (1979) simulated the effects of nutrition on various organs in different species and concluded that they were adequate bases for mechanistic and dynamic representations of the growth process. The concepts and equation forms used by Baldwin and Black (1979) served as the starting point for the beef cattle growth model of Oltjen *et al.* (1986), now called the DGM (Davis Growth Model). The DGM is dynamic, thus differential equations were integrated to estimate gain (or loss) of DNA and body protein. Body fat gain is estimated from the difference between energy intake and energy used for maintenance and protein gain. Since the model requires initial estimates of whole-body DNA, protein and fat, empirical relationships between these and animal weight, mature size and condition score were developed to set beginning values for model implementation. Later, additional research was conducted to fill in gaps of knowledge regarding cell number and sizes in various tissues of the bovine (DiMarco *et al.*, 1987; Sainz and Bentley, 1997), an example of the interplay between modeling and experimentation that was the hallmark of Baldwin's research.

The DGM has been used widely in beef cattle management. For example, this model, coupled with frame, weight and ultrasound measurements of cattle entering the feedlot, is being used to sort cattle into more uniform outcome groups (Sainz and Oltjen, 1994; Sainz *et al.*, 1995, 2009). The DGM has also been reparameterized for *Bos indicus* cattle (Sainz *et al.*, 2006), and served as the basis for a practical management tool for the Australian beef industry (Oddy *et al.*, 2008). Despite its simplicity and practical applicability, the DGM's representation of fundamental mechanisms has also enabled its use in research, for example, to evaluate hypotheses regarding variations in feed efficiency (Cruz *et al.*, 2007), and as the basis for a sheep growth model (Oltjen *et al.*, 2010). More recent versions of the DGM include the introduction of variable maintenance functions (Oltjen and Sainz 2001; Oltjen *et al.*, 2006) and distribution of body fat among different depots (Sainz and Hasting, 2000; McPhee *et al.*, 2008). A study (Garcia *et al.*, 2008) comparing the DGM with a model of cattle growth and body composition based upon explicit representation of protein and fat metabolism (Hoch and Agabriel, 2004) demonstrated that each model adequately described accretion of protein and fat, but the DGM was superior under discontinuous feeding and with different diet qualities.

Many other models of animal growth have been developed and are in use today for practically every livestock species. These models were developed for different purposes and using different approaches, and so may or may not derive directly from the growth modeling work of Baldwin and coworkers. Indirectly, of course, the notion that mathematical models can help us solve research and practical problems had no greater champion. In his words (Baldwin, 1995):

> In our view, dynamic models are absolutely essential to economic evaluations of risks associated with current decisions because these clearly influence subsequent performance. When a proven, dynamic, mechanistic model becomes available to students and practitioners ... to enable cause-and-effect analyses of underlying reasons for the observed responses, we will have created a superior instrument for both the teaching and application of our science. This is the challenge we pose.

References

Baldwin, R. L. and J. L. Black, 1979. Simulation of the effects of nutritional and physiological status on the growth of mammalian tissues: description and evaluation of a computer program. Animal Research Laboratories Technical Paper 6. Commonwealth Science and Industrial Research Organization, Melbourne, Australia.

Baldwin, R. L., 1995. Modelling Ruminant Digestion and Metabolism. Chapman & Hall Ltd. Scientific, London.

Cruz, G. D., J. W. Oltjen and R. D. Sainz, 2007. Evaluation of hypotheses about residual feed intake in beef steers using the Davis Growth Model. Can. J. Anim. Sci. 87, 649.

DiMarco, O. N., R. L. Baldwin and C.C. Calvert, 1987. Relative contributions of hyperplasia and hypertrophy to growth in cattle. J. Anim. Sci. 65, 150-157.

Enesco, M. and C. P. Leblond, 1962. Increase in cell number as a factor in the growth of the organs and tissues of the young male rat. J. Embryol. 10, 530-562.

Garcia, F., R. D. Sainz, J. Agabriel, L. G. Barioni, and J. W. Oltjen, 2008. Comparative analysis of two dynamic mechanistic models of beef cattle growth. Anim. Feed Sci. Technol. 143, 220-241.

Hoch, T. and J. Agabriel, 2004. A mechanistic dynamic model to estimate beef cattle growth and body composition: 1. Model description. Agric. Syst. 81, 1-15.

McPhee, M. J., J. W. Oltjen, J. G. Fadel, D. Perry, and R. D. Sainz, 2008. Development and evaluation of empirical equations to interconvert between twelfth-rib fat and kidney, pelvic, and heart fat respective fat weights and to predict initial conditions of fat deposition models for beef cattle. J. Anim. Sci. 86, 1984-1995.

Oddy, V. H., R.C. Dobos, M. J. McPhee, W. McKiernan, J. W. Oltjen and R. D. Sainz, 2008. A new tool to predict beef cattle fatness in the field. Can. J. Anim. Sci. 88, 735.

Oltjen, J. W. and R. D. Sainz, 2001. Alternate forms for heat production in ruminant growth and composition models. In: Chwalibog, A. and K. Jakobsen (editors), Energy Metabolism in Animals. Wageningen Pers, Wageningen, The Netherlands, EAAP Pub. 103, 39-42.

Oltjen, J. W., A. C. Bywater, R. L. Baldwin and W. N. Garrett, 1986. Development of a dynamic model of beef cattle growth and composition. J. Anim. Sci. 62, 86-97.

Oltjen, J. W., A. Cannas, A., A. S. Atzori, L. O. Tedeschi, R. D. Sainz and D. G. Fox, 2010. Integration of the Small Ruminant Nutrition System and of the UC Davis sheep growth model for improved predictions. In: Crovetto, G. M. (editor), Energy and Protein Metabolism and Nutrition. EAAP Pub. 127, 553-554.

Oltjen, J. W., R. D. Sainz, A. B. Pleasants, T. K. Soboleva, and V. H. Oddy. 2006. Representation of fat and protein gain at low levels of growth and improved prediction of variable maintenance requirement in a ruminant growth and composition model. In: Kebreab, E., J. Dijkstra, A. Bannink, W. J. J. Gerrits, and J. France (editors), Nutrient Digestion and Utilization in Farm Animals: Modelling Approaches. CAB International, Wallingford, U.K., pp. 144-169.

Robinson, D. W., 1971. Cellular basis for changes in body composition. J. Anim. Sci. 33, 416-420.

Sainz, R. D. and B. E. Bentley, 1997. Visceral organ mass and cellularity in growth-restricted and refed beef steers. J. Anim. Sci. 75, 1229-1236.

Sainz, R. D. and J. W. Oltjen, 1994. Improving uniformity of feeder steers using ultrasound and computer modelling. Proc. Am. Soc. Anim. Sci. Western Section 45, 179-181.

Sainz, R. D., and E. Hasting. 2000. Simulation of the development of adipose tissue in beef cattle. In: McNamara, J. P., J. France, and D. E. Beever (editors), Modelling Nutrient Utilization in Farm Animals. CAB International, Wallingford, U.K., pp. 175-182.

Sainz, R. D., Barioni, L. G., Paulino, P. V., Valadares Filho, S. C., and Oltjen, J. W. 2006. Growth patterns of Nellore vs. British beef cattle breeds assessed using a dynamic, mechanistic model of cattle growth and composition. In: Kebreab, E., J. Dijkstra, A. Bannink, W. J. J. Gerrits, and J. France (editors), Nutrient Digestion and Utilization in Farm Animals: Modelling Approaches. CAB International, Wallingford, UK., pp. 160-170.

Sainz, R. D., J. G. Smith, I. Garnett and Y. B. Lee, 1995. Use of ultrasound and computer modeling to predict days on feed and improve beef carcass uniformity. Proc. Am. Soc. Anim. Sci. Western Section 46, 148-150.

Sainz, R. D., L. G. Barioni, F. R. C. Araujo, and J.W. Oltjen· 2009. Development and evaluation of a tool to sort feedlot beef cattle into uniform outcome groups. Can. J. Anim. Sci. 89, 533.

Advances in rumen microbiology

M.R. Murphy
Department of Animal Science, University of Illinois, Champaign, IL, USA; mrmurphy@illinois.edu

With the publication of his MSc thesis at Michigan State University in 1958, R. Lee Baldwin began contributing to advances in rumen microbiology by reporting on an 'Investigation of the oxidation-reduction potential of rumen contents.' His Ph.D. dissertation there in 1962 examined the conversion of lactate to propionate 'in a natural biological system,' bovine rumen fluid. Those results were detailed in Journal of Bacteriology articles published in 1962 and 1963. The next year Baldwin presented a paper at the Second International Symposium on the Physiology of Digestion in the Ruminant. His topic was pathways of carbohydrate metabolism in the rumen; however, he also called attention to the fact that very few data were available 'regarding the quantitative contributions of different metabolic pathways or microorganisms, and the specific factors which control the growth and metabolic activity of the microorganisms in the rumen.'

By the time that the Third International Symposium on the Physiology of Digestion and Metabolism in the Ruminant occurred five years later (1969), Baldwin was able to demonstrate how a relatively new technique, computer simulation, could be used to model ruminant systems. Identified advantages of this approach included: the use of familiar mathematical formulations, applicability to specific experimental data, and flexibility and capacity to represent metabolic and regulatory mechanisms at various levels. This modeling effort focused on energetic relationships in the formation and utilization of fermentation end-products in the rumen. Notably, it was both quantitative and dynamic; i.e. based on differential equations representing specific metabolic reactions.

Rumen fermentation characteristics were considered to be largely determined by the microbial species present but representing each species explicitly in a model of rumen function was likely neither feasible nor necessary. What was essential was that species characteristics and functions be considered, and represented implicitly in the model. To that end, three composite microbial groups were defined: the cellulolytic complex and methanogenic species, amylolytic and associated species, and species utilizing a broad range of substrates. Another important element of the approach was that the validity of the model was tested by simulating data collected in experiments that were not considered in its construction. The conclusion of this article also extended the exercise by addressing philosophical concerns related to mathematical modeling and computer simulation. The point was that, properly applied, these approaches could contribute to our understanding of biological and physiological systems which are both complex and dynamic.

These results were then utilized, along with a separate analysis of energy metabolism in anaerobes, to develop two input-output balance models based on systems of simultaneous linear equations. The first examined relationships among elementary compositions of microbial cells (bacteria and protozoa), digestible feed nutrients, and fermentation products. Another calculated balances for biosynthetic and fermentation equations for rumen microbiota.

Because relative proportions of fermentation products differed depending on the type of substrate, stoichiometric and precursor-product relationships used in previous studies of rumen fermentation balance then had to be extended to provide more detailed evaluations of nutrient digestive patterns. This involved the computation of metabolic flux and stoichiometric parameters for soluble sugars, starch, cellulose, hemicellulose, and protein using statistical methods. Expected products of rumen digestion were then predicted quantitatively and compared to results observed in a sample problem. This framework was adapted to estimate these relationships for a wider range of diets and input-output data, results of which were then utilized in later models of rumen function.

The next major development was a dynamic model of ruminant digestion for evaluation of factors affecting nutritive value. This effort integrated aspects of earlier models with recent experiments in his laboratory on the dynamics of fermentation and microbial growth, and results of animal studies from the literature and by a coauthor. Representations of rumen function remained similar in subsequent models; however, the models incorporated many other elements.

Recent advances and future directions discussed in, and inspired by, Baldwin's work on rumen microbiology are many but only a few are mentioned here. One is that data for estimating stoichiometric coefficients of substrate fermentation in the rumen for animals fed a range of diets at various intakes remain limiting. Specifically, many more measurements of volatile fatty acid rates are needed; perhaps a way forward is determination using stable isotope techniques (e.g. ^{13}C-propionate) or estimation via in vitro fermentation methods.

Another potentially fruitful topic was identified early but has received little attention; i.e. the interaction between comminution of feed and digesta particles by chewing during eating and rumination, and the kinetics of microbial digestion and passage from the reticulorumen. It has been demonstrated, at least for two forages, that synergism exists between animal and microbial effects: mastication during eating enhances microbial fermentation, which increases the effectiveness of comminution during rumination. Obviously such effects need to be quantified for more forages and mixed diets to better understand the dynamics of this interaction.

The structure of microbial populations in the rumen and interactions among species has received increased attention. Of particular interest is how bacterial communities in the rumen are affected by changes in diet and physiological state. Challenges include quantification of their metabolic impact and assessment of the level of detail needed to adequately describe rumen function. Basically, how many 'composite microbial groups' are justified? Some work suggests that only a few classes of fibrolytic bacteria and species are needed to explain much of the variation associated with substrate solubilization and utilization.

Author index